Handbook of

ANALOG CIRCUIT DESIGN

Handbook of ANALOG CIRCUIT DESIGN

Dennis L. Feucht

ACADEMIC PRESS, INC.
Harcourt Brace Jovanovich, Publishers
San Diego New York Boston
London Sydney Tokyo Toronto

This book is printed on acid-free paper. ∞

Academic Press, Inc.
San Diego, California 92101

United Kingdom Edition published by
Academic Press Limited
24–28 Oval Road, London NW1 7DX

Library of Congress Cataloging-in-Publication Data

Feucht, Dennis.
Handbook of analog circuit design / Dennis L. Feucht.
p. cm.
ISBN 0-12-254240-1 (alk. paper)
1. Linear integrated circuits--Design and construction.
I. Title
TK7874.F47 1990
621.381'5--dc20

89-28080
CIP

Printed in the United States of America
90 91 92 93 9 8 7 6 5 4 3 2 1

This book is dedicated to my mathematics and electronics teachers
at Benson Polytechnic High School:

Carl Bryson, mathematics

Vearl Overton, mathematics

Earl Schrader, electronics

Contents

3. Feedback Circuits 52

4. Multiple-Path Amplifiers 94

Preface

Solid-state electronics has been a familiar technology now for about a quarter century. Yet some circuit ideas, like the transresistance method of finding amplifier gain or identifying resonances above an amplifier's bandwidth, are so simple and intuitively appealing that I have been puzzled as to why they are not generally better known in the industry. Fortunately, I began encountering them in my earlier days at Tektronix, but could not find them in engineering textbooks. In writing this book, very few original circuit ideas are presented. Somewhere in the literature, they can be found in some form. My motivation has been to present them from a perspective from which they can be remembered and applied — even quantitatively — with little effort.

The behavior of most circuits is determined most easily by computer simulation. What circuit simulators do not provide is knowledge of what to compute. The creative aspect of circuit design and analysis must be performed by the circuit designer, and this aspect of design is emphasized here. Two kinds of reasoning in electronics seem to be most closely related to creative circuits intuition:

1. Geometric reasoning — a kind of visual or graphic reasoning that applies to the topology (component interconnection) of circuit diagrams and to graphs (such as reactance charts)
2. Causal reasoning — the kind of reasoning that most appeals to our sense of understanding of mechanisms; when we can trace a sequence of causes for circuit behavior, we feel we understand how the circuit works

These two kinds of reasoning combine when we try to understand a circuit by causally thinking our way through the circuit diagram. Answers obtained by *inspection* lie at the root of the quest.

It is possible to write down accurate design equations (based on the given circuit representations) by inspection of the circuit topology. Usually a circuit can be analyzed several ways, but most approaches will not provide a simple insight into how the circuit works or give a perspective that can be applied generally to circuits of its class. Here, I attempt to make explicit these simple but powerful ways of envisioning analog circuits.

Analog circuit design is a very broad discipline. This book recapitulates some of my experience as an instrument design engineer and research engineer. It emphasizes instrumentation and control circuits. It focuses less on narrow-band, communications, opto-electronic, power, and, of course, purely digital circuits. Although some special-function circuits are investigated, the objective is to provide a foundation in general techniques and a familiarity with some particular circuits.

To this end, the book divides conceptually into three parts. The first part, Chapters 1–4, focuses on the transresistance method and feedback concepts. Topological analysis is emphasized by restriction to low-frequency ac (quasistatic) circuits. Dynamic response is then emphasized in Chapters 5–8, covering frequency compensation and frequency-dependent design. Building upon the previous two sections, Chapters 9–12 investigate precision and high-performance amplifier techniques followed by a wide variety of mainly nonlinear, commonly used signal generating and processing circuits. Finally, there are sampled-data circuits—namely, D/A, A/D, S/H, and switched-capacitor circuits.

This book is somewhat unusual in the following ways. Many equations may be seen while thumbing through it. Engineers, scientists, and technicians who actually practice circuit design are not always proficient mathematicians. Consequently, I have put in more intermediate steps and more circuit-oriented interpretation of mathematical results than is typically found in circuits literature. The emphasis is on intuitively appealing ideas, quantified by mathematical development.

Secondly, electronics terminology contains many synonyms or associated expressions, and I have parenthetically added them here and there so that the reader who is familiar with one term can associate it with the term I use instead. Sometimes synonyms are exchanged frequently, as they are outside of books.

Finally, the level of sophistication of concepts varies. For example, the first four chapters are appropriate for an undergraduate circuits course. But I have encountered enough engineers in industry who do not know these concepts that it is worth their inclusion. After twenty years of electronics, I do not put myself beyond discovery of those new elemental insights into circuits that delighted me in my teenage years. And it is my hope that something in this book will also fascinate you.

Dennis L. Feucht

Acknowledgments

This book was made possible largely because of the work of others, especially many good engineers with whom I worked at Tektronix, Inc. In particular, I am indebted to:

Bruce Hofer, a founder of Audio Precision and designer of Tek 7000-series time-bases and 5000-series audio equipment, a rare source of engineering inspiration, who informally was my mentor in college and in the early days at Tek, and a good friend.

Carl Battjes, a major contributor to oscilloscope vertical amplifier design, who founded and teaches the Amplifier Frequency and Transient Response (AFTR) course at Tektronix, a source of many important amplifier ideas that are not generally well known.

Laudie Doubrava, who has investigated power-supply distribution, grounding, and EMI problems and provided useful insights into their solutions.

Art Metz, for his clever contributions to a wide number of amplifier designs.

Jim Woo, for his general wisdom on vertical amplifier design.

Ian Getreu, from whom I learned about transistors.

Michael Freiling, an artificial intelligence researcher in Tektronix Laboratories, whose work in knowledge representation influenced my broader understanding of electronics.

Fred Beckett, Tim Sauerwein, Wayne Kelsoe, Cal Diller, Marv LaVoie, Lee Jalovec, George Ermini, Jim Geddes, Carl Hollingsworth, Chuck Barrows, Dick Hung, Carl Matson, Don Hall, Phil Crosby, Keith Ericson, Keith Lofstrom, Alan Plunkett, and Neldon Wagner also have contributed to my development as an engineer and consequently to this book. And most of all, I am indebted to the Creator of our universe, who made electronics possible.

Any errors or weaknesses in this book, however, are my own.

CHAPTER 1

Introduction

1.1 The Organization of Electronics

This book is about the design of analog circuits and systems. Electronic systems can be described by means of a multilevel hierarchy of concepts. At the most concrete level are the physical circuits themselves, usually represented by a schematic diagram or netlist; these are structural descriptions of the circuit. From these, various electrical (and thermal or mechanical) behaviors are deduced through a causal theory of circuits; when applied to a circuit, a causal description of the circuit results. At the next (more abstract) level of description, these causes are explained in terms of a functional (or teleological) theory that leads to a functional description. So we have three levels of description in electronics:

- A *structural* description of a circuit describes what it *is*.
- A *causal* description of a circuit describes what it *does*.
- A *functional* description of a circuit describes what it *is for*.

Each of these descriptions may be complex enough to require hierarchical organization. For example, a structural description of a system consisting of hundreds of parts is too unwieldy to deal with directly. *Systems* are often organized into *subsystems*, usually described by a block diagram. These subsystems consist of *circuits*, which in turn are composed of *components*. It is common for electronic systems to be structurally described by this kind of four-level hierarchy.

Structural descriptions are often presented in a way that makes the causal and functional descriptions explicit. Block diagrams not only show which parts are grouped together but also represent various subsystem functions that help to show the overall function of the system.

1.2 An Analysis of Circuit Analyses

Circuit analyses can be categorized (Fig. 1.1) as static (or dc) and dynamic (or ac). *Static analysis* involves constant or parametric circuit quantities around which variations occur. These fixed quantities define the circuit operating point or bias. If no variations are present, the state of the circuit is *quiescent.*

Dynamic analysis reveals circuit behavior in terms of changing (dynamic) quantities. Behavioral descriptions are given in either the time domain or the frequency domain. Circuits that are not frequency-dependent can be analyzed entirely using real numbers; this is called *low-frequency* or (from thermodynamics usage) *quasistatic* analysis. Complex analysis includes the effects of reactive components (inductors and capacitors) and characterizes circuit behavior in terms of a complex frequency s. In the complex-frequency domain, both time and frequency responses are characterized.

These kinds of analyses can be done on linearized functions (around an operating point), which is called *incremental* or *small-signal* analysis. More

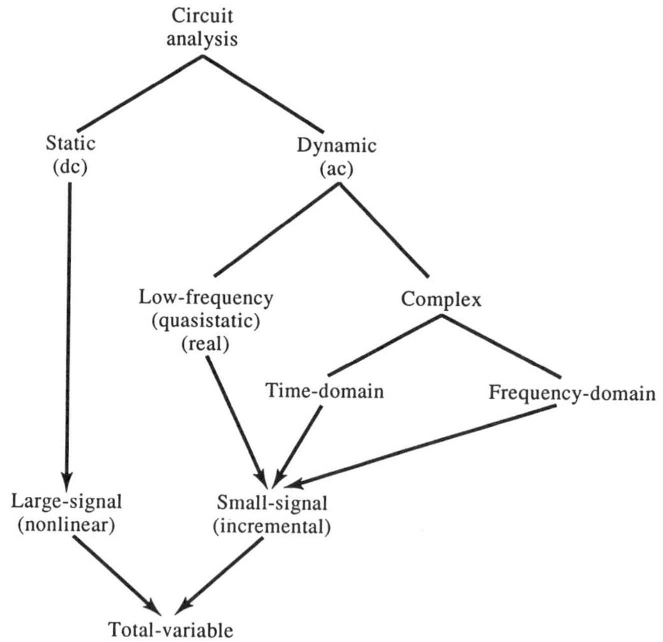

FIG. 1.1 Categories of circuit analysis. Synonyms are in parentheses with dominant expression first. Total-variable analysis combines a dc operating point with a linearized incremental analysis.

generally, analysis of nonlinear circuits involves *total-variable* or *large-signal* analysis.

This book uses common terminology and symbols found elsewhere in circuits literature:

- Static (dc) quantities are in uppercase with (optional) uppercase subscripts (such as V_S).
- Dynamic (small-signal ac) quantities are in lowercase with (optional) lowercase subscripts (such as v_{be}).
- Total-variable (dc + small-signal ac) quantities are in lowercase with (optional) uppercase subscripts (such as v_{BE}).
- Complex (frequency-domain) quantities are in uppercase with lowercase subscripts, if any (such as V_s).

Occasionally, there are exceptions for well-established usage (such as β_o for static transistor current gain).

1.3 The Nature of Design

Design is a kind of creative activity that begins with a definition of the problem to be solved or specification of the device to be built. Solving the problem or specifying the device in enough detail to build it is the goal of the designer. Usually, more than one alternative solution or design is possible. Sometimes these alternatives are already known, and the problem consists mainly of adapting a known general solution to a particular application. This is "standard engineering" practice. Other problems have no known solution and require a search or novel adaptation of existing solutions to similar problems. This is "state of the art" engineering and is sometimes called "research and development" or R&D.

When a solution is found, it is then refined and specified for use. If the problem must be solved many times, the device that solves the problem is manufactured. The process of creating devices and specifying them for manufacture is often called *new product introduction* in the electronics industry. Typically it consists of the following steps or phases:

1. Concept phase: Clarify the idea for a new product with a one-page description of it and a quickly built functional prototype device (or model) that demonstrates the product idea. This phase is completed after *product* or *project approval.*
2. Design phase: Specify the performance parameters of the new product and design the product more carefully to meet the specifications. Build a few models of this design, characterize their performance by testing, and refine

the design to meet specifications where performance is inadequate. This phase is completed upon *design completion.*

3. Evaluation phase: Evaluate the design more extensively by building several units of the product with the materials and processes that will be used in manufacturing, and then testing these units rigorously in the laboratory for performance under all anticipated operating conditions. Design refinement proceeds until the design meets the specification or until the models cannot be modified further and continue to embody the design. This phase is completed at *prototype release.*

4. Verification phase: Verify that the design meets performance and reliability specifications under the conditions of its intended (end) use. (This is called *field testing.*) Build a statistically significant number of units and subject them to extensive use. Refinements to the design in this phase should be minimal and testing maximal. This phase is completed when all documentation that specifies the design for manufacture is acceptable at *engineering release* (or, to manufacturing personnel, *manufacturing acceptance*).

After this, the manufacture of a batch of product units using the design documentation (a *pilot* or *preproduction run*) is carried out by manufacturing personnel to test the documented design for production flaws. Engineers may be required to correct these flaws.

Analog circuit design consists mostly of analysis. The performance constraints of circuits are uncovered by their analysis, and design is largely a matter of achieving a desired function within given constraints. Therefore, a significant aspect of design skill is the ability to understand how circuit constraints affect desired function. Some analytic techniques, especially those best executed by computer, give the designer little insight into the relationship between circuit structure and function. The techniques developed here are intended to provide the kind of insight into circuit function that readily leads to construction of mathematical descriptions of its constraints.

Besides mastering analytic methods, a designer must become familiar with a large number of circuits.

Finally, a designer must understand something of the activity of design itself. Software engineering is more explicit about this than electronics engineering. Electronic designs are performed top-down, using the four-level hierarchy in Section 1.1. In R&D projects, not enough is known about the detailed levels to proceed purely top-down, so experimentation with details is necessary. When the details are adequately understood, the systems-level design can then be clarified. Complexity is handled in electronics (as in software design) by use of modular functions or subsystems. (In electronics, a *module* is a physical, not only functional, distinct subsystem.) An entire subsystem can be defined purely in terms of its interactions with other subsystems. Instead of passing objects, parameters, or pointers to data structures, one must make electrical connections between input and output ports. Just as parameter passing must be done according to a protocol, electrical connections between modules must take into account impedance matching, dynamic range, and loading effects.

Reference

Johan de Kleer, "How Circuits Work," in *Qualitative Reasoning about Physical Systems*, Daniel Bobrow, ed., Bradford Books, MIT Press, 1985, pp. 205–280.

CHAPTER 2

Basic Amplifier Circuits

2.1 Active Device Models

Solid-state devices are nonlinear. Because of this it is not easy to analyze them without developing linear approximations to their behavioral models. With a linear model, the powerful techniques of linear analysis can be applied to circuits.

Nonlinear device models are linearized by selecting an *operating point* for the device. This is a point on a curve of the model at which a linear approximation to the curve is constructed. This approximation is valid as long as the excursions from the operating point are small; it is an *incremental* or *small-signal model.* In contrast, the *large-signal* (total-variable) *model* is the exact (nonlinear) model since operation anywhere along its curve is valid.

To illustrate this, consider the voltage–current (v–i) relationship for a diode,

$$i = I_S(e^{v/V_T} - 1) \tag{2.1}$$

plotted in Fig. 2.1. V_T is the thermal voltage, defined as

$$V_T = \frac{kT}{q} \tag{2.2}$$

where k is Boltzmann's constant, q the electron charge, and T the absolute temperature (in degrees Kelvin). $V_T \cong 26$ mV at a temperature of 298 K (25°C). For a fixed operating point, $Q = (V, I)$, the static or dc resistance of the diode is

$$R|_{(V,I)} = \frac{V}{I} \cong \frac{V_T \ln(i/I_S)}{i}, \qquad i \gg I_S \tag{2.3}$$

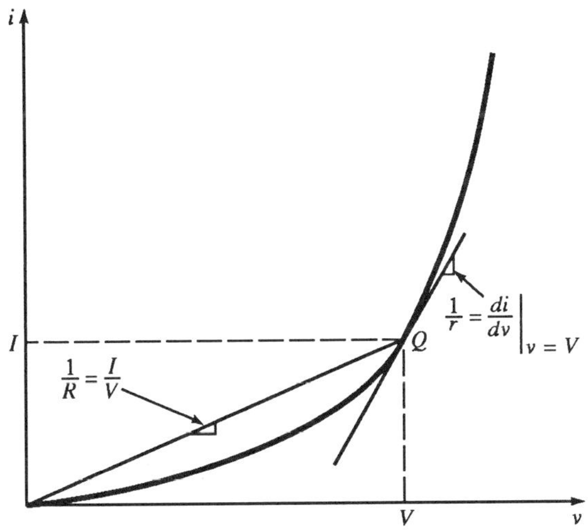

FIG. 2.1 The v–i characteristic of a diode, showing the difference between static large-signal resistance R and dynamic small-signal resistance r.

This is shown as 1/slope of the line from the origin to Q in Fig. 2.1. For a typical diode operating at 1 mA, $Q = (0.6\text{ V}, 1\text{ mA})$ and

$$R = \frac{0.6\text{ V}}{1\text{ mA}} = 600\ \Omega$$

For small excursions around Q, R varies slightly. A linear approximation to $i(v)$ at Q is a line tangent to the curve. Its 1/slope is

$$r\big|_{(V,I)} = \left.\frac{dv}{di}\right|_{I} \cong \frac{V_T}{i} \cong \left.\frac{\Delta V}{\Delta I}\right|_{(V,I)}, \qquad i \gg I_S \tag{2.4}$$

This is the dynamic or incremental ac resistance of the diode at Q. A small change in v results in a small change in i of about $\Delta V/r$. For 1 mA operation, $r = 26\text{ mV}/1\text{ mA} = 26\ \Omega$ and is considerably less than the static resistance. For a linear device such as a resistor, small- and large-signal behavior is identical for both small and large variations in variables.

For a bipolar junction transistor (BJT), the dynamic resistance of the base–emitter junction under forward bias is

$$r_e = \frac{dv_{BE}}{di_E} = \frac{v_{be}}{i_e} \tag{2.5}$$

BJT transconductance g_m can be expressed as a transresistance,

$$r_m = \frac{1}{g_m} = \frac{dv_{BE}}{di_C} = \frac{v_{be}}{i_c} \tag{2.6}$$

For the BJT, static current gain is defined as

$$\beta_o = \frac{I_C}{I_B} \tag{2.7}$$

and dynamic current gain as $\beta = i_c / i_b$. Then, combining equations (2.5) and (2.7) with $i_e = i_c + i_b$, we obtain

$$r_e = \frac{v_{be}}{i_e} = \frac{v_{be}}{(\beta + 1) i_b} \tag{2.8}$$

and

$$r_m = \frac{v_{be}}{i_c} = \frac{v_{be}}{\beta i_b} \tag{2.9}$$

Substituting (2.9) into (2.8) as r_m gives

$$r_e = \left(\frac{\beta}{\beta + 1}\right) r_m = \alpha \cdot r_m \tag{2.10}$$

where

$$\alpha = \frac{i_c}{i_e} = \frac{\beta}{\beta + 1} \tag{2.11}$$

Since β and r_e are incremental BJT parameters, they can be used to construct a simple incremental model, the *T model* (Fig. 2.2a). A more complete incremental model can be derived from the Ebers–Moll or Gummel–Poon large-signal models, the hybrid-π model (Fig. 2.2b). We will make use of this model when studying frequency response. For FETs, the simple incremental model of Fig. 2.2c will also be of use.

The T-model dynamic emitter resistance r_e is related to the hybrid-π model dynamic base resistance r_π. It is one of the more interesting transistor relationships. Since both r_e and r_π are across the same nodes—base and emitter—it would seem at first that they must be the same resistance. They differ, however, in the connection of the collector current source; in the hybrid-π model, it is connected to the emitter whereas in the T model, it is connected to the base. Consequently, both base and collector current flow through r_e, and only base current flows through r_π. By definition,

$$r_\pi = \frac{v_{be}}{i_b} \tag{2.12}$$

Since $i_e = (\beta + 1) i_b$, v_{be} causes $(\beta + 1)$ times as much current to flow through r_e as r_π. Since $(\beta + 1)$ times as much current flows in the emitter as in the base for the same applied voltage, the resistance in the base side of the base-emitter loop can be transformed into an equivalent emitter resistance by the β transform:

$$r_E = \frac{R_B}{(\beta + 1)} \tag{2.13a}$$

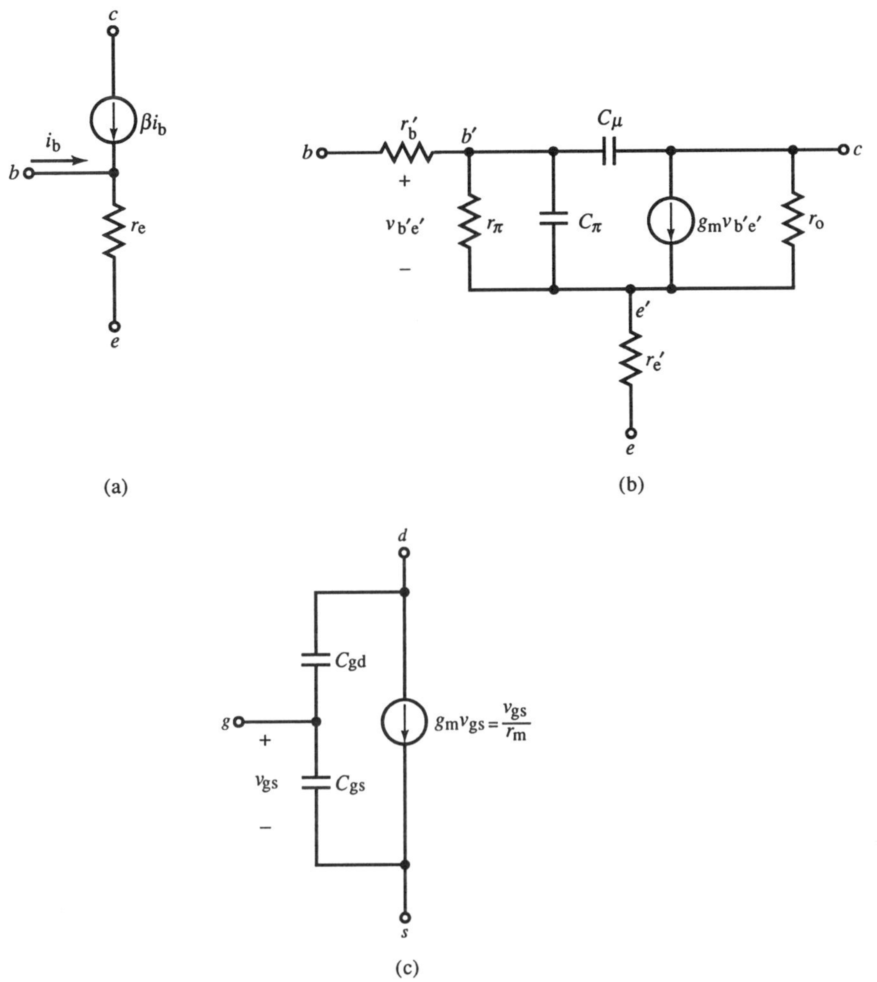

FIG. 2.2 Transistor models: (a) low-frequency, small-signal T model; (b) small-signal, hybrid-π model; and (c) small-signal FET model.

Similarly, emitter resistance can be transformed to an equivalent base resistance:

$$r_B = (\beta + 1) R_E \tag{2.13b}$$

This transform is extremely useful in transistor circuit analysis. It lets us place all resistances on either the base or emitter side of a circuit loop containing the base–emitter junction. This results in elimination from the analysis of one of the variables i_b or i_e.

2.2 Basic Amplifier Configurations

A single transistor can be configured as an amplifier in three ways. When viewed as a two-port network, an amplifier has an input port and an output port, or four terminals in all. Since a transistor is a three-terminal device, one of the terminals must be common to both input and output circuits, resulting in the three basic amplifier configurations for a single BJT:

Common emitter (CE)
Common base (CB)
Common collector (or emitter-follower) (CC)

Equivalently, for FETs:

Common source (CS)
Common gate (CG)
Common drain (or source-follower) (CD)

In Fig. 2.3, typical amplifier configurations are shown along with their incremental models. Since these models involve only incremental changes, the total variables for input and output voltages,

$$v_I = V_{BB} + v_i, \qquad v_O = V_{CC} + v_o$$

are replaced by incremental variables v_i and v_o, respectively. A complete circuit analysis involves both large- and small-signal analyses:

large-signal (dc) analysis $\Rightarrow$ operating point (or bias)

small-signal (ac) analysis $\Rightarrow$ dynamic resistances, gain

Although the emphasis here is on small-signal analysis, large-signal analysis must be examined first because dynamic quantities such as r_e depend on it. For example, r_e depends on i_E in (2.5); r_e is approximately constant when $i_e \ll I_E$ and, like (2.4), is

$$r_e \cong \frac{V_T}{|I_E|} \tag{2.14}$$

Of major interest in the dc analysis, based on dc circuit quantities, is not only the operating point but also its stability. With significant operating point change, incremental parameters can vary too widely, resulting in unacceptable performance. This is caused by *temperature drift* and changing values of aging components. The sensitivity of the operating point to component value variations is important in this analysis.

FIG. 2.3 Three basic BJT configurations and equivalent circuit models: (a) common emitter (CE), (b) common base (CB), and (c) common collector or emitter-follower (CC).

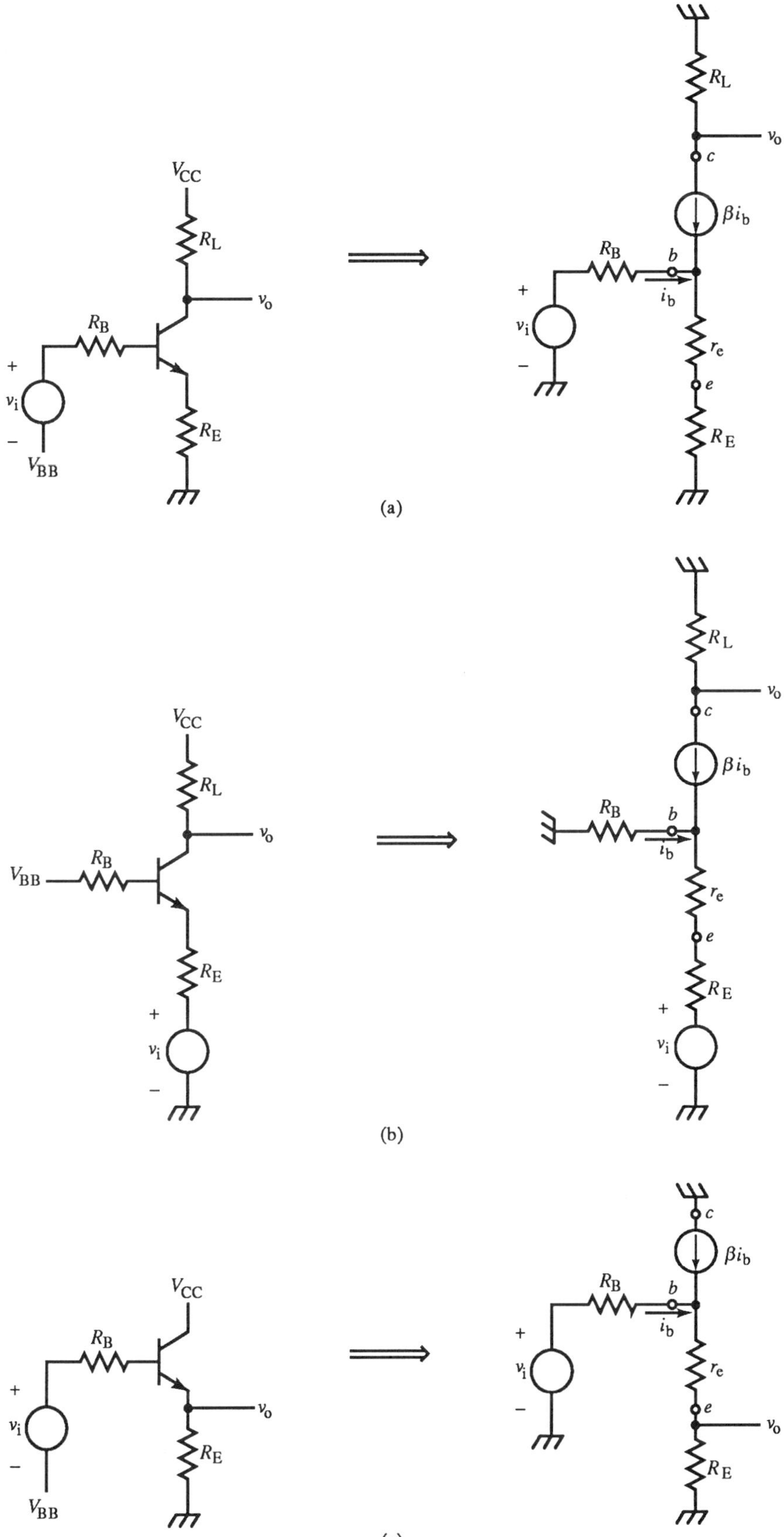

V_{CC}
R_L
v_o
R_B
v_i
V_{BB}
R_E
c
βi_b
b
i_b
r_e
e
(a)
(b)
(c)

2.3 Basic Amplifier Analysis Procedure

The quantities of common interest about an amplifier are its amplification (or gain) and its input and output resistances. For single-input, single-output circuits, several kinds of incremental gain are of interest:

$$\text{power gain} = A_p = \frac{P_{out}}{P_{in}} = \frac{v_o i_o}{v_i i_i} \tag{2.15}$$

$$\text{voltage gain} = A_v = \frac{v_o}{v_i} \tag{2.16}$$

$$\text{current gain} = A_i = \frac{i_o}{i_i} \tag{2.17}$$

$$\text{transconductance} = G_m = \frac{i_o}{v_i} \tag{2.18}$$

$$\text{transresistance} = R_m = \frac{v_o}{i_i} \tag{2.19}$$

The three transistor configurations can be analyzed using a common procedure. More complex circuits can also be analyzed by the same procedure by decomposing them into the three basic configurations.

The procedure is based on recognition of two circuit loops or nodes, one relating to the input and the other to the output. In Fig. 2.3a, the input loop consists of v_i, R_B, r_e, and R_E. The currents that flow in this loop, i_b and i_e, are caused by the input voltage source v_i. Similarly, the output loop consists of R_L, the βi_b current source, r_e, and R_E. The associated currents are i_c and i_e. Since this is a CE circuit, i_e is common to both input and output loops and is the key to relating input to output.

The procedure—the *transresistance method*—is as follows:

1. Refer all input circuit quantities to a common terminal by use of the β transform. Calculate a variable common to both input and output circuits.
2. Calculate the output in terms of the common variable and output circuit quantities.

The effect of using this procedure is to work forward mathematically from the input source to the output. Consider again the CE amplifier of Fig. 2.3a. By referring R_B to the emitter side of the circuit using the β transform, we calculate i_e as

$$i_e = \frac{v_i}{R_B/(\beta+1) + r_e + R_E} \tag{2.20}$$

The output quantity v_o is

$$v_o = -R_L \cdot i_c = -R_L \cdot \alpha \cdot i_e \tag{2.21}$$

Substituting the variable common to both input and output, i_e, into (2.21), we obtain

$$\text{CE} \qquad A_v = \frac{v_o}{v_i} = -\alpha \frac{R_L}{R_B/(\beta+1) + r_e + R_E} \tag{2.22}$$

This can be interpreted as the ratio of two resistances through which the common current i_e (adjusted by α) flows. The numerator is the resistance across which the common current develops the output voltage, and the denominator is the *transresistance*, the resistance across which the input source voltage develops the common current. For a more accurate result, the i_e-to-i_c factor α must be included and the sign of the gain deduced from the circuit topology. The essence of the method is to develop the following relationships in the order:

$$x_i \Rightarrow x_{\text{common}} \Rightarrow x_o \tag{2.23}$$

For the CE, this amounts to: $v_i \Rightarrow i_e \Rightarrow i_c \Rightarrow v_o$.

An alternative derivation based on the same approach is to refer the resistances r_e and R_E on the emitter side to the base and calculate i_b as the common variable. Then the form of A_v is

$$A_v = -\frac{\beta R_L}{R_B + (\beta+1)(r_e + R_E)} \tag{2.24}$$

If $(\beta+1)$ is factored from the denominator (thus transforming this resistance to an emitter-referred transresistance), the result is the same as (2.22), after (2.11) is used; the common variable is i_b instead.

2.4 Common-Base and Common-Collector Amplifier Analyses

The CB amplifier of Fig. 2.3b can be analyzed by first using the β transform to refer R_B to the emitter circuit. Then the emitter current generated by v_i and the transresistance is

$$i_e = -\frac{v_i}{R_B/(\beta+1) + r_e + R_E} \tag{2.25}$$

The collector current is $i_c = \alpha i_e$, and output voltage

$$v_o = -i_c \cdot R_L \tag{2.26}$$

Combining these equations gives the voltage gain:

$$\text{CB} \qquad A_v = \alpha \frac{R_L}{R_B/(\beta+1) + r_e + R_E} \tag{2.27}$$

For the CC amplifier of Fig. 2.3c, the order of variables is

$$v_i \Rightarrow i_e \Rightarrow v_o \tag{2.28}$$

The transresistance is

$$r_M = \frac{R_B}{\beta+1} + r_e + R_E \tag{2.29}$$

Since $i_e = v_i / r_M$ and $v_o = i_e \cdot R_E$, the voltage gain is

$$\text{CC} \qquad A_v = \frac{R_E}{R_B/(\beta+1) + r_e + R_E} \tag{2.30}$$

For both the CB and CC, A_v is a ratio of resistances (adjusted by α and polarity). For the CC, (2.30) has an intuitive interpretation as a voltage divider with input v_i and output v_o. The top resistance of the divider is $R_B/(\beta+1) + r_e$, and the bottom resistor is R_E.

2.5 Dynamic Input and Output Resistances

Besides voltage gain, the dynamic input and output resistances, r_{in} and r_{out}, can be found using the β transform. For the CE, r_{in} is a resistance referred to the base side of the input loop and is

$$r_{in}(\text{CE}) = \frac{v_i}{i_i} = \frac{v_i}{i_b} = R_B + (\beta+1)(r_e + R_E) \tag{2.31}$$

The base-side resistances are equivalently emitter-side resistances $(\beta+1)$ times larger. This results in a relatively high input resistance when R_E is large.

For the CB and CC circuits, using similar analysis we obtain

$$r_{in}(\text{CB}) = \frac{v_i}{i_i} = \frac{v_i}{i_e} = R_E + r_e + \frac{R_B}{(\beta+1)} \tag{2.32}$$

$$r_{in}(\text{CC}) = \frac{v_i}{i_i} = \frac{v_i}{i_b} = R_B + (\beta+1)(r_e + R_E) \tag{2.33}$$

Both CE and CC have the same r_{in} whereas $r_{in}(\text{CB})$ is $1/(\beta+1)$ times less than that of the CE and CC.

The output resistance of the CE and CB configurations are

$$r_{out}(\text{CE}) = r_{out}(\text{CB}) = R_L \tag{2.34}$$

For the CC,

$$r_{out}(\text{CC}) = R_E \parallel \left(r_e + \frac{R_B}{\beta+1}\right) \tag{2.35}$$

where $\parallel$ means "in parallel with" and indicates parallel resistances. Because the rightmost resistance is relatively small, $r_{out}(\text{CC})$ is small.

Example 2.1 CE Amplifier

Assume the BJT of Fig. E2.1 has the following parameters:

$$\beta + 1 = 100$$

$$I_S = 10^{-6}\ \text{A}$$

The saturation current I_S determines the v–i relationship of the base-emitter junction. For 2.7 mA,

$$V_{BE} \cong V_T \ln\left(\frac{I_E}{I_S}\right), \qquad I_E \gg I_S \tag{E1}$$

and $V_{BE} = 0.80$ V. This gives us a place to start for the static analysis.

First, the emitter circuit can be simplified by Thévenin's theorem. The result is a 688 Ω resistance and −3.75 V source. Next, we need to find I_E to determine r_e. The maximum I_E is the current that flows with a collector-emitter short, or

$$\max I_E = \frac{12\ \text{V} - (-3.75\ \text{V})}{1.0\ \text{k}\Omega + 688\ \Omega} \cong 9\ \text{mA}$$

We can estimate I_E by assuming $V_{BE} = 0.8$ V. Then,

$$I_E \cong \frac{3.75\ \text{V} - 0.8\ \text{V}}{688\ \Omega} = 4.3\ \text{mA}$$

Then (using E1) $V_{BE} = 0.81$ V. I_E can be recalculated using this more refined value for V_{BE}. After only two iterations, the numbers converge to

$$V_{BE} = 0.81\ \text{V}, \qquad I_E = 4.27\ \text{A}$$

Because V_{BE} is logarithmically related to I_E, it is relatively insensitive to I_E variation. (This is why convergence was rapid.)

Now, solving for r_e gives

$$r_e = \frac{26\ \text{mV}}{4.3\ \text{mA}} = 6.1\ \Omega$$

The transresistance method can now be applied (2.22):

$$\frac{v_o}{v_i} = -(0.99)\frac{1.0\ \text{k}\Omega}{6\ \Omega + 688\ \Omega} = -1.43$$

Input resistance is

$$r_{in} = (\beta + 1)[r_e + R_E] = (100)[6\ \Omega + 688\ \Omega] = 69.4\ \text{k}\Omega$$

and output resistance is $r_{out} = 1.0$ kΩ.

These results agree with those of the SPICE circuit simulation to the two significant digits of the manual calculations. The simulation uses the

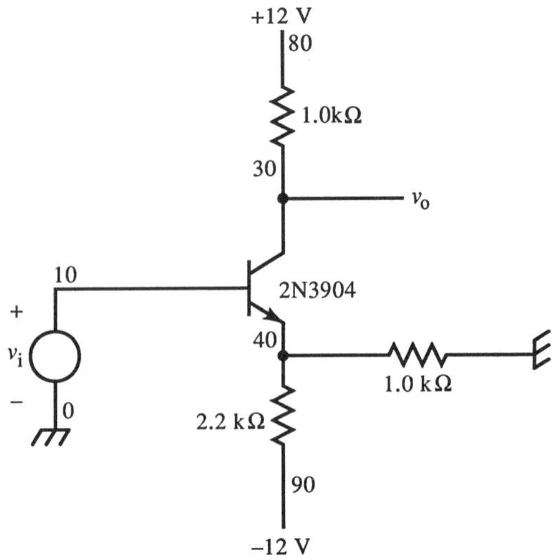


```
E2.1 CE Amplifier

.OPT NOMOD OPTS NOPAGE
.DC VI -0.25 0.25 0.05

.TF V(30) VI

VI 10 0 DC 0V
VCC 80 0 DC 12
VEE 90 0 DC -12

RE1 40 90 2.2K
RE2 40  0 1.0K
RL  80 30 1.0K

Q1 30 10 40 BJT1
.MODEL BJT1 NPN (BF=99)

.END

NODE    VOLTAGE
(30) 7.7686  (40) -.8115

VOLTAGE SOURCE CURRENTS
NAME          CURRENT

VI           -4.274E-05
VCC          -4.231E-03
VEE           5.086E-03

TOTAL POWER DISSIPATION  1.12E-01 WATTS

V(30)/VI = -1.427E+00
INPUT RESISTANCE AT VI = 6.936E+04
OUTPUT RESISTANCE AT V(30) = 1.000E+03
```

```
E2.1A CE Amplifier with 2N3904 model

.OPT NOMOD OPTS NOPAGE
.DC VI -0.25 0.25 0.05

.TF V(30) VI

VI 10 0 DC 0V
VCC 80 0 DC 12
VEE 90 0 DC -12

RE1 40 90 2.2K
RE2 40  0 1.0K
RE  80 30 1.0K

Q1 30 10 40 BJT1
.MODEL BJT1 NPN (BF=150 IS=1E-16
  VA=110 RB=15 RE=2)

.END

SMALL SIGNAL BIAS SOLUTION
TEMPERATURE = 27.000 DEG C

NODE    VOLTAGE
(30) 7.7627  (40) -.8187

VOLTAGE SOURCE CURRENTS
NAME          CURRENT

VI           -2.639E-05
VCC          -4.237E-03
VEE           5.082E-03

TOTAL POWER DISSIPATION  1.12E-01 WATTS

V(30)/VI = -1.427E+00
INPUT RESISTANCE AT VI = 1.060E+05
OUTPUT RESISTANCE AT V(30) = 9.995E+02
```

FIG. E2.1

same idealized T model of the analysis in this example. How do the results compare for a more realistic BJT model? The parameters of a 2N3904 were used and the simulation rerun. The results show good agreement except for r_{in}. A typical 2N3904 β of 150 is about 50% larger than the value of 99 used, and the discrepancy between the r_{in} values is also 50%. Therefore, the simple T model can produce accurate (typically <1% error) results.

Example 2.2 CC Amplifier

The emitter-follower of Fig. E2.2 has a voltage divider at its output. Assuming the same BJT parameters and using a dc analysis as in Example

```
E2.2 CC Amplifier

.OPT NOMOD OPTS NOPAGE
.DC VI -0.25 0.25 0.05

.TF V(50) VI

VI 10 0 DC 0V
VCC 80 0 DC 12
VEE 90 0 DC -12

RB  10 20 10K
RE1 40 50 160
RE2 50 90 2.0K
RC  80 30 1.2K

Q1 30 20 40 BJT1
.MODEL BJT1 NPN (BF=99)

.END

SMALL SIGNAL BIAS SOLUTION     TEMPERATURE = 27.000 DEG C

NODE   VOLTAGE
(20) -.4949  (30) 6.1206  (40) -1.3102  (50) -2.1020

VOLTAGE SOURCE CURRENTS
NAME          CURRENT

VI            -4.949E-05
VCC           -4.899E-03
VEE            4.949E-03

TOTAL POWER DISSIPATION  1.18E-01 WATTS

V(50)/VI = 8.829E-01
INPUT RESISTANCE AT VI = 2.265E+05
OUTPUT RESISTANCE AT V(50) = 2.342E+02
```

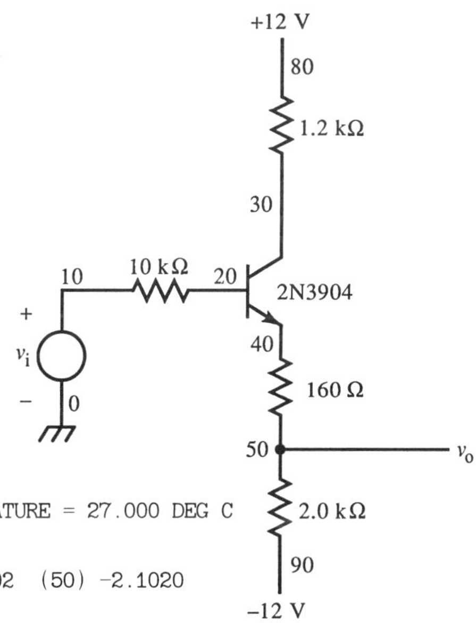

FIG. E2.2

2.1, we get $V_{BE} = 0.82$ V and $I_E = 4.95$ mA. Consequently, $V_C = 6.1$ V, but this does not affect the small-signal amplifier parameters (using the T model). Next, $r_e = 5.2\ \Omega$, and the ac parameters of interest are

$$\frac{v_o}{v_i} = \frac{2.0\ \text{k}\Omega}{2.0\ \text{k}\Omega + 160\ \Omega + 5\ \Omega + 10\ \text{k}\Omega/100} = 0.883$$

$$r_{in} = 10\ \text{k}\Omega + (100)(5\ \Omega + 160\ \Omega + 2.0\ \text{k}\Omega) = 227\ \text{k}\Omega$$

$$r_{out} = 2.0\ \text{k}\Omega \parallel \left(160\ \Omega + 5\ \Omega + \frac{10\ \text{k}\Omega}{100}\right) = 234\ \Omega$$

These compare to three digits with the SPICE results. Except for arithmetic round-off, there is no difference between these results. A more accurate calculation of the operating point is necessary to produce a more accurate value of r_e, however.

The input and output resistances of the three configurations can be summarized:

configuration	input resistance	output resistance
CE	large	medium
CB	small	medium
CC	large	small

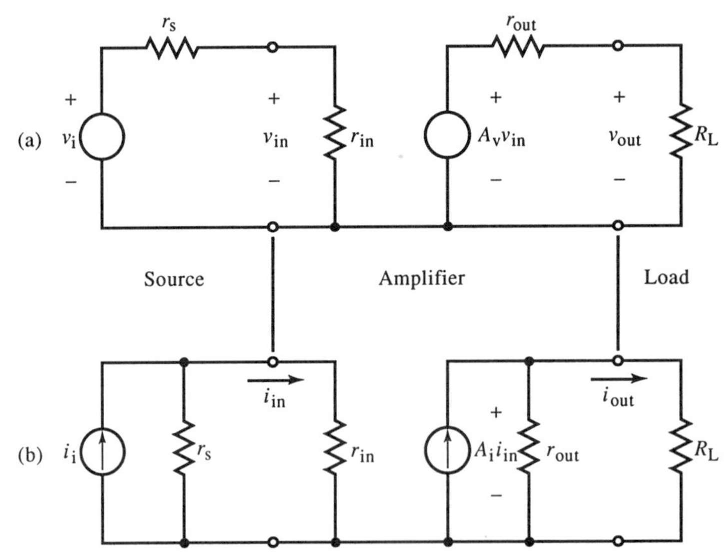

FIG. 2.4 Amplifier input and output loading for (a) voltage and (b) current amplifiers.

The large input resistances of the CE and CC cause them to appear as open circuits to the voltage sources driving them. In Fig. 2.3, the internal (Thévenin equivalent) resistances of the sources are omitted, but actual circuits have a nonzero resistance. This source resistance forms a voltage divider with the input resistance of the amplifier circuit causing attenuation of v_i (Fig. 2.4a). If the voltage source resistance r_s is variable or unknown, the attenuation of the divider and the overall voltage gain will be too. To avoid this, the ideal amplifier input resistance is infinite, so $v_{in} = v_i$ independent of r_s.

Similarly, for the voltage divider at the output, formed by the nonzero amplifier output resistance and the load resistance, $A_v \cdot v_{in} = v_{out}$ and is independent of R_L when $r_{out} = 0$. The ideal voltage amplifier therefore has infinite r_{in} and zero r_{out}. In actual amplifier circuits, the input and output dividers must be taken into account when calculating the voltage gain.

For a current amplifier, current dividers at input and output similarly affect the current gain unless $r_{in} = 0$ and $r_{out} \to \infty$, the conditions for an ideal current amplifier. Considering the four basic amplifier types, the ideal terminal resistances are as tabulated:

ideal amplifier type	input resistance	output resistance
voltage	infinite	zero
transconductance	infinite	infinite
current	zero	infinite
transresistance	zero	zero

When these ideal properties are compared with the three configurations, the following optimal matches can be made:

ideal amplifier type	optimal configuration(s)
voltage	CC, CE
transconductance	CE
current	CB, CE
transresistance	CB

This table shows that none of the configurations is ideal. Although the CC resistances approach the ideal, it has a maximum voltage gain of only one. Similarly, the CB has a good resistance match but also has a maximum current gain of one. In these cases, the CE is the best choice because it provides useful voltage and current gain. It also is optimal for transconductance amplification because its resistances match best. The CB is best for transresistance amplifiers for the same reason. Overall, the CE is the most versatile configuration and is used the most in practice. Later, we will see that when these basic configurations are combined in pairs, the resulting two-transistor configurations exceed the basic configurations in approaching the ideal.

2.6 Bipolar-Junction Transistor Output Resistance

The simple BJT T model used in the preceding sections will now be extended by considering the output resistance r_o (Fig. 2.5). In the Ebers–Moll three (EM3) model (of which the T model is a simplification), r_o is defined as

$$r_o = \left.\frac{V_A + |V_{BC}|}{|I_C|}\right|_{v_{be}=0} \tag{2.36}$$

where V_A is the *Early voltage.* This relationship is represented graphically by the collector family of curves as displayed by a curve tracer and shown idealized, in Fig. 2.5b. Since v_{be} is zero, r_e can be neglected and r_b' considered part of R_B. Then, from Fig. 2.5a,

$$v_c = R_E(i_c + i_b) + r_o(i_c - \beta i_b) \tag{2.37a}$$

$$-i_b R_B = R_E(i_c + i_b) \tag{2.37b}$$

Solving (2.37) for r_c results in

$$r_c = \frac{v_c}{i_c} = R_E \parallel R_B + r_o\left(1 + \beta \cdot \underbrace{\frac{R_E}{R_B + R_E}}_{\text{fraction of } i_c \text{ into } R_B}\right) \tag{2.38}$$

This expression for r_c can be understood in terms of the circuit of Fig. 2.5. As indicated in (2.38), the current divider formed by R_E and R_B determines the

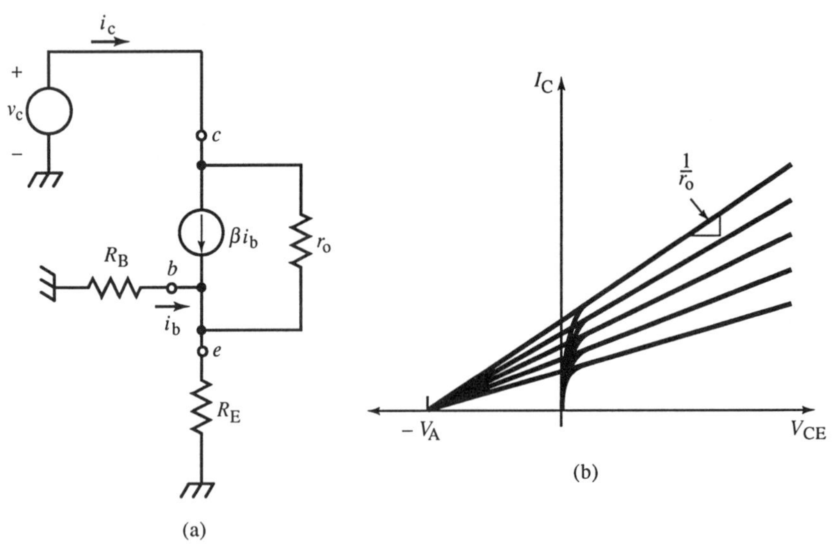

FIG. 2.5 BJT collector output resistance can be determined by circuit model (a); (b) the Early voltage V_A of the BJT is independent of I_C, and V_{CE} and can be used to calculate $r_o(I_C, V_{CE})$.

fraction of i_c that flows into R_B. When R_B is much larger than R_E (or the base terminal approaches an open circuit), then

$$r_c\big|_{R_B \to \infty} = R_E + r_o \tag{2.39}$$

and for a shorted base,

$$r_c\big|_{R_B = 0} = (\beta + 1) r_o \tag{2.40}$$

Thus, r_c increases as R_B decreases or R_E increases. Summarizing,

$$R_B \to \infty \Rightarrow r_c = r_o + R_E \tag{2.41a}$$

$$R_B = 0 \Rightarrow r_c = (\beta + 1) r_o \tag{2.41b}$$

$$R_E \to \infty \Rightarrow r_c = R_B + (\beta + 1) r_o \tag{2.41c}$$

$$R_E = 0 \Rightarrow r_c = r_o \tag{2.41d}$$

To envision r_c, begin at the b–e node where R_B and R_E are in parallel. This is accounted for in (2.38) by the first term. This parallel resistance is in series with r_o and the current source βi_b, accounted for by the second term of (2.38). If βi_b has no effect, then r_o is in series with $R_E \parallel R_B$, and the second term is only r_o. In this case, $\beta i_b = 0$ when $i_b = 0$. This occurs when R_B is infinite (2.41a).

When all of the current through r_o flows in the base ($R_B = 0$), then both βi_b and the current through r_o, i_o, flow in the collector, so

$$i_c = \beta i_b + i_o \tag{2.42}$$

With the base shorted to ground, no current flows in R_E and

$$i_c = -i_b \tag{2.43}$$

Current flowing out of the base is opposite in polarity to the indicated direction for the βi_b current source and causes βi_b to flow toward the collector terminal. Consequently, βi_b contributes to i_o and adds to i_c, flowing down through r_o and out the base terminal. The effect is that most of i_o comes from βi_b instead of being supplied as i_c. Because v_c causes a current of $v_c / r_o = i_o$, most of this current is supplied internal to the transistor as βi_b. Substituting i_b of (2.43) into (2.42) and solving for i_c gives

$$i_c = \frac{i_o}{\beta + 1} = \frac{v_c / r_o}{\beta + 1} = \frac{v_c}{(\beta + 1) r_o} \tag{2.44}$$

The collector resistance v_c / i_c is $(\beta + 1) r_o$, as (2.41b) indicates. The main insight here is

> Current through r_o that flows in the base causes r_o to appear $(\beta + 1)$ times larger from the collector

This is what (2.38) suggests: The fraction of i_c that becomes base current causes r_o to be multiplied by β. The influence of r_o on amplifier performance can be significant because the collector node can affect collector current. The collector current is no longer isolated from the input circuit that causes it.

2.7 The Effect of a Base–Emitter Shunt Resistance

An analysis similar to that for output resistance can be applied to a base-emitter shunt resistance (Fig. 2.6). The equivalent resistance for both emitter (a, b) and base (c, d) can be derived from the equivalent small-signal circuits.

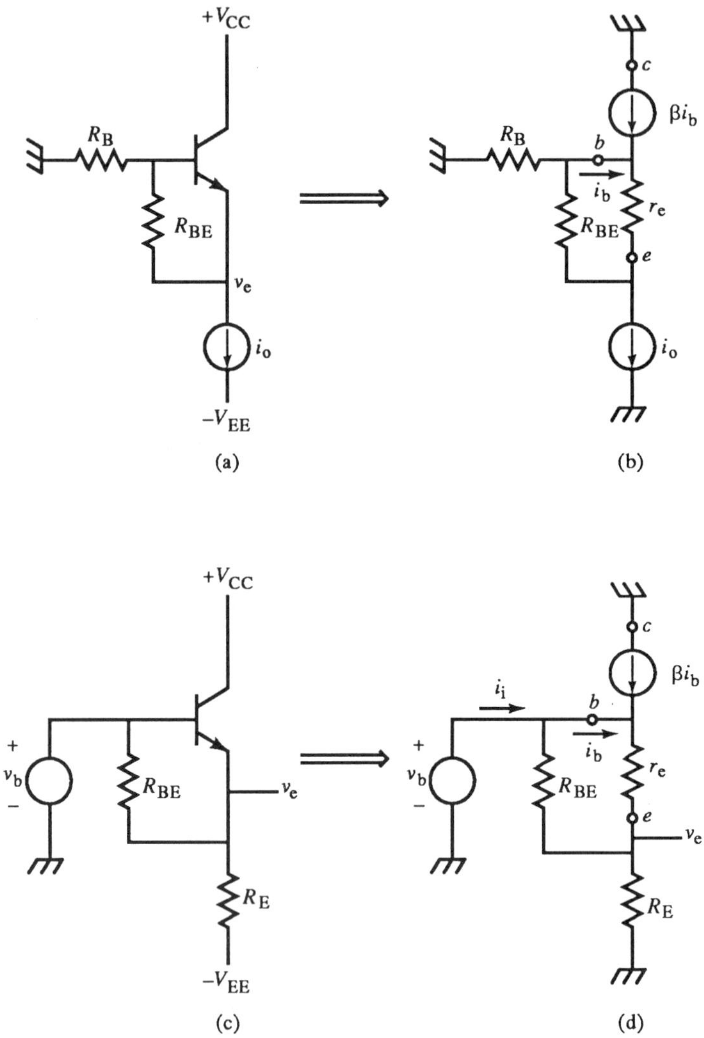

FIG. 2.6 The effect of shunt base-emitter resistance R_{BE} on resistance: (a) at emitter, modeled in (b), and (c) at base, modeled in (d).

Beginning with the emitter side first, we find the nodal equations for base and emitter to be

$$\frac{v_b}{R_B}+\frac{v_b-v_e}{R_{BE}}+\underset{\substack{\uparrow \\ i_b}}{\frac{v_b-v_e}{(\beta+1)r_e}}=0 \qquad \text{base node} \tag{2.45}$$

$$\frac{v_b-v_e}{R_{BE}}+\frac{v_b-v_e}{r_e}=i_o \qquad \text{emitter node} \tag{2.46}$$

Solving for emitter resistance from (2.46) gives

$$\frac{v_e}{i_o}=\frac{v_e}{\left(\dfrac{v_b-v_e}{R_{BE}}+\dfrac{v_b-v_e}{r_e}\right)}=\frac{v_e}{(v_b-v_e)\left[\dfrac{1}{R_{BE}}+\dfrac{1}{r_e}\right]} \tag{2.47}$$

Solving for v_b in (2.45) gives

$$v_b\left[\frac{1}{R_B}+\frac{1}{R_{BE}}+\frac{1}{(\beta+1)r_e}\right]=v_e\left[\frac{1}{R_{BE}}+\frac{1}{(\beta+1)r_e}\right] \tag{2.48a}$$

$$v_b=v_e\frac{R_B}{R_B+R_{BE}\parallel(\beta+1)r_e} \tag{2.48b}$$

Substituting for v_b in (2.47) and solving yields

$$\frac{v_e}{i_o}=R_{BE}\parallel r_e+\frac{R_B}{\beta+1}\left(\frac{R_{BE}}{r_e+R_{BE}}\right)+R_B\left(\frac{r_e}{r_e+R_{BE}}\right) \tag{2.49}$$

↑	↑
fraction of i_o through r_e causing R_B to be $R_{BE}/(\beta+1)$ at *e*	fraction of i_o through R_{BE} that also flows through R_B

This result for emitter resistance with a shunt R_{BE} can be understood in terms of the β transform. The first term is the parallel resistance of R_{BE} and r_e on the emitter side of the base-emitter loop. The drive current i_o divides between the emitter current and the shunt *R*. The fraction of i_o that is emitter current results in base current that flows through R_B and causes R_B to appear $1/(\beta+1)$ times smaller from the emitter. This fraction is determined by the current-divider factor in the second term of (2.49). The third term accounts for the fraction of i_o that flows through *R*. Since the β transform does not apply to it, this current flows through R_B without being scaled down. This result is intuitively appealing since it can be constructed by use of the β transform and inspection of the circuit.

The current gain of this circuit, also derived from the basic circuit equations, is

$$A_i = \frac{i_c}{i_o} = \frac{\beta i_b}{i_o} = \alpha\left(\frac{R_{BE}}{R_{BE}+r_e}\right) \tag{2.50}$$

Without R_{BE}, the current gain would be α, but a fraction of i_o is lost to R_{BE}. The remaining fraction that is emitter current is expressed by the current divider factor in (2.50).

The circuit of Fig. 2.6 can also be analyzed from the base side (Figs. 2.6c,d). Writing the nodal equations at base and emitter and solving for v_e results in

$$v_e = v_b \cdot \frac{R_E}{R_E + R_{BE} \parallel r_e} \tag{2.51}$$

Substituting this into the emitter nodal equation gives

$$\frac{v_b}{i_i} = R_{BE} \parallel (\beta+1)r_e + \underset{\substack{\uparrow \\ \text{due to } i_b}}{(\beta+1)R_E\left(\frac{R_{BE}}{R_{BE}+(\beta+1)r_e}\right)} + \underset{\substack{\uparrow \\ \text{due to } i \\ \text{through } R_{BE}}}{R_E\left(\frac{(\beta+1)r_e}{R_{BE}+(\beta+1)r_e}\right)} \tag{2.52}$$

This result is similar to (2.49) in form, as might be expected. The first term is the parallel combination of R_{BE} and r_e from the base side. The second term is due to the β transform effect of the base current according to the fraction of input current i_i that flows in the base. The shunt-current fraction of i_i that flows through R contributes the third term, where R_E appears unscaled by $\beta+1$.

For the circuit of Fig. 2.6d, the transconductance is

$$\frac{i_c}{v_b} = \frac{\beta i_b}{v_b} = \alpha \underset{\substack{\uparrow \\ \text{fraction of} \\ \text{current in} \\ R_E \text{ that} \\ \text{is } i_e}}{\left(\frac{R_{BE}}{R_{BE}+r_e}\right)} \underset{\substack{\uparrow \\ v_b \text{ times} \\ \text{this is the} \\ \text{current} \\ \text{in } R_E}}{\left(\frac{1}{R_E + R_{BE} \parallel r_e}\right)} \tag{2.53}$$

Equations (2.49, 2.50) and (2.50, 2.51) can be better understood by considering the extremes of R_{BE}. For $R_{BE}=0$, no current flows through r_e and no $\beta+1$ scaling occurs. In this case, a passive resistor network results, and the second term in (2.49) and (2.52) is zero. When $R_{BE}\to\infty$, the analysis using the β transform applies completely, and the third terms of (2.49) and (2.52) are eliminated.

The circuit of Fig. 2.6c has the useful property that the input resistance is higher than if R_{BE} were returned to ground instead of the emitter. R_{BE} is *bootstrapped* since its bottom terminal voltage follows the top terminal. This causes v_R across it to be less than the v_b that would be across a grounded R_{BE}, and R_{BE} appears to be much larger than its actual value. To show this, consider first that $R_{BE} \parallel (\beta+1)r_e$ (not R_{BE} alone) is the bootstrapped resistance. The equivalence factor for this resistance can be found by expressing (2.52) as

$$\text{BJT} \qquad r_{in} = [R_{BE} \parallel (\beta+1)r_e]\left[1+\alpha \cdot \frac{R_E}{r_e}\right] + R_E \tag{2.54}$$

The interpretation of this expression is that $R_{BE} \parallel (\beta+1)r_e$ has an equivalent value $(1+\alpha R_E/r_e)$ larger than its actual value.

This result or that of (2.52) can be modified for a FET from that of a BJT by allowing $\beta \to \infty$ (or $\alpha = 1$) since FET $i_g = 0$ would be equivalent to setting i_b to zero. From (2.52),

$$\frac{v_g}{i_i} = R_{BE} + R_S \cdot \frac{R_{BE}}{r_m} + R_S \tag{2.55}$$

Here it is apparent that for infinite β, $r_e \to r_m$ and R_E replaces R_S. This result can be given physical meaning by factoring R_{BE} from the first two terms on the right. Then the equivalent R_{BE} due to bootstrapping is

$$\text{FET} \qquad R_{equiv} = R_{BE} \cdot \left[1+\frac{R_S}{r_m}\right] \tag{2.56}$$

The equivalence factor is the reciprocal of the divider formed by r_m and R_S and is identical in form to the second factor of (2.54) for BJTs. The difference is that for a FET, only R_{BE} is bootstrapped (since $(\beta+1)r_e \to \infty$). For a FET with $R_{BE} = 100\ \text{k}\Omega$, $r_m = 100\ \Omega$, and $R_S = 1\ \text{k}\Omega$; then $R_{equiv} \cong 1\ \text{M}\Omega$.

2.8 The Cascade Amplifier

Now that we have examined several aspects of single-transistor amplifier circuits, we will investigate two- and three-transistor combinations. The three basic configurations can be combined in various ways to produce circuits with useful properties. These properties are not found in either of the combined configurations alone, just as molecules have properties that their constituent atoms do not exhibit. Consequently, these circuits can be considered basic building blocks in themselves.

The most common combination of multiple-transistor amplifiers is the *cascade amplifier*, which consists of two CE amplifiers, the output of one driving the input of the other (Fig. 2.7). Each of these CE amplifiers is called a *stage* of amplification; any unit of a sequence of consecutive amplifiers is a stage. This amplifier can be analyzed by the transresistance method. The only additional complication results from the interconnection of the two stages. If the base of the second stage is disconnected from the collector of the first

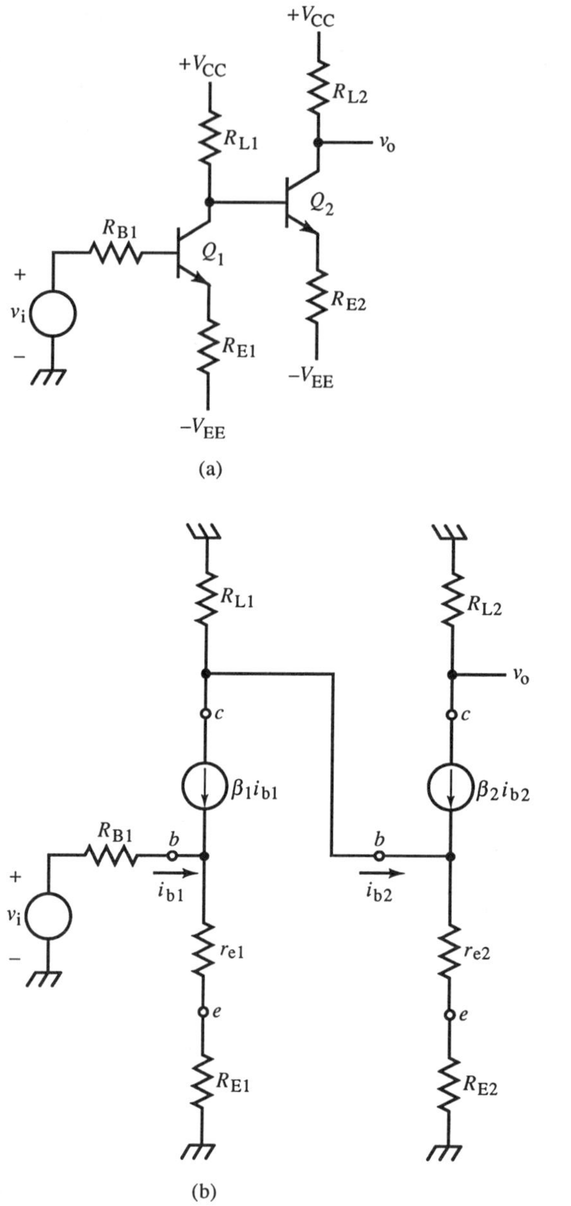

FIG. 2.7 The cascade amplifier (a) and its small-signal model (b).

stage, the gain of the first stage is

$$A_{v1} = -\alpha_1 \frac{R_{L1}}{R_{B1}/(\beta_1 + 1) + r_{e1} + R_{E1}} \tag{2.57}$$

The output of the first stage can be represented as a Thévenin equivalent circuit and connected to the input of the second stage. The voltage source is $A_{v1} \cdot v_i$

with a Thévenin resistance of R_{L1}. The calculation of the second stage gain is similar to that of the first stage, resulting in a total gain of

$$
\begin{aligned}
A_v &= A_{v1} \cdot A_{v2} \\
&= -\alpha_1 \frac{R_{L1}}{R_{B1}/(\beta_1+1)+r_{e1}+R_{E1}} \cdot -\alpha_2 \frac{R_{L2}}{R_{L1}/(\beta_2+1)+r_{e2}+R_{E2}}
\end{aligned} \tag{2.58}
$$

An alternative view of the interaction of the stages is to consider the collector of the first stage to be loaded by the input resistance of the second stage so that the collector load resistance is

$$R_{C1} = R_{L1} \parallel (\beta_2+1)(r_{e2}+R_{E2}) \tag{2.59}$$

Then the gain formula for the second stage does not include base resistance since it is already taken into account in the first-stage collector resistance. With R_{C1} in the first-stage gain formula, the output is at the base terminal of the second stage, not at the Thévenin equivalent voltage source as in the first approach. The second approach is explicit in the gain formula when it is expressed as

$$A_v = -\alpha_1 \frac{R_{L1} \parallel (\beta_2+1)(r_{e2}+R_{E2})}{R_{B1}/(\beta_1+1)+r_{e1}+R_{E2}} \cdot -\alpha_2 \frac{R_{L2}}{r_{e2}+R_{E2}} \tag{2.60}$$

Equations (2.58) and (2.60) are equivalent but have different algebraic forms. Two alternative views of stage interaction result. Often, the most difficult step in gaining insight into a new circuit is in expressing equations resulting from standard circuit analysis in a form that makes clear an equivalent circuit topology that leads to a simple explanation of the circuit.

The cascade amplifier stage interaction phenomenon occurs often and can be dealt with in general form by analyzing the cascade attenuator shown in Fig. 2.8. The Thévenin equivalent circuit and loaded-divider approaches, as used above, achieve the same result. For the Thévenin circuit approach,

$$\frac{v_o}{v_i} = \left(\frac{R_2}{R_1+R_2}\right)\left(\frac{R_4}{R_1 \parallel R_2+R_3+R_4}\right) \tag{2.61}$$

and for the loaded-divider approach,

$$\frac{v_o}{v_i} = \frac{R_2 \parallel (R_3+R_4)}{R_2 \parallel (R_3+R_4)+R_1}\left(\frac{R_3}{R_3+R_4}\right) \tag{2.62}$$

As in the case of the cascade amplifier, (2.61) and (2.62) are equivalent.

2.9 The Cascode Amplifier

A *cascode amplifier* is a CE stage followed by a CB stage (Fig. 2.9). Because the output is CB, r_{out} is high, and r_{in} is also high due to a CE input stage. This results in a good transconductance amplifier with higher output resistance

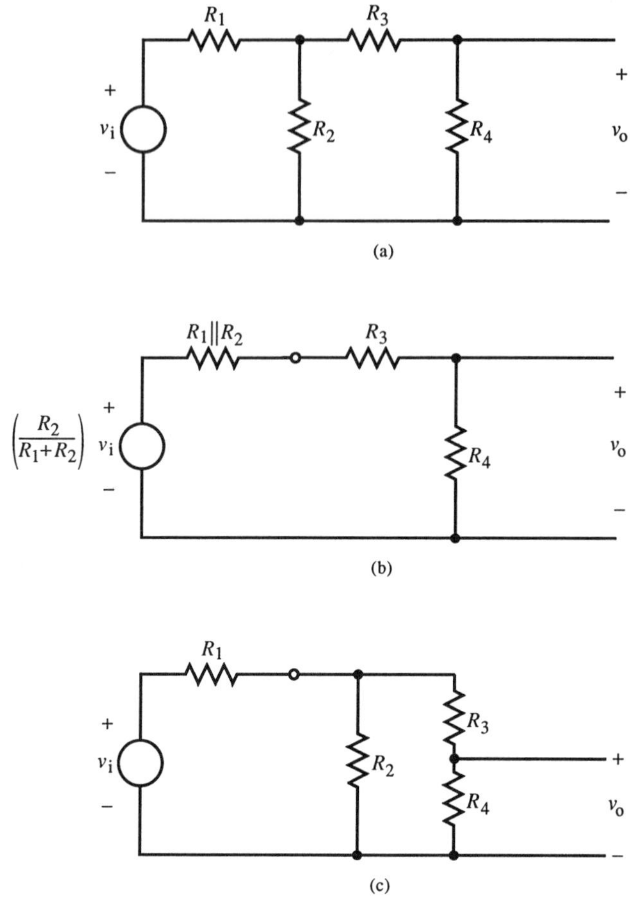

FIG. 2.8 The loaded divider (a) can be solved by (b) use of Thévenin's theorem or (c) the loaded divider method.

than a CE amplifier alone. When the cascode amplifier is analyzed as we have done with previous circuits, the voltage gain is

$$A_v = G_m \cdot R_L = -\alpha_1 \cdot \alpha_2 \frac{R_L}{R_B/(\beta+1) + r_{e1} + R_E} \tag{2.63}$$

The $\alpha_1\alpha_2$ factor is due to loss of base current in both transistors; otherwise, the analysis holds no surprises. When the CB transistor r_o is taken into account, the numerator of (2.63) is modified so that

$$A_v = -\alpha_1\alpha_2 \cdot \frac{R_L \parallel [(\beta_2+1)r_{o2}]}{R_B/(\beta_1+1) + r_{e1} + R_E} \tag{2.64}$$

Because of the CB output stage, not only is the output resistance higher but so is the voltage gain. The input resistance is that of a CE amplifier.

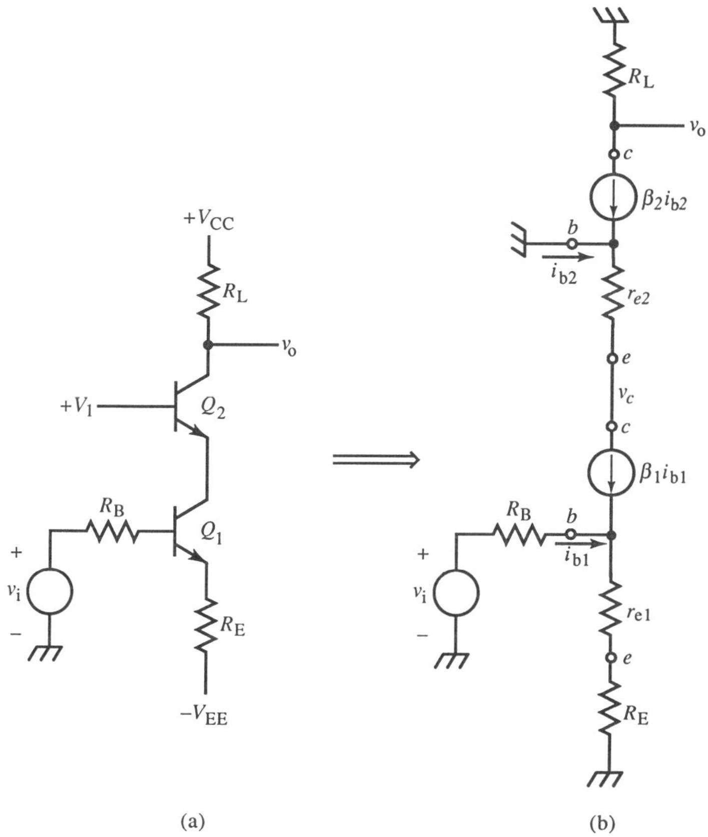

FIG. 2.9 The cascode amplifier (a) and its small-signal model (b).

Example 2.3 Cascode Amplifier

A typical cascode amplifier is shown in Fig. E2.3. Let us use a conservative β of 99. The dc calculation of Q_1 emitter current yields 600 μA. The emitter current of Q_2 is the collector current of Q_1, or 594 μA. Then the dynamic emitter resistances are

$$r_{e1} = 43.3\ \Omega, \qquad r_{e2} = 43.8\ \Omega$$

Applying (2.63), we obtain a voltage gain of $A_v = -0.26$. This gain compares well with the SPICE results. The output resistance (assuming infinite r_{o2}) is $R_L = 1.0\ \text{k}\Omega$. The input resistance is

$$r_{in} = R_B + (\beta_1 + 1)(r_{e1} + R_E) = 374.3\ \text{k}\Omega$$

```
E2.3 Cascode Amplifier
.OPT NOMOD OPTS NOPAGE
.DC VI -0.25 0.25 0.05
.OP
.TF V(60) VI
VI 10 0 DC 0V
VCC 80 0 DC 12
VEE 90 0 DC -3
RI  10 20 10K
RE  30 90 3.6K
RB1 80 50 1.0K
RB2 50  0 3.3K
RL  80 60 1.0K
Q1 40 20 30 BJT1
Q2 60 50 40 BJT1
.MODEL BJT1 NPN (BF=99)
.END
```

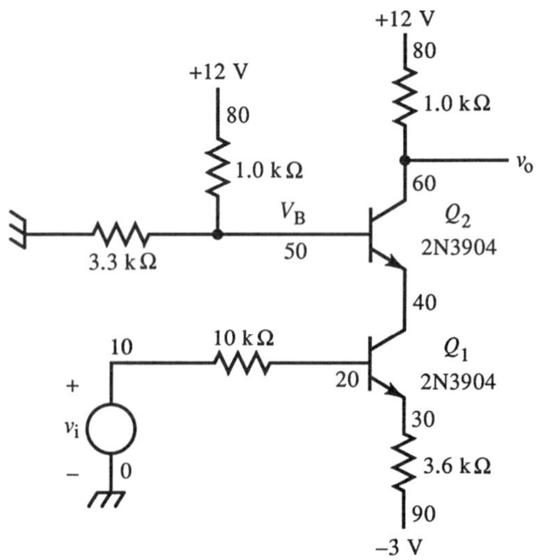


```
SMALL SIGNAL BIAS SOLUTION      TEMPERATURE = 27.000 DEG C

NODE    VOLTAGE
(20) -.0605   (30) -.8215   (40) 8.4440   (50) 9.2047
(60) 11.4070

BIPOLAR JUNCTION TRANSISTORS

NAME          Q1           Q2
MODEL         BJT1         BJT1
IB            6.05E-06     5.99E-06
IC            5.99E-04     5.93E-04
VBE           7.61E-01     7.61E-01
VBC          -8.50E+00    -2.20E+00
VCE           9.27E+00     2.96E+00
BETADC        9.90E+01     9.90E+01
GM            2.32E-02     2.29E-02

V(60)/VI = -2.619E-01
INPUT RESISTANCE AT VI = 3.743E+05
OUTPUT RESISTANCE AT V(60) = 1.000E+03
```

FIG. E2.3

The SPICE output agrees. This particular example of a cascode amplifier does not have a useful gain (>1) but can function as a dc voltage shifter. If we reduce R_E, the gain (magnitude) is increased, but the dc emitter current also increases and input resistance decreases. The difficulty here is partly due to the values of the available power supplies. For a smaller R_E, $-V_{EE}$ must also be made smaller for the same bias current. But a decreasing V_{EE} makes the bias current more sensitive to V_{BE1}. To achieve both a stable operating point and higher gain, use a large V_{EE} for stable bias current and construct a Thévenin equivalent source by placing another resistor from the emitter to ground. The two emitter resistors give the freedom needed to choose both a Thévenin equivalent supply voltage and an emitter resistance.

Example 2.4 Complementary Cascode Amplifier

A variation on the cascode amplifier of Fig. E2.3 is the *complementary cascode* shown in Fig. E2.4. The output transistor Q_2 is of opposite polarity

```
E2.4 Complementary Cascode Amplifier
.OPT NOMOD OPTS NOPAGE
.DC VI -0.25 0.25 0.05
.OP
.TF V(50) VI
VI 10 0 DC 0V
VCC 80 0 DC 12
VEE 90 0 DC -12
RE1 20 90 3.74K
RB1 80 40 1.00K
RB2 40  0 1.00K
RO  30 80 866
RL  50 90 4.02K
Q1 30 10 20 BJT1
Q2 50 40 30 BJT2
.MODEL BJT1 NPN (BF=99)
.MODEL BJT2 PNP (BF=99)
.END
```

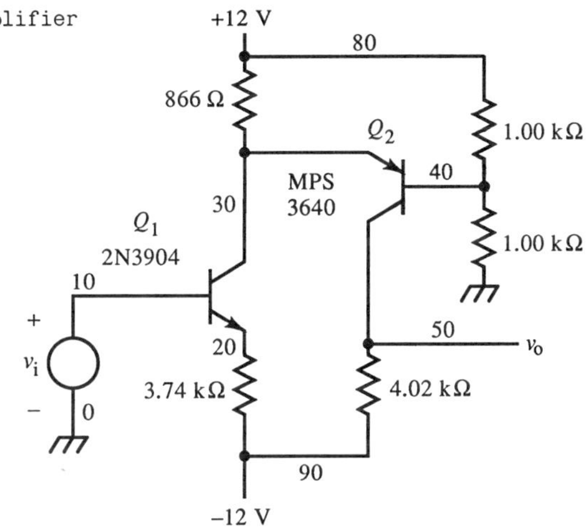


```
SMALL SIGNAL BIAS SOLUTION    TEMPERATURE = 27.000 DEG C

NODE   VOLTAGE
(20) -.8023  (30) 6.8176  (40) 6.0151  (50) .0196

BIPOLAR JUNCTION TRANSISTORS

NAME         Q1          Q2
MODEL        BJT1        BJT2
IB           2.99E-05   -3.02E-05
IC           2.96E-03   -2.99E-03
VBE          8.02E-01   -8.03E-01
VBC         -6.82E+00    6.00E+00
VCE          7.62E+00   -6.80E+00
BETADC       9.90E+01    9.90E+01
GM           1.15E-01    1.16E-01

V(50)/VI = -1.035E+00
INPUT RESISTANCE AT VI = 3.749E+05
OUTPUT RESISTANCE AT V(50) = 4.020E+03
```

FIG. E2.4

to Q_1. The gain formula of (2.59) does not exactly apply since some current is lost in the 866 Ω biasing resistor that shunts the emitter of Q_2. A dc solution for I_{E1} is 3.0 mA. Then, $r_{e1} = 8.7\ \Omega$. For Q_2, node 40 is driven by a Thévenin equivalent circuit of 6 V and 500 Ω. Solving for the dc solution of the base-emitter circuit of Q_2 is simplified by referring the 500 Ω equivalent base resistance to the emitter (as 5 Ω) and offsetting the 12 V supply by $(866\ \Omega)(I_{C1}) = 2.60$ V. Then the familiar diode-resistance circuit can be iteratively solved. This produces $I_{E2} = 3.05$ mA and $r_{o2} = 8.5\ \Omega$. Knowing the dynamic emitter resistances, we can find the voltage gain. For Q_1, the transresistance is 3.75 kΩ. While we are at it,

$$r_{in} = (\beta + 1)(r_{e1} + R_E) = 374.9\ \text{k}\Omega$$

The collector current of Q_1 is $\alpha_1(v_i/3.75\ \text{k}\Omega)$. For $\beta = 99$, $\alpha = 0.99$. Next, the 866 Ω resistor forms a current divider with the emitter circuit of Q_2, and

$$i_{e2} = \left(\frac{866\ \Omega}{866\ \Omega + r_{e2} + 500\ \Omega/(\beta_2 + 1)}\right)\frac{0.99}{3.75\ \text{k}\Omega} \cdot v_i = \frac{0.975}{3.75\ \text{k}\Omega} \cdot v_i$$

Then $-i_{c2} \cdot R_L = v_o$. Combining this with $\alpha_2 = 0.99$ and i_{e2} gives the gain:

$$A_v = -1.03$$

and this agrees with the SPICE gain. The dc output voltage is, according to SPICE, about 20 mV, or about zero volts, the same as the input. This amplifier is an inverting, nonoffsetting ×(−1) buffer. Its output resistance of R_L is high for a voltage output amplifier. The addition of a CC stage and dc modification (to correct for V_{BE} of the CC) would result in a more acceptable inverting buffer amplifier.

2.10 The Darlington (or Compound) Amplifier

Another two-transistor amplifier with useful properties is the *Darlington* or *compound amplifier* (sometimes called the Darlington configuration), shown in Fig. 2.10. It consists of a CE stage emitter driving a CE stage, except that the collectors are connected. This is a three-terminal amplifier that resembles a single transistor with improved properties.

This circuit presents a challenge in using the transresistance method. Notice that it has a single-input loop containing the transresistance but generates two output currents in parallel (from the two collectors). This results in a two-term expression for the voltage gain, one term per transistor. For the input transistor,

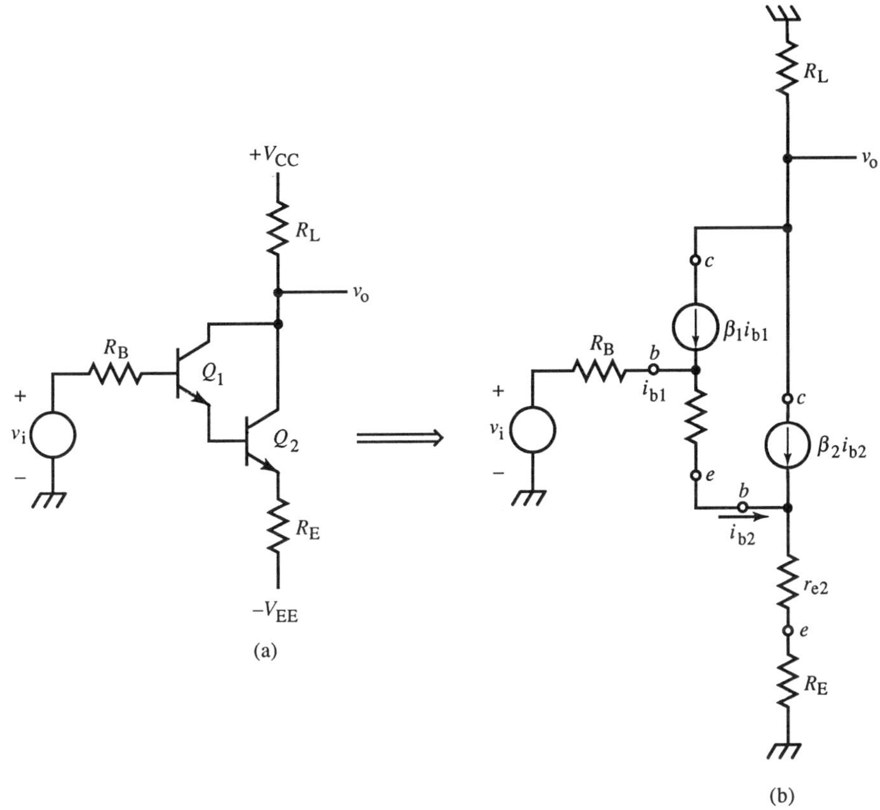

FIG. 2.10 The Darlington configuration (a) and small-signal model (b).

the voltage gain is

$$A_{v1} = -\alpha_1 \frac{R_L}{(\beta_2+1)(r_{e2}+R_E)+r_{e1}+R_B/(\beta_1+1)} \tag{2.65}$$

The first term of the denominator is the external emitter resistance of the input transistor and is the input resistance of the driven BJT. The second term in the voltage gain is due to the driven transistor:

$$A_{v2} = -\alpha_2 \frac{R_L}{R_E+r_{e2}+r_{e1}/(\beta_2+1)+R_B/(\beta_2+1)(\beta_1+1)} \tag{2.66}$$

For the driven BJT, the emitter resistance of the input transistor is its base resistance and is divided by (β_2+1). The sum of A_{v1} and A_{v2} is the Darlington voltage gain:

$$A_v \cong -\frac{R_L}{R_E+r_{e2}+r_{e1}/(\beta_2+1)+R_B/(\beta_1+1)(\beta_2+1)} \qquad \beta \gg 1 \tag{2.67}$$

This gain is slightly less that that of a CE amplifier. Its advantage is its input resistance

$$r_{\text{in}} = R_{\text{B}} + (\beta_1+1)[r_{\text{e1}} + (\beta_2+1)(r_{\text{e2}} + R_{\text{E}})]$$
$$\cong (\beta+1)^2[r_{\text{e2}} + R_{\text{E}}], \qquad \beta_1 = \beta_2 = \beta \tag{2.68}$$

which is larger than that of the CE by a factor of $(\beta+1)$. A Darlington amplifier makes a good input stage for voltage and transconductance amplifiers because of its high input resistance. When transistor r_o is taken into account, output resistance involves two parallel collector resistances with relatively low r_c, especially for the driven transistor.

Example 2.5 Darlington Amplifier

Figure E2.5 shows a Darlington amplifier with a shunt emitter resistor on Q_2 terminating at a common bootstrapping resistor of 10 kΩ at node 50. The dc analysis follows a development similar to the ac analysis. Therefore, the dc solution will be taken from the SPICE model and only the ac solution worked out. The emitter resistances of the BJTs are

$$r_{\text{e1}} = 156\ \Omega, \qquad r_{\text{e2}} = 31.9\ \Omega$$

Because the input of Q_2 is bootstrapped, the input resistance to Q_2 (from the base of Q_2, including the 10 kΩ resistor between nodes 20 and 50) can be found by using (2.52). Substituting values gives

$$\begin{aligned} r_{\text{in2}} &= 10\ \text{k}\Omega \parallel (100)(31.9\ \Omega + 1.0\ \text{k}\Omega) \\ &\quad + (100)(10\ \text{k}\Omega)\left[\frac{10\ \text{k}\Omega}{10\ \text{k}\Omega + (100)(31.9\ \Omega + 1.0\ \text{k}\Omega)}\right] \\ &\quad + 10\ \text{k}\Omega\left[\frac{(100)(31.9\ \Omega + 1.0\ \text{k}\Omega)}{10\ \text{k}\Omega + (100)(31.9\ \Omega + 1.0\ \text{k}\Omega)}\right] \\ &= 9116.53\ \Omega + 88347\ \Omega + 9116.43\ \Omega = 106580\ \Omega \end{aligned}$$

Now (2.65) is used to solve for the gain due to Q_1:

$$A_{\text{v1}} = -\alpha_1 \cdot \frac{R_{\text{L}}}{r_{\text{e1}} + r_{\text{in2}}} = -0.0473$$

The gain of Q_2 is found as follows:

$$A_{\text{v2}} = \underset{\substack{\uparrow \\ (2.53)}}{\frac{v_{\text{b2}}}{v_{\text{i}}}} \cdot \frac{i_{\text{c2}}}{v_{\text{b2}}} \cdot (-R_{\text{L}}) = \left(\frac{r_{\text{in2}}}{r_{\text{e1}} + r_{\text{in2}}}\right)\frac{i_{\text{c2}}}{v_{\text{b2}}} \cdot (-R_{\text{L}})$$

Again, we use another bootstrapping equation, (2.53), to find the trans-

```
E2.5 Darlington Amplifier
.OPT NOMOD OPTS NOPAGE
.DC VI -0.25 0.25 0.05
.OP
.TF V(30) VI
VI 10 0 DC 0V
VCC 80 0 DC 12
VEE 90 0 DC -12
RE1 20 50 10K
RE2 40 50 1.0K
RE3 50 90 10K
RL  80 30 5.1K
Q1 30 10 20 BJT1
Q2 30 20 40 BJT1
.MODEL BJT1 NPN (BF=99)
.END
SMALL SIGNAL BIAS SOLUTION     TEMPERATURE = 27.000 DEG C
NODE    VOLTAGE
(20) -.7275  (30) 7.0652  (40) -1.4961  (50) -2.3073
BIPOLAR JUNCTION TRANSISTORS
NAME            Q1            Q2
MODEL           BJT1          BJT1
IB              1.66E-06      8.11E-06
IC              1.64E-04      8.03E-04
VBE             7.28E-01      7.69E-01
VBC            -7.07E+00     -7.79E+00
VCE             7.79E+00      8.56E+00
BETADC          9.90E+01      9.90E+01
GM              6.35E-03      3.11E-02
V(30)/VI = -4.652E-01
INPUT RESISTANCE AT VI = 1.067E+07
OUTPUT RESISTANCE AT V(30) = 5.100E+03
```

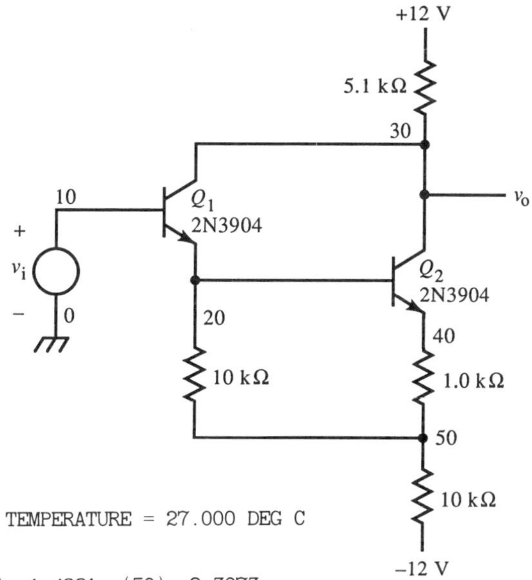

FIG. E2.5

conductance of the Q_2 stage. Substituting values into this equation, we get

$$A_{v2} = -(0.09985)(0.4185) = -0.4179$$

Thus the total gain is

$$A_v = A_{v1} + A_{v2} = -0.4652$$

The SPICE result agrees. Finally, the input resistance is

$$r_{in} = (\beta_1 + 1)(r_{e1} + r_{in2}) = (100)(156\ \Omega + 106580\ \Omega) = 10.67\ \text{M}\Omega$$

With a gain of less than one, this amplifier is not very useful. The gain can be easily increased (see Example 2.3). The input resistance, however, is very useful for a voltage-input amplifier.

2.11 The Differential (Emitter-Coupled) Amplifier

The previously considered amplifiers were *single-ended*: The input and output share a common (ground) node. Amplifiers with port voltages for which neither terminal is grounded are *differential amplifiers*. Usually, an amplifier with differential input (and differential or single-ended output) is called a differential amplifier (or *diff-amp* for short) and has an output of

$$v_o = A_v \cdot (v_{i2} - v_{i1}) \tag{2.69}$$

for a voltage amplifier. Some amplifiers have single-ended inputs and differential outputs. A differential output voltage is

$$\text{differential} \quad v_o = v_{o2} - v_{o1}$$

A differential amplifier can be built from two CE amplifiers that share R_o (Fig. 2.11). Because their emitters are coupled, they are sometimes called emitter-coupled amplifiers. To achieve true differential amplification, the circuit must be symmetric so that the gains of each input to the output are the same. The output voltage for a general two-input voltage amplifier is

$$v_o = A_{v2} v_{i2} - A_{v1} v_{i1} \tag{2.70}$$

The condition for a differential amplifier is that $A_{v2} = A_{v1}$. The voltage gain of the diff-amp of Fig. 2.11 can be found by the transresistance method and superposition. Because of its symmetric topology, we need only find A_{v1} and rewrite it for A_{v2} since it will have the same form:

$$A_{v1} = A_{v1+} - A_{v1-} = \frac{v_{o2}}{v_{i1}} - \frac{v_{o1}}{v_{i1}} \tag{2.71}$$

Beginning with A_{v1}, we have

$$A_{v1-} = -\alpha_1 \frac{R_{L1}}{R_{B1}/(\beta_1+1) + r_{e1} + R_{E1} + R_o \parallel [R_{E2} + r_{e2} + (R_{B2}/(\beta_2+1))]} \tag{2.72}$$

A_{v1+} is somewhat more complicated in that it involves the input transistor, operating as a CC, driving the output transistor as a CB with a cascaded attenuator in the emitter circuit. With the loaded-divider approach, the gain can be factored into two gains, using center node with voltage v_e as a splitting point:

$$A_{v1+} = \frac{v_{o2}}{v_{i1}} = \frac{v_e}{v_{i1}} \cdot \frac{v_{o2}}{v_e} \tag{2.73}$$

Defining

$$R_1 = \frac{R_{B1}}{\beta_1 + 1} + r_{e1} + R_{E1} \tag{2.74}$$

$$R_2 = \frac{R_{B2}}{\beta_2 + 1} + r_{e2} + R_{E2} \tag{2.75}$$

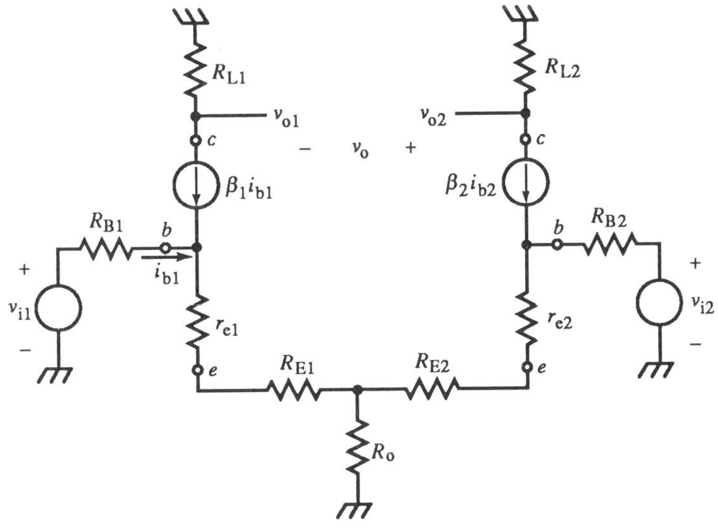

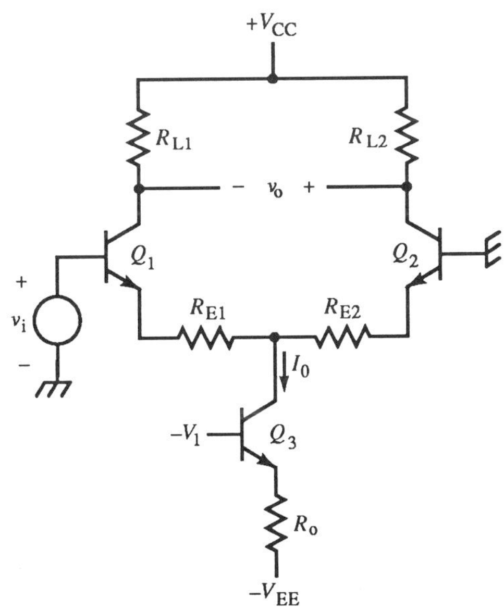

FIG. 2.11 The differential or emitter-coupled amplifier, modeled in (a). If we replace R_o with a CB current source, implemented in (b), the gain from either input to both outputs is equal.

then,

$$A_{v1+} = \frac{R_o \parallel R_2}{R_o \parallel R_2 + R_1}\left(\alpha_2 \cdot \frac{R_{L2}}{R_2}\right) \tag{2.76}$$

The first factor of (2.76) is the loaded divider; multiplied by v_{i1}, it produces v_e. The remaining factor is the voltage gain of the output transistor. Because calculation of v_e took R_2 into account, it is the input voltage to R_2 when calculating gain.

The Thévenin circuit approach breaks R_{E2} at R_o and solves for the gain from v_{i1} to v_e. Because the loading of R_2 is neglected, a Thévenin equivalent circuit must then drive R_2. The alternative expression for A_{v1+} is

$$A_{v1+} = \left(\frac{R_o}{R_1 + R_o}\right) \cdot \left(\alpha_2 \frac{R_{L2}}{R_1 \parallel R_o + R_2}\right) \tag{2.77}$$

The Thévenin resistance appears in the transresistance in the second factor as $R_1 \parallel R_o$. Equations (2.76) and (2.77) are equivalent. The total gain A_{v1}, according to (2.71), is

$$\begin{aligned} A_{v1} &= \alpha_2 \frac{R_o \parallel R_2}{R_o \parallel R_2 + R_1}\left(\frac{R_{L2}}{R_2}\right) + \alpha_1 \frac{R_{L1}}{R_1 + R_o \parallel R_2} \\ &= \alpha_2 \cdot \frac{R_{L2}}{R_o \parallel R_2 + R_1}\left(\frac{R_o}{R_o + R_2}\right) + \alpha_1 \cdot \frac{R_{L1}}{R_o \parallel R_2 + R_1} \end{aligned} \tag{2.78}$$

So far, we have calculated the gain for a single-input diff-amp in that v_{i2} is shorted (by superposition). Before we complete the derivation of the total differential gain, notice that with $v_{i2} = 0$, this is a single-ended input, differential output amplifier. This circuit is common and useful; vertical input amplifiers of oscilloscopes use this as an input stage from the probe. To produce a balanced differential output ($v_{o1} = -v_{o2}$), the magnitudes of the gains to both outputs must be equal. The conditions required (that the terms of (2.78) be equal) are

$$\alpha_1 = \alpha_2 \tag{2.79a}$$

$$R_{L1} = R_{L2} \tag{2.79b}$$

$$\frac{R_o}{R_o + R_2} = 1 \tag{2.79c}$$

Condition (2.79a) requires matched transistors. For high-β transistors, this condition is not critical and is easily met. Condition (2.79c) can be satisfied either by letting $R_2 = 0$ or $R_o \to \infty$. Since the first alternative is not physically realizable (because $r_{e2} > 0$), a finite R_o causes an imbalance. R_o is often replaced by a current source, satisfying the condition. In practice, this can be the collector of another transistor generating the current I_0 (Fig. 2.11b).

Returning to the full diff-amp gain derivation, by symmetry of the circuit topology, the gains to both outputs due to v_{i2} (with v_{i1} shorted) have the same form but with corresponding components from the other side of the circuit:

$$A_{v2} = \alpha_1 \cdot \frac{R_{L1}}{R_o \parallel R_1 + R_2}\left(\frac{R_o}{R_o + R_1}\right) + \alpha_2 \cdot \frac{R_{L2}}{R_o \parallel R_1 + R_2} \tag{2.80}$$

The total gain is

$$A_v = \frac{v_o}{v_i} = \frac{v_{o2} - v_{o1}}{v_{i2} - v_{i1}} = \frac{A_{v2}v_{i2} - A_{v1}v_{i1}}{v_{i2} - v_{i1}} \tag{2.81}$$

The condition for differential amplification is that

$$A_{v2} = A_{v1} \tag{2.82}$$

This condition can be met in two ways:

$$A_{v1+} = -A_{v2-}, \qquad A_{v2+} = -A_{v1-} \qquad \text{(antisymmetric)} \tag{2.83}$$

$$A_{v1+} = A_{v2+}, \qquad A_{v1-} = A_{v2-} \qquad \text{(symmetric)} \tag{2.84}$$

The first approach leads to the circuit component conditions:

$$R_1 = R_2 = R \tag{2.85a}$$

$$R_o \to \infty \tag{2.85b}$$

and the second to the conditions

$$\alpha_1 = \alpha_2 = \alpha \tag{2.86a}$$

$$R_{L1} = R_{L2} = R_L \tag{2.86b}$$

$$R_1 = R_2 = R \tag{2.86c}$$

The symmetric conditions (2.84) require that the gains from the two inputs be the same to their corresponding inverting and noninverting outputs, as shown graphically in Fig. 2.12a. The antisymmetric conditions (2.83) are illustrated in Fig. 2.12b). Here, the gains from the two inputs to a given output must be equal. Neither approach to differential amplification necessarily satisfies the conditions for balance (2.79).

Either way, the differential gain is

$$A_v = 2A_{v1} = 2A_{v2} \tag{2.87}$$

When the circuit is differential and balanced, components on corresponding sides are equal, and the gain reduces to

$$A_v = 2\alpha\frac{R_L}{R} \tag{2.88}$$

With symmetric circuit topology and $v_{i1} = -v_{i2}$, $v_e = 0$ because the two superimposed inputs at v_e are equal and opposite. In this case, the v_e node is a virtual ground, and the gains of each side of the amplifier can be calculated under this assumption. Consequently, the transresistance is R on each side,

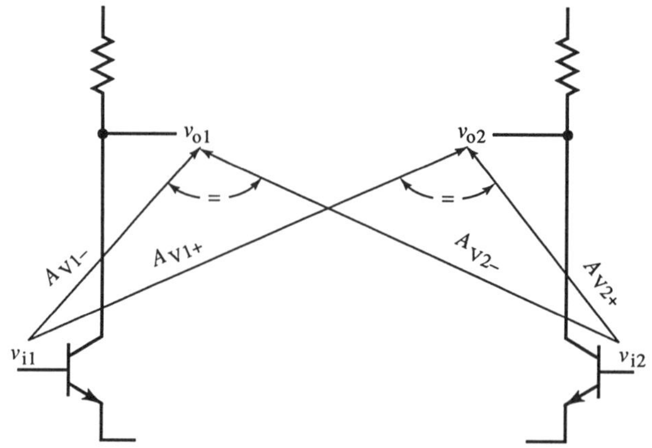

(a)

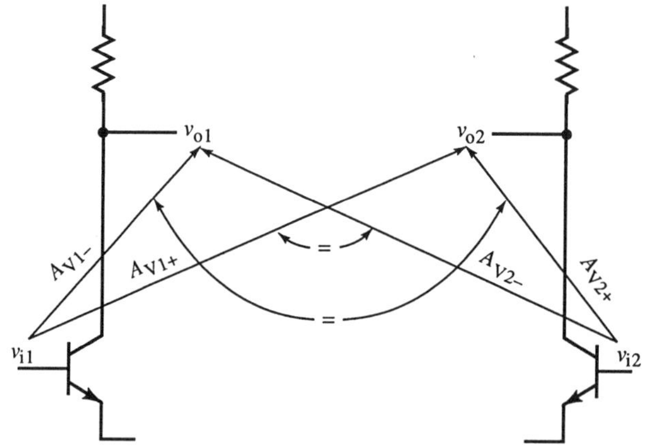

(b)

FIG. 2.12 Diff-amp symmetry can be achieved under (a) symmetric or (b) antisymmetric conditions.

and the gain for each side (v_{o1}/v_{i1} and v_{o2}/v_{i2}) is $\alpha R_L/R$ in magnitude. For differential outputs, according to (2.87), the gain is twice that of a single side.

The *common-mode rejection ratio* (CMRR) is the measure of how differential an amplifier is. It is the change in output when both inputs are changed the same amount and is defined as

$$\text{CMRR} = \frac{v_o/(v_{i2} - v_{i1})}{v_o/(v_{i2} + v_{i1})} \tag{2.89}$$

The numerator is the differential-mode gain of the amplifier; the denominator is the common-mode gain. Two arbitrary inputs, v_{i1} and v_{i2}, can be combined into differential-mode v_d and common-mode v_c voltages:

$$v_d = \frac{v_{i2} - v_{i1}}{2} \tag{2.90a}$$

$$v_c = \frac{v_{i2} + v_{i1}}{2} \tag{2.90b}$$

A purely differential input occurs when $v_{i2} = -v_{i1}$. Then $v_d = v_{i2}$ and $v_c = 0$. When $v_{i2} = v_{i1}$, the input is purely common mode and $v_c = v_{i2}$ whereas $v_d = 0$. An infinite CMRR is ideal because there the amplifier only amplifies the differential-mode voltage. CMRR is a measure of how well the conditions of (2.83) or (2.84) are met.

Example 2.6 Differential Amplifier with CC Output

Figure E2.6 is a differential-amplifier stage buffered by an emitter-follower. Since both bases of the diff-amp are at the same dc voltage (0 V), and their emitter resistors are the same value, it is reasonable to assume that their currents are equal. Assume they both conduct $(12\text{ V} - 0.8\text{ V})/(220\ \Omega + 1.8\text{ k}\Omega \times 2) \cong 3\text{ mA}$. Then

$$I_{E1} = I_{E2} = \frac{12\text{ V} - (0.8\text{ V} + (3\text{ mA})(220\ \Omega))}{220\ \Omega + 2(1.8\text{ k}\Omega)} = 2.76\text{ mA}$$

Using this current, we find that V_{BE} agrees with the assumed value of 0.80 V and that the bias point has converged. Then $r_{e1} = r_{e2} = 9.6\ \Omega \cong 10\ \Omega$. Continuing the dc analysis at the output collector, we have

$$V_{C2} = 12\text{ V}\left(\frac{470\text{ k}\Omega}{470\text{ k}\Omega + 2.0\text{ k}\Omega}\right) - (2.7\text{ mA})(2.0\text{ k}\Omega \parallel 470\text{ k}\Omega) = 6.6\text{ V}$$

and

$$I_{E3} \cong \frac{6.6\text{ V} + 12\text{ V} - 0.8\text{ V}}{4.7\text{ k}\Omega} = 3.8\text{ mA}$$

Then $r_{e3} = 6.8\ \Omega$. Solving for the usual amplifier parameters gives

$$\frac{v_o}{v_i} = \left(\frac{1.8\text{ k}\Omega}{1.8\text{ k}\Omega + 220\ \Omega + 10\ \Omega}\right) \times \left((0.99)\frac{2.0\text{ k}\Omega}{10\ \Omega + 220\ \Omega + 1.8\text{ k}\Omega \parallel (220\ \Omega + 10\ \Omega)}\right) \times \left(\frac{4.7\text{ k}\Omega}{4.7\text{ k}\Omega + 7\ \Omega + 2\text{ k}\Omega/100}\right) = (0.887)(4.56)(0.994) = 4.02$$

```
E2.6 Differential Amplifier with CC Output
.OPT NOMOD OPTS NOPAGE
.DC VI -0.25 0.25 0.05
.OP
.TF V(70) VI
VI 10 0 DC 0V
VCC 80 0 DC 12
VEE 90 0 DC -12
RE1 30 40 220
RE2 50 40 220
RO  40 90 1.8K
RL  80 60 2.0K
RE3 70 90 4.7K
Q1 80 10 30 BJT1
Q2 60  0 50 BJT1
Q3 80 60 70 BJT1
.MODEL BJT1 NPN (BF=99)
.END
```

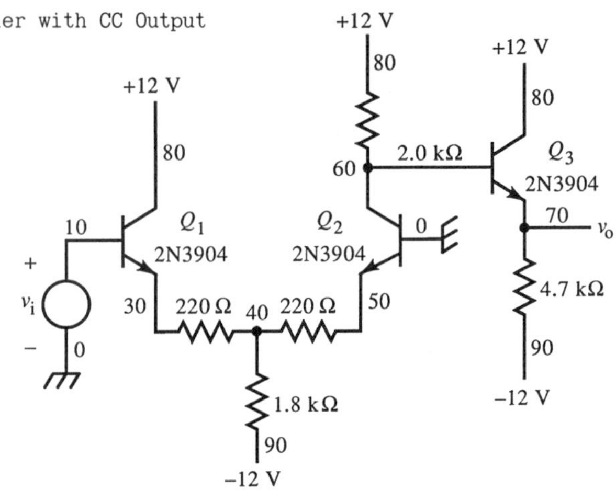


```
SMALL SIGNAL BIAS SOLUTION    TEMPERATURE = 27.000 DEG C

NODE   VOLTAGE
(30) -.8018  (40) -1.4467  (50) -.8018
(60) 6.1220  (70) 5.3143

BIPOLAR JUNCTION TRANSISTORS

NAME         Q1          Q2          Q3
MODEL        BJT1        BJT1        BJT1
IB           2.93E-05    2.93E-05    3.68E-05
IC           2.90E-03    2.90E-03    3.65E-03
VBE          8.02E-01    8.02E-01    8.08E-01
VBC         -1.20E+01   -6.12E+00   -5.88E+00
VCE          1.28E+01    6.92E+00    6.69E+00
BETADC       9.90E+01    9.90E+01    9.90E+01
GM           1.12E-01    1.12E-01    1.41E-01

V(70)/VI = 4.045E+00
INPUT RESISTANCE AT VI = 4.318E+04
OUTPUT RESISTANCE AT V(70) = 2.686E+01
```

FIG. E2.6

$$r_{in} = (100)[10\ \Omega + 220\ \Omega + 1.8\ \text{k}\Omega \parallel (220\ \Omega + 10\ \Omega)] = 43\ \text{k}\Omega$$

$$r_{out} = 4.7\ \text{k}\Omega \parallel (7\ \Omega + 2\ \text{k}\Omega/100) = 27\ \Omega$$

Again, compared with the simulation, these results are right on.

2.12 Current Mirrors

A circuit that supplies a current of the same polarity and magnitude as its input current is a *current mirror*. A Widlar current mirror (after Bob Widlar)

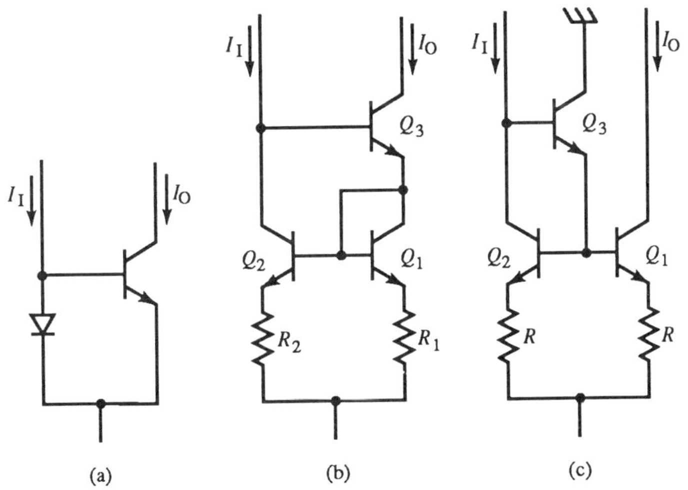

FIG. 2.13 Active current sources: (a) Widlar, (b) Wilson, and (c) common IC source with CC.

is shown in Fig. 2.13a. Input current I_I flows through the diode, creating a voltage that is also V_{BE} of the transistor. If the diode and b–e junctions are matched (that is, have the same v–i function), then the resulting emitter current equals I_I. Consequently, the output current I_O is equal to αI_I. For a typical $\alpha \cong 1$, I_O is a replication of I_I. This circuit is useful, for example, in supplying I_O to a diff-amp. If the emitter and anode are connected to $-V_{EE}$, the current can be set by a resistor from the input to ground. If the voltage across the resistor is much larger than the diode voltage, I_I is largely determined by the resistor value.

The basic current mirror of Fig. 2.13a can be improved by adding another transistor to compensate for base current lost to the transistor. If we take into account I_B, then

$$I_O = \alpha(I_I - I_B) = \alpha\left(\frac{I_I - I_O}{\beta}\right) \tag{2.91}$$

Solving for the current gain gives

$$\frac{I_O}{I_I} = \frac{\beta}{\beta+2} \cong \alpha, \qquad \beta \gg 1 \tag{2.92}$$

Three improvements have been made in Fig. 2.13b. First, the diode has been made out of a similar transistor, Q_1, by connecting the base and collector. This kind of diode is often used in integrated circuits to achieve the best match of two pn junctions. Second, the transistor Q_3 has been added to compensate for α loss, now occurring in Q_2. Third, to further reduce current-gain error, emitter resistors have been added.

The effect of Q_3 is to divert I_B amount of current from I_I. Q_3 emitter current is then $(\beta+1)I_B$. From this current, I_B is diverted into the base of Q_2.

This loss of base current to Q_2 was compensated by the diversion of Q_3 base current. Though it is better than the simple current mirror of Fig. 2.13a, the compensation is not perfect, even with matched junctions. If we assume that the *b*-*e* junctions are matched and $R_1 = R_2$, since the bases of Q_1 and Q_2 are connected, the same voltage occurs across identical branches. Thus $I_{E1} = I_{E2}$ and

$$I_{E3} = (\beta_3 + 1) I_{B3} = I_{B2} + I_{E1} \tag{2.93}$$

Solving for I_{E3} gives

$$I_{E3} = \frac{I_{E2}}{\beta_2 + 1} + I_{E2} = I_{E2}\left(1 + \frac{1}{\beta_2 + 1}\right) \tag{2.94}$$

Also,

$$I_I = \alpha_2 I_{E2} + I_{B2} \tag{2.95}$$

Since $I_O = \beta_3 I_{B3}$, the current gain can be found from (2.94) and (2.95) and is

$$\frac{I_O}{I_I} = \frac{\beta_1\beta_2 + 2\beta_3}{\beta_1\beta_2 + 2\beta_2 + 2} \tag{2.96}$$

For $\beta_1 = \beta_2 = \beta_3 = \beta$,

$$\frac{I_O}{I_I} = \frac{\beta^2 + 2\beta}{\beta^2 + 2\beta + 2} \tag{2.97}$$

This current gain is a closer approximation to unity than that of Fig. 2.13a; it is tabulated here for various values of β:

β	I_O/I_I	$\beta/(\beta+2)$
1	0.60	0.33
2	0.80	0.50
10	0.98	0.83
50	0.999	0.96
100	0.9998	0.98

For $\beta = 100$, I_O/I_I is 100 times better than $\beta/(\beta+2)$ and 40 times better at $\beta = 50$.

This analysis assumes perfect matching of Q_1 and Q_2. In practice, the effect of mismatch tends to be minimized by R_1 and R_2 if the voltage dropped across them is much greater than the *b*-*e* junction voltages of the transistors. Since resistors can be matched much better than transistors and can be made very stable, the current mirror emitter currents can be determined dominantly by the emitter resistors. In integrated current mirrors, transistor matching can be very good, and the additional voltage drop of the emitter resistors can be minimized, giving the circuitry connected to the mirror a wider voltage range. This current mirror is called the Wilson current mirror after its inventor, George Wilson.

Another three-BJT mirror, well-suited for integrated circuit (IC) layout, is shown in Fig. 2.13c. For matched junctions, $I_{E1} = I_{E2}$ and

$$I_I - \frac{[I_{E2}/(\beta_2+1) + I_O/\beta_1]}{\beta_3+1} = \alpha_2 I_{E2} \tag{2.98}$$

where the numerator of the second term on the left is I_{E3}. Using the substitution $I_O/\alpha = I_{E2}$ yields the current gain:

$$\frac{I_O}{I_I} = \frac{\beta_1\beta_2\beta_3 + \beta_1\beta_2 + \beta_1\beta_3 + \beta_1}{\beta_1\beta_2\beta_3 + \beta_1\beta_2 + \beta_2\beta_3 + \beta_1 + 2\beta_2 + 2} \tag{2.99}$$

For $\beta_1 = \beta_2 = \beta_3 = \beta$, then

$$\frac{I_O}{I_I} = \frac{\beta^2 + \beta}{\beta^2 + \beta + 2} \tag{2.100}$$

The terms in the numerator and denominator differ only at the constant term, resulting in accurate current mirroring. The current gain versus β for several values is given in the following table, along with the gain values for the Wilson mirror (Fig. 2.13b):

β	$(I_O/I_I)_b$	I_O/I_I
1	0.60	0.50
2	0.80	0.78
10	0.98	0.98
50	0.99923	0.99922
100	0.99980	0.99980

For $\beta \gg 1$, the mirrors of Figs. 2.13b,c have almost identical current gains. Although Q_3 of Figs. 2.13b,c recirculate their base currents, taken from I_I, back to Q_2, only Q_3 of Fig. 2.13c provides base current for Q_1 and Q_2 and carries no output current.

Sometimes a minimum component current mirror is preferred to minimize silicon or circuit-board area. The mirror of Fig. 2.14 uses a compensating base resistor for Q_1. R_B compensates for loss of I_{B2} from I_I and for α_2. For β-matched transistors and base-emitter junction areas of Q_1 and Q_2, ratioed so that $V_{BE1} = V_{BE2}$, then the desired gain is $I_O/I_I = R_{E1}/R_{E2} = A_I$. From the circuit, $I_{E1} = I_I - I_O/\beta$; and from the area-ratio constraint,

$$I_{E1}R_{E1} + I_{B1}R_B = I_{E2}R_{E2} \tag{2.101}$$

Furthermore, $I_O/\alpha = I_{E2}$. Solving these equations for R_B, we finally obtain

$$R_B = \frac{(A_I+1)R_{E1}}{\alpha(1 - A_I/\beta)} \cong (\beta+1)R_{E1}\left(\frac{A_I}{\beta - A_I}\right) \tag{2.102}$$

This result gives the compensating value of R_B and is more easily derivable in the following way. The loss in voltage across R_{E1} due to the loss of I_{B2} is

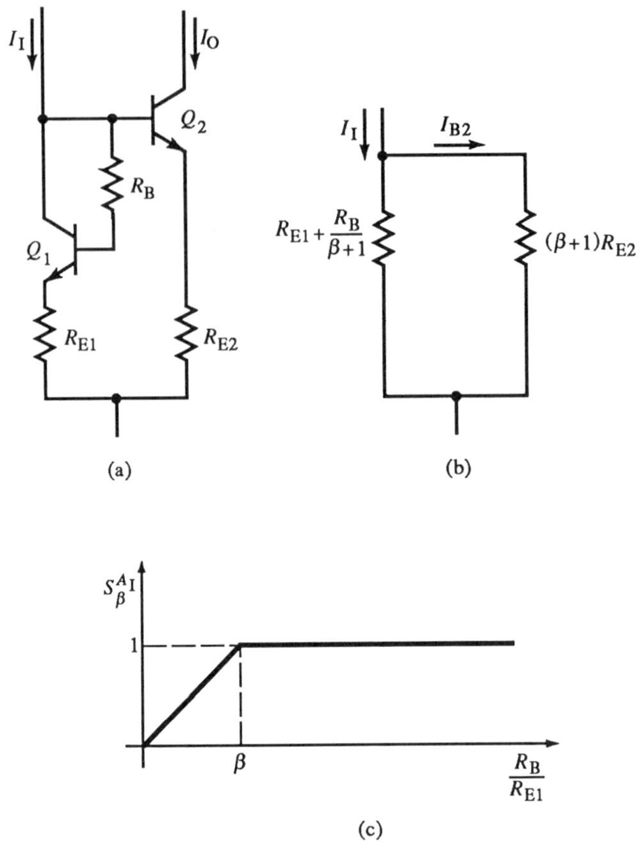

FIG. 2.14 An R_B-compensated current source (a), equivalent resistance of Q_1 (b), and the sensitivity of the current gain to β with the ratio of R_B to R_{E1} (c).

$R_{E1}I_O/\beta$ and is compensated by the drop across R_B. Then,

$$\frac{I_O}{\beta}\cdot R_{E1}=\frac{(I_I-I_O)}{\beta}\left(\frac{R_B}{\beta+1}\right) \tag{2.103}$$

Rearranging gives

$$\frac{I_O}{\beta}\left(R_{E1}+\frac{R_B}{\beta+1}\right)=I_I\left(\frac{R_B}{\beta+1}\right) \tag{2.103a}$$

Because $A_I=I_O/I_I$,

$$A_IR_{E1}=\frac{\beta R_B}{\beta+1}-\frac{A_IR_B}{\beta+1}=\frac{(\beta-A_I)R_B}{\beta+1} \tag{2.104}$$

Solving for R_B, we obtain the approximate expression of (2.104). This simpler derivation does not take into account the α_2 loss, only the loss due to I_{B2}.

This analysis was based on Kirchhoff's laws and led to much algebraic manipulation in deriving R_B. By using the β transform, we can simplify the analysis by using the equivalent input circuit shown in Fig. 2.14b. Since $I_O = \beta I_{B2}$, I_O/I_I can be expressed largely as a current divider, where

$$A_I = \frac{I_O}{I_I} = \beta \cdot \frac{R_{E1} + R_B/(\beta+1)}{R_{E1} + R_B/(\beta+1) + (\beta+1)R_{E2}} \tag{2.105}$$

Under the constraint that

$$A_I = \frac{R_{E1}}{R_{E2}} \tag{2.106}$$

the numerator and denominator of (2.105) can be divided by R_{E1} and A_I can be substituted for the resistor ratio of (2.106). Solving the resulting equation for R_B results in the exact expression of (2.102).

R_B was chosen according to the constraint of (2.106). This constraint simplifies the selection of emitter resistor values. The current gain is nevertheless sensitive to β variation, and R_B can instead be chosen for minimum β sensitivity, or for

$$\min \frac{\partial}{\partial \beta}\left(\frac{I_O}{I_I}\right)$$

For $\beta \gg 1$, I_O/I_I can be rewritten as

$$\frac{I_O}{I_I} \cong \beta \cdot \frac{\beta R_{E1} + R_B}{\beta R_{E1} + R_B + \beta^2 R_{E2}} = \beta \cdot \frac{a}{a + \beta^2 R_{E2}} = \beta \cdot f(\beta) \tag{2.107}$$

Then

$$\frac{\partial}{\partial \beta}\left(\frac{I_O}{I_I}\right) = \frac{\partial}{\partial \beta} \beta \cdot f(\beta) = f + \beta \cdot \frac{\partial f}{\partial \beta} \tag{2.108}$$

$$\frac{\partial f}{\partial \beta} = \frac{-a(R_{E1} + 2\beta R_{E2})}{(a + \beta^2 R_{E2})^2} + \frac{R_{E1}}{(a + \beta^2 R_{E2})} \tag{2.109}$$

Substituting (2.109) into (2.108) yields

$$\frac{\partial f}{\partial \beta}\left(\frac{I_O}{I_I}\right) = \frac{a^2 + a\beta^2 R_{E2} - \beta a R_{E1} - 2a\beta^2 R_{E2} + a\beta R_{E1} + \beta^3 R_{E1} R_{E2}}{(a + \beta R_{E2})^2} \tag{2.110}$$

To obtain R_E at minimum sensitivity to β, (2.110) is set to zero and the numerator solved for R_B. The result is

$$R_B = -\beta\left(R_{E1} - \frac{\beta}{2} \cdot R_{E2}\right) \pm \frac{\beta}{2}\sqrt{\beta R_{E2}(\beta R_{E2} - 4R_{E1})} \tag{2.111}$$

Under the previous constraint of (2.106), the sensitivity of A_I with respect to β is

$$S_{\beta}^{A_I} = \frac{\partial A_I / A_I}{\partial \beta / \beta} = \left(\frac{\beta}{A_I}\right)\frac{\partial A_I}{\partial \beta} \tag{2.112}$$

where A_I is (2.105) with R_{E2} expressed in terms of R_{E1} and A_I. With R_B given by (2.102) and substituting into (2.112), we get

$$S_{\beta}^{A_I} = k^2 \cdot \frac{\beta}{(k-1)\beta^2 + [(k+1)(k-1)-1]\beta - (k+1)}, \qquad k = \frac{R_B}{R_{E1}},$$

$$\cong \frac{1}{\beta/k+1}, \qquad \beta, k \gg 1 \tag{2.113}$$

The approximation to S is asymptotic with unity (Fig. 2.14c). For $k > \beta$, A_I varies directly with β. This result suggests that k be kept less than β to reduce gain sensitivity to β. This is not always possible when attempting to satisfy (2.106) as well. Or, the circuit has gain limits for acceptable gain sensitivity to β.

2.13 Matched Transistor Buffers and Complementary Combinations

A simple but elegant circuit (Fig. 2.15a) consists of a pair of matched JFETs. The lower JFET acts as a current source for the upper source-follower. The beauty of this circuit is that the lower transistor sinks a particular amount of current (I_{DSS}) with $V_{GS} = 0$ and that, with negligible loss of current to the load, I_{DSS} also flows through the upper JFET, resulting in the same V_{GS} of zero volts (because they are matched). This voltage amplifier of unity gain (or

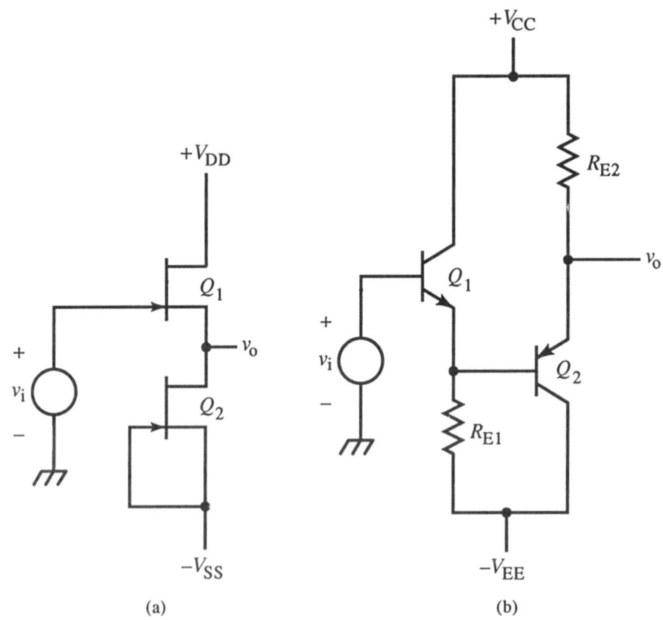

FIG. 2.15 Voltage buffer (×1 voltage amplifier) circuits: (a) matched FET buffer and (b) complementary CC buffer.

buffer) consequently has zero voltage offset. This is desirable because the purpose of a buffer is to provide a voltage source at a much reduced output resistance than the input voltage (from a higher-resistance source). A simple emitter or source-follower would cause an offset due to an undetermined V_{BE} or V_{GS} and introduce a voltage error at the output. A BJT circuit based on the same general idea is shown in Fig. 2.15b. Here, the V_{BE} offsets of opposite polarity CC BJTs cancel to the extent that their currents are equal (for matched junctions). This offset match is more difficult than with the JFET circuit in that the devices are of opposite polarity. This circuit is nevertheless quite useful for acceptable offsets of typically less than 50 mV.

The NPN/PNP pair of Fig. 2.16 functions like a PNP with current gain from the output NPN. This circuit is commonly used in the output stage of amplifiers so that only power NPN transistors need be used (and is sometimes called a "complementary PNP" circuit). It can also be used, as in Fig. 2.16b, to source current. Although the output is from an emitter, the base resistance is large (r_c of the PNP), resulting in an acceptably large output resistance to pass as a current source in applications in which the driven node is of relatively low resistance.

Some complementary pairs are regenerative and form latching circuits. Thyristors are a class of four-layer (PNPN or NPNP) devices that are used as high-power switches and also are formed as parasitic elements in ICs that have multiple n and p layers (such as CMOS or biMOS circuits that have a tendency to latch if their inputs exceed the supply voltages). A common thyristor, the *silicon-controlled rectifier* (SCR), is shown in Fig. 2.17a along

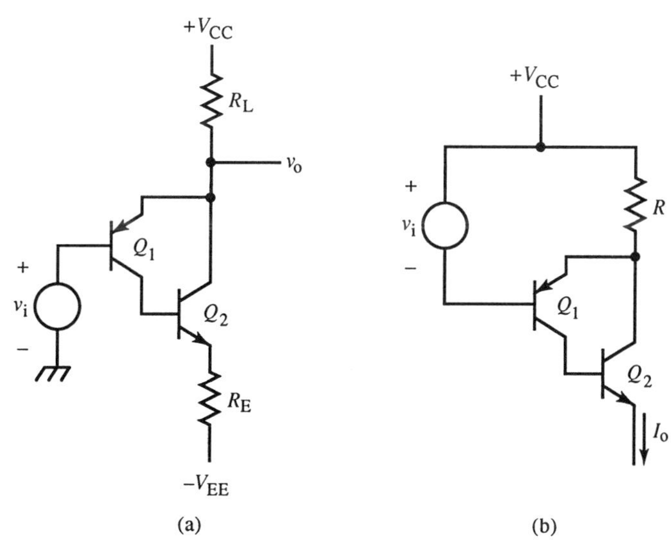

FIG. 2.16 Uses for a complementary PNP circuit: (a) PNP equivalent CC amplifier and (b) current source.

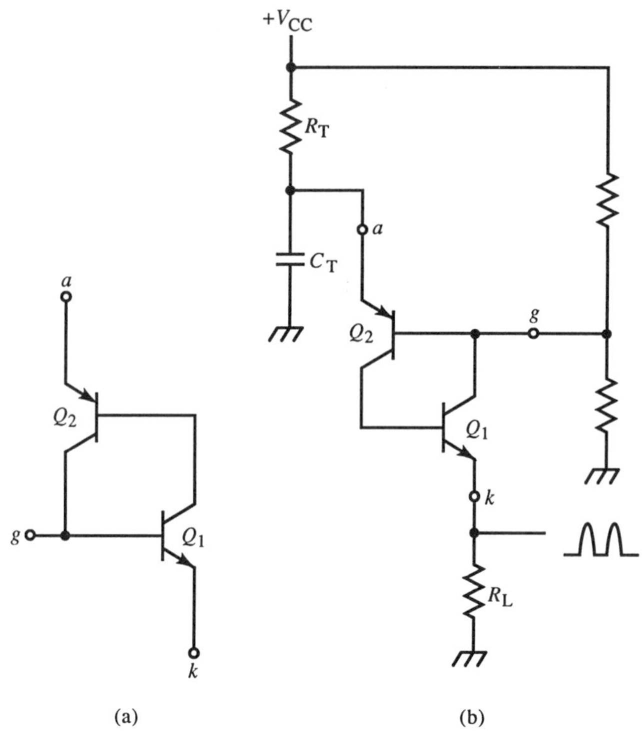

FIG. 2.17 Thyristor equivalent BJT circuits: (a) SCR and (b) PUT.

with a variant, the *programmable unijunction transistor* (PUT). For either device, transistor collector currents supply base current to the other transistor, causing regenerative action. SCRs cannot be turned off by the gate; the anode-to-cathode voltage must reverse, causing cessation of conduction. SCRs are used in power conversion, and PUTs are useful devices for constructing simple oscillators and programmable timers.

2.14 Closure

We have examined a variety of amplifier circuits that have one to three transistors. More complicated "building-blocks" will be introduced later. As additional transistors are added, complexity grows to the point that a multilevel or hierarchical organization is needed. Multiple circuits are combined to form complete subsystems, which in turn are combined with other subsystems to form the final system. We can deal with complexity at these various levels in the same way. An op-amp (introduced in the next chapter) contains many circuits but, like the transistor, can be modeled as a single device with a simple functional description. In this chapter, we developed a "library" of basic circuits that can be used to develop a library of basic subsystems, in the same way that commonly used computer routines can be joined to form more complicated routines.

This circuit discussion was based on simplifying assumptions that must now be examined. We assumed that a circuit input was independent of the output, that there was no *feedback*. Furthermore, although both static and dynamic quantities were introduced, we omitted reactive components such as capacitors and inductors. These components require the use of complex numbers to describe their behavior and lead to discussion of transient and frequency response.

References

E. James Angelo, Jr. *Electronics: BJTs, FETs, and Microcircuits*, McGraw-Hill, 1969. Ch. 12 and 13.

David Cheng, *Analysis of Linear Systems*, Addison-Wesley, 1959. Ch. 9.

R. D. Middlebrook, *Differential Amplifiers*, Wiley, 1963.

CHAPTER 3

Feedback Circuits

3.1 Basic Feedback Topology

An amplifier with an input that includes some of its output is a *feedback* amplifier. This kind of amplifier can improve control of the output. If we compare the actual output with the desired output (represented by the input), an error can be constructed and used to correct the output. Feedback is analogous to recursion in mathematics and to iterative loops that branch backward in software.

The block diagram representation of a classical feedback system is shown in Fig. 3.1a. The input R is summed with output C through feedback gain H or HC. The summation is defined such that HC is subtracted from R, resulting in the error E. The forward gain G amplifies E, resulting in output C. The algebra represented by the block diagram is

$$E = R - HC \tag{3.1}$$

$$C = GE \tag{3.2}$$

Solving (3.1) and (3.2) for the overall gain of the feedback amplifier, the *closed-loop gain*, results in

$$T = \frac{C}{R} = \frac{G}{1+GH} \tag{3.3}$$

The input and output quantities R and C are arbitrary but are usually voltages or currents in circuits. The total gain around the feedback loop GH is the *loop gain.*

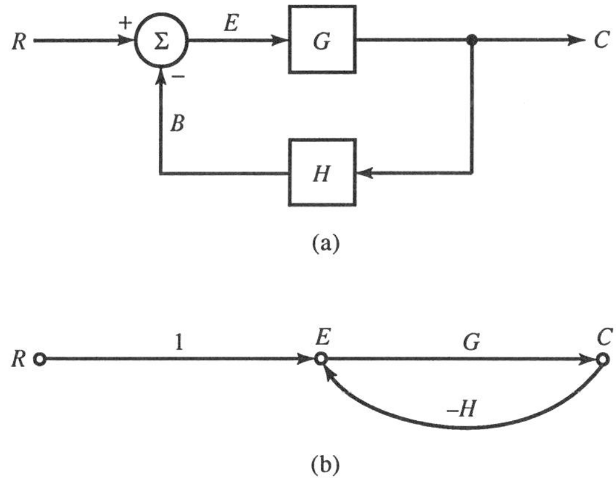

FIG. 3.1 The classical feedback system represented as (a) a block diagram or (b) a signal-flow graph.

An alternative representation for the same feedback circuit is the *flow graph*, shown in Fig. 3.1b. By convention, nodes represent quantities such as R, E, and C, and directed arcs or paths between nodes represent transmittances (gains or attenuations). When multiple arcs enter a node, they add. The output of an arc is the product of its input node quantity and the arc gain, written along it. Block diagrams are often associated with control theory and flow graphs have traditionally been used in network theory. They are equivalent representations but block diagrams are somewhat easier to read and flow graphs easier to draw compactly. Both representations will be used here. *Transmittance* in network analysis is *transfer function* in control theory; it is the ratio of output to input quantities. (In network theory, G is usually called a, and H is called f.)

3.2 Identification of Forward and Feedback Paths

One of the difficulties of feedback circuit analysis is in identifying the forward (G) and feedback (H) paths, the summer Σ, and the output pick-off point. Circuits often consist of components connected in messy topologies that are hard to compare with the simple feedback forms of Fig. 3.1. We need to develop a view of feedback circuits that makes their decomposition into G and H blocks intuitive yet exacting. This will be done inductively; examples will be analyzed so that a more general perspective can emerge. After G and H are known, the techniques of control theory can be applied. In this chapter, we will constrain our investigations to real, dynamic analysis.

Before we attempt to identify G, H, and Σ in examples of feedback circuits, some general principles can be deduced. First, from (3.1) it is clear that R

and B $(=HC)$ must be compatible quantities; they must have the same units to be summed. Of course, E must also have the same units. If R is a voltage, then E and B must also be voltages. Beginning with the input and output quantities, we will find that the kind of transfer functions (or gains) that G and H are will follow from this analysis of units. G has the units of C/E, and because the units of E and R are the same, G has the units of the overall amplifier C/R. Since $H = E/C$, its units are the inverse of G. For example, if G is a transconductance, H is a transresistance; if G is unitless, H is also unitless.

Second, since E is the summation of R and B, the kind of circuit topology that realizes this summation depends on the kind of quantity being summed. If voltages are summed, then look for a loop, because, by Kirchhoff's voltage law (KVL), voltages sum in loops. For current summation, by Kirchhoff's current law (KCL), look for a node. Because voltages in loops are summed in series, voltage summation is often called *series feedback*. Similarly, current summation at a node is *shunt feedback*. Sometimes the summing loop or node corresponding to Σ is not obvious, and some circuit transformations are required to reveal it.

Finally, the block-diagram pick-off or *sampling* point for the output will be realized in circuit form as a loop for currents and a node for voltages. For sampling, currents flow in loops and current sampling is *series sampling*, whereas voltage sampling at the output node is *shunt sampling*.

Most feedback circuits are more complex than the simple form of Fig. 3.1. A gain block can precede the input or follow the sampling point to the output. In these cases, the input does not directly sum with the feedback, and the output is not the fed-back quantity (see Figs. 3.2a,b, respectively). There-

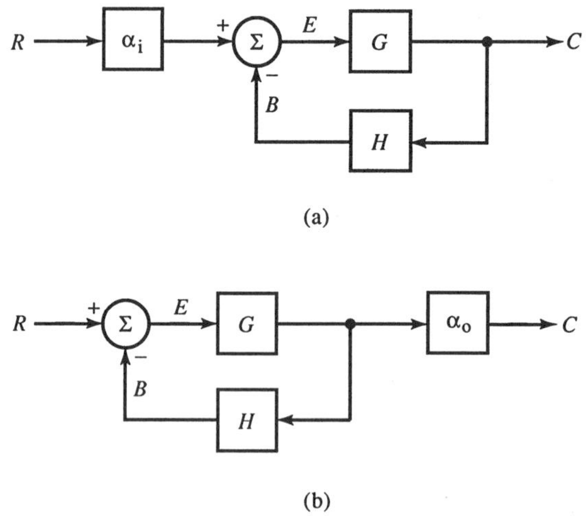

FIG. 3.2 The classical feedback system with additional input (a) and output (b) transfer functions.

fore, our first task in the general analysis is to identify the summing and sampling quantities and their corresponding loops or nodes.

Generally, the choice of summing and sampling quantities is arbitrary. Figure 3.3 shows that choosing the sampling anywhere in the feedback loop results in a correct overall transfer function T as long as accounting is made for G_2, the gain block between the output and sampling point. This flow graph begins (Fig. 3.3a) with a feedback circuit in which α_i precedes the summing circuit. Then

$$E = \alpha_i R - HC' \tag{3.4}$$

The output C is shown buried somewhere in the assumed forward path G, where

$$G = G_1 \cdot G_2 \tag{3.5}$$

The output is along this path at $C = G_1 \cdot E$. A further gain is required to get to C', where the assumed feedback path H begins.

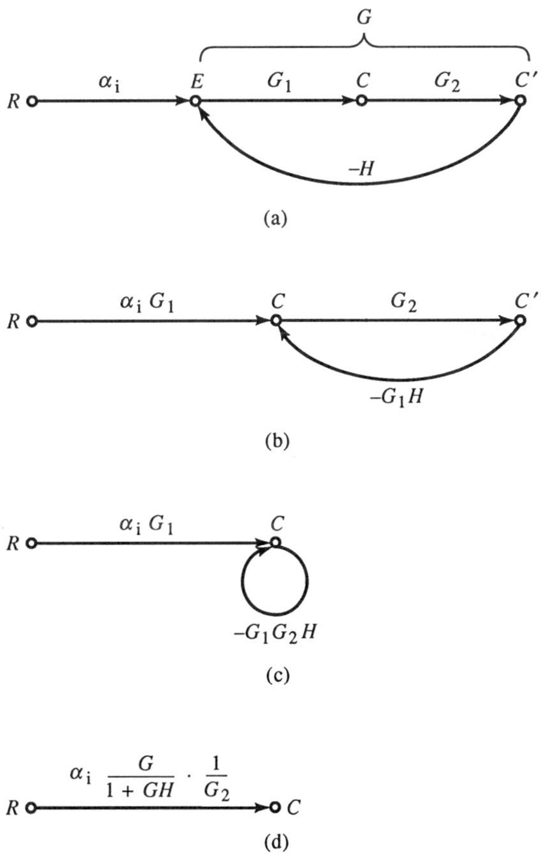

FIG. 3.3 Choosing the sampling anywhere within the feedback loop, such as at C' instead of C results in a correct transfer function where G_2 becomes part of the feedback path.

Flow-graph reduction is shown in Figs. 3.3a–d. Each step in the reduction illustrates a basic reduction rule. The feedback path is moved from E to C in Fig. 3.3b, and the G_1 path is eliminated. Using (3.4), we find this is algebraically equivalent to

$$C = G_1 E = G_1(\alpha_i R - HC') = G_1 \alpha_i R - G_1 HC' \tag{3.6}$$

Eliminating the G_1 path is equivalent to distributing G_1 over the terms in E. Eliminating C' by combining G_2 and $-G_1 H$ results in Fig. 3.3c with a direct feedback path to the same node. Writing the expression for C from Fig. 3.3c we have

$$C = \alpha_i G_1 R - G_1 G_2 HC = \frac{\alpha_i G_1 R}{1-(-G_1 G_2 H)} \tag{3.7}$$

The last expression is the solution for C. The closed-loop gain T is

$$T = \frac{C}{R} = \alpha_i \cdot \frac{G}{1+GH} \cdot \frac{1}{G_2} \tag{3.8}$$

The significance of this result is that if a sampled quantity is chosen that is not the actual output quantity, as long as it is within the loop the correct expression for T still results. Because G_2 is actually part of the feedback path from C, the total feedback path is $G_2 H$, and (3.8) can be expressed as

$$T = \alpha_i \cdot \frac{G_1}{1+G_1(G_2 H)} \tag{3.8a}$$

The forward path is G_1.

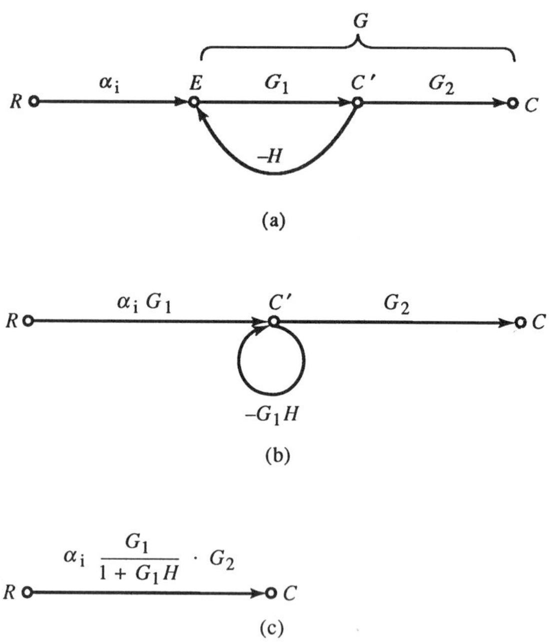

FIG. 3.4 Choosing the sampling at C' instead of at the output C results in a correct transfer function with $\alpha_o = G_2$.

For the case in which the sampling precedes the output, as in Fig. 3.2b, α_o must be introduced. In flow-graph form, this is shown in Fig. 3.4. The reduction of the flow-graph results in

$$T = \alpha_i \cdot \frac{G_1}{1 + G_1 H} \cdot G_2 \tag{3.9}$$

In this case, $\alpha_o = G_2$, and the forward path gain is G_1. As long as the sampled quantity is chosen within the loop and (3.5) holds, G_1 and G_2 can be arbitrarily chosen.

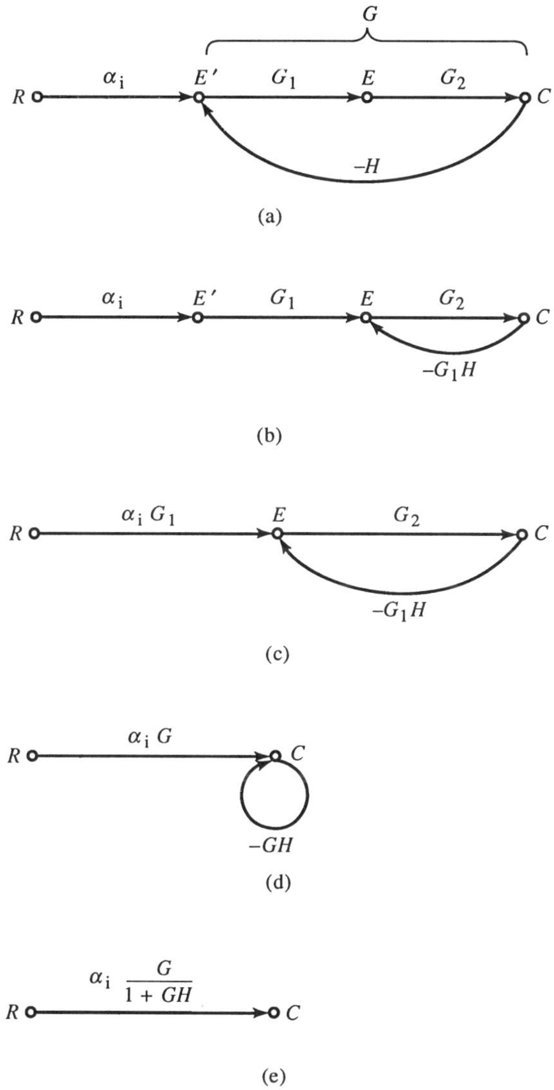

FIG. 3.5 Choosing the error summing anywhere in the forward path, such as at E instead of E', results in a correct transfer function. The choice of E' eliminates the common factor of G_1 from the branches into E.

A similar situation for choice of E is shown in Fig. 3.5. Choosing E anywhere in the forward path results in a correct expression for T. Suppose E is chosen as the summing circuit instead of E'. Since E is in the forward path, T is correctly derived (Fig. 3.5e). The flow-graph summing node closest to the input is usually preferred because the common factor G_1 in $\alpha_i G_1$ and $-G_1 H$ is eliminated from each transmittance. The common factor suggests that input and feedback paths are sharing a common path G_1, and that a different summing circuit closer to the input can be chosen.

3.3 Operational Amplifier Configurations

The circuits closest to ideal block or flow-graph representation are operational amplifier (op-amp) circuits. An amplifier is *operational* when its behavior is dominated by feedback path components instead of forward path gain. A feedback path can be as simple as a divider composed of resistors. Generally, resistive dividers are more stable and precise than amplifier gain. If the forward path gain is made as large as possible (that is, $G \to \infty$), then

$$\lim_{G \to \infty} \frac{G}{1 + GH} = \frac{1}{H} \tag{3.10}$$

and G is no longer a determinant of closed-loop gain, only H.

Integrated circuit (IC) op-amps are commonly used in analog signal processing. *Op-amps*, by definition, have a differential input and single-ended output with (open-loop) gain K so that

$$v_o = K(v_+ - v_-) \tag{3.11}$$

where v_+ is the noninverting $(+)$ input voltage and v_- is the inverting $(-)$ input voltage. An ideal op-amp has infinite input resistance, zero output resistance, and infinite gain.

Figure 3.6 shows the two basic op-amp configurations. In the noninverting configuration (Fig. 3.6a), Σ is the differential op-amp input, where

$$v_+ = v_i, \qquad v_- = Hv_o, \qquad E = v_+ - v_- = v_i - Hv_o \tag{3.12}$$

From (3.1) and (3.2), we can create general expressions for G and H. In (3.1), if R is set to zero, then by superposition

$$H = -\frac{E}{C}\bigg|_{R=0} \tag{3.13}$$

and from (3.2), with no feedback from H,

$$G = \frac{C}{R}\bigg|_{B=0} \tag{3.14}$$

Since $R = v_i$, a voltage, then E is a voltage. $C = v_o$ and H denotes a voltage

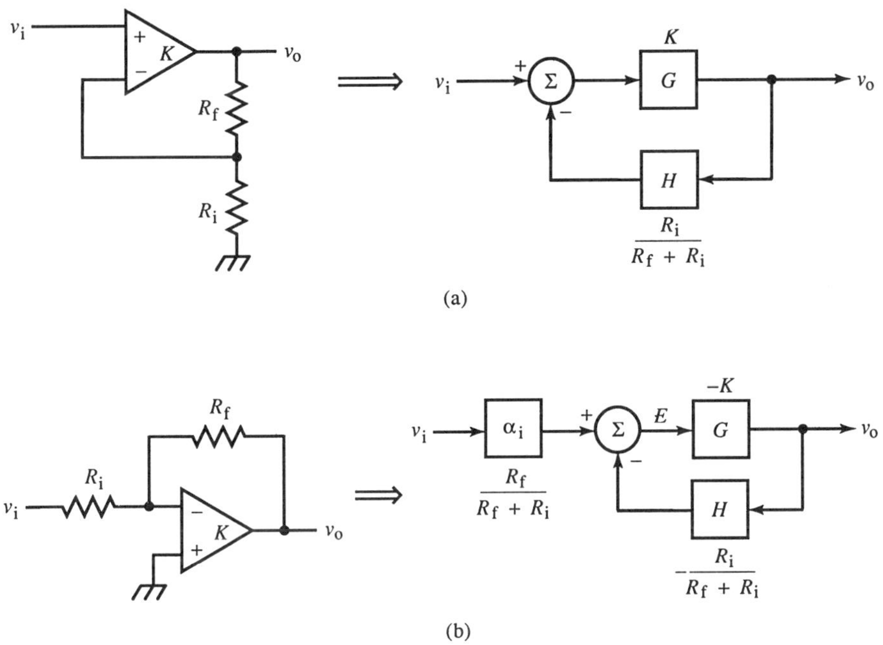

FIG. 3.6 Basic op-amp configurations: (a) noninverting, (b) inverting, and their block diagrams.

gain (unitless). From (3.13),

$$H = -\frac{-v_-}{v_o} = \frac{v_-}{v_o} = \frac{R_i}{R_f + R_i} \tag{3.15}$$

H is the voltage divider attenuation from v_o to v_-, given by (3.15).

The forward path gain is found from (3.14) by opening the loop so that no feedback is introduced into Σ. This is done by setting B $(=v_-)$ to zero:

$$G = \frac{v_o}{v_i} = \frac{K(v_i)}{v_i} = K \tag{3.16}$$

Since $K \to \infty$, we can use (3.10) to find the closed-loop gain. It is

$$\text{noninverting op-amp} \qquad A_v|_{K\to\infty} = \frac{R_f + R_i}{R_i} = 1 + \frac{R_f}{R_i} \tag{3.17}$$

A finite-gain expression for the closed-loop voltage gain follows from (3.3) and is

$$\text{noninverting op-amp} \qquad A_v = \frac{K}{1 + K(R_i/(R_f + R_i))} \tag{3.18}$$

In the case of finite K, E is nonzero. A typical open-loop gain K of IC op-amps is 10^5. Because $E \cong 0$ for proper op-amp function, if E is represented by a circuit node voltage, it is a virtual ground because the feedback characteristic of the op-amp keeps it from changing significantly.

For the noninverting op-amp configuration, it is not difficult to identify the sampling node v_o or Σ. For the inverting configuration (Fig. 3.6b), Σ is more difficult to identify but must involve the v_- node because the input and feedback path components both connect to it. Since $v_+ = 0$, let $E = v_-$. The v_- node is thus a virtual ground. A further complication is that v_i does not add directly to v_-; there is an input voltage divider between them. The divider attenuation is

$$\alpha_i = \frac{R_f}{R_f + R_i} \tag{3.19}$$

The value of α_i of the block diagram of Fig. 3.6b does not involve v_o since α_i is not in the feedback loop. Consequently, since an ideal op-amp has zero output resistance, the attenuation from v_i to v_- can be found by setting v_o to zero, and (3.19) results. α_i is v_-/v_i when $v_o = 0$, so only the α_i branch (and not the feedback branch, Hv_o) contributes to E.

Similarly, for H, the same components R_f and R_i form a divider in the reverse direction. In this case, from (3.15), $R = v_i$ is set to zero, and H is

$$H = -\frac{v_-}{v_o}\bigg|_{v_i=0} = -\frac{R_i}{R_f + R_i} \tag{3.20}$$

In effect, what we have done to find v_- is to apply the superposition theorem when we found α_i and H. Both v_i and v_o contribute to v_- through the divider components, and they sum to v_- by superposition. Voltage summation must occur in a loop, but here it occurs where two loops intersect. Both v_i and v_o sources can be Thévenized to produce the equivalent loop in which the summation Σ occurs. In this case, Σ is not obvious from the circuit topology.

The last block to be determined is G. It is $-K$ of the op-amp. In this case, $G < 0$ because the gain is from the inverting input. Combining these blocks, the closed-loop voltage gain is

$$\text{inverting op-amp} \qquad A_v = \frac{R_f}{R_f + R_i} \cdot \frac{-K}{1 + (-K)(-R_i/(R_f + R_i))} \tag{3.21}$$

The simplified gain expression for $K \to \infty$ is found by applying (3.10):

$$\text{inverting op-amp} \qquad A_v|_{K \to \infty} = -\frac{R_f}{R_i} \tag{3.22}$$

This simplified gain expression can be explained in terms of the circuit. With infinite K, v_- is a virtual ground, and we can let $v_- = 0$. Then the input current is v_i/R_i, and it must all flow through R_f, creating an output voltage of

$$v_o = -R_f\left(\frac{v_i}{R_i}\right) \tag{3.23}$$

From (3.23), (3.22) readily follows. This explanation does not easily fit into the previous approach, but it does suggest an alternative analysis.

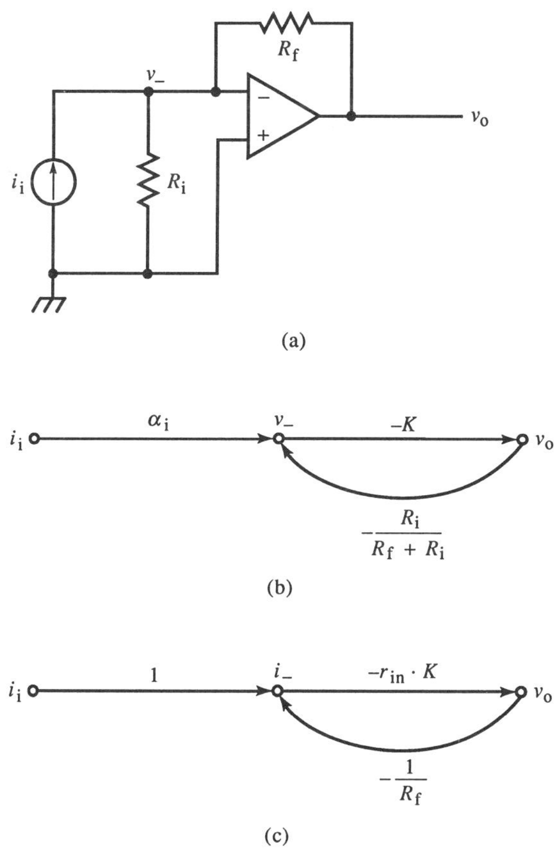

FIG. 3.7 The inverting op-amp configuration with a Norton equivalent input source (a). Choosing $E = v_-$ results in the flow graph of (b) whereas $E = i_-$ results in (c). Both are correct representations of (a).

Suppose that v_i and R_i are replaced by a Norton equivalent circuit, as shown in Fig. 3.7a. Here, $i_i = v_i / R_i$. What results is a transresistance amplifier with $r_m = v_o / i_i$. The blocks can be identified as follows:

$$\alpha_i = \left.\frac{v_-}{i_i}\right|_{B=0} = r_{in} = R_i \parallel R_f \tag{3.24}$$

$$G = \left.\frac{v_o}{v_i}\right|_{B=0} = -K \tag{3.25}$$

$$H = -\left.\frac{v_-}{v_o}\right|_{v_i=0} = -\frac{R_i}{R_f + R_i} \tag{3.26}$$

The closed-loop transresistance is α_i times (3.3) and is

$$r_m = \alpha_i \frac{G}{1 + GH} = \frac{R_f R_i}{R_f + R_i} \cdot \frac{-K}{1 + (-K)(-R_i/(R_f + R_i))} \tag{3.27}$$

For infinite K, this reduces to

$$r_m|_{K\to\infty} = -R_f \tag{3.28}$$

The gain of this transresistance amplifier is independent of R_i because it is from v_- to ground. For v_- as a virtual ground, no current flows through R_i. Its contribution to α_i cancels its contribution to the reverse voltage divider of H.

The voltage gain, A_v, of (3.21) can be derived using (3.27) and $i_i = v_i/R_i$:

$$A_v = r_m \cdot \frac{i_i}{v_i} = \frac{r_m}{R_i} \tag{3.29}$$

When the substitutions shown in (3.29) are made, (3.21) results. This shows that there is not necessarily a unique choice of blocks for closed-loop gain derivation.

To further illustrate this point, consider the choice of E in Fig. 3.7c. Instead of an error voltage v_-, an error current i_- is assumed. This choice of E most closely approaches the alternative explanation based on currents flowing into and out of a virtual ground at the v_- node. For this alternative, $\alpha_i = 1$; the input current contributes directly to $E = i_-$. G is a transresistance v_o/i_-. The op-amp responds to an input voltage of v_- at its inverting input, and i_- develops v_- across r_{in}. This voltage is multiplied by $-K$ of the op-amp to produce v_o. Therefore,

$$G = -r_{in}K \tag{3.30}$$

Finally, H must be a transconductance. With i_i set to zero, B is the current generated by v_o that flows into the inverting op-amp node at voltage v_-. This feedback current is $(v_o - v_-)/R_f$. Because this expression contains v_-, it is not too useful in determining feedback current B. Notice that R_f and v_o form a Thévenin equivalent circuit connected to the summing node. If we change it to a Norton equivalent circuit, R_f and R_i are in parallel (which is r_{in}), and the two current sources i_i and v_o/R_f are in parallel. Then,

$$H = -\frac{v_o}{i_B} = -\frac{1}{R_f} \tag{3.31}$$

Combining (3.30) and (3.31) with (3.3) gives the resulting transresistance:

$$r_m = \frac{v_o}{i_i} = \frac{G}{1+GH} = \frac{-r_{in}K}{1+(-r_{in}K)(-1/R_f)} \tag{3.32}$$

Substituting for r_{in} from (3.24) and rearranging (3.32) results in (3.27). We have solved for the closed-loop gain of the inverting op-amp configuration in three ways. What is mainly required to produce the correct closed-loop gain is consistency, making sure that blocks connect correctly and that sampling and summing circuits are in the loop.

3.4 Feedback Effects on Input and Output Resistance

The input resistance of a voltage summing amplifier is increased due to negative feedback. In Fig. 3.1a, let R and E be voltages and the circuit represented by this block diagram have an open-loop input resistance R_i, due to the input resistance of G shunting the output resistance of H (Fig. 3.8a). Without feedback, $E = R = v_i$, and the input voltage is applied to R_i directly. The same input voltage is largely canceled by feedback, resulting in a much smaller error voltage applied to R_i. In effect, the input resistance is increased because the same input voltage, v_i produces a current in R_i that is E/R_i. The ratio of closed-loop to open-loop input resistance is the ratio E/R. This ratio can be derived from (3.1) and (3.2) as follows:

$$\frac{E}{R} = 1 - H \cdot \frac{C}{R} = 1 - \frac{GH}{1+GH} = \frac{1}{1+GH} \tag{3.33}$$

The closed-loop reduction through R_i makes the input resistance effectively larger by $(1+GH)$, or

$$\text{voltage input} \qquad r_{in}(\text{cl}) = \frac{v_i}{i_i} = \frac{E+B}{i_i} = \frac{E(1+GH)}{i_i} = R_i(1+GH) \tag{3.34}$$

For the noninverting op-amp circuit of Fig. 3.6a, the effective op-amp differential input resistance is made larger by feedback. Since v_- tracks v_+ as feedback, the differential voltage across the input terminals is the error voltage. This small voltage produces a much smaller current in the op-amp input resistance than v_i alone would produce (with v_- grounded). In effect, the op-amp input resistance is bootstrapped by the feedback voltage. For the inverting configuration (Fig. 3.6b), since v_- is the error voltage and is small, op-amp input current is also much smaller than it would be with v_i alone applied to v_-. Thus, the resistance across which the error voltage is developed is effectively larger with feedback.

A similar argument applies to an amplifier with an error current. In this case, input resistance is reduced by $(1+GH)$. The input current (rather than voltage) is reduced by feedback current. The same input current results in a

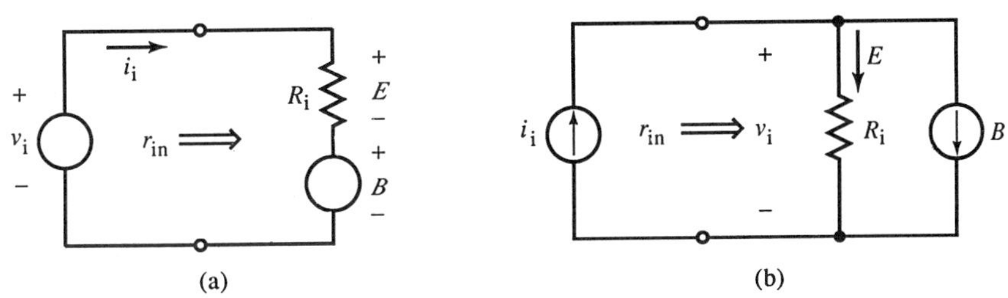

FIG. 3.8 Feeback effects on input resistance for (a) voltage and (b) current inputs.

reduced voltage across the input resistance due to feedback current cancellation (Fig. 3.8b). The resulting voltage across R_i produced by the error current is $1/(1+GH)$ times smaller than the voltage that the input current alone would produce. A smaller voltage resulting from the same input current means that the effective resistance is smaller:

$$\text{current input} \qquad r_{in}(\text{cl}) = \frac{v_i}{i_i} = \frac{v_i}{E+B} = \frac{v_i}{E(1+GH)} = \frac{R_i}{1+GH} \tag{3.35}$$

For the circuit of Fig. 3.7, open-loop input resistance is $R_i \parallel R_f$. With negative feedback, the resulting error current produces an error voltage of v_- that is much reduced from the voltage that i_i alone would produce across r_{in}. The closed-loop input resistance is effectively $(R_i \parallel R_f)/(1+GH)$.

The effect of negative feedback on output resistance can be analyzed by representing the output as a Thévenin equivalent voltage source with internal voltage of v and open-loop output resistance of R_o. An output voltage of v_o is produced by a current i_o applied to the output. With feedback, output voltage error due to the drop across R_o is corrected. The output voltage is (with no input signal applied):

$$v_o = v + i_o R_o = -GHv_o + i_o R_o = \frac{i_o R_o}{1+GH} \tag{3.36}$$

The effective closed-loop output resistance v_o/i_o is

$$\text{voltage output} \qquad r_{out}(\text{cl}) = \frac{R_o}{1+GH} \tag{3.37}$$

Similarly, for current output amplifiers, r_{out} is increased by feedback. If the output is represented by a Norton equivalent circuit with current source i, then internal shunt resistance R_o reduces output current. The output resistance can be found by applying a voltage v_o to the output. The resulting output current is

$$i_o = i + \frac{v_o}{R_o} = (-GH)i_o + \frac{v_o}{R_o} = \frac{v_o}{R_o(1+GH)} \tag{3.38}$$

The effective closed-loop output resistance is

$$\text{current output} \qquad r_{out}(\text{cl}) = R_o(1+GH) \tag{3.39}$$

Feedback has advantages for both input and output resistance. For the four cases considered, the effect of feedback is toward the ideal. We have considered the effect of input resistance at the error node or loop and output resistance at the sampled node or loop. For voltage summing, the resistance across the error voltage is increased; for current summing, the resistance through which the error current flows is decreased. If other gain blocks separate these nodes or loops from input or output, their effect on resistance must also be considered.

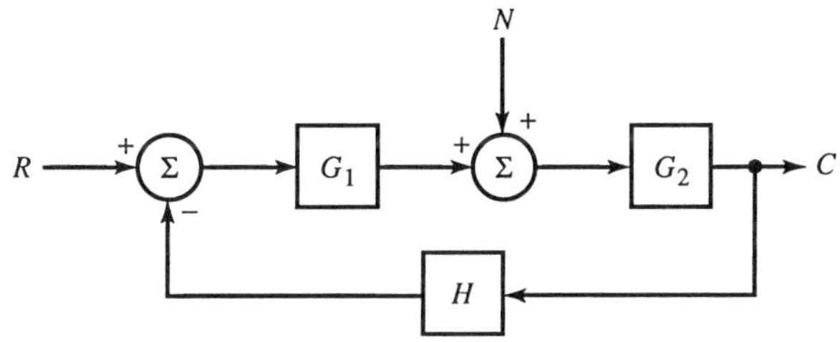

FIG. 3.9 A noise source N is injected into the feedback system forward path. Feedback rejects noise to the extent that N is injected toward the output C.

3.5 Noise Rejection by Feedback

Feedback increases the immunity of a circuit to noise, which is any undesirable electrical disturbance to the circuit. Figure 3.9 shows the classical feedback topology with the addition of noise N injected into the forward path. Noise rejection ability will be defined as the *signal-to-noise ratio* (SNR), the ratio of signal to noise at the output, or

$$\mathrm{SNR} = \frac{C/R}{C/N} \tag{3.40}$$

For the open-loop case,

$$\frac{C}{R} = G_1 G_2 = G, \qquad \frac{C}{N} = G_2 \tag{3.41}$$

Then open-loop SNR is

$$\mathrm{SNR}_{\mathrm{ol}} = \frac{G_1 G_2}{G_2} = G_1 \tag{3.42}$$

With feedback, the ratios are

$$\frac{C}{R} = \frac{G}{1+GH}, \qquad \frac{C}{N} = \frac{G_2}{1+GH} \tag{3.43}$$

The closed-loop SNR is

$$\mathrm{SNR}_{\mathrm{cl}} = G_1 \tag{3.44}$$

The open- and closed-loop SNRs are the same, suggesting no advantage to feedback. However, for the same input, the open-loop output is much larger than the closed-loop output for $G \gg 1$. Comparing C/R for open and closed loop, the open-loop gain is $(1+GH)$ times larger. For the same C/R (signal gain), the open-loop G must be $1/(1+GH)$ that of the closed-loop amplifier, or

$$G_{\mathrm{ol}} = \frac{G_{\mathrm{cl}}}{1+G_{\mathrm{cl}}H} \tag{3.45}$$

The ratio of open- to closed-loop SNRs shows the advantage of feedback:

$$\frac{\text{SNR}_{\text{cl}}}{\text{SNR}_{\text{ol}}}=\frac{G_{1\text{cl}}}{G_{1\text{ol}}}=\frac{G_{\text{cl}}/G_{2\text{cl}}}{G_{\text{ol}}/G_{2\text{ol}}}=\left(\frac{G_{2\text{ol}}}{G_{2\text{cl}}}\right)(1+G_{\text{cl}}H) \tag{3.46}$$

When $G_{2\text{cl}}=G_{2\text{ol}}$, the familiar $(1+GH)$ factor reappears as the advantage of feedback on SNR. This feature of op-amp circuits leads to better rejection of noise from the power supply by stages following a sufficiently large G_1. The closer to the input that N is injected (that is, the smaller proportion of G that G_1 is), the less advantage feedback has. In the extreme, noise injected at the input summer is indistinguishable from signal. Noise at the output is rejected by a factor of $(1+GH)$.

3.6 Reduction of Nonlinearity with Feedback

A further benefit of feedback is the linearization of nonlinear forward path gain blocks. Assuming the classical feedback topology of Fig. 3.1, let $G=K+\varepsilon$, where K is a fixed gain and ε represents the nonlinear terms of G; ε varies with E or C. The closed-loop gain T is

$$T=\frac{(K+\varepsilon)}{1+(K+\varepsilon)H} \tag{3.47}$$

For $K \gg \varepsilon$, $1+(K+\varepsilon)H \cong 1+KH$, and T can be separated into linear and nonlinear terms:

$$T \cong \frac{K}{1+KH}+\frac{\varepsilon}{1+KH}, \qquad K \gg \varepsilon \tag{3.48}$$

The second term is the nonlinear closed-loop gain. The open-loop nonlinearity has been reduced by $(1+KH)$.

In all of the improvements brought about by feedback, the improvement factor has been $(1+GH)$. The improvements investigated here were the sensitivity of the closed-loop gain to the open-loop gain, input and output resistances, noise rejection, and linearization of nonlinear open-loop forward-path gain.

3.7 Miller's Theorem

The inverting op-amp configuration has a resistor connected from output to inverting input. This is not uncommon for feedback amplifiers and can be generalized as shown in Fig. 3.10a. The inverting voltage amplifier has a gain of $-K$ with input quantities of v_i and i_i. The output voltage is v_o. The equivalent input resistance can be found as follows. First,

$$v_0=-K\cdot v_i \tag{3.49}$$

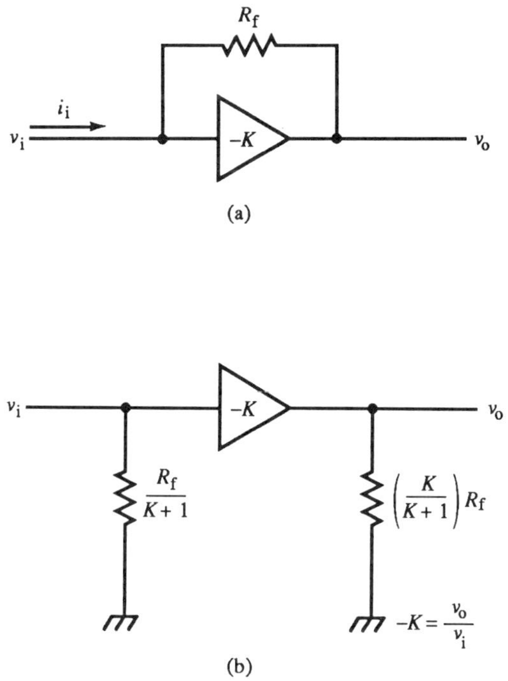

FIG. 3.10 Miller's theorem produces equivalent circuit (b) from the inverting voltage amplifier shunted by resistance R_f in (a).

For the input node, applying KCL we obtain

$$\frac{v_i - v_o}{R_f} - i_i = 0 \tag{3.50}$$

Substituting for v_o from (3.49) gives

$$\frac{v_i - (-Kv_i)}{R_f} = \frac{(1+K)v_i}{R_f} = i_i \tag{3.51}$$

Rearranging (3.51) for input resistance, we have

$$\text{Miller's theorem} \qquad r_{in} = \frac{R_f}{1+K} \tag{3.52}$$

This result is in conformance with (3.35) and also follows from it as a special case. It is given separately here because it appears repeatedly when working with CE amplifier stages with collector-to-base feedback.

For an amplifier with output resistance, the equivalent shunt contribution due to R_f can be found similarly, or from (3.39):

$$r_{out} = \frac{v_o}{-i_i} = \frac{(-K)v_i}{-v_i(1+K)/R_f} = \left(\frac{K}{1+K}\right)R_f \tag{3.53}$$

From the output, R_f appears to be slightly less than its actual value for large K. From the input, R_f appears to be $1/(1+K)$ times its actual value, causing input resistance to be much reduced and providing a low-resistance path for i_i. For infinite K, the input node is a virtual ground, as for the inverting op-amp. The equivalent circuit resulting from Miller's theorem is shown in Fig. 3.10b.

3.8 An Inverting Feedback Voltage Amplifier

The first discrete transistor feedback amplifier we will analyze using the techniques already developed is shown in Fig. 3.11. To simplify analysis, an ideal unity-gain buffer follows the collector of Q_1, but otherwise could be an emitter-follower. This circuit poses a challenge in identifying the input summer Σ. Since the input is v_i, a voltage, the error quantity must also be a voltage (without a quantity-transforming α_i). Voltages are summed around loops, according to KVL. Since the error voltage must be in a loop that includes the feedback, the loop containing R_f (the only component providing feedback) must include the error voltage. It is appealing to let the error voltage be v_b since it meets these requirements. Then both v_i and v_o contribute to the loop containing R_f and R_b. This error loop is shown in Fig. 3.11b and is similar to that for the inverting op-amp configuration—two loops combine, resulting in error voltage v_b across part of the loop.

What is not apparent is what to do with the input resistance of Q_1: $r_{in} = (\beta+1)(r_e + R_E)$. It turns the two voltage sources with their series resistances into a loaded divider similar to that of the cascade amplifier (Fig. 2.8). As in that case, a decision must be made about what to do with r_{in}. If it is included in calculation of v_b, then when the gain of G is found, the external base resistance is not included in the gain formula because it was taken into account in finding v_b. Alternatively, by solving for a Thévenin equivalent v_E in series with $R_B \parallel R_f$ (Fig. 3.11c), this resistance is included in the gain formula of Q_1 as external base resistance.

We first analyze the circuit by letting r_{in} load the input. The contribution of v_i to v_b is through the divider formed by R_B in series with $r_{in} \parallel R_f$. This divider has a transfer function of $v_b/v_i = \alpha_i$:

$$\alpha_i = \frac{R_f \parallel r_{in}}{R_f \parallel r_{in} + R_B} \tag{3.54}$$

A similar divider defines H:

$$H = -\left.\frac{v_o}{v_b}\right|_{v_i=0} = -\frac{R_B \parallel r_{in}}{R_B \parallel r_{in} + R_f} \tag{3.55}$$

Since r_{in} is taken into account in α_i and H, v_b is the actual base voltage with

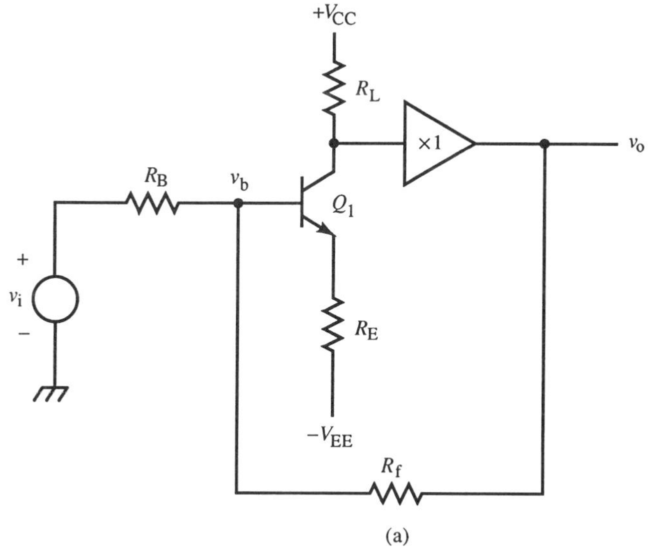

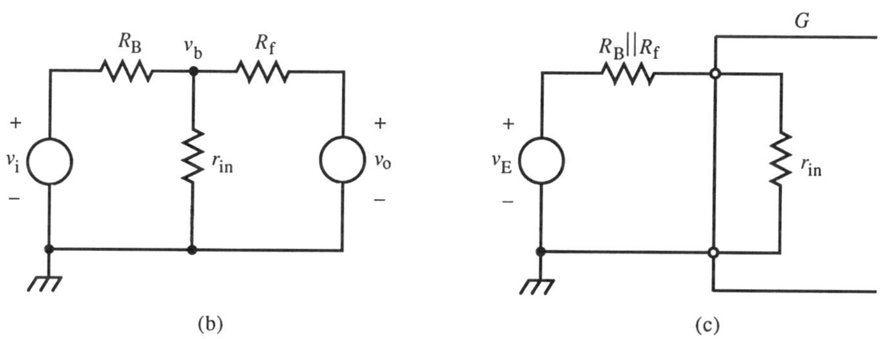

FIG. 3.11 (a) A typical inverting voltage feedback amplifier. (b) For $E = v_b$, v_i and v_o sum by superposition across r_{in}. (c) For $E = v_E$, the Thévenin equivalent voltage source is a combination of input and fedback output.

external base resistance taken into account. Therefore, G does not include its effect and is

$$G = \frac{v_o}{v_b} = -\alpha \cdot \frac{R_L}{r_e + R_E} \tag{3.56}$$

The closed-loop voltage gain is found by substituting (3.54)-(3.56) into (3.3) and is

$$A_v = \alpha_i \cdot \frac{G}{1+GH} = -\alpha \cdot \frac{R_L}{r_e + R_E + R_B/(\beta+1) + (r_e + R_E + \alpha R_L)(R_B/R_f)} \tag{3.57}$$

This gain expression is similar to that of nonfeedback amplifier stages but has no transresistance interpretation; it is merely a simplified form of the feedback formula in terms of component values and Q_1 parameters. It is useful for calculating the closed-loop gain but offers little of the insight into feedback characteristics of individual block transfer functions.

An alternative derivation of (3.57) is based on incorporation of external base resistance into G. In this approach, E is different; the error voltage is no longer v_b but is the Thévenin equivalent voltage from the voltage divider formed by the two sources with the base of Q_1 open. It is

$$v_E = \left(\frac{R_f}{R_f + R_n}\right) v_i + \left(\frac{R_B}{R_f + R_B}\right) v_o \tag{3.58}$$

This Thévenin error voltage is in series with a Thévenin resistance of $R_B \parallel R_f$ that forms a divider with r_{in} (Fig. 3.11c). Then,

$$v_b = \left(\frac{r_{in}}{r_{in} + r_B \parallel R_f}\right) v_E \tag{3.59}$$

It is apparent from (3.58) that α_i is different from the previous analysis and is

$$\alpha_i = \left.\frac{v_E}{v_i}\right|_{v_o = 0} = \frac{R_f}{R_f + R_B} \tag{3.60}$$

From (3.58), H is

$$H = -\left.\frac{v_E}{v_o}\right|_{v_i = 0} = -\frac{R_B}{R_f + R_B} \tag{3.61}$$

Finally, G is

$$G = \left.\frac{v_o}{v_E}\right|_{B = 0} = -\alpha \cdot \frac{R_L}{r_e + R_E + (R_B \parallel R_f)/(\beta + 1)} \tag{3.62}$$

Combining (3.60)-(3.62) into (3.3) and rearranging gives (3.57). This circuit demonstrates a multiplicity of valid choices for the error voltage. In the loaded-divider case, the error is an actual node voltage v_b whereas in the second analysis it is the Thévenin voltage v_E. Both approaches produce the correct closed-loop gain.

3.9 Input and Output Loading

If the buffer of the inverting amplifier (Fig. 3.11) were replaced by a CC stage, it would be included in G, and its output resistance would form a voltage divider with R_f. The divider would be part of G since the divider output is v_o. The CC stage gain depends on the resistance at the output node. But what is this resistance? At the output, R_f is in series with $R_B \parallel r_{in}$. This appears to be the CC load, but it is not. Output loading due to H is R_f only because v_b is an actual voltage calculated by taking into account the loading of R_f. Similarly, the loading of the output on the input suggests that, from the v_b

node, R_f is in series with the CC emitter and its β-transformed base resistance R_L and is

$$R_f + r_e + \frac{R_L}{\beta + 1}$$

But likewise, this is not the loading on the input. The CC output resistance would not be added to R_f when determining external base resistance in (3.62) because it would already have been taken into account in the output divider in G.

The situation is similar to the input loading problem. If the loading of r_{in} is taken into account when calculating v_b, then the base resistance $R_B \parallel R_f$ should not be accounted for when calculating the gain of the Q_1 stage; it was already accounted for when calculating v_b. But keeping track of this accounting is not always easy, and a more formal basis for it is needed when determining the input and output loading on G. If the loading interactions between G and H are applied to G, then when H is determined, loading will already have been accounted for. To clarify this, we consider the general problem of determining loading on the input and output of G.

To approach the loading interactions of G and H generally, we model both of these blocks as two-port networks. Each will have either Thévenin or Norton equivalent circuits representing input and output. Both input and output (voltage or current) sources are controlled by either the two-port voltage or current of the opposite port.

We now establish constraints on what these source dependencies must be so that loading can be found. Because these are controlled sources, equivalent resistance cannot be found by shorting or opening them because the sources themselves can affect the resistance. If the source value varies with a terminal quantity, it appears to have a finite resistance. For example, a Thévenin resistance cannot be found by shorting the controlled voltage source in series with it. (See the derivation of Miller's theorem. The output voltage is controlled by the input voltage, affecting the input resistance.) Also, the two-port sources do not correspond to actual sources in the circuit topology. Shorting or opening them does not necessarily correspond to a short or open in the actual circuit. Their controlling variables, however, are actual circuit quantities. For example, if a two-port equivalent voltage source is $k_i v_o$, it can be nulled by shorting the v_o node of the actual circuit.

First consider the loading on the output of G caused by the input to H. The input equivalent circuit to H could be either a Norton or Thévenin circuit. In either case, its source is dependent on either the H output port voltage or current. If voltage, then shorting the H output port causes the port voltage to be zero, and the dependent source driving the H input has zero output. This removes the effect of the source and leaves the H input resistance alone loading the output of G. The output loading due to H has been isolated. The dual case is a source dependent on the H output current. An open H output port causes this current to be zero, nulling the source at the input of *H* and leaving only the *H* input resistance.

In either case, the *H* input source is dependent on an *H* output quantity that is set to zero by either shorting a node or opening a loop at the output port of *H*. If the error quantity is a voltage, then it is summed around a loop that includes the *H* output port. Opening the port (thus breaking the loop) causes the error loop current to be set to zero. The *H* input source is made dependent on this loop current to be removed by opening the error voltage summing loop. Similarly, if the error quantity is a current, input and feedback currents are summed at a node. The *H* input source is made dependent on the voltage at this node (to ground) so that when it is shorted, the *H* input source is removed. This can be summarized by the following conditions (Fig. 3.12).

Output loading conditions:

- Voltage feedforward through *H* is nulled when the error summing node is shorted, causing the output of *G* to be loaded by the *H* input. (If $r_{in} = 0$, open the input to *G* to assure that error current $= 0$.)
- Current feedforward through *H* is nulled when the error summing loop is opened, causing the output of *G* to be loaded by the *H* input. (If r_{in} of *G* is infinite, short the input to *G* to assure that error voltage $= 0$.)

Since the error quantity is arbitrary within the feedback loop, the loading is also arbitrary but is dependent on the choice of error.

A simplifying constraint upon the two-port output sources of *H* involves the error circuit. For voltage summation, input and feedback voltages must sum around a loop. To insure that these voltages are being summed in the same loop, the same current i_{loop}, must flow through all loop elements. When the output of *H* is represented as a Thévenin equivalent circuit (Fig. 3.12a), the Thévenin voltage source and resistance are in series with the loop, and i_{loop} flows through both. (A Norton circuit would have created another loop.) In the case of an error current, summation occurs at a node (Fig. 3.12b). To insure that input and feedback currents are summed at the same node, the same voltage v_{node} must be across each source at the summing node. The output of *H* is thus a Norton equivalent circuit. Both current source and equivalent resistance are across v_{node}.

This constraint on choice of equivalent circuit is not necessary because whatever representation is used of the port, the port voltage and current are still the same. With this constraint, the summing circuit is easier to identify because an additional loop or node has been avoided. The identification of *H* is also made easier because the *H* output source quantity is the same kind as the error.

Now consider input loading due to the output of *H*. If the sampled quantity to be fed back is a voltage, it must be at the sampled node (to ground) and

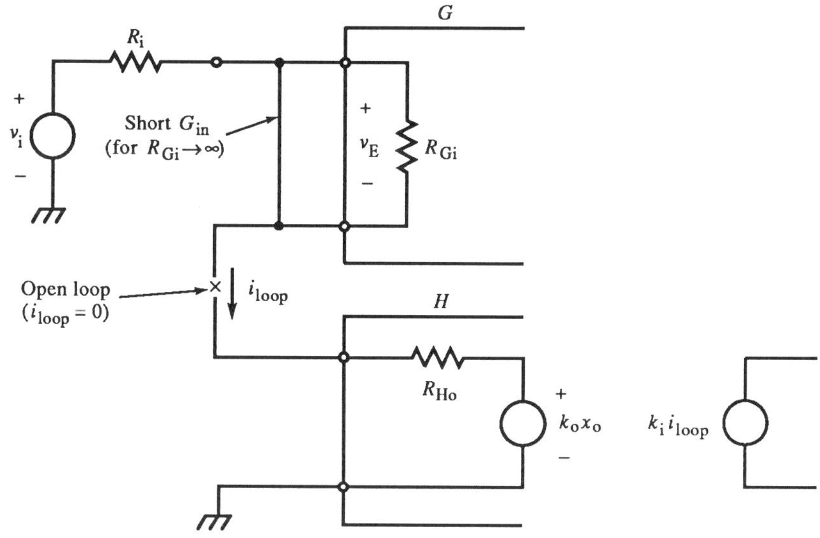

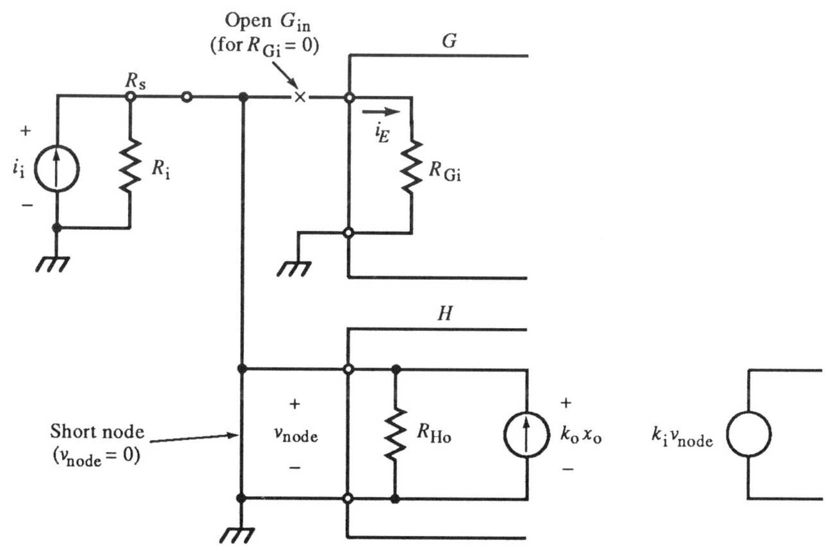

FIG. 3.12 Loading on the output of G is found by nulling the H input source, which is driven by an H output quantity. For an error voltage, E is summed around a loop (a) which, when opened, causes E and $k_i i_{loop}$ to be zero. For an error current, E is summed at a node (b) which, when shorted, causes E and $k_i v_{node}$ to be zero.

across the H input. It is nulled ($B = 0$) when the sampled node is shorted. Then the output loading of H on the input can be found. For sampling a current, feedback is nulled when the loop containing the current is opened. The second simplifying constraint on H, therefore, is that its output source be dependent on the sampled quantity. Then the following conditions can be applied to find input loading.

Input loading conditions:

- Voltage feedback through H is nulled ($B = 0$) when the output sampling noded is shorted, causing the input of G to be loaded by the H output.
- Current feedback through H is nulled ($B = 0$) when the output sampling loop is opened, causing the input of G to be loaded by the H output.

The reason is the same as for output loading. If the H output source is set to zero, what appears across the port is the H output loading resistance.

Simplifying constraints on H as a two-port network:

- The H input source must be controlled by the H output quantity that, when set to zero, makes the error (input to G) zero.
- The H output source must be controlled by the G output quantity.

These constraints apply to the given loading conditions. When the conditions are evaluated for voltage and current cases, simple loading rules result.

Loading rules:

- For input loading, short H input if G output is a voltage; open H input if G output is a current.
- For output loading, short H output if error is a current; open H output if error is a voltage.

To illustrate the use of two-port equivalent models, the passive network of Fig. 3.13a will be analyzed using the two-port model of Fig. 3.13b. We need to determine the two-port parameters R_a, R_b, a, and b. For the resistances,

$$R_a|_{v_b=0} = R_f,\ R_b|_{v_a=0} = R_f \tag{3.63}$$

These resistances are found by setting their series sources to zero. For R_a this requires that $v_b = 0$. This is accomplished in Fig. 3.13a by shorting the node of v_b. The resulting resistance loading the v_a node is R_f. The method is the same for R_b.

For the transmittances a and b,

$$a = \left.\frac{v_a}{v_b}\right|_{i_a=0} = 1, \qquad b = \left.\frac{v_b}{v_a}\right|_{i_b=0} = 1 \tag{3.64}$$

By opening the ports, we can find the (open-circuit) Thévenin equivalent

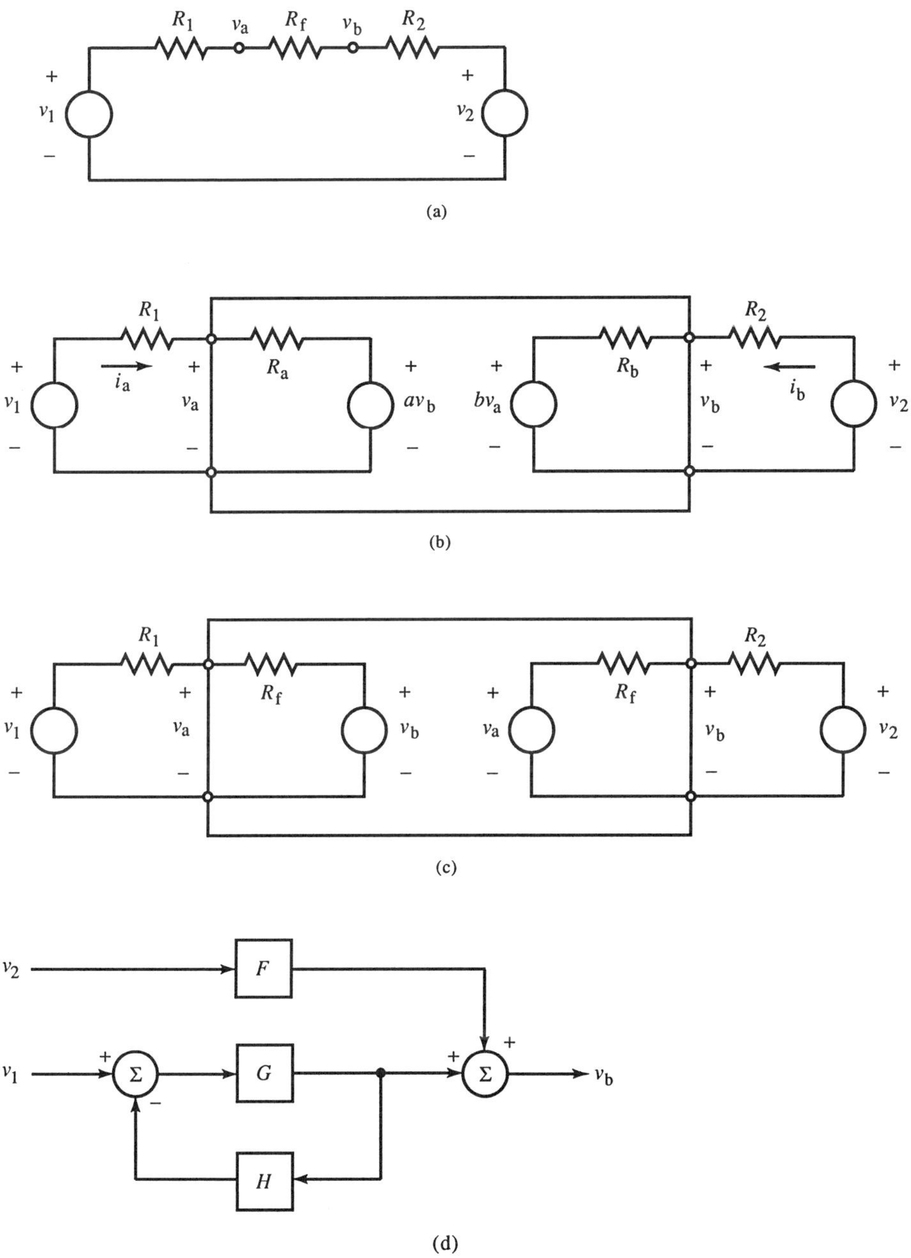

FIG. 3.13 A simple network (a) with two-port equivalent circuit (b). The four parameters of the two-port are determined in (c). The block diagram of (c) is given in (d).

voltages in terms of the controlling variable. The resulting two-port equivalent circuit is shown in Fig. 3.13c. The two sides of the circuit have been isolated from each other except through controlled sources. When this network is solved for v_a and v_b, the result is

$$v_a = \frac{R_2 + R_f}{R_1 + R_2 + R_f} \cdot v_1 + \frac{R_1}{R_1 + R_2 + R_f} \cdot v_2 \tag{3.65}$$

$$v_b = \frac{R_2}{R_1 + R_2 + R_f} \cdot v_1 + \frac{R_1 + R_f}{R_1 + R_2 + R_f} \cdot v_2 \tag{3.66}$$

Of curious note, these equations can be cast into feedback form (Fig. 3.13d). For v_b,

$$G = \frac{R_2}{R_2 + R_f} \cdot \frac{R_f}{R_1 + R_f} \quad H = -\frac{R_1}{R_1 + R_f} \quad F = \frac{(R_f/(R_2 + R_f))}{1 - (R_1/(R_1 + R_f))(R_2/(R_2 + R_f))}$$

A similar formulation is possible for v_a. Although this simple network could be solved for v_a and v_b more easily by using the voltage divider formula, the two-port approach is illustrated by it.

The inverting amplifier of Fig. 3.11 is the second example. The two-port network for H is shown in Fig. 3.14a. Since $E = v_b$, the error voltage is directly across (rather than in series with) the H output. When it is shorted, E is directly nulled. The H input source is thus set to zero, and R_f loads the output.

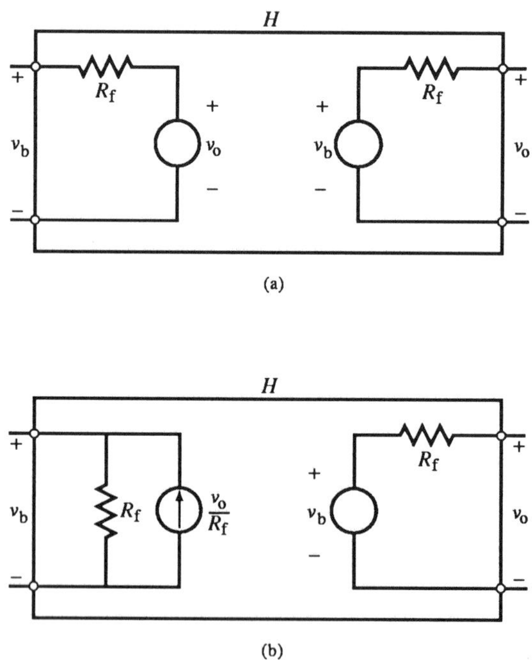

FIG. 3.14 A two-port equivalent network for H of the inverting amplifier of Fig. 3.11 for (a) $E = v_b$ and (b) $E = i_b$.

The sampled quantity is v_o. The H output source must be controlled by it. Shorting it allows the input loading resistance R_f to be found.

If E is chosen as i_b instead, the resulting H network is shown in Fig. 3.14b. The H output is a Norton circuit, so the current source directly connects to the current summing node. The H output port voltage is still v_b. When shorted, i_b is diverted through the short, ensuring that $E = 0$. This causes the H input voltage source, controlled by v_b, to be set to zero, and the input loading on the G output can be found. To find the loading of the H output on the G input, the G output quantity v_o is shorted. Since it is the H input voltage, it controls the H output current source and sets it to zero. This leaves R_f as the H output loading on the G input.

Example 3.1 Inverting Feedback Amplifier

Figure E3.1 shows an inverting feedback amplifier similar to that of Fig. 3.11. After dc analysis[1] (and assuming $\beta = 99$, $I_S = 10^{-16}$ A), the dynamic emitter resistances are

$$r_{e1} = 230.93\ \Omega; \qquad r_{e2} = 18.61\ \Omega$$

A few other incremental resistances are needed. The input can be Théven-ized, combining R_B and R_P into r_s; the input voltage source has an attenuation of α_s:

$$r_s = R_B \parallel R_P = 861.11\ \Omega, \qquad \alpha_s = \frac{R_P}{R_P + R_B} = 0.8611$$

Let $E = v_b = v(20)$ and $C = v_{e2} = v(40)$. The input resistance of Q_1 is

$$r_{Gi} = (\beta_1 + 1)(r_{e1} + R_{E1}) = (100)(230.9\ \Omega + 330\ \Omega) = 56.093\ \text{k}\Omega$$

Then,

$$\alpha_i = \alpha_s \cdot \frac{r_{Gi} \parallel R_f}{r_{Gi} \parallel R_f + r_s} = 0.78179$$

$$G = -\alpha_1 \cdot \frac{R_L}{r_{e1} + R_E} \cdot \frac{R_f \parallel (R_{E2} + R_{E3})}{R_f \parallel (R_{E2} + R_{E3}) + r_{e2} + R_L/(\beta_2 + 1)} = -36.201$$

[In Chapter 4, we learn that there is another forward path through R_f to v_{e2}. It adds to G because it is in parallel with it:

$$G_2 = \frac{(R_{E2} + R_{E3}) \parallel (r_{e2} + R_L/(\beta_2 + 1))}{(R_{E2} + R_{E3}) \parallel (r_{e2} + R_L/(\beta_2 + 1) + R_f)} = 2.225 \times 10^{-2}$$

When this is added to G, $G = -36.179$, a difference of 0.06%.]

1. This analysis can be avoided for ac problem-solving exercises by using the dc bias solution from the circuit simulation data.

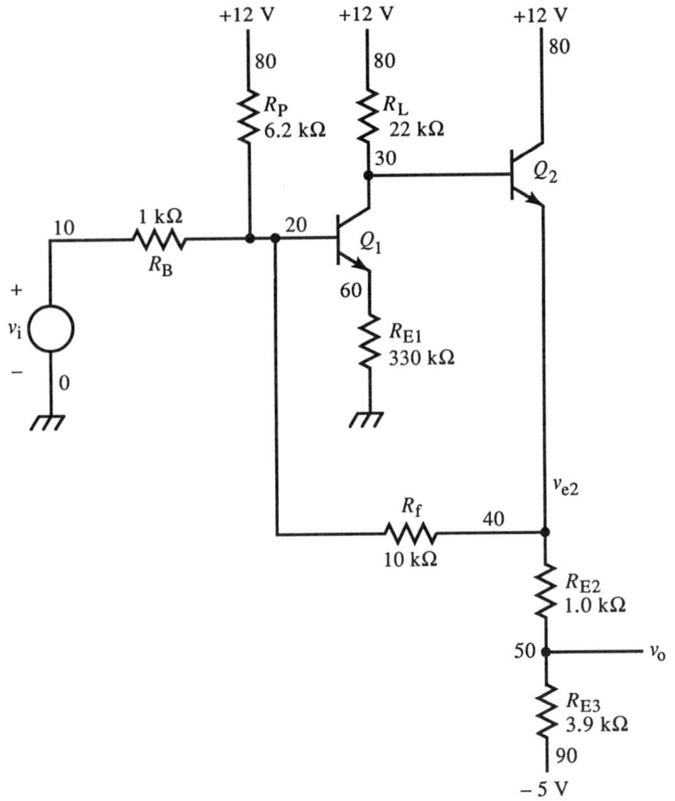


```
E3.1 Inverting 2-BJT Feedback
     Amplifier

.OPT NOMOD OPTS NOPAGE
.OP

.DC VI -2V 2V 0.05V
.TF V(50) VI

VCC 80 0 DC 5V
VEE 90 0 DC -5V
VI  10 0 DC 0V

RB  10 20 1 K
RP  80 20 6.2K
RL  30 80 22K
RF  20 40 10K
RE  60  0 330
RE1 40 50 1K
RE2 50 90 3.9K

Q1 30 20 60 BJT1
Q2 80 30 40 BJT1
.MODEL BJT1 NPN (BF=99)

.END

NODE VOLTAGE
(20) .7545  (30) 2.2458
(40) 1.4634  (50) .1443
(60) .0371

OPERATING POINT INFORMATION
TEMPERATURE=   27.000 DEG C

BIPOLAR JUNCTION TRANSISTORS

NAME     Q1          Q2
MODEL    BJT1        BJT1
IB       1.12E-06    1.39E-05
IC       1.11E-04    1.38E-03
VBE      7.17E-01    7.82E-01
VBC     -1.49E+00   -2.75E+00
VCE      2.21E+00    3.54E+00
BETADC   9.90E+01    9.90E+01
GM       4.30E-03    5.32E-02

V(50)/VI=-5.883E+00
INPUT RESISTANCE AT VI=1.256E+03
OUTPUT RESISTANCE AT V(50)=8.327E+02
```

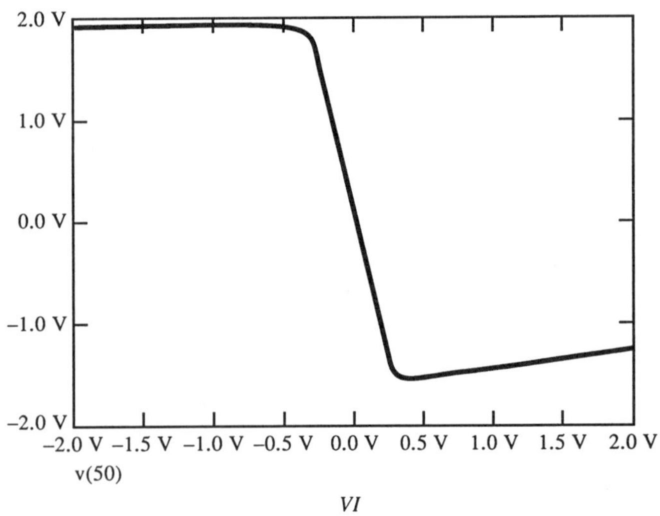

FIG. E3.1

The remaining transmittances are

$$H = -\frac{r_s \parallel r_{Gi}}{r_s \parallel r_{Gi} + R_f} = -7.8179 \times 10^{-2}$$

$$\alpha_o = \frac{R_{E3}}{R_{E2} + R_{E3}} = 0.79592$$

Putting the transmittances together, we obtain a voltage gain of

$$\frac{v_o}{v_i} = \alpha_i \cdot \frac{G}{1+GH} \cdot \alpha_o = -5.8802$$

where $1 + GH = 3.8284$. The input and output resistances are

$$r_{in} = R_B + R_P \parallel r_{Gi} \left\| \frac{R_f}{1+(-G)} \right. = 1.2566 \text{ k}\Omega$$

$$r_{out} = \left\{ \frac{(R_B \parallel R_P \parallel r_{Gi} + R_f) \parallel [r_{e2} + R_L/(\beta_2+1)] \parallel (R_{E2} + R_{E3})}{1+GH} + R_{E2} \right\} \Bigg\| R_{E3}$$

$$= 832.3\ \Omega$$

The SPICE simulation verifies these numbers to three digits, the given simulation convergence accuracy. The graph of $v_o = v(50)$ versus v_i shows feedforward through R_f outside the linear range of the active path (through the transistor) when its gain is zero.

3.10 A Noninverting Feedback Amplifier

The next feedback amplifier example is shown in Fig. 3.15a. It feeds back via R_f to the emitter of the CE input transistor Q_1. Since both input and output quantities are voltages, this is a voltage amplifier. To simplify analysis, the output of the CE stage is buffered by an ideal $\times(-1)$ amplifier. (This could be implemented as a PNP CE stage.)

We will analyze this circuit for three different choices of error E. The first choice is preferred because it is simplest to reduce to gain and summing blocks.

From Fig. 3.15a, let

$$R_E = R_f \parallel R_i \qquad \text{and} \qquad r_s = r_e + \frac{R_B}{\beta+1} \tag{3.67}$$

The summing loop must include both v_i and v_o terms (Fig. 3.15b). The feedback

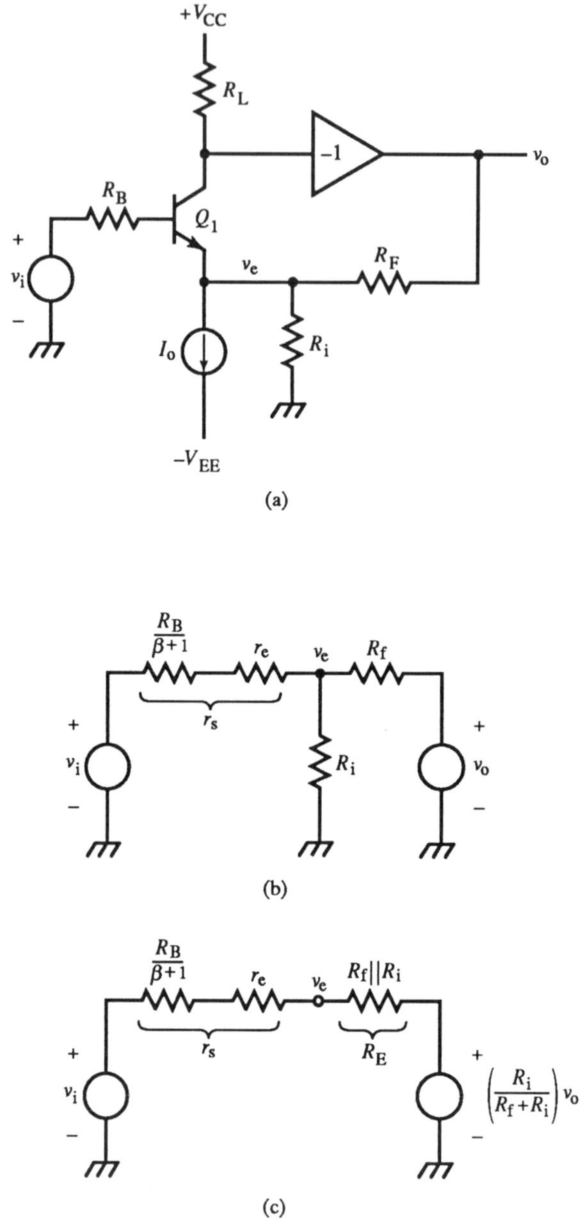

FIG. 3.15 A typical noninverting voltage feedback amplifier (a) showing details of the loaded divider summing loop (b). By Thévenizing the right loop of (b), the summing loop becomes explicit (c). E is the Thévenin voltage source coefficient.

from v_o is Thévenized with a series resistance of R_E and Thévenin voltage source of

$$v_{fb} = \left(\frac{R_i}{R_f + R_i}\right) v_o \tag{3.68}$$

The output of this equivalent circuit is v_e. The loop continues (Fig. 3.15c) and includes r_s and v_i. The Thévenin equivalent circuit of Fig. 3.15b is shown in Fig. 3.15c.

The error is chosen to be

$$E = v_i - v_{fb} \tag{3.69}$$

Since the input quantity v_i adds directly to E, $\alpha_i = 1$. From (3.69),

$$H = -\frac{E}{v_o}\bigg|_{v_i=0} = \left(\frac{R_i}{R_f + R_i}\right) \tag{3.70}$$

and the forward path is

$$G = \frac{v_o}{E}\bigg|_{B=0} = \alpha\frac{R_L}{r_s + R_E} \tag{3.71}$$

This results in a closed-loop voltage gain of

$$A_v = \frac{\alpha(R_L/(r_s + R_E))}{1 + [\alpha(R_L/(r_s + R_E))][R_i/(R_f + R_i)]} \tag{3.72}$$

Now that the circuit has been analyzed by a straightforward analysis, we will investigate alternatives to show that feedback circuits can be solved several ways. Examination of two alternatives should clarify some of the subtler aspects of the approach.

First, notice that in the preceding solution of A_v, because v_{fb} was chosen as the term relating v_o to E in (3.69), the gain of G was calculated to include R_E in the transresistance of Q_1. If v_e were chosen for E instead of (3.69), R_E would have been taken into account in the expression for v_e, and G would have only r_s in its denominator. In this case, the transresistance of G would be around an input loop from v_i to v_e instead of from v_i to v_{fb}.

Taking this approach, we begin by noting that both v_i and v_o sum at the emitter of Q_1. The loop that contains E is shown in Fig. 3.15c. An expression for the emitter voltage v_e can be constructed by superposition:

$$v_e = \underbrace{\left(\frac{R_E}{r_s + R_E}\right)}_{\uparrow\ \alpha_i} v_i + \underbrace{\left(\frac{r_s}{r_s + R_E}\right)\left(\frac{R_i}{R_f + R_i}\right)}_{\uparrow\ -H} v_o \tag{3.73}$$

This time, let $E = v_e$. It qualifies because it contains terms for both v_i and v_o. The coefficients of these quantities are expressions for blocks α_i and $-H$, as designated in (3.73). These blocks are

$$\alpha_i = \frac{R_E}{r_s + R_E}, \qquad H = -\left(\frac{r_s}{r_s + R_E}\right)\left(\frac{R_i}{R_f + R_i}\right) \tag{3.74}$$

Now an unusual step must be taken to produce G. We need two equivalent expressions, which are

$$G = -\alpha\frac{R_L}{r_s} = \alpha\frac{R_L}{R_E} \tag{3.75}$$

These expressions for G are equal because the voltage developed across R_E is the same as that developed across r_s except that they are inverted in polarity (Fig. 3.15a). If v_E increases, the voltage across R_E increases and causes an increase in I_E. This same voltage change across r_s causes a decrease in I_E out of the emitter of Q_1, inverting the polarity of the change in I_E. Hence the minus sign for the first expression of (3.75).

Combining the gain expressions gives the closed-loop gain as

$$A_v = \left(\frac{R_E}{r_s + R_E}\right) \cdot \frac{\alpha(R_L/R_E)}{1 + \alpha(R_L/r_s)[r_s/(r_s + R_E)] \cdot [R_i/(R_f + R_i)]} \tag{3.76}$$

This is equivalent to (3.72) after the α_i factor is multiplied by the numerator. Although the result is the same, the derivation is less obvious due to the need for two expressions for G.

The final approach uses a different choice for E in that v_e is used as the output-related term and is subtracted from the input v_i:

$$E = v_i - v_e \tag{3.77}$$

E can be expanded as we did in (3.73):

$$E = v_i - \left[\left(\frac{R_E}{r_s + R_E}\right)v_i + \left(\frac{r_s}{r_s + R_E}\right)\left(\frac{R_i}{R_f + R_i}\right)v_o\right] \tag{3.77a}$$

This reduces to

$$E = \left(\frac{r_s}{r_s + R_E}\right)\left(v_i + \left[\frac{R_i}{R_f + R_i}\right]v_o\right) \tag{3.77b}$$

From (3.77b), α_i and H are

$$\alpha_i = \left(\frac{r_s}{r_s + R_E}\right), \qquad H = \left(\frac{r_s}{r_s + R_E}\right)\left(\frac{R_i}{R_f + R_i}\right) \tag{3.78}$$

Since $H = -(v_i - v_e)/v_o$, solving (3.77) for $-E/v_o$ with $v_i = 0$ yields a negative expression that is then negated to produce a positive H of v_e/v_o.

The gain of G is from E to v_o, and G is calculated based on the difference in (3.77), or

$$G = \alpha\frac{R_L}{r_s} \tag{3.79}$$

Combining (3.78) and (3.79) with (3.3) and rearranging gives (3.72). This choice of E does not require two gain expressions for G and is somewhat simpler conceptually. In both solutions, E is expressed as a linear combination of v_i and v_o.

These solutions were constructed by inspection of the loops and nodes of the circuit without use of the two-port constraints. The analysis took place at a higher level than basic circuit laws, however, since the transresistance method and identification of voltage or current dividers make it possible to write the block gains directly from inspection of the circuit. This approach is intuitively simple as long as the H loading is obvious and G and H are easy to identify. Otherwise, the two-port approach would be applied (Fig. 3.16).

To set the error voltage to zero, the loop around which it is summed must be opened. The loop current is i_e; opening the loop at the emitter of Q_1 sets E to zero. Since $i_e = 0$ nulls E, the H input source must be controlled by i_e. Also, since v_o is the sampled output of G, the H output source must be controlled by it. The H output is the result of Thévenizing involving R_i, R_f, and v_o. When v_o is shorted, the G input loading is $R_i \parallel R_f$. When i_e is nulled by opening the emitter, the G output loading is $R_f + R_i$. The H output source voltage is the voltage due to v_o with zero port current ($i_e = 0$); it is

$$\left(\frac{R_i}{R_f + R_i}\right) v_o$$

due to the divider action of R_f and R_i. The H input source voltage appears at the H input when the H input port current is zero. When the port is opened (disconnecting it from v_o), the resulting voltage is due to i_e flowing through R_i, or $R_i i_e$.

As for the virtues of this feedback amplifier, two transistors provide a gain determined by R_f and R_i for large G. With two BJTs, G can be made larger by allowing the output inverting buffer to have a gain magnitude $\gg 1$. A typical

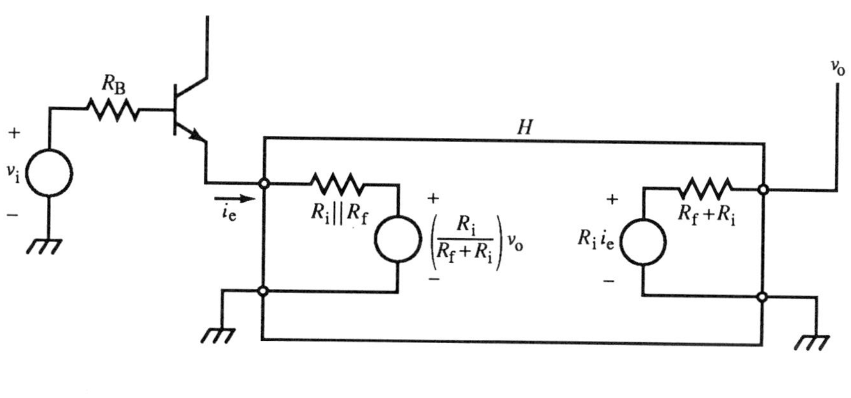

FIG. 3.16 Two-port equivalent circuit for H for the noninverting feedback amplifier.

achievable gain for G is 500. For a closed-loop gain $\ll 500$, the gain is set predominantly by the external resistors. This circuit is a simple discrete implementation similar to the noninverting op-amp configuration.

Example 3.2 Noninverting Feedback Amplifier

Figure E3.2 is a noninverting feedback amplifier with an idealized $\times(-1)$ buffer, similar to the one in Fig. 3.15. Assuming the dc analysis from the simulation data, we obtain $r_e = 32.73\ \Omega$. Then the three quantities of interest are

$$\frac{v_o}{v_i} = \frac{G}{1+GH} = \frac{3.2106}{1+(3.2106)(7.5630)\times 10^{-2}} = 2.5833$$

$$r_{in} = (\beta+1)(r_e + R_i \parallel R_f)(1+GH) = 314.25\ \text{k}\Omega$$

$$r_{out} = 0\ \Omega$$

The SPICE data are

$$\frac{v_o}{v_i} = 2.583; \qquad r_{in} = 314.2\ \text{k}\Omega; \qquad r_{out} = 0\ \Omega$$

Also, from bias point calculation,

$$I_E = 0.7907\ \text{mA}; \qquad V_{BE} = -0.7679\ \text{V}; \qquad V_O = 0.4192\ \text{V}$$

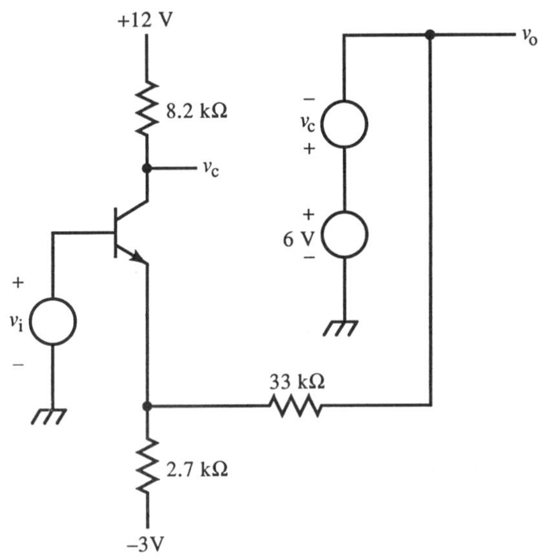

FIG. E3.2

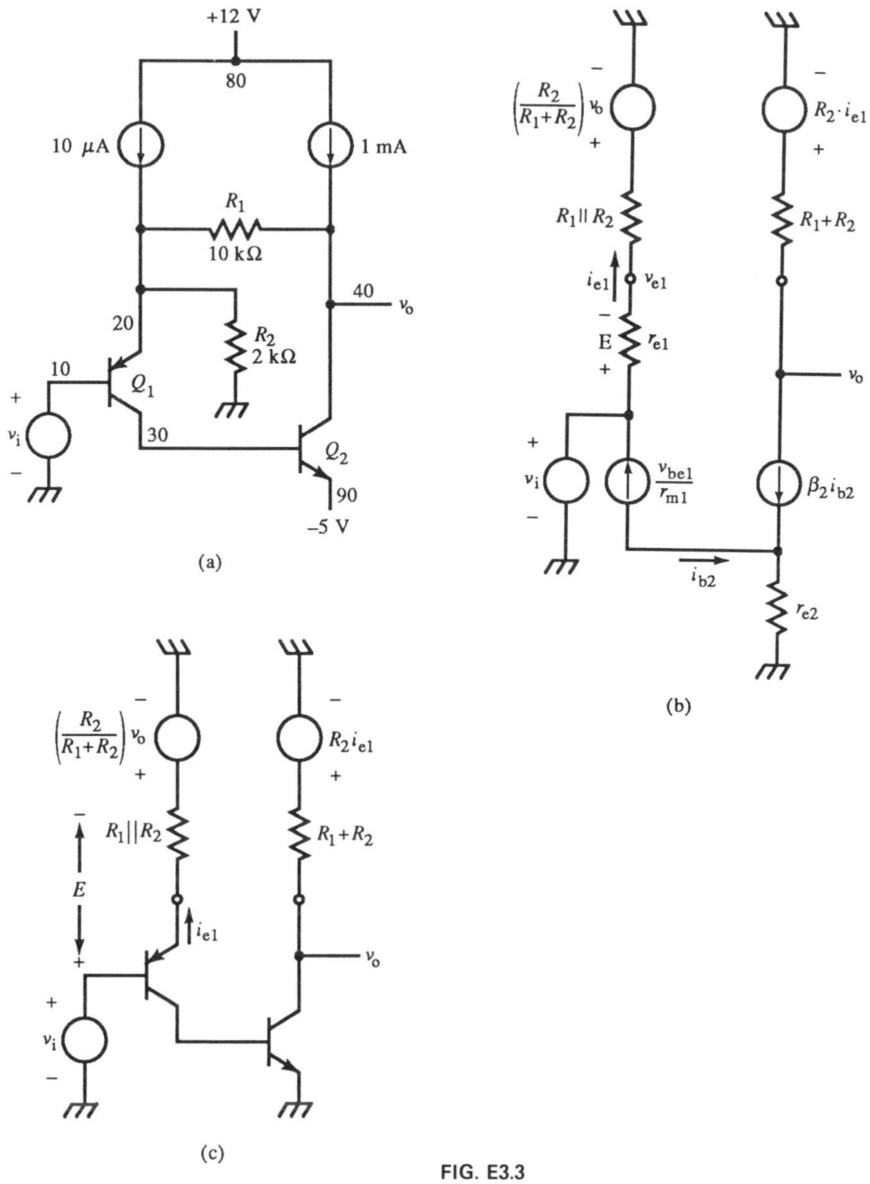

FIG. E3.3

Example 3.3 Noninverting BJT Feedback Amplifier

The circuit of Fig. E3.3a is analyzed by first choosing the error E and then determining the loaded equivalent circuit (Fig. E3.3b). Let $E = v_{be1}$. As in Section 3.9, the choice of a Thévenin equivalent output port for H results in a single summing loop. The port current is i_{e1}. When the emitter is opened, $i_{e1}=0$, and the H output voltage is

$$v_{fb}=\left(\frac{R_2}{R_1+R_2}\right)v_o=0.167\cdot v_o$$

From the emitter of Q_1, R_1 and R_2 are in parallel. The Thévenin resistance of the H output port is $2\text{ k}\Omega \parallel 10\text{ k}\Omega=1667\ \Omega$.

The H input port samples the output voltage v_o. The feedforward path through H is represented by a Thévenin equivalent circuit at the H input. The contribution to v_o through H is the voltage at v_o due to the path through the feedback network and is dependent on the H output port current i_{e1}. It is $i_{e1}\cdot R_2$. The resistance of the port is found by nulling i_{e1}. This is accomplished by opening the Q_1 emitter. The resistance is R_1+R_2.

We will analyze the circuit for

$$E=v_i-v_{e1}=v_{be1}=v_i-\left[\left(\frac{R_2}{R_1+R_2}\right)\left(\frac{r_{e1}}{r_{e1}+R_1\parallel R_2}\right)v_o+\left(\frac{R_1\parallel R_2}{r_{e1}+R_1\parallel R_2}\right)v_i\right]$$

and $v_o=GE$. When E is simplified, α_i and H are the coefficients of v_i and v_o, respectively. They are

$$\alpha_i=\left(\frac{r_{e1}}{r_{e1}+R_1\parallel R_2}\right),\qquad H=-\left.\frac{v_{be1}}{v_o}\right|_{G\text{off}}=\left(\frac{R_2}{R_1+R_2}\right)\left(\frac{r_{e1}}{r_{e1}+R_1\parallel R_2}\right)$$

Now we must determine G. It has two paths, the active path G_1 through the transistors and the passive path G_2 through H. G is

$$G=G_1+G_2=\left.\frac{v_o}{v_{be1}}\right|_{H,G_2\text{off}}+\left.\frac{v_o}{v_{be1}}\right|_{H,G_1\text{off}}=\beta_2\cdot\frac{R_1+R_2}{r_{m1}}+\frac{R_2}{r_{e1}}$$

Finally, this is combined in

$$A_v=\alpha_i\cdot\frac{G}{1+GH}$$

for the closed-loop gain. A numerical solution begins with a dc solution, in which

$$I_{E1}=6.87\ \mu\text{A},\qquad I_{E2}=680\ \mu\text{A}$$

$$r_{e1}=3.765\text{ k}\Omega,\qquad r_{m1}=3.803\text{ k}\Omega,\qquad r_{e2}=38.04\ \Omega$$

Substituting these and the circuit element values into the above equations, we get

$$G_1=312.4,\qquad G_2=0.5312,\qquad G=312.9$$

Also,

$$H = (0.1667)(0.6932) = 0.1155, \qquad \alpha_i = 0.6932$$

Putting it all together, we get

$$A_v = 5.838$$

The circuit is solved more easily by choosing

$$E = v_i - v_{fb}$$

as shown in Fig. E3.3c. Since this E is also a voltage, the loading of G is the same as we found before. But now

$$E = 1 \cdot v_i - \left(\frac{R_2}{R_1 + R_2}\right) v_o$$

and $\alpha_i = 1$ whereas

$$H = \left(\frac{R_2}{R_1 + R_2}\right) = 0.1667$$

G has two paths, as before:

$$G_1 = \left.\frac{v_o}{E}\right|_{H,G_2\text{off}} = \alpha_1 \cdot \frac{R_1 + R_2}{r_{e1} + R_1 \parallel R_2} \cdot \beta_2 = 216.5$$

$$G_2 = \left.\frac{v_o}{E}\right|_{H,G_1\text{off}} = \frac{R_2 \cdot i_{e1}}{i_{e1}(r_{e1} + R_1 \parallel R_2)} = \frac{R_2}{r_{e1} + R_1 \parallel R_2} = 0.368$$

Then $G = G_1 + G_2 = 216.9$. The closed-loop gain is 5.838, the same as before. The SPICE simulation confirms the result.

3.11 A Noninverting Voltage Feedback Amplifier with Output Block

In the previous examples, feedback was sampled directly at the output. This is not always the case, however. Another block between the sampling circuit and output must then be introduced just as α_i was required for similar situations at the input. This block will be labelled α_o. Both α_i and α_o can be included in an expression for E:

$$E = \alpha_i R - \frac{HC}{\alpha_o} \tag{3.80}$$

The amplifier of Fig. 3.15 can be modified by adding a PNP BJT, Q_2 (Fig. 3.17a). Its block diagram is Fig. 3.17b. Here, C/α_o, the sampled voltage, is

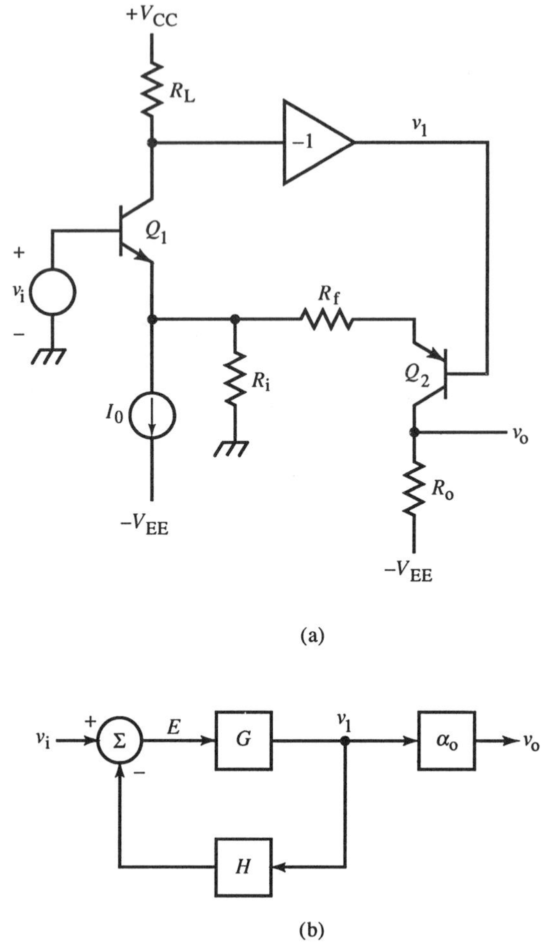

FIG. 3.17 A noninverting voltage amplifier (a) with α_o. Feedback is from base to emitter of Q_2 whereas output path is through Q_2 to collector. Block diagram of (a) is shown in (b).

labeled v_1. Choosing sampled quantities is like choosing error quantities; within the loop, the choice is arbitrary. The sampling point is preferably chosen as far forward in the loop as possible to minimize common expressions in α_o and H.

Let E be that of (3.69). Then v_{fb} is somewhat different due to Q_2. Writing v_{fb} in the complete error expression gives

$$E = v_i - \left(\frac{R_i}{R_i + R_f + r_{e2}}\right) v_1 \tag{3.81}$$

For $r_{e2} \ll R_f + R_i$, the two resistors dominate the denominator of H, so variations in r_{e2} are negligible. G is the same as in (3.71) except that now $R_E =$

$R_i \parallel (R_f + r_{e2})$ and

$$G = \alpha_1 \frac{R_L}{r_{e1} + R_i \parallel (r_{e2} + R_f)} \tag{3.82}$$

The final block to be determined is α_o:

$$\alpha_o = \left.\frac{v_o}{v_1}\right|_{E=0} = -\alpha_2 \cdot \frac{R_o}{r_{e2} + R_f + R_i} \tag{3.83}$$

This is the gain of Q_2 from v_1 to v_o, with the loading of H included. (E is a voltage, so the summing loop is opened.) The closed-loop gain can then be constructed by substituting the expressions for the blocks into

$$A_v = \frac{G}{1 + GH} \cdot \alpha_o \tag{3.84}$$

The resulting expression is unwieldy and not intuitively beneficial. For complicated feedback amplifiers, more insight is gained into amplifier behavior from the expressions for the blocks themselves.

An inverting voltage amplifier with the same block diagram is shown in Fig. 3.18. The error voltage is chosen to be

$$E = v_i - v_{fb} = v_i - R_3 i_{c2} \tag{3.85}$$

The error loop is shown in Fig. 3.18b. The collector current source of Q_2 is shunted by R_3. This Norton source can be transformed into a Thévenin source in which the voltage source is $R_3 i_{c2}$. The sampled quantity is i_{c2}. G is consequently a transconductance amplifier with gain

$$\begin{aligned} G = \left.\frac{i_{c2}}{E}\right|_{v_{fb}=0} &= -\alpha_1 \frac{R_1}{r_{e1} + R_3} \cdot \alpha_2 \frac{-1}{R_2 + r_{e2} + R_1/(\beta_2 + 1)} \\ &= \alpha_1 \cdot \frac{R_1}{r_{M1}} \cdot \alpha_2 \cdot \frac{1}{r_{M2}} \end{aligned} \tag{3.86}$$

Consequently, H must be a transresistance and is

$$H = -\left.\frac{E}{i_{c2}}\right|_{v_i=0} = -(-R_3) = R_3 \tag{3.87}$$

Finally, α_o is

$$\alpha_o = \frac{v_o}{i_{c2}} = \frac{v_o}{i_{e2}} \cdot \frac{i_{e2}}{i_{c2}} = -R_2 \cdot \frac{1}{\alpha_2} \tag{3.88}$$

These blocks are combined for the closed-loop voltage gain according to (3.84), resulting in

$$A_v = -\frac{\alpha_1 (R_1/r_{M1})(R_2/r_{M2})}{1 + \alpha_1 (R_1/r_{M1}) \alpha_2 (R_3/r_{M2})} \tag{3.89}$$

To demonstrate an alternative derivation, α_o can be eliminated from the

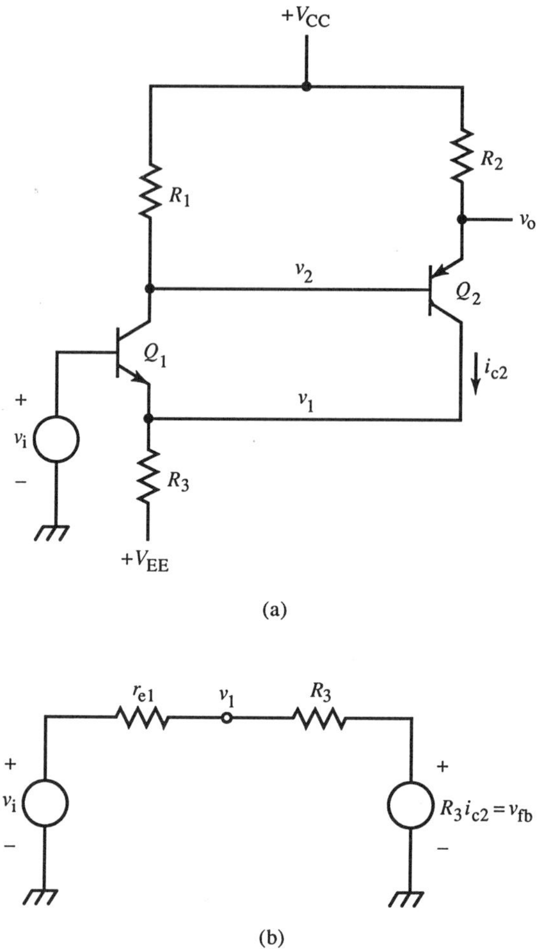

FIG 3.18 An inverting feedback amplifier with α_o (a) and error-voltage summing loop (b).

feedback topology by choosing the sampled node to be v_o instead. For this choice, the blocks are composed somewhat differently. With the same E, G and H become voltage amplifiers:

$$G = -\alpha_1 \cdot \frac{R_1}{r_{M1}} \cdot \frac{R_2}{r_{M2}}, \qquad H = -\frac{R_3}{R_2} \cdot \alpha_2 \tag{3.90}$$

Substituting these into (3.3) yields (3.89). For this circuit, this latter choice of sampled quantity is better since it eliminates the redundant path of i_{c2} to v_o in H and α_o.

This two-transistor amplifier not only provides a gain determined by R_2 and R_3 but also provides a low-resistance source at the output with voltage translation from the input. Unless R_3 is very small, the input resistance is large, approaching the ideal input–output requirements for a voltage amplifier.

A limitation is that all of the voltage gain must be realized in the first stage (Q_1). Thus, R_1 must be large relative to R_3. Since R_1 is loaded by the large r_{in}(CC), a large gain can be achieved from Q_1. This amplifier achieves much functional capability from its five components. Because V_{EE} constrains the value of R_3 for biasing, an additional resistor from the emitter of Q_1 to ground could be required.

3.12 Field-Effect Transistor Buffer Amplifier

Another amplifier similar to that of Fig. 3.18 is a common buffer amplifier that uses feedback to reduce gain error. The goal is to achieve an accurate ×1 gain from an ideal voltage amplifier so that a high-resistance voltage source can be transformed into a low-resistance source capable of supplying varying amounts of load current at the input voltage. The circuit of Fig. 3.19 provides this capability with few components. The input transistor is chosen to be a JFET for high static resistance. A BJT could be used instead if the base current caused negligible dc offset across R_G. The current source at the drain simplifies the analysis and can be implemented as a large resistor or a PNP collector current supply. Since it is constant, it enters into the dynamic analysis as an open circuit.

Let us choose the error voltage to be

$$E = v_i - v_o \tag{3.91}$$

Then $E = v_{gs}$. Since, in the FET model (see Fig. 2.2c), v_{gs} develops i_d across

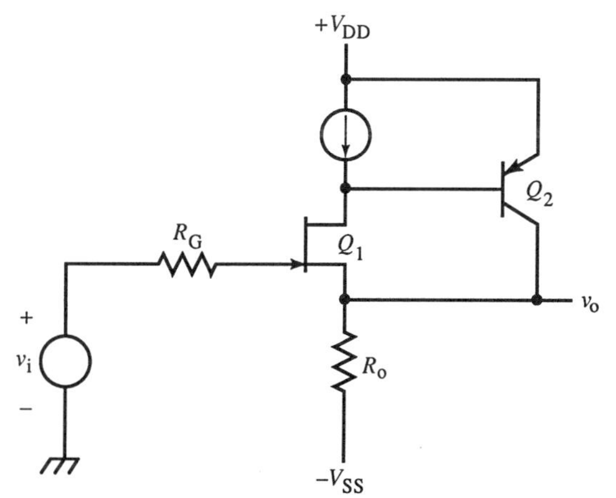

FIG. 3.19 A feedback voltage buffer with FET input. The voltage gain of Q_2 increases the loop gain and decreases gain deviation from unity.

r_m, then

$$G = \left.\frac{v_o}{E}\right|_{B=0} = \frac{v_o}{v_{gs}} = \beta\frac{R_o}{r_m} \tag{3.92}$$

Since both input and output are directly in E, $H = 1$. To find G, we set $B = 0$ as usual. This means that the feedback contribution to E must be eliminated to calculate forward path gain. In this case, we set v_o to zero. But since v_o is the numerator of G, it must remain free to vary. Therefore, to conceptually satisfy (3.92), another v_o source is separately connected to provide feedback, and it is set to zero. Think of it this way. If the loop is broken at the output and the fedback v_o is v_o', then v_o' is set to zero while calculating G. For calculating H, $v_o' = v_o$. The closed-loop gain of this buffer is

$$A_v = \frac{R_o}{(r_m/\beta) + R_o} \tag{3.93}$$

For a typical JFET, $r_m = 100\ \Omega$. If $R_o = 1\ \mathrm{k}\Omega$ and $\beta = 100$, then A_v is 0.9990, or approximately 10 bits of accuracy (in A/D converter terms).

The previous examples illustrate the idea that feedback analysis is a very fluid activity in that a given circuit can be analyzed several ways, all consistent with the basic concept. Whenever a circuit can be cast in the feedback form, it can be viewed as having feedback. Even the CD (or CC) configuration can be analyzed from a feedback perspective. The FET buffer amplifier is a CD stage with an additional transistor that increases loop gain. But even without Q_2 of Fig. 3.19, feedback analysis can be applied. Let $E = v_i - v_o$ as before. Then $G = v_o/E = R_o/r_m$ and $H = 1$. The closed-loop gain is

$$A_v = \frac{v_o}{v_i} = \frac{G}{1+GH} = \frac{R_o/r_m}{1+(R_o/r_m)(1)} = \frac{R_o}{r_m + R_o} \tag{3.94}$$

This result is what the transresistance method would yield. (See (2.30) with $R_B = 0$, $R_o = R_E$, and $r_e = r_m$.)

3.13 Closure

We have seen that a major difficulty in analyzing feedback circuits is in relating their flow-graph representations to their circuit diagrams. The major difficulty is in identifying error summing and feedback sampling. Furthermore, the G and H blocks load each other so that loading interactions must be accounted for. We approached loading by using two-port models of G and H to derive some general rules that are simple and intuitive. (No memorization of various kinds of two-port parameters were required.) Fortunately, as long as the summing and sampling quantities are chosen within the feedback loop, closed-loop analysis can be performed. This allows various choices for E and C.

Various feedback amplifier examples were investigated with multiple derivations to cultivate the art of choosing them well.

We are not yet finished with feedback amplifiers. In Chapter 2, we analyzed a BJT circuit with a shunt base-emitter resistance. Two other possibilities are shunt resistances from collector to emitter (r_o) and collector to base. We will study these topologies further in the next chapter.

References

Roberto Saucedo and Earl E. Schiring, *Introduction to Continuous and Digital Control Systems*, Macmillan, 1968, Ch. 1 and 6.

Paul E. Gray and Campbell Searle, *Electronic Principles: Physics, Models, and Circuits*, Wiley, 1969, Ch. 18.

Paul R. Gray and Robert Meyer, *Analysis and Design of Analog Integrated Circuits*, Wiley, 1984, pp. 478–525.

CHAPTER 4

Multiple-Path Amplifiers

4.1 The Reduction Theorem

The β transform presented in Chapter 2 greatly simplifies open-loop amplifier circuit analysis and makes the transresistance method possible. It can transform feedback circuits into open-loop equivalents. For example, the CC or CD amplifier has a feedback interpretation (Section 3.12) but can be β-transformed so that the voltage gain takes the form of a resistive divider, (2.30).

In this chapter, we examine circuits with more complex topologies. It is common for feedback amplifiers to have a significant forward transmittance through the feedback path. This results in multiple parallel (or shunt) forward paths. Shunt c–e or c–b resistance causes bilateral signal flow with a combination of feedback and multiple forward paths.

Before taking the feedback approach to analyze shunt-feedback circuits, we will examine some network theorems that are useful for simplifying these circuits. Analytic techniques adaptable to intuitive use are based on powerful, general circuit theorems. The β transform is half of a more general theorem, the *reduction theorem.* It has two forms:

$$\text{current form} \Rightarrow \beta \text{ transform}$$

$$\text{voltage form} \Rightarrow \mu \text{ transform}$$

These forms are duals. In Fig. 4.1, two networks, represented by blocks, share a common port with a controlled source between them. In the current-source case (Fig. 4.1a), network N1 could be a BJT base circuit, in which i is the

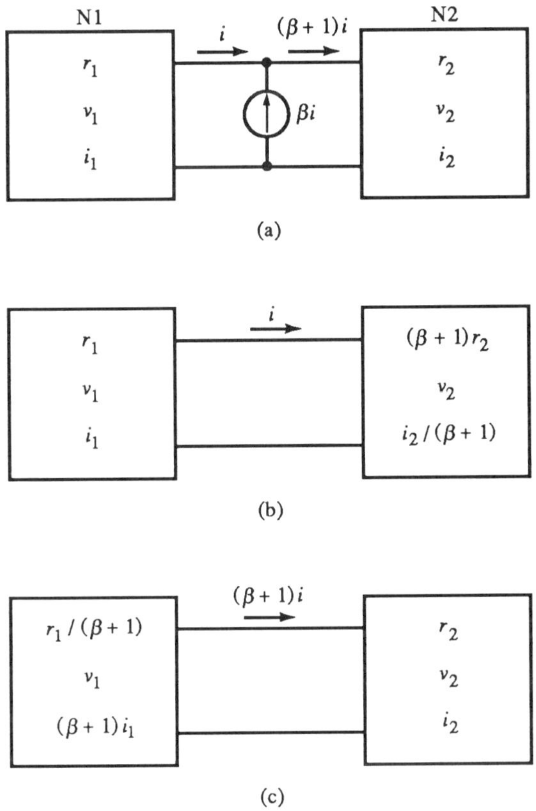

FIG. 4.1 The current form of the reduction theorem. For networks N_1 and N_2 with CCCS βi shunting their ports (a), a β-transformed N_2 (b) or N_1 (c) is equivalent to (a).

base current. Then network N2 is the emitter circuit, and the current source that shunts the common port is a BJT collector current source.

Wherever Fig. 4.1a applies, two equivalent circuits are possible (Figs. 4.1b,c). These correspond, respectively, to equivalent base and emitter circuits for a BJT. In Fig. 4.1b, N2 is transformed using $\beta + 1$; in (c), N1 is transformed instead. All voltages, currents, and resistances in the transformed network are affected as shown.

Figure 4.2 displays the dual of Fig. 4.1 with its corresponding dual transform, the μ transform. It applies to circuits with a voltage gain because μ is a voltage gain. This transform is used extensively in the modeling of vacuum tubes and applies especially to FET's because of their low drain resistance. It enables us to avoid use of feedback analysis in shunt-feedback circuits by transforming them into circuits most easily analyzed open-loop.

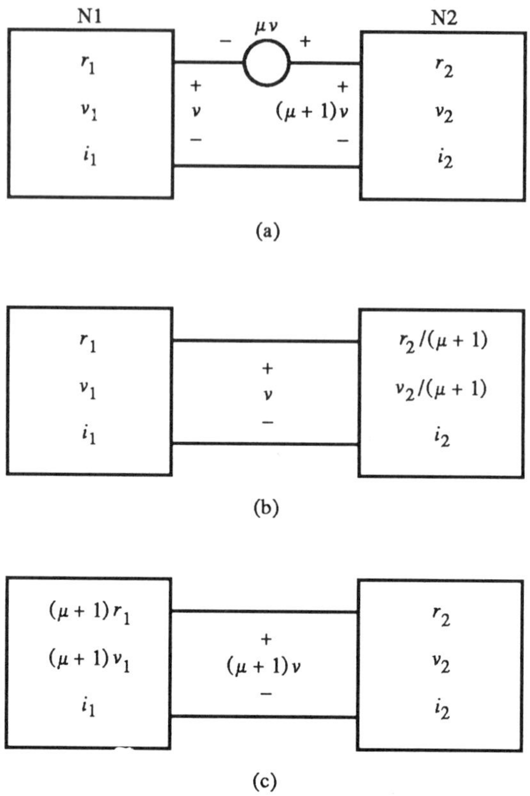

FIG. 4.2 The voltage form of the reduction theorem. For networks N_1 and N_2 with VCVS μv in series with their ports (a), a μ-transformed N_2 (b) or N_1 (c) is equivalent to (a).

4.2 μ Transform of Bipolar-Junction Transistor and Field-Effect Transistor T Models

The μ transform cannot be applied directly to circuits using the T model because the transform is based on a controlled voltage source. The T model in Fig. 4.3a is shunted by r_o. We will use this familiar model later when feedback analysis is applied to multipath circuits. For now, it must be transformed into a model with a controlled voltage source. This can be done by first referring r_e to the base as r_π (using the β transform). Then r_o shunts the controlled current source and forms a Norton equivalent circuit with it (Fig. 4.3b). The Norton circuit can be converted to a Thévenin equivalent (Fig. 4.3c) by noting that

$$\beta i_b = \alpha i_e = \alpha\left(\frac{v_{be}}{r_e}\right) = \frac{v_{be}}{r_m} \tag{4.1}$$

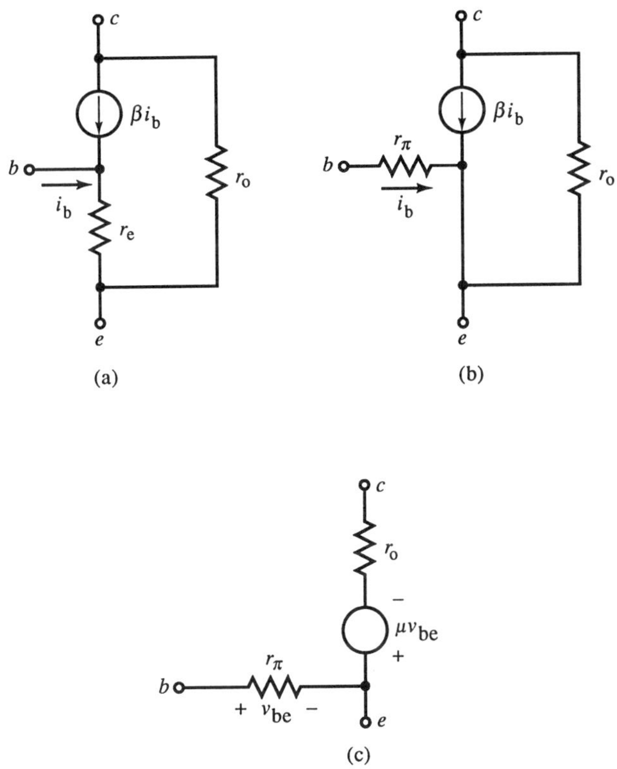

FIG. 4.3 Transformation of the BJT T model with r_o (a) to a Norton equivalent low-frequency hybrid-π model (b) to a Thévenized form (c).

This current is converted to a Thévenin voltage by multiplying by the series resistance r_o, resulting in

$$r_o\left(\frac{v_{be}}{r_m}\right) = \left(\frac{r_o}{r_m}\right)v_{be} = \mu \cdot v_{be} \tag{4.2}$$

More precisely, the definition of μ for the BJT model is

$$\mu \equiv -\frac{v_{ce}}{v_{be}}\bigg|_{i_c=0} \tag{4.3}$$

The condition that i_c be zero allows v_{ce} to be the voltage of the controlled source alone, without additional drop across r_o. Furthermore,

$$\left(\frac{\mu}{\mu+1}\right) = \frac{r_o}{r_o+r_m} = \frac{-v_{ce}/v_{be}|_{i_c=0}}{1-(v_{ce}/v_{be})|_{i_c=0}} = \frac{-v_{ce}}{v_{be}-v_{ce}}\bigg|_{i_c=0}$$

$$= \frac{v_{ce}}{v_{cb}}\bigg|_{i_c=0} = \frac{v_{ec}}{v_{bc}}\bigg|_{i_c=0} \tag{4.4}$$

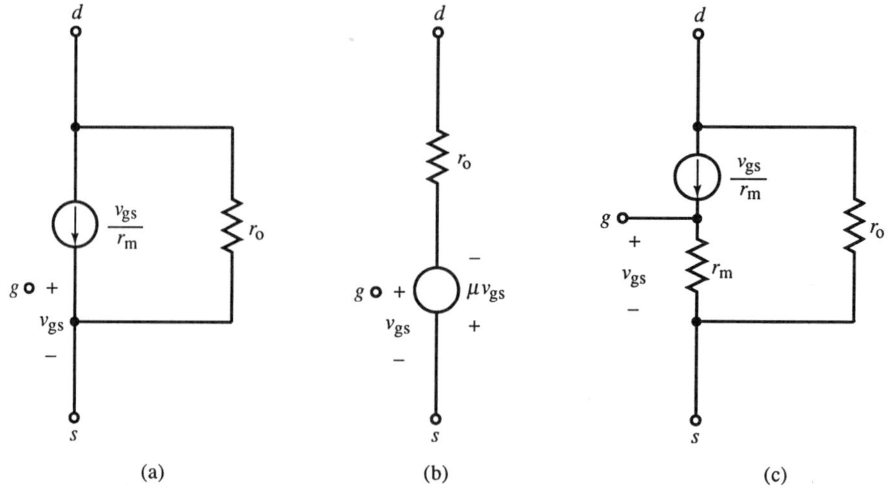

FIG. 4.4 Transformation of low-frequency FET model with r_o (a) to Thévenized form (b). A FET T model, equivalent to (a), is shown in (c).

For $i_c = 0$ (and collector resistance $R_C < \infty$), then $v_c = 0$ and

$$\nu = \left(\frac{\mu}{\mu + 1}\right) = \frac{v_e}{v_b} \tag{4.5}$$

This relationship appears often in circuit analyses and is designated by ν, the counterpart of $\alpha = \beta/(\beta + 1)$.

A similar model transformation for the FET model of Fig. 2.2c (without capacitances) begins with the model of Fig. 4.4a where r_o is added. It immediately converts to the Thévenin equivalent form of Fig. 4.4b. An alternative equivalent model is shown in Fig. 4.4c, in which the gate is connected to the current source and r_m is added. This is a FET T model. The gate current i_g remains zero because all i_g must flow through r_m. Its resulting voltage drop affects v_{gs}, and since the current source is controlled by v_{gs}, a change in drain current equal to i_g is injected into the gate node; or, more simply, since the voltage across r_m is v_{gs}, the current that must be flowing in r_m is v_{gs}/r_m. But this is the amount of current injected into the gate node by the drain current source. By KCL, i_g must be zero.

The definition of μ applied to this FET model is substantially the same as the BJT model. The relationship between the BJT and FET models is simple; if r_e of the BJT model is replaced by r_m, the FET model results.

BJT to FET T model conversion:

$$r_e \Rightarrow r_m, \quad e \Rightarrow s, \quad b \Rightarrow g, \quad c \Rightarrow d, \quad v_{be} \Rightarrow v_{gs} \tag{4.6}$$

Applying this conversion to (4.3) results in the FET version of μ:

$$\text{FET } \mu \equiv -\left.\frac{v_{ds}}{v_{gs}}\right|_{i_d=0} \tag{4.7}$$

Because the r_o for FETs is typically much lower than for BJTs, the use of transistor models that include r_o is more common for FETs. We now analyze some basic FET circuits using the extended model before applying it to BJT circuits.

4.3 Common-Gate Amplifier with r_o

Figure 4.5a shows a CG amplifier, drawn so that the reduction theorem can be easily applied to it. The gate is at ground and is the common terminal of the two networks shown in boxes. Network N1 is the source circuit, and N2 is the drain circuit. The FET model of Fig. 4.4b is between N1 and N2. To

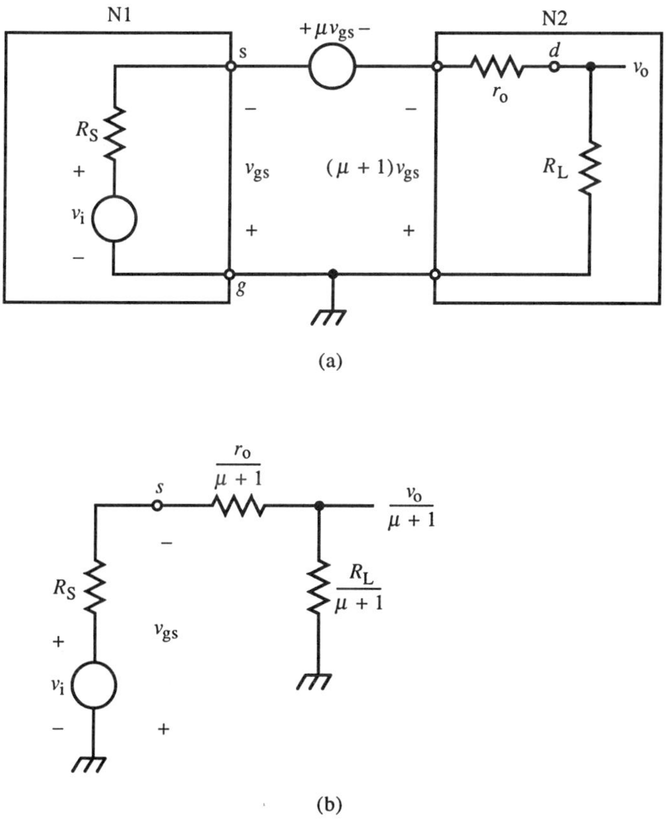

FIG. 4.5 The CG amplifier, drawn to make the application of the μ transform obvious (a), and the resulting model after transformation (b).

make this circuit correspond to Fig. 4.2a, r_o must be included in N2. The result of transforming the drain circuit (N2) is shown in Fig. 4.5b. The drain circuit has been referred to the source side. The output voltage v_o across R_L is also transformed to $v_o/(\mu+1)$. This transformed circuit is now a voltage divider between input v_i and output $v_o/(\mu+1)$:

$$\frac{v_o}{\mu+1} = \frac{R_L/(\mu+1)}{R_S + r_o/(\mu+1) + R_L/(\mu+1)} \cdot v_i \tag{4.8}$$

The voltage gain is thus

$$\text{CG} \quad A_v = \frac{R_L}{R_S + (R_L + r_o)/(\mu+1)} \tag{4.9}$$

This result is reminiscent of the transresistance method but uses the μ instead of the β transform. It demonstrates the voltage form of the transresistance method. The denominator of (4.9) can be interpreted as amplifier transresistance r_M. The resistance in the drain contributes to r_M and appears smaller by $1/(\mu+1)$ when referred to the source side of the FET. The β transform involves base and emitter networks; the μ transform involves the drain (or collector) and source (or emitter) circuits instead.

The input resistance r_{in} can be seen directly in Fig. 4.5b to be

$$\text{CG} \quad r_{in} = \frac{r_o + R_L}{\mu+1} + R_S \tag{4.10}$$

The CG output resistance can be found by μ-transforming the source circuit. In this case, the resistance of the source referred to the drain is $(\mu+1)$ times larger, so

$$\text{CG} \quad r_{out} = R_L \parallel [r_o + (\mu+1)R_S] \tag{4.11}$$

4.4 Common-Source Amplifier with r_o

A CS amplifier is shown in Fig. 4.6a. The voltage-source FET model makes KVL analysis easy since there is only one loop. The needed equations are:

$$v_s = i_s R_S$$

$$v_o = -i_s R_L$$

$$i_s[R_S + r_o + R_L] = \mu v_{gs} = \mu v_g - \mu R_S i_s$$

Solving for A_v gives

$$\frac{v_o}{v_g} = -\frac{R_L}{r_m + (R_L/\mu) + ((\mu+1)/\mu)R_S} \tag{4.12}$$

Although this gain is a ratio of resistances, the terms in the denominator

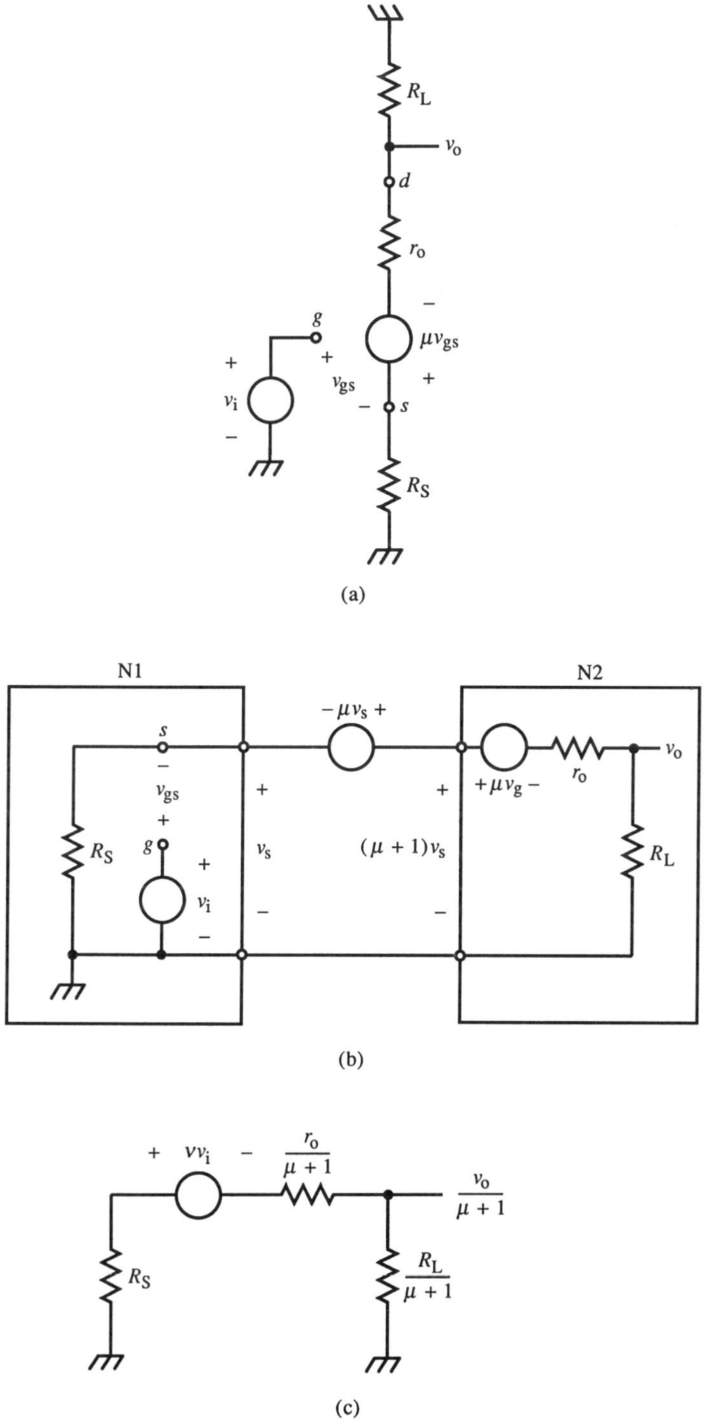

FIG. 4.6 The CS amplifier model (a), redrawn for μ transformation (b) and after transformation of N_2 and simplification (c).

involving μ do not have a simple interpretation in terms of the μ transform and circuit topology. By factoring $(\mu+1)/\mu$ out of the denominator of (4.12), we obtain two factors containing v_s:

$$\text{CS} \quad A_v = -\underbrace{\left(\frac{\mu}{\mu+1}\right)}_{(v_s/v_g)} \underbrace{\frac{R_L}{R_S+(r_o+R_L)/(\mu+1)}}_{(v_o/v_s)} \tag{4.13}$$

The first factor, ν, is the gate-to-source gain. The second factor is the same as the CG A_v. Its denominator can be interpreted as r_M, keeping in mind that it is v_s (not v_g) across r_M that generates i_s. Consequently, the voltage form of the transresistance method is based on finding r_M across v_s and then (if needed) relating v_s to v_g through ν:

$$r_M = \frac{v_s}{i_s} \tag{4.14}$$

and

$$v_s = \left(\frac{\mu}{\mu+1}\right) v_g = \nu \cdot v_g \tag{4.15}$$

The gain expression of (4.12) was found using basic circuit laws, not by applying the μ transform directly to the circuit topology. For the CS, it is not as obvious as for the CG since the gate is not common to both source and drain circuits (Fig. 4.6a). In Fig. 4.6b, it is redrawn so that application of the μ transform is explicit. Because the port voltage is chosen to be v_s, the drain voltage source μv_{gs} is split into two sources so that the first is dependent upon v_s. The remaining source, μv_g, becomes part of the drain network and is transformed along with it. When the μ transform is applied to the drain circuit, Fig. 4.6c results. The transformed voltage across R_L is

$$\frac{v_o}{\mu+1} = -\nu \cdot \frac{v_i}{r_M} \cdot \left(\frac{R_L}{\mu+1}\right) \tag{4.16}$$

Solving for the voltage gain gives

$$\text{CS} \quad A_v = -\nu \cdot \frac{R_L}{R_S+(r_o+R_L)/(\mu+1)} = -\nu \cdot \frac{R_L}{R_S+r_s+R_L/(\mu+1)} \tag{4.17}$$

The expression $r_o/(\mu+1)$ has been expressed as

$$r_s = \frac{r_o}{\mu+1} = \frac{r_o}{\mu} \cdot \left(\frac{\mu}{\mu+1}\right) = \nu \cdot r_m \tag{4.18}$$

When r_o is referred to the source, it transforms to r_s, the FET analog of r_e, in that both are related to r_m by dual factors, α and ν. Although α expresses a current loss due to base current, ν expresses a voltage loss due to v_{gs}; μ and β are duals, as are $\nu = \mu/(\mu+1)$ and $\alpha = \beta/(\beta+1)$.

The input resistance of the CS amplifier is infinite. The output resistance is the same as the CG since the source circuit referred to the drain is the same for both.

4.5 Common-Drain Amplifier with r_o

The last of the three basic FET configurations is the CD or source-follower (Fig. 4.7). Applying the voltage form of the transresistance method, we find r_M by determining the resistance across which the source voltage generates the source current i_s. The μ transform is required to refer the resistance on the drain side of the FET voltage source to the source side; as before, it is

$$r_s + \frac{R_D}{\mu + 1}$$

This resistance, when referred to the source circuit, is in series with R_S. The total transresistance is thus

$$r_M = R_S + r_s + \frac{R_D}{\mu + 1}$$

The source current generated by v_s across r_M develops an output voltage across R_S. Since the voltage gain from the gate is desired, ν must be included as a

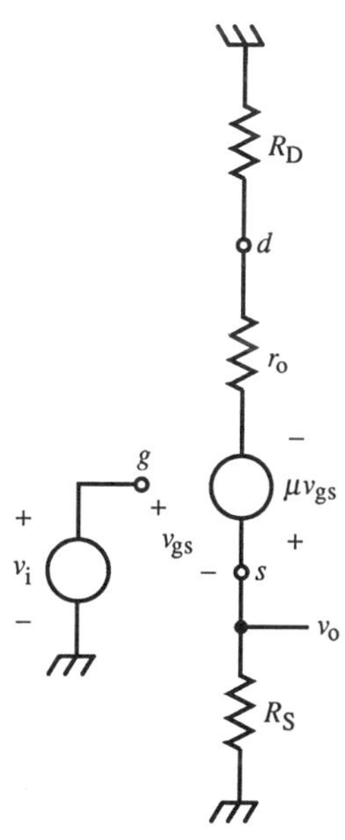

FIG. 4.7 Model of a generalized CD amplifier with r_o.

factor:

$$\text{CD} \quad A_v = \nu \cdot \frac{R_S}{R_S + r_s + (R_D/\mu + 1)} \tag{4.19}$$

This gain is more general than a pure CD amplifier since it includes a resistance in the drain R_D.

The input resistance of the CD is infinite, and output resistance is

$$\text{CD} \quad r_{out} = R_S \parallel \left(r_s + \frac{R_D}{\mu + 1} \right) \tag{4.20}$$

4.6 Field-Effect Transistor Cascode Amplifier with r_o

The voltage form of the transresistance method extends directly to multiple-transistor amplifier stages. The FET cascode amplifier model (Fig. 4.8) has a

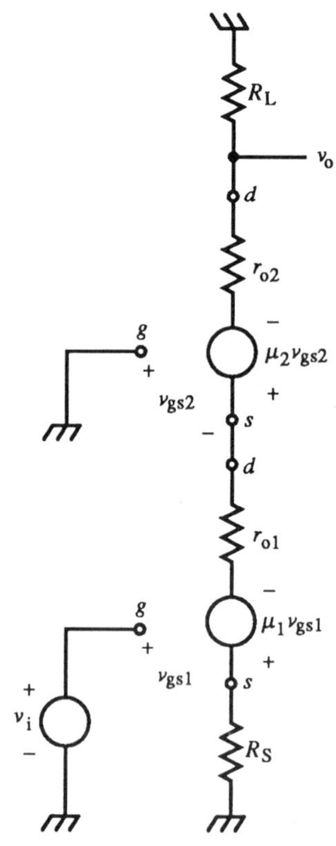

FIG. 4.8 Model of the FET cascode amplifier taking r_o into account.

voltage gain of

$$\text{cascode} \quad A_v = -\nu_1 \cdot \frac{R_L}{R_S + \dfrac{r_{o1} + ((r_{o2} + R_L)/(\mu_2 + 1))}{\mu_1 + 1}} \tag{4.21a}$$

$$\text{cascode} \quad A_v = -\nu_1 \cdot \frac{R_L}{R_S + r_{s1} + r_{s2}/(\mu_1 + 1) + \dfrac{(R_L/(\mu_2 + 1))}{\mu_1 + 1}} \tag{4.21b}$$

This can be interpreted (and also constructed) by inspection of the circuit diagram. The input voltage v_i at the gate of the CS produces v_s via ν_1. The CS r_M is R_S in series with the drain resistance, referred to the source. Drain resistance is r_{o1} in series with the CG drain circuit referred to its source, or $(r_{o2} + R_L)/(\mu_2 + 1)$. When these resistances are referred to the CS source, the denominator of (4.21), r_M, results. The source current develops the output voltage, v_o, over R_L (in the numerator) and is an inverting output. A_v can be written as (4.21b) using the definition of r_s in (4.18)—as r_o referred to the source circuit.

The output resistance of the cascode stage can be found using the same approach; it is

$$\text{cascode} \quad r_{out} = R_L \parallel [r_{o2} + (\mu_2 + 1)(r_{o1} + (\mu_1 + 1)R_S)] \tag{4.22}$$

In this case, the μ transform is used to refer source resistances to the drain circuit.

4.7 Common-Base Amplifier with r_o

The application of the voltage form of the transresistance method to BJT amplifiers adds the complication of r_π (Fig. 4.9). It forms an additional loop or node when compared to the CG circuit. This complication does not significantly affect the approach. The circuit model is redrawn in Fig. 4.9b to make the application of the μ transform explicit. After the drain circuit is referred to the source side (Fig. 4.9c), the divider formed by R_E and r_π is Thévenized (Fig. 4.9d). The voltage gain can then be found by solving the voltage divider:

$$\text{CB} \quad A_v = \underset{\substack{\uparrow \\ (v_e/v_i)}}{\left(\frac{r_\pi}{r_\pi + R_E}\right)} \underset{\substack{\uparrow \\ (v_o/v_e)}}{\frac{R_L}{r_\pi \parallel R_E + (R_L + r_o)/(\mu + 1)}} \tag{4.23}$$

This gain expression has two additional complications over that of the CG (4.9). At the emitter, R_E is now shunted by r_π. This affects r_M in the second factor of (4.23). The first factor accounts for the divider formed by r_π with R_E.

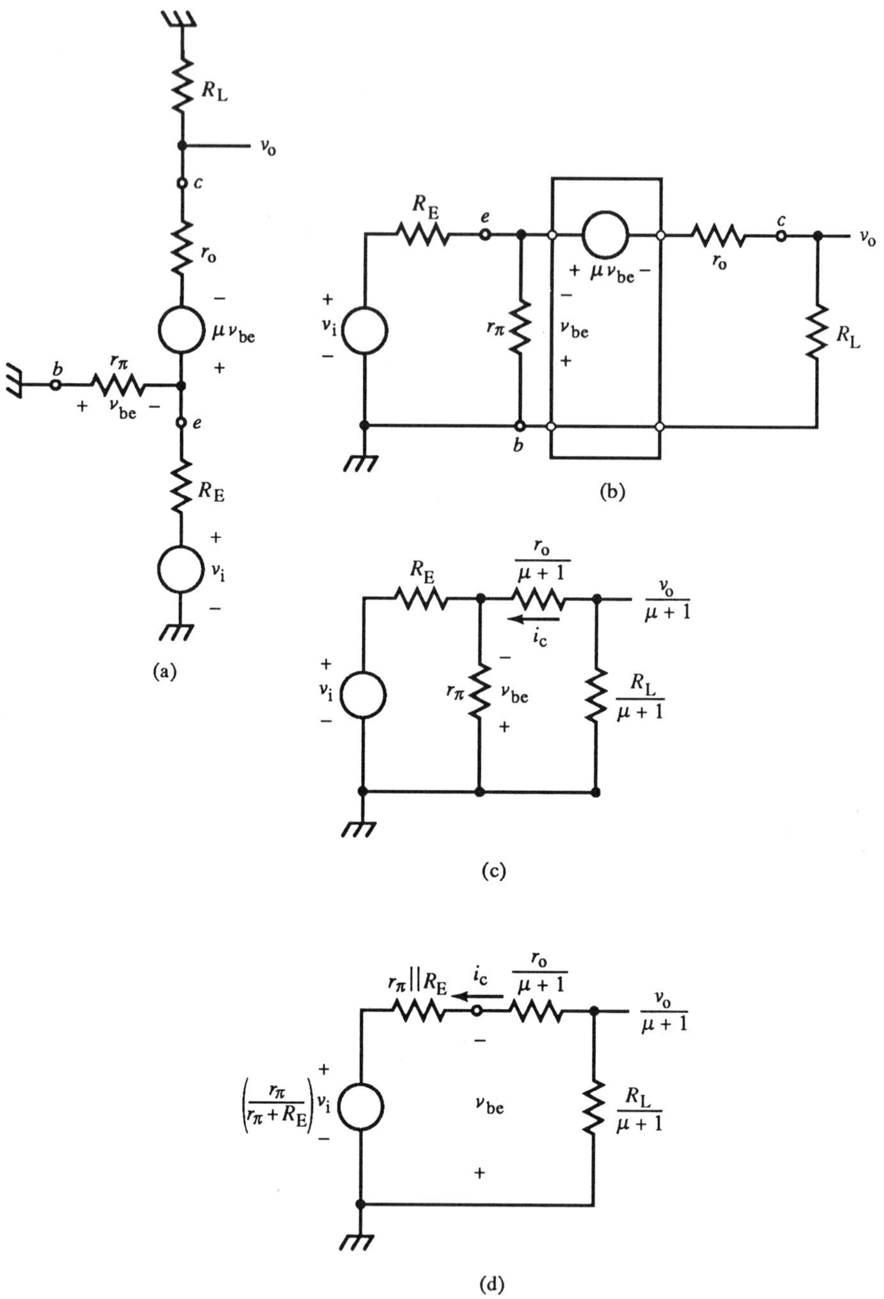

FIG. 4.9 CB amplifier model with r_o (a), drawn for μ transform (b), transformed (c), and simplified (d).

An alternative formulation of A_v regards r_π and R_E as forming a current divider with a transmittance of (i_c/i_e):

$$\text{CB} \quad A_v = \left[\frac{r_\pi}{r_\pi + (r_o + R_L)/(\mu+1)}\right] \cdot \frac{R_L}{R_E + r_\pi \parallel [(r_o + R_L)/(\mu+1)]}$$

$$= \left(\frac{i_o}{i_e}\right)\underset{\substack{\uparrow \\ 1/r_{in}}}{\left(\frac{i_e}{v_i}\right)}\underset{\substack{\uparrow \\ R_L}}{\left(\frac{v_o}{i_c}\right)} \tag{4.24}$$

For (4.24), i_e is the common quantity of the transresistance method. The input v_i generates i_e across the input resistance r_{in}, which acts as r_M, the denominator of the second factor:

$$\text{CB} \quad r_{in} = R_E + r_\pi \parallel \left(\frac{r_o + R_L}{\mu+1}\right) \tag{4.25}$$

Some of i_e is lost to the base, leaving i_c, and is accounted for by the first factor of (4.24). The output voltage is then developed across R_L by i_c. In this formulation, both voltage and current forms of the transresistance method are present. The μ transform refers the collector resistances to the emitter; the voltage form is applied. The (i_c/i_e) factor, however, is a circuit-dependent α characteristic of gain equations resulting from the current form. In contrast, (4.23) has a purely voltage form interpretation. Since it is easier to apply only one form, (4.23) is preferred in most cases.

The CB output resistance is found by applying the μ transform to the emitter circuit:

$$\text{CB} \quad r_{out} = R_L \parallel [r_o + (\mu+1)(r_\pi \parallel R_E)] = R_L \parallel r_c \tag{4.26}$$

The μ-transformed expression for the collector resistance r_c has been derived before as (2.38). There, an equivalent formula was derived and given a β-transform interpretation. (To derive r_c of (4.26) from (2.38), substitute μr_π for βr_o and let $R_B = r_\pi$.)

4.8 Common-Collector and Common-Emitter Amplifiers with r_o

The CC or emitter-follower is shown in Fig. 4.10a, with simplified equivalent circuit in (b). This is a generalized CC amplifier in that collector resistance is included. Following the approach of Section 4.4, the gain is

$$\text{CC} \quad A_v = \underbrace{\left[\frac{R_E}{R_E + r_\pi}\right]\left[\frac{(R_C + r_o)/(\mu+1)}{R_E \parallel r_\pi + (R_C + r_o)/(\mu+1)}\right]}_{\text{passive path}} + \underbrace{\nu\left[\frac{R_E \parallel r_\pi}{R_E \parallel r_\pi + (R_C + r_o)/(\mu+1)}\right]}_{\text{active path}} \tag{4.27}$$

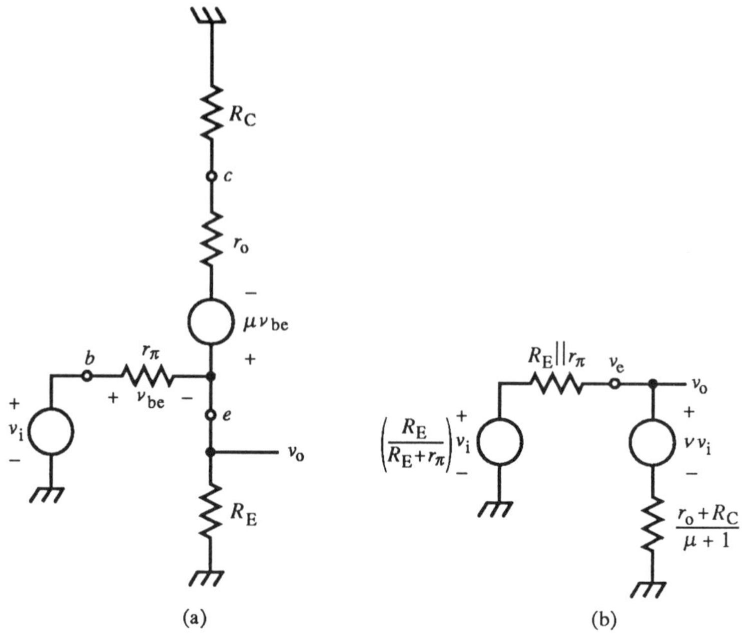

FIG. 4.10 CC amplifier model with r_o (a). The μ transform is applied and the resulting circuit simplified (b).

The CC has two gain paths, an active path due to the gain of the transistor and a passive path due to a finite r_π. The first factor of both terms of (4.27) is (v_e/v_i). For the active path, the second factor is a ratio of load resistance $R_E \| r_\pi$ over r_M. The second factor of the passive path term is a voltage divider gain due to the drop across the collector resistance, referred to the emitter. This is a loaded divider with

$$R_E \left\| \frac{(r_o + R_C)}{(\mu + 1)} \right.$$

The passive path gain can be rewritten to make this explicit:

$$\frac{R_E \| [(r_o + R_C)/(\mu + 1)]}{R_E \| [(r_o + R_C)/(\mu + 1)] + r_\pi}$$

Figure 4.11 shows the CE circuit model and its successive modifications leading to Fig. 4.11e. Again, the μ transform reduces this circuit to a voltage divider. r_π creates a loaded divider (Fig. 4.11c) that is Thévenized in Fig. 4.11d. The voltage source νv_i is combined with v_i in Fig. 4.11e, from which a

---

FIG. 4.11 CE amplifier model with r_o (a), redrawn for application of μ transform (b); resulting circuit after output network is transformed (c), input circuit Thévenized (d), and further reduced (e).

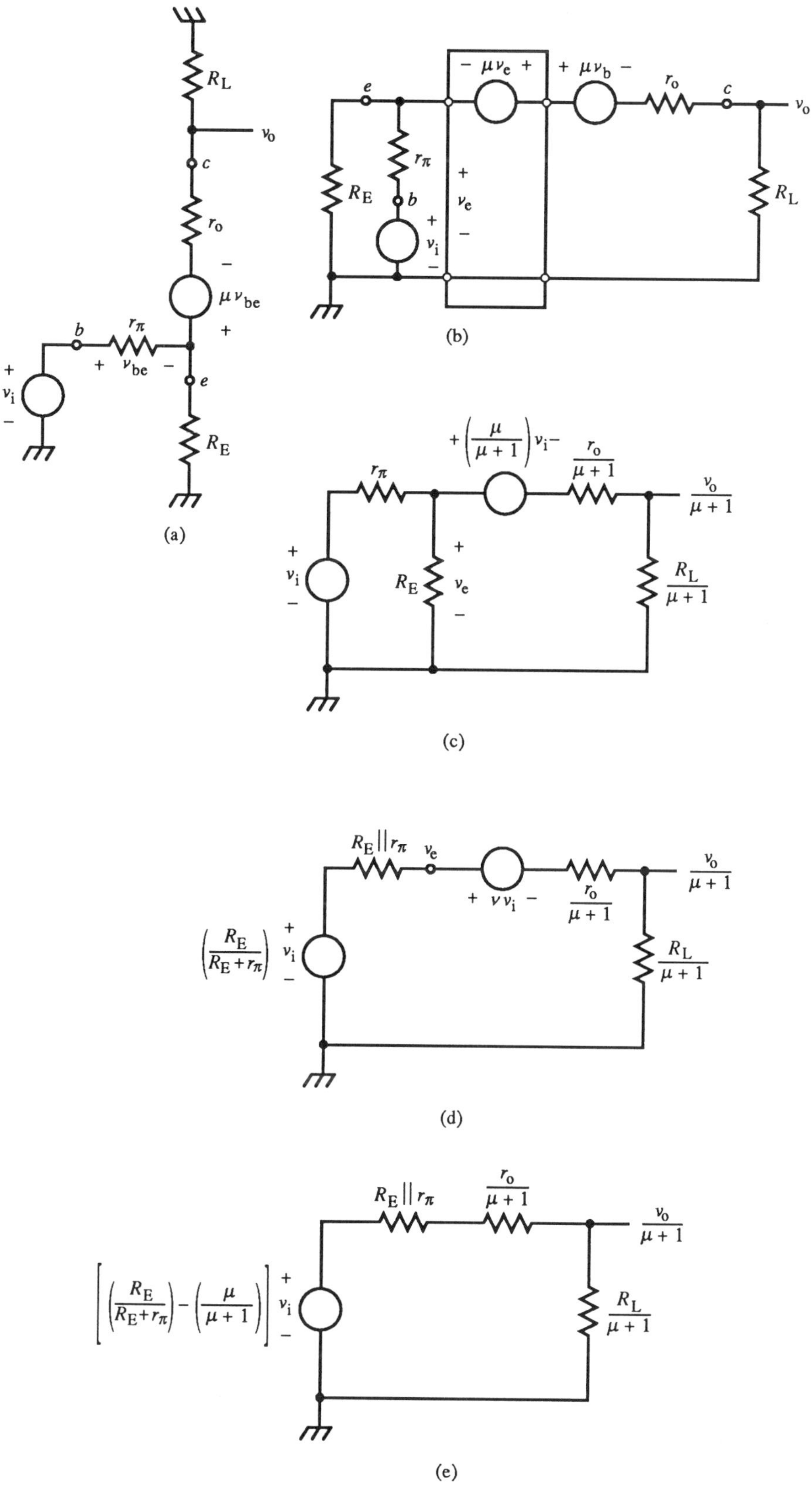

R_L
v_o
c
r_o
μ v_be
b
r_π
v_be
e
v_i
R_E
(a)
e
− μ v_e +
+ μ v_b −
r_o
c
v_o
r_π
R_E
b
v_e
v_i
R_L
(b)
(μ/(μ + 1)) v_i
r_π
r_o/(μ + 1)
v_o/(μ + 1)
v_i
R_E
v_e
R_L/(μ + 1)
(c)
R_E || r_π
v_e
ν v_i
r_o/(μ + 1)
v_o/(μ + 1)
(R_E/(R_E + r_π)) v_i
R_L/(μ + 1)
(d)
R_E || r_π
r_o/(μ + 1)
v_o/(μ + 1)
[(R_E/(R_E + r_π)) − (μ/(μ + 1))] v_i
R_L/(μ + 1)
(e)

voltage gain expression can be written:

$$\text{CE} \quad A_v = \Bigg(\underset{\substack{\uparrow \\ (v_e/v_i) \\ \text{active} \\ \text{path}}}{-\nu} + \underset{\substack{\uparrow \\ (v_e/v_i) \\ \text{passive} \\ \text{path}}}{\frac{R_E}{R_E + r_\pi}} \Bigg) \cdot \underset{\substack{\uparrow \\ (v_o/v_e)}}{\frac{R_L}{(R_L + r_o)/(\mu + 1) + r_\pi \parallel R_E}} \tag{4.28}$$

The second factor of (4.28) is the load resistance over r_M, as with the CB amplifier. The novelty is in the first factor. The first term, $-\nu$, is the μ-transform base-to-emitter voltage gain due to the active device μ amplification; it expresses the gain due to the active forward path. The CS A_v contained only this term (negated). With the CE, the second term is added due (once again) to r_π. This term represents a voltage divider formed from r_π and R_E and expresses the gain of a passive path from input to output. This term gives the passive gain from v_i to v_e. The voltage component of v_e due to the passive path is then amplified along with the active path component by the second factor of (4.28). Since the passive path gain is noninverting, it decreases the overall (inverting) gain somewhat.

The output resistance can be obtained by direct application of the μ transform to the input side of the circuit. Then

$$\text{CE} \quad r_{out} = R_L \parallel [r_o + (\mu + 1)(r_\pi \parallel R_E)] \tag{4.29}$$

The input resistance r_{in} of the CE can be found by redrawing Fig. 4.11c as shown in Fig. 4.12a. The right side is Thévenized in Fig. 4.12b. Since the voltage source on the right is controlled by v_i, it affects r_{in}. Resorting to basic circuit analysis, we can solve for the input resistance; it is

$$r_{in} = \frac{v_i}{i_i} = \frac{v_i}{\dfrac{v_i - \nu\left(\dfrac{R_E}{R_E + (R_L + r_o)/(\mu + 1)}\right) v_i}{r_\pi + R_E \parallel (R_L + r_o)/(\mu + 1)}}$$

$$\text{CE} \quad r_{in} = r_\pi \left[\mu \left(\frac{R_E}{R_E + R_L + r_o} \right) + 1 \right] + R_E \parallel (R_L + r_o) \tag{4.30}$$

This expression is not immediately apparent from the circuit topology, as previous circuit expressions were, and reveals limits to the extent a topology-oriented approach can take. Substituting βr_o for μr_π gives an alternative β-transformlike expression. It is left to the reader to find a topology-oriented explanation for these expressions of r_{in}.

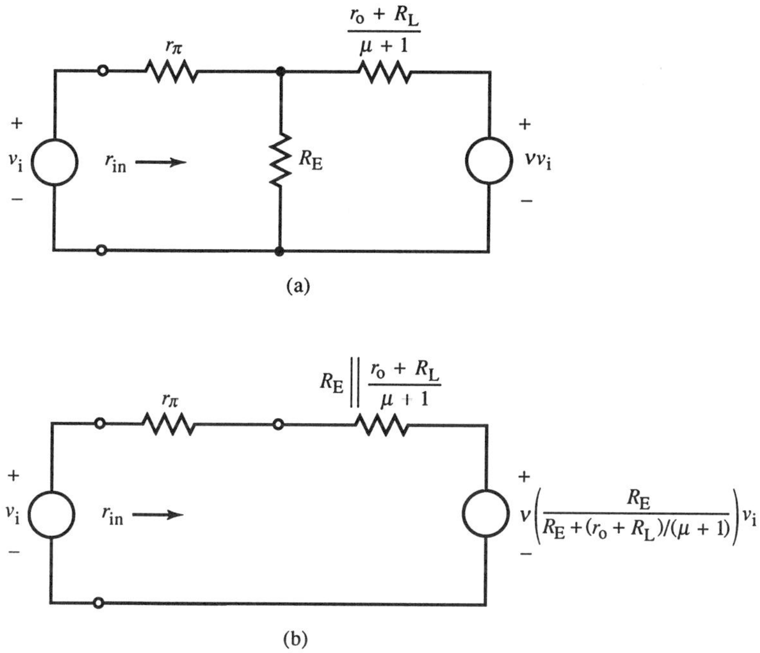

FIG. 4.12 The output resistance of circuit of Fig. 4.28a, Fig. 4.28c in (a) and Thévenized in (b).

4.9 Some Circuit Transformations

For circuits with more than three branch or loop equations, finding algebraic solutions can be tedious. In these situations, the following formulas are often useful. They will be used in Section 4.10.

Thévenin and Norton circuits for loaded dividers are shown in Figs. 4.13a,b. For the voltage divider in (a),

$$v = v_{\mathrm{i}}\left(\frac{a \parallel c}{a \parallel c + b}\right), \qquad i = \frac{v}{c} = \left(\frac{a \parallel c}{a \parallel c + b}\right)\left(\frac{1}{c}\right) v_{\mathrm{i}} \tag{4.31}$$

and for the current divider in (b),

$$i = \left(\frac{a \parallel b}{a \parallel b + c}\right)\left(\frac{v_{\mathrm{i}}}{b}\right) \tag{4.32}$$

Loaded dividers often appear, and it is useful to be able to reverse the loading, as the following formulas allow:

$$a \parallel b + c = (a \parallel c + b)\left(\frac{a + c}{a + b}\right) \tag{4.33}$$

$$\frac{a \parallel c}{a \parallel c + b} = \frac{a \parallel b}{a \parallel b + c} \cdot \frac{c}{b} \tag{4.34}$$

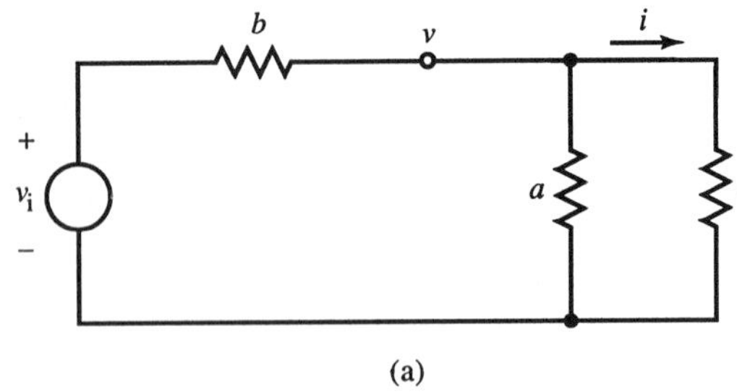

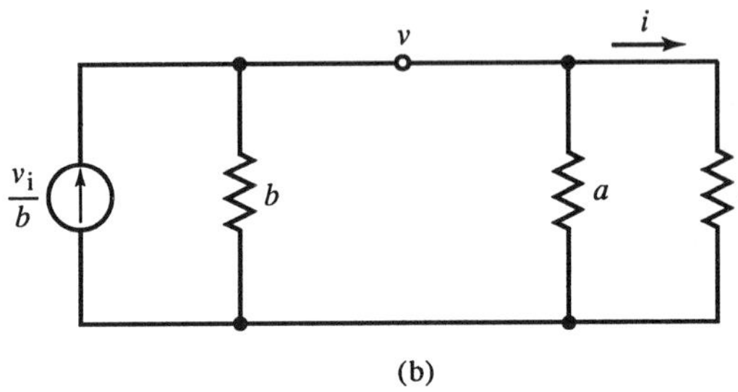

FIG. 4.13 Equivalent loaded voltage (a) and current (b) dividers.

It is also handy to note that

$$\frac{a \parallel b}{a} = \frac{b}{a+b} \tag{4.35}$$

The manipulation of expressions involving the $\parallel$ operation are made easier by the following properties:

$$\text{associative property of } \parallel: \quad (a \parallel b) \parallel c = a \parallel (b \parallel c) \tag{4.36}$$

$$\text{distributive property of } \times \text{ over } \parallel: \quad ab \parallel ac = a(b \parallel c) \tag{4.37}$$

$$\text{commutative property of } \parallel: \quad a \parallel b = b \parallel a \tag{4.38}$$

$$\frac{a \parallel b}{c \parallel d} = \frac{a \parallel b}{c} + \frac{a \parallel b}{d} \tag{4.39}$$

An alternative to algebraic manipulation is the manipulation of circuit models. Two circuit transforms that can be used to separate circuits into two more independent circuits are shown in Figs. 4.14 and 4.15. In Fig. 4.14a, a current source is replaced by two sources with the same current in series. This change introduces an additional node c between the two sources. This is useful, for example, in transforming a loop with a floating current source into two separate loops with ground-referenced current sources (Fig. 4.14b).

The dual of this transformation is shown in Fig. 4.15, where a voltage source is replaced by two parallel sources of the same voltage. This transformation is useful in separating two branches, giving each its own source, as in

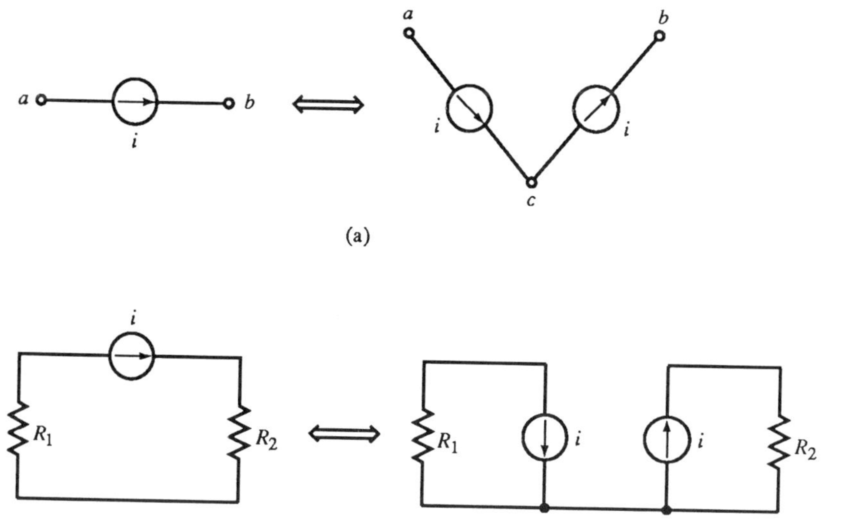

(a)

(b)

FIG. 4.14 The current source-shifting transformation (a) introduces an additional node *c*. A floating source (b) can be referenced to a common node of two loops, isolating the resistors.

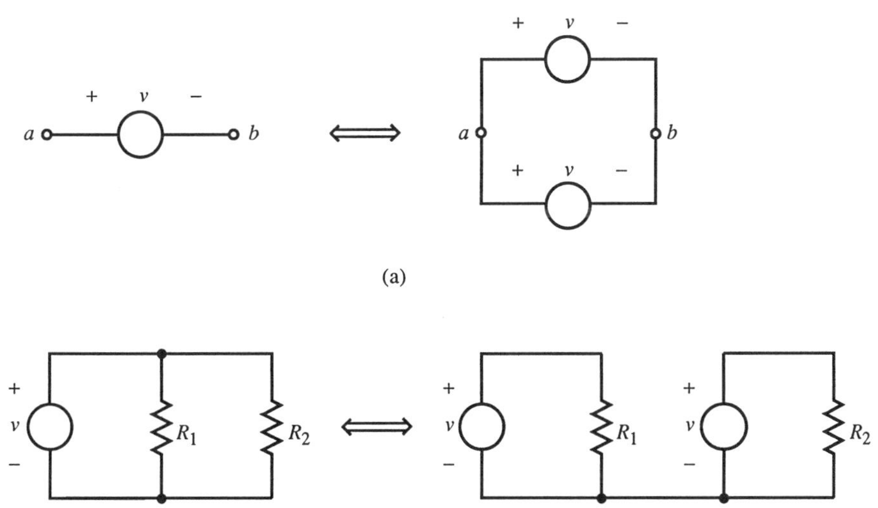

(a)

(b)

FIG. 4.15 The voltage source-shifting transformation (a) introduces an additional branch. Shunt resistances can be referenced to a common node while isolated in separate loops (b).

Fig. 4.15b. These transformations are voltage and current *source-shifting* transformations.

Finally, the *substitution theorem* applies to controlled sources as shown in Fig. 4.16. It too has voltage and current dual forms. In Fig. 4.16a, a voltage-controlled current source (VCCS) of current v/r has a terminal voltage of v. Because it is controlled by the voltage across its terminals, it acts as a resistance of r. Similarly, the current-controlled voltage source (CCVS) of Fig. 4.16b has a terminal voltage of ri with current i. It also is equivalent to a resistance of r.

To demonstrate source shifting and the substitution theorem, the CS with r_o is modeled in Fig. 4.17a. Current-source shifting is applied, resulting in Fig. 4.17b. This circuit is also modified by splitting v_{gs}/r_m into two sources, v_g/r_m and v_s/r_m. Since the current source v_s/r_m is across v_s, the substitution theorem can be applied, resulting in r_m in Fig. 4.17c. Successive applications of Norton and Thévenin conversions then reduce the circuit to an equivalent form, from which the gain given by (4.17) readily follows.

4.10 Feedback Analysis of Multipath Transistor Amplifiers

Now that we have developed methods for avoiding feedback analysis using the reduction theorem, we shall investigate how to extend the results of Chapter

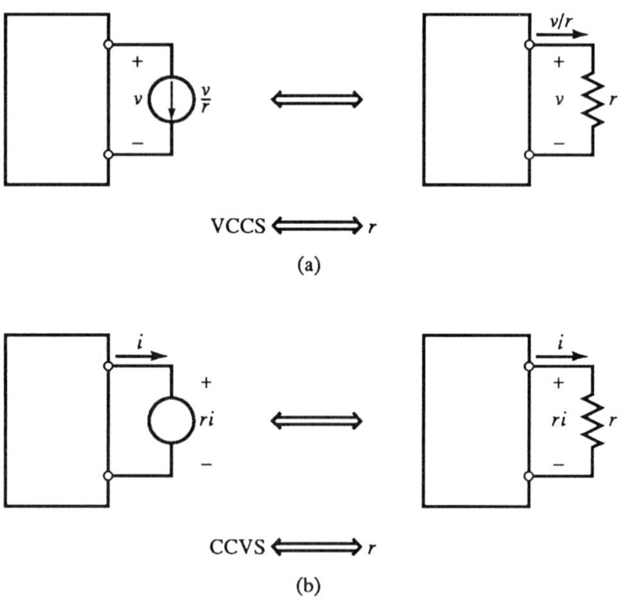

FIG. 4.16 The substitution theorem for VCCS (a) and CCVS (b). A source controlled by its other terminal quantity is equivalent to a resistance.

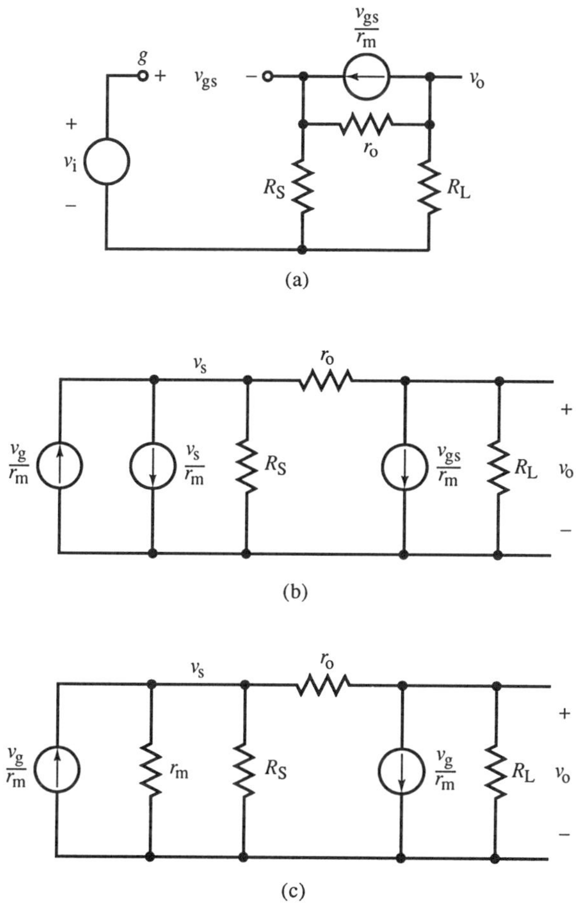

FIG. 4.17 The CS with r_o (a) after source-shifting the collector current source and then splitting it into two sources (b), and after applying the substitution theorem to current source v_s/r_m (c).

3 to multipath topologies involving a feedback loop. In particular, we will restrict, for now, the topologies to those with a single *feedforward* path *F*.

The two topologies of interest are shown in Fig. 4.18. Each of these has the basic feedback loop embedded within it. They differ in that the topology of Fig. 4.18b isolates the feedforward path from the feedback loop with a ×1 transmittance from C to v_o. This is functionally a ×1 buffer amplifier. In the topology of Fig. 4.18a, the feedforward path injects its signal into the feedback loop at the output. When this flow graph is reduced, the voltage gain and error are

$$\frac{v_o}{v_i} = \frac{\alpha_i G + F}{1 + GH} \tag{4.40}$$

$$E = \frac{\alpha_i + FH}{1 + GH} \tag{4.41}$$

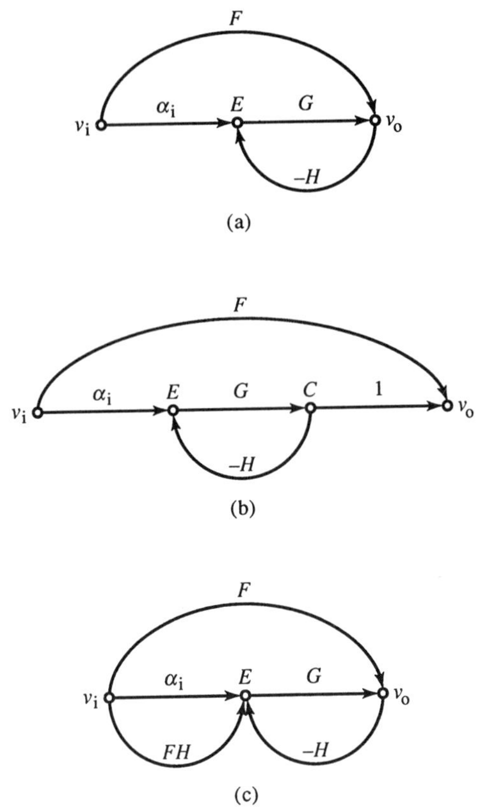

FIG. 4.18 Flow graphs for feedback analysis of transistor configurations with r_o: (a) common multipath topology with feedforward path F injecting into output of feedback loop; (b) isolated feedforward topology. When (b) is reconstructed in the form of (a), an extra path FH results from v_i to E (c) that cancels $F(-H)$ through the path $v_i \rightarrow v_o \rightarrow E$.

For the topology with the isolated feedforward path:

$$\frac{v_o}{v_i} = \alpha_i \cdot \frac{G}{1+GH} + F \tag{4.42}$$

$$E = \frac{\alpha_i}{1+GH} \tag{4.43}$$

Notice that in (4.42) F adds to the output without involving the loop, whereas in (4.40) its output contributes to E via H, and the FH term is also divided by $1+GH$ of the loop.

The relationship between the two topologies can be made explicit by writing (4.42) as

$$\frac{v_o}{v_i} = \frac{\alpha_i G + F}{1+GH} + \frac{(FH)G}{1+GH} \tag{4.40a}$$

where the first term is (4.40) and the second adds transmittance FH from v_i

to E. This gives the equivalent topology of Fig. 4.18b shown in (c). Isolating F from C at v_o is equivalent to adding the FH branch parallel to α_i. What this branch adds to E exactly cancels the contribution to E from the path from v_i through F and $-H$.

To find F from circuit topology, the G path must be nulled. Then F can be found from

$$F = \left.\frac{v_o}{v_i}\right|_{GE=0} \tag{4.44}$$

If we set GE to zero, then v_o is due to F alone. In the isolated feedforward topology of Fig. 4.18b, if possible, C can be set to zero as an alternative to E. Similarly, because the F branch contributes to v_o, finding G must have the additional condition (besides $B=0$) that $Fv_i=0$. These conditions are necessary in envisioning the transmittances from the topology.

This approach to feedback circuit analysis is summarized in the following steps:

1. Identify signal flow paths and summing and sampling points (i.e., the flow graph) from the circuit topology.
2. Use the transresistance method and divider formulas to find path transmittances by inspection.
3. Write gain expression by use of flow-graph reduction.

Feedback analysis can be applied only if forward and feedback paths are separately identified. Then their interactions can be reduced to loading and two-port source transmittances. If path interactions are not apparent, it is always possible to fall back on basic circuit laws; KCL, KVL, and ΩL (Ohm's law). The disadvantage in this is the difficulty of expressing equations so that the path transmittances in them are explicit.

Falling back to the basic laws is not always the end of intuitive elegance. More algebraic calculation is involved, but this too can be neatly minimized by an orderly procedure based on building flow graphs from circuit equations and then reducing the graphs. This method will be called *flow-graph analysis*:

1. Apply KVL, KCL, and ΩL to produce circuit equations.
2. Construct a flow graph from these equations.
3. Reduce the flow graph.

A good approach to solving extremely difficult circuits is:

1. Make simplifying assumptions and apply feedback analysis.
2. Apply flow-graph analysis.
3. Use the results of step 1 to guide the formulation of equations from step 2 so that the flow paths become obvious.

An example of this approach is to solve a FET equivalent of a given BJT circuit and then attempt to construct the more complicated BJT expressions, guided by the FET results.

4.11 Feedback Analysis of the Common-Base Amplifier

We shall now apply both flow graph and two-port feedback analysis to the BJT configurations with r_o since (as we shall see) all the signal paths are not trivially obvious. The corresponding FET circuits follow from the BJT-to-FET transformation (4.6).

We begin with the simplest configuration, the CB amplifier in Fig. 4.19a. Applying KCL to the emitter node e, we obtain

$$\text{at } e, \qquad \frac{v_e - v_i}{R_E} + \frac{v_e}{r_e} + \frac{v_e - v_c}{r_o} = 0 \tag{4.45}$$

This can be rewritten as

$$v_e = \left(\frac{r_e \parallel r_o}{r_e \parallel r_o + R_E}\right) v_i + \left(\frac{R_E \parallel r_e}{R_E \parallel r_e + r_o}\right) v_c \tag{4.46}$$

Similarly, for the collector node c, we obtain

$$\text{at } c, \qquad \frac{v_c}{R_C} + \frac{v_c - v_e}{r_o} + \frac{v_{be}}{r_m} = 0 \tag{4.47}$$

which reduces to

$$v_e = \left(\frac{r_m \parallel r_o}{R_C \parallel r_o}\right) v_c \tag{4.48}$$

Equations (4.46) and (4.48) have the general form:

$$\text{at } e, \qquad E = av_i + bv_c \tag{4.49}$$

$$\text{at } c, \qquad E = dv_c \Rightarrow v_c = \left(\frac{1}{d}\right) E \tag{4.50}$$

where

$$E = v_e = -v_{be} \tag{4.51}$$

The form of (4.49) and (4.50) is similar to the basic feedback equations (3.1) and (3.2), respectively. These equations are represented as a flow graph in Fig. 4.19b.

From the flow graph, α_i, G, and H can be identified as

$$\alpha_i = a \qquad G = \left(\frac{1}{d}\right), \qquad H = -b \tag{4.52}$$

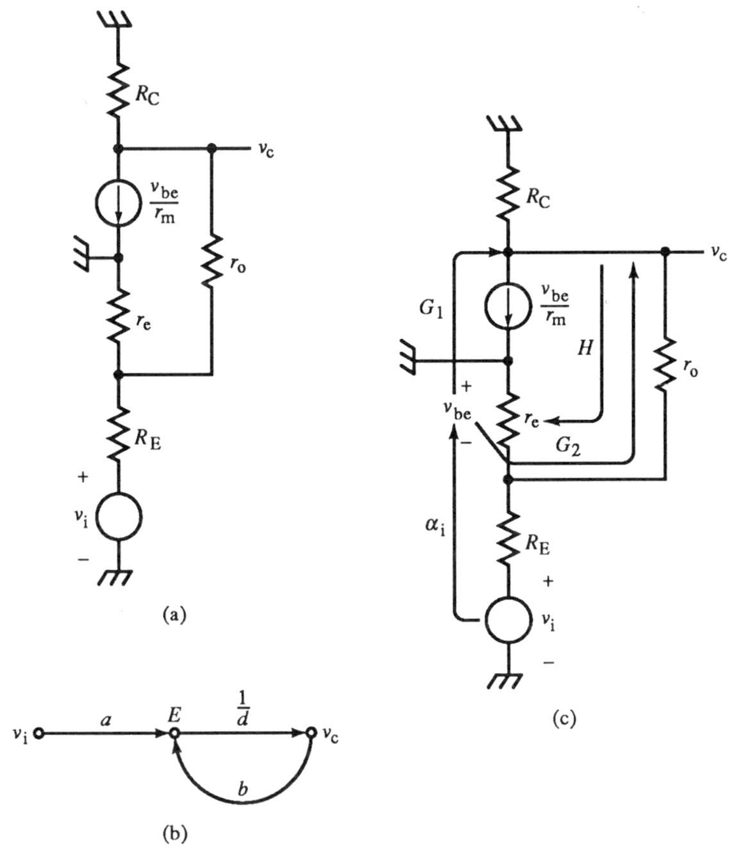

FIG. 4.19 Feedback analysis of the CB amplifier (a) and flow graph (b) with transmittance paths shown in (c). Notice that the forward path G consists of two paths: G_1, an active path, and G_2, a passive path.

When two-port loading rules from feedback analysis are applied, α_i is the voltage divider attenuation from v_i to v_e. R_E forms a divider with $r_e \parallel r_o$. From (4.46), H is

$$-\left(\frac{R_E \parallel r_e}{R_E \parallel r_e + r_o}\right)$$

Finally, G requires further decomposition:

$$G = \frac{1}{d} = \frac{R_C \parallel r_o}{r_m \parallel r_o} = \frac{R_C \parallel r_o}{r_m} + \frac{R_C}{R_C + r_o} = G_1 + G_2 \tag{4.53}$$

G can be interpreted as having two signal paths, G_1 and G_2. G_1 has a transresistance interpretation and is the gain from v_{be} to v_c. This is the active path. G_2 is a passive path attenuation from v_e to v_c, formed by divider resistances r_o and R_C. Both paths are from E to v_c; therefore, these parallel paths both contribute to G.

In this analysis, the circuit equations led to the identification of four signal paths (Fig. 4.19c). In Fig. 4.19b, these paths interact to form a familiar feedback topology. Combining α_i, G, and H into the feedback formula, the CB closed-loop gain is

$$\text{CB} \quad A_v = \alpha_i \cdot \frac{G}{1+GH} = \frac{a/d}{1-(b/d)} \tag{4.54}$$

The input and output resistances can be derived by making use of the gain calculations. The input resistance can be found by Nortonizing v_i and R_E. Then the closed-loop emitter resistance is

$$r_E = \frac{v_e}{i_i} = \frac{E}{i_i} = \frac{[\alpha_E i_i/(1+GH)]}{i_i} = \frac{\alpha_E}{1+GH} = \frac{r_e \parallel r_o \parallel R_E}{1+GH} \tag{4.55}$$

where

$$\alpha_E = \left.\frac{v_e}{i_i}\right|_{B=0} = r_e \parallel r_o \parallel R_E \tag{4.56}$$

and $i_i = v_i/R_E$. Then r_{in} can be put in the following form:

$$\text{CB} \quad r_{in} = R_E + \left(\frac{r_e \parallel r_o}{1+GH}\right) \Bigg\| \left(\frac{R_E}{GH}\right) \cong R_E + \left(\frac{r_e \parallel r_o}{1+GH}\right), \qquad R_E \gg r_e \parallel r_o \tag{4.57}$$

The feedback in Section 3.4 was assumed negative so that $1+GH>1$. Here, $0<1+GH<1$; the feedback is positive but less than 1. This causes some reversals of effect. For r_{in}, R_E is in series with the closed-loop resistance of the emitter node. By inspection, the resistance there is due to r_e (grounded at the base) in parallel with the resistance to the output, r_o. The closed-loop resistance is $r_e \parallel r_o$, divided by $1+GH$, or (4.57). The exact expression includes R_E/GH in parallel with this resistance and is due to interaction of the feedback loop with R_E.

More directly, r_{out} is

$$\text{CB} \quad r_{out} = \frac{R_C \parallel r_o}{1+GH} = \frac{R_C \parallel r_o}{1-b/d} \tag{4.58}$$

The loop gain for this amplifier is less than 1 because the feedback due to r_o is positive. Although typically $G>1$, $-H \ll 1$ so that $|GH|<1$, and the circuit does not oscillate. But the effect of positive feedback is to work against the benefits of negative feedback. In most circuits of this kind, where r_o is due to transistor base width or channel length modulation, r_o is much larger than external circuit resistances and has little effect on circuit performance. For accurate calculations, however, it is among the dominant second-order effects to be accounted for.

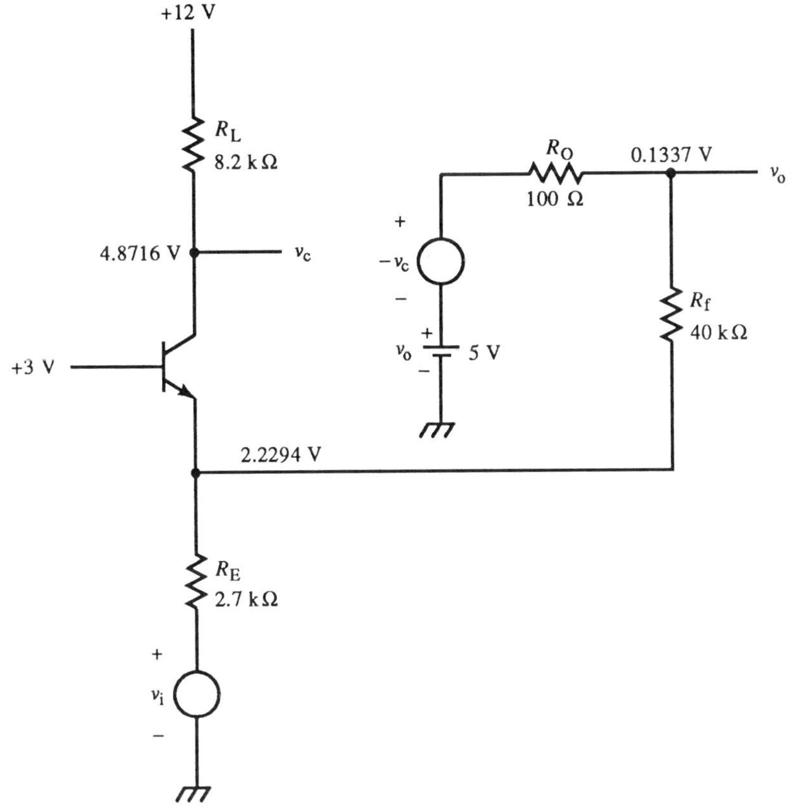

FIG. E4.1

Example 4.1 Inverting CB Feedback Amplifier

The amplifier of Fig. E4.1 is a ×(−1)-buffered CB stage with feedback through R_f. (The default BJT model is being used here: $\beta = 99$, $I_S = 10^{-16}$ A.) From SPICE, we have the following data:

$I_E = 0.8781$ mA, $V_E = 2.2294$ V, $V_C = 4.8716$ V, $V_O = 0.1337$ V

and

$$\frac{v_o}{v_i} = -2.470, \qquad r_{in} = 2.724\ \text{k}\Omega, \qquad r_{out} = 83.12\ \Omega$$

From these data, $r_e = 29.46\ \Omega$ and $r_e \parallel R_f = 29.44\ \Omega$. Then, applying feedback analysis with $E = v_{be}$, we have

$$G_1 = -\alpha \cdot \frac{R_L}{r_e}\left(\frac{R_f}{R_O + R_f}\right) = -274.78$$

$$G_2 = \frac{R_O}{R_O + R_f} = 2.4938 \times 10^{-3} \cong 0$$

Then $G = G_1 + G_2 = -274.78$. Furthermore,

$$H = -\frac{R_E \parallel r_e}{R_f + R_E \parallel r_e} = -7.2826 \times 10^{-4}$$

$$\alpha_i = \frac{R_f \parallel r_e}{R_f \parallel r_e + R_E} = 1.0789 \times 10^{-2}$$

Combining these transmittances yields

$$\frac{v_o}{v_i} = \alpha_i \cdot \frac{G}{1+GH} = -2.4703$$

$$r_{in} = R_E + r_e \left\| \frac{R_f}{1+(-G)} \right. = 2.7245\ \text{k}\Omega$$

$$r_{out} = \frac{R_O \parallel (R_f + R_E \parallel r_e)}{1+GH} = 83.118\ \Omega$$

Example 4.2 CB BJT Amplifier with R_{CE}

The circuit in Fig. E4.2 is a CB amplifier with a fixed external collector-emitter resistance R_{CE}. This resistor is a simplified form of r_o since it is independent of I_E. To analyze this circuit, the approach given in Section 4.10 for difficult circuits will be taken. If we solve the same circuit without R_{CE}, then if R_{CE} has a minor effect on the circuit, the simplified analysis gives us an approximation by which to evaluate more complicated solutions. The dc solution for the simplified circuit is

$$I_E = 0.826\ \text{mA}, \qquad V_E = 2.231\ \text{V}, \qquad V_C = 5.292\ \text{V}$$

(The default BJT model is being used here: $\beta = 99$, $I_S = 10^{-16}$ A.) Then the ac solution follows:

$$r_e = 31.3\ \Omega, \qquad v_c/v_i = 2.97, \qquad r_{in} = 2.73\ \text{k}\Omega, \qquad r_{out} = 8.2\ \text{k}\Omega$$

SPICE simulation results for the operating point are

$$I_E = 0.7511\ \text{mA}, \qquad I_{RCE} = 76.075\ \mu\text{A}, \qquad V_E = 2.2355\ \text{V}, \qquad V_C = 5.2785\ \text{V}$$

The dynamic parameters are

$$r_e = 34.456\ \Omega, \qquad r_m = 34.804\ \Omega, \qquad r_\pi = 3.4456\ \text{k}\Omega, \qquad \mu = 1149.3$$

The simulation results are

$$\frac{v_c}{v_i} = 2.955, \qquad r_{in} = 2.741\ \text{k}\Omega, \qquad r_{out} = 8.162\ \text{k}\Omega$$

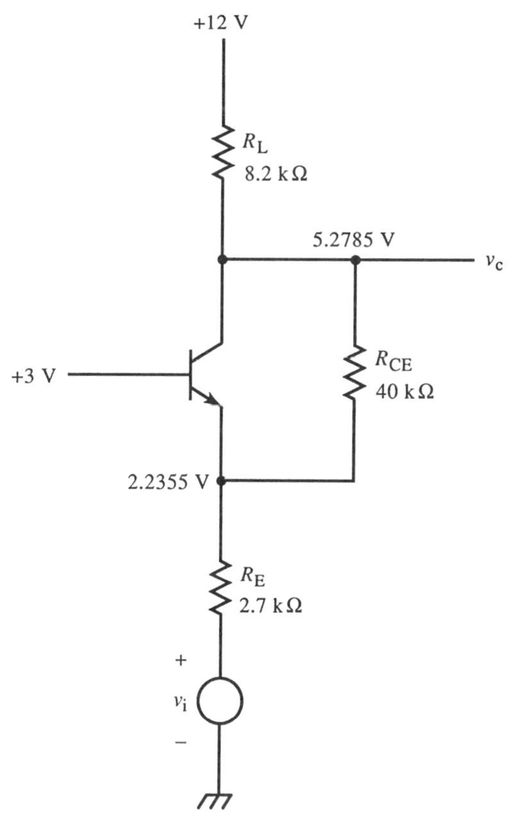

FIG. E4.2

With these data, we can now apply various methods to find the three parameters of interest and verify the results. The first solution is based on (4.54), derived from KCL, and involves simple substitution:

$$\frac{v_o}{v_i} = \frac{a}{d-b} = \frac{1.2591 \times 10^{-2}}{-8.4982 \times 10^{-4} + 5.1100 \times 10^{-3}} = 2.9556$$

From this calculation, $1 + GH = 1 - b/d = 0.83370$. The resistances are

$$r_{in} \cong \frac{r_e \parallel r_o}{1+GH} + R_E = 2.7413 \text{ k}\Omega$$

$$r_{out} = \frac{R_C \parallel r_o}{1+GH} = \frac{6.8050 \text{ k}\Omega}{0.83370} = 8.1624 \text{ k}\Omega$$

Next, we shall solve the circuit using the μ transform and (4.23), (4.25), and (4.26):

$$\frac{v_c}{v_i} = (0.56066)(5.2710) = 2.9552$$

$$r_{in} = 2.7\ \text{k}\Omega + 41.399\ \Omega = 2.7414\ \text{k}\Omega$$

$$r_{out} = 8.2\ \text{k}\Omega \parallel (40\ \text{k}\Omega + 1.7413\ \text{M}\Omega) = 8.1624\ \text{k}\Omega$$

Finally, we can use Miller's theorem to find r_{in}. First,

$$K = -\frac{v_c}{E} = -\frac{1}{d} = -195.69$$

then,

$$r_{in} = R_E + r_e \left\| \frac{R_{CE}}{1+K} \right. = 2.7\ \text{k}\Omega + (34.46\ \Omega) \parallel (40\ \text{k}\Omega/(-194.7) = 2.7414\ \text{k}\Omega$$

4.12 Feedback Analysis of the Common-Emitter Amplifier

For the CE of Fig. 4.20a, KCL is applied at emitter and collector:

$$\text{at } e, \qquad v_e = \left(\frac{R_E \parallel r_o}{R_e \parallel r_o + r_e}\right) v_i + \left(\frac{R_E \parallel r_e}{R_E \parallel r_e + r_o}\right) v_c \tag{4.59}$$

$$\text{at } c, \qquad v_e = \left(\frac{r_o}{r_o + r_m}\right) v_b + \left(\frac{r_m \parallel r_o}{R_C \parallel r_o}\right) v_c \tag{4.60}$$

Let $E = v_{be}$. These equations can be made explicit in E by negating them and adding v_b. The result is

$$v_{be} = \left(\frac{r_e}{R_E \parallel r_o + r_e}\right) v_i - \left(\frac{R_E \parallel r_e}{R_E \parallel r_e + r_o}\right) v_c \tag{4.61}$$

$$v_{be} = \left(\frac{r_m}{r_o + r_m}\right) v_i - \left(\frac{r_m \parallel r_o}{R_C \parallel r_o}\right) v_c \tag{4.62}$$

The more general forms of these equations are, respectively:

$$E = av_i + bv_c \tag{4.63}$$

$$E = cv_i + dv_c \Rightarrow v_c = \left(\frac{1}{d}\right) E - \left(\frac{c}{d}\right) v_i \tag{4.64}$$

Compared with (4.49) and (4.50), the CE has the additional term with coefficient c. These equations are represented by the flow graph of Fig. 4.20b. The identifiable paths are

$$\alpha_i = a, \qquad H = -b, \qquad G = \left(\frac{1}{d}\right), \qquad F = -\left(\frac{c}{d}\right) \tag{4.65}$$

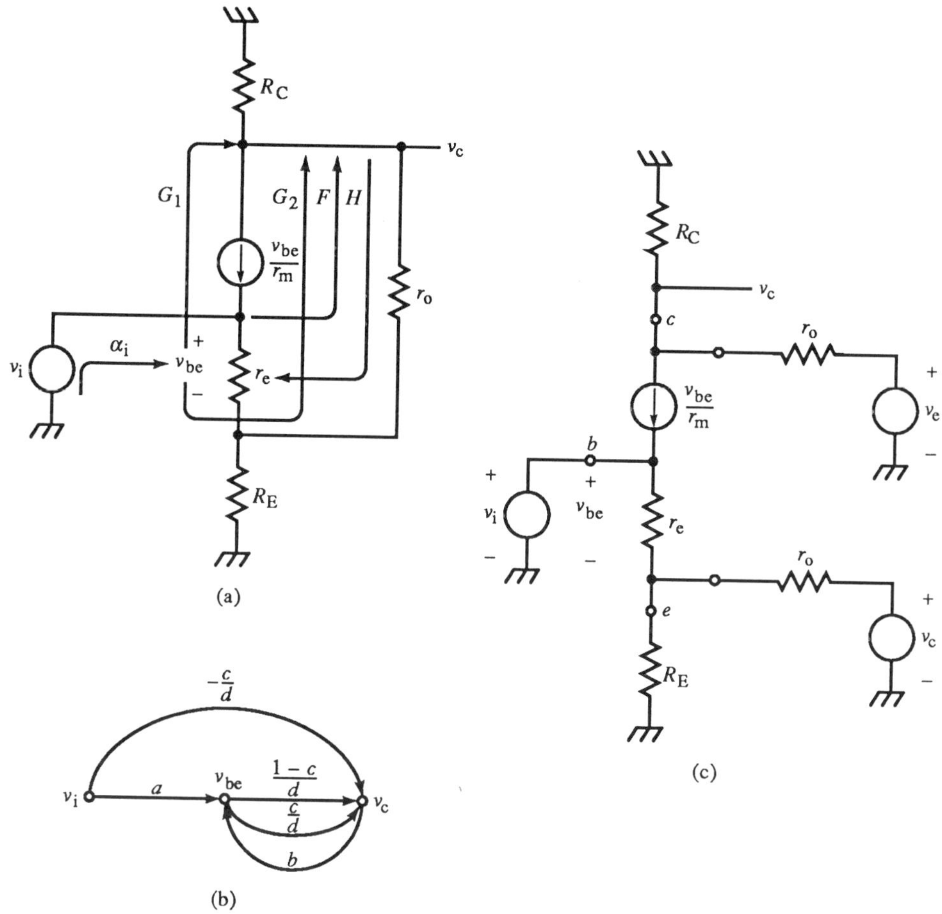

FIG. 4.20 Feedback analysis of the CE amplifier with r_o (a), showing the five paths of flow graph (b). A two-port equivalent circuit is shown in (c).

G is the same as for the CB but is negative, as expected for a CE. (The CE E is also the negated CB E.) G represents the same two paths, G_1 and G_2, as for the CB. H is a voltage divider from v_c to $v_e = -E$ (and is the negated CB H). α_i is the voltage divider attenuation from v_i to v_{be}. Because of the additional term in v_c, there is a feedforward path F from v_i to v_c in the CE that is absent in the CB. F can be expressed (using formulas from Section 4.9) as

$$F = -\frac{c}{d} = \left(\frac{R_C \parallel r_o}{r_m \parallel r_o}\right)\left(\frac{r_m}{r_o + r_m}\right) = \left(\frac{R_C}{R_C + r_o}\right) \tag{4.66}$$

The path of F can be traced from this expression. It is the attenuation of the passive divider from emitter to collector through r_o. F is v_c/v_i, where v_i is at the base (not the emitter). To find F, as we did with G and H in Section 4.10, the path through G must be nulled to allow only the signal through F to affect

v_c. This can be accomplished in this case by setting E to zero, or

$$F = \left.\frac{v_c}{v_i}\right|_{E=0} \tag{4.67}$$

Then for $E = v_{be} = 0$, then $v_b = v_e$ and $F = v_c/v_e$, as in (4.66). Figure 4.20a shows the five signal paths. The two additional paths, G_2 and F, are a result of the r_o branch to the collector. They differ in that G_2 is the path from v_{be} whereas F is from the input v_i. For F, an increase in base voltage causes an increase in emitter voltage. This increase is transmitted through the divider to the collector uninverted. For $E = v_{be}$, an increasing E causes a decreasing v_e. Consequently, G_2 is inverted. E and v_i do not follow identical paths to the output since v_i (through F) can affect v_c even though the effect of G_2 on v_c is due to E.

A two-port equivalent of H is shown in Fig. 4.20c. To meet the constraints on H (from Section 3.9), the H input source v_e must be expressed in terms of v_{be} (using a divider formula). Since this is possible, this approach also leads to a solution. We now examine this aspect further.

Since the choice of E is arbitrary within the feedback loop, Fig. 4.21 shows the solution with

$$E = v_i - \left(\frac{R_E}{R_E + r_o}\right) v_c \tag{4.68}$$

The two-port equivalent CE circuit is shown in Fig. 4.21a. The flow graph (Fig. 4.21b) can be derived from Fig. 4.20b by moving a forward through the E node. Paths out of E become multiplied by a and paths into E divided by a, as shown. The paths are identified as

$$\alpha_i = 1 \tag{4.69a}$$

$$G = \frac{a}{d} = \frac{(1-c)}{d} \cdot a + \frac{c}{d} \cdot a = G_1 + G_2 \tag{4.69b}$$

where

$$G_1 = -\alpha \cdot \frac{R_C \parallel r_o}{R_E \parallel r_o + r_e}, \tag{4.69c}$$

$$G_2 = -\left(\frac{r_e}{R_E \parallel r_o + r_e}\right)\left(\frac{R_C}{R_C + r_o}\right) \tag{4.69d}$$

$$H = -\frac{b}{a} = \frac{R_E}{R_E + r_o} \tag{4.69e}$$

$$F = -\frac{c}{d} = \left(\frac{R_C \parallel r_o}{r_m \parallel r_o}\right)\left[1 - \frac{r_o}{r_m + r_o}\right] = \frac{R_C}{R_C + r_o} \tag{4.69f}$$

This choice of E places it across the transresistance r_M of the transistor forward

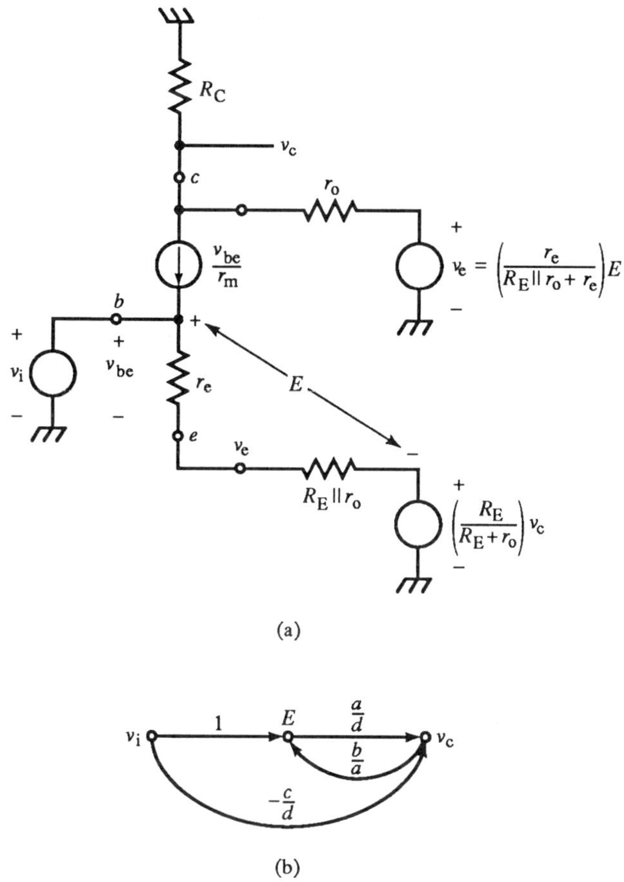

FIG. 4.21 Feedback analysis of the CE amplifier with r_o, with two-port circuit for H and E as shown in (a); flow graph (b).

path gain G_1. v_i is directly in the loop containing E, so that $\alpha_i = 1$. Similarly, the divider formula of the Thévenin feedback source is H.

G_2 is the passive forward path through r_o and consists of two voltage divider factors in (4.69d). The first is the divider from E to v_e with output across v_{be}. The second is from v_e to v_c. To find G_2, then, $G_1 E$, B, and $F \cdot v_i$ must be nulled. The G_1 path is nulled by disconnecting the BJT collector current source. B is set to zero when the H output source is set to zero. Nulling the signal through F is not easy because it shares essentially the same path as G_2. In this case, instead of attempting to null Fv_i, we subtracted it from the derived expression for G_2. We must find

$$v_c = G_2 E + Fv_i|_{B,G_1=0} = (G_2\alpha_i + F)v_i, \qquad B, G_1 = 0 \tag{4.70}$$

Then,

$$G_2 = \left.\frac{\left(\frac{v_c}{v_i}\right) - F}{\alpha_i}\right|_{B,G_1=0} = \left(\frac{v_e}{v_i}\right)\left(\frac{v_c}{v_e}\right) - F$$
$$= \left(\frac{R_E \parallel r_o}{R_E \parallel r_o + r_e}\right)\left(\frac{R_C}{R_C + r_o}\right) - \left(\frac{R_C}{R_C + r_o}\right) = -\frac{r_e}{R_E \parallel r_o + r_e} \cdot \frac{R_C}{R_C + r_o} \tag{4.71}$$

This derivation of G_2 shows that when paths overlap significantly, isolating the path being found may require an approach other than nulling sources. Opening paths and subtracting the effects of intertwined paths are also options in meeting the transmittance constraints. Use of a different flow graph, such as Fig. 4.18b instead of (a), can "untangle" paths. The basic idea when finding a path transmittance is to eliminate contributions from other paths to the output node of the path being found.

Example 4.3 CE Amplifier with R_{CE}

Figure E4.3 is a CE amplifier with the default BJT model: $\beta = 99$, $I_S = 10^{-16}$ A. From SPICE, the dc values are

$$I_E = 0.6775 \text{ mA}, \qquad V_E = -0.7639 \text{ V}, \qquad V_C = 5.2644 \text{ V}$$

The simulation ac solution is

$$\frac{v_c}{v_i} = -2.946, \qquad r_{in} = 216.4 \text{ k}\Omega, \qquad r_{out} = 8.160 \text{ k}\Omega$$

From these data, we can calculate the following:

$$r_e = 38.2\ \Omega, \qquad r_m = 38.585\ \Omega, \qquad r_\pi = 3.82 \text{ k}\Omega, \qquad \mu + 1 = 1037.7$$

With $R_{CE} \to \infty$, $v_c/v_i \cong -2.96$. Applying Blackman's formula (Section 4.18) for r_{in}, we get $r_{in} = 216.40$ kΩ. Applying (4.28) through (4.30) gives

$$\frac{v_c}{c_i} = (-0.58493)(5.0358) = -2.9456$$

$$r_{in} = (3.82 \text{ k}\Omega)(55.992) + (2.5568 \text{ k}\Omega) = 216.45 \text{ k}\Omega$$

$$r_{out} = 8.2 \text{ k}\Omega \parallel (1037.7)(1.5819 \text{ k}\Omega + 40 \text{ k}\Omega) = 8.1602 \text{ k}\Omega$$

Next, we use feedback analysis to find a solution. Following Section 4.12, let $E = v_{be}$. Then the transmittances are

$$\alpha_i = 1.4878 \times 10^{-2}, \qquad G = -176.53, \qquad H = 9.4079 \times 10^{-4}, \qquad F = 0.17012$$

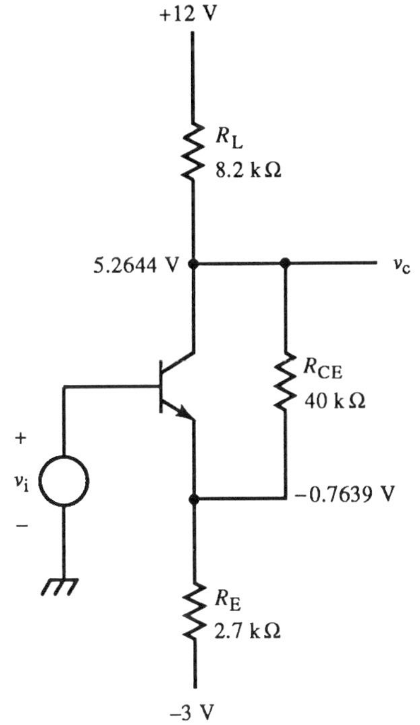

FIG. E4.3

Then $1+GH=0.83392$ and

$$\frac{v_c}{v_i}=\frac{\alpha_i G}{1+GH}+\frac{F}{1+GH}=-3.1496+0.20401=-2.9456$$

$$r_{out}=\frac{R_C \parallel R_{CE}}{1+GH}=8.1602\text{ k}\Omega$$

Another feedback solution is based on (4.68):

$$E=v_i-\left(\frac{R_E}{R_E+R_{CE}}\right)v_c$$

The transmittances are

$$\alpha_i=1, \qquad G=G_1+G_2=-2.6240-2.5312\times 10^{-3}=-2.6265,$$

$$H=-6.3232\times 10^{-2}, \qquad F=0.17012$$

Then $1+GH=0.83392$ and $v_c/v_i=-2.9456$. Furthermore,

$$r_{in}=(1+GH)(\beta+1)(r_e+R_E \parallel (R_{CE}+R_C))=216.40\ \text{k}\Omega$$

$$r_{out}=\frac{R_C \parallel R_{CE}}{1+GH}=8.1602\ \text{k}\Omega$$

These results are further confirmed by the μ transform. From Section 4.8,

$$\frac{v_c}{v_i}=-2.9456, \qquad r_{in}=216.45\ \text{k}\Omega,$$

$$r_{out}=R_C \parallel [(\mu+1)(r_\pi \parallel R_E)+R_{CE}]=8.1602\ \text{k}\Omega$$

4.13 Feedback Analysis of the Common-Collector Amplifier

The last BJT configuration is the CC, shown in Fig. 4.22a. If we base the CC analysis on the results of the CE, then E is chosen to be v_{be}. Combining (4.63)

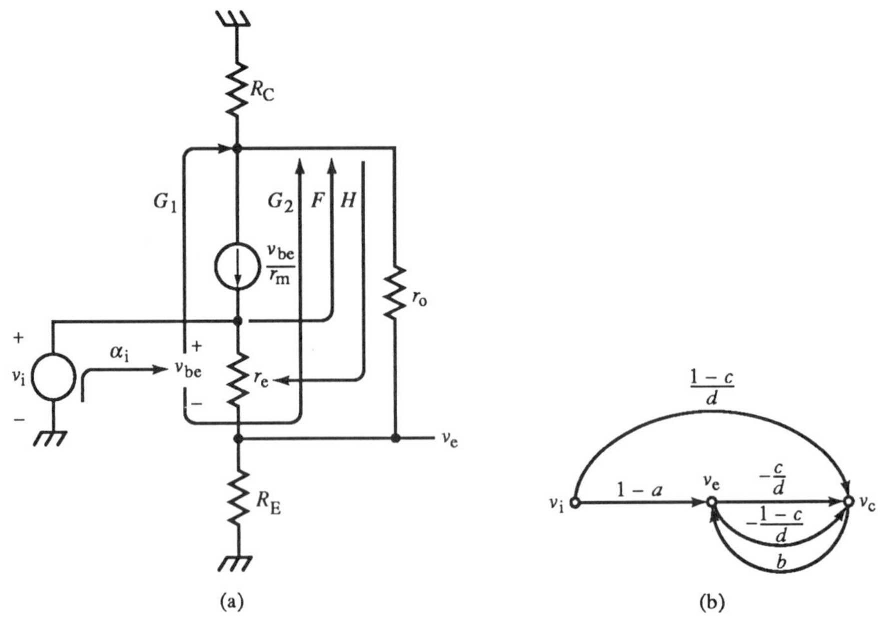

FIG. 4.22 Feedback analysis of the CC amplifier with r_o, showing the five paths (a) with flow-graph of Fig. 4.20b; flow graph (b) with explicit v_e.

and (4.64) we obtain

$$E = v_{be} = av_i + b\left[\left(\frac{1}{d}\right)E - \left(\frac{c}{d}\right)v_i\right] = \frac{a-(bc/d)}{1-(b/d)} \cdot v_i \tag{4.72}$$

The CC output, v_e, is related to E by $v_e = v_i - E$. Dividing by v_i and substituting,

$$\frac{v_e}{v_i} = 1 - \frac{E}{v_i} = 1 - \frac{a-(bc/d)}{1-(b/d)} = \frac{(1-a)-(b/d)(1-c)}{1-(b/d)} \tag{4.73}$$

The paths, identified on the topological model of Fig. 4.22, correspond to paths of the CE flow graph, Fig. 4.20b. In Fig. 4.22b, (4.73) is represented as a flow graph with v_e made explicit. The paths are:

$$1-a = \frac{R_E \parallel r_o}{R_E \parallel r_o + r_e}; \qquad -\frac{1}{d} = -\frac{c}{d} - \frac{(1-c)}{d} = \frac{R_C \parallel r_o}{r_m} + \frac{R_C}{R_C + r_o}$$

$$-b = \frac{R_E \parallel r_e}{R_E \parallel r_e + r_o}; \qquad \frac{1-c}{d} = -\frac{R_C \parallel r_o}{r_m} \tag{4.74}$$

4.14 Inverting Op-Amp with Output Resistance

A feedback approach will now be taken to the inverting op-amp with op-amp output resistance of R_o (Fig. 4.23a). A nonzero R_o provides another forward path to v_o through R_f. Following the approach in the previous sections, let us assume that the circuit can be represented by the flow graph in Fig. 4.23b, modified accordingly (this time as a block diagram) in Fig. 4.23c. Applying feedback analysis, let $E = v_-$. Then

$$v_1 = -Kv_- \tag{4.75}$$

$$E = v_- = \alpha_i v_i - Hv_o \tag{4.76}$$

$$\alpha_i = \frac{R_f + R_o}{R_f + R_i + R_o} \tag{4.77}$$

$$G = -K\left(\frac{R_f + R_i}{R_f + R_i + R_o}\right) \tag{4.78}$$

$$H = -\frac{R_i}{R_f + R_i} \tag{4.79}$$

The transfer function from Fig. 4.23b is

$$A_v = \frac{v_o}{v_i} = \alpha_i \cdot \frac{G}{1+GH} + F \tag{4.80}$$

The feedforward path F must also be determined:

$$F = \left.\frac{v_o}{v_i}\right|_{C=0} = \frac{R_o}{R_o + R_f + R_i} \tag{4.81}$$

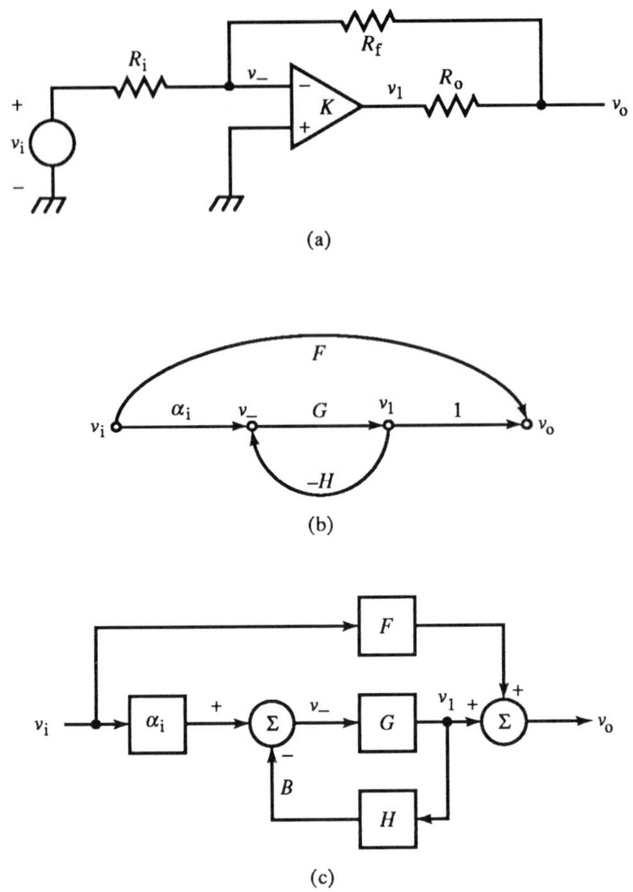

FIG. 4.23 Inverting op-amp with output resistance (a), flow graph (b), and equivalent block diagram (c).

Substituting (4.77)-(4.79) and (4.81) into (4.80) and simplifying, the closed-loop gain is

$$A_v = \left(\frac{R_f + R_i}{R_f + R_o + R_i}\right)\left(\frac{-K(R_f + R_o)}{R_f + (1+K)R_i}\right) + \frac{R_o}{R_o + R_f + R_i} \tag{4.82}$$

Taking the limit of A_v as $K \to \infty$, the familiar $-R_f/R_i$ results, as in (3.22). This shows that R_o does not affect the closed-loop gain with sufficiently large K.

This op-amp circuit can also be used to demonstrate an alternative gain derivation based on a different flow graph, that of Fig. 4.24b with the two-port equivalent circuit in (a). This choice of topology has no feedforward though G still has two parallel paths. The difference that this flow graph makes in the analysis is that the output of G is v_o, not v_1 as in Fig. 4.23b. The

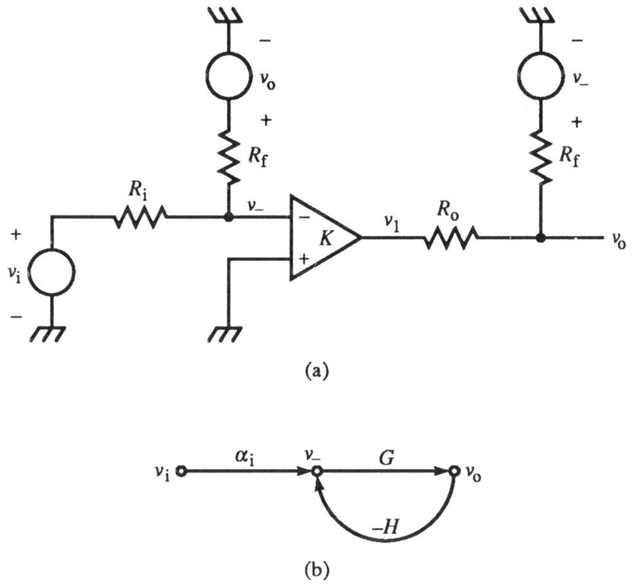

FIG. 4.24 Two-port feedback equivalent circuit of Fig. 4.23a where output of G is v_o instead of v_1 (a), and its flow graph with no feedforward path (b).

transmittances, found by the usual feedback analysis, are

$$\alpha_i = \frac{R_f}{R_f + R_i} \tag{4.83}$$

$$G = G_1 + G_2 = -K\left(\frac{R_f}{R_f + R_o}\right) + \left(\frac{R_o}{R_f + R_o}\right) \tag{4.84}$$

$$H = -\frac{R_i}{R_f + R_i} \tag{4.85}$$

Substituting these transmittances into the feedback formula,

$$\frac{v_o}{v_i} = \alpha_i \cdot \frac{G}{1 + GH} \tag{4.86}$$

results in an equivalent equation to (4.82). The gain expression of (4.86) can be verified by applying KCL at v_- and v_o. From the resulting equations, the three transmittances can be readily extracted.

The implementation of the circuits modeled in Figs. 3.10, 3.14, and 3.16 has a finite output resistance and a feedforward path. Similar analyses are applicable to them.

We have examined in some detail the effects of r_o or similar shunt resistance around the active path of single transistors or amplifiers. Usually the contribution of the extra forward path is negligible, as for the preceding op-amp circuit.

In precision amplifier stages such as the differential input stage of an op-amp, finite r_o can contribute significantly to imbalance in the collector (or drain) load resistance. A differential cascode input stage reduces this problem because the CB (or CG) output transistors have a maximum output resistance due to r_o. Another influence of a feedforward path is to cause the step response to begin with a momentary inversion before responding with the expected polarity of step. This is due to a faster passive forward path than the active inverting path. This dynamic phenomenon is called *preshoot.*

4.15 Feedback Analysis of the Shunt-Feedback Amplifier

The shunt-feedback amplifier is similar to the inverting op-amp but it has limited gain. A typical one-transistor shunt-feedback amplifier is shown in Fig. 4.25. R_f shunts the transistor from collector to base. This is the third and last

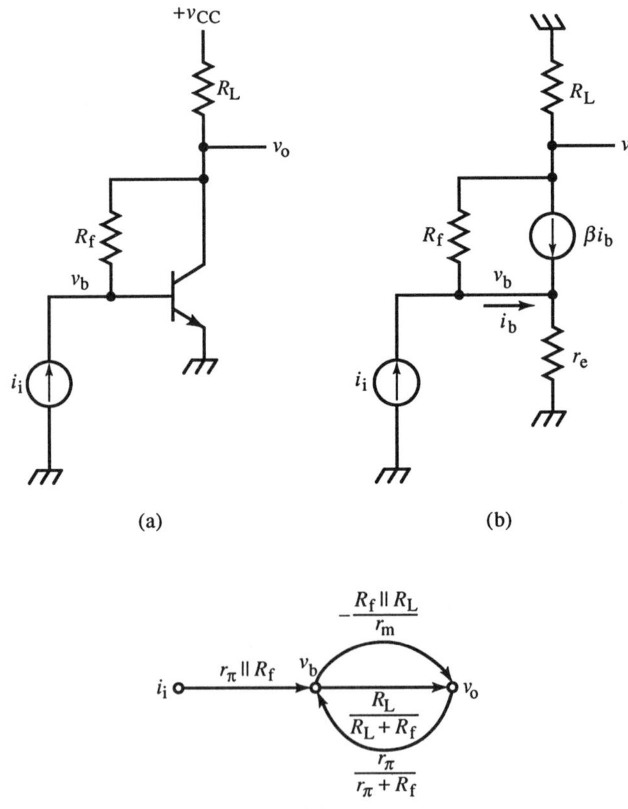

FIG. 4.25 The shunt-feedback amplifier (a), equivalent circuit model (b), and flow graph with $E = v_b$ (c).

possible shunting configuration for a transistor and is another basic kind of amplifier stage. Unlike the base-emitter or collector-emitter (r_o) shunts, this one is quite useful. We shall analyze it from several perspectives.

This is a transresistance amplifier because the input is a current i_i. Applying KCL at base and collector, we obtain

$$\text{at } b, \quad v_b = (r_\pi \parallel R_f)i_i + \left(\frac{r_\pi}{r_\pi + R_f}\right)v_o \tag{4.87}$$

$$\text{at } c, \quad v_o = \left[\frac{1}{R_f} - \frac{1}{r_m}\right](R_L \parallel R_f)\cdot v_b = \left[\frac{R_L}{R_L + R_f} - \frac{R_f \parallel R_L}{r_m}\right]v_b \tag{4.88}$$

These equations can be written more compactly and directly in terms of transmittances of Fig. 4.25c with $E = v_b$:

$$v_b = \alpha_i i_i - Hv_o, \qquad v_o = (G_1 + G_2)v_b \tag{4.89}$$

The closed-loop gain is

$$\frac{v_o}{i_i} = \alpha_i \cdot \frac{G}{1+GH} = -\frac{(r_\pi \parallel R_f)[R_L/(R_f + R_L)][(R_f/r_m) - 1]}{1 + (r_\pi \parallel R_f)[R_L/(R_f + R_L)][(R_f/r_m) - 1](1/R_f)} \tag{4.90}$$

This derivation can be used to construct the results for a similar amplifier with R_i in shunt with i_i. This makes the input a Norton circuit, which can be converted to a Thévenin equivalent with

$$v_i = R_i \cdot i_i$$

resulting in a voltage amplifier. These equations are easily modified to account for R_i by replacing each occurrence of r_π with $r_\pi \parallel R_i$, since R_i shunts r_π.

Now consider some simplifications of (4.90). If R_L is replaced by a current source,

$$\left.\frac{v_o}{i_i}\right|_{R_L \to \infty} = -\alpha R_f + r_e \tag{4.91}$$

For BJT $\beta \to \infty$,

$$\left.\frac{v_o}{i_i}\right|_{\beta \to \infty} = (-R_f + r_e)\left(\frac{R_L}{R_L + r_e}\right) = -R_f\left(\frac{R_L}{R_L + r_e}\right) + r_e \parallel R_L \tag{4.92}$$

With both of the above assumptions,

$$\left.\frac{v_o}{i_i}\right|_{R_L, \beta \to \infty} = -R_f + r_e \tag{4.93}$$

A two-port feedback analysis with $E = i_b$ is based on Fig. 4.26, in which H is identified as the two-port equivalent circuit. Here,

$$E = i_b = \alpha_i i_i + \left(\frac{1}{R_f + r_\pi}\right)v_o \tag{4.94}$$

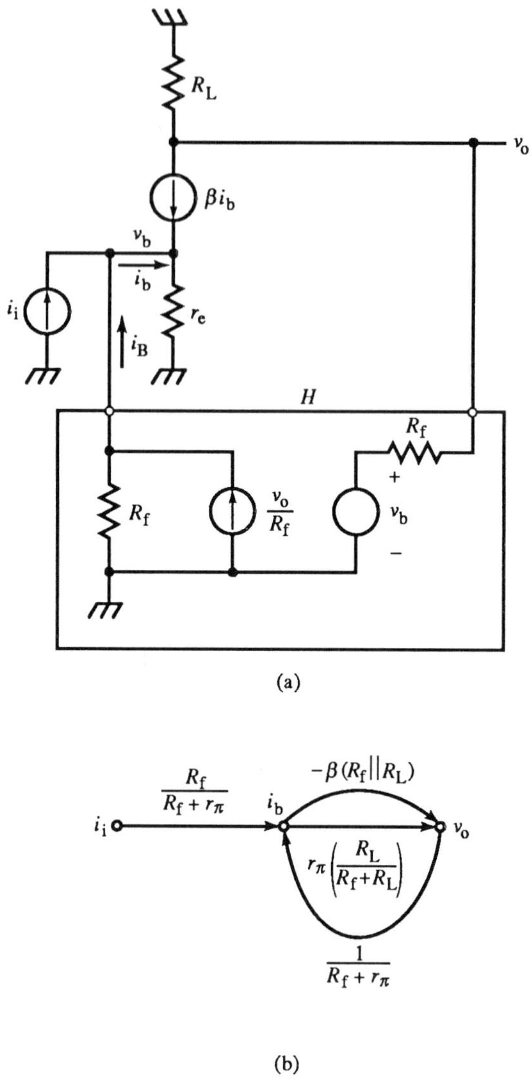

FIG. 4.26 Two-port H circuit for shunt-feedback amplifier (a) and flow graph (b).

The transmittances can be derived from the previous calculations by shifting r_π forward, out of α_i in Fig. 4.25c. The previous $E = v_b$ is then divided by r_π, making $E = i_b$. Then r_π multiplies G and divides H (see Fig. 4.26b). The closed-loop gain is the same as (4.90). From inspection of Fig. 4.26, α_i is now an input current divider:

$$\alpha_i = \frac{R_f}{R_f + r_\pi} = \text{fraction of } i_i \text{ through } r_\pi \ (= i_b) \tag{4.95}$$

G has two paths, an active G_1 path and a passive G_2 path:

$$G = G_1 + G_2 = -\beta(R_f \parallel R_L) + r_\pi \cdot \left(\frac{R_L}{R_f + R_L}\right) \tag{4.96}$$

The first term is i_b, multiplied by β to become i_c. This current develops v_o across the collector resistance $R_f \parallel R_L$ and is negative. The second term is the passive path from the base through R_f to the collector. The H input source of Fig. 4.26 is v_b. Since E is i_b, then $i_b \cdot r_\pi$, or v_b, is divided by R_f and R_L. The second term G_2 is thus $(v_b/i_b)(v_o/v_b)$.

If $R_L \ll R_f$, then $G_2 \cong 0$ and, for $\beta \to \infty$,

$$\left.\frac{v_o}{i_i}\right|_{\substack{\beta\to\infty \\ R_L \ll R_f}} \cong -R_f \tag{4.97}$$

This simple formula is the approximate transresistance for the single-BJT shunt-feedback amplifier for implementations in which R_f is large. The greatest error is usually due to finite β, causing r_π to excessively shunt R_f.

The input resistance, easy to find by using the feedback approach, is

$$r_{in} = \frac{v_b}{i_i} = \frac{r_\pi \parallel R_f}{1 + GH} = \frac{r_\pi \parallel R_f}{1 + [r_\pi/(r_\pi + R_f)][R_L/(R_f + R_L)][(R_f/r_m) - 1]} \tag{4.98}$$

This can be checked by resorting to the basic feedback equations for this circuit:

$$v_b = (r_\pi \parallel R_f)i_i - Hv_o \tag{4.99a}$$

$$v_o = Gv_b \tag{4.99b}$$

Then

$$v_b = (r_\pi \parallel R_f)i_i - GHv_b$$

Solving for v_b/i_i yields

$$r_{in} = \frac{r_\pi \parallel R_f}{1 + GH} = \frac{r_\pi \parallel R_f}{1 - G[r_\pi/(r_\pi + R_f)]} = r_\pi \parallel \left(\frac{R_f}{1 - G}\right) \tag{4.100}$$

The last expression is cast in the form of Miller's theorem, showing that it too could have been applied to find r_{in}. R_f is across an amplifier of gain v_o/v_b and is reduced by $1/(1 - G)$ times its value. This effective resistance is shunted by r_π. If R_i is involved, r_{in} is easily modified by replacing r_π by the parallel combination, as for (4.90).

A KCL solution for r_{out} follows from

$$i_o = \frac{v_o}{R_L} + \frac{v_o - v_b}{R_f} + \frac{v_b}{r_m}, \qquad v_b = \left(\frac{r_\pi}{R_f + r_\pi}\right)v_o \tag{4.101}$$

and is v_o/i_o:

$$r_{out} = R_L \parallel \frac{R_f + r_\pi}{\beta + 1} \tag{4.102}$$

For a current-source load, $R_L \to \infty$ and

$$r_{out}|_{R_L \to \infty} = \frac{R_f + r_\pi}{\beta + 1} \tag{4.103}$$

This result has a topological interpretation. A change in ouput voltage v_o causes a current in R_f that flows entirely into the base. This current is

$$i_b = \frac{v_o}{R_f + r_\pi}$$

The total current resulting from v_o is this current plus the collector current, or

$$i_o = i_b + i_c = (\beta + 1)i_b = (\beta + 1)\frac{v_o}{R_f + r_\pi} \tag{4.104}$$

Solving for v_o/i_o, we again obtain (4.103).

Miller's theorem cannot be used to find r_{out} because v_o is not a voltage source; R_L is part of the internal resistance.

Example 4.4 Shunt-Feedback BJT Amplifier

Figure E4.4 has $V_{BE} = 0.8085$ V, $I_E = 3.81$ mA, and $V_C = 1.1896$ V. (The default BJT model is being used here: $\beta = 99$, $I_S = 10^{-16}$ A.) Then $r_e =$

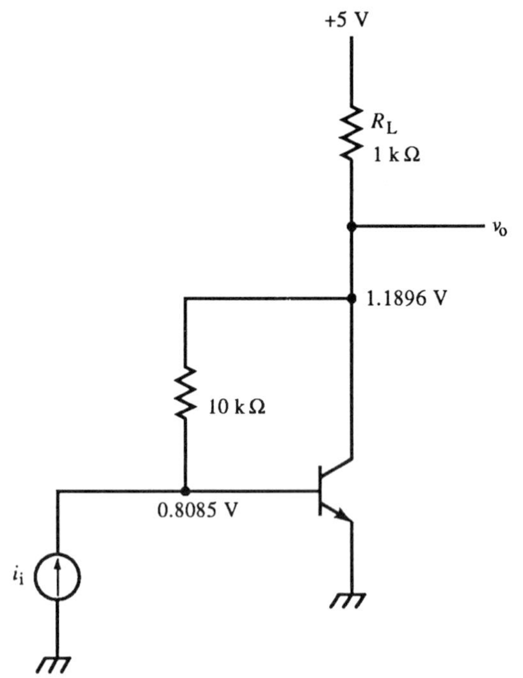

FIG. E4.4

6.79 Ω, $r_m = 6.858\ \Omega$, and $r_\pi = 678.92\ \Omega$. Feedback analysis, with $E = v_{be}$, yields

$$G = -\frac{909.09\ \Omega}{6.858\ \Omega} + 9.0909 \times 10^{-2} = -132.47$$

$$H = -\frac{r_\pi}{r_\pi + R_f} = -6.3576 \times 10^{-2}$$

so that $1 + GH = 9.4220$ and, from (4.90),

$$\frac{v_o}{i_i} = -8.9386\ \text{k}\Omega$$

The resistances are

$$r_{out} = \frac{R_L \parallel R_f}{1 + GH} = 96.486\ \Omega$$

Alternatively, from (4.102),

$$r_{out} = R_L \left\| \left(\frac{R_f + r_\pi}{\beta + 1} \right) \right. = 96.486\ \Omega$$

Finally,

$$r_{in} = r_\pi \left\| \left(\frac{R_f}{1 - G} \right) \right. = 67.476\ \Omega$$

To check these results, the SPICE simulation produced

$$\frac{v_o}{i_i} = -8.939\ \text{k}\Omega, \qquad r_{in} = 67.44\ \Omega, \qquad r_{out} = 96.48\ \Omega$$

4.16 Shunt-Feedback Amplifier Analysis: Substitution Theorem

Another way to find shunt-feedback amplifier output resistance is to use the substitution theorem to find the effective resistance of the collector path. Since i_c is controlled by v_o, it can be expressed as a resistance r_c:

$$r_c = \frac{r_m}{(v_b/v_o)} = \frac{r_m}{(r_\pi/(R_f + r_\pi))} = \frac{R_f + r_\pi}{\beta} \tag{4.105}$$

The resistance shunting r_c through R_f is $R_f + r_\pi$. Combining these in parallel results in (4.103). In (4.105), v_b/v_o is $-H$ rather than $1/G$ because i_i is nulled.

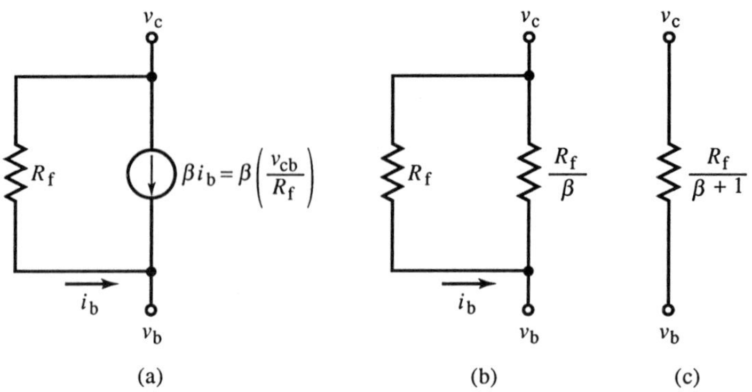

FIG. 4.27 Shunt-feedback amplifier analysis using the substitution theorem. When all of i_b flows through R_f, then R_f and the transistor current source form a Norton circuit (a). The current source is controlled by its terminal voltage and can be converted to a resistance (b) and merged with R_f (c).

The substitution theorem can be applied in a more general way to the combination of R_f and the βi_b current source, as in Fig. 4.27. Since

$$i_b = \frac{v_c - v_b}{R_f} = \frac{v_{cb}}{R_f}$$

the current source becomes $\beta v_{cb}/R_f$. The source is across v_{cb} and is also dependent on it, making the substitution theorem applicable. In Fig. 4.27b, the current source is replaced by a resistance of R_f/β. When this is combined with R_f, the result is a single resistance of $R_f/(\beta+1)$.

This equivalent circuit makes shunt-feedback amplifier resistance analysis much simpler than feedback analysis. Figure 4.28 shows the shunt-feedback amplifier equivalent circuit for finding r_{out}. The circuit has been generalized slightly by including external emitter resistance R_E. This resistance always adds to r_e and can be lumped with it. With this equivalent circuit, output resistance reduces to divider formulas and parallel resistances:

$$r_{out} = R_L \left\| \left(\frac{R_f}{\beta+1} + R_E + r_e \right) \right. \tag{4.106}$$

Input resistance cannot be found as easily because i_i is injected at the base. The resulting i_b is not from the R_f branch alone as assumed in the equivalent circuit.

Another direct application of the shunt-feedback equivalent circuit is to find r_{in} of an emitter-driven shunt-feedback amplifier, shown in Fig. 4.28b and modeled in (c). Here $i_f = i_b$, and the shunt-feedback equivalent is exact. The

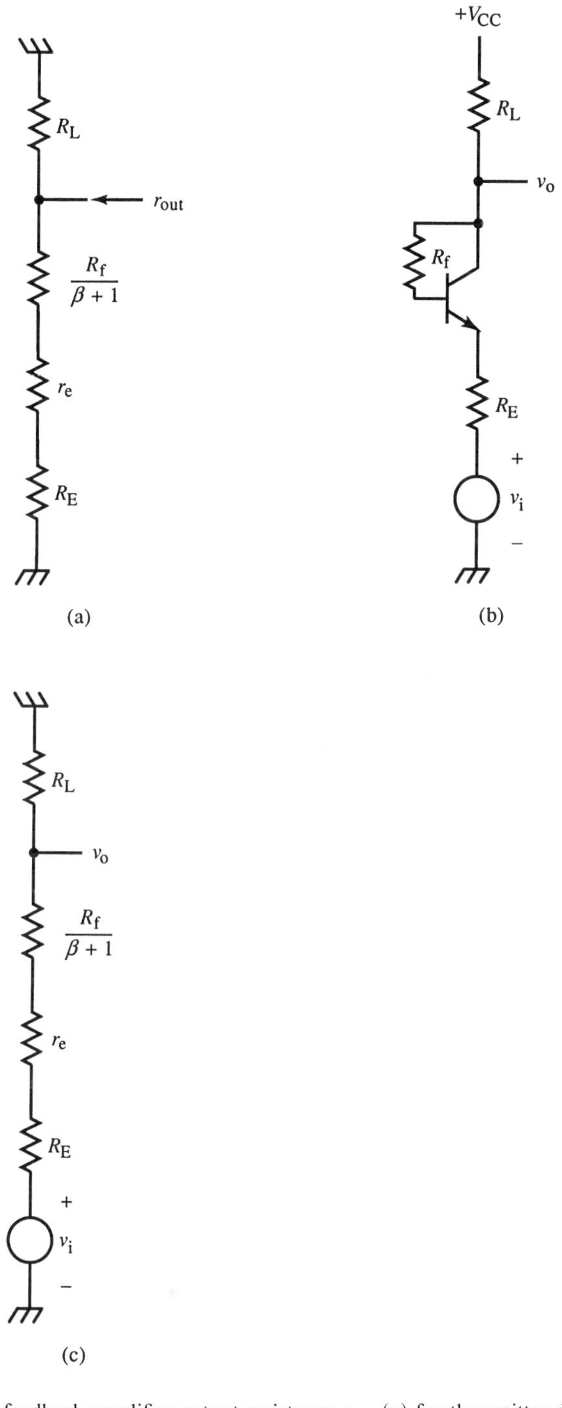

FIG. 4.28 Shunt-feedback amplifier output resistance r_{out} (a) for the emitter-driven circuit (b) can be found from its equivalent circuit model (c).

input and output resistances and voltage gain of this amplifier are

$$r_{in} = R_L + \frac{R_f}{\beta+1} + r_e + R_E \tag{4.107}$$

$$r_{out} = R_L \,\Big\|\, \left[\frac{R_f}{\beta+1} + r_e + R_E\right] \tag{4.108}$$

$$\frac{v_o}{v_i} = \frac{R_L}{R_L + R_f/(\beta+1) + r_e + R_E} \tag{4.109}$$

This amplifier is of limited use for amplification; it has a voltage gain of less than 1. It can be used as a voltage translator for meeting biasing conditions.

Example 4.5 CB Shunt-Feedback Amplifier

Figure E4.5 is an instance of Fig. 4.28. For $\beta = 99$, $I_S = 10^{-16}$ A; then, from SPICE

$V_E = 0.1671$ V, $V_B = 0.9571$ V, $V_C = 1.1428$ V, $I_E = 1.86$ mA

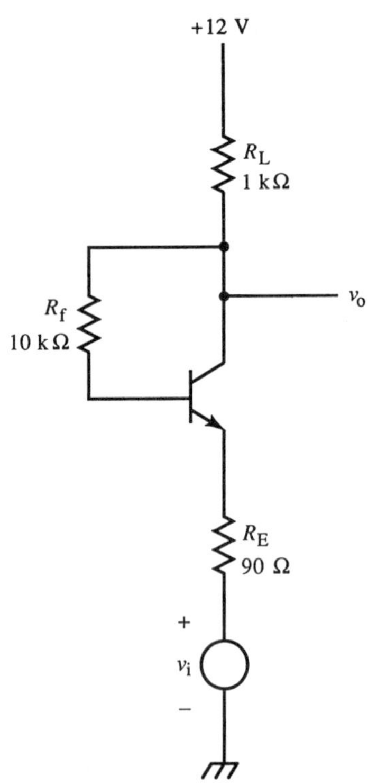

FIG. E4.5

Then $r_e = 13.91\ \Omega$ and

$$\frac{v_o}{v_i} = \frac{R_L}{R_L + R_f/(\beta+1) + r_e + R_E} = 0.83063$$

$$r_{in} = R_L + \frac{R_f}{\beta+1} + r_e + R_E = 1.2039\ \text{k}\Omega$$

$$r_{out} = R_L \,\Bigg\|\, \left[\frac{R_f}{\beta+1} + r_e + R_E\right] = 169.37\ \Omega$$

The SPICE values for the above parameters are

$$\frac{v_o}{v_i} = 0.8306, \qquad r_{in} = 1.204\ \text{k}\Omega, \qquad r_{out} = 169.4\ \Omega$$

4.17 An Idealized Shunt-Feedback Amplifier

A more general form of shunt-feedback amplifier is shown in Fig. 4.29. A transconductance amplifier replaces the BJT, and R_i is included. Solving for the usual incremental quantities using feedback analysis based on the flow graph of Fig. 4.29b, we obtain the transmittances:

$$\alpha_i = R_f \parallel R_i \tag{4.110a}$$

$$G = \frac{R_f \parallel R_L}{(1/G_M)} + \frac{R_L}{R_f + R_L} \tag{4.110b}$$

$$H = -\frac{R_i}{R_f + R_i} \tag{4.110c}$$

Then if we combine these transmittances, the transresistance is

$$\frac{v_o}{i_i} = \alpha_i \cdot \frac{G}{1+GH} = (R_f \parallel R_i)\frac{\dfrac{R_f \parallel R_L}{R_f \parallel (1/G_M)}}{1 - \dfrac{R_f \parallel R_L}{R_f \parallel (1/G_M)} \cdot \dfrac{R_i}{R_f + R_i}} \tag{4.111}$$

Removing R_L simplifies (4.111) somewhat:

$$\left.\frac{v_o}{i_i}\right|_{R_L \to \infty} = (R_f \parallel R_i)\frac{\dfrac{R_f + 1/G_M}{1/G_M}}{1 - \dfrac{R_f + 1/G_M}{1/G_M} \cdot \dfrac{R_i}{R_f + R_i}} \tag{4.112}$$

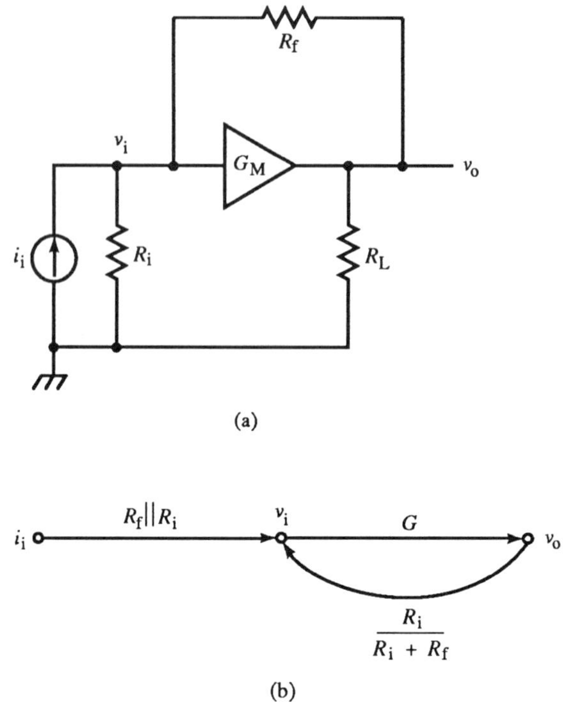

FIG. 4.29 A shunt-feedback amplifier with ideal transconductance amplifier (a) and flow graph (b).

If R_i is removed instead, then

$$\left.\frac{v_o}{i_i}\right|_{R_i\to\infty} = R_f \cdot \frac{\dfrac{R_f \parallel R_L}{R_f \parallel (1/G_M)}}{1 - \dfrac{R_f \parallel R_L}{R_f \parallel (1/G_M)}} \tag{4.113}$$

Finally, when both R_L and R_i are removed, the transresistance reduces to the simple form of

$$\left.\frac{v_o}{i_i}\right|_{R_i, R_L\to\infty} = -\left(R_f + \frac{1}{G_M}\right) \tag{4.114}$$

This expression is similar to (4.93). When R_i is set to r_π and $1/G_M$ to r_m, then the transconductance amplifier is equivalent to the BJT shunt-feedback amplifier.

The input resistance can be found from the feedback equations:

$$v_i = \alpha_i i_i - H v_o, \qquad v_o = G v_i \tag{4.115}$$

Substituting v_o from the second equation into the first and solving, we obtain

$$r_{in} = \frac{v_i}{i_i} = \frac{\alpha_i}{1 + GH} = \frac{R_f \parallel R_i}{1 + GH} \tag{4.116}$$

An alternative solution that uses recursion begins with the first equation of (4.115). It is divided by i_i:

$$r_{in} = \frac{v_i}{i_i} = \alpha_i - H\left(\frac{v_o}{i_i}\right) = \alpha_i - H \cdot \frac{v_o}{v_i} \cdot \frac{v_i}{i_i}$$

$$= \alpha_i - GH \cdot r_{in} = \frac{\alpha_i}{1+GH}$$

Writing H in terms of circuit component values, we can reformulate r_{in} as

$$r_{in} = \frac{R_f \parallel R_i}{1 - G[R_i/(R_f + R_i)]} = \frac{R_i R_f}{(1-G)R_i + R_f} = R_i \,\Bigg\|\, \frac{R_f}{1-G} \tag{4.117}$$

This result suggests that we can apply Miller's theorem as an alternative approach to finding r_{in}.

Output resistance is derived as follows, applying KCL to the output node:

$$i_o = \frac{v_o}{R_L} + \frac{v_o - v_i}{R_f} - G_M v_i, \qquad v_i = \left(\frac{R_i}{R_f + R_i}\right) v_o \tag{4.118}$$

Solving for v_o / i_o gives

$$r_{out} = R_L \,\Bigg\|\, \frac{R_f + R_i}{1 - R_i/(1/G_M)} \tag{4.119}$$

This form of r_{out} is similar to (4.102) when $R_i \to r_\pi$ and $1/G_M \to -r_m$. (G_M is positive as defined in Fig. 4.29a.)

Since the feedback output quantity v_o is across r_{out}, feedback analysis applies directly, resulting in

$$r_{out} = \frac{R_f \parallel R_L}{1+GH} \tag{4.120}$$

Finally, for a voltage amplifier version, the Norton equivalent circuit formed by i_i and R_i can be Thévenized so that $v_i = i_i R_i$. The voltage gain is then v_o / v_i. This transformation of Fig. 4.29a is easy to make by changing α_i and the input node of the flow graph (Fig. 4.29b). The new transmittance is

$$\alpha_i = \frac{R_i}{R_f + R_i}; \qquad \text{input node is } v_i$$

The voltage gain is

$$\frac{v_o}{v_i} = \frac{(4.111)}{R_i} = \frac{R_f}{R_f + R_i} \cdot \frac{\dfrac{R_f \parallel R_L}{R_f \parallel (1/G_M)}}{1 - \dfrac{R_f \parallel R_L}{R_f \parallel (1/G_M)} \cdot \dfrac{R_i}{R_f + R_i}} \tag{4.121}$$

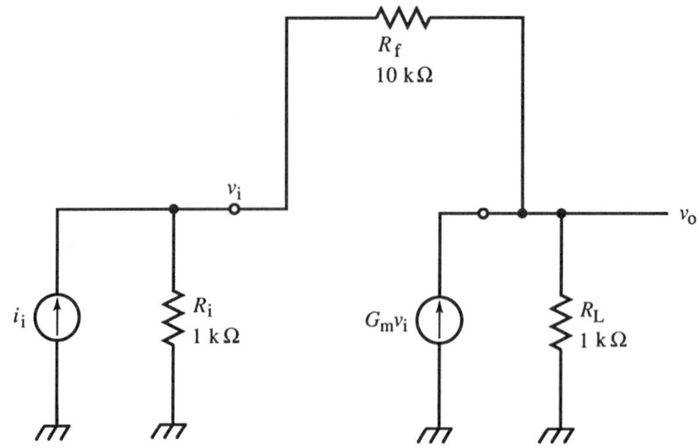

FIG. E4.6

Example 4.6 Transconductance Amplifier

Figure E4.6 is a transconductance amplifier with a forward path transconductance of $G_m = -10$ mS $= -1/100\ \Omega$. The SPICE results are

$$\frac{v_o}{v_i} = -4.500\ \text{k}\Omega, \qquad r_{in} = 500.0\ \Omega, \qquad r_{out} = 500.0\ \Omega$$

For $1/G_m = +100\ \Omega$, this positive feedback amplifier has

$$\frac{v_o}{v_i} = -50.50\ \text{k}\Omega, \qquad r_{in} = 5.500\ \text{k}\Omega, \qquad r_{out} = 5.500\ \text{k}\Omega$$

These results agree with those derived from Section 4.17.

4.18 Feedback Circuit Resistances

The effects of feedback on circuit resistances were introduced in Section 3.4. The situations considered were general but not general enough. In the previous sections, it was difficult to derive r_{in} for the CB, and no attempt was made for the CE and CC. Miller's theorem provides r_{out}, but under the condition that the amplifier output be a voltage source (no resistance). This is very limiting, as we have seen. Usually it was necessary to resort to KCL. The results of feedback analysis, mainly transmittances, are often useless in finding resistances, for several reasons.

First, from Section 3.4, feedback results can be applied directly to determine the closed-loop resistance across an error voltage or in series with an error current. But these error quantities are often not the resistances of interest. In the three basic BJT configurations, input and output resistances were not directly associated with the error quantity. For example, for the CE and CC, for $E = v_{be}$, we can immediately determine that r_e is $r_e \times (1 + GH)$ with feedback. But r_{in} involves r_e in series with the rest of the emitter circuit. How is its resistance affected by feedback? It is usually not obvious.

Second, most of the equivalent circuit methods used to find transmittances do not preserve circuit resistances. Two-port, Thévenin, and Norton equivalent circuits and divider formulas do not preserve resistances. For example, a voltage divider consisting of two 1 kΩ resistors has a transmittance of 0.5, $r_{in} = 2$ kΩ, and $r_{out} = 500$ Ω (when the input is driven by a voltage source). But $r_{out} \neq 0.5 r_{in}$; the transmittances do not apply to resistances as they do to voltages or currents.

The reduction theorem is resistance-preserving. Consequently, we are able to apply the β transform directly to circuit topology to find resistances. Earlier in this chapter, we did the same with the μ transform. The reduction theorem reduces circuits to a form that makes resistances available by topological inspection (by appealing to causal and topological reasoning, or intuition). Therefore, feedback analysis is usually not a good approach to resistance determination whereas the reduction theorem is. Miller's theorem can also be used, but like feedback analysis it is limited in its application. No single method is generally best; judgment is required, based on the particular circuit and what aspects of it are desired to be made explicit.

An early method for finding resistances was published by R. B. Blackman in 1943, but it lay dormant for decades and is not found in many circuit textbooks. In recent years, R. D. Middlebrook of CalTech and Sol Rosenstark of New Jersey Institute of Technology have been reviving it. Blackman developed a simple formula for calculating resistances in feedback circuits that is also based on inspection of the topology. A feature of this method is that only loop gain is needed; no decisions about input or output feedback quantities are required. Another advantage is that it can use loop gain results from feedback analysis. Its disadvantage is that it is not as intuitive to use as the reduction theorem because the answer results from substituting these inspected values into a formula that, in itself, is not easily envisioned in terms of circuit topology. But it is easy to apply and minimizes calculation. Consequently, it will be developed here.

Figure 4.30 represents a feedback circuit with loop gain $-T$ and a port with terminal voltage v driven by a current source i. We want to find the closed-loop resistance at this port. Within the feedback loop, choose a convenient point where it can be opened so that two flow graph nodes, x_i and x_o, are created. We have done this before when finding GH; the gain from E through G and H back to E again (or $-B/E = -GH$) is represented here by

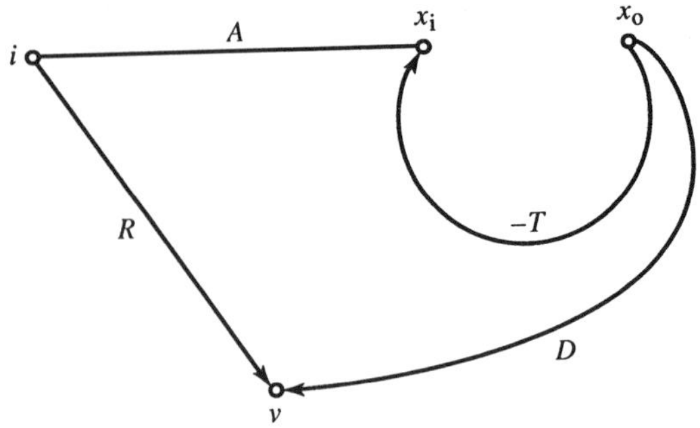

FIG. 4.30 Blackman's formula is derived from this flow graph, where $-T$ is the loop gain. The loop is opened between x_i and x_o so that when $i=0$, $-T = x_i/x_o$.

$-T$. To derive Blackman's formula, it is not necessary to choose an E, but instead to pick a point within the loop where it can be opened so that loop gain under different conditions can be derived. Loop gain is x_i/x_o. The x quantities can be either voltages or currents somewhere in the loop. The simplicity of this approach is that G, H, and E need not be identified, only loop gains.

The flow graph of Fig. 4.30 can be expressed algebraically as

$$x_i = Ai - Tx_o \tag{4.121a}$$

$$v = Ri + Dx_o \tag{4.121b}$$

For a closed-loop amplifier, $x_o = x_i = x$, and

$$x = Ai - Tx = \frac{A}{1+T} \cdot i, \qquad v = Ri + D\left[\frac{A}{1+T}\right]i$$

Solving for closed-loop terminal resistance gives

$$\left.\frac{v}{i}\right|_{\text{cl}} = \left.\frac{v}{i}\right|_{x_o=x_i} = R + \frac{DA}{1+T} = R \cdot \frac{1+[T+(DA/R)]}{1+T} \tag{4.122}$$

This is the resistance we are seeking, but to use it in this form requires that we know the transmittances it contains. This can be an onerous task. An ingenious simplification is made by finding the topological meaning of the subexpressions in (4.122).

Let us find the open-loop resistance. To obtain an open-loop circuit, set $x_o = 0$. Then from (4.121b),

$$r_{\text{ol}} = \left.\frac{v}{i}\right|_{\text{ol}} = \left.\frac{v}{i}\right|_{x_o=0} = R \tag{4.123}$$

From this, we know that R in (4.122) is the terminal resistance when the loop is opened.

Next, consider the expressions for loop gain that result from both open- and short-circuiting the port. For an open-circuited port, $i = 0$. Substituting for i in (4.121) and solving for loop gain, we obtain

$$T_{oc} = \left.\frac{x_i}{x_o}\right|_{oc} = \left.\frac{x_i}{x_o}\right|_{i=0} = -T \tag{4.124}$$

The denominator of (4.122) can be expressed in terms of the open-circuit loop gain as $1 - T_{oc}$. Finally, for the short-circuit loop gain, set $v = 0$ and solve for T_{sc} from (4.121):

$$T_{sc} = \left.\frac{x_i}{x_o}\right|_{sc} = \left.\frac{x_i}{x_o}\right|_{v=0} = -\left[\frac{AD}{R} + T\right] \tag{4.125}$$

Interestingly enough, this matches the numerator of (4.122), so it can be expressed as $1 - T_{sc}$. When these expressions are substituted into (4.122), *Blackman's resistance formula* results:

$$r_{cl} = r_{ol} \cdot \frac{1 - T_{sc}}{1 - T_{oc}} \tag{4.126}$$

To find closed-loop resistance at an arbitrary port in a feedback amplifier:

1. Open the feedback loop and find the port resistance r_{ol}.
2. Open the port and find the closed-loop gain T_{oc}.
3. Short the port and find the closed-loop gain T_{sc}.
4. Substitute these results into Blackman's formula for r_{cl}.

Blackman's formula can be applied to feedback amplifiers to find input and output resistances. For the pathological cases of the CE and CC with r_o (Sections 4.12 and 4.13), r_{in} is easily found. From the input port, the topology of the CC is the same as the CE and r_{in} is identical:

$$r_{in} = (\beta + 1)[r_e + R_E \parallel (r_o + R_C)] \cdot \frac{1 - (b/d)}{1 - 0}$$

or

$$\text{CE, CC} \quad r_{in} = (\beta + 1)[r_e + R_E \parallel (r_o + R_C)]\left[1 - \left(\frac{R_E \parallel r_e}{R_E \parallel r_e + r_o}\right)\left(\frac{R_C \parallel r_o}{r_m \parallel r_o}\right)\right] \tag{4.127}$$

Finally, let us find r_{in} for the CB with r_o (Section 4.11):

$$\text{CB} \quad r_{in} = [R_E + r_e \parallel (r_o + R_L)] \cdot \frac{1 - b/d}{1 - (1/d)[r_e/(r_e + r_o)]}$$

For the CB, neither T_{sc} nor T_{oc} is zero.

4.19 The Asymptotic Gain Method

In close connection with Blackman's resistance method is the *asymptotic gain method* for finding feedback circuit gain. It has been developed extensively by R. D. Middlebrook of CalTech. Here the equivalence of the asymptotic gain method and the signal flow graph feedback method we have been using will be shown. In doing so, the idea of the asymptotic gain method will be made clear.

Consider again the feedback topology of Fig. 4.18b, expressed as Fig. 4.31. From Section 4.10, we know that

$$\frac{x_o}{x_i} = \frac{\alpha_i G}{1 + GH} + F \tag{4.128}$$

where the path transmittances are

$$\alpha_i = \left.\frac{E}{x_i}\right|_{-HC=0}, \qquad G = \left.\frac{x_o}{E}\right|_{Fx_i=0}, \qquad -H = \left.\frac{E}{C}\right|_{\alpha_i x_i=0}, \qquad F = \left.\frac{x_o}{x_i}\right|_{C=0} \tag{4.129}$$

The asymptotic gain formula is

$$\frac{x_o}{x_i} = G_\infty \cdot \frac{T}{1+T} + G_0 \cdot \frac{1}{1+T} \tag{4.130}$$

where

$$G_\infty = \left.\frac{\alpha_i G}{1+GH}\right|_{GH\to\infty} + F = \frac{\alpha_i}{H} + F = \frac{\alpha_i G}{T} + F, \qquad T = GH; \qquad G_0 = F \tag{4.131}$$

Substituting for G_∞ and G_0 into (4.130) gives

$$\frac{x_o}{x_i} = \left(\frac{\alpha_i G}{T} + F\right) \cdot \frac{T}{1+T} + F \cdot \frac{1}{1+T} = \frac{\alpha_i G}{1+GH} + F \tag{4.132}$$

The significance of (4.130) is that it reformulates (4.128) in a form that makes another method explicit. By finding G_∞ and G_0 from circuit maneuvers, the results are substituted into (4.130). The method is similar to Blackman's formula: Find some circuit quantities by imposing constraints on the circuit and then substitute these results into a simple formula. G_∞ is x_o/x_i with infinite loop gain. This is not unfamiliar; we analyzed what happens to op-amp circuits when op-amp gain is infinite. Discrete transistor amplifiers can be analyzed

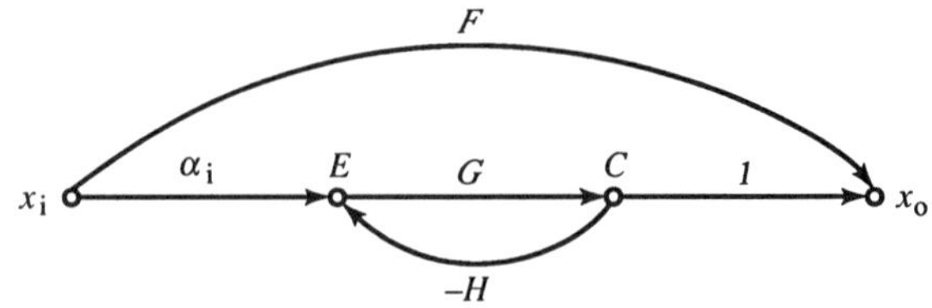

FIG. 4.31 The asymptotic gain method is derived from this flow-graph topology.

similarly; the result is the feedforward path added to $\alpha_i(1/H)$. G_0 is x_o/x_i with zero loop gain, which is the feedforward path F.

4.20 The Cascode and Differential Shunt-Feedback Amplifiers

The basic shunt-feedback amplifier can be combined with other elemental circuits such as the cascode or differential amplifiers. Figure 4.32 shows a shunt-feedback cascode amplifier with current-source load, modeled in Fig. 4.32b. First, i_{c2} must flow through R_2. Thus,

$$i_{R1} = i_{c2} = i_{c1}$$

and

$$i_i = i_{bi} + i_{R1} = i_{b1} + i_{c1} = (\beta_1 + 1)i_{b1} = (\beta_1 + 1)\frac{v_{b1}}{r_{\pi 1}}$$

Then

$$v_{b1} = r_{e1} \cdot i_i$$

The output voltage is

$$v_o = v_{b1} - i_{c1}R_1 - i_{c2}R_2$$

Substituting, we obtain

$$v_o = r_{e1}i_i - i_{c1}R_1 - \alpha_2 i_{c1}R_2 = r_{e1}i_i - \alpha_1 i_i(R_1 + \alpha_2 R_2)$$

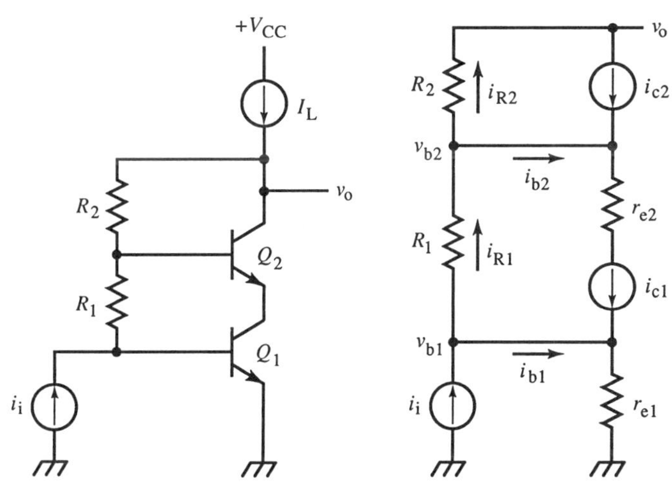

FIG. 4.32 The shunt-feedback cascode amplifier (a) and equivalent circuit (b).

The transresistance is

$$\frac{v_o}{i_i} = -\alpha_1 R_1 - \alpha_1 \alpha_2 R_2 + r_{e1} \tag{4.133}$$

This result is similar to (4.91), where R_f corresponds approximately to $R_1 + R_2$. Since i_{c2} suffers from loss of current to both Q_1 and Q_2 bases, both α_1 and α_2 are factors of R_2 in (4.133).

A differential shunt-feedback amplifier is shown in Fig. 4.33. The results of (4.107) apply in finding the equivalent emitter resistance of one side of the differential pair due to the other side. With one input open, the other input

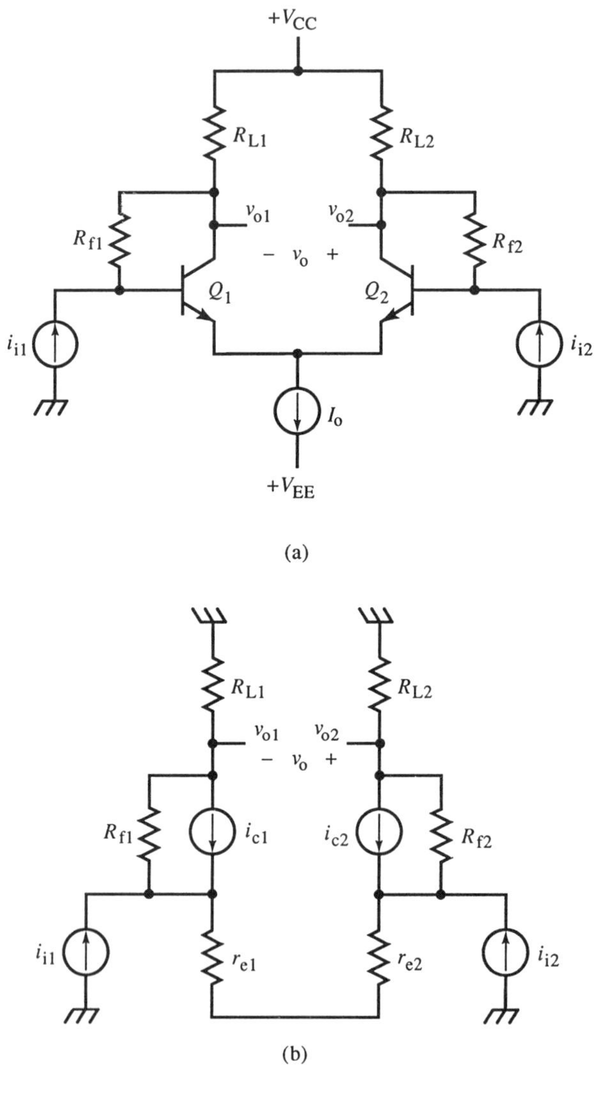

FIG. 4.33 The differential shunt-feedback amplifier (a) and equivalent circuit (b).

appears as a high resistance, but when both inputs are driven differentially, the emitter currents caused by each input are equal and opposite for a symmetrical circuit.

Example 4.7 Differential Shunt-Feedback Amplifier

Figure E4.7, using the default BJT model, has (reduced) SPICE program and output as shown. The transmittance follows directly from substitution

```
E4.7 Shunt Differential Feedback
     Amplifier
.OPT NOMODS OPTS NOPAGE
.OP
.DC II -0.10mA 0.10mA 10uA
.TF V(40,20) II

VCC 80 0 DC 12V
IO  30 0 DC 2mA
II 50 10 DC 0A

RL1 80 20 2.7K
RL2 80 40 2.7K
RF1 20 10 10K
RF2 40 50 10K

Q1 20 10 30 BJT1
Q2 40 50 30 BJT1
.MODEL BJT1 NPN (BF=99)

.END

NODE     VOLTAGE
(10) 9.2000
(20) 9.30000
(30) 8.4261
(40) 9.3000
(50) 9.1000

OPERATING POINT INFORMATION
TEMPERATURE=   27.000 DEG C

BIPOLAR JUNCTION TRANSISTORS
NAME      Q1          Q2
MODEL     BJT1        BJT1
IB        1,00E-05    1.00E-05
IC        9.90E-04    9.90E-04
VBE       7.74E-01    7.74E-01
VBC      -1.00E-01   -1.00E-01
VCE       8.74E-01    8.74E-01
BETADC    9.90E+01    9.90E+01
GM        3.83E-02    3.83E-02

V(40,20)/II=1.887E+04
INPUT RESISTANCE AT II=2.325E+02
OUTPUT RESISTANCE AT V(40,20)=2.405E+02
```

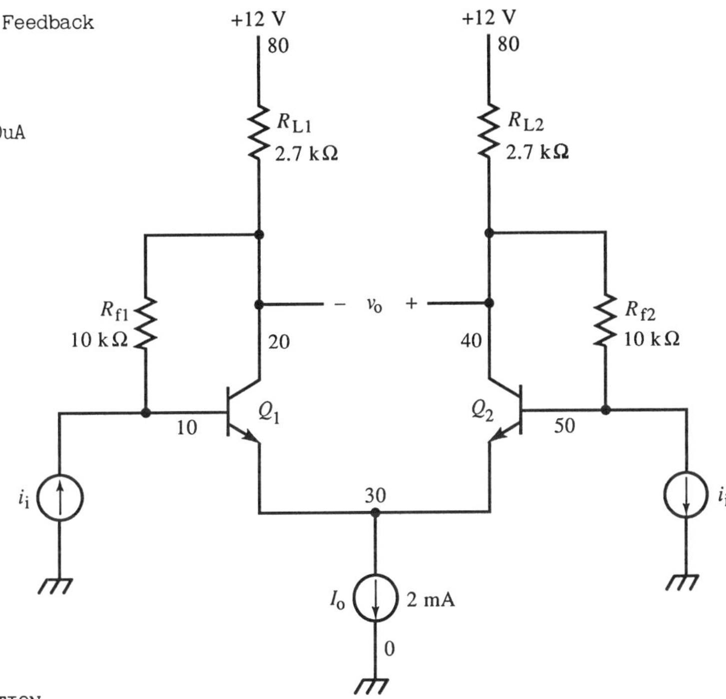

FIG. E4.7

of circuit element values into (4.90) and doubling the result to account for both sides of the circuit.

4.21 The Emitter-Coupled Feedback Amplifier

A common multipath topology involving both cascade and emitter coupling is shown in Fig. 4.34a with small-signal equivalent circuit shown in (b), two-port equivalent for H in (c), and flow graph in (d). Let $E = i_{e1}$. The error-summing node is at the output port of H. $F = 0$ since all paths from v_i to v_o are along either G_1 (cascade path) or G_2 (emitter-coupled path). The transmittances are

$$\alpha_i = \left.\frac{i_{e1}}{v_i}\right|_{i_{e2}=0} = \frac{1}{r_{e1} + R_E}, \qquad \alpha_o = \frac{v_o}{i_{e2}} = -\alpha_2 R_{L2}$$

$$G = \left.\frac{i_{e2}}{i_{e1}}\right|_{Hi_{e2}=0} = G_1 + G_2 = \frac{\alpha_1 R_{L1}}{r_{e2} + R_E} - \frac{R_E}{r_{e2} + R_E}$$

$$H = -\left.\frac{i_{e1}}{i_{e2}}\right|_{v_i=0} = \frac{R_E}{r_{e1} + R_E}$$

Since the input and output quantities are voltages and the error quantity is a current, α_i is a conductance and α_o a resistance. Other choices for E are possible, of course. The voltage gain is

$$A_v = \alpha_i \cdot \frac{G}{1 + GH} \cdot \alpha_o \tag{4.134}$$

This amplifier can be implemented by replacing the $\times(-1)$ amplifier with a PNP CE stage or NPN CE stage with negative voltage offset. Unlike previous amplifiers with multiple forward paths (for G), the passive path (G_2) has a significant transmittance and cannot be ignored.

Example 4.8 Emitter-Coupled Feedback Amplifier

Figure E4.8 feedback analysis assumes $E = i_{e1}$. Then dc analysis produces the following values:

$V_{C1} = 6.7112$ V, $V_E = -0.7773$ V, $V_O = 9.6345$ V, $V_{B2} = -0.0012$ V

$I_{E1} = 1.14$ mA, $I_{E2} = 1.09$ mA

FIG. 4.34 An emitter-coupled cascade amplifier (a) and model (b), two-port equivalent circuit for H (c), and flow graph (d). This amplifier has significant transmittance in its passive forward path.

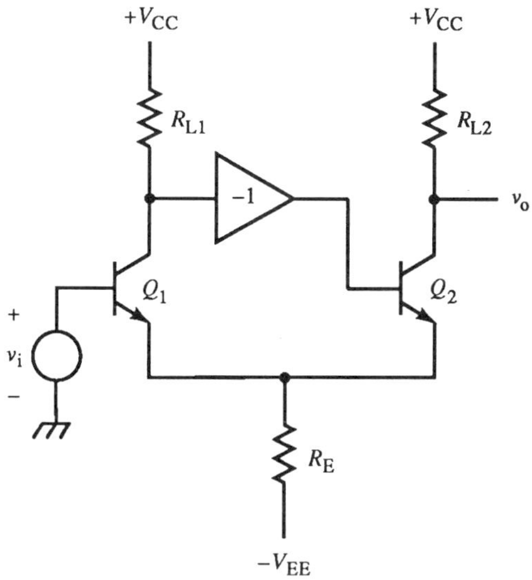

+V_CC
+V_CC
R_L1
R_L2
−1
v_o
Q_1
Q_2
+
v_i
−
R_E
−V_EE

(a)

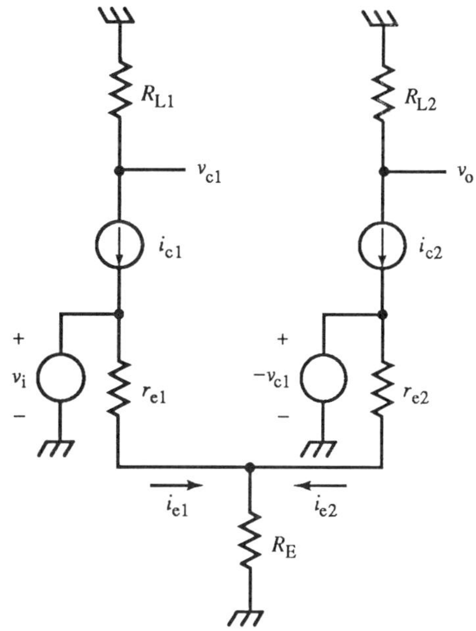

R_L1
R_L2
v_c1
v_o
i_c1
i_c2
+
v_i
−
r_e1
+
−v_c1
−
r_e2
i_e1
i_e2
R_E

(b)

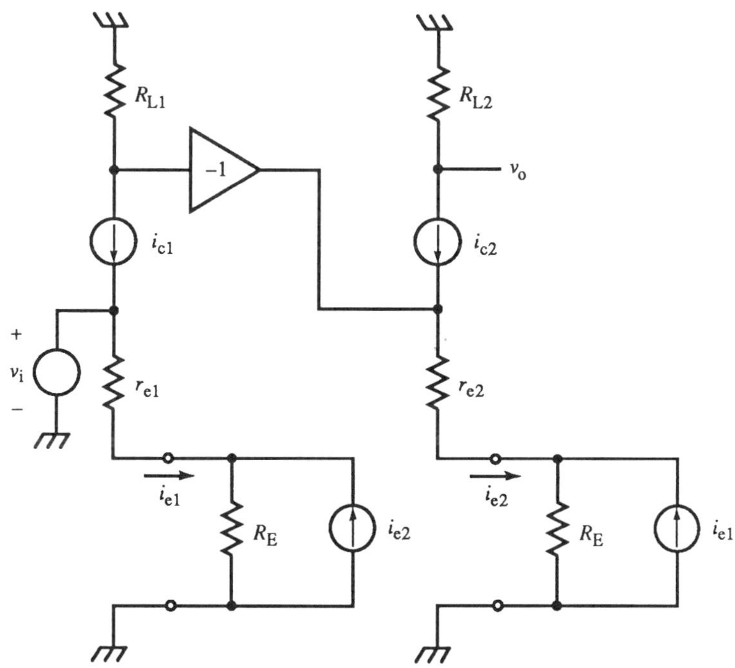

R_L1
R_L2
−1
v_o
i_c1
i_c2
+
v_i
−
r_e1
r_e2
i_e1
R_E
i_e2
i_e2
R_E
i_e1

(c)

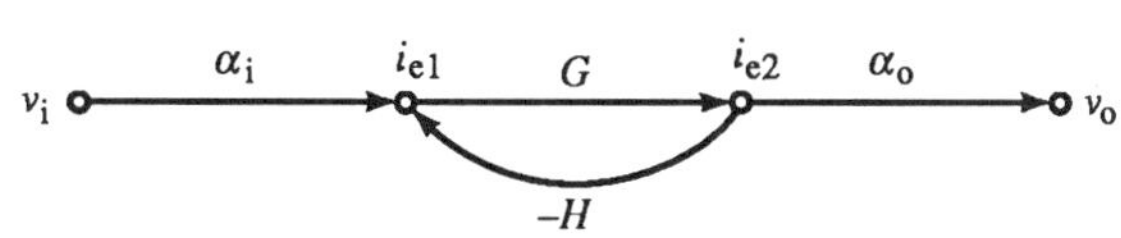

v_i
α_i
i_e1
G
i_e2
α_o
v_o
−H

(d)

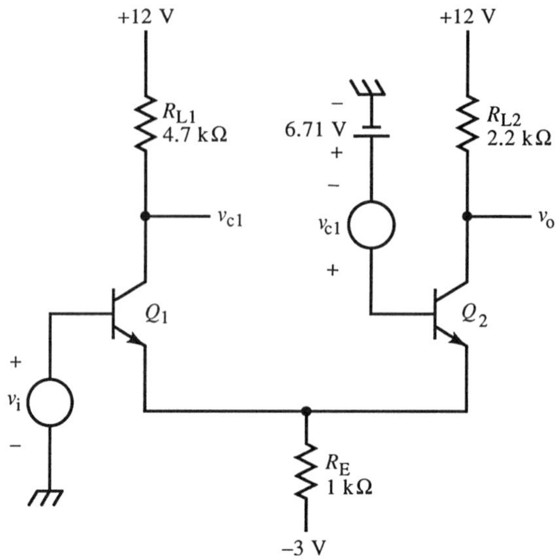

FIG. E4.8

From the current values,

$$r_{e1} = 22.69\ \Omega, \qquad r_{e2} = 23.73\ \Omega$$

The transmittances are

$$\alpha_i = 9.7781 \times 10^{-4}\ \text{S}, \qquad H = 0.97781,$$

$$G_1 = 4.5451, \qquad G_2 = -0.97682$$

$$G = G_1 + G_2 = 3.5683, \qquad \alpha_o = -2.178\ \text{k}\Omega, \qquad 1 + GH = 4.4892$$

Then

$$\frac{v_o}{v_i} = \alpha_i \cdot \frac{G}{1+GH} \cdot \alpha_o = -1.6928$$

$$r_{in} = (1+GH)[(\beta_1+1)(r_{e1}+R_E)] = 459.1\ \text{k}\Omega$$

$$r_{out} = 2.2\ \text{k}\Omega$$

4.22 Closure

We have found that multiple paths through amplifiers are common and that several methods are usually applicable for finding a circuit quantity of interest. We shall examine multipath amplifiers again when frequency response is of primary interest.

Example 4.9 Inverting Feedback Amplifier

The Fig. E4.9 dc bias solution is given in the SPICE simulation. The transistor parameters are for 2N930 transistors. From these values we obtain

$$r_{e1} = 244\ \Omega, \qquad r_{e2} = 26.6\ \Omega$$

```
E4.9 Inverting Feedback Amplifier
.OPT NOMOD OPTS NOPAGE
.OP
.DC VI -2V 2V 0.05V
.TF V(70 VI
VCC 80 0 DC 16V
VI  10 0 DC 1.3484V
RB  10 30 2.2K
RF  30 70 220K
RL1 80 40 100K
RE1 50 0 7.5K
RE2 60  0 100
RL2 80 70 15K
Q1 40 30 50 BJT1
Q2 70 50 60 BJT1
* 2N930
.MODEL BJT1 NPN (BF=170 IS=1E14)
.END
```

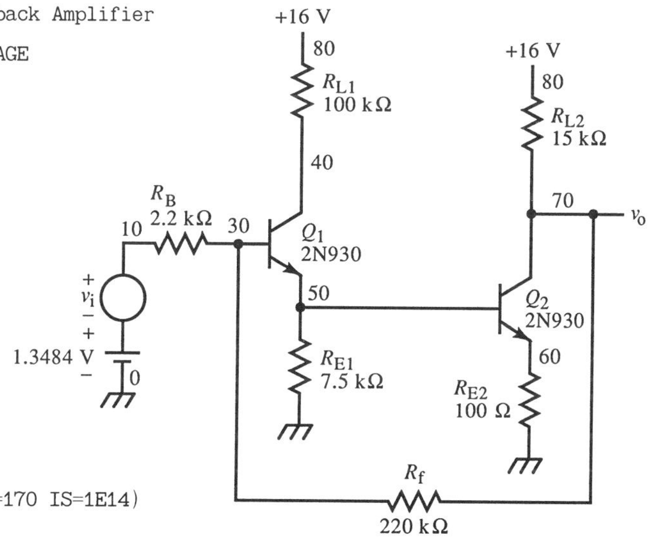


```
NODE    VOLTAGE
(10) 1.3484  (30) 1.3484  (40) 5.4728  (50) .7515
(60 .0973  (70 1.4837

BIPOLAR JUNCTION TRANSISTORS

NAME     Q1          Q2
MODEL    BJT1        BJT1
IB        6.19E-07    5.69E-06
IC        1.05E-04    9.67E-04
VBE       5.97E-01    6.54E-01
VBC      -4.12E+00   -7.32E-01
VCE       4.72E+00    1.39E+00
BETADC    1.70E+02    1.70E+02
GM        4.07E-03    3.74E-02

V(70)/VI=-5.106E+01
INPUT RESISTANCE AT VI=4.260E+03
OUTPUT RESISTANCE AT V(70)=6.873E+03
```

FIG. E4.9

Furthermore, the base input resistance of Q_1 is

$$r_i = (\beta_1 + 1)[r_{e1} + R_{E1} \parallel (\beta_2 + 1)(r_{e2} + R_{E2})]$$

$$= (171)(244\ \Omega + 5.5700\ \text{k}\Omega) = 994.2\ \text{k}\Omega$$

Additionally,

$$r_i \parallel R_f = 180.1\ \text{k}\Omega, \qquad r_i \parallel 2.2\ \text{k}\Omega = 2.1951\ \text{k}\Omega$$

Let $E = v_{b1} = v(30)$. Then,

$$\alpha_i = \frac{180.1\ \text{k}\Omega}{180.1\ \text{k}\Omega + 2.2\ \text{k}\Omega} = 0.9879$$

$$G_1 = \left(\frac{5.5700\ \text{k}\Omega}{5.570\ \text{k}\Omega + 244\ \Omega}\right)\left(-\frac{15\ \text{k}\Omega \parallel 220\ \text{k}\Omega}{26.6\ \Omega + 100\ \Omega} \cdot \frac{170}{171}\right)$$

$$= (0.9580)(-110.3) = -105.7$$

$$G_2 = \frac{15\ \text{k}\Omega}{15\ \text{k}\Omega + 220\ \text{k}\Omega} = 6.3830 \times 10^{-2} \cong 0$$

$$G = -105.6$$

$$H = -9.8793 \times 10^{-3} = \frac{-1}{101.2}$$

$$1 + GH = 2.0433$$

Then

$$\frac{v_o}{v_i} = \alpha_i \cdot \frac{G}{1 + GH} = -51.057$$

$$r_{out} = \frac{15\ \text{k}\Omega \parallel 220\ \text{k}\Omega}{1 + GH} = 6.8726\ \text{k}\Omega$$

$$r_{in} = [2.2\ \text{k}\Omega + 180.1\ \text{k}\Omega] \cdot \frac{1 + (GH)|_{v_i=0}}{1 + (GH)|_{i_i=0}} = 182.3\ \text{k}\Omega \cdot \frac{2.0433}{87.466}$$

$$= (182.3\ \text{k}\Omega)(2.3360 \times 10^{-2}) = 4.2586\ \text{k}\Omega$$

Note that in using Blackman's formula to find r_{in} that G remains the same for both numerator and denominator, that the numerator is the $1 + GH$ calculated previously, and that what is different in the denominator is H:

$$H|_{i_i=0} = \frac{r_i}{r_i + R_f} = -0.81881$$

The SPICE simulation results are

$$\frac{v_o}{v_i} = -51.06, \qquad r_{in} = 4.260\ \text{k}\Omega, \qquad r_{out} = 6.873\ \text{k}\Omega$$

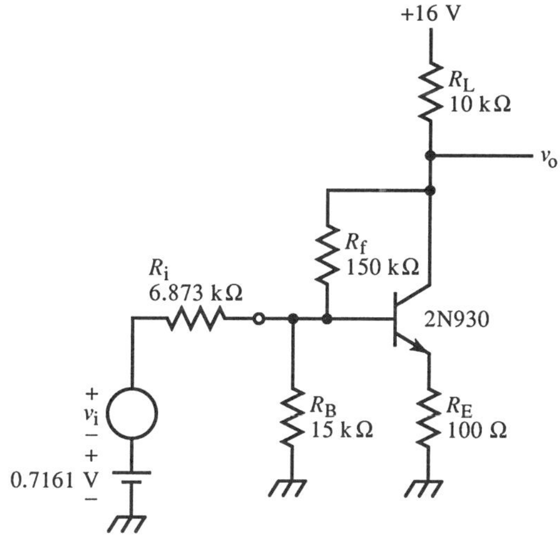

FIG. E.4.10

Example 4.10 Shunt-Feedback Voltage Amplifier

The Fig. E4.10 analysis follows from Section 4.15. The 2N930 BJT is used here with $\beta = 170$ and $I_S = 10^{-14}$ A. The dc analysis yields

$$I_E = 411\ \mu\text{A}, \qquad V_{BE} = 0.646\text{ V}, \qquad V_C = 8.4941\text{ V}, \qquad V_E = 70.3\text{ mV}$$

The rather exacting value of R_i happens to be the output resistance of Fig. E4.9. (In Fig. E4.11, they are connected.) This R_i accounts for interstage loading. The feedback calculations are based on $E = v_b$, and the formulas of Section 4.15 apply directly.

$$r_M = \frac{r_e + R_E}{\alpha} = r_m + \frac{R_E}{\alpha} = 37\ \Omega + \frac{100\ \Omega}{0.99415} = 137.63\ \Omega$$

$$G = -\left(\frac{10\text{ k}\Omega}{160\text{ k}\Omega}\right)\left(\frac{150\text{ k}\Omega}{137.63\ \Omega} - 1\right) = -68.057$$

Let

$$r_s = R_i \parallel R_B \parallel (\beta+1)(r_e + R_E) = 6.873\text{ k}\Omega \parallel 9.1400\text{ k}\Omega = 3.9230\text{ k}\Omega$$

Then,

$$H = -\frac{r_s}{r_s + R_f} = -\frac{3.9230\text{ k}\Omega}{3.9230\text{ k}\Omega + 150\text{ k}\Omega} = -2.5487 \times 10^{-2}$$

From this, $1 + GH = 2.7346$ and $r_s \parallel R_f = 3.8230$ kΩ. Then,

$$\frac{v_o}{v_i} = \left(\frac{r_e \parallel R_f}{R_i}\right)\left(\frac{G}{1+GH}\right) = \left(\frac{3.8230\text{ k}\Omega}{6.873\text{ k}\Omega}\right)(-24.888) = -13.843$$

Using Blackman's formula for r_{in}, we obtain

$$r_{in} = [R_i + R_B \parallel (\beta + 1)(r_e + R_E) \parallel R_f] \frac{1 + (GH)|_{sc}}{1 + (GH)|_{oc}}$$

$$= 15.488 \text{ k}\Omega \frac{2.7346}{4.9088} = 8.6280 \text{ k}\Omega$$

For $(GH)|_{oc} = (GH)|_{i_i=0}$, G is the same as before, but H is

$$H|_{oc} = -\frac{9.1400 \text{ k}\Omega}{9.1400 \text{ k}\Omega + 150 \text{ k}\Omega} = -5.7434 \times 10^{-2}$$

The output resistance is

$$r_{out} = \frac{R_L \parallel R_f}{1 + GH} = \frac{9.3750 \text{ k}\Omega}{2.7346} = 3.4283 \text{ k}\Omega$$

Example 4.11 Audiotape Playback Amplifier

In Fig. E4.11 we combine Figs. E4.9 and E4.10 into a typical discrete BJT tape playback amplifier design. From the analysis of Figs. E4.9 and E4.10, the voltage gain is

$$A_v(\text{E4.9}) \cdot A_v(\text{E4.10}) = (-51.06)(-13.84) = 706.7$$

A SPICE ac simulation was performed on Fig. E4.11. Although we have not studied frequency response yet, the quasistatic parameters are closely approximated by the ac results at high frequencies. At 100 kHz, $A_v = v(90)/v(10) = 706.8$, showing good agreement. Additionally, from the simulation, $A_{v1} = v(70)/v(10) = 10.38$ and $r_{in} = v(10)/i(R_B) = 19.19$ kΩ at 100 kHz. The discrepancy between A_v (E4.9) and A_{v1} is due to how interstage loading is handled. In combining Figs. E4.9 and E4.10, the Thévenin equivalent of Fig. E4.9 was used to drive Fig. E4.10 instead of the output of Fig. E4.9 loaded by Fig. E4.10. The loaded Fig. E4.9 has a gain of A_{v1} and can be calculated using feedback analysis with a loaded $R_L' = 15 \text{ k}\Omega \parallel r_{in}(Q_3) = 1.563$ kΩ. Then the gain of the shunt-feedback output stage is its open-loop $G = -68.057$. By accounting for the loading of $r_{in}(Q_3)$ on Fig. E4.9, the actual output voltage of Fig. E4.9 is derived, and this is the base voltage of Fig. E4.10. Since v_{b3} is known (with Q_3-stage feedback taken into account in the loading), the remaining transmittance is from the base of Q_3 to its collector, or G. The overall gain is then

$$A_v = (-10.38)(-68.057) = 706.4$$

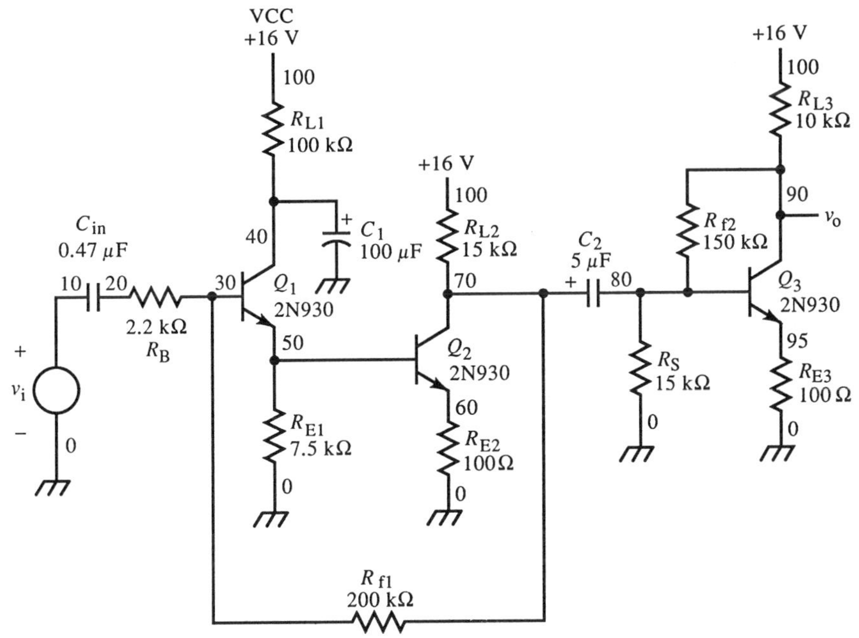

FIG. E.4.11

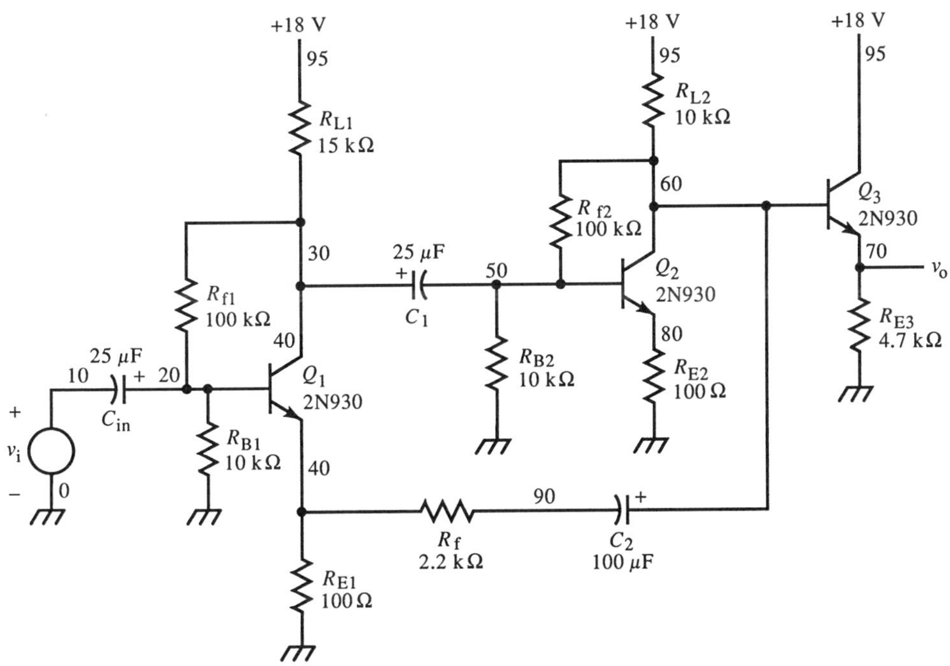

FIG. E4.12

Example 4.12 Audio Preamplifier with Noninverting Feedback

The circuit in Fig. E4.12 is an audio preamplifier design using similar gain stages as Fig. E4.11 but employing a noninverting feedback amplifier similar to that of Section 3.10. It is followed by an emitter-follower to provide a low-resistance (voltage source) output. The amplifier has high input resistance due to feedback, as desired for a voltage amplifier (see Section 2.5). SPICE simulation produced the following voltage gains:

$$\frac{v(30)}{v(10)} = 1.512 \text{ at } 100\text{ kHz}, \qquad \frac{v(70)}{v(10)} = 21.09 \text{ at } 100\text{ kHz}$$

References

E. James Angelo, Jr., *Electronics: BJTs, FETs, and Microcircuits*, McGraw-Hill, 1969. Ch. 12.

R. D. Middlebrook, "Design-Oriented Circuit Analysis and Measurement Techniques" Short Course, Power Electronics Group, California Institute of Technology, Pasadena, California.

R. D. Middlebrook, "Measurement of loop gain in feedback systems," *Int. J. Electronics*, 1975, Vol. 38, No. 4. pp. 485–512.

Sol Rosenstark, *Feedback Amplifier Principles*, Macmillan, 1986. pp. 11–24.

Vasil Uzunoglu, "Feedback analysis can be speeded," *Electronic Design* 19, 13 Sept. 1970. pp. 84–88.

CHAPTER 5

Transient and Frequency Response

The previous circuit analyses were low-frequency (real) and did not take into account inductance and capacitance. The reactive effects of these circuit elements[1] require complex analysis, and circuit behavior can be expressed either as functions of time (*time-domain* analysis) or frequency (*frequency-domain* analysis). These domains are united by expressing circuit response in terms of a complex frequency.

5.1 Reactive Circuit Elements

What makes circuit behavior different with reactances - inductances and capacitances - is that they can store energy and release it later. Their behavior depends on the rate of change of signals applied to them. The definitions of inductance L and capacitance C are

$$L \text{ definition} \qquad L \equiv \frac{d\lambda}{di} \tag{5.1a}$$

$$C \text{ definition} \qquad C \equiv \frac{dq}{dv} \tag{5.1b}$$

where λ is the magnetic flux linkage and q the electric charge. Since

$$v = \frac{d\lambda}{dt} \quad \text{and} \quad i = \frac{dq}{dt} \tag{5.2}$$

1. A circuit *element* is an idealization of a circuit component or other parasitic (distributed-parameter) effect equivalent to a component.

then

$$L\ v\text{-}i\ \text{relation} \qquad v = \frac{d\lambda}{di} \cdot \frac{di}{dt} = L\frac{di}{dt} \tag{5.3a}$$

$$C\ v\text{-}i\ \text{relation} \qquad i = \frac{dq}{dv} \cdot \frac{dv}{dt} = C\frac{dv}{dt} \tag{5.3b}$$

L and C differ by an interchange of v and i; they are duals. Besides the definition and v–i relation for L and C, the basic expression based on geometry is also important:

$$L = \frac{\mu A}{l} \tag{5.4a}$$

$$C = \frac{\varepsilon A}{l} \tag{5.4b}$$

where A is the coil loop or capacitor plate area, l the coil length or capacitor plate separation, μ the relative permeability of the material inside the coil and ε is the relative permittivity (or *dielectric constant*) of the material between the plates. Not all inductors or capacitors are constructed of solenoidal coils or parallel plates, but the form of (5.4) is generally correct. It is of interest to note that conductance can be expressed similarly:

$$G = \frac{\sigma A}{l} \tag{5.4c}$$

where σ is conductivity of the conductive material. Equation (5.4c) is the reciprocal of the more common formula for resistance:

$$R = \frac{\rho l}{A} \tag{5.5}$$

where ρ is resistivity $= 1/\sigma$, and $R = 1/G$.

The energy W stored by an inductor or capacitor can be derived by noting that power P is the rate of energy flow,

$$P = \frac{dW}{dt} \tag{5.6}$$

and that power is, by definition,

$$\text{Watt's law} \qquad P = v \cdot i \tag{5.7}$$

Combining (5.6) and (5.7) and solving for W, we have

$$W = \int vi\, dt \tag{5.8}$$

Substituting into (5.8) from (5.2) gives

$$\text{inductor energy,} \qquad W = \int \frac{d\lambda}{di} \cdot i\, dt = \int i\, d\lambda \tag{5.9a}$$

$$\text{capacitor energy,} \qquad W = \int \frac{dq}{dv} \cdot v\, dt = \int v\, dq \tag{5.9b}$$

For linear elements, (5.1) can be simplified to

$$\text{linear } L, \qquad \lambda = Li \tag{5.10a}$$

$$\text{linear } C, \qquad q = Cv \tag{5.10b}$$

Substituting these expressions into (5.9), the energy stored in a linear L or C is

$$W = \int i\, d(Li) = L \int i\, di = \tfrac{1}{2} \cdot Li^2 \tag{5.11a}$$

$$W = \int v\, d(Cv) = C \int v\, dv = \tfrac{1}{2} \cdot Cv^2 \tag{5.11b}$$

The energy stored in an inductor is proportional to the square of the current through it; for a capacitor, it is the square of the voltage. The proportionality constants are half L and C.

The v–i relations for L and C can also be expressed as

$$v = \frac{1}{L} \int_c^t i\, d\tau, \qquad i = \frac{1}{C} \int_c^t v\, d\tau \tag{5.12}$$

where τ is a dummy variable of integration and t is time.

5.2 First-Order Time-Domain Transient Response

Using the relations for L and C from Section 5.1, we can find the time-domain response (that is, the response as a function of time; time is the independent variable) of circuits such as that of Fig. 5.1a. This is a simple RC integrator or low-pass filter. The response can be found using KCL at the output node:

$$\frac{v_o - v_i}{R} + C\frac{dv_o}{dt} = 0 \tag{5.13}$$

or, arranging (5.13) into the form of a standard differential equation,

$$\frac{dv_o}{dt} + \left(\frac{1}{RC}\right) v_o = \left(\frac{1}{RC}\right) v_i \tag{5.13a}$$

This is an ordinary linear, constant-coefficient differential equation and can

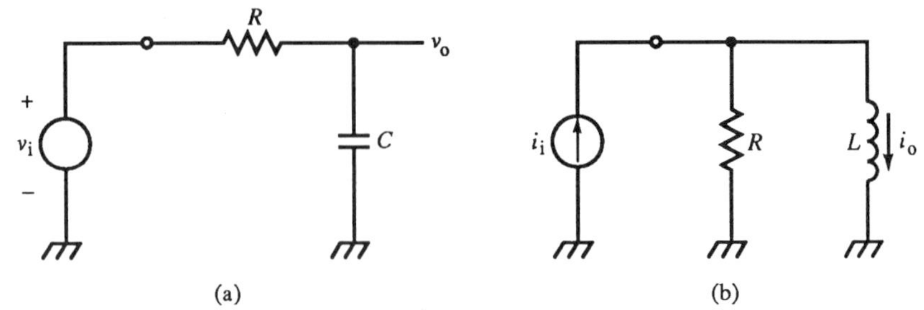

FIG. 5.1 First-order circuits: (a) *RC* integrator or low-pass filter and (b) *RL* dual of *RC* integrator.

be solved by using the substitution

$$v_o = Ve^{st} \tag{5.14}$$

where s is a variable. Notice that v_i, the input function driving the circuit, appears only on the right side of (5.13a). If a specific function is substituted for $v_i(t)$, the output time response can be found for that input. The output response also depends on the characteristic of the circuit itself, as represented by the left side of (5.13a). By setting the input function to zero and solving for v_o, we obtain the *natural response* of the circuit,

$$\frac{dv_o}{dt} + \left(\frac{1}{RC}\right) v_o = 0 \tag{5.15}$$

Substituting (5.14) gives an algebraic equation:

$$V\left(s + \frac{1}{RC}\right) e^{st} = 0 \tag{5.16}$$

Since the exponential cannot be zero and $V \neq 0$, the first factor must be zero, or

$$s + \frac{1}{RC} = 0 \tag{5.17}$$

This algebraic equation in s is the *characteristic equation* of (5.15) and of the circuit. Solving for s gives

$$s = -\frac{1}{RC} \tag{5.18}$$

Finally, the natural response is

$$v_o(t) = Ve^{-(1/RC)t} \tag{5.19}$$

If the capacitor has an initial charge of $v_o(t) = V_0$ and $v_i = 0$, then

$$V_0 = Ve^0 = V$$

and

$$v_o = V_0\, e^{-(1/RC)t} \tag{5.20}$$

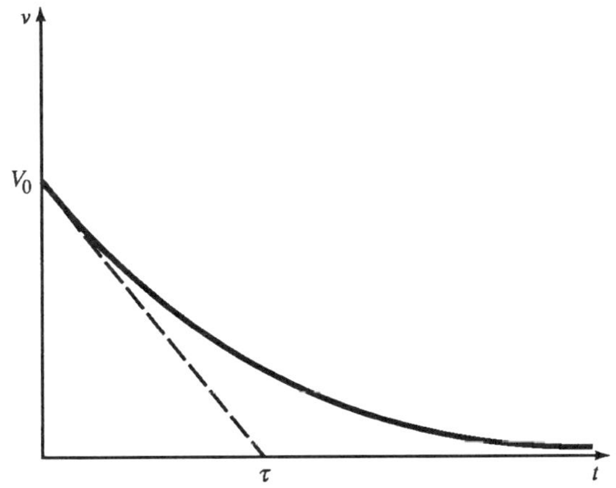

FIG. 5.2 First-order response: exponential decay. The initial slope at initial voltage V_0, when projected to the t-axis, intercepts at the time constant τ.

The response shown in Fig. 5.2 is very common and is called *exponential decay.* The output voltage decays, asymptotically approaching zero at infinite time. This is typical of the natural response of circuits. Given an initial energy in reactive elements, this energy is eventually dissipated by any resistive elements. Consequently, circuit response to a nonzero initial condition, the natural response, is also called the *transient response.*

The time scale of the decay is measured as the *time constant*, the value of $-1/s$ in (5.18), which is

$$\tau = RC \tag{5.21}$$

The initial slope of $v_o(t)$ is projected to the t-axis in Fig. 5.2 and intercepts the axis at τ. At $t = \tau$, $v_o(\tau)/V_0 = e^{-1} \cong 36.7\%$. After 5τ, v_o is within 1% of zero.

The RL circuit of Fig. 5.1b is the dual of (a), and its time constant is $\tau = L/R$. For $i_i = 0$ and an initial inductor current of I_0, $i_o(t)$ is about 37% of its final value after τ. Replacing V_0 with I_0 in Fig. 5.2 results in the response of $i_o(t)$.

5.3 Complex Poles and the Complex-Frequency Domain

For circuits with more than one reactive element, the differential equations describing them can also be solved with the substitution (5.14). The complication is in solving the characteristic equation. Its degree is equal to the number of reactive circuit elements. After its roots are found, the time-domain response is obtained by substitution. For multiple roots, linear combinations of (5.14) are required.

The roots of the characteristic equation are called *poles*. For single-pole response, s is real. For two or more poles, s can be real, imaginary, or complex. In general, s is a complex variable. Complex numbers can be expressed in rectangular or polar form:

$$\text{rectangular form,} \qquad s = \sigma + j\omega \tag{5.22}$$

$$\text{polar form,} \qquad s = \|s\|\, e^{j\phi} \tag{5.23}$$

In rectangular form, s is the sum of real and imaginary numbers (where $j = \sqrt{-1}$). In polar form, s is expressed by a magnitude and phase angle. The polar and rectangular forms are related through Euler's formula,

$$e^{j\phi} = \cos\phi + j\sin\phi \tag{5.24}$$

and the Pythagorean theorem,

$$\|s\| = \sqrt{\sigma^2 + \omega^2} \tag{5.25}$$

The rectangular components are expressed in terms of the polar components as

$$\sigma = \|s\| \cos\phi \tag{5.26}$$

$$\omega = \|s\| \sin\phi \tag{5.27}$$

Therefore,

$$s = \sigma + j\omega = \|s\| \cos\phi + j\|s\| \sin\phi = \|s\|(\cos\phi + j\sin\phi) = \|s\|\, e^{j\phi}$$

Dividing (5.27) by (5.26) and solving for ϕ gives

$$\phi = \tan^{-1}\left\{\frac{\omega}{\sigma}\right\} \tag{5.28}$$

With these expressions we can now examine the general form of a quadratic characteristic equation for

$$s = (-\alpha, \pm\omega_{\mathrm{d}}) = -\alpha \pm j\omega_{\mathrm{d}} \tag{5.29}$$

which is

$$\begin{aligned} s^2 + 2\zeta\omega_{\mathrm{n}} s + \omega_{\mathrm{n}}^2 &= (s + \alpha + j\omega_{\mathrm{d}})(s + \alpha - j\omega_{\mathrm{d}}) = 0 \\ &= s^2 + 2\alpha s + (\alpha^2 + \omega_{\mathrm{d}}^2) \end{aligned} \tag{5.30}$$

Equating terms, we get

$$\alpha = \zeta\omega_{\mathrm{n}} \tag{5.31}$$

$$\omega_{\mathrm{n}}^2 = \alpha^2 + \omega_{\mathrm{d}}^2 \tag{5.32}$$

The general solution of (5.30) is

$$s_{1,2} = -\zeta\omega_{\mathrm{n}} \pm j\omega_{\mathrm{n}}\sqrt{1 - \zeta^2} \tag{5.33}$$

where

$$\omega_d = \omega_n\sqrt{1-\zeta^2} \tag{5.34}$$

The quantity ω_n is called the *natural frequency*, α the *damping factor*, and ζ the *damping ratio.* From (5.31) and (5.32), ζ can be directly related to the pole angle ϕ as

$$\zeta = \cos\phi \tag{5.35}$$

We can express (5.33) in polar form by using (5.25), (5.28) and $\tan\vartheta = -\tan(180° - \vartheta) = -\tan\phi$:

$$s_{1,2} = \omega_n\, e^{\pm j\phi} \tag{5.36}$$

This pole pair is shown in Fig. 5.3. Complex poles always occur in conjugate pairs, as shown. In the complex plane of s (*s-plane* or *s-domain*), poles are represented by × marks.

The units of s must be frequency. Since s is complex, it is a complex frequency. The s-domain is called the *complex-frequency domain.*

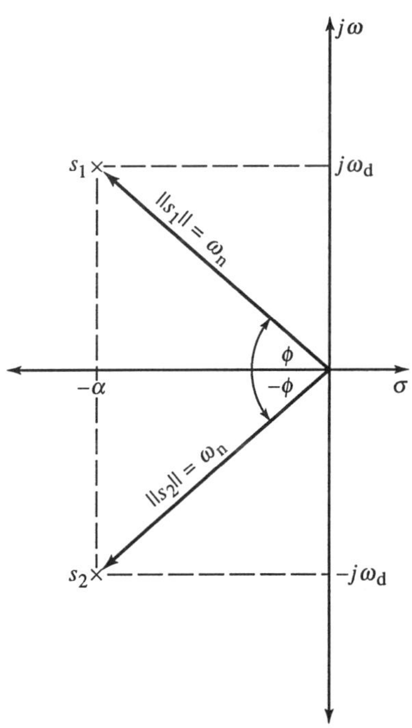

FIG. 5.3 General complex pole pair in the complex-frequency domain. The pole angles are $\pm\phi$ (pole angle is ϕ), magnitude is the natural frequency ω_n, real component is $-\alpha$, and imaginary components are $\pm j\omega_d$. The poles are always complex conjugates and are always symmetrical about the σ-axis.

5.4 Second-Order Time-Domain Response: *RLC* Circuit

The series *RLC* circuit of Fig. 5.4 is a typical circuit with a second-order differential equation (and second-degree characteristic equation). Using KCL, its differential equation is

$$\frac{d^2 i}{dt^2} + \left(\frac{R}{L}\right)\frac{di}{dt} + \left(\frac{1}{LC}\right) i = \frac{1}{L} \cdot \frac{di_i}{dt} \tag{5.37}$$

The transient response is found when $i_i(t) = 0$. Solving for the characteristic equation using $i = e^{st}$, it is

$$s^2 + \left(\frac{R}{L}\right) s + \left(\frac{1}{LC}\right) = 0 \tag{5.38}$$

Using the quadratic formula to solve for the poles in s, we obtain

$$s_{1,2} = -\frac{R}{2L} \pm j\sqrt{\frac{1}{LC} - \left(\frac{R}{2L}\right)^2} \tag{5.39}$$

These poles are of the form

$$s_{1,2} = -\alpha \pm j\omega_d \tag{5.40}$$

where

$$\alpha = \frac{R}{2L}, \qquad \omega_d = \sqrt{\frac{1}{LC} - \left(\frac{R}{2L}\right)^2} \tag{5.40a}$$

Each pole contributes a solution to (5.37) when substituted into e^{st}. Because the response is the solution to a linear differential equation, then by superposition, the independent pole solutions can be combined linearly to form a complete solution.

The transient response of a second-order circuit depends on the value of its elements. For the *RLC* circuit of Fig. 5.4, the poles are real when $j\omega_d$ is real:

$$\left(\frac{R}{2L}\right)^2 - \frac{1}{LC} \geq 0$$

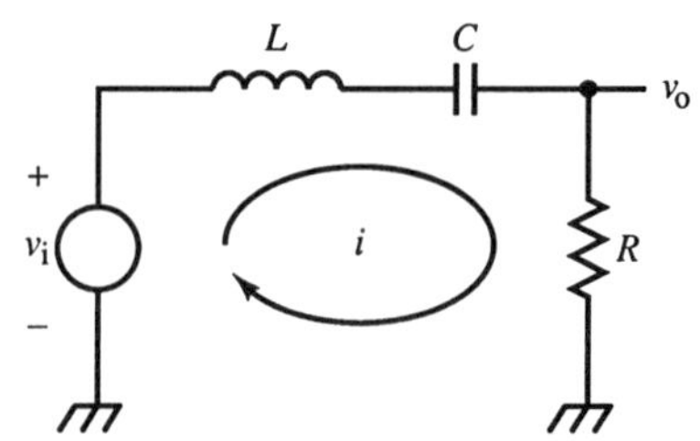

FIG. 5.4 A second-order *RLC* circuit.

or

$$R \geq 2\sqrt{\frac{L}{C}} \tag{5.41}$$

When distinct poles are real, response is exponential, and the natural frequency ω_n equals the real frequency α. Distinct real poles are located at $s = -\alpha \pm \omega_d$.

A special case of real poles is when they are equal (or *repeated*). Their solutions cannot be combined by superposition to produce a response because they are not independent. In this case, the general form of the pole solutions is, for n poles,

$$\frac{t^{n+1}}{(n-1)!} \cdot e^{-\alpha t} \tag{5.42}$$

For the RLC circuit, $n = 2$, and the response is

$$i(t) = I_1\, e^{-\alpha t} + I_2 t\, e^{-\alpha t} \tag{5.43}$$

For imaginary poles, $\omega_n = \omega_d$, $\alpha = 0$, and the natural response is a sinusoid. This is the case of oscillators and is a conditionally stable response.

The last case to consider is that of complex poles. The solution is

$$i(t) = c_1\, e^{(-\alpha + j\omega_d)t} + c_2\, e^{(-\alpha - j\omega_d)t} \tag{5.44}$$

This can be written as

$$\begin{aligned} i(t) &= e^{-\alpha t}(c_1 \cos \omega_d t + jc_1 \sin \omega_d t + c_2 \cos \omega_d t - jc_2 \sin \omega_d t) \\ &= e^{-\alpha t}[(c_1 + c_2) \cos \omega_d t + j(c_1 - c_2) \sin \omega_d t] \end{aligned} \tag{5.45}$$

If c_1 and c_2 are complex conjugate constants, then the coefficients in (5.45) are real. Let

$$I_1 = c_1 + c_2, \qquad I_2 = j(c_1 - c_2) \tag{5.46}$$

Then the response is

$$i(t) = e^{-\alpha t}(I_1 \cos \omega_d t + I_2 \sin \omega_d t) \tag{5.47}$$

with real I_1 and I_2 determined by initial conditions. This can also be expressed as a single sinusoid with a phase angle ϑ:

$$i(t) = I\, e^{-\alpha t} \sin(\omega_d t + \vartheta) \tag{5.48}$$

where

$$I = \sqrt{I_1^2 + I_2^2}, \qquad \vartheta = \tan^{-1}\left(\frac{I_1}{I_2}\right) \tag{5.49}$$

This response is a damped sinusoid (Fig. 5.5). For left half-plane poles, the sinusoid decays with time due to the decaying exponential factor. This factor is the *envelope* of the sinusoid and is shown as a dotted line. For right half-plane poles, the response is unstable; it is an exponentially growing sinusoid. The

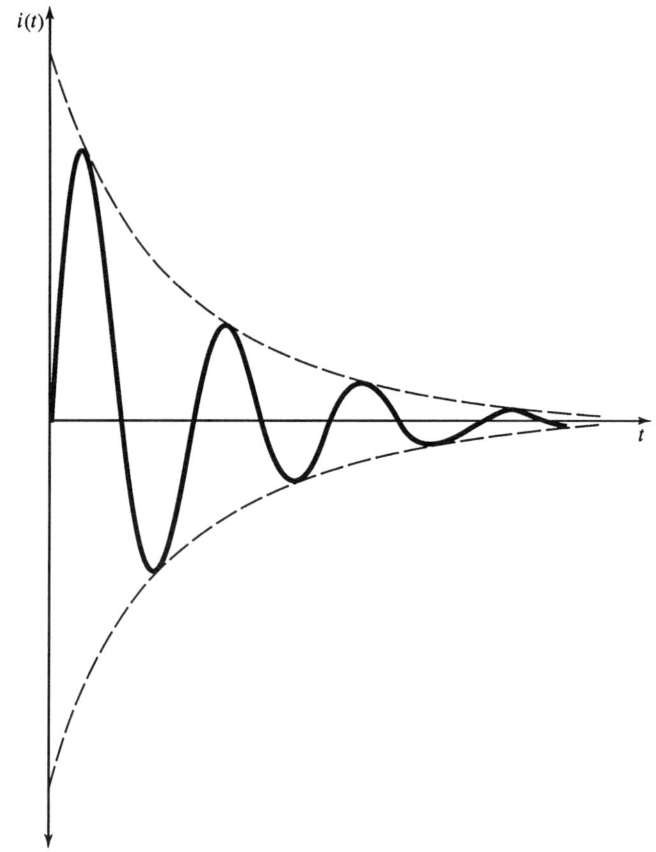

FIG. 5.5 Damped sinusoidal response of second-order circuit. The envelope (dotted curves) of the sinusoid is exponential decay, dependent upon α. The damped frequency is ω_d.

limiting of actual circuits causes nonlinear limiting of such a response and distorts the sine wave.

The pole angle ϕ and its related parameter ζ most explicitly express the kind of response a circuit will have. The pole angle is

$$\phi = \tan^{-1}\left(\frac{\omega_d}{\alpha}\right) \tag{5.50}$$

As ζ decreases, poles move toward each other and then split off the real axis, increasing in pole angle (Fig. 5.6). As the poles leave the real axis, the time-domain response begins to show a sinusoid, with noticeable "ringing." The larger the pole angle, the more sinusoidal cycles occur before being damped out by the decaying exponential factor. We have, then, the following

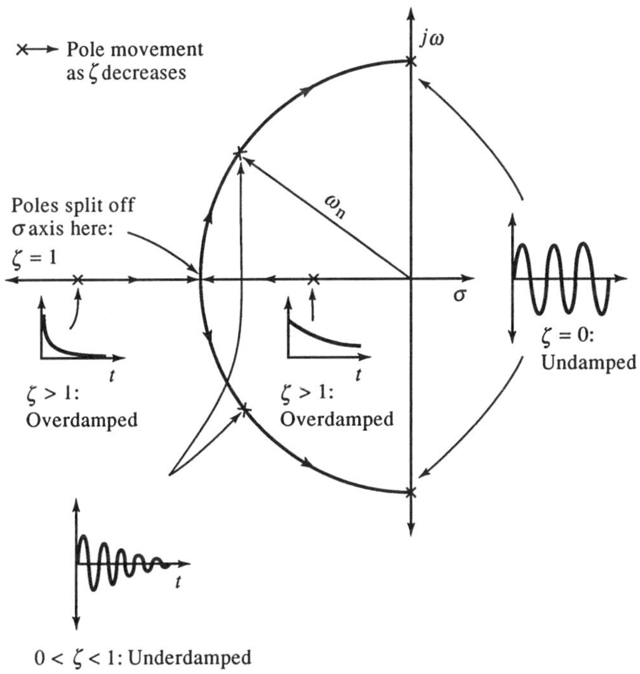

FIG. 5.6 Responses for various pole locations. As ζ decreases, the imaginary component gains in dominance, and sinusoidal response becomes more prominent.

categories of response:

$\zeta > 1$	overdamped response	poles real and distinct
$\zeta = 1$	critically damped response	two equal real poles
$0 < \zeta < 1$	underdamped response	complex pole pair
$\zeta = 0$	undamped response	poles imaginary

As ζ decreases below unity, the poles move in a circular arc of radius ω_n and increasing pole angle until, at $\zeta = 0$, they are located on the $j\omega$ axis at $\pm j\omega_d = \pm j\omega_n$. The frequency ω_d, the *damped frequency*, is less than the natural frequency ω_n when $\zeta > 0$, that is, when damping exists due to R (5.34).

The real component of s, $-\alpha$, is related to the exponential factor in the response. It has units of 1/time (or frequency) and is the reciprocal of the time constant of the exponential factor. For large α, the exponential response is fast. For poles on the $j\omega$ axis, $\alpha = 0$ and no exponential decay occurs.

The RLC circuit has

$$\alpha = \frac{1}{2\tau} = \frac{R}{2L} \tag{5.51}$$

$$\omega_n = \frac{1}{\sqrt{LC}} \tag{5.52}$$

and

$$\zeta = \frac{\alpha}{\omega_n} = \frac{R}{2\sqrt{L/C}} \tag{5.53}$$

The *RLC* circuit is critically damped at $\zeta = 1$. Solving (5.53) for R, we obtain the equality of (5.41), Define

$$Z_n = \sqrt{\frac{L}{C}} \tag{5.54}$$

This is the *characteristic impedance* of the *LC* elements. L and C also determine ω_n (5.52). A combination of elements (L and C or negative elements) can give rise to an underdamped response or a *resonance.* The behavior of resonant circuits exhibiting sinusoidal response is oscillatory. Resonance occurs at a frequency of ω_n (or ω_o) and a characteristic impedance of Z_n (or Z_o). Circuits with more than two reactive elements have more than one resonance (or resonant mode). These circuits can have more than one complex pole pair.

Resonance is either *series resonance* or *parallel resonance.* The previous *RLC* circuit is series resonant since L and C are in series with R. For critical damping of a series resonance, R must be equal to the sum of the (equal) reactances of L and C at resonance, or $2Z_n$. A circuit with R, L, and C in parallel is parallel resonant, and for critical damping, once again, R must equal the combined reactance of L and C or be half of Z_n.

5.5 Forced Response and Transfer Functions in the *s*-Domain

We have seen that the natural or transient response of a circuit is due to initial nonzero energy storage in reactive elements. At $t > 0$, the circuit responds to this energy without external input sources. With resistive elements in the circuit, this initial energy is dissipated and eventually goes to zero. For a circuit quantity (a voltage or current) x, then

$$\lim_{t\to\infty} x_{tr} = 0$$

When the circuit is driven by a source, it continues to respond indefinitely to the source. This is the *forced* or *steady-state response.* It continues after the transient response has decayed and is

$$x_{ss} = \lim_{t\to\infty} x$$

For linear circuits, the total response is the superposition of the transient and steady-state response, or

$$x(t) = x_{tr}(t) + x_{ss}(t) \tag{5.55}$$

The transient response can be found, as we did in Section 5.4, by solving the circuit differential equations for zero input. This (homogeneous) solution can then be used to find the (particular or complementary) solution with a nonzero input, resulting in the total response.

For linear circuits, input sinusoids always result in output sinusoids. A differential circuit equation describing an output quantity x_o and input x_i is, in general,

$$D_1(x_o) = N_1(x_i) \tag{5.56}$$

The transient response is found by setting $N_1 = 0$ and $x_o = X_o(s)\, e^{st}$, where X_o is a complex parameter of s (and constant in t). Then,

$$D_1(X_o\, e^{st}) = 0 \tag{5.57}$$

We have seen that the exponential can be factored out of D, leaving the characteristic equation $D(s) = 0$. This factorization is

$$D_1(X_o\, e^{st}) = D(s) \cdot X_o\, e^{st} = D(s) \cdot x_o \tag{5.58}$$

If we let x_i similarly be a complex exponential, then we have a similar factorization on the right side of (5.56), or

$$N_1(x_i)\big|_{x_i = X_i e^{st}} = N(s) \cdot x_i \tag{5.59}$$

Substituting (5.58) and (5.59) into (5.56) and solving for the output/input ratio, or *transfer function*, results in

$$\frac{X_o}{X_i} = \frac{N(s)}{D(s)}, \qquad x_i, x_o = X(e^{st}) \tag{5.60}$$

This is an extremely important result. Whenever the input is a complex exponential function, the output will also be, modified by $N(s)/D(s)$. Sinusoidal input is a special case in which $s = j\omega$. Equation (5.60) is a very general result because complex exponentials can be summed in a Fourier series to create arbitrary functions. Therefore, in (5.60) we have the key to finding the transfer functions of linear circuits in general, expressed in s.

N and D can be factored into the *canonical form* of a transfer function:

$$\frac{N(s)}{D(s)} = K \cdot \frac{(s+z_1)(s+z_2)\cdots(s+z_m)}{(s+p_1)(s+p_2)\cdots(s+p_n)}, \qquad m \le n \tag{5.61}$$

The roots of $D(s)$ are poles, $-p_i$, and the roots $-z_i$ of $N(s)$ are called *zeros* because N/D is zero at $s = -z_i$. (The word *pole* fits the idea that since the poles make N/D infinite, a plot of $N(s)/D(s)$ typically looks like a tent with poles holding it up at the *poles*.) Poles and zeros are called *critical frequencies*. For actual (causal) circuits, the degree of N does not exceed that of D, or $m \le n$. K is a constant, but it is not the dc transmittance. Equation (5.61) can be expressed in *normalized form* by factoring out z_i and p_i. The factors are normalized to unity at $s = 0$, and the new constant K is the dc transmittance.

$$\frac{N(s)}{D(s)} = K \cdot \frac{(s/z_1+1)(s/z_2+1)\cdots(s/z_m+1)}{(s/p_1+1)(s/p_2+1)\cdots(s/p_n+1)}, \qquad m \le n \tag{5.62}$$

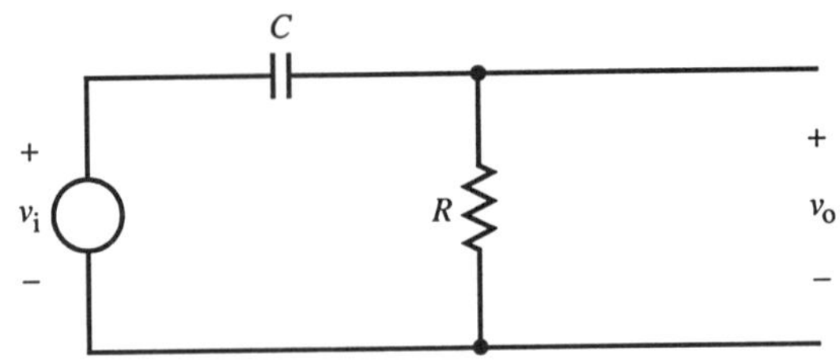

FIG. 5.7 *RC* differentiator.

To illustrate these general concepts by example, consider the RC differentiator circuit of Fig. 5.7. Using KCL at the output node, we obtain

$$C\frac{d(v_o - v_i)}{dt} + \frac{v_o}{R} = 0 \tag{5.63}$$

This can be rearranged as

$$\frac{dv_o}{dt} + \left(\frac{1}{RC}\right) v_o = \frac{dv_i}{dt} \tag{5.64}$$

The right side of the equation describes the effect of the input and is related to the steady-state response. The left side, as we have seen, characterizes the transient response. Letting both v_o and v_i be complex exponentials and factoring, we obtain

$$M(s) = \frac{N(s)}{D(s)} = \frac{s}{s + (1/RC)} = RC \cdot \frac{s}{sRC + 1} \tag{5.65}$$

There is a zero at the origin ($s = 0$) and a pole at $-1/RC$. $D(s)$ is the same as that of (5.17), also a first-order RC circuit. Since the poles characterize the transient response, it is identical for this RC circuit. The numerator describes the transformation of the input. Since

$$\frac{d}{dt} e^{st} = s\, e^{st} \tag{5.66}$$

s acts as a differentiation operator for complex exponentials and, more generally, for functions that can be described in terms of complex exponentials. In (5.65), $N(s) = s$ can be interpreted as a differentiation of the input function to the circuit, hence the name *RC differentiator.*

Applying this method of finding the transfer function in s to the RC integrator of Fig. 5.1a, described by (5.13a), we obtain

$$M(s) = \frac{1}{RC} \cdot \frac{1}{s + (1/RC)} = \frac{1}{sRC + 1} \tag{5.67}$$

5.6 The Laplace Transform

The *RC* integrator and differentiator were described by one differential equation resulting from application of KCL in the time domain. For more complicated circuits, we must solve a system of differential equations. The total solution requires substitution of a particular function for the input. We saw in Section 5.5 that for complex exponential inputs, the response is also a complex exponential, and the transfer function in the *s*-domain can be found. Since Fourier series descriptions of arbitrary functions can be given in terms of complex exponentials, this is a generally useful result. But what remains is to determine the time-domain response of a circuit described by transfer function $M(s)$ for a given input. Specifically, we need a way of converting a function $x(t)$ to its equivalent in the *s*-domain, $X(s)$. Then we can multiply $X_i(s)$ by $M(s)$ to get the resulting output $X_o(s)$:

$$X_o(s) = M(s) \cdot X_i(s) \tag{5.68}$$

Finally, if we can transform $X_o(s)$ to the time domain, we have the desired result, $x_o(t)$.

The Fourier series is limited to periodic waveforms but in the limit becomes the Fourier integral. This integral is a transform from the time domain to the $j\omega$ domain. By adding a real component σ to $j\omega$, we have an extended transform in *s*, the *Laplace transform*[2], defined as

$$\mathscr{L}\{f(t)\} \equiv \int_0^\infty f(t)\, e^{-st}\, dt \tag{5.69}$$

Some transformed functions are the following:

$f(t)$	$F(s) = \mathscr{L}\{f(t)\}$	
$\delta(t)$	1	(5.70)
$u(t)$	$\frac{1}{s}$	(5.71)
$e^{-\alpha t}$	$\frac{1}{s+\alpha}$	(5.72)
$\sin \omega_d t$	$\frac{\omega_d}{s^2+\omega_d^2}$	(5.73)
$\cos \omega_d t$	$\frac{s}{s^2+\omega_d^2}$	(5.74)
$e^{-\alpha t} \sin \omega_d t$	$\frac{\omega_d}{(s+\alpha)^2+\omega_d^2}$	(5.75)
t^n	$\frac{n!}{s^{n+1}}$	(5.76)

2. This is the commonly used *one-sided* Laplace transform since the lower bound on integration is zero, not $-\infty$. For causal systems (and all analog circuits are causal), no response is possible before an input at $t = 0$ is applied; the integral is zero before $t = 0$, and consequently we can set the lower integration bound at zero.

These are among the functions useful as inputs to circuits for characterizing their responses. The unit impulse function $\delta(t)$ is defined as

$$\delta(t) \equiv \begin{cases} 0, & t \neq 0 \\ \infty, & t = 0 \end{cases} \tag{5.77}$$

Although $\delta(0)$ is infinite, $\int \delta \, dt = 1$. At $t = 0^+$, the circuit responds as it would to nonzero initial conditions with the transient response. In practice, it is not easy to generate δ, and a step input is used to characterize response instead. The unit step function $u(t)$ is defined as

$$u(t) = \begin{cases} 0, & t < 0 \\ 1, & t > 0 \end{cases} \tag{5.78}$$

Notice that $u(t)$ is not periodic. Step functions are commonly approximated as a periodic function - a square wave - with a relatively long period.

The Laplace transform of operations such as differentiation and integration can also be taken:

t-domain	s-domain		
$Kf(t)$	$KF(s)$	scale invariance	(5.79)
$\sum f_i(t)$	$\sum F_i(s)$	superposition	(5.80)
$\dfrac{df(t)}{dt}$	$sF(s) - f(0^+)$		(5.81)
$\dfrac{d^2f(t)}{dt^2}$	$s^2F(s) - sf(0^+) - \dfrac{df(0^+)}{dt}$		(5.82)
$\int_c^t f(\tau)\, d\tau$	$\dfrac{F(s)}{s}$		(5.83)
$e^{-\alpha t}f(t)$	$F(s+\alpha)$		(5.84)
$t^n f(t)$	$(-1)^n \cdot \dfrac{d^nF(s)}{ds^n}$		(5.85)
$u(t-\tau)f(t-\tau)$	$e^{-\tau s}F(s)$	shifting theorem	(5.86)
$\int_c^t f(\tau)g(t-\tau)\, d\tau$	$F(s)G(s)$	convolution	(5.87)
$\lim_{t\to\infty} f(t)$	$\lim_{s\to 0} sF(s)$	final value theorem	(5.88)
$\lim_{t\to 0} f(t) = f(0^+)$	$\lim_{s\to\infty} sF(s)$	initial value theorem	(5.89)

Equations (5.79) and (5.80) affirm the linearity of the Laplace transform. The transform of a derivative is consistent with (5.66) and the use of s as a differentiation operator in the s-domain in Section 5.5. The initial condition $f(0^+)$ is part of the transformed derivative. The shifting theorem expresses the effect of shifting f in time by a delay of τ. The convolution integral offers an alternative to the inverse Laplace transform for finding $x_o(t)$ from (5.68) but is usually not as easy to use.

The Laplace transform can be applied to the v–i relations of R, L, and C to find their s-domain impedances. They can then be used in basic circuit analysis. KCL, KVL, and ΩL can be applied directly in the s-domain with no need for intermediate steps involving differential equations. For R, C, and L, we have

$$\mathscr{L}\{v\} = \mathscr{L}\{Ri\} = R\mathscr{L}\{i\} \Rightarrow Z_R(s) = \frac{\mathscr{L}\{v\}}{\mathscr{L}\{i\}} = \frac{V(s)}{I(s)} = R \tag{5.90}$$

$$\mathscr{L}\{v\} = \mathscr{L}\left\{\frac{1}{C}\int i\,dt\right\} = \frac{1}{C}\cdot\frac{\mathscr{L}\{i\}}{s} \Rightarrow Z_C(s) = \frac{\mathscr{L}\{v\}}{\mathscr{L}\{i\}} = \frac{1}{sC} \tag{5.91}$$

$$\mathscr{L}\{i\} = \mathscr{L}\left\{\frac{1}{L}\int v\,dt\right\} = \frac{1}{L}\cdot\frac{\mathscr{L}\{v\}}{s} \Rightarrow Z_L(s) = \frac{\mathscr{L}\{v\}}{\mathscr{L}\{i\}} = sL \tag{5.92}$$

Initial conditions for L and C in the s-domain can be accounted for by Laplace transforming the time-domain expressions for L and C with initial conditions, or

$$i_L(t) = \frac{1}{L}\int v_L(t)\,dt + i_L(0^+) \overset{\mathscr{L}}{\Rightarrow} I_L(s) = \frac{V_L(s)}{sL} + \frac{i_L(0^+)}{s} \tag{5.93}$$

$$v_C(t) = \frac{1}{C}\int i_C(t)\,dt + v_C(0^+) \overset{\mathscr{L}}{\Rightarrow} V_C(s) = \frac{I_C(s)}{sC} + \frac{v_C(0^+)}{s} \tag{5.94}$$

The s-domain equivalent circuit for L with initial current $i_L(0^+)$ is a current source of $i_L(0^+)/s$ in parallel with L. The s-domain equivalent circuit for C with initial voltage $v_C(0^+)$ is a voltage source of $v_C(0^+)/s$ in series with $1/sC$ (Fig. 5.8a). By transforming the derivative form of the L and C v–i relations, we obtain an equivalent circuit that accounts for initial conditions (Fig. 5.8b).

By working directly in s, we can find the transfer function of the RC integrator and differentiator from inspection, by treating them as voltage dividers. For the RC integrator,

$$M(s) = \frac{V_o(s)}{V_i(s)} = \frac{1/sC}{R + 1/sC} = \frac{1}{sRC + 1} \tag{5.95}$$

and for the RC differentiator,

$$M(s) = \frac{V_o(s)}{V_i(s)} = \frac{R}{R + 1/sC} = \frac{sRC}{sRC + 1} \tag{5.96}$$

By writing circuit equations directly in the s-domain, we avoid the need to either transform or solve differential equations.

Example 5.1 Series *RC* Circuit

Figure E5.1 is a passive circuit with series RC divider impedances. The transfer function is found by directly writing out the voltage divider

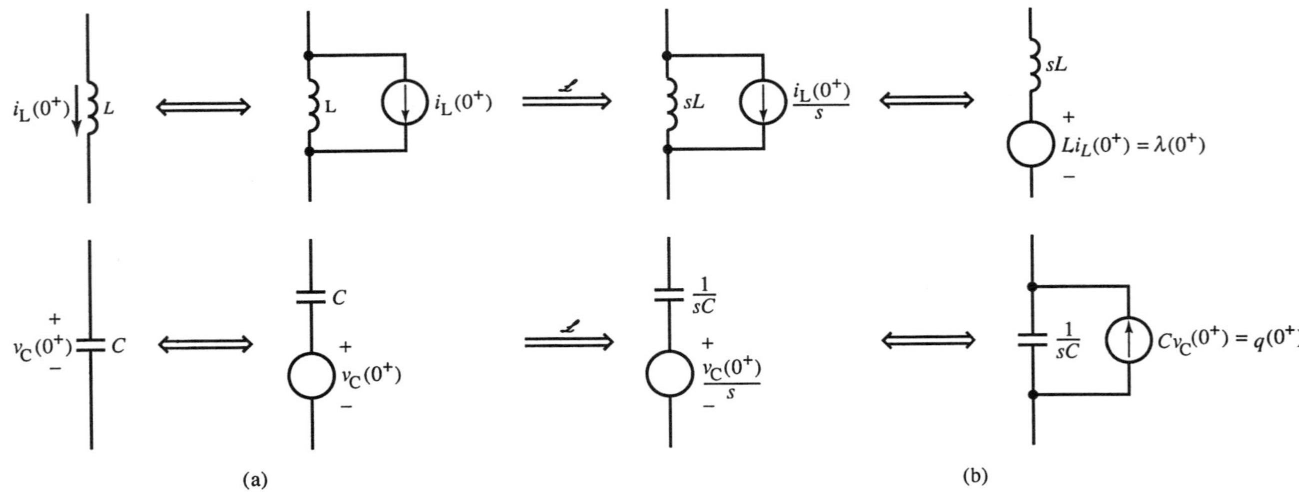

FIG. 5.8 Equivalent circuits in the s-domain for elements with initial energy (a) and alternative circuits (b).

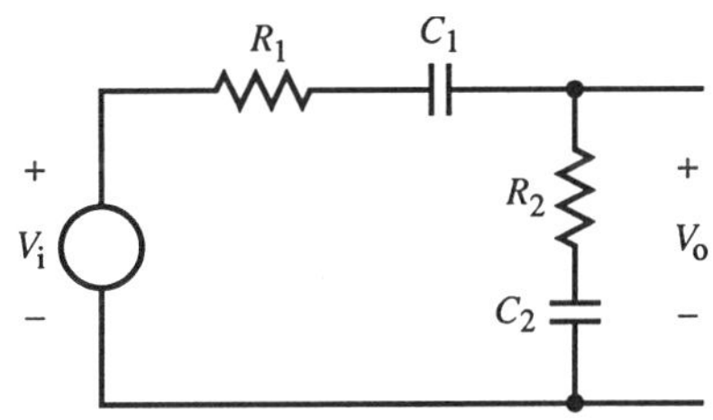

FIG. E5.1

formula using s-domain impedances. Both impedances are of form $R+1/sC$. The transfer function is

$$\frac{V_o(s)}{V_i(s)}=\left(\frac{C_1}{C_1+C_2}\right)\frac{sR_2C_2+1}{s(R_1+R_2)(C_1 \parallel C_2)+1}, \qquad C_1 \parallel C_2=\frac{C_1C_2}{C_1+C_2}$$

Example 5.2 Wien-Bridge Filter

Figure E5.2a is a filter topology used as the feedback path of the Wien-bridge oscillator. It is another voltage divider, for which the transfer function is

$$\frac{V_o(s)}{V_i(s)}=\frac{sR_2C_2}{s^2R_1R_2C_1C_2+s[R_1C_1+R_2(C_1+C_2)]+1}$$

For a Wien-bridge filter, $R_1=R_2=R$ and $C_1=C_2=C$. Then,

$$\text{Wien-bridge,} \qquad \frac{V_o(s)}{V_i(s)}=\frac{sRC}{s^2(RC)^2+s[3RC]+1}$$

For this filter, $\omega_n=1/RC$ and $\zeta=1.5$.

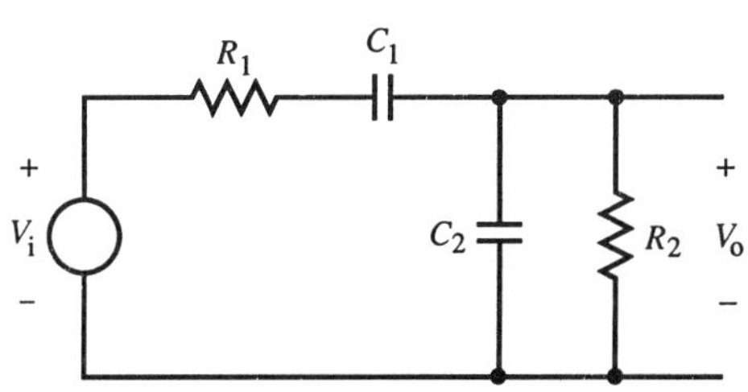

(a)

FIG. E5.2(a)

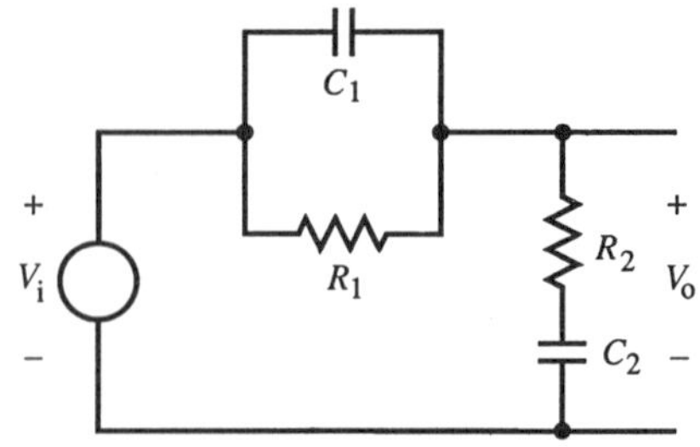

FIG. E5.3

Example 5.3 Inverse of Wien-Bridge Filter

In Fig. E5.3, the divider topology of Fig. E5.2a is inverted. The transfer function is

$$\frac{V_o(s)}{V_i(s)} = \frac{(sR_1C_1+1)(sR_2C_2+1)}{s^2R_1R_2C_1C_2+s[R_1(C_1+C_2)+R_2C_2]+1}$$

For the Wien-bridge conditions, this reduces to

$$\frac{(sRC+1)^2}{s^2(RC)^2+s[3RC]+1}$$

The poles are in the same place as in Example 5.2, but instead of a zero at the origin, repeated zeros appear at $1/RC$.

Example 5.4 Shunt-Series *RC* Circuit

Figure E5.4 is another example of an *RC* circuit. It has a terminal impedance of

$$Z = R_1 \cdot \frac{sR_2C_2+1}{s^2[R_1C_1R_2C_2]+s[R_1C_1+(R_1+R_2)C_2]+1}$$

This circuit is sometimes the external emitter network of CE amplifiers,

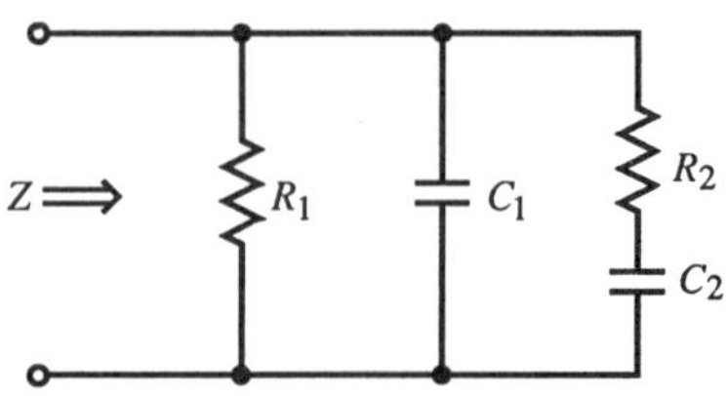

FIG. E5.4

in which R_1 is the emitter resistor, C_1 provides bypassing, and R_2C_2, frequency compensation.

Example 5.5 Crystal Equivalent Circuit

Figure E5.5 shows the equivalent circuit of a quartz crystal. The terminal impedance is

$$Z_{\text{xtal}} = \frac{1}{C_p + C_s} \cdot \frac{s^2LC_s + sRC_s + 1}{s[s^2L(C_s \parallel C_p) + sR(C_s \parallel C_p) + 1]}$$

where (as usual), $\parallel$ is an algebraic operator not a topological descriptor; C_s and C_p are in series in the quadratic pole.

Resonance occurs when the phase is zero. The phases of numerator and denominator are

$$\phi_N = \frac{\omega RC_s}{1 - \omega^2 LC_s}, \qquad \phi_D = \frac{\omega - \omega^3 L(C_s \parallel C_p)}{-\omega^2 R(C_s \parallel C_p)}$$

Then $\angle Z(s) = \phi_N - \phi_D$. Setting this to zero and simplifying, we must solve for resonant frequency ω_r in

$$\omega_r^4 L^2 C_s (C_s \parallel C_p) - \omega_r^2 [R^2 C_s (C_s \parallel C_p) + L(C_s \parallel C_p) + LC_s] + 1 = 0$$

Solving for ω_r^2 we obtain

$$\omega_r^2 = \tfrac{1}{2}(\omega_s^2 + \omega_p^2) \pm \tfrac{1}{2}\sqrt{(\omega_s - \omega_p)^2}$$

where $\omega_s^2 = 1/LC_s$ and $\omega_p^2 = 1/L(C_s \parallel C_p)$. The two resonant frequencies are at the series resonance ω_s and parallel resonance ω_p of the crystal

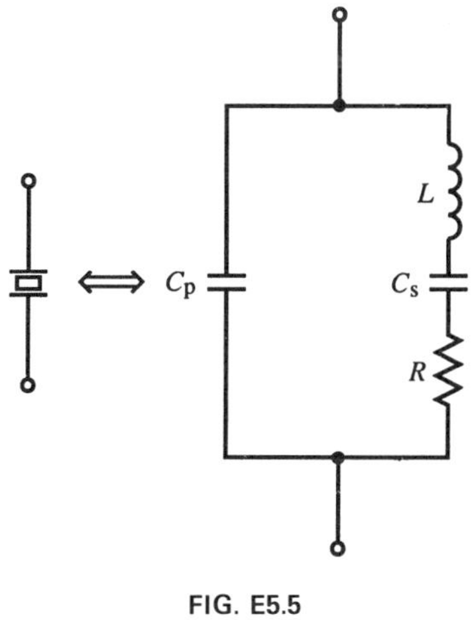

FIG. E5.5

5.7 Time-Domain Response to a Unit Step Function

To find the time-domain response to a given input function, we can take the inverse Laplace transform $\mathscr{L}^{-1}$ of $X_o(s)$ in (5.68) (or apply the convolution integral). After $X_o(s)$ is in the form of a known transform, it is inverse Laplace transformed to produce the time-domain response. Since $M(s)$ is a rational function, partial-fraction expansion is the usual method of expressing X_o in terms that can be inverse transformed.

The Laplace-transformed impulse function, when multiplied by $M(s)$, yields the s-domain transient response. Because it is such a difficult function to generate and observe, the step function is the dominant alternative. It is approximated in practice by a square wave with a period much longer than the duration of significant transient response (and thereby is effectively aperiodic). Various characteristics of circuit response to the step are of interest and all are time related. This approach to circuit characterization is *time-domain analysis.*

Transfer functions represent the complex dynamic behavior of circuits but are an abstraction of actual circuit behavior. The response of a circuit under controlled conditions produces features that characterize the circuit. We now investigate the characterization of circuits by their time-domain response to a unit step input, $u(t)$. The time-domain response can be determined by multiplying the transfer function by the Laplace transform of $u(t)$, or $1/s$, and inverse transforming the result.

The RC integrator response is calculated as

$$v_o(t) = \mathscr{L}^{-1}\left\{\frac{1}{sRC+1}\cdot\frac{1}{s}\right\} \tag{5.97}$$

The s-domain expression is partial-fraction expanded to

$$\frac{A}{s}+\frac{B}{sRC+1}=\frac{1}{s}-\frac{RC}{sRC+1} \tag{5.98}$$

This inverse transforms, using (5.71) and (5.72), to

$$v_o(t) = u(t) - e^{-t/RC} = 1 - e^{-t/RC}, \qquad t>0 \tag{5.99}$$

which is plotted against t/RC in Fig. 5.9, curve a.

For the RC differentiator,

$$v_o(t) = \mathscr{L}^{-1}\left\{\frac{RCs}{sRC+1}\cdot\frac{1}{s}\right\} = \mathscr{L}^{-1}\left\{\frac{1}{s+(1/RC)}\right\} = e^{-t/RC} \tag{5.100}$$

which is shown in 5.9, curve b, with time scaled in time constants.

The response of a circuit with complex poles is demonstrated by the RLC circuit of Fig. 5.10. Its transfer function can be written by treating it as a

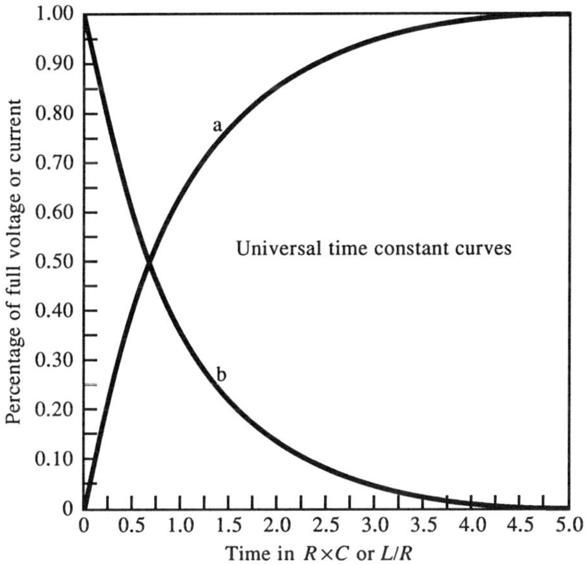

FIG. 5.9 First-order responses of (a) *RC* integrator and (b) *RC* differentiator. These curves apply to any first-order circuit and allow rapid determination of fractional decay versus time constant.

voltage divider. Then

$$\frac{V_o(s)}{V_i(s)} = \frac{(1/sC) \parallel R}{(1/sC) \parallel R + sL} = \frac{1}{LC} \cdot \frac{1}{s^2 + s(1/RC) + (1/LC)}$$

$$= \frac{1}{s^2 LC + s(L/R) + 1} \tag{5.101}$$

where

$$K = 1, \qquad \omega_n = \frac{1}{\sqrt{LC}}, \qquad \alpha = \frac{1}{2RC} \tag{5.102}$$

For $V_i(s) = 1/s$, the step response of the *RLC* circuit is

$$v_{step}(t) = \mathscr{L}^{-1}\left\{\frac{1}{s} \cdot \frac{1}{s^2 LC + s(L/R) + 1}\right\} \tag{5.103}$$

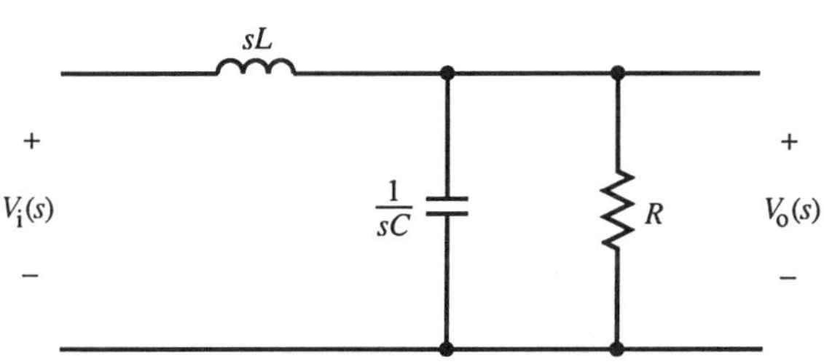

FIG. 5.10 *RLC* circuit with quadratic pole factor.

The quadratic factor is of the form

$$\frac{N(s)}{s^2+2\alpha s+\omega_n^2}=\frac{1}{\omega_n^2}\cdot\frac{N(s)}{(s^2/\omega_n^2)+(2\alpha/\omega_n^2)s+1} \tag{5.104}$$

The denominator can be factored into

$$\frac{N(s)}{(s+\alpha+j\omega_d)(s+\alpha-j\omega_d)}$$

$$=\frac{1}{\omega_n^2}\cdot\frac{N(s)}{[(s/\omega_n)+(\alpha/\omega_n)s+j(\omega_d/\omega_n)][(s/\omega_n)+(\alpha/\omega_n)s-j(\omega_d/\omega_n)]} \tag{5.105}$$

This can be expressed as a partial fraction expansion:

$$\frac{N(s)}{(s-p)(s-p^*)}=\frac{A^*}{(s-p)}+\frac{A}{(s-p^*)} \tag{5.106}$$

where X^* is the complex conjugate of X and

$$p=-\alpha+j\omega_d \tag{5.107}$$

This form can be shown to be valid by letting

$$\begin{cases} A=a+jb \\ N(s)=cs+d \end{cases}$$

This is the most general form N can take, with its degree one less than the denominator. [If (5.106) is a transfer function of a circuit with zero magnitude at infinite frequency, the fraction must be less than 1, or $m<n$ of (5.61).] Then the partial-fraction expansion coefficients are

$$a=\frac{c}{2},\qquad b=\frac{-\alpha c+d}{2\omega_d} \tag{5.108}$$

The time-domain response of the right side of (5.106) is found by making use of the polar form of A. Substituting $A=\|A\|\,e^{j\vartheta}$ and $A^*=\|A\|\,e^{-j\vartheta}$, we obtain

$$\mathscr{L}^{-1}\left\{\frac{A^*}{s-p}+\frac{A}{s-p^*}\right\}=A^*\,e^{pt}+A\,e^{p^*t} \tag{5.109a}$$

$$=\|A\|\,e^{-\alpha t}(e^{-j\vartheta}\,e^{j\omega_d t}+e^{j\vartheta}\,e^{-j\omega_d t})$$

$$=\|A\|\,e^{-\alpha t}(e^{j(\omega_d t-\vartheta)}+e^{-j(\omega_d t-\vartheta)})$$

$$=2\|A\|\,e^{-\alpha t}\cos(\omega_d t-\vartheta) \tag{5.109b}$$

Therefore, the general transform involving complex pole pairs is

$$\frac{\|A\|\,e^{j\vartheta}}{s+\alpha+j\omega_d}+\frac{\|A\|\,e^{-j\vartheta}}{s+\alpha-j\omega_d}\overset{\mathscr{L}^{-1}}{\Longrightarrow}2\|A\|\,e^{-\alpha t}\cos(\omega_d t-\vartheta) \tag{5.110}$$

Returning to (5.103), its partial fraction expansion is

$$\frac{1}{s}\cdot\frac{1}{s^2LC+s(L/R)+1}=\frac{1}{LC}\left\{\frac{A}{s}+\frac{B^*}{s-p}+\frac{B}{s-p^*}\right\} \tag{5.111}$$

where $p=-\alpha+j\omega_{\rm d}$. Solving for the numerators of (5.111) gives

$$A=\frac{1}{\omega_{\rm n}^2},\qquad B=-\frac{1}{2\omega_{\rm n}^2}\left[1+j\left(\frac{\alpha}{\omega_{\rm d}}\right)\right]=-\frac{1}{2\omega_{\rm n}^2}\cdot\frac{1}{\cos\gamma}\cdot e^{j\gamma},\qquad \gamma=\tan^{-1}\left\{\frac{\alpha}{\omega_{\rm d}}\right\} \tag{5.112}$$

Inverse transforming (5.111), using (5.71) and (5.110), gives

$$v_{\rm step}(t)=1-\frac{1}{\sin\gamma}\cdot e^{-\alpha t}\cos(\omega_{\rm d}t-\gamma),\qquad \gamma=\tan^{-1}\left\{\frac{\alpha}{\omega_{\rm d}}\right\} \tag{5.113}$$

With the trigonometric relation

$$\tan\vartheta=\frac{1}{\tan(90^\circ-\vartheta)}$$

(5.113) becomes

$$v_{\rm step}(t)=1-\frac{1}{\sin\phi}\cdot e^{-\alpha t}\sin(\omega_{\rm d}t+\phi),\qquad \phi=\tan^{-1}\left\{\frac{\omega_{\rm d}}{\alpha}\right\} \tag{5.114}$$

The factor $1/\sin\phi$ can be expressed as

$$\frac{1}{\sin\phi}=\sqrt{1+\left(\frac{\alpha}{\omega_{\rm d}}\right)^2}=[1-\zeta^2]^{-1/2} \tag{5.115}$$

In circuit element values, for the *RLC* circuit of Fig. 5.10,

$$\zeta=\frac{Z_{\rm n}}{2R}=\frac{\sqrt{L/C}}{2R} \tag{5.116}$$

An alternative approach to the inverse Laplace transformation of (5.104) is to complete the square for the quadratic denominator:

$$s^2+2\alpha s+\omega_{\rm n}^2=(s+\alpha)^2-(\alpha^2-\omega_{\rm n}^2)=(s+\alpha)^2+\omega_{\rm d}^2 \tag{5.117}$$

Then, for $N(s)=cs+d$, (5.104) becomes

$$\begin{aligned}\frac{N(s)}{s^2+2\alpha s+\omega_{\rm n}^2}&=\frac{cs}{(s+\alpha)^2+\omega_{\rm d}^2}+\frac{d}{(s+\alpha)^2+\omega_{\rm d}^2}\\&=c\cdot\frac{(s+\alpha)}{(s+\alpha)^2+\omega_{\rm d}^2}-\left(\frac{\alpha c}{\omega_{\rm d}}\right)\frac{\omega_{\rm d}}{(s+\alpha)^2+\omega_{\rm d}^2}+\left(\frac{d}{\omega_{\rm d}}\right)\frac{\omega_{\rm d}}{(s+\alpha)^2+\omega_{\rm d}^2}\end{aligned} \tag{5.118}$$

$$\frac{N(s)}{s^2+2\alpha s+\omega_{\rm n}^2}=c\cdot\frac{(s+\alpha)}{(s+\alpha)^2+\omega_{\rm d}^2}+\left(\frac{d-\alpha c}{\omega_{\rm d}}\right)\frac{\omega_{\rm d}}{(s+\alpha)^2+\omega_{\rm d}^2} \tag{5.119}$$

Using (5.75) and a similar extension of (5.74), we obtain

$$\mathscr{L}^{-1}(5.119) = c\,e^{-\alpha t}\cos\omega_d t + \left(\frac{d-\alpha c}{\omega_d}\right)e^{-\alpha t}\sin\omega_d t \tag{5.120}$$

For the sum of a sine and cosine,

$$a\cos\vartheta + b\sin\vartheta = \sqrt{a^2+b^2}\,\sin\left[\vartheta + \tan^{-1}\left(\frac{a}{b}\right)\right] \tag{5.121}$$

Equation (5.121) can be applied to (5.120) to express it as a single sinusoid. After some manipulation,

$$\frac{cs+d}{s^2+2\alpha s+\omega_n^2} \overset{\mathscr{L}^{-1}}{\Longrightarrow} \frac{c}{\sin\phi}\cdot e^{-\alpha t}\sin(\omega_d t-\phi), \qquad \phi = \tan^{-1}\left(\frac{c\omega_d}{\alpha c-d}\right) \tag{5.122}$$

If we apply this method to (5.103), the partial-fraction expansion is

$$\frac{1}{s} - \frac{s+2\alpha}{s^2+2\alpha s+\omega_n^2} \tag{5.123}$$

From this expression, $c = 1$ and $d = 2\alpha$. Substituting into (5.122) for the quadratic term yields

$$\mathscr{L}^{-1}(5.123) = 1 - \frac{1}{\sin\phi}\cdot e^{-\alpha t}\sin(\omega_d t+\phi), \qquad \phi = \tan^{-1}\left\{\frac{\omega_d}{\alpha}\right\} \tag{5.124}$$

This is the same result as (5.114).

For the case of repeated real poles (critical damping),

$$v_{\text{step}}(t) = 1-(1+\alpha t)\,e^{-\alpha t} \tag{5.125}$$

and for distinct real poles, since $\zeta > 1$, ω_d is imaginary and

$$p_{1,2} = -\alpha \mp \omega_d = -\alpha \mp \omega_n\sqrt{\zeta^2-1} \tag{5.126}$$

These are real roots. The step response for them is

$$v_{\text{step}}(t) = 1 - \left(\frac{p_1}{p_1-p_2}\,e^{p_2 t} - \frac{p_2}{p_1-p_2}\,e^{p_1 t}\right) \tag{5.127}$$

Example 5.6 Magnetic Deflection Yoke Coil Circuit

Figure E5.6 shows a simplified CRT deflection circuit. The deflection yoke consists of horizontal and vertical deflection coils that magnetically deflect the CRT electron beam. A yoke coil has significant series resistance and intrawinding capacitance, modeled as shown. If $i_i(t)$ is a ramp function (producing a horizontal or vertical sweep needed for raster scanning of the CRT screen by the electron beam), then it can be expressed

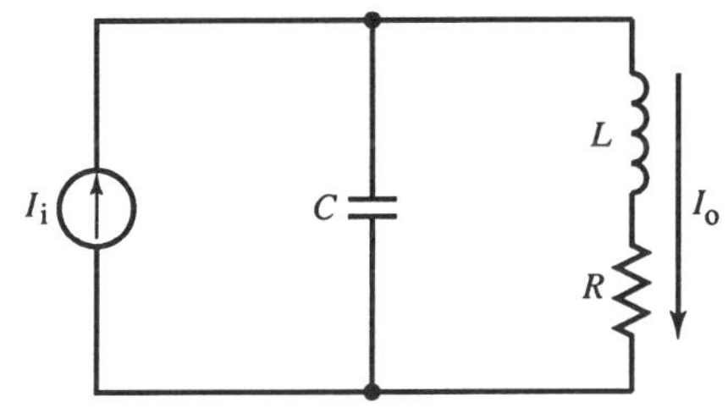

FIG. E5.6

as

$$i_i(t) = \left(\frac{I}{T}\right) t = mt$$

where I is the peak ramp current and T the ramp duration (or the period of an ideal sawtooth function). The output current $i_o(t)$ is the current that flows through L, creating the deflection field. Our goal is to find a general expression in s for $I_o(s)$ and also to find the time-domain response.

The current divider formula is used here and yields

$$\frac{I_o(s)}{I_i(s)} = \frac{1}{s^2LC + sRC + 1}$$

For $I_i(s) = m/s^2$,

$$I_o(s) = \frac{1}{s^2LC + sRC + 1} \cdot \frac{m}{s^2}$$

and $\omega_n^2 = 1/LC$, $\alpha = \zeta\omega_n = 1/(2L/R)$. I_o can be written by completing the square of the quadratic pole and expressing I_o as a partial-fraction expansion. Because of multiple roots at $s = 0$, it is necessary to take the derivative of the partial-fraction equation to find the coefficient k for the k/s term. Then,

$$I_o(s) = \frac{(2\zeta/\omega_n) \cdot s + m(4\zeta^2 - 1)}{(s+\alpha)^2 + \omega_d^2} + \frac{m}{s^2} - \frac{2\zeta m/\omega_n}{s}$$

Using (5.122) to perform $\mathscr{L}^{-1}$ on the first term, we find that

$$i_o(t) = \left(\frac{2\zeta}{\omega_n \sin\phi}\right) e^{-\alpha t} \sin(\omega_d t - \phi) + m\left(t - \frac{2\zeta}{\omega_n}\right)$$

$i_o(t)$ is a ramp delayed by $2\zeta/\omega_n$. Superimposed on this ramp is a decaying sinusoid, the first term. When the response of the horizontal deflector is too underdamped, the resulting ringing causes the picture on the left side of the CRT screen to show an alternating compaction and expansion until the sinusoid dies out.

5.8 Circuit Characterization in the Time Domain

With the derivations of the previous section, we have both first- and second-order responses to a unit step function. Higher-order responses are combinations of first- and second-order responses. Most circuits can be separated into lower-order circuits and analyzed individually. Transfer function numerators and denominators can be factored into first- and second-order factors that are separated by partial-fraction expansion.

In Section 5.4, the effect of ζ on the response was examined. For a step response, we are interested in the amount above the step amplitude that the response reaches for complex poles. For accurate reproduction of a step, this *overshoot* should be minimal. We also want to avoid the other extreme of a highly overdamped response. The larger ζ becomes, the longer it takes for the response to approach its asymptotic value. In other words, the *risetime* (or for a negative step, *falltime*) is excessive. An obvious compromise is at critical damping, when $\zeta = 1$. This value of ζ is seldom chosen for wideband amplifiers because a much faster step can be achieved for a small amount of overshoot.

When the step response overshoots, its peak occurs at time t_p. This time is derived by taking the derivative of (5.114), setting it to zero, and solving for t. The derivative is

$$\frac{dv_{\text{step}}(t)}{dt} = \frac{\alpha}{\sin \phi} \cdot e^{-\alpha t} \sin(\omega_d t + \phi) - \omega_n e^{-\alpha t} \cos(\omega_d t + \phi) \tag{5.128}$$

This reduces to

$$\tan(\omega_d t + \phi) = \tan \phi$$

which requires that

$$\omega_d t = k\pi, \qquad k = 0, 1, 2, \ldots$$

The peak occurs at $k = 1$ for a damped response and is

$$t_p = \frac{\pi}{\omega_d} \tag{5.129}$$

The peak at this time is, from (5.114),

$$v_{\text{step}}(t_p) = 1 + e^{-\pi/\tan \phi} \tag{5.130}$$

Since the input is a unit step, (5.130) is the fractional peak. The *overshoot* is defined as

$$M_p = e^{-\pi/\tan \phi} \tag{5.131}$$

M_p is related to ζ through $\tan \phi$ and (5.35), or

$$\tan \phi = \frac{\sqrt{1-\zeta^2}}{\zeta} \tag{5.132}$$

Overshoots for several angles are tabulated:

ϕ	M_p (%)	ζ
0° (0 rad)	0	1
30° $\left(\frac{\pi}{6}\text{ rad}\right)$	0.433	0.866
45° $\left(\frac{\pi}{4}\text{ rad}\right)$	4.321	0.707
60° $\left(\frac{\pi}{3}\text{ rad}\right)$	16.30	0.500

To find a compromise optimum between overshoot and risetime, we need to define risetime. Furthermore, a third criterion is the time it takes the step to settle to its steady-state value. This is the *settling time* t_s; its definition depends on the amount of settling that is adequate for the application. If we define M_s to be the fractional settling amplitude, then

$$M_s \equiv e^{-\alpha t_s} \tag{5.133}$$

Solving for the settling time, we obtain

$$t_s = -\frac{1}{\alpha} \cdot \ln M_s = -\frac{\lg M_s}{\alpha \lg e} = -\frac{\ln 2}{\alpha} \cdot \lg M_s \cong \frac{0.7}{\alpha} \cdot \lg\left(\frac{1}{M_s}\right) \tag{5.134}$$

where $\lg \equiv \log_2$ and $\lg(1/M_s)$ is the number of bits of resolution of settling. For a second-order response, $\alpha = 1/(2\tau)$ and t_s further reduces to

$$t_s \cong 1.4 \cdot \tau \cdot \lg\left(\frac{1}{M_s}\right) \tag{5.135}$$

A few values of t_s are given in the following table:

$t_s(\tau)$	$\lg(1/M_s)$ (bits)
5.5	4
8.3	6
11.1	8
13.9	10
16.6	12
19.4	14

When we observe an underdamped step, we can make an estimate of ϕ or ζ based on the number of cycles of oscillation N_s until the waveform settles. The oscillation frequency is ω_d. Then,

$$N_s = \frac{\omega_d t_s}{2\pi} = \frac{\omega_d}{2\pi}\left(-\frac{1}{\alpha} \ln M_s\right) = \ln 2 \cdot \frac{\tan\phi}{2\pi} \cdot \lg\left(\frac{1}{M_s}\right)$$

$$\cong (0.11) \tan\phi \cdot \lg\left(\frac{1}{M_s}\right) \tag{5.136}$$

Some values of N_s for 8 bits of resolution (approximately the resolution of viewing an oscilloscope trace) are the following:

N_s	ϕ	ζ
0	0°	1
0.5	30°	0.866
0.9	45°	0.707
1.5	60°	0.500
3.3	75°	0.259
8.8	84°	0.100

Finally, the risetime t_r could be defined as settling time or as a given number of time constants. The most commonly used definition is the time it takes the step to change from 10 to 90% of its final value. This definition is general in that it does not assume a particular kind of response. For a first-order system, from (5.99), we have

$$t_r \equiv t_{90\%} - t_{10\%} = -\tau \ln(1-0.9) + \tau \ln(1-0.1) = \tau \ln(9) \cong 2.2\tau \quad (5.137)$$

This risetime formula holds approximately for complex poles with small pole angles.

5.9 The *s*-Plane Frequency Response of Transfer Functions

An alternative to characterization of circuits by their response to a step input is to use a sinusoid (or "sinewave"). Unlike the step, this is periodic and characterized by amplitude and phase, neither of which is time dependent. By exciting the circuit input with sinewaves over a range (or *band*) of frequencies, we can determine the amplitude and phase as a function of frequency. This is the *frequency response* of the circuit; it is steady-state sinusoidal response. In practice, approximation of this procedure is accomplished by sweeping slowly enough through a band of frequencies to let the transient response die out. This approach to circuit characterization is *frequency-domain analysis*.

Time- and frequency-domain analyses reveal different aspects of the complex-frequency domain. A cross section of the transfer function $M(s)$ along the $j\omega$ axis is a function of ω only and is the frequency response $M(j\omega)$. The magnitude and phase of M can be found for a particular $j\omega_1$ by substituting into $M(j\omega)$. The magnitude and angle of the resulting complex number is the amplitude and phase of the frequency response at ω_1. This can be done graphically on the s-plane. (Zeros are marked on s-plane plots by an open circle.) $M(j\omega_1)$ can be calculated from the graph by first finding the length and angle of each vector. Then,

$$\|M(j\omega_1)\| = K \cdot \frac{\prod_{\text{zeros}} (\text{zero vector magnitudes})}{\prod_{\text{poles}} (\text{pole vector magnitudes})} \tag{5.138a}$$

$$\angle M(j\omega_1) = \sum_{\text{zeros}} (\text{zero vector angles}) - \sum_{\text{poles}} (\text{pole vector angles}) \tag{5.138b}$$

For example, Fig. 5.11 shows an $M(s)$ with three poles and one zero. If $K = 100$, then the frequency response at $\omega = 1$ is

$$\|M(j1)\| = (100) \cdot \frac{(2.24)}{(6.08)(4.12)(5.00)} = 1.78$$

Each of the numbers in the fraction is the contribution of a critical frequency. For example, the contribution of p_2 is

$$\|[j1 - (-4 + j2)]\| = \|4 - j1\| = \sqrt{(4)^2 + (-1)^2} \cong 4.12$$

The phase angle of M at $j1$ is

$$\angle M(j1) = 26.57° - (9.46° - 14.04° + 36.87°) = -5.73°$$

where, for example, the angle of p_1 is

$$\angle(j1 - (-6)) = \angle(6 + j1) = \tan^{-1}\left\{\frac{1}{6}\right\} \cong 9.46°$$

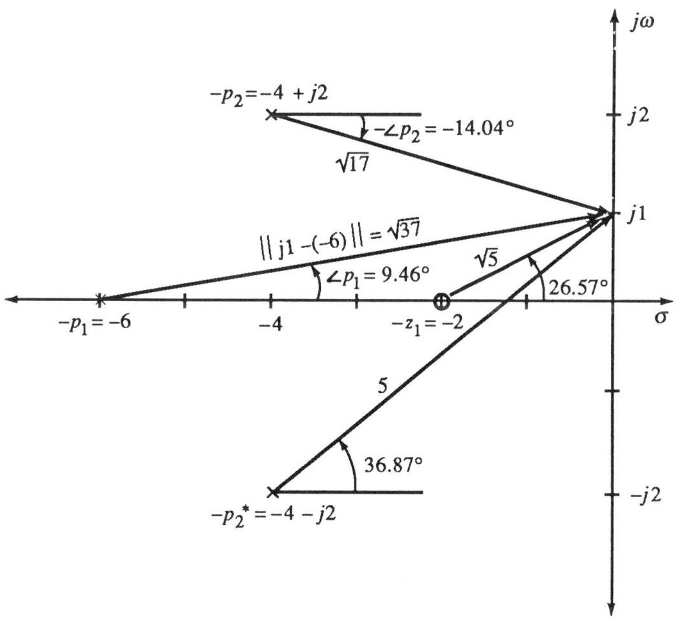

FIG. 5.11 Transfer function magnitude and phase can be calculated directly from the s-plane for a given input frequency, $j\omega_d$ (here $j1$). Magnitude is the product of zero vector lengths divided by the product of pole vector lengths. Phase is the sum of zero angles minus the sum of pole angles.

5.10 Graphical Representation of Frequency Response

Several other graphical methods of frequency response representation are found in circuits and control systems literature. Some of the more common are the following:

1. Bode plot. Two graphs, one of amplitude and the other of phase, against frequency. The amplitude graph is of log–log scale; the phase-angle graph is semilog (log frequency).
2. Reactance chart. Similar to a Bode plot; a log–log plot of impedance magnitude, on which divider-type transfer functions can be constructed directly.
3. Nyquist diagram (complex polar plot). Graph of $M(j\omega)$ with frequency as a parameter, on the complex plane with axes $j\mathcal{Im}\{M(j\omega)\}$ and $\mathcal{Re}\{M(j\omega)\}$. For feedback systems the loop gain GH is plotted.
4. Root-locus plot. Feedback amplifier plot on the s-plane of the movement of the closed-loop poles with dc gain K.
5. Nichols chart. Plot on rectangular coordinates of magnitude versus phase (the *gain-phase plane*) or open-loop response of a feedback amplifier, with superimposed contours of constant magnitude and phase of $M(j\omega)$ of the resulting closed-loop magnitude and phase [for $M(j\omega)=G/(1+G)$].
6. Hall chart. Complex plot of $\mathcal{Im}\{G(j\omega)\}$ versus $\mathcal{Re}\{G(j\omega)\}$ for feedback system with superimposed contours of constant $\|M(j\omega)\|$ loci [for $M = G/(1+G)$].

We will survey the construction of Bode plots and later make extensive use of them. For a transfer function in normalized form, the rational expression in s is unity at $s=0$, so the constant K must be the dc transmittance. A transfer function $M(s)$ having one real pole p, evaluated at $j\omega$, is

$$\frac{1}{(j\omega/p+1)}=\frac{1}{\sqrt{(\omega/p)^2+1}}\cdot e^{-\tan^{-1}(\omega/p)} \tag{5.139}$$

The log–log plot of the magnitude of a real pole is

$$\log\left\{\frac{1}{\sqrt{(\omega/p)^2+1}}\right\}=-\frac{1}{2}\log\left[\left(\frac{\omega}{p}\right)^2+1\right] \tag{5.140}$$

For $\omega/p \ll 1$, (5.140) becomes

$$-\frac{1}{2}\log\left[\left(\frac{\omega}{p}\right)^2+1\right]\Bigg|_{\omega/p\ll 1}\cong \log(1)=0 \tag{5.141}$$

For $\omega/p \gg 1$,

$$-\frac{1}{2}\log\left[\left(\frac{\omega}{p}\right)^2+1\right]\Bigg|_{\omega/p\gg 1}\cong -(\log\omega-\log p)\cong -\log\omega \tag{5.142}$$

For $\omega \gg p$, $\log \omega - \log p \cong \log \omega$. Equations (5.141) and (5.142) are the piecewise linear *asymptotic approximations* of the exact pole magnitude. An *ideal Bode plot* can easily be constructed from these straight-line approximations, as shown in Fig. 5.12. The graph is flat until it reaches a frequency of p, where the slope changes to -1 (in log-log coordinates). Hence, p, the frequency of the pole, is called the *corner* or *break frequency*.

The relationship between the real part of a pole or zero and Bode plot break frequencies is that the break frequency corresponds to $-1/\sigma$ for negative poles or zeros. The convention is used here of indicating pole location on Bode plots as p instead of $-p$. Since Bode plot frequencies are always positive, no confusion should result.

Often, magnitude is scaled in decibels (dB). For voltage, $\|A\|_{dB} \equiv 20 \log_{10}\|A\|$. (Notice that decibels, like radians, is a pseudounit, a scaling transformation.) A real pole "rolls off" (that is, decreases in magnitude) with a slope of -1 on a log-log plot or -20 dB/dec $\cong -6$ dB/oct. (A decade (dec) is a 10 to 1 frequency range; an octave (oct) is a 2 to 1 range.)

The error in the asymptotic approximation is greatest at the break frequency, where the actual curve is at $\sqrt{2}/2 \cong 0.707$ or -3 dB.

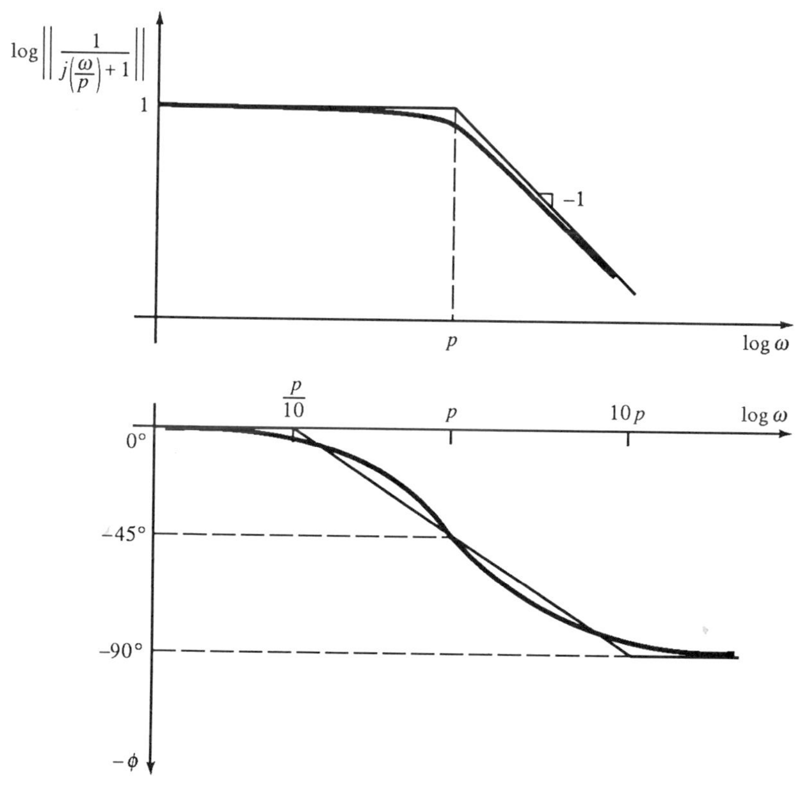

FIG. 5.12 Bode plot asymptotic approximations to real pole p, magnitude and phase.

The phase-angle plot is also subject to asymptotic approximation. On a semilog plot, the exact phase is

$$\phi = -\tan^{-1}\left(\frac{\omega}{p}\right) \tag{5.143}$$

For $\omega/p \ll 1$,

$$-\tan^{-1}\left(\frac{\omega}{p}\right)\bigg|_{\omega/p \ll 1} \cong 0 \tag{5.144}$$

and for $\omega/p \gg 1$,

$$-\tan^{-1}\left(\frac{\omega}{p}\right)\bigg|_{\omega/p \gg 1} \cong \tan^{-1}\{\infty\} = -\frac{\pi}{2} = -90^\circ \tag{5.145}$$

At $\omega = p$, $\phi = -\pi/4 = -45^\circ$. The slope of ϕ at p is

$$\frac{d}{d(\log \omega)}\left\{-\tan^{-1}\left(\frac{\omega}{p}\right)\right\}\bigg|_{\omega=p} = \frac{-1}{1+(\omega/p)^2} \cdot \ln(10) \cdot \frac{\omega}{p}\bigg|_{\omega=p} = -\frac{\ln 10}{2} \tag{5.146}$$

A line tangent to ϕ at p intercepts the asymptotes at 0° and −90° ($-\pi/2$) at frequencies logarithmically symmetric about p, at ap and p/a:

$$\frac{-\pi/2 - 0}{\log(ap) - \log(p/a)} = -\frac{\pi}{2} \cdot \ln(10) \tag{5.147}$$

Then $a = e^{\pi/2} \cong 4.81$.[3] This linear approximation to ϕ does not minimize the maximum error. Instead, when $a = 10$, the maximum error is less than 6°. The phase plot for a single pole is shown in Fig. 5.12.

Linear approximations of Bode plots for other cases are shown in Fig. 5.13. Since frequency response analysis is linear, these elemental plots can be combined linearly to produce the total response plot. For complex critical frequencies, decreasing ζ increases the magnitude peak and the slope of the phase near the break frequency.

The maximum magnitude, M_m, for underdamped response occurs at frequency ω_m. This is derived by setting the derivative of $M(j\omega)$ to zero and solving. For a quadratic pole factor,

$$\frac{d}{d\omega}\left\|\frac{1}{[j(\omega/\omega_n)]^2 + j2\zeta(\omega/\omega_n) + 1}\right\| = \frac{d}{d\omega}\left(\frac{1}{\sqrt{[1-(\omega/\omega_n)^2]^2 + (2\zeta\omega/\omega_n)^2}}\right)$$

When set to zero, ω is

$$\omega_m = \omega_n\sqrt{1-2\zeta^2}, \qquad \zeta < \frac{\sqrt{2}}{2} \tag{5.148}$$

and

$$M_m = \frac{1}{2\zeta\sqrt{1-\zeta^2}}, \qquad \zeta < \frac{\sqrt{2}}{2} \tag{5.149}$$

3. Bruce Hofer has observed that the frequency ratio, $e^{\pi/2} = j^{-j}$.

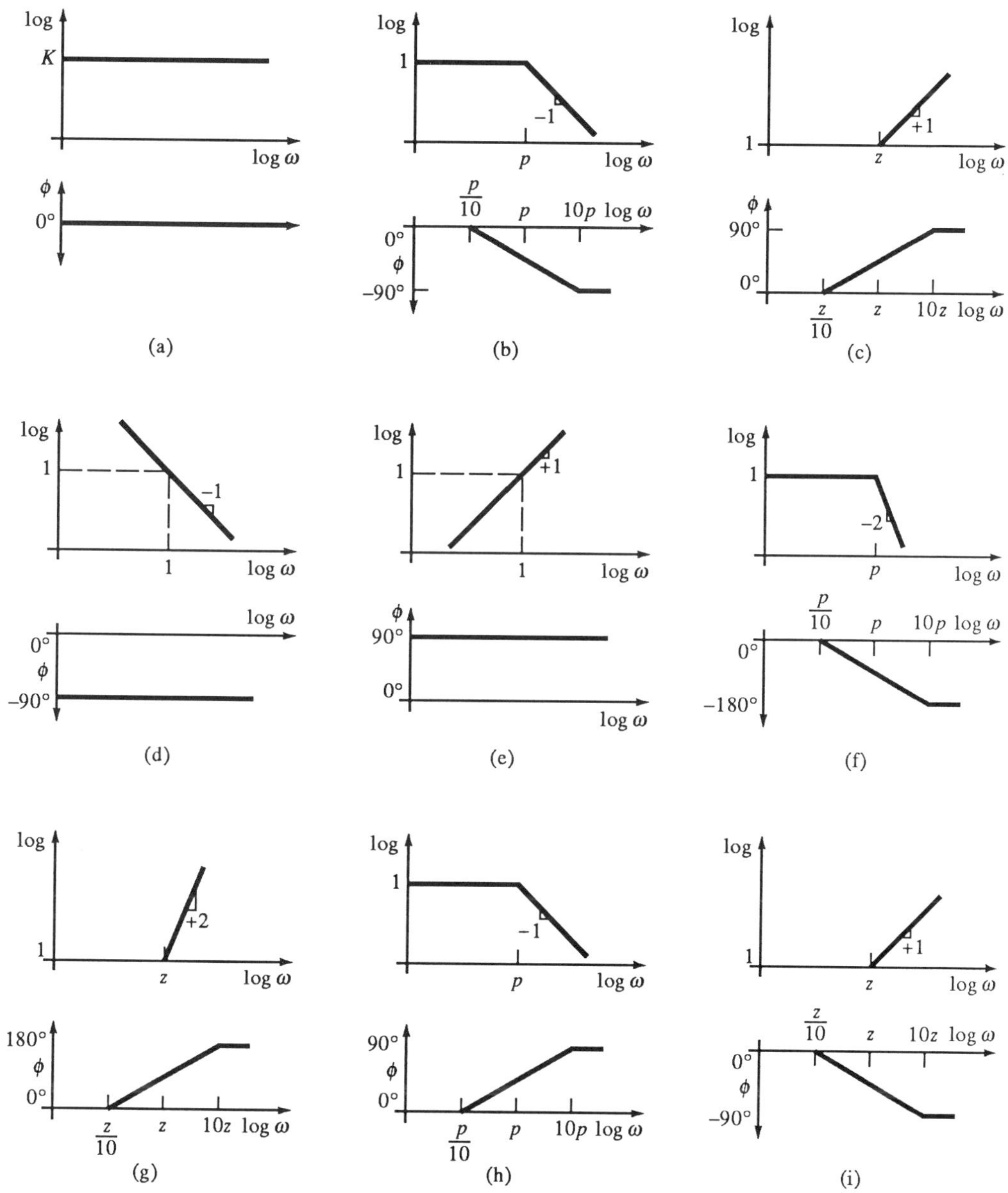

FIG. 5.13 Bode plot elements for (a) constant K, (b) real left half-plane pole p, (c) real left half-plane zero z, (d) pole at origin, (e) zero at origin, (f) complex left half-plane pole pair, (g) complex left half-plane zero pair, (h) real right half-plane pole p, and (i) real right half-plane zero z.

For $\zeta \ll 1$, M_m varies inversely with ζ. Construction of an approximate Bode plot for $\zeta < 1$ is aided by a few points around the break frequency. Besides the peak at ω_m, the magnitude at ω_n is

$$M(j\omega_n) = 2\zeta \tag{5.150}$$

M crosses unity at

$$\omega|_{M=1} = \sqrt{2} \cdot \omega_m \tag{5.151}$$

The frequency at which M is $1/\sqrt{2}$ (or -3 dB) down from unity is defined as the *bandwidth*. This definition, like that for risetime, does not assume a particular kind of response. The bandwidth of a single pole M is its break frequency. For a quadratic pole, we set $\|M\| = 1/\sqrt{2}$ and solve for ω. This simplifies to

$$\left[1 - \left(\frac{\omega}{\omega_n}\right)^2\right]^2 + \left(\frac{2\zeta\omega}{\omega_n}\right)^2 = 2 \tag{5.152}$$

and solving for ω (the larger root) yields

$$\omega_{bw} = \omega_n \sqrt{1 - 2\zeta^2 + \sqrt{4\zeta^4 - 4\zeta^2 + 2}} \tag{5.153}$$

Example 5.7 Phase-Lag Circuit with Capacitive Output Loading

Figure E5.7 is another passive RC circuit; it is the phase-lag circuit of Fig. 5.19a with capacitive output loading. It was constructed of 1% tolerance metal film resistors; 1 nF, 1% mica and 0.1 μF, 2% plastic film capacitors. A 6 V peak-peak sinewave for V_i was applied at various frequencies, and the corresponding peak-peak voltages at V_o were recorded as follows:

frequency	$\|V_o\|$	$\|V_o/V_i\|$	frequency	$\|V_o\|$	$\|V_o/V_i\|$
10 Hz	6.00 V	1.00	5.0 kHz	0.20 V	0.033
20	6.00	1.00	10	0.11	0.018
50	5.95	0.99	20	0.075	0.013
100	5.10	0.85	50	0.064	0.011
200	3.74	0.62	100	0.064	0.011
500	1.80	0.30	200	0.064	0.011
1.0 kHz	0.96	0.16	500	0.060	0.010
2.0	0.47	0.078	1.0 MHz	0.050	0.008
			2.0	0.035	0.006

From these data, asymptotic approximation of pole and zero frequencies results in

$$p_1 = 150 \text{ Hz}, \qquad z = 15.5 \text{ kHz}, \qquad p_2 = 1.6 \text{ MHz}$$

The transfer function is

$$\frac{V_o(s)}{V_i(s)} = \frac{sR_2C_2 + 1}{s^2[R_1C_1R_2C_2] + s[R_1(C_1 + C_2) + R_2C_2] + 1}$$

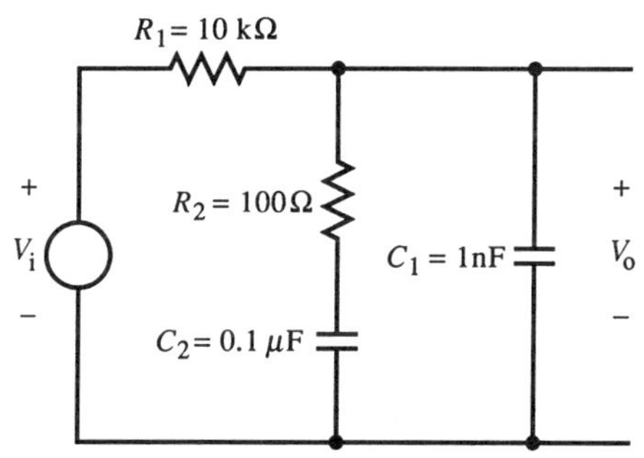

FIG. E5.7

When the circuit values are substituted, the exact theoretical poles and zero are

$$p_1 = 156\ \text{Hz}, \qquad z = 15.9\ \text{kHz}, \qquad p_2 = 1.62\ \text{MHz}$$

The error is largely a matter of parts tolerances, accurate asymptote plotting, and chart reading.

5.11 Loci of Quadratic Poles

The quadratic equation

$$as^2 + bs + 1 = 0 \tag{5.154}$$

has complex roots at

$$s = -\frac{b}{2a} \pm j\sqrt{\frac{1}{a} - \left(\frac{b}{2a}\right)^2} \tag{5.155}$$

For complex poles,

$$\sigma = -\left(\frac{b}{2a}\right) \tag{5.156a}$$

$$\omega^2 = \frac{1}{a} - \left(\frac{b}{2a}\right)^2 \tag{5.156b}$$

and

$$\alpha = \frac{b}{2a}, \qquad \omega_n = \frac{1}{\sqrt{a}}, \qquad \zeta = \frac{b}{2\sqrt{a}} \tag{5.157}$$

The coefficients a and b are composed of circuit values when (5.154) is the characteristic equation of a transfer function. Often, a circuit element value appears in only one coefficient. This allows control of the poles by varying the value of the element until desired pole placement is achieved. We now consider three cases in which constraints are placed on the loci of the poles in the s-plane.

5.11.1 a = constant, b is parameter

Let b be the parameter. Solving for b in (5.156a) gives

$$b = -2a\sigma$$

and substituting into (5.156b) gives

$$\omega^2 + \sigma^2 = \frac{1}{a} \tag{5.158}$$

This equation describes the locus of the poles for constant a as a circle with radius $\sqrt{1/a}$ and centered at the origin (Fig. 5.14).

5.11.2 b = constant, a is parameter

From (5.156a), $1/a = -2\sigma/b$. Substituting into (5.156b) gives

$$\omega^2 = -\frac{2\sigma}{b} - \left(\frac{b}{2}\right)^2\left(-\frac{2\sigma}{b}\right)^2 \Rightarrow \omega^2 + \left(\sigma^2 + \frac{2}{b}\sigma\right) = 0$$

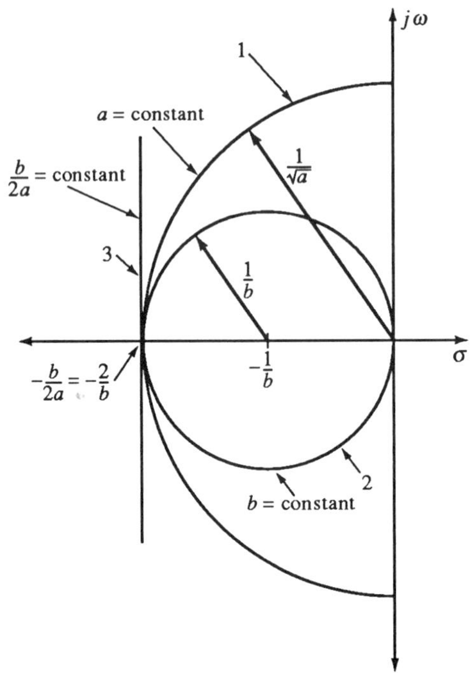

FIG. 5.14 Loci of $as^2 + bs + 1$ for (1) constant a, b parameter; (2) constant b, a parameter; and (3) $b/2a$ constant.

Adding $(1/b)^2$ to both sides and completing the square, we obtain

$$\omega^2 + \left(\sigma + \frac{1}{b}\right)^2 = \left(\frac{1}{b}\right)^2 \tag{5.159}$$

This equation describes the locus as a circle centered at $-1/b$ with a radius of $1/b$ (Fig. 5.14).

5.11.3 $b/2a$ = constant

Since $-\sigma = b/2a$, the locus is a vertical line at

$$\sigma = -\frac{b}{2a} \tag{5.160}$$

This locus is also shown in Fig. 5.14.

By including the root locus for real roots, we plot the total loci (Fig. 5.15). The + and − marks on the loci associate a locus with the positive or negative second term in (5.155). With these precalculated loci, we can predict pole movement based on variation of circuit element values.

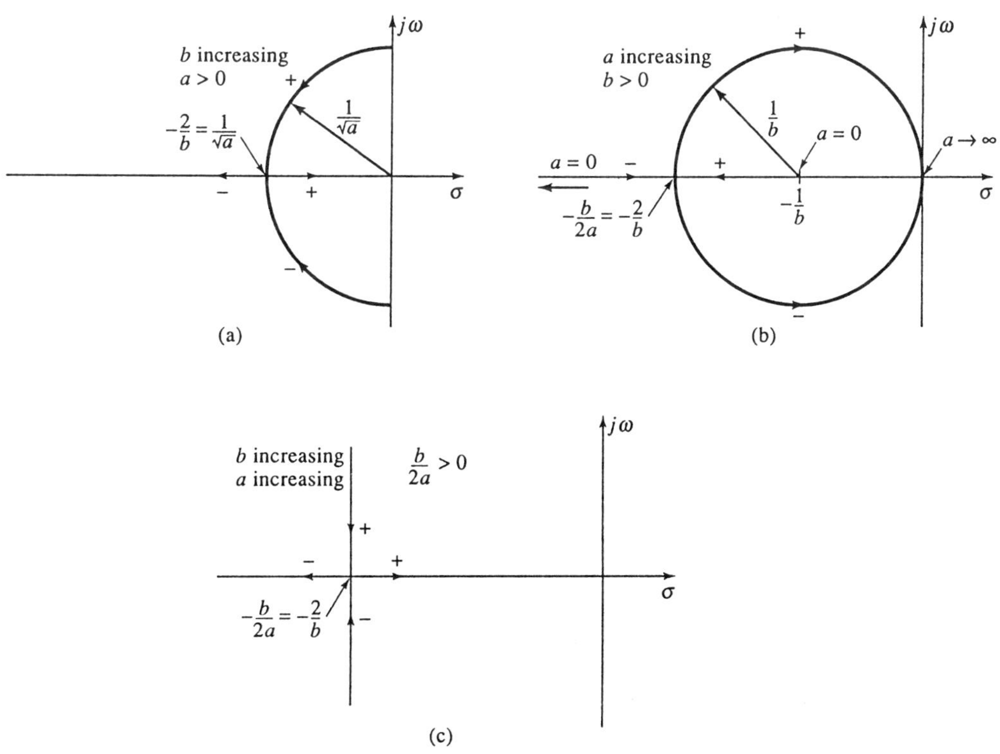

FIG. 5.15 Real and complex loci for (a) constant a, (b) constant b, and constant $b/2a$.

5.12 Optimization of Time-Domain and Frequency-Domain Response

For accurate step response, the major criteria are minimum risetime and overshoot. For wideband amplifiers, the usual criterion of performance in the frequency domain is constant magnitude (or *flat response*) out to a maximum bandwidth. A wider bandwidth can be achieved at the expense of greater M_m (or *peaking*). In the time domain, a faster risetime is achievable if more overshoot is allowed. As ζ and t_r decrease, ω_{bw}, M_m, and M_p all increase.

Optimization of time and frequency response requires identification of relationships between the two domains. For an amplifier with one pole (or with a dominant pole approximation), the relationship between risetime and bandwidth is

$$t_r \cong 2.2\tau = \frac{2.2}{2\pi f_{bw}} \cong \frac{0.35}{f_{bw}} \tag{5.161}$$

This relationship is approximate for complex poles with $\zeta \cong 0.7$.

Of particular interest are the pole placements for ϕ of 30°, 45°, and 60°. A complex pole pair at 45° gives a maximally flat amplitude (MFA) response over frequency (also called a *Butterworth response*). For $\phi = 30°$, the response has maximally flat envelope or group delay (MFED) (or *Bessel* or *Thompson response*). The major characteristics of these responses are as follows:

characteristic	critical damping	MFED	MFA	
ϕ	0°	30°	45°	60°
ζ	1.000	0.866	0.707	0.500
$\left(\frac{\omega_d}{\omega_n}\right)$	0	0.500	0.707	0.866
$\left(\frac{\omega_{bw}}{\omega_n}\right)$	0.644	0.786	1.000	1.272
M_m	—	—	1.000	1.155
$\left(\frac{\omega_m}{\omega_n}\right)$	—	—	0.000	0.707
$t_r \cdot \omega_n$	$t_r\alpha = 3.36$	2.73	2.15	1.64
M_p	0%	0.433%	4.32%	16.3%

From (5.149), $M_m = 1$ when $\zeta = \sqrt{2}/2 \cong 0.707$. At this value of ζ, M_m is at the onset of peaking.

Phase delay is defined as

$$\text{phase delay,} \qquad \tau_p \equiv -\frac{\phi}{\omega} \tag{5.162}$$

This is the delay time of a sinusoid at frequency ω with a phase lag of ϕ. If phase angle decreases linearly with frequency, each frequency component of a signal maintains its alignment in time with the others and no waveform distortion occurs. A related quantity, *envelope* or *group delay*, is defined as

$$\tau_g \equiv -\frac{d\phi}{d\omega} \tag{5.163}$$

Group delay characterizes amplitude distortion in the time domain. If all frequency components of a signal are delayed the same amount, they remain aligned in time, and the waveshape remains unchanged. If not, components of different frequencies are shifted in time, resulting in waveform distortion. A pole angle of 30° results in a second-order Bessel response, with maximally flat group delay. For a quadratic pole factor, the phase is

$$\phi = -\tan^{-1}\left\{\frac{2\zeta(\omega/\omega_n)}{1-(\omega/\omega_n)^2}\right\} \tag{5.164}$$

From (5.163), the group delay is

$$\text{quadratic pole group delay} = \left(\frac{2\zeta}{\omega_n}\right)\frac{(\omega/\omega_n)^2+1}{(\omega/\omega_n)^4+2[2\zeta^2-1](\omega/\omega_n)^2+1} \tag{5.165}$$

where the denominator can be factored into poles of $(\omega/\omega_n)^2$ located at

$$1-2\zeta^2 \pm 2\zeta\sqrt{\zeta^2-1}$$

We can find ζ for the MFED as we found M_m in Section 5.10. To find the maximum τ_g, set $d\tau_g/d\omega = 0$ and solve for ω. It is more convenient to find $(\omega/\omega_n)^2$ after the derivative is taken, and it is

$$\max \tau_g \text{ at } \left(\frac{\omega}{\omega_n}\right)^2 = -1 \pm 2\sqrt{1-\zeta^2}, \qquad 0 \le \zeta \le 1 \tag{5.166}$$

Substituting this into (5.165), we get the maximum τ_g:

$$\max \tau_g = \left(\frac{2\zeta}{\omega_n}\right)\frac{\pm\sqrt{1-\zeta^2}}{4(1+\zeta^2)[1\mp\sqrt{1-\zeta^2})} \tag{5.167}$$

For MFED response, the maximum τ_g must equal τ_g at $\omega = 0$, or $2\zeta/\omega_n$. Setting (5.167) equal to this and solving for ζ, we get $\zeta = \sqrt{3}/2$ and a 30° pole angle.

This subject has been developed into the electronic circuits specialty of *filter circuits.* Higher-order responses are often characterized according to optimal parameters. Butterworth filters have maximally flat amplitude response; Bessel filters have maximally flat group delay; Chebyshev filters optimize the trade off between amplitude ripple and sharp amplitude roll-off (or cutoff) with no ripple in the frequency response above the cutoff frequency; and elliptic (or Cauer) filters have the maximum (or "sharpest") cutoff for a given order of filter but have ripple above the cutoff frequency.

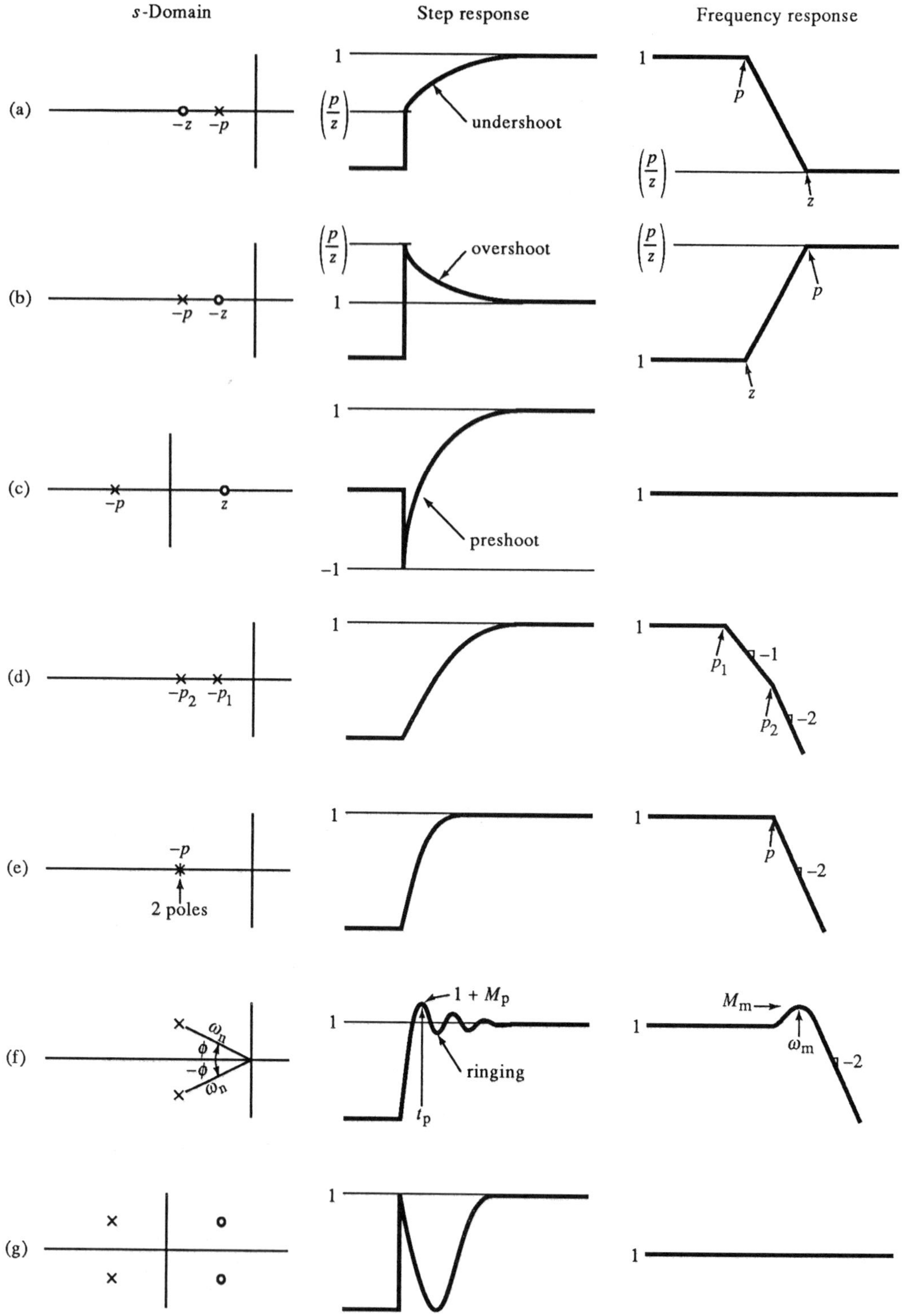

FIG. 5.16 Relationship of s-domain, step response, and frequency response representations as commonly encountered. Cases (c) and (g) are those of all-pass circuits.

The responses of common transfer functions are shown in Fig. 5.16 in three representations: s-domain pole-zero locations, time-domain step response, and frequency response. Right half-plane zeros cause a preshoot in the step response. The frequency response in Fig. 5.16g is independent of frequency and is an instance of an *all-pass* filter. It is not a Bessel filter, however, because the distorted step response indicates much phase distortion. Amplifier designs are often a trade off between conflicting transient and frequency response performance.

Example 5.8 Parallel Resonant Circuit

The *RLC* circuit of Fig. E5.8 provides a way of generating the response of a quadratic pole. Its transfer function is

$$\frac{V_o(s)}{V_i(s)} = \frac{1}{s^2LC + s(L/R) + 1}$$

When we set $L = C = 1$, ω_n is normalized to unity, and

$$\zeta = \frac{1}{2R}\sqrt{\frac{L}{C}} = \frac{1}{2R}$$

or $R = 1/(2\zeta)$. A SPICE simulation produced the response to a unit step input for the following:

R (Ω)	ζ	ϕ (deg)
1.000	0.500	60
0.707	0.707	45
0.577	0.866	30
0.500	1.000	0

The step response, frequency response (amplitude and phase), and group delay are plotted from SPICE. Notice that for $\zeta = 0.707$, the amplitude remains flat to the highest frequency without peaking (MFA) and that the group delay for $\zeta = 0.866$ similarly remains flat longest without peaking.

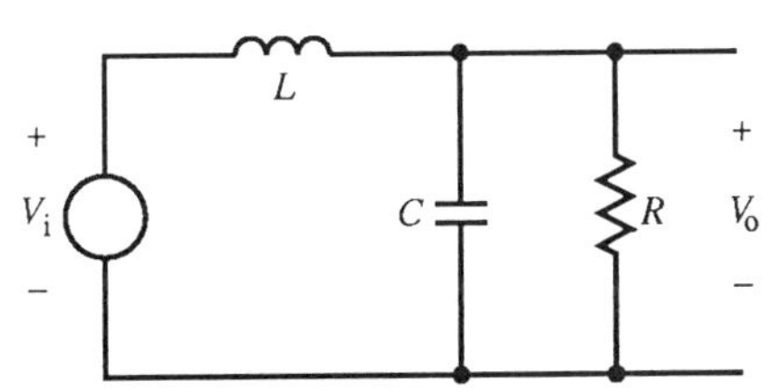

FIG. E5.8 (Figure continues.)

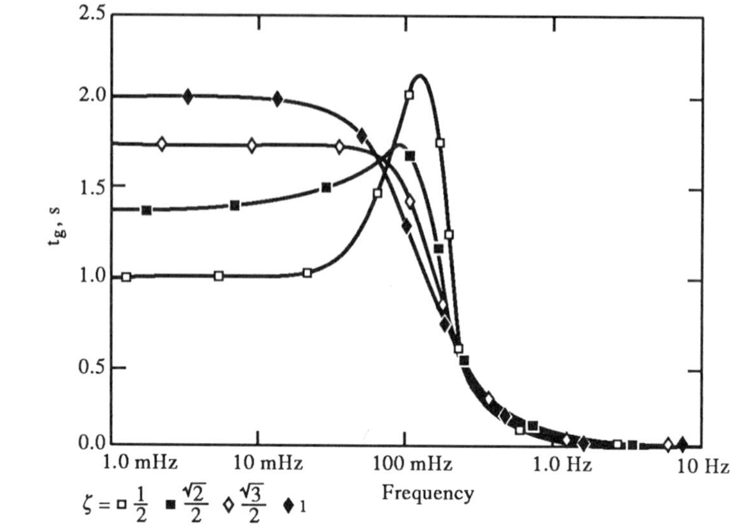

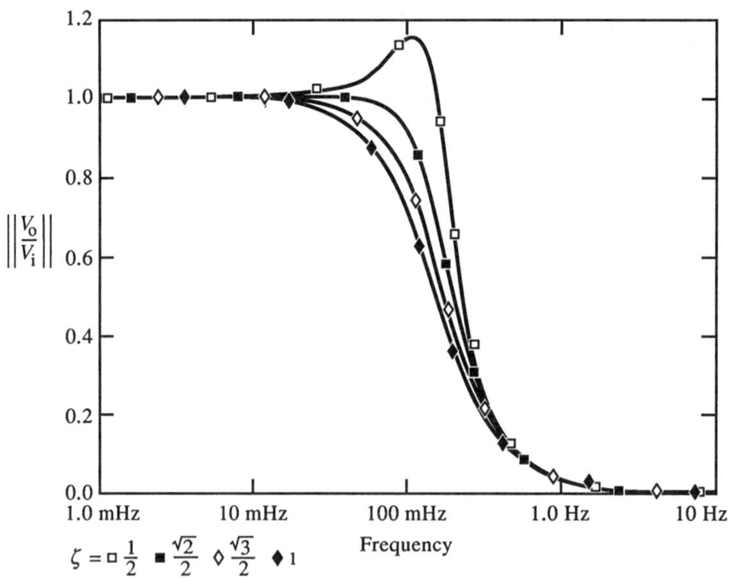

FIG. E5.8 (continued)

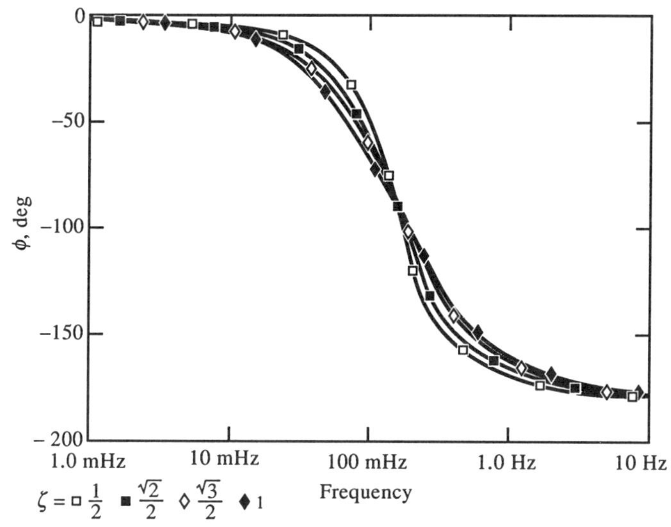

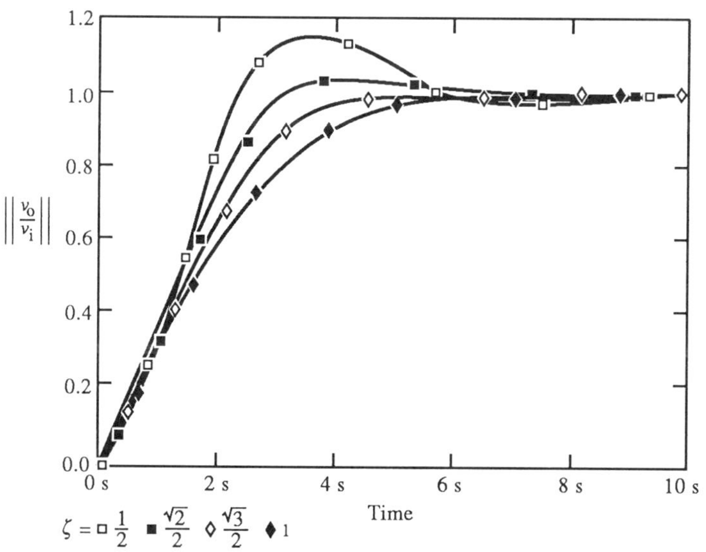

FIG. E5.8 (continued)

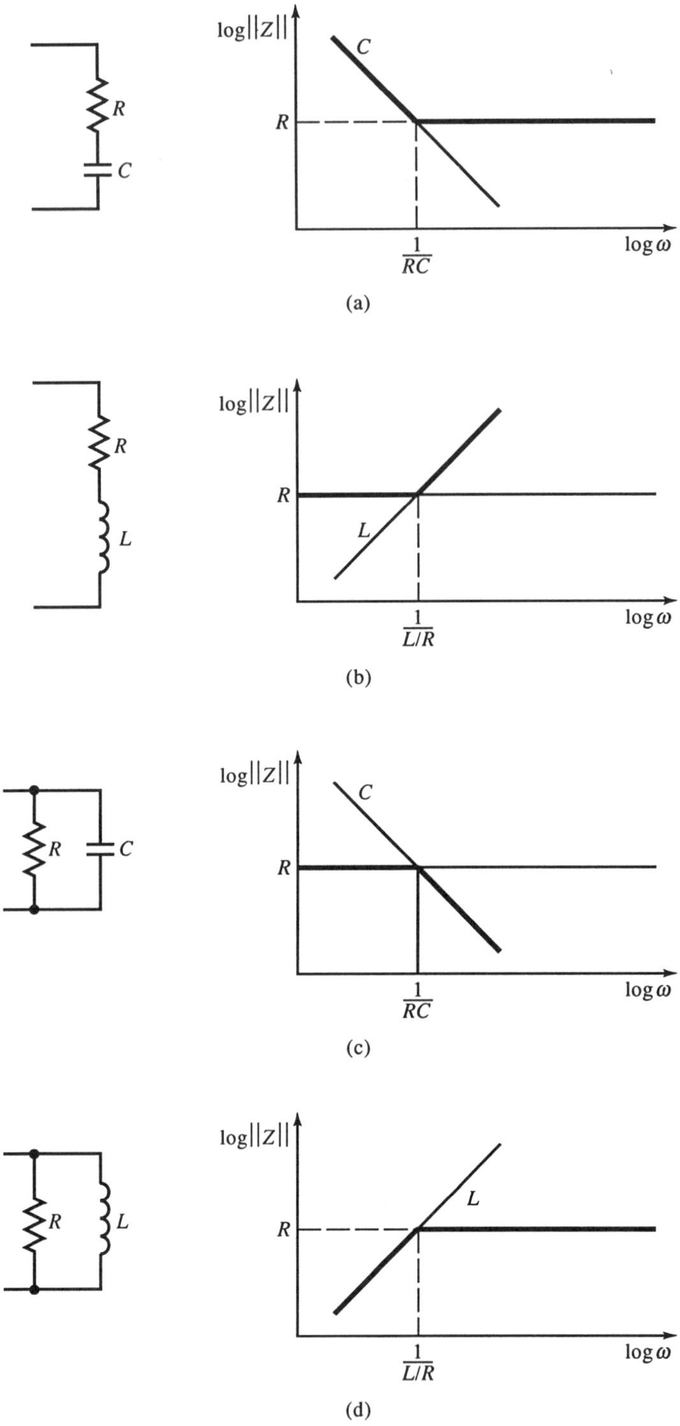

FIG. 5.17 Elemental combinations of elements found in passive networks: (a) series RC, (b) series RL, (c) shunt RC, and (d) shunt RL. The reactance charts are given for each, with pole or zero at the circuit time constant.

5.13 Reactance Chart Transfer Functions of Passive Circuits

The reactance chart is a powerful aid for graphically determining the magnitudes of the transfer functions of passive divider circuits. Combinations of *RL* and *RC* are plotted in Fig. 5.17. Asymptotic approximations similar to those of Bode plots are used to construct $\|Z(j\omega)\|$. The impedance of more complicated circuits can be built from these basic combinations.

These impedance plots follow directly from the behavior of the elements. For the series *RC*, at low frequencies the reactance of *C* dominates the series *RC* combination; but at high frequencies, the impedance is dominated by *R*. The frequency of equal impedance magnitudes is at the pole frequency $1/RC$, where the impedance plot shows an asymptotic break. The other basic combinations are analyzed similarly. The break frequency in all cases is $1/\tau$, where τ, the time constant, is *RC* for *RC* combinations and L/R for *RL* combinations.

Transfer functions of voltage dividers are constructed by first plotting the divider input impedance magnitude $\|Z_{in}\|$ and the impedance $\|Z_L\|$, across which the output voltage is developed. Then $\|M(j\omega)\|$ is plotted as $\|Z_L\|/\|Z_{in}\|$ by visually dividing the two impedances. The inverse of $\|Z_{in}\|$ for a linear segment is a segment with the opposite slope. Division is accomplished by subtraction (since the reactance chart is a log–log graph).

Figure 5.18 shows examples of a transfer function constructed with a reactance chart for the *RC* integrator and differentiator. When the plots of $\|Z_{in}\|$ and $\|Z_L\|$ track in slope, $\|V_o/V_i\|$ is constant (flat, zero-slope plot). For the integrator, $\|Z_L\|$ and $\|Z_{in}\|$ track in slope until $\omega = 1/RC$, where $\|Z_{in}\|$ becomes flat, causing the transfer function to roll off with $\|Z_L\|$. For the *RC* differentiator, it is $\|Z_{in}\|$ that slopes below $\omega = 1/RC$ and causes $\|V_o/V_i\|$ to slope in the opposite direction. The *s*-domain transfer functions are given with the circuit for comparison.

The basic combinations of Fig. 5.17 are part of the dividers of Fig. 5.19. In Fig. 5.19, the impedance magnitudes for Z_{in} and Z_L are plotted. $\|Z_{in}\|$ has a break frequency where the line for *C* intersects the line for $R_1 + R_2$ at $\omega = 1/(R_1 + R_2)C$. $\|Z_L\|$ rolls off with *C* until it reaches R_2, where it breaks and is flat. This break frequency is at $\omega = 1/R_2C$.

Below $\omega_p = 1/(R_1 + R_2)C$, $\|Z_{in}\|$ and $\|Z_L\|$ track, and $\|V_o/V_i\|$ is flat. Along this segment, the input and load impedances are equal, and the transmittance is 1. At ω_p, $\|Z_{in}\|$ becomes flat while $\|Z_L\|$ continues to roll off (with a slope of -1). This causes $\|V_o/V_i\|$ to roll off until $\|Z_L\|$ breaks at $\omega_2 = 1/R_2C$. Above ω_2, both $\|Z_L\|$ and $\|Z_{in}\|$ are flat, and $\|V_o/V_i\|$ is flat at the ratio of $\|Z_L\|/\|Z_{in}\|$, or $R_2/(R_1 + R_2)$. As on Bode plots, the phase of V_o/V_i decreases (or lags) whenever the magnitude decreases (or rolls off). This circuit causes a phase lag for frequencies between ω_p and ω_z.

In Fig. 5.19b, we encounter a parallel *RC* for Z_L. Parallel combinations introduce a slight complication in identifying a break frequency location. First,

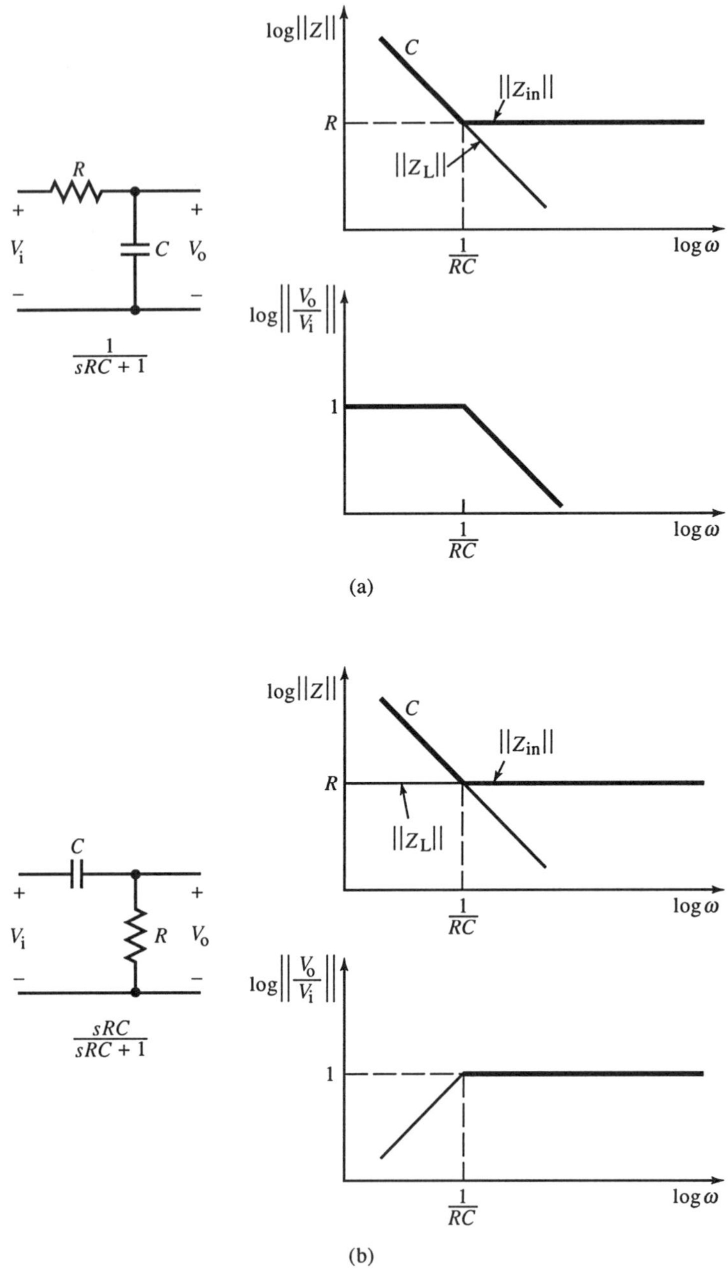

FIG. 5.18 Reactance chart analysis of transfer functions of (a) *RC* integrator and (b) *RC* differentiator.

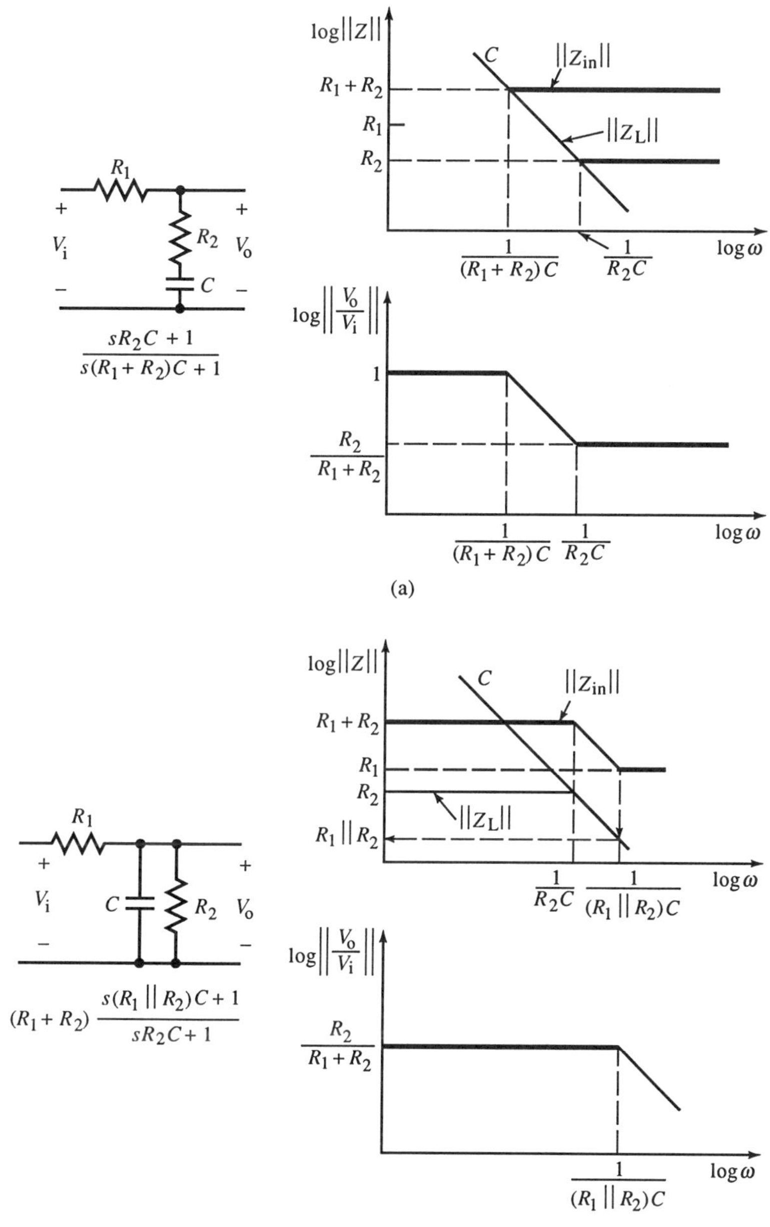

FIG. 5.19 Reactance chart analysis of commonly encountered passive *RC* circuits: (a) phase-lag compensator, (b) capacitively loaded divider, (c) *RC* differentiator with source resistance, and (d) phase-lead compensator. (Figure continues.)

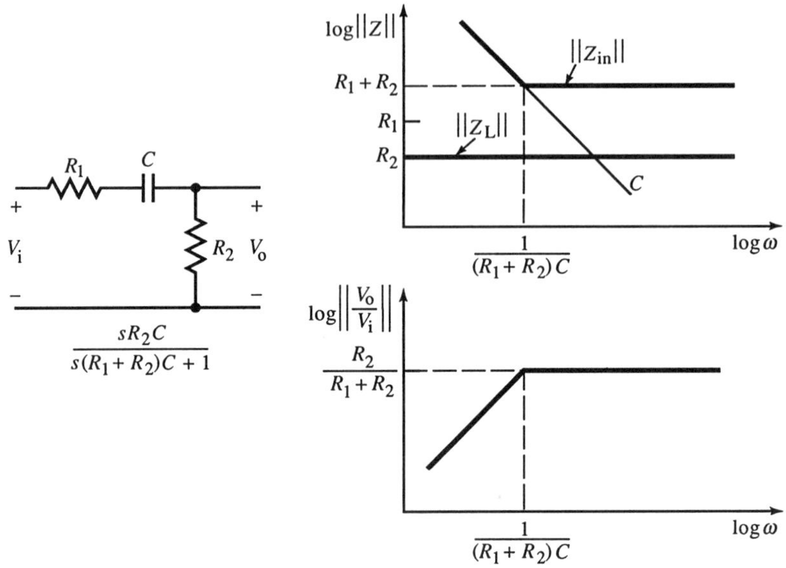

(c)

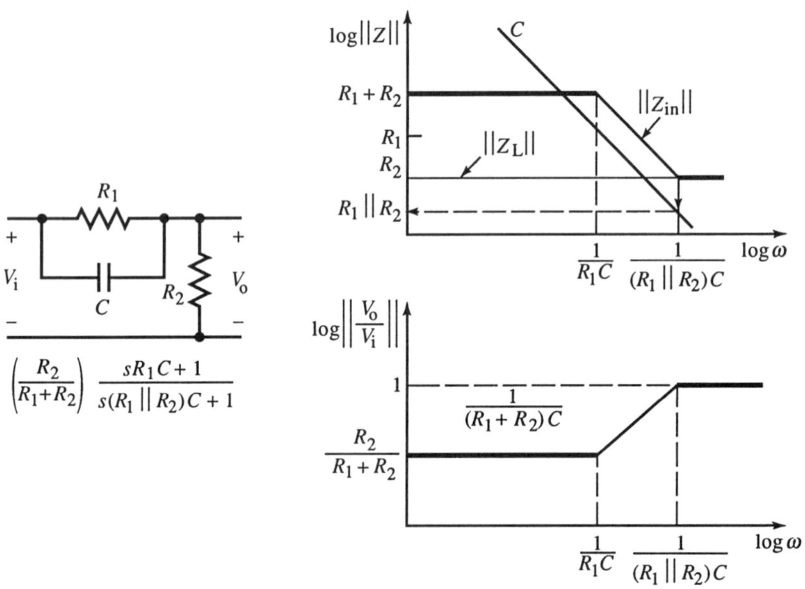

(d)

FIG. 5.19 (continued)

$\|Z_L\|$ is dominated by R_2 until $\omega_p = 1/R_2C$, where the shunt reactance of C dominates and $\|Z_L\|$ rolls off with C, as shown. $\|Z_{in}\|$ is $\|Z_L\|$ added to R_1. On the reactance plot, this is accomplished by shifting $\|Z_L\|$ upward until, at low frequencies, $\|Z_{in}\| = R_1 + R_2$. The break frequency due to $\|Z_L\|$ is present in $\|Z_{in}\|$, which rolls off until it reaches R_1, where it again breaks flat. This frequency is found by going down vertically to the curve for C and reading the resistance from the vertical axis. It is $R_1 \parallel R_2$. The associated break frequency is therefore at $\omega_z = 1/(R_1 \parallel R_2)C$.

That the equivalent resistance of ω_z is $R_1 \parallel R_2$ can be demonstrated by taking into account the log-log scaling of the reactance chart axes. The impedance of C, $\|X_C\|$, rolls off at a (log-log) slope of -1 as does $\|Z_{in}\|$ between ω_p and ω_z. By calculating the slopes of $\|X_C\|$ and $\|Z_{in}\|$ between ω_p and ω_z and equating, we obtain

$$\frac{\Delta \log\|Z\|}{\Delta \log \omega} = \frac{\log R_1 - \log(R_1 + R_2)}{\log \omega_z - \log \omega_p} = \frac{\log\|X_C(\omega_z)\| - \log\|X_C(\omega_p)\|}{\log \omega_z - \log \omega_p} \tag{5.168}$$

At ω_z, $\|X_C\|$ equals the equivalent resistance we are seeking. Reducing (5.168), we obtain

$$\log\|X_C(\omega_z)\| = \log\|X_C(\omega_p)\| + \log R_1 - \log(R_1 + R_2) \tag{5.169}$$

At ω_p, $\|X_C(\omega_p)\| = R_2$. Substituting and simplifying gives

$$\log\|X_C(\omega_z)\| = \log R_2 + \log R_1 - \log(R_1 + R_2) = \log\left(\frac{R_1 R_2}{R_1 + R_2}\right) = \log(R_1 \parallel R_2) \tag{5.170}$$

Therefore, the value of resistance that is read off the graph where ω_z intersects the line for C is $R_1 \parallel R_2$.

For $\|V_o/V_i\|$, $\|Z_{in}\|$ and $\|Z_L\|$ are flat to ω_p and $\|V_o/V_i\|$ is also flat with a value of $R_2/(R_1 + R_2)$. At ω_p, both $\|Z_{in}\|$ and $\|Z_L\|$ roll off, maintaining a flat $\|V_o/V_i\|$ until ω_z, where $\|Z_{in}\|$ flattens. Since $\|Z_L\|$ continues to roll off, so does $\|V_o/V_i\|$. Consequently, $\|V_o/V_i\|$ has a pole at ω_z.

Further examples are shown in Figs. 5.19c,d. The circuit in (d) is a phase-lead circuit since it causes an increase in phase while $\|V_o/V_i\|$ is increasing. This circuit also demonstrates how to handle a parallel resistance in the time constant.

Example 5.2 (*continued*) Wien-Bridge Filter

The reactance chart transfer function (see Fig. E5.2b) rolls up to $1/RC$, and then rolls off. Whenever a slope change of 2 or more occurs in a narrow band (less than a decade) on a reactance chart, the asymptotic

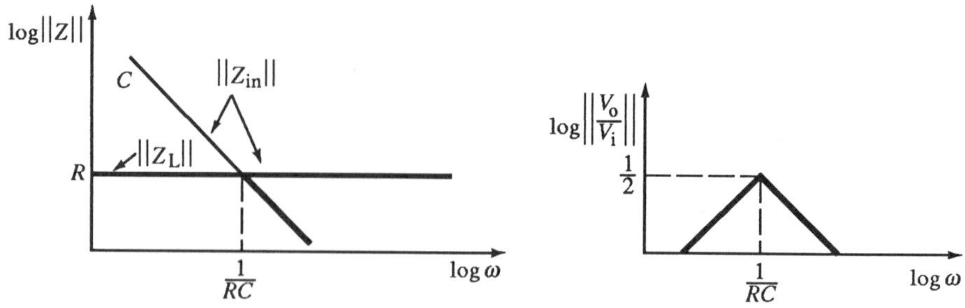

(b)

FIG. E5.2(b)

approximation is degraded. The actual poles are at

$$p_{1,2} = -\frac{1}{2RC}(3 \mp \sqrt{5}) \cong \begin{cases} -0.38/RC \\ -2.62/RC \end{cases}$$

These poles are centered at $-1/RC$ on a log ω scale.

Example 5.3 (*continued*) Inverse of Wien-Bridge Filter

The reactance chart method produces a flat transfer function of 1 through $1/RC$. $\|Z_L\|$ dominates $\|Z_{in}\|$ below $1/RC$ by C and above by R, making $\|Z_{in}\| = \|Z_L\|$. If C_2 is made large, the transfer function approaches that of Fig. 5.19d, with a zero at $1/R_1C_1$ and a pole at $1/(R_1 \| R_2)C_1$. At low frequencies it is $R_2/(R_1+R_2)$.

Example 5.7 (*continued*) Phase-Lag Circuit with Capacitive Output Loading

Reactance chart determination of the transfer function is simplified by the wide separation of break frequencies. The low-frequency pole occurs at about $1/R_1C_2$, or 159 Hz.

The zero is at $1/R_2C_2$, or 15.9 kHz and the high-frequency pole is at $1/R_2C_1$ or 1.59 MHz. These frequencies are not exact but are derived from the reactance chart. For example, the exact high-frequency pole is 1.6232 MHz. This difference is insignificant if the accuracy of a graphical method is considered adequate.

Example 5.9 Cascaded *RC* Integrators

Two *RC* integrators are cascaded in Fig. E5.9a to form a passive filter. Its transfer function can be found using the reactance chart. The first step is to find $V_1(s)$ by loading the first stage with the second. Then, at the input port, $\|Z_{in}\|$ is constructed on the reactance chart by beginning with the graph of R_2C_2. It follows C_2 until it interesects R_2 at $\omega_z = 1/R_2C_2$. Then C_1 shunts this impedance, with C_1 and C_2 in parallel at low frequencies ($C_p = C_1 + C_2$). The combined Z_1 decreases along C_p until it reaches the break frequency ω_z. It then flattens, following the R_2C_2 curve, but at a lower resistance. The curve again breaks where C_1 dominates, at $\omega_p = 1/R_2C_s$ (where C_s is the series combination of C_1 and C_2), and $\|Z_1\|$ rolls off.

The situation here is similar to that of Fig. 5.19d, in which a curve is shifted from its original location by the addition of another impedance. In Fig. 5.19d, the R_2C curve was shifted upward when the series resistance R_1 was added to it. This caused the capacitive roll-off of $\|Z_{in}\|$ to be shifted to the right so that its break frequency at R_2 was at the same frequency as C when combined with $R_1 \parallel R_2$. A similar effect occurs in this example, except that it is due to the addition of shunt C instead of series R. Because C_p dominates Z_1 at low frequencies instead of C_2, it reaches ω_z at a resistance of

$$R_2\left(\frac{C_2}{C_1 + C_2}\right)$$

When C_1 dominates Z_1 at ω_p, the break in $\|Z_1\|$ occurs where this resistance intersects C_1. As the upward arrow in Fig. E5.9b shows, the capacitance that would result when combined with R_2 is C_s. To construct $\|Z_{in}\|$, R_1 is added to $\|Z_1\|$; it shifts the graph upward. Then the transfer function with first stage output V_i can be constructed from $\|Z_{in}\|$ and $\|Z_i\|$. Similar construction for the second stage (R_2C_2) and a combination of reactance chart transfer functions produces the desired transfer function (magnitude).

The expression for $Z_1(s)$ is found by writing a voltage divider formula from Fig. E5.9a. It is

$$Z_1(s) = \frac{sR_2C_2 + 1}{sC_p(sR_2C_s + 1)}; \qquad C_p = C_1 + C_2, \qquad C_s = \frac{C_1C_2}{C_1 + C_2} = C_1 \parallel C_2$$

This expression is consistent with the reactance graph of Fig. E5.9b.

The first stage transfer function is

$$\frac{V_1}{V_i} = \frac{sR_2C_2 + 1}{s^2[R_1R_2C_1C_2] + s[R_1C_p + R_2C_2] + 1}$$

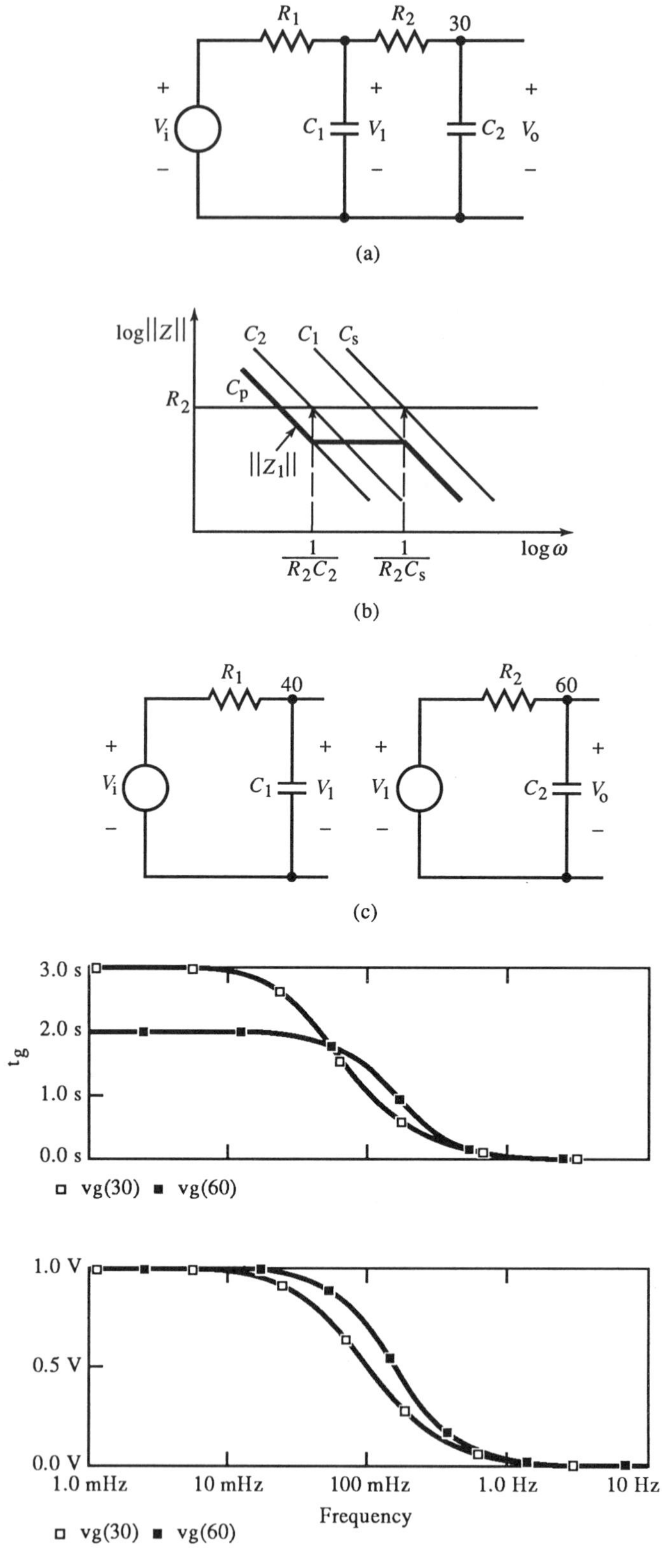

FIG. E5.9 (Figure continues.)

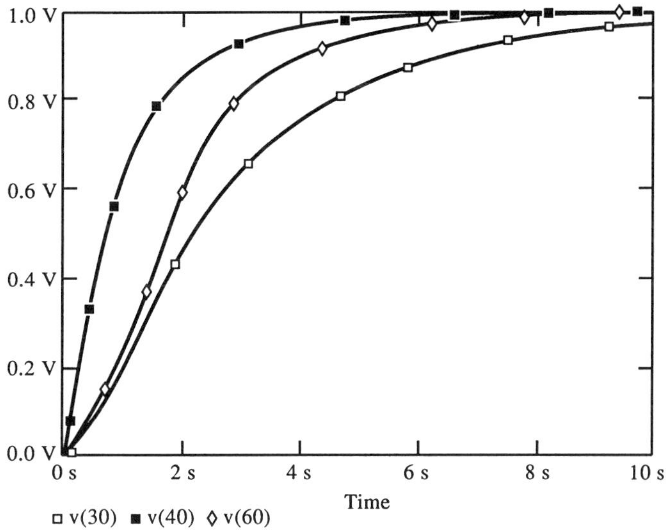

FIG. E5.9 (continued)

and the overall transfer function is

$$\frac{V_o}{V_i} = \frac{V_1}{V_i} \cdot \frac{1}{sR_2C_2+1} = \frac{1}{s^2[R_1R_2C_1C_2] + s[R_1C_p + R_2C_2] + 1}$$

It is of interest to note that a double pole at $1/RC$ does *not* occur when the resistors and capacitors are of the same values. Under the conditions

$$R_1 = R_2 = R, \qquad C_1 = C_2 = C \Rightarrow \zeta = 1.5, \qquad p_{1,2} = \begin{cases} -0.382/RC \\ -2.618/RC \end{cases}$$

This attempt to design a two-pole filter at $1/RC$ fails because the second-stage loading causes the poles to shift. To achieve a two-pole roll-off at a specified frequency, circuit values must be chosen to make the denominator of the transfer function a perfect square. The minimum value of $\zeta = 1$ is approached when the second stage loading is minimized by making $R_2 \gg R_1$ and $C_2 \ll C_1$. For $R_1 = R_2$ or $C_1 = C_2$, minimum $\zeta = \sqrt{2}$.

An alternative two-pole filter without interstage loading in Fig. E5.9c has a voltage buffer separating the stages. The frequency response simulations of (a) and (c) show the difference. For Fig. E5.9c, both poles are at $1/RC$ for these conditions.

5.14 Closure

This survey of linear dynamic response is the foundation for analysis of active circuits with reactive elements. We shall return to the amplifiers of previous chapters and extend their analyses to the complex-frequency domain, using the methods presented here. In practical circuits, the assumption of linearity applies for small-signal amplifiers. The extensive analysis done here of second-order circuits does not readily apply to higher-order circuits, so the formulas for t_p, M_p, M_m, and ζ may not be valid when zeros or additional poles are present.

References

Norman Balabanian and Theodore Bickart, *Electrical Network Theory*, Wiley. Ch. 6, pp. 392–431.

David K. Cheng, *Analysis of Linear Systems*, Addison-Wesley, 1959.

Someshwar Gupta and Lawrence Hasdorff, *Fundamentals of Automatic Control*, Wiley, 1970.

S. Madhu, *Linear Circuit Analysis*, Prentice-Hall, 1988.

James W. Nilsson, *Electric Circuits*, Addison-Wesley, 1986.

Joseph Pettit and Malcolm McWhorter, *Electronic Amplifier Circuits: Theory and Design*, McGraw-Hill, 1961.

James K. Roberge, *Operational Amplifiers: Theory and Practice*, Wiley, 1975.

Roberto Saucedo and Earl Schiring, *Introduction to Continuous and Digital Control Systems*, Macmillan, 1968.

Michael A. Street, *Passive A.C. Circuit Analysis Using Reactance-Bode Charts*, Bonneville Power Administration, Portland, Oregon.

Michael Street, "Simplify ac circuit analysis with Reactance-Bode charts," *EDN*, 5 May 1976. pp. 83–89.

CHAPTER 6

Dynamic Response Compensation

Dynamic response compensation techniques are applied to achieve desirable transient or frequency response of amplifier circuits. In Chapter 5, the response of linear circuits in both the time and frequency domain was investigated. Here, methods for achieving desired response are developed.

6.1 Passive Compensation: Voltage Divider

The familiar resistive voltage divider of Fig. 6.1a illustrates the idea of compensation. When a capacitive load C_2 shunts R_2, the step response is overdamped and bandwidth is reduced. To compensate for C_2, C_1 is added in parallel with R_1. The transfer function of this divider is

$$\frac{V_o(s)}{V_i(s)} = \left(\frac{R_2}{R_1 + R_2}\right) \cdot \frac{sR_1C_1 + 1}{s(R_1 \parallel R_2)(C_1 + C_2) + 1} \tag{6.1}$$

The addition of C_2 introduces a finite zero and makes $N(s)$ and $D(s)$ of the same degree in s, a condition for an all-pass filter. When the pole and zero are equated, the (all-pass) compensation condition is

$$R_1C_1 = R_2C_2 \tag{6.2}$$

A similar technique can be used with the current divider dual of Fig. 6.1, in which series load inductance is compensated by placing series inductance in the other branch of the divider. To compensate, the L/R time constants of the two branches are set equal.

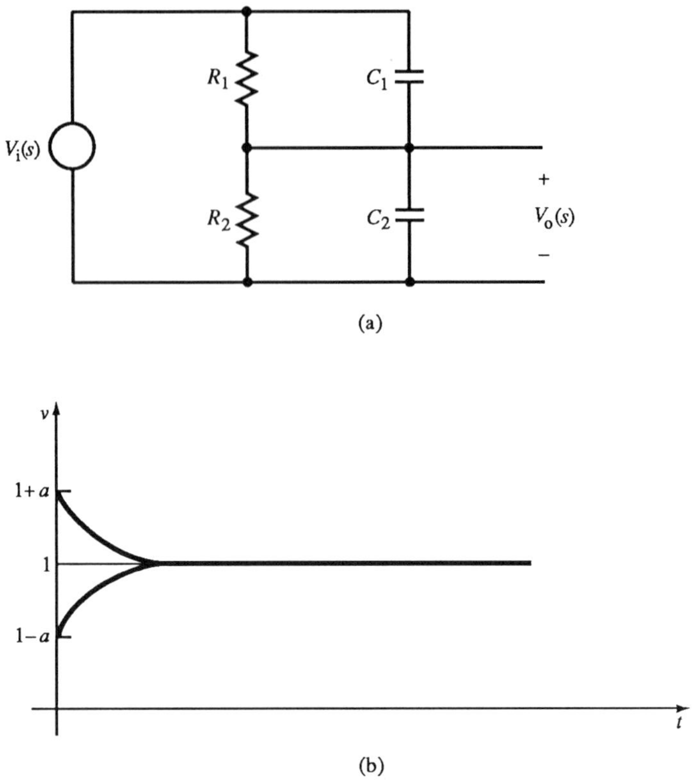

FIG. 6.1 A compensated voltage divider (a) and possible step responses (b).

Now suppose that this divider, or a circuit with a similar transfer function, is not properly compensated and has a step response like that of Fig. 6.1b, in which the fractional overshoot (or undershoot) is a. This time response is

$$\mathscr{L}^{-1}\left\{\frac{s\tau_z+1}{s\tau_P+1}\cdot\frac{1}{s}\right\}=\mathscr{L}^{-1}\left\{\frac{\tau_z}{s\tau_P+1}+\frac{1}{s(s\tau_P+1)}\right\}$$

$$=\mathscr{L}^{-1}\left\{\frac{\tau_z}{s\tau_P+1}+\frac{1}{s}+\frac{-\tau_P}{s\tau_P+1}\right\}=1+\left[\frac{\tau_z}{\tau_P}-1\right]e^{-t/\tau_P} \tag{6.3}$$

At $t=0$, the exponential is 1; its coefficient in (6.3) therefore is a, and the relationship between the pole and zero time constant is

$$\tau_z=(1+a)\tau_P \tag{6.4}$$

An additional cascaded compensation network with a pole time constant $\tau_{pc}=\tau_z$ and a zero time constant $\tau_{zc}=\tau_p$ results in a flat frequency and step response. The value of τ_p can be estimated by observing the transient decay of the step response. The settling time is 4 to 5 times τ_p as observed on an oscilloscope. With this estimate for τ_p and from measurement of a from the step response, τ_z can be calculated from (6.4).

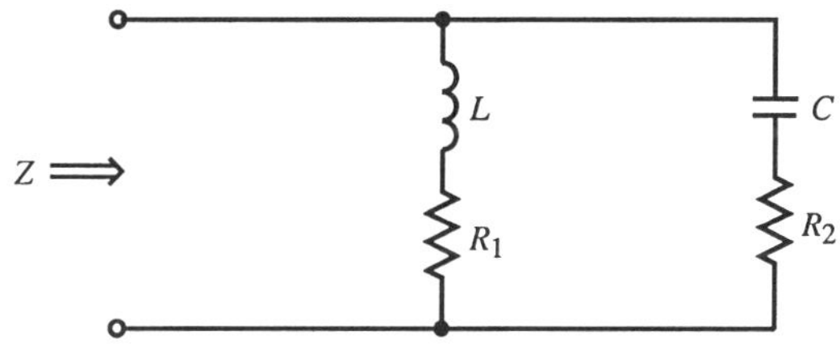

FIG. E6.1

Example 6.1 Shunt-Series All-Pass Circuit

Figure E6.1 has a terminal impedance of

$$Z = R_1 \cdot \frac{(sL/R_1+1)(sR_2C+1)}{s^2LC+s(R_1+R_2)C+1}$$

$$= R_1 \cdot \frac{s^2(LCR_2/R_1)+s(L/R_1+R_2C)+1}{s^2LC+s(R_1+R_2)C+1}$$

Z has two poles and two zeros. If the poles and zeros cancel, the input resistance is merely R_1 and is independent of frequency. This is achieved when

$$LC\left(\frac{R_2}{R_1}\right) = LC, \qquad \frac{L}{R_1}+R_2C = (R_1+R_2)C$$

or

$$R_1 = R_2 = R, \qquad \frac{L}{R} = RC$$

This circuit suggests frequency compensation schemes. A series RC can be compensated with a series RL and vice versa.

Example 6.2 Series-Shunt All-Pass Circuit

The dual of Fig. E6.1 is Fig. E6.2, in which

$$Z = \frac{sL}{sL/R_1+1} + \frac{R_2}{sR_2C+1} = R_2 \cdot \frac{s^2LC+s[(L/R_1)+(L/R_2)]+1}{s^2[LCR_2/R_1]+s[L/R_1+R_2C]+1}$$

The all-pass conditions are found by equating $N(s)$ and $D(s)$ and equating coefficients:

$$\frac{LCR_2}{R_1} = LC, \qquad \frac{L}{R_1}+\frac{L}{R_2} = \frac{L}{R_1}+R_2C$$

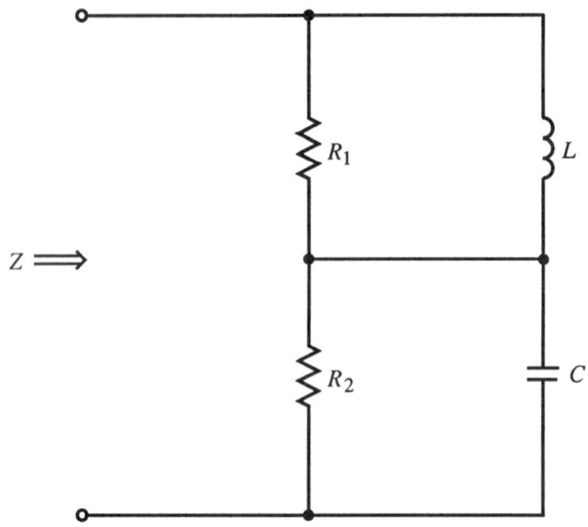

FIG. E6.2

This reduces to the all-pass conditions:

$$R_1 = R_2 = R, \qquad \frac{L}{R} = RC$$

Notice that they are the same as for Fig. E6.1. For both, $R = Z_n$.

6.2 Op-Amp Transfer Functions from Reactance Charts

We now return to op-amp circuits and apply the reactance chart method (of Section 5.13) to find their transfer functions. With this capability, we can more easily attend to their compensation. We begin by considering the voltage gain of inverting op-amps:

$$\frac{V_o(s)}{V_i(s)} = -\frac{Z_f(s)}{Z_i(s)} \tag{6.5}$$

This is an s-domain extension of (3.22) and assumes an infinite op-amp bandwidth and gain. Under these simplifying conditions, the op-amp integrator and differentiator of Fig. 6.2 have gains of

$$\text{op-amp integrator,} \qquad \frac{V_o(s)}{V_i(s)} = -\frac{1}{sRC} \tag{6.6}$$

$$\text{op-amp differentiatior,} \qquad \frac{V_o(s)}{V_i(s)} = -sRC \tag{6.7}$$

The transfer functions are shown on Bode plots of Fig. 6.2b and (d).

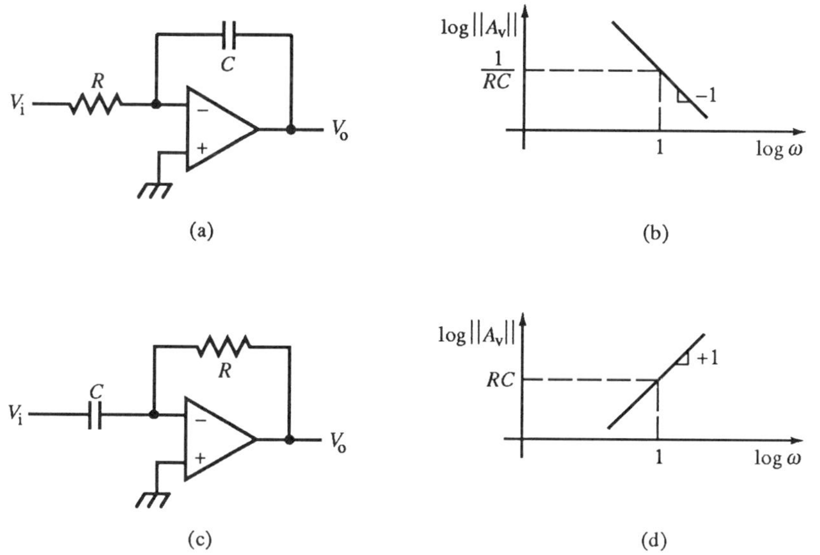

FIG. 6.2 Op-amp integrator (a) and its Bode plot (b); op-amp differentiator (c) and Bode plot (d).

For the op-amp integrator, a finite-gain op-amp cannot supply adequate gain as the input frequency approaches zero. At dc, the op-amp circuit is open-loop and subject to dc drift from offset errors. To stabilize the closed-loop gain (at some high value at a low frequency), the feedback capacitor is shunted by a large resistor (Fig. 6.3a). The dc gain is then $-R_f/R_i$ and the output, though not exactly the integral, is predictable and stable.

The reactance plots for $\|Z_f\|$ and $\|Z_i\|$ are shown in Fig. 6.3b. The ratio, $\|Z_f\|/\|Z_i\|$, is the magnitude of the gain $\|A_v\|$. At frequencies below $\omega_p = 1/R_fC$, C is effectively an open circuit, and the gain is determined by the resistors. Above ω_p, C dominates R_f, and integration occurs; the -1 slope (single-pole roll-off) is characteristic of time-domain integration.

The op-amp differentiator has similar limitations but at high frequencies. To limit high-frequency gain, R_i is added in series with C (Fig. 6.3c). The differentiator is dc-stable because of the resistive feedback. Above $\omega_p = 1/R_iC$, the circuit fails to accurately differentiate, and the gain is determined by the resistors. The transfer function plot is derived from the reactance chart as before.

The reactance chart method is not limited to these simpler examples. Figure 6.4 shows more involved circuits. In (a), $\|Z_f\| = 1/\omega C_f$, and $\|Z_i\|$ is shown on the $\|Z\|$ plot. As we saw in Section 5.13, the addition of R_s to the R_iC_i plot shifts it upward to $R_i + R_s$ at dc. This $\|Z_i\|$ plot rolls off and intersects R_s at a break frequency that, if it were caused by C_i, would be due to an equivalent resistance of $R_s \parallel R_i$. This is shown by the dotted lines with arrows. The upward-shifted $\|Z_i\|$ plot rolls off at a capacitive value less than C_i. Since the circuit has no capacitance of this value, the zero of $\|Z_i\|$ is referred to the

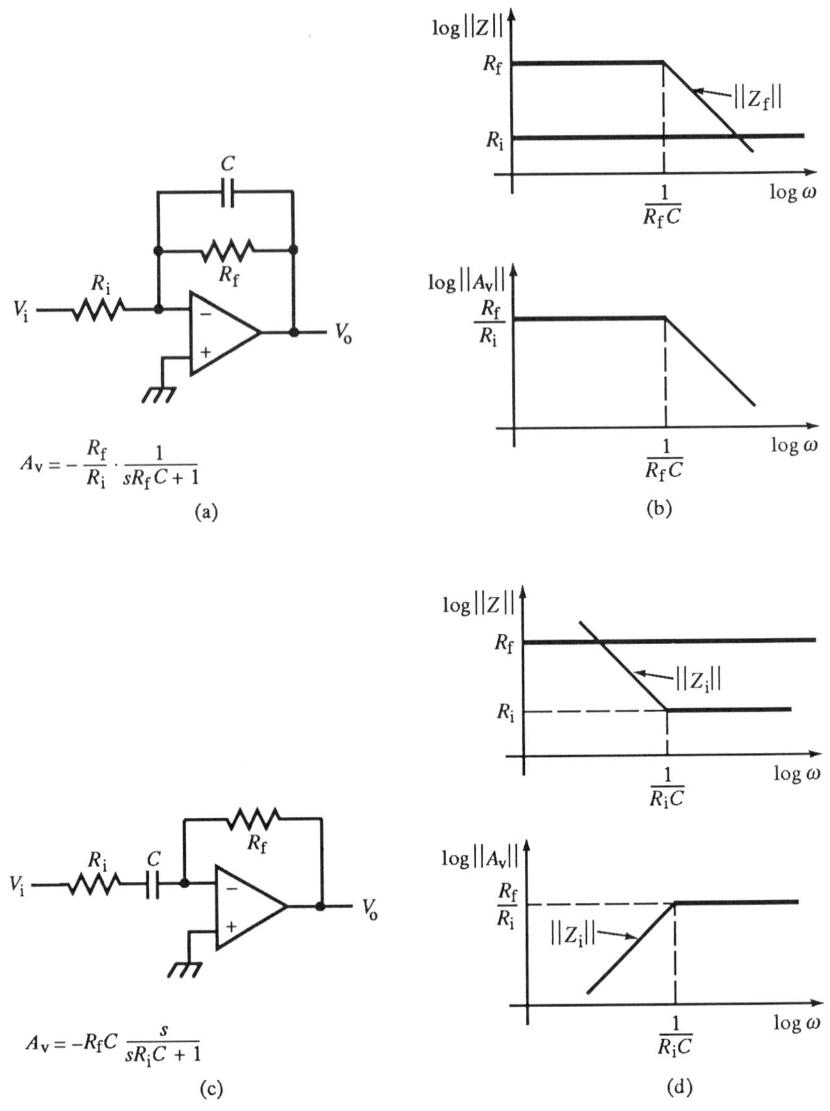

FIG. 6.3 Op-amp integrator (a) with finite dc gain and Bode plot derived from reactance chart (b); op-amp differentiator with finite high-frequency gain (c) with reactance charge and Bode plot (d).

C_i curve so that its resulting expression is readily interpretable in terms of the circuit topology. The $\|A_v\|$ plot follows, as in previous examples, from the plots of $\|Z_f\|$ and $\|Z_i\|$.

The op-amp circuit of Fig. 6.4b does not have a unique transfer function plot but depends on the relative values of its poles and zeros. The reactance chart method is limited in generality (compared with the s-domain transfer function A_v) because only one case can be plotted. All possible orderings of

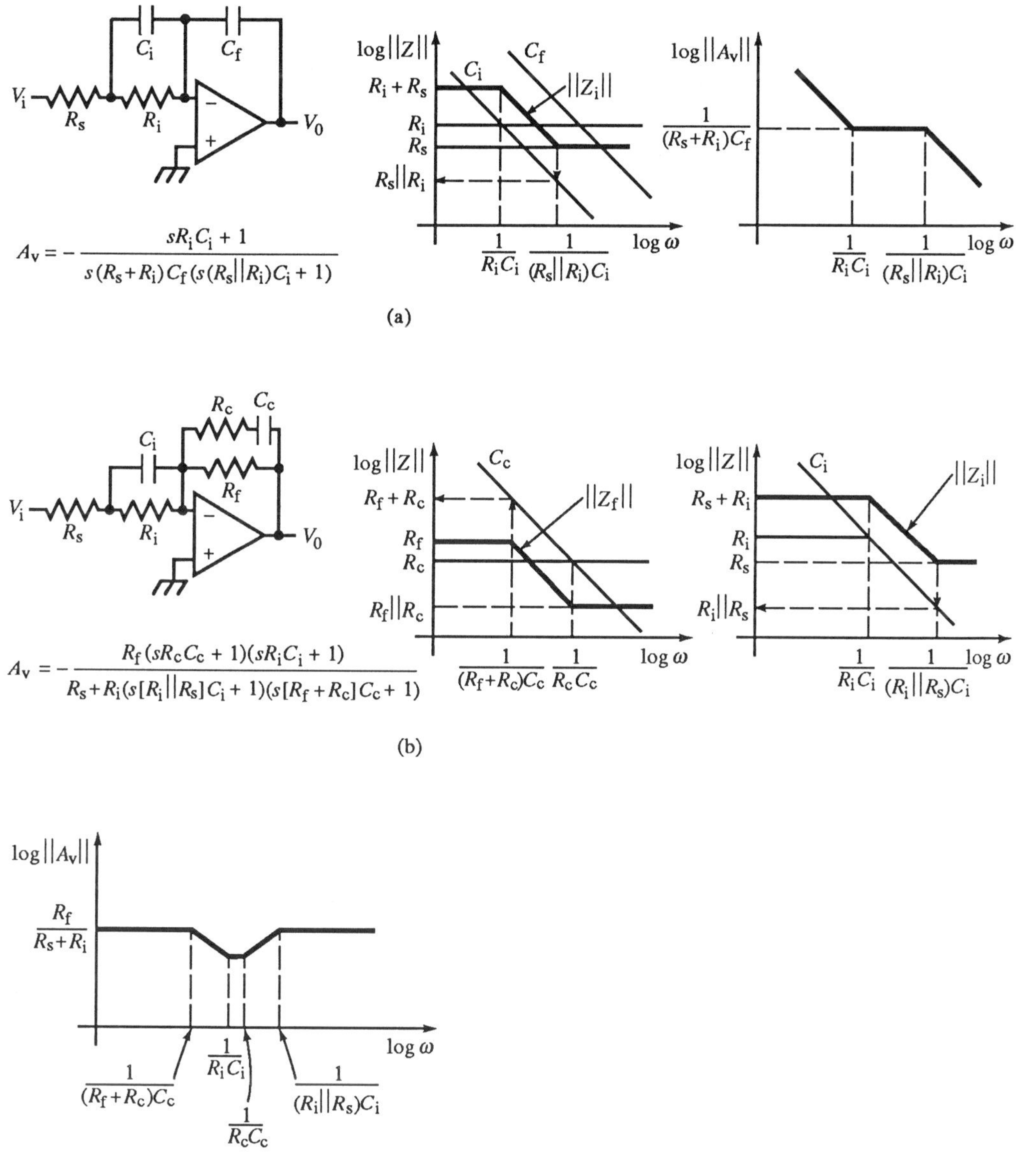

FIG. 6.4 More complicated op-amp circuits with Bode plots derived from reactance charts.

pole and zero values have to be considered by generating separate plots. In practice, the relative (if not actual) values of the elements are known because they are determined by the functional requirements of the circuit. In Fig. 6.4b, $R_f > R_c$, $R_i > R_s$, and the ordering of poles and zeros shown in (c) is assumed.

The reactance plot of $\|Z_f\|$ is shifted from the plot of R_cC_c because $\|Z_f\| = R_f \| R_c$ at high frequencies. At dc, $\|Z_f\|$ must be R_f. The zero of $\|Z_f\|$

is set by the R_cC_c plot, and $\|Z_f\|$ has a -1 slope between resistances of $R_f \parallel R_c$ and R_f. This slope represents a capacitance greater than C_c but not an actual circuit element value. Therefore, the break frequency at R_f is found by referring the resistance to the C_c plot (the dotted line with arrow pointing upward). The resistance at C_c is $R_f + R_c$, and the pole of $\|Z_f\|$ is at $1/(R_f + R_c)C_c$. This technique of scaling the impedance at a given frequency by referring to a reactive circuit element (such as C_c here) to find the associated resistance is also used to find $\|Z_i\|$.

When $\|Z_f\|$ and $\|Z_i\|$ are combined to form $\|A_v\|$, the transfer function in Fig. 6.4c results. Again, this frequency response is not unique but depends upon the placement of poles and zeros. Some ordering limitations are imposed by basic circuits laws. The pole at $1/(R_i \parallel R_s)C_i$ must always be higher in frequency than the zero at $1/R_iC_i$, and the zero at $1/R_cC_c$ must be greater than the pole at $1/(R_f + R_c)C_c$. Furthermore, depending on circuit values, complex poles and zeros are possible for the circuit in Fig. 6.4b, and the reactance chart asymptotic approximations may not be adequate for lightly damped response.

Noninverting op-amp frequency response is determined with the reactance chart method in the same way that passive dividers were treated in Section 5.13. The difference is that for the op-amp, the closed-loop response is the reciprocal of the divider H, or

$$\|A_v\| = \frac{\|Z_f + Z_i\|}{\|Z_i\|} = \frac{\|Z_{\mathrm{Hin}}\|}{\|Z_i\|} \tag{6.8}$$

where Z_{Hin} is the impedance of the feedback network from the op-amp output. On a reactance chart, $\|Z_{\mathrm{Hin}}\|$ is plotted by adding $\|Z_f\|$ and $\|Z_i\|$ on the chart. Since asymptotic approximations are used,

$$\log\|Z_1 + Z_2\| = \log\sqrt{\|Z_1\|^2 + \|Z_2\|^2} = \tfrac{1}{2}\cdot\log(\|Z_1\|^2 + \|Z_2\|^2)$$
$$= \begin{cases} \log\|Z_1\|, & \|Z_1\| \gg \|Z_2\| \\ \log\|Z_2\|, & \|Z_2\| \gg \|Z_1\| \end{cases} \tag{6.9}$$

Consequently, $\|Z_1 + Z_2\| = \|Z_1\| + \|Z_2\|$ under the constraints of (6.9) and reactance chart impedance magnitudes can be combined by addition of individual impedance magnitudes.

6.3 Feedback Circuit Response Representation

The feedback techniques of Chapter 3 derived closed-loop response from loop gain. The closed-loop gain $A_v(s)$ is also determined from the loop gain $GH(s)$. Feedback in the s-domain is the subject of control theory, found typically in control and circuits textbooks, and will not be systematically developed here.

developed here. Instead, basic aspects of amplifier stability and good dynamic response are explained, leading to methods for compensation of amplifiers that have undesirable responses.

Of the representations of $A_v(s)$, the Bode, polar (or Nyquist), and root-locus plots are the most commonly used. Bode plots are already familiar and present the frequency and phase response. Polar plots of the imaginary ($j\omega$-axis) and real (σ-axis) components of GH with ω as the parameter are an alternative representation in polar form. For each of these representations, closed-loop performance is determined by the loop-gain characteristics.

Root-locus diagrams are s-plane plots of the loci of closed-loop poles with open-loop gain K as a parameter. As K increases from zero, the closed-loop poles begin at open-loop poles and proceed toward open-loop zeros (some of which may be at infinity). When these poles leave the left half-plane, the feedback circuit becomes unstable. The pole loci can be found by setting the denominator of $A_v(s)$ to zero. Then,

$$1 + G(s)H(s) = 0$$

or $GH = -1 = 1e^{\pm\pi}$. In polar form, the locus conditions are

$$\|GH\| = 1, \qquad \angle(GH) = \pm 180° \tag{6.10}$$

Locating the loci in the s-plane is simplified by root-locus rules. These rules are constraints imposed on the location of the closed-loop poles by (6.10). Some of the more commonly used (and easily remembered) rules are the following:

1. The root loci start at the poles of GH (for $K = 0$).
2. The root-loci terminate at the zeros of GH.
3. There are as many separate root loci as poles of GH.
4. The loci are symmetrical about the real axis.
5. The root loci are on the real axis to the left of an odd number of real poles and zeros of GH.
6. The sum of the closed-loop poles is constant. (The centroid of the loci remains constant.)

Other rules can be constructed from (6.10).

The Bode and root-locus plots for an amplifier with a frequency-independent H and a single, real pole $-p$ are shown in Fig. 6.5. The amplifier gain is

$$G(s) = \frac{K}{s/p + 1} \tag{6.11}$$

The closed-loop gain for positive K and H is then

$$A_v(s) = \frac{G}{1 + GH} = \left(\frac{K}{KH + 1}\right)\frac{1}{s/(KH + 1)p + 1} \tag{6.12}$$

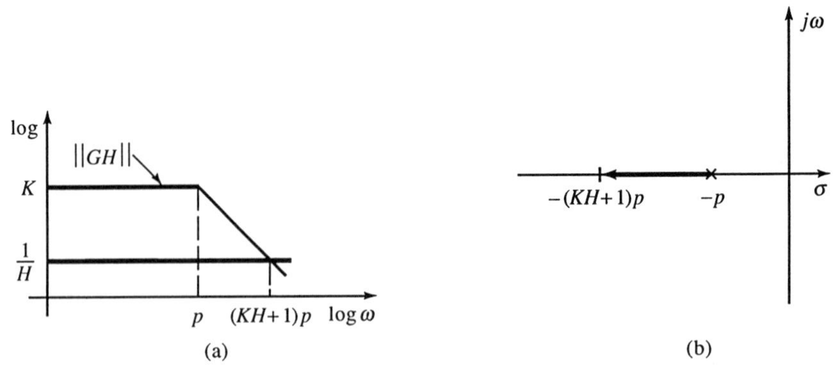

FIG. 6.5 Bode plot (a) of single-pole feedback amplifier with root locus (b). The open-loop pole at p increases in frequency by the dc loop gain $1+KH$.

The closed-loop response is also that of a single, real pole, but at the frequency of $\omega_{bw}=(KH+1)p$. The bandwidth has been extended by $KH+1$. This response is unconditionally stable. [Whenever steady-state frequency response ($j\omega$-axis response) is related to pole locations in s, it is assumed that the positive value of the real component of the pole location is used in relation to the steady-state frequency. To be precise, $\omega_{bw}=(KH+1)|-p|$ for real poles. Since frequency response involves only positive frequencies, and $p>0$ for negative poles, no confusion should result.] The root-locus plot is shown in Fig. 6.5b. The open-loop pole at $-p$ moves toward and terminates at the closed-loop pole $-(KH+1)p$.

Next, consider an amplifier with two poles:

$$G(s)=\frac{K}{(s/p_1+1)(s/p_2+1)} \tag{6.13}$$

For H constant with frequency, the closed-loop response is

$$A_v(s)=\left(\frac{KH}{KH+1}\right)\cdot\frac{1}{s^2/(KH+1)\omega_n^2+2\zeta s/(KH+1)\omega_n+1} \tag{6.14}$$

A_v is also a quadratic pole response. The closed-loop parameters are

$$\omega_{nc}=\omega_n\sqrt{KH+1}=\sqrt{p_1p_2(KH+1)} \tag{6.15}$$

and

$$\zeta_c=\frac{\zeta}{\sqrt{KH+1}}=\frac{p_1+p_2}{2\omega_{nc}} \tag{6.16}$$

For complex poles, both pole angle and magnitude depend on the dc loop gain, as did the single-pole response. That is why dc loop gain is the parameter of closed-loop pole movement for root-locus plots. For both first- and second-order loop gain, stability is unconditional. Response can become unacceptably

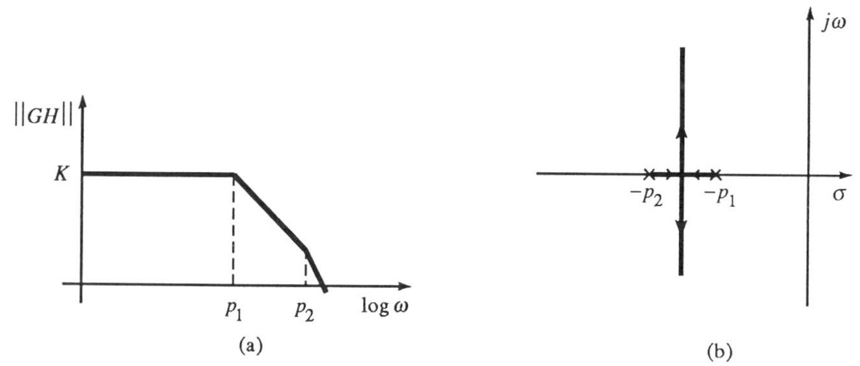

FIG. 6.6 A two-pole feedback amplifier Bode plot (a) and root locus (b). As K increases, the poles become complex.

underdamped for excessive loop gain in (6.14), but the poles remain in the left half-plane. The Bode magnitude and root-locus plots are shown for second-order loop gain in Fig. 6.6.

The Bode plot of $\|G\|$ and $\|1/H\|$, for G of (6.11) and constant H, is shown in Fig. 6.7. Because the magnitude axis is logarithmic, the difference between the $\|G\|$ and $1/H$ plots is the loop gain. That is,

$$\log\|G\| - \log(1/H) = \log\|GH\| \tag{6.17}$$

These Bode plots are an alternative to calculation and plotting of $\|GH\|$ to determine response characteristics. We need only plot $\|G\|$ and $1/H$ separately and then use $1/H$ as the unity-gain axis. This applies also for $\|H(j\omega)\|$. In Fig. 6.7, the open- and closed-loop gains intersect at ω_{bw} of (6.12). $\|A_v\|$ rolls off with $\|GH\|$ above this closed-loop bandwidth.

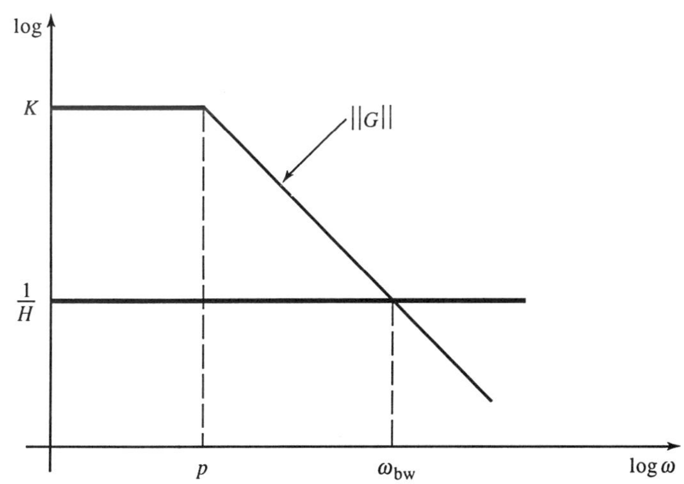

FIG. 6.7 The $1/H$ curve can be used as the unity-gain "axis" for analyzing loop gain.

The closed-loop bandwidth can be calculated from Fig. 6.7. The dc gain magnitude of G is K, and since $1/H$ is constant, the difference between them is KH on a Bode plot. The slope of $\|G\|$ due to the pole at p is -1. Since the ω axis is also logarithmic, a logarithmic frequency difference is a ratio, and $\omega_{\text{bw}}/p = KH + 1$. The bandwidth is then

$$\omega_{\text{bw}} = (KH + 1) \cdot p \tag{6.18}$$

and is the same as for the plot of $\|GH\|$ in Fig. 6.5a.

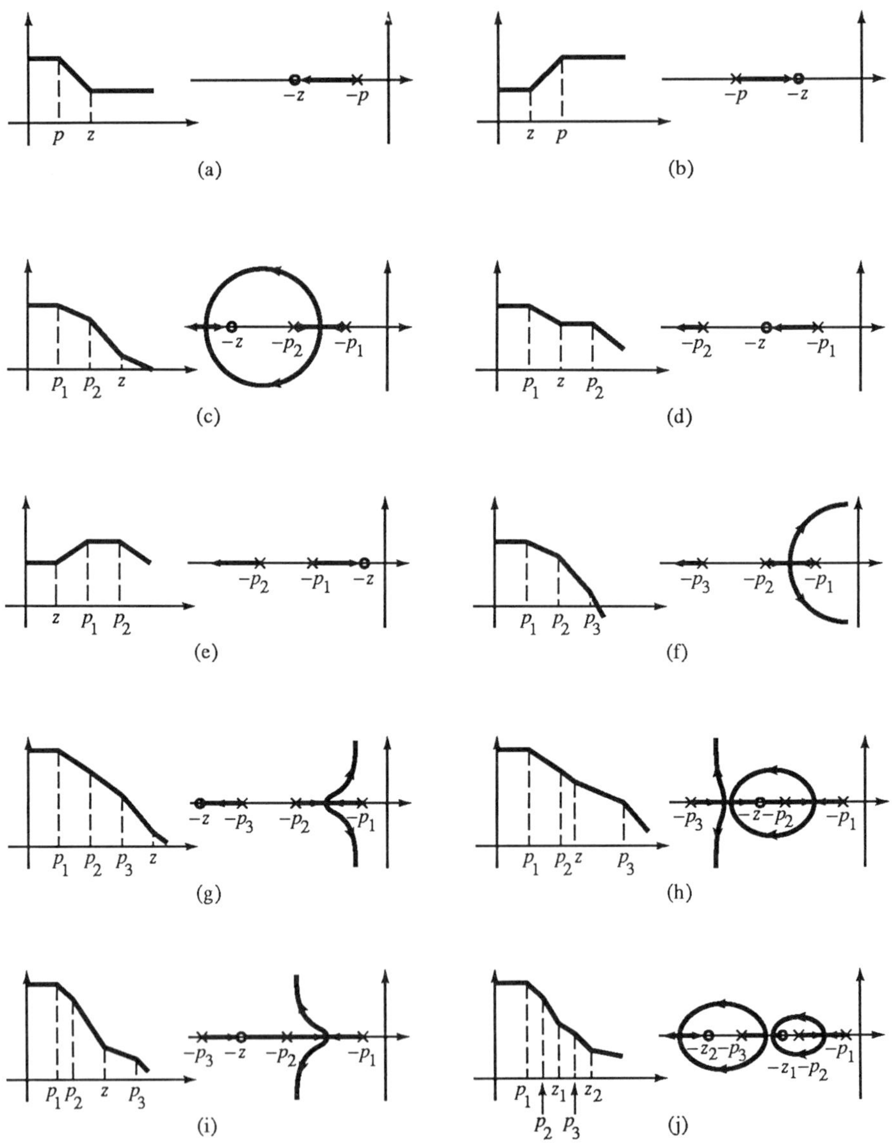

FIG. 6.8 Several common pole-zero configurations: Bode and root-locus plots.

Figure 6.8 shows some Bode and root-locus plots for circuits with up to three poles and two zeros. Bode plots show the gain–phase relationship with frequency directly and are most useful for compensating fixed-gain amplifiers. Root-locus plots show the closed-loop poles in the s-plane and how these poles vary with loop gain. For circuits with three or more poles, the closed-loop poles can leave the left half-plane with increasing K. The addition of zeros tends to "bend" the loci back from the $j\omega$-axis. This effect is a basis for response compensation.

6.4 Feedback Circuit Stability

Circuits with no right half-plane (RHP) poles or zeros are *minimum-phase* circuits. Most circuits are of this kind. The stability of a minimum-phase circuit can be determined from a Bode plot. When loop gain, $G(j\omega)H(j\omega) \leq -1$ (or $G(-H) \geq 1$), the feedback is in phase with (and thus reinforces) the error input with a loop gain magnitude ≥ 1, enough to sustain oscillation. In other words, the phase lead or lag around the GH loop is large enough to invert the signal and cause it to come back to the input *in phase.* This is positive feedback. When $GH(j\omega) = -1$, then $\|GH\| = 1$ and $\phi = \pm 180°$. On a Bode plot, when ϕ crosses $-180°$, stability requires that $\|GH\| < 1$. Or, when $\|GH\| \geq 1$, $-180° < \phi < 180°$ for stability. This stability condition is called the *Nyquist criterion.*

For minimum-phase circuits, stability can be determined from a polar plot of $GH(j\omega)$ by observing whether GH *encloses* the $(-1, j0)$ point. By traversing GH as ω goes from 0^+ to infinity, if $(-1, j0)$ remains to the left of the curve, it is not enclosed and the circuit is stable (that is, has no closed-loop RHP poles). Figure 6.9 shows some examples of nonenclosing curves.

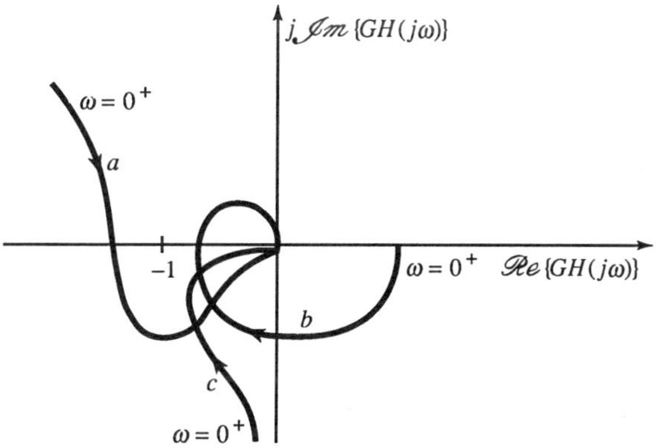

FIG. 6.9 Polar plots that do not enclose $-1 + j0$.

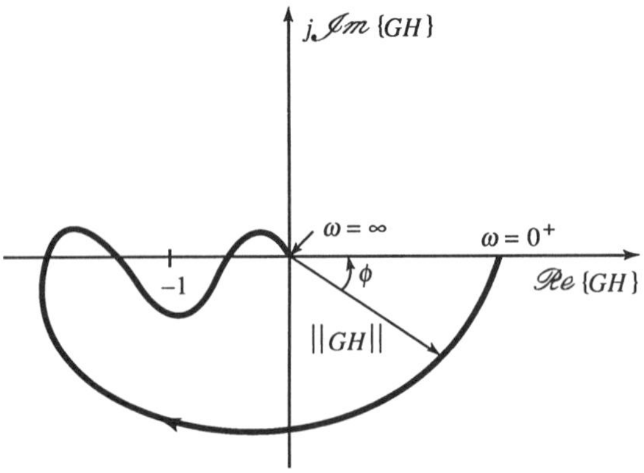

FIG. 6.10 Polar plot of a conditionally stable system.

Stability is not as easy to determine for *nonminimum-phase* circuits, those with right half-plane poles or zeros. Circuits with RHP zeros can be *conditionally stable* within a loop-gain range. For minimum-phase circuits, a decrease in dc loop gain K increases the relative stability. But for a conditionally stable circuit, a decrease in gain can decrease stability instead. The reason for this can be seen graphically in Fig. 6.10. The plot of $GH(j\omega)$ extends above $\phi = -180°$ with a magnitude exceeding unity. As magnitude decreases, the phase reverts to the stable side of $-180°$ (to quadrant III) and skirts around $(-1, j0)$, not enclosing it. The phase again lags beyond $-180°$ at a loop-gain magnitude of less than unity. Because the plot crosses $-180°$ on both sides of -1, too great an increase or decrease of K could cause it to enclose $(-1, j0)$.

Figure 6.11 shows a typical nonminimum-phase circuit polar plot in which GH *encircles* points A and B but *encloses* only A. The complete locus is needed to see the encirclements and includes GH for $\omega = 0^-$ to $\omega = -\infty$. The negative frequency range locus of GH is symmetric with the positive range locus relative

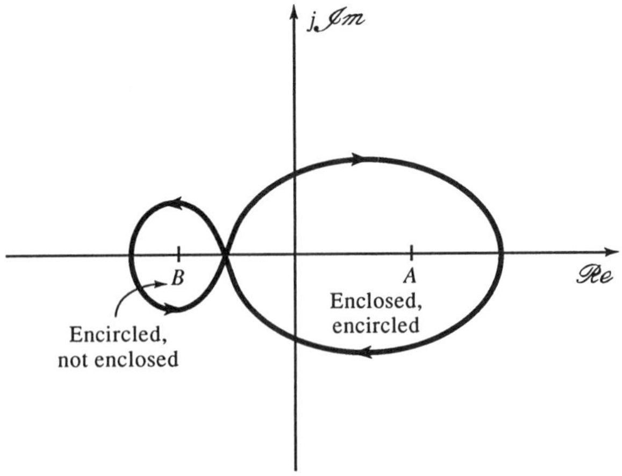

FIG. 6.11 Encirclement versus enclosure of points on a polar plot.

to the real axis. The GH locus in the s-plane closes at infinity (that is, from $\omega = -\infty$ to $\omega = +\infty$) with a counterclockwise path at infinity, enclosing the stable LHP. The Nyquist criterion must be generalized to include the nonminimum-phase case. The number of RHP poles must be zero for stability; their number is

$$\begin{aligned}&\text{number of closed-loop RHP poles}\\ &\quad = \text{number of poles of } GH \text{ in RHP}\\ &\qquad - \text{net number of counterclockwise encirclements of } (-1, j0) \text{ by } GH\end{aligned} \tag{6.19}$$

For nonminimum-phase circuits, stability cannot be determined by enclosure; the Nyquist criterion, (6.19), requires encirclements instead. From (6.19), for stability the net number of encirclements of $(-1, j0)$ must equal the number of positive poles of GH.

Bode plots cover only the positive frequency range of GH and, for nonminimum-phase circuits, are likely to be misleading. But for minimum-phase circuits, stability and (to some extent) major response characteristics can be readily determined from them. Since most circuits are minimum-phase, we can usually use Bode plots.

Relative stability is measured by gain and phase margins. The *gain margin* is the difference between unity and the gain at $\phi = -180°$. The *phase margin* (PM) is the difference between the phase at unity gain and $-180°$ and is the amount of additional phase lag that will make the circuit unstable. Although second-order circuits are unconditionally stable, phase margin still describes relative stability whereas gain margin is infinite. Therefore, phase margin is usually more meaningful in circuits than gain margin.

Gain and phase margins are related to second-order response parameters such as ζ, M_p, and M_m. As the margins decrease, the closed-loop damping ratio ζ_c decreases, and M_{pc} and M_{mc} increase. For second order feedback circuits with no finite zeros, the relationship between PM and ζ_c is approximately

$$\zeta_c \cong \frac{\text{PM}}{100}, \quad \text{PM in deg}, \quad 0 < \zeta_c < 0.7, \quad 0 < \text{PM} < 64°, \quad \text{2nd-order} \tag{6.20}$$

Since overshoot is a function of ζ_c [by combining (5.131) and (5.132)], PM can be expressed in terms of overshoot:

$$M_{pc} \cong 75 - \text{PM}, \qquad M_{pc} \text{ in \%}, \quad \text{PM in deg}, \quad \text{PM} > 20°, \quad \text{2nd-order} \tag{6.21}$$

From (5.149), M_{mc} is also a function of ζ_c and can be related similarly to PM. Since pole angle is $\cos^{-1} \zeta_c$, a phase margin of 50° corresponds to a pole angle of 60° and an overshoot of 25%. This is greater than the 16% of an open-loop second-order circuit (see Section 5.8). The exact relationship between PM and ζ_c is found by choosing

$$G(s) = \frac{1}{s(s/\omega_{nc} + 2\zeta_c/\omega_{nc})}, \qquad H = 1$$

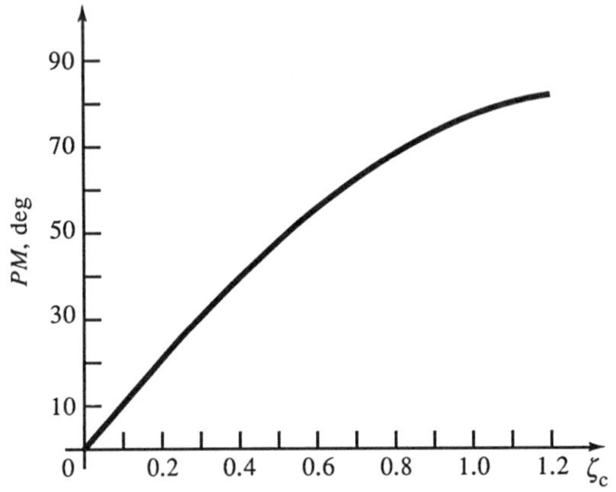

ζ_c	PM, deg
0.00000000	0.00000000
0.50000000	5.7247926
0.10000000	11.420619
0.15000000	17.057280
0.20000000	22.602284
0.25000000	28.020176
0.30000000	33.272491
0.35000000	38.318598
0.40000000	43.117598
0.45000000	47.631221
0.50000000	51.827292
0.55000000	55.683028
0.60000000	59.187267
0.65000000	62.341037
0.70000000	65.156393
0.75000000	67.653970
0.80000000	69.859999
0.85000000	71.803450
0.90000000	73.513722
0.95000000	75.019028
1.0000000	76.345415
1.0500000	77.516297
1.1000000	78.552331
1.1500000	79.471514
1.2000000	80.289394

FIG. 6.12 Phase margin as a function of damping ratio for a second-order system.

This choice results in a closed-loop transfer function with only a quadratic pole factor. Solve for the unity-gain crossover frequency ω_T. Then solve for the phase margin, and substitute ω_T. The result, in radians, is

$$\mathrm{PM} = \frac{\pi}{2} - \tan^{-1}\left\{\frac{1}{2\zeta_c\sqrt{2\zeta_c^2 + \sqrt{4\zeta_c^4 + 1}}}\right\} \tag{6.22}$$

Based on this result, the error of (6.20) is calculated to be less than ±5 degrees. Figure 6.12 is a plot of this function, converted to units of degrees.

Example 6.3 Two-Pole Feedback Amplifier Stability

A feedback amplifier has two poles in *G*, none in *H*, and has a closed-loop step response that has 45% overshoot. The dc loop gain is 20. What is its phase margin and damping ratio?

The loop gain has only a quadratic pole factor, so the previous formulas apply. (If *G* has zeros or *H* has poles, they become zeros of the closed-loop gain, and the quadratic-pole analysis does not apply.) The closed-loop transfer function is similar to (6.14), where KH is the dc loop gain. For overshoot, ζ_c is calculated from (5.131) (or found in the table in Section 5.8) and is $\zeta_c \cong 0.25$. From (6.16), the open-loop ζ is

$$\zeta = \sqrt{KH+1} \cdot \zeta_c = \sqrt{21} \cdot (0.25) = 1.13$$

ζ can also be calculated by using the approximations of (6.20) and (6.21) with (6.16):

$$\zeta \cong \frac{(75 - M_{pc})}{100}\sqrt{KH+1} = 1.37$$

This overdamped open-loop response becomes underdamped when the loop is closed. From (6.20), PM $\cong 100\zeta_c = 25°$.

Example 6.4 Transimpedance Amplifier with Input Capacitance

The amplifier of Fig. E6.4 consists of a voltage amplifier with voltage gain $G(s)$. The feedback blocks are

$$\alpha_i = R \,\Big\|\, \frac{1}{sC} = \frac{R}{sRC+1}, \qquad H = -\frac{1}{sRC+1}$$

The closed-loop transimpedance is

$$R_m(s) = \alpha_i(s)\frac{G(s)}{1+G(s)H(s)} = \frac{R}{sRC+1} \cdot \frac{G(s)}{1 - G(s)/(sRC+1)}$$

$$= R \cdot \frac{G(s)}{sRC + 1 - G(s)}$$

For a single-pole amplifier,

$$G(s) = \frac{-K}{s\tau_G + 1}$$

and

$$R_m(s) = -R\left(\frac{K}{K+1}\right)\frac{1}{s^2[RC\tau_G/(K+1)] + s[(RC+\tau_G)/(K+1)] + 1}$$

To find PM, because of the zero in $G/(1+GH)$ due to H, we cannot use the second-order approximations even though $R_m(s)$ has no finite zeros. Instead we apply a more general approach using the Bode plot.

For $K = 999$, $R = 1\ \text{M}\Omega$, $\tau_G = 159\ \mu\text{s}$ ($p_G = 1$ kHz), and $C = 5$ pF, then at dc, $R_m(0) = -999\ \text{k}\Omega$, and the closed-loop poles are at

$$(1\ \text{kHz})(1000) = 1\ \text{MHz}, \qquad \frac{1}{2\pi RC}(1000) = (31.8\ \text{kHz})(1000)$$

$$= 31.8\ \text{MHz}$$

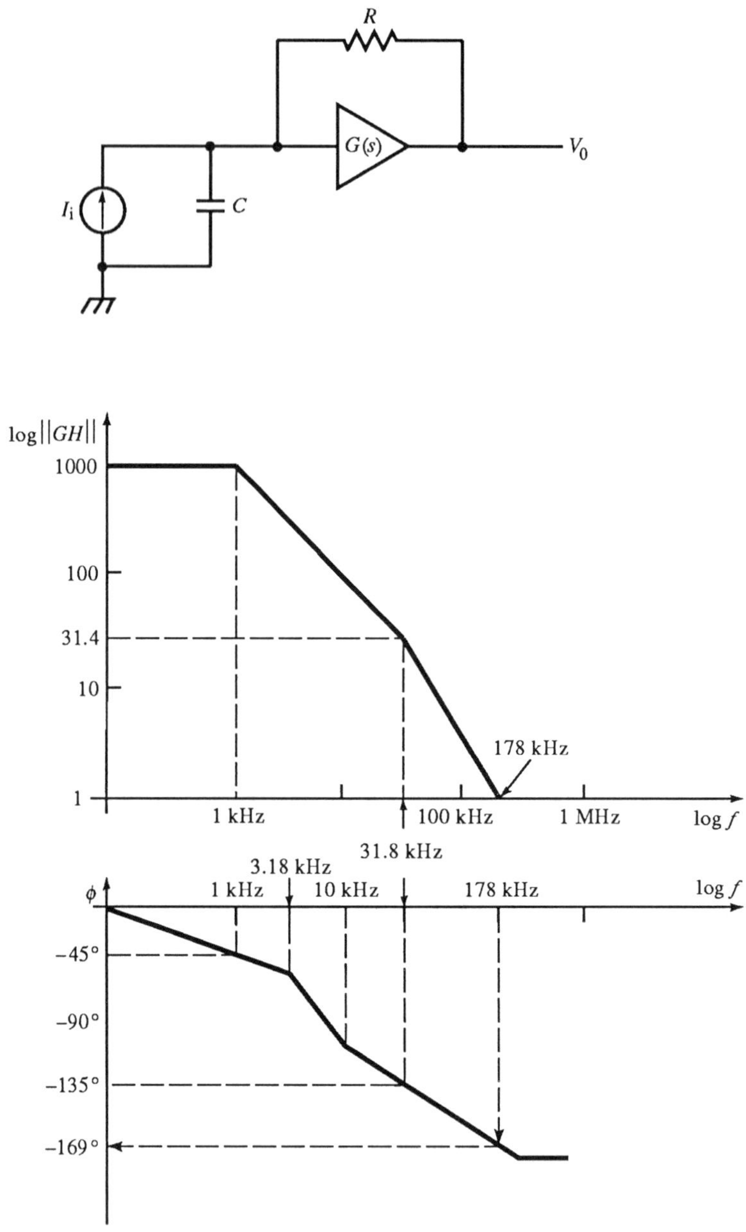

R
G(s)
V_0
I_i
C
log||GH||
1000
100
31.4
10
1
178 kHz
1 kHz
100 kHz
1 MHz
log f
31.8 kHz
3.18 kHz
ϕ
1 kHz
10 kHz
178 kHz
log f
-45°
-90°
-135°
-169°

FIG. E6.4

These poles are not too close, but the dc loop gain is high. The damping ratio is low:

$$\zeta_c = \frac{b}{2\sqrt{a}} = 0.092$$

From the ideal Bode plot, the PM is about 11°, which is nearly unstable. PM is found as follows. First, we find the gain at the higher pole. It is 31.8 kHz/1 kHz = 31.8 times less than the first pole. The gain slope between poles is −1, so the ratio of gains is also 31.8. The open-loop gain at 31.8 kHz is thus 1000/31.8 = 31.4. The magnitude then decreases at a slope of −2 and crosses unity gain at

$$f_T = (31.8\text{ kHz})\sqrt{31.4} = (31.8\text{ kHz})(5.60) = 178\text{ kHz}$$

We now know f_T and proceed to plot the phase. The phase lags of the two poles overlap. The 1 kHz pole phase range extends from 100 Hz to 10 kHz, and the 31.8 kHz pole range is from 3.18 kHz to 318 kHz. In the overlap (between 3.18 kHz and 10 kHz), the phase slope is twice that due to a single pole. For a single pole, phase changes −90° in two decades, for a −45°/dec slope. In the overlap, it is −90°/dec. At the higher pole, $\phi = -135°$. To find the additional phase lag to f_T, we calculate the number of decades and multiply by the phase slope, and then add −135°:

$$\log\left(\frac{178\text{ kHz}}{31.8\text{ kHz}}\right)(-45°/\text{dec}) - 135° = -169°$$

Then PM = −169° − (−180°) = 11°. The result from the exact Bode plot is also 11°.

Most feedback circuits have more than two poles and are capable of instability. Feedback circuit compensation relies on an intuitive understanding of how pole and zero placement affects stability. Although optimal compensator design techniques exist, they are rarely the most expedient, cost-effective, or reliable ways to compensate most feedback circuits.

Consider a loop gain with n poles at frequency $\omega = p$. The Bode plots of magnitude and phase are shown in Fig. 6.13. The magnitude rolls off at p with a slope of $-n$. The phase lags by $-45° \cdot n$ at p and rolls off at $-45° \cdot n/\text{dec}$. The frequency at which the asymptotic approximation for phase crosses −180° is

$$\omega_\phi = p \cdot 10^{(4-n)/n} \tag{6.23}$$

Similarly for the unity-gain crossover frequency ω_T of the magnitude with dc gain of K:

$$\omega_T = p \cdot K^{1/n} \tag{6.24}$$

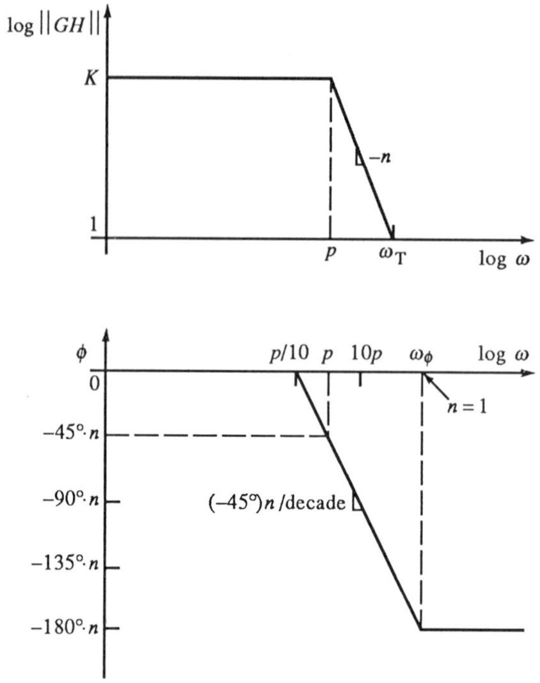

FIG. 6.13 Bode plot for n poles at p. Phase margin rapidly decreases as number of poles increases.

For stability, $\omega_T < \omega_\phi$ or $\omega_\phi - \omega_T > 0$. Subtracting $\log \omega_T$ from $\log \omega_\phi$ gives

$$\log \omega_\phi - \log \omega_T = \frac{1}{n} \log\left(\frac{10^{4-n}}{K}\right) > 0 \tag{6.25}$$

As n increases, ϕ rolls off toward $-180°$ faster than $\|GH\|$ does toward unity. As the frequency difference of (6.25) approaches zero, the maximum K for stability is approximately

$$K < 10^{4-n} \tag{6.26}$$

For $n = 4$, $K < 1$, which is hardly a useful feedback circuit. If the poles are complex, the situation is worse: ϕ rolls off even faster. Frequency plots for a three-pole circuit are shown in Fig. 6.8f.

Now consider the effect on stability of separating the poles. For a two-pole A_v, (6.14), ζ_c increases with open-loop ζ. ζ can be expressed in p_1 and p_2 by multiplying the factors of (6.13). The coefficients yield

$$\omega_n = \sqrt{p_1 p_2}, \qquad \zeta = \frac{1}{2} \cdot \frac{p_1 + p_2}{\sqrt{p_1 p_2}} \tag{6.27}$$

ω_n is the geometric mean of the two poles and lies midway between them on a Bode plot. Relate the poles by a constant γ:

$$p_1 = \gamma p_2, \qquad \gamma \geq 0 \tag{6.28}$$

Then

$$\zeta = \frac{(\gamma + 1)}{2\sqrt{\gamma}} \tag{6.29}$$

Minimum ζ is 1 when $\gamma = 1$ for real poles. For maximum pole separation of $\gamma = 0$ or ∞, $\zeta = \infty$. For the two-pole case, maximum pole separation increases stability.

A root-locus plot of two real poles, maximally separated, shows that they must travel a maximum distance along the real axis before meeting and becoming complex. This can be generalized from inspection of Fig. 6.8f for three poles. A heuristic stability rule suggested by these observations is

- Pole separation increases stability.

6.5 Compensation Techniques

Compensation is often necessitated by circuit imperfections. Parasitic circuit elements, unavoidable reactive input and output loading, and undesirable amplifier frequency response are the major reasons. Some of these are shown in Fig. 6.14. The power-supply leads to the op-amp terminals contribute series inductance (L_1 and L_2). Stray capacitance from the supply terminals to the inputs is significant if appreciable high-frequency ac voltage is present at the supply terminals. The op-amp inputs have some internal capacitance to ground, causing a shunt RC with R_i. The op-amp output is an equivalent shunt RL in series with a voltage source. The inductance is due to gain roll-off above the op-amp bandwidth (to be studied in the next chapter). This output impedance can resonate with a capacitive load. Furthermore, the op-amp usually has several poles. All of this amounts to a "naturally occurring" unstable circuit requiring response compensation.

A clue to compensation comes from studying Fig. 6.8, in which we see that multiple poles cause instability (with sufficient gain), as in Fig. 6.8f. The inclusion of a zero in the loop gain causes poles that would head to the right to be "pulled back" from their course toward the $j\omega$ axis. Root-locus rule 6 from Section 6.3 is intuitively powerful for envisioning where the poles of separate loci will move. They maintain a fixed centroid on the real axis so that pole movement, say, to the left, is accompanied by corresponding pole movement to the right (as in Fig. 6.8f). When a zero terminates the movement of a pole to the left (as in Fig. 6.8g), the poles moving right also cease moving in that direction. Depending on the order of poles and zero, various loci occur but always act according to rule 6 (as seen in Fig. 6.8h–j). Adding LHP zeros to the loop gain enables the response to be compensated. So another heuristic stability rule is

- LHP zeros tend to increase stability.

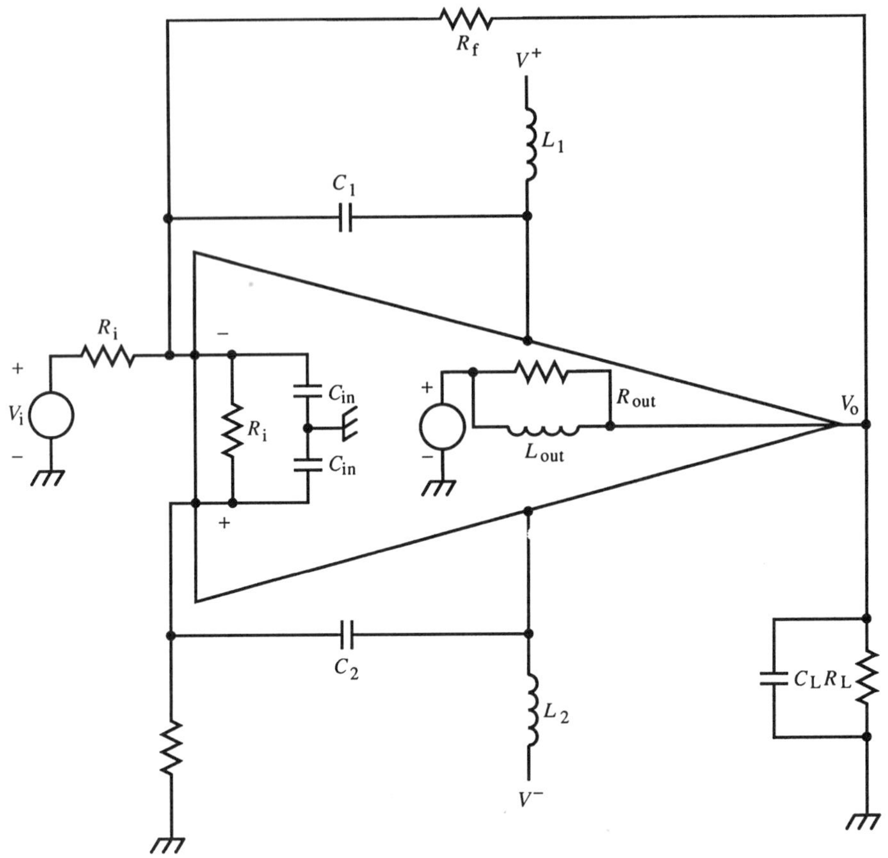

FIG. 6.14 An inverting op-amp circuit with parasitic elements that degrade performance or stability.

We now consider how to apply these heuristic guidelines more specifically as compensation techniques.

Pole-zero cancellation places the compensator zero on an offending pole of *GH*. If the compensator pole is far removed from its zero, then the offending pole is effectively shifted far away. Pole-zero cancellation is demonstrated in Fig. 6.15a.

Phase-lead compensation places the zero near ω_ϕ, where $\phi = -180°$. This prolongs a stable phase while magnitude continues to roll off toward unity. The compensator pole is an implementation side-effect that must be put somewhere. Since the zero is placed where phase lead is needed, the pole should be placed at a higher frequency, beyond ω_T, where the additional phase lag it contributes will occur beyond where the magnitude crosses unity. Phase-lead compensation is demonstrated in Fig. 6.15b. Because phase-lead compensation occurs at high frequencies, it mainly affects the transient response.

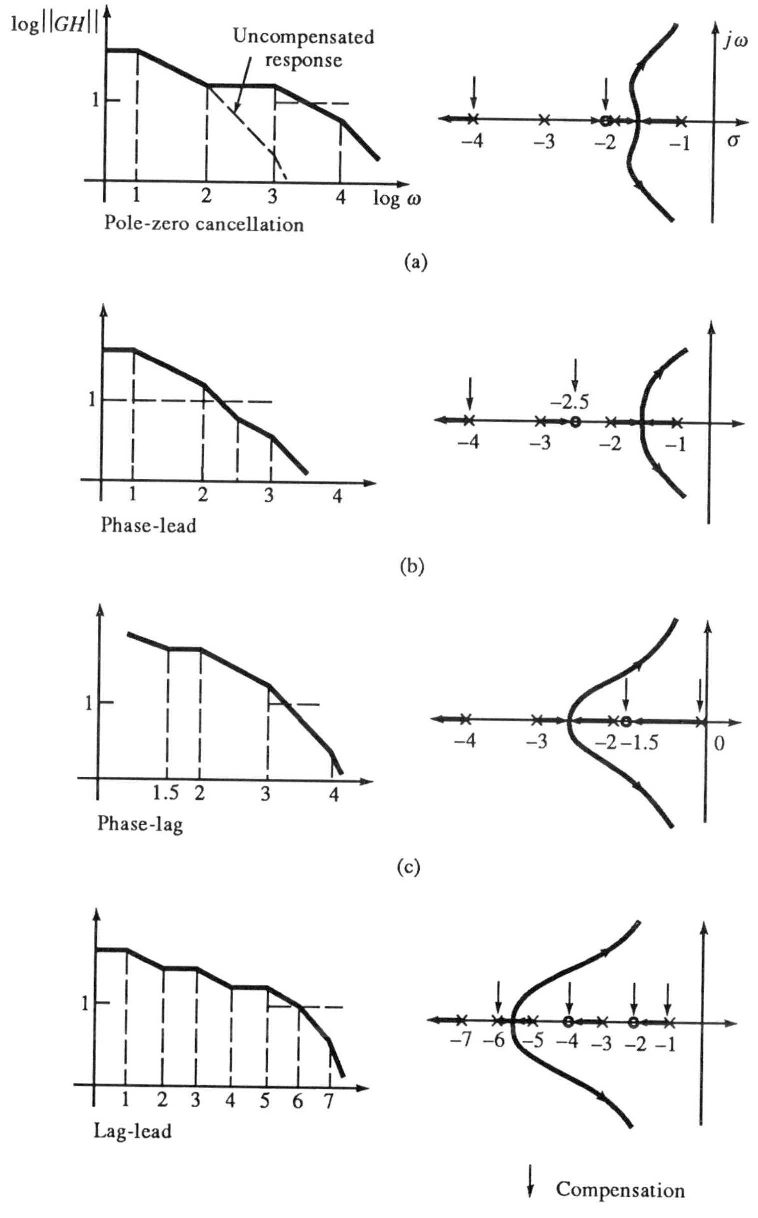

FIG. 6.15 Some first-order compensation techniques: (a) pole-zero cancellation, (b) phase-lead, (c) phase-lag, and (d) lag-lead compensation. The compensator poles and zeros are indicated by the downward arrows.

Phase-lag compensation places the compensator pole at a lower frequency than the zero. The idea is to introduce the pole at a frequency below the poles of the loop, where the magnitude is flat. By decreasing the magnitude while the phase lag is still small and then correcting it with a zero, we can reduce the magnitude while contributing little phase lag. This technique allows higher dc loop gain and consequently smaller steady-state error.

Phase-lag compensation is demonstrated in Fig. 6.15c. It mainly affects low-frequency response error since the compensating pole and zero are placed at low frequencies, relatively near dc. The step response can have a long-lasting exponential decay (or "tail") (see Fig. 5.15a) before settling to the steady-state value. Whenever a low-frequency pole and zero are meant to cancel but are misaligned, a *dipole* is created with a time-domain response that shows a long-lasting exponential.

Lag-lead compensation is a combination of lag and lead compensation, in which two poles and two zeros are introduced into the loop (Fig. 6.15d). Both techniques may be required to stabilize amplifiers with many close poles.

Pole separation can itself be a technique. If the poles are far enough apart, the magnitude, starting from the lower-frequency poles, has enough frequency range to decrease to unity gain before excessive phase lag accumulates. An important instance of pole separation is *dominant-pole compensation*, in which one pole is placed at a frequency much lower than the others (and thus dominates the response). Another pole separation technique is *pole-splitting*, in which a low-frequency zero is introduced to pull an adjacent pole toward it; all the while the next higher frequency pole increases in frequency. The effect is just the opposite of what is usually expected on a root-locus plot; the poles separate instead of moving toward each other.

One of the simplest of all compensation techniques is *dc loop gain reduction*. This may not be desirable in many applications due to its reduction in the beneficial effects of feedback. But for circuits with abundant loop gain (such as many op-amp circuits), this can be an attractive technique.

Although these techniques are usually sufficient to achieve desirable response, combinations of them may be necessary for highly unstable amplifiers. In addition to stabilization of the loop with compensators, stages in GH can be individually compensated. Sometimes a transistor causes an oscillation and must be stabilized before overall loop response can even be considered. Therefore, good design practice is to start with an evaluation of stage responses before considering loop response.

The techniques described in this section have various realizations in analog circuitry. But a technique and its various realizations (and how to design them) are different considerations, just as filter types (Butterworth, Bessel, etc.) have corresponding circuit realizations (state-variable, negative impedance converter, Sallen–Key, etc.). The limitations on circuit topology can affect the choice of technique (bottom-up design) though ideally the nature of the problem determines the best choice of technique (top-down design). We now turn our attention to various ways that these compensation techniques can be realized as analog circuits.

6.6 Compensator Design: Compensating with Zeros in *H*

First, we consider how zeros can be realized and at what frequencies they should be placed. Realizable circuits have no fewer poles than zeros. This complicates compensation because we also must be careful where the added poles are placed. If the pole is less than the zero, the response of Fig. 6.8a results; if the zero is less than the pole, Fig. 6.8b results.

Some passive compensator circuit realizations are shown in Fig. 6.16. (The first two are the same as in Figs. 5.18d, a, respectively.) For the phase-lead (Fig. 6.16a) and phase-lag (Fig. 6.16b) compensators, the separation of pole and zero depends on the ratio of R_2 to $R_1 + R_2$. For effective compensation, this separation must be significant; therefore, the values of R_1 and R_2 must be significantly different. The lag-lead compensator of Fig. 6.16c has the following transfer function:

$$\text{lag-lead compensator} \Rightarrow \frac{(sR_1C_1+1)(sR_2C_2+1)}{s^2R_1C_1R_2C_2+s(R_1C_1+R_1C_2+R_2C_2)+1} \tag{6.30}$$

From Fig. 6.16c, the conditions on pole and zero placement are

$$p_1 < z_1 \ll z_2 < p_2 \tag{6.31}$$

and $R_1C_1 \ll R_2C_2$. The wide separation of these critical frequencies is desirable. By choosing the separation of the zeros, we can determine the pole and zero pair separations. A trade off between these separations must be based on the particular amplifier requirements.

These are not the only compensator realizations. The uncompensated amplifier topology affects choice of design, especially if the compensator can be synthesized from part of the given topology. We now examine some particular amplifier compensations.

The op-amp of Fig. 6.17 has one op-amp pole p in $G(s)$. Another pole is due to the input capacitance C_i. If the poles are too close, compensation may be needed. Inserting a phase-lead compensator in cascade with either the input or output of the op-amp is undesirable because it decreases input resistance or increases output resistance. Since H consists of a voltage divider, it can be modified to form a phase-lead compensator. The topology of H is familiar; it is an uncompensated voltage divider. A compensation capacitor C_f is placed in parallel with R_f. Then

$$H = -\left(\frac{R_i}{R_f+R_i}\right)\frac{sR_fC_f+1}{s(R_f \parallel R_i)(C_f+C_i)+1} \tag{6.32}$$

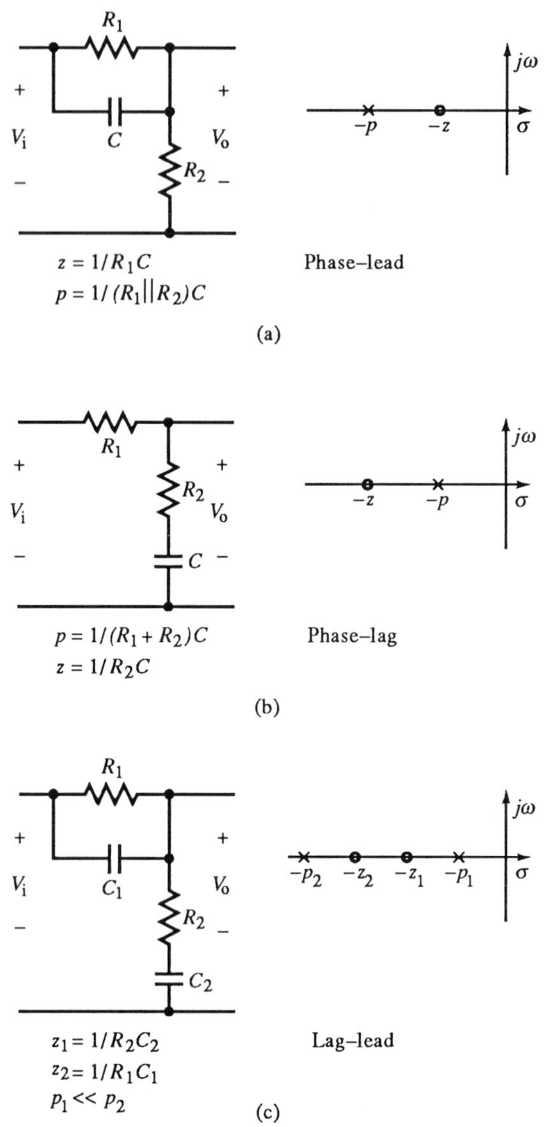

FIG. 6.16 Passive first-order compensators: (a) phase-lead, (b) phase-lag, and (c) lag-lead.

Now H is equivalent to (6.1), the compensated divider formula. If we set pole and zero equal, H becomes an all-pass network, and the pole due to C_i is cancelled. In this case,

$$C_f = \left(\frac{R_f}{R_i}\right) C_i \tag{6.33}$$

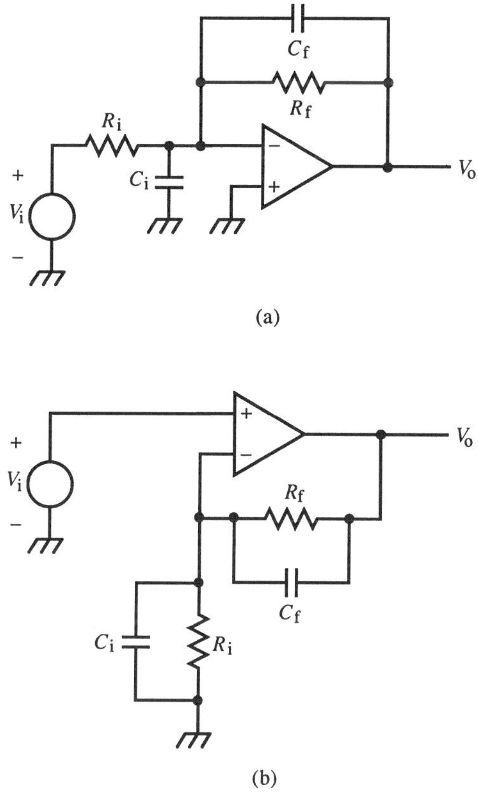

FIG. 6.17 Phase-lead compensation of op-amp input capacitance with shunt feedback capacitor C_f in H for (a) inverting and (b) noninverting configurations.

Example 6.5 Op-Amp Input Capacitance Compensation

The amplifiers of Fig. 6.17 have values of $R_i = 47\ \text{k}\Omega$, $R_f = 220\ \text{k}\Omega$, $C_i = 100\ \text{pF}$, $K = 100\ \text{k}$, $p = 10$ Hz. The uncompensated loop has an op-amp pole at 10 Hz and a pole due to C_i at $1/(220\ \text{k}\Omega \parallel 47\ \text{k}\Omega)(100\ \text{pF})$ or 41.1 kHz (Fig. E6.5a). From the Bode plot of GH (Fig. E6.5b), $f_T \cong 80$ kHz and PM $\cong 26°$. (We cannot apply (6.20) or (6.21) here because H has a pole. Poles of H become zeros of the closed-loop gain.)

The addition of C_f creates a zero at $1/(220\ \text{k}\Omega)C_f$. According to (6.33), $C_f = 470$ pF. The compensated response has a single pole at 10 Hz.

We could have chosen to cancel p with the zero instead. The compensated response (Fig. E6.5a) has a maximum phase lag of −90°, leaving a 90° PM. If p were cancelled, C_f would be 72.3 nF, and the magnitude plot would be flat to the compensator pole, now at only 56.8 Hz, with single-pole roll-off from this pole. This alternative provides greater loop bandwidth. For the inverting op-amp, the larger C_f causes a lower-frequency pole in α_i, reducing closed-loop bandwidth considerably.

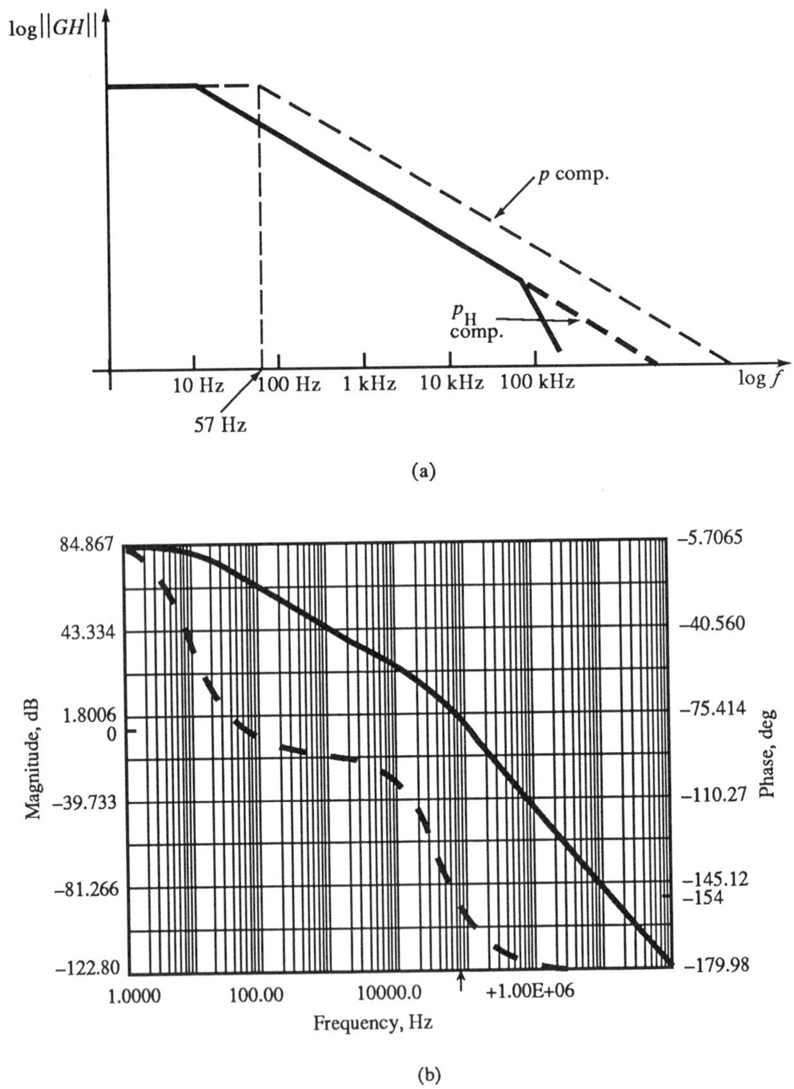

FIG. E6.5 (a) 10-Hz pole cancellation strategy. (b) Open-loop response. (c) Closed-loop response. (d) Closed-loop step response.

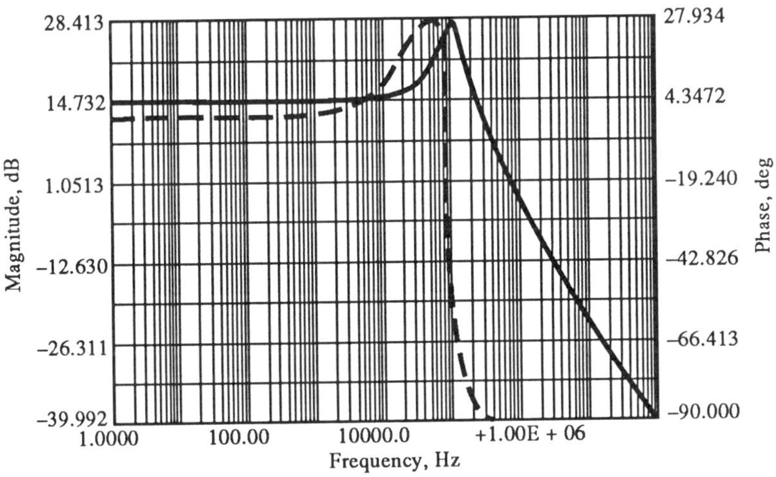

(c)

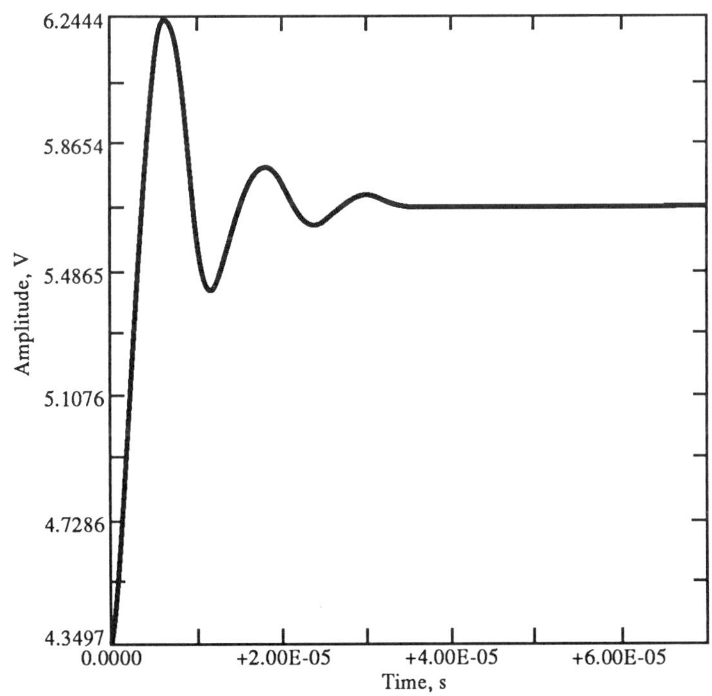

(d)

FIG. E6.5 (continued)

The maximum amount of phase lead that a phase-lead compensator introduces into a loop depends on the separation of its pole and zero. From the asymptotic approximation for phase, both pole and zero linearly affect phase for one decade on each side of them. If they are separated by two decades, the zero achieves a full 90° of phase lead before the pole begins to take effect. Consequently,

$$\text{maximum phase lead} = \begin{cases} (45^\circ)\log\left(\dfrac{p}{z}\right), & 1 \le \left(\dfrac{p}{z}\right) \le 100 \\ 90^\circ, & \left(\dfrac{p}{z}\right) > 100 \end{cases} \tag{6.34}$$

The frequency of maximum phase lead is at $\sqrt{pz}$ or about $p/10$, where the pole begins to cancel the effect of the zero. The frequency range over which phase-lead compensation occurs is

$$\text{phase-lead frequency range} = \begin{cases} \log(p/z)\ \text{dec}, & 1 \le p/z \le 100 \\ 2\ \text{dec}, & p/z > 100 \end{cases} \tag{6.35}$$

Example 6.6 Op-Amp Phase-Lead Compensation

The op-amp of Fig. E6.6a has a gain of 2.2M and poles at 100 Hz and 1 MHz (Fig. E6.6b). The Bode plot of GH (without C_f) shows that phase lag approaches $-180°$, causing oscillatory response. C_f is introduced to phase-lead compensate the loop.

First, Bode plot gain at the second pole frequency and the unity-gain frequency f_T are of interest. On a log–log plot, a line with slope n relates changes in magnitude and frequency according to

$$\left(\frac{y_2}{y_1}\right) = \left(\frac{f_2}{f_1}\right)^n \tag{E1}$$

The loop gain is $2.2\text{M}/11 = 200\text{k}$. The pole separation of p_1 and p_2 is 1 MHz/100 Hz or 4 decades. With a -1 slope, the loop gain is 4 decades reduced, or $200\ \text{k}/10^4 = 20$. For f_T, the slope is -2, and

$$f_T = (1\ \text{MHz})\sqrt{20/1} \cong 4.47\ \text{MHz}$$

The phase plot (Fig. E6.6b, dashed curve) shows a phase lag of $-169°$ at f_T, and PM $= 11°$. At p_2, the phase lag is $-135°$, and the phase rolls off at $-45°$/dec. At 4.47 MHz, or an additional log(4.47 MHz/1 MHz) = 0.65 dec, this is an additional 29°, or a total of $-164°$. This linear approximation is $+5°$ in error. The phase plot value results in $\zeta \cong 0.11$, a pole angle of about 84°, and a step-response overshoot M_p of 71%. This response is too underdamped for most applications.

The maximum phase lead that can be introduced is

$$(45°)\log\left(\frac{p}{z}\right) = (45°)\log\left[\frac{(R_f \parallel R_i)C_f}{R_f C_f}\right] = (45°)\log\left(\frac{R_f}{R_i}+1\right) \quad \text{(E2)}$$

or $(45°)(\log 11) \cong 47°$. The phase lead of the zero acts over a frequency range of $\log(11) \cong 1.04$ dec. If the high end of this range is placed at the compensated f_T, or f_{Tc}, then phase lag is held constant from $f_{Tc}/(p/z)$ to f_{Tc}. This placement of phase-lead range is accomplished by noting that p begins to affect phase at $p/10$. So we set

$$f_{Tc} = \frac{p}{10}, \qquad z = \frac{10 f_{Tc}}{(p/z)} \quad \text{(E3)}$$

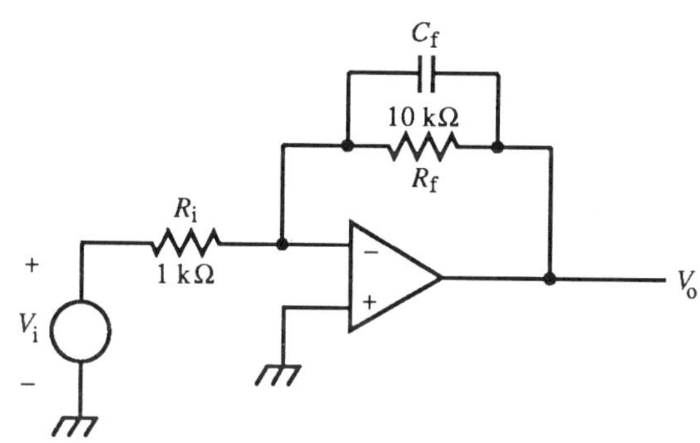

(a)

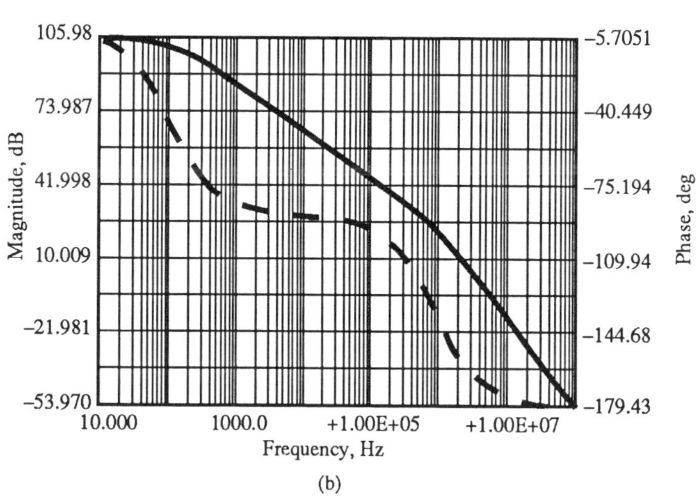

(b)

FIG. E6.6 (a) Circuit diagram. (b) Open-loop response. (c) Closed-loop response. (d) Response compensator. (Figure continues.)

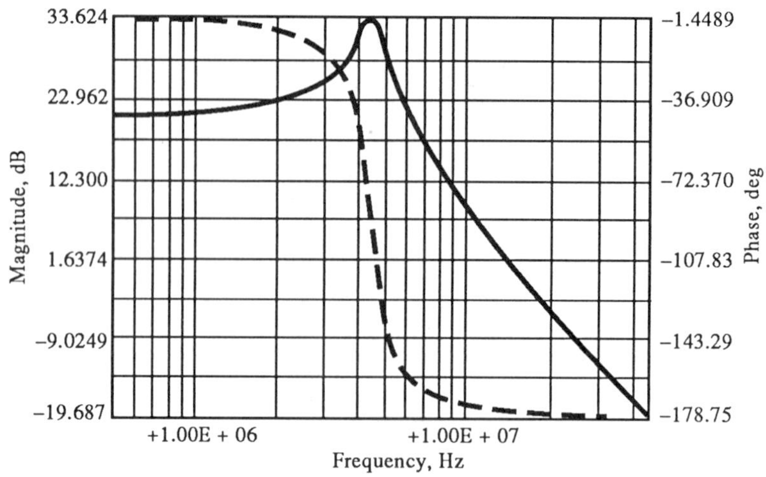

(c)

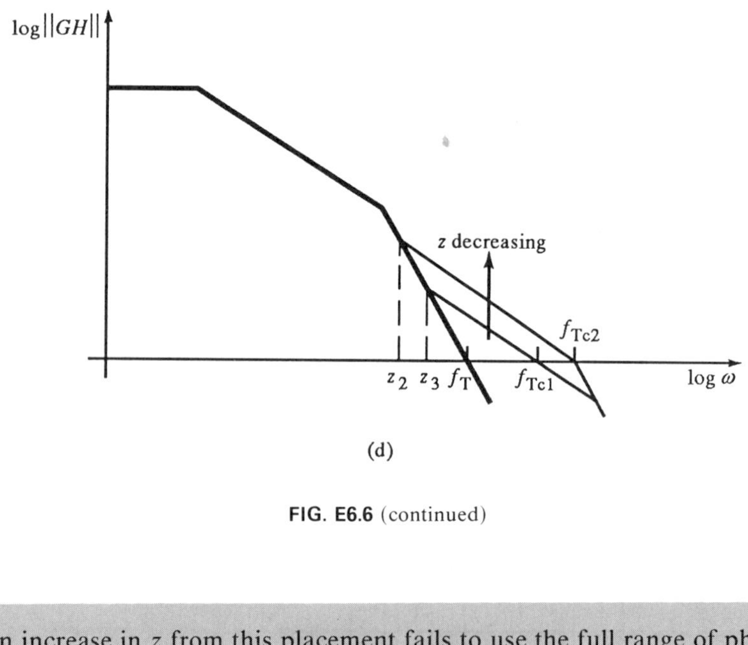

(d)

FIG. E6.6 (continued)

An increase in z from this placement fails to use the full range of phase lead. A decrease in z increases f_{Tc} until the break to a -2 slope occurs at f_{Tc2} (Fig. E6.6d). Then f_{Tc} remains fixed as z continues to decrease. When a decreasing z increases f_{Tc}, the magnitude slope at f_{Tc} is -1, and phase is decreasing. For maximum PM, phase should begin to decrease again at f_{Tc}. The phase-lead range is then placed with the high end at f_{Tc}.

We still must relate (E3) to the uncompensated plot. At z the two plots roll off at their respective slopes to f_{T} and f_{Tc}. Since their gain change is the same, their locations depend on the slope differences.

Consequently, using (E1) gives

$$\left(\frac{f_{\mathrm{Tc}}}{z}\right)^{-1} = \left(\frac{f_{\mathrm{T}}}{z}\right)^{-2} \Rightarrow f_{\mathrm{T}} = \sqrt{f_{\mathrm{Tc}} z}$$

Combining with (E3), we obtain

$$f_{\mathrm{T}} = \sqrt{\frac{pz}{10}}, \qquad z = f_{\mathrm{T}}\sqrt{\frac{10}{p/z}}, \qquad p = f_{\mathrm{T}}\sqrt{10\left(\frac{p}{z}\right)} \tag{E4}$$

For the example, $f_{\mathrm{T}} - 4.47$ MHz and $p/z = 11$. Then

$$z = 4.26\ \mathrm{MHz}, \qquad p = 11z = 46.9\ \mathrm{MHz}, \qquad f_{\mathrm{Tc}} = \frac{p}{10} = 4.69\ \mathrm{MHz}$$

At $z/10 = 426$ kHz, the phase is

$$\log\left(\frac{426\ \mathrm{kHz}}{1\ \mathrm{MHz}}\right)(45°) - 135° = -118°$$

an improvement of about 46° (as calculated from (E2)) over the uncompensated amplifier. The compensated PM $\simeq$ 57°.

Phase-lag compensation of the amplifiers of Fig. 6.17 can be implemented in H by connecting a series RC between the op-amp input terminals (Fig. 6.18). Given the two-pole op-amp of (6.13),

$$\begin{aligned} H(s) &= -\left(\frac{R_{\mathrm{i}}}{R_{\mathrm{f}} + R_{\mathrm{i}}}\right)\frac{sR_{\mathrm{c}}C_{\mathrm{c}} + 1}{s^2[R_{\mathrm{c}}C_{\mathrm{c}}R_{\mathrm{p}}C_{\mathrm{i}}] + s[R_{\mathrm{p}}(C_{\mathrm{c}} + C_{\mathrm{i}}) + R_{\mathrm{c}}C_{\mathrm{c}}] + 1} \\ &= -\left(\frac{R_{\mathrm{i}}}{R_{\mathrm{f}} + R_{\mathrm{i}}}\right) \\ &\quad \times \frac{sR_{\mathrm{c}}C_{\mathrm{c}} + 1}{(sR_{\mathrm{p}}C_{\mathrm{c}} + 1)[sR_{\mathrm{c}}C_{\mathrm{i}} + (1 + C_{\mathrm{i}}/C_{\mathrm{c}} + R_{\mathrm{c}}/R_{\mathrm{p}} - R_{\mathrm{c}}C_{\mathrm{i}}/R_{\mathrm{p}}C_{\mathrm{c}})] - (C_{\mathrm{i}}/C_{\mathrm{c}} + R_{\mathrm{c}}/R_{\mathrm{p}} - R_{\mathrm{c}}C_{\mathrm{i}}/R_{\mathrm{p}}C_{\mathrm{c}})} \end{aligned} \tag{6.36}$$

where $R_{\mathrm{p}} = R_{\mathrm{f}} \parallel R_{\mathrm{i}}$. For $R_{\mathrm{c}} \ll R_{\mathrm{p}}$ and $C_{\mathrm{c}} \gg C_{\mathrm{i}}$,

$$H(s) \cong -\left(\frac{R_{\mathrm{i}}}{R_{\mathrm{f}} + R_{\mathrm{i}}}\right)\frac{sR_{\mathrm{c}}C_{\mathrm{c}} + 1}{(sR_{\mathrm{p}}C_{\mathrm{c}} + 1)(sR_{\mathrm{c}}C_{\mathrm{i}} + 1)} \tag{6.37}$$

The effect of this compensation is to add a pole and zero. Because $R_{\mathrm{c}} \ll R_{\mathrm{p}}$, the pole frequency, $1/R_{\mathrm{p}}C_{\mathrm{c}}$, is less than the zero frequency, $1/R_{\mathrm{c}}C_{\mathrm{c}}$. Since $C_{\mathrm{c}} \gg C_{\mathrm{i}}$, this zero is less than the second pole $1/R_{\mathrm{c}}C_{\mathrm{i}}$. This results in a pole and zero ordering of

$$p < \frac{1}{R_{\mathrm{p}}C_{\mathrm{c}}} < \frac{1}{R_{\mathrm{c}}C_{\mathrm{c}}} < \frac{1}{R_{\mathrm{p}}C_{\mathrm{i}}} < \frac{1}{R_{\mathrm{c}}C_{\mathrm{i}}} \tag{6.38}$$

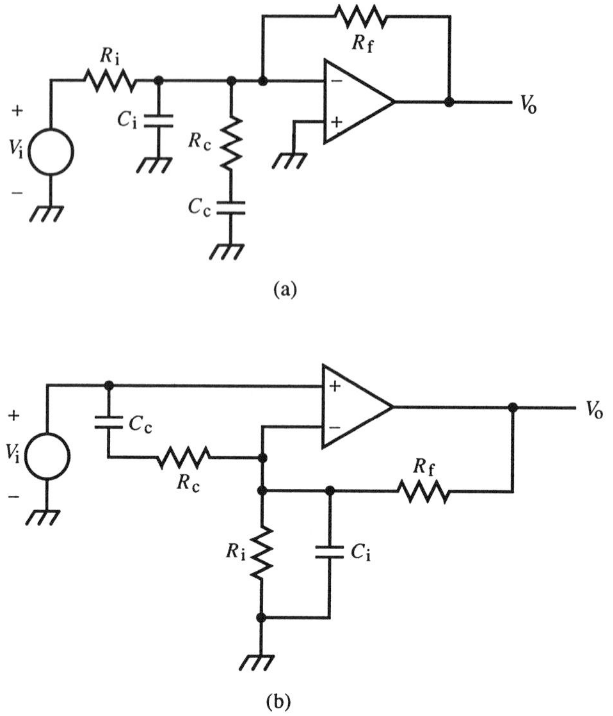

FIG. 6.18 Phase-lag compensation of op-amp at input by R_c and C_c for (a) inverting and (b) noninverting configurations.

The effect of the compensation network is shown in Fig. 6.19, which demonstrates phase-lag compensation. Above $1/R_pC_c$, the magnitude decreases at a steeper slope at small phase angles. Then, as ω approaches unity gain, the zero is introduced (at $1/R_cC_c$) to reduce the phase angle slope and increase phase margin.

Another way to add a zero to H when a feedback capacitor is used for compensation is to add a resistor in series with it. In Fig. 6.4b, the effect of R_c is to add a zero at $1/R_cC_c$ and move the pole at $1/R_fC_c$ down in frequency to $1/(R_f + R_c)C_c$.

Example 6.7 Phase-Lag Compensation

We want to phase-lag compensate the amplifier of Fig. E6.7 for maximum PM. The op-amp has poles at $p_1 = 100$ kHz and $p_2 = 1$ MHz and a gain of 220 k. A phase-lag compensator is realized by adding a series RC from the inverting input of the op-amp to ground with elements R_c and

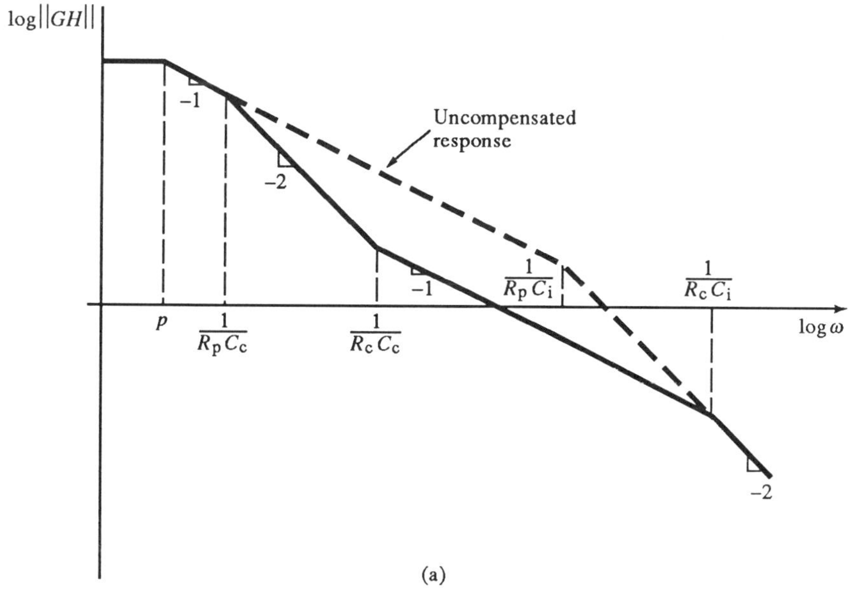

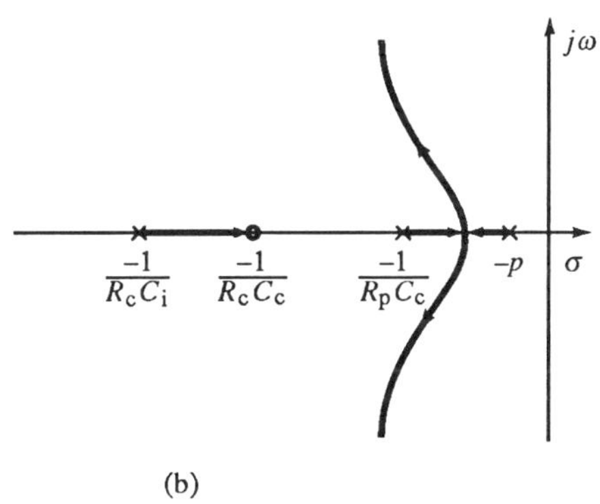

FIG. 6.19 Bode plot (a) and root-locus plot (b) showing the effects of phase-lag compensation. The gain is decreased at low frequencies where the phase lag is small. The phase is restored with a zero, but at a decreased gain.

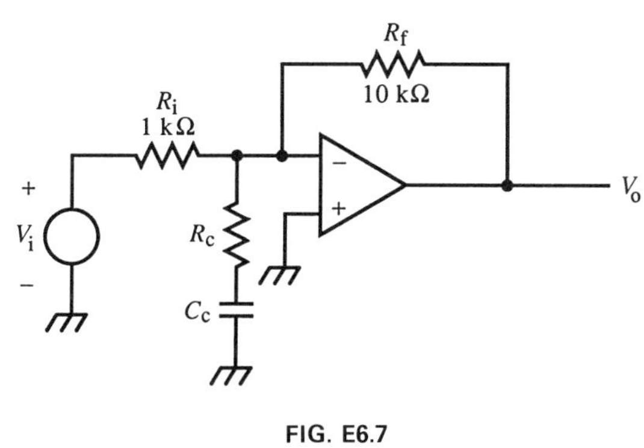

FIG. E6.7

C_c. Then

$$H(s) = -\left(\frac{R_i}{R_f + R_i}\right)\frac{sR_cC_c + 1}{s[R_c + (R_f \parallel R_i)]C_c + 1} \quad \text{(E1)}$$

Substituting values, $H(0) = 1/11 = 0.0909$ and $GH(0) = 20$ k. To achieve as much PM as possible, the pole and zero of H must be widely separated. The ratio of the H zero to pole is

$$\frac{z}{p} = 1 + \frac{R_f \parallel R_i}{R_c} \quad \text{(E2)}$$

and is large when $R_c \ll R_f \parallel R_i$. Since R_f and R_i are given, we can both place and separate z and p by selecting R_c and C_c. If the high end of the phase range of z is placed two decades below p_1, then phase is restored to zero at $p_1/10$, but the gain will have rolled off by a decade to 2 k. By placing z at 1 kHz, $R_cC_c = 159\ \mu$s. For a full 90° of phase lag, $z/p = 100$. Substituting into (E2) gives $R_c = 9.18\ \Omega$. The closest 5% resistor value is 9.1 Ω. C_c is 159 μs/9.18 Ω = 17 μF, or a 5% tolerance value of 18 μF.

We estimate the compensated PM as follows. The gain at p_1 is 200, and ϕ must be −45°. The gain rolls off another decade to p_2, where it is 20. Then $f_{Tc} = \sqrt{20}(1\text{ MHz}) = 4.47$ MHz. This is log(4.47), or 0.65, decades above p_2. At p_2, $\phi = -90°$. At −90°/dec, the additional phase lag at f_{Tc} is 59° or −149° total, and the PM = 31°.

Example 6.4 (*continued*) Transimpedance Amplifier with Phase-Lead Compensation

In Example 6.4, ζ was unacceptably low. We can apply phase-lead compensation in H by shunting R with a compensation capacitor C_c.

The open-loop poles are at frequencies of $1/RC$ and $1/\tau_G = p_G$. The addition of C_c shifts the pole in H to $p_H = 1/R(C + C_c)$ and adds a zero at $z = 1/RC_c$. With phase-lead compensation, $z > p_H$. The root-locus plot is Fig. 6.8c. The complex pole locus bends left in a circle and rejoins the real axis above the zero. Since the dc loop gain is given, pole placement for the desired response largely depends on the placement of z.

The addition of C_c results in a closed-loop transimpedance of

$$\frac{V_o}{I_i} = -\left(\frac{K}{K+1}\right)\frac{sR(C + C_c) + 1}{s^2[\tau_G R(C + C_c)/(K+1)] + s[\tau_G/(K+1) + RC/(K+1) + RC_c] + 1} \quad \text{(E1)}$$

The closed-loop ζ of the poles can be derived as

$$\zeta = \frac{b}{2\sqrt{a}} = \frac{1}{2} \cdot \frac{(RC + \tau_G)/(K+1) + RC_c}{\sqrt{\tau_G R(C + C_c)/(K+1)}} \tag{E2}$$

For an MFED pole placement, $\zeta = \sqrt{3}/2$. Substituting into (E2) and solving for C_c gives 2.3 pF.

6.7 Compensator Design: Reducing dc Loop Gain

In Section 6.6, phase-lag compensation allowed dc gain to remain at K, the uncompensated value, because of C_c. In effect, phase-lag compensation reduces loop gain except at low frequencies. The simpler technique of reducing K does not require C_c and does not appreciably degrade dc performance for amplifiers (such as op-amps) with high loop gains. Neither does it introduce compensation poles and zeros that must be readjusted when closed-loop gain is adjusted. The Bode magnitude plot is shifted downward. With C_c shorted in Fig. 6.18, A_v of the inverting op-amp in (a) is not affected by R_c but the loop gain is. From Section 3.3, the inverting op-amp has input attenuation $\alpha_i = 1 + H$; both α_i and H are affected by R_c such that A_v remains unchanged. Consequently, GH can be adjusted by adjusting H with R_c without affecting closed-loop gain. Reduction of K is accomplished by R_c for the inverting op-amp (Fig. 6.20a).

The noninverting op-amp configuration can also be compensated by R_c. To achieve the same result, R_c must be placed across the op-amp inputs (Fig. 6.20b). The flow graph for this topology has a transmittance of α_i in front of the feedback loop. The feedback equations are

$$V_o = KE, \qquad E = \alpha_i V_i - HV_o = V_i - V_- \tag{6.39}$$

The transmittances are

$$\alpha_i = \frac{R_c}{R_i \parallel R_f + R_c} \tag{6.40}$$

$$H = \frac{R_c \parallel R_i}{R_c \parallel R_i + R_f} \tag{6.41}$$

$$G = K \tag{6.42}$$

Combining (6.39)–(6.42) gives the closed-loop gain:

$$A_v = \left(\frac{R_f + R_i}{R_i}\right) \frac{[K/(1+K)][R_c \parallel R_i]}{R_c \parallel R_i + R_f/(1+K)} = \left(\frac{R_f + R_i}{R_i}\right)\Bigg|_{K \to \infty} \tag{6.43}$$

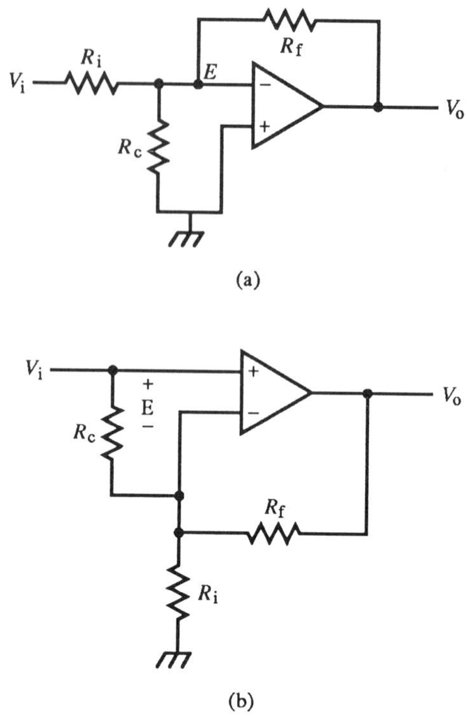

FIG. 6.20 Compensation through reduction of dc loop gain by R_c for the (a) inverting and (b) noninverting op-amp. The closed-loop gain remains unaffected by R_c since it affects α_i and H equally.

This is the familiar noninverting op-amp gain formula when $K \to \infty$. A_v is not affected by R_c, but the loop gain is. Since $GH = KH$, the effect on loop gain is to attenuate H by R_c shunting R_i. An apparent disadvantage of this topology is that R_c reduces the input resistance. But the effect is minimal with large K since R_c is across E, a small voltage, and is bootstrapped.

R_i and R_f can be generalized to impedances. The closed-loop gain is unaffected by R_c whereas the dc loop gain is reduced. This shifts the Bode magnitude plot downward, causing it to cross unity gain at a lower frequency, where the phase lag is less.

6.8 Compensator Design: Pole Separation and Parameter Variation

One of the simplest compensation techniques is pole separation by *dominant-pole compensation.* If one pole is introduced into a feedback loop at a much lower frequency than the other loop poles, it causes the gain to roll off at

−20 dB/dec (−1 slope) over a large frequency range until the next pole is encountered. If the range is large enough, the unity-gain frequency is less than the remaining poles, so that their influence is insignificant. An existing pole can often be reduced in frequency by modifying the value of its associated circuit elements.

Another pole separation technique, *pole-splitting*, is commonly applied by placing a capacitor from input to output of an inverting amplifier stage with an output pole. Figure 6.21a shows a feedback amplifier with dominant forward-path pole at $-p_G$ and a transfer function of

$$G = -K \cdot \frac{1}{s/p_G + 1} \tag{6.44}$$

The feedback path is that of an RC differentiator. The resulting loop gain is

$$GH = \frac{K}{s/p_G + 1} \cdot \frac{sRC}{sRC + 1} \tag{6.45}$$

The loop-gain root locus is plotted in Fig. 6.21b. Because of the zero at the origin, the pole at $-p_G$ migrates to the right while the pole at $-1/RC$ increases in frequency. This is pole-splitting. The poles split apart instead of coming

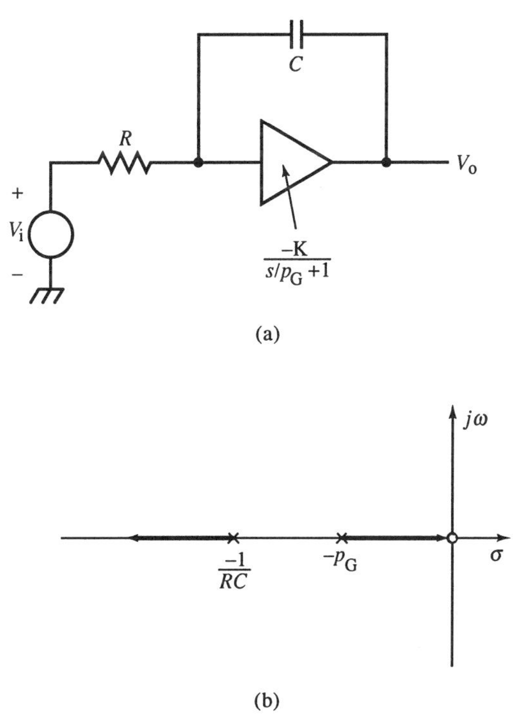

FIG. 6.21 Pole-splitting due to zero at origin in H due to C (a). Instead of moving together, the poles separate (b) so that p_G becomes dominant.

together and thereby achieve pole separation. The closed-loop transfer function is

$$\frac{V_o(s)}{V_i(s)} = -K \cdot \frac{1}{s^2(RC/p_G) + s[(K+1)RC + 1/p_G] + 1} \tag{6.46}$$

Because of the Miller effect, evident in the linear term of the denominator, the integrator pole moves away from p_G as K increases, separating the poles.

We now analyze the shunt capacitive realization of pole-splitting in more detail. Figure 6.22a shows an inverting amplifier stage with its flow graph, (b).

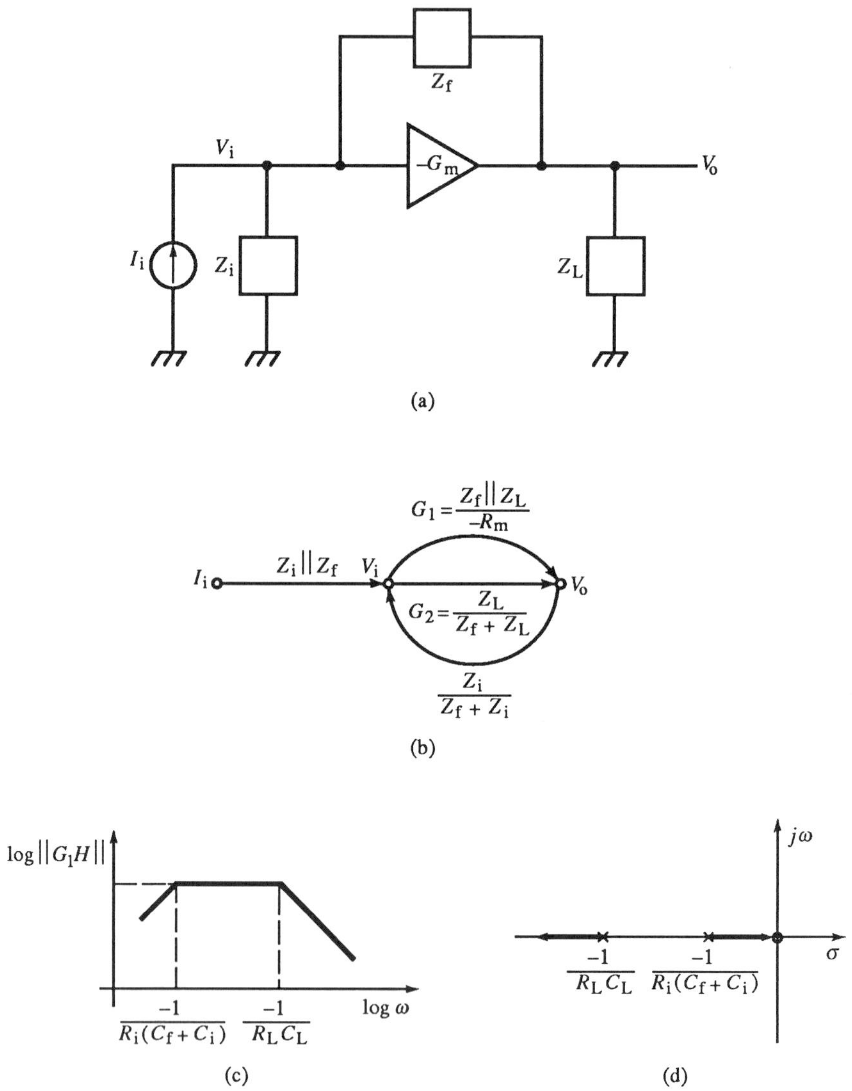

FIG. 6.22 A general transimpedance amplifier (a), its flow graph (b), Bode magnitude plot of loop gain (c), and root-locus plot (d).

The active forward path is a transadmittance amplifier with a gain of $-G_m = -1/R_m < 0$. The input and output are both loaded by general impedances Z_i and Z_L, and the transadmittance amplifier is shunted by a feedback impedance Z_f. From the flow graph, the closed-loop transimpedance is

$$\frac{V_o}{I_i} = (Z_f \parallel Z_i) \frac{(Z_f \parallel Z_L)/(Z_f \parallel -R_m)}{1+\underset{\substack{\uparrow \\ G}}{[(Z_f \parallel Z_L)/(Z_f \parallel -R_m)]}\underset{\substack{\uparrow \\ -H}}{[Z_i/(Z_f+Z_i)]}} \tag{6.47}$$

and Z_{in} is

$$Z_{in} = Z_i \left\| \frac{Z_f}{1-G} \right. \tag{6.48}$$

For the simpler case of no loading, Z_i and Z_L are removed, and the transimpedance is

$$\left.\frac{V_o}{I_i}\right|_{Z_i, Z_L \to \infty} = -Z_f + R_m \tag{6.49}$$

Since the output quantity is a voltage, a low-impedance output is desirable (to approximate a voltage source). In this case, $Z_L \ll Z_f$, and G_2, the passive path in G, is negligible. Then

$$G_1 = \frac{Z_f \parallel Z_L}{-R_m} \cong \frac{Z_L}{-R_m} \tag{6.50}$$

and

$$G_1 H = \left(\frac{Z_L}{R_m}\right)\left(\frac{Z_i}{Z_f + Z_i}\right) \tag{6.51}$$

To make the circuit more specific, let Z_L contribute a single pole due to a parallel RC load:

$$Z_L = \frac{R_L}{sR_L C_L + 1} \tag{6.52}$$

Similarly, let the input loading of the G_m amplifier be a parallel RC,

$$Z_i = \frac{R_i}{sR_i C_i + 1} \tag{6.53}$$

and let the feedback impedance be a capacitance,

$$Z_f = \frac{1}{sC_f} \tag{6.54}$$

These choices of impedances simulate a CE (or CS) stage with collector-base (or drain-gate) capacitance. Substituting these impedance expressions into (6.51), we obtain

$$G_1 H = \frac{R_L}{R_m} \cdot \frac{sR_i C_f}{[sR_i(C_f + C_i) + 1](sR_L C_L + 1)} \tag{6.55}$$

The frequency response and root loci for (6.55) are plotted in Figs. 6.22c, d. The zero at the origin splits the poles. Because the zero is not positive, there is no danger of instability with too much dc loop gain G_m. As gain increases, however, the lower-frequency pole decreases in frequency, and the bandwidth is correspondingly decreased.

We assumed that G_2, the passive forward path through Z_f, was negligible. If we extend this analysis to include it, we get some interesting and important results. The complete G is

$$G = G_1 + G_2 = -\frac{Z_f \parallel Z_L}{R_m} + \frac{Z_L}{Z_f + Z_L} = \frac{Z_f \parallel Z_L}{Z_f \parallel -R_m} \tag{6.56}$$

Specifically, G_1 and G_2 are

$$G_1 = -\frac{R_L}{R_m} \cdot \frac{1}{sR_L(C_f + C_L) + 1} \tag{6.57}$$

$$G_2 = \frac{Z_L}{Z_f + Z_L} = \frac{sR_L C_f}{sR_L(C_f + C_L) + 1} \tag{6.58}$$

Then G becomes

$$G = -\frac{R_L}{R_m} \cdot \frac{-sR_m C_f + 1}{sR_L(C_f + C_L) + 1} \tag{6.59}$$

This more complete expression for G has an additional RHP zero at $+1/R_m C_f$. The loop gain is

$$GH = \frac{R_L}{R_m} \cdot \frac{(-sR_m C_f + 1)(sR_i C_f)}{[sR_L(C_f + C_L) + 1][sR_i(C_f + C_i) + 1]} \tag{6.60}$$

The root locus of (6.60) is not directly obtainable as before because the RHP zero varies with R_m, and R_L/R_m is the dc loop gain. Root-locus plotting is based on fixed open-loop poles and zeros that are independent of dc gain. In (6.60), both elements that affect dc loop gain also affect a pole and zero. This situation suggests limits on the applicability of the root-locus technique for circuits.

We consequently proceed directly to the closed-loop voltage gain V_o/V_i:

$$\frac{V_o}{V_i} = -\frac{R_L}{R_m} \cdot \frac{(-sR_m C_f + 1)[sR_i(C_f + C_i) + 1]}{s^2 a + sb + 1} \tag{6.61}$$

where

$$a = R_i R_L(C_i C_f + C_L C_f + C_i C_L) \tag{6.61a}$$

$$b = R_i C_f\left(1 + \frac{R_L}{R_m}\right) + R_i C_i + R_L C_f + R_L C_L \tag{6.61b}$$

Finally, the closed-loop transimpedance is

$$\frac{V_o}{I_i} = Z_f \parallel Z_i \cdot \frac{V_o}{V_i} = Z_f(-H) \cdot \frac{V_o}{V_i} = -R_i \cdot \frac{R_L}{R_m} \cdot \frac{-sR_m C_f + 1}{s^2 a + sb + 1} \tag{6.62}$$

This expression has two left half-plane (LHP) poles and one RHP zero. Its characteristic equation is the same as (6.61).

The Miller effect is evident in the linear coefficient b in the first term of (6.61b), where the Miller capacitance C_f is multiplied by the voltage gain plus 1. As R_m decreases, b increases while a remains constant. This causes the poles to move apart with decreasing R_m. It also causes the RHP zero to increase in frequency. The LHP zero of (6.61) remains fixed as R_m varies. The movement of the poles and zero of (6.61) with decreasing R_m (or increasing G_m) is shown in Fig. 6.23. The poles move to zero and $-\infty$; the RHP zero goes to $+\infty$.

A root-locus plot with a circuit element as parameter is a *root-contour plot.* If KGH can be reformulated as $X \cdot F(s)$, where $F(s)$ is independent of X, then X is a constant relative to F and can be varied in the root-locus equation, $KGH = XF = -1$. The root-locus rules then apply, and the pole loci are mapped as a parameter of X. If a formulation compatible with the root-locus technique is not feasible, then movement of the poles and zeros of the closed-loop transfer function with variation of a circuit element can still be investigated. This *closed-loop parameter variation* technique is often quite useful in determining the effect of a circuit element on the dynamic response.

If C_f is made the parameter instead of R_m, the movement of the poles with C_f can be plotted. As C_f increases in (6.62), the RHP zero decreases toward zero. For the poles, C_f is in both a and b. We can estimate pole location from extreme values of C_f. When $C_f = 0$, the poles are located at $-1/R_iC_i$ and $-1/R_LC_L$. These are also the open-loop poles for $C_f = 0$. As C_f increases slightly from zero, the poles decrease until C_f dominates (that is, $C_f \gg C_i, C_L$). Then the quadratic pole factor becomes

$$s^2C_f[R_iR_L(C_i + C_L)] + sC_f\left(R_i\left(1 + \frac{R_L}{R_m}\right) + R_L\right) + 1$$

In solving for the poles, we find that the $b/2a$ term is independent of C_f. Also the $(C_i + C_L)$ factor in a is not found in b. If it is varied (as long as C_f continues to dominate), the pole loci move together and form a complex circular arc

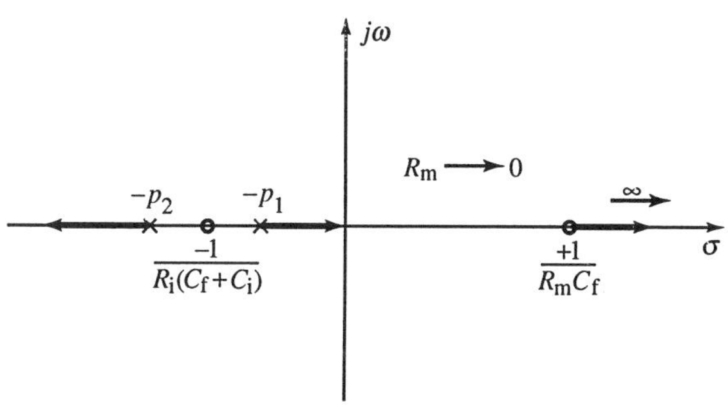

FIG. 6.23 Closed-loop parameter-variation plot (not root locus) for Fig. 6.22 when $Z_f = 1/sC_f$, as R_m decreases.

centered at the origin. (See Section 5.11 for a description of root loci due to parameter variation.)

As $C_f \to \infty$ in (6.62), both poles and zero move toward the origin. One pole and the RHP zero reach the origin and cancel, leaving a single pole at

$$\frac{-1}{(R_L \parallel R_i \parallel R_m)(C_i + C_L)}$$

An infinite C_f is a short between input and output such that $V_o = V_i$. The substitution theorem applies to the transadmittance current source, transforming it into a resistance across V_i of value R_m. All circuit components are in parallel. Therefore, a very large C_f sufficiently couples input and output so that their separate poles are merged effectively into one.

From the open-loop gain, (6.60), if $C_L \gg C_f$, then C_f has negligible effect on the open-loop pole at $-1/R_L(C_f + C_L)$; it remains relatively fixed as C_f increases. For the other open-loop pole, if C_i is not much greater than C_f, it moves appreciably to the right. Under these conditions, variation in C_f causes pole separation.

We now analyze the closed-loop transadmittance of Fig. 6.22a for other choices of circuit impedances. Consider new circuit conditions, in which Z_i is removed and the load is only capacitive:

$$Z_L = \frac{1}{sC_L}, \qquad Z_i \to \infty, \qquad Z_f = \frac{1}{sC_f} \tag{6.63}$$

Then

$$\frac{V_o}{I_i} = \frac{1}{sC_f} \cdot \frac{-sR_m C_f + 1}{sR_m C_L + 1} \tag{6.64}$$

G has the RHP zero and a pole at the origin with coefficient $R_m(C_f + C_L)$. $H = -1$ and $GH = -G$. As R_m decreases, closed-loop pole and zero separate (Fig. 6.24a). The pole and zero can be adjusted independently by varying C_L or C_f, respectively.

If we add a resistive Z_i to this circuit, the modified conditions are

$$Z_L = \frac{1}{sC_L}, \qquad Z_i = R_i, \qquad Z_f = \frac{1}{sC_f} \tag{6.65}$$

Then

$$G = -\frac{-sR_m C_f + 1}{sR_m(C_f + C_L)} \tag{6.66}$$

and

$$H = -\frac{R_i}{1/sC_f + R_i} = -\frac{sR_i C_f}{sR_i C_f + 1} \tag{6.67}$$

The closed-loop transimpedance is

$$\frac{V_o}{I_i} = \frac{-R_i}{R_m(C_f + C_L) + R_i C_f} \cdot \frac{-sR_m C_f + 1}{s(s\{[R_i(C_f \parallel C_L)] \parallel R_m C_L\} + 1)} \tag{6.68}$$

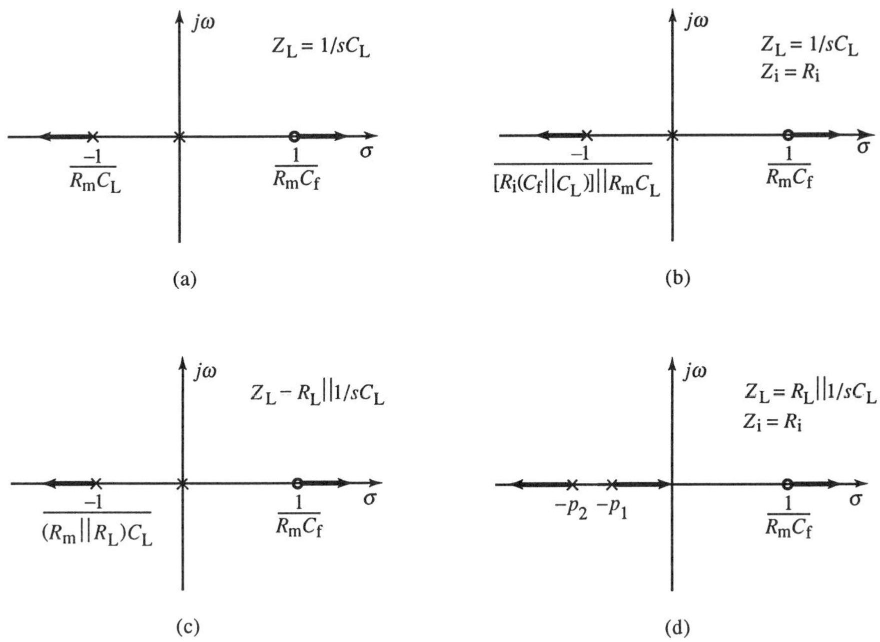

FIG. 6.24 Closed-loop parameter variation plots for Fig. 6.22a for specific impedance topologies as R_m decreases. Differences in (a)–(d) reflect the changes in impedances for Z_i and Z_L.

The closed-loop poles and zero are plotted in Fig. 6.24b. The dc factor, a pole, and the RHP zero vary with R_m. Again, pole and zero move outward, away from the origin. This circuit acts as an integrator due to the fixed pole at the origin. The effect of adding R_i is only to shift the non-zero pole.

R_i is now removed and R_L added. The conditions are:

$$Z_L = R_L \left\| \frac{1}{sC_L} = \frac{R_L}{sR_L C_L + 1}, \qquad Z_i \to \infty, \qquad Z_f = \frac{1}{sC_f} \right. \tag{6.69}$$

For these conditions, $H = -1$, and the open-loop gain is

$$GH = \frac{R_L}{R_m} \cdot \frac{-sR_m C_f + 1}{sR_L(C_f + C_L) + 1} \tag{6.70}$$

The closed-loop transimpedance is

$$\frac{V_o}{I_i} = -\left(\frac{R_L}{R_L + R_m}\right)\frac{1}{sC_f} \cdot \frac{-sR_m C_f + 1}{s(R_m \parallel R_L)C_L + 1} \tag{6.71}$$

Again, the s-plane situation is similar to the previous two cases, with the pole location modified due to R_L.

Finally, consider the addition of a resistive R_i to the circuit of (6.71). The conditions are then

$$Z_L = R_L \left\| \frac{1}{sC_L} = \frac{R_L}{sR_L C_L + 1}, \qquad Z_i = R_i, \qquad Z_f = \frac{1}{sC_f} \right. \tag{6.72}$$

The closed-loop transimpedance is

$$\frac{V_o}{I_i} = -R_i \cdot \frac{R_L}{R_m} \cdot \frac{-sR_m C_f + 1}{s^2(R_L R_i C_L C_f) + s[R_L(C_f + C_L) + R_i C_f(1 + R_L/R_m)] + 1} \tag{6.73}$$

The effect of resistances at both input and output is to move the low-frequency pole off the origin. The circuit no longer acts as a pure integrator. When R_m decreases, the poles split in the usual way; one goes to the origin and the other to $-\infty$. Again, this locus of poles is due to variation in the linear-term coefficient of the denominator of (6.73). The quadratic coefficient remains constant with R_m.

This extended analysis of the circuit of Fig. 6.22 demonstrates the conditions for pole-splitting due to variation in dc loop gain G_m and in C_f. The limitation of the root-locus technique was largely overcome by closed-loop parameter variation. This circuit is representative of CE and CS amplifiers and wideband amplifiers in general.

Example 6.8 Transimpedance Amplifier Pole-Splitting

The amplifier of Fig. E6.8 is an idealized form of inverting transistor amplifier with a feedback capacitance. Let

$$R_L = 1\ \text{k}\Omega, \quad R_m = 100\ \Omega, \quad R_i = 10\ \text{k}\Omega, \quad C_f = 10\ \text{pF},$$
$$C_i = 10\ \text{pF}, \quad C_L = 90\ \text{pF}$$

From (6.62), we solve for the zero and poles. They are

$$z = 159\ \text{MHz}, \quad p_{1,2} = 124\ \text{kHz},\ 10.8\ \text{MHz}$$

The Bode plot confirms the pole and zero values.

Since b remains constant with R_m, the poles should move toward each other as R_m is increased. Calculating again for $R_m = 1\ \text{k}\Omega$, we obtain

$$z = 15.9\ \text{MHz}, \quad p_{1,2} = 461\ \text{kHz},\ 2.89\ \text{MHz}$$

As predicted, the poles are now closer. With further increase in R_m, they will eventually become complex.

6.9 Two-Pole Compensation

High-performance feedback amplifiers require high loop gain over a wide frequency range. Unfortunately, some of the simpler compensation techniques —notably, dominant-pole compensation—are successful because they reduce

```
E6.8 Transresistance Amplifier
.OPT NOMOD OPTS NOPAGE
.OP
.DC II -2uA 0.2uA 0.05uA
.TF V(20) II
.AC DEC 100 10K 1G
.TRAN 10N 10U

II 0 10 AC 10uA PULSE (0 10uA)
GM 0 20  10 0  -0.01S

RL  20 0   1K
CL  20 0   90pF
CF  20 10  10pF
RI  10 0   10K
CI  10 0   10pF

.PROBE
.END

V(20)/II = -1.000E+05
INPUT RESISTANCE AT
   II = 1.000E+04
OUTPUT RESISTANCE AT
   V(20) = 1.000E+03
```

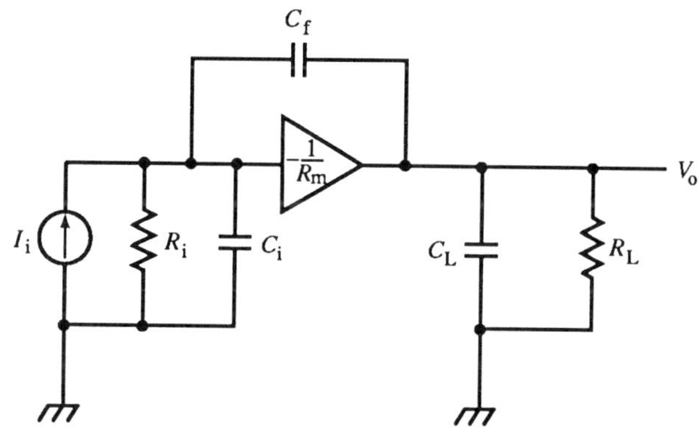

(a)

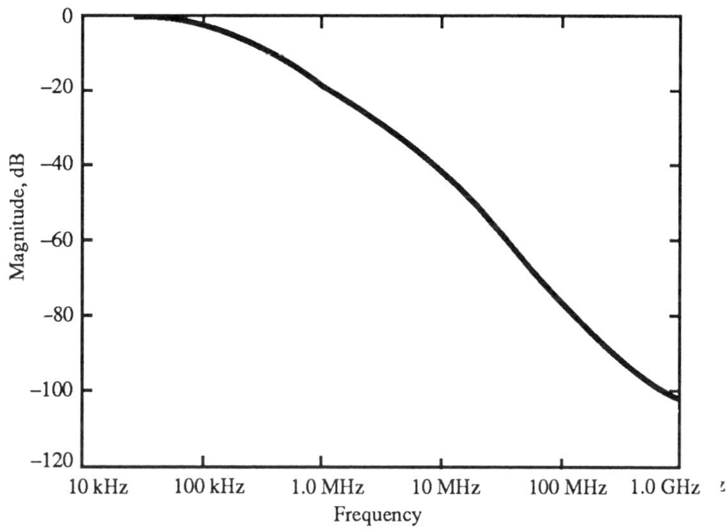

(b)

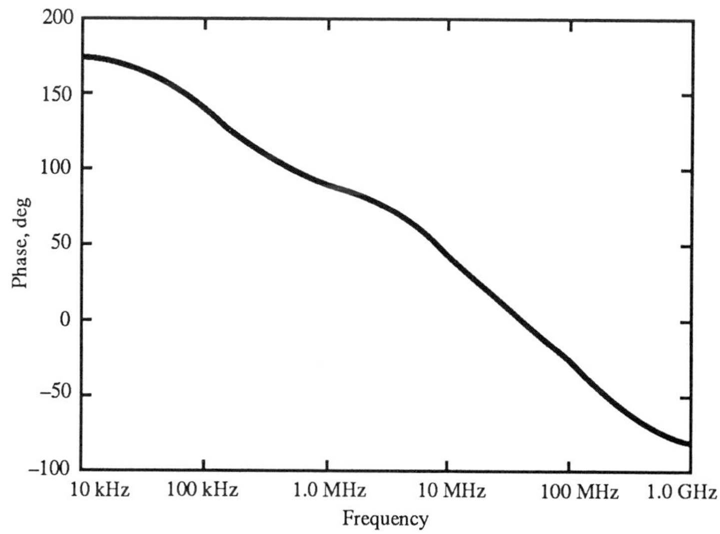

(c)

FIG. E6.8 (a) Circuit and listing. (b) Magnitude response. (c) Phase response.

gain appreciably at higher frequencies. The *two-pole compensation* technique sustains high gain to a higher break frequency, where it then rolls off at −40 dB/dec (−2 slope) followed by a zero that restores the magnitude to that of dominant-pole compensation. The difference is shown in Fig. 6.25b, with a typical two-pole compensator feedback realization in (a). The high loop gain is extended from the dominant-pole break frequency of p_d to p, where two poles reside.

Most op-amps have three stages of gain:

1. a differential (transconductance) amplifier input stage
2. a frequency-compensated high-gain stage
3. a current-gain output stage

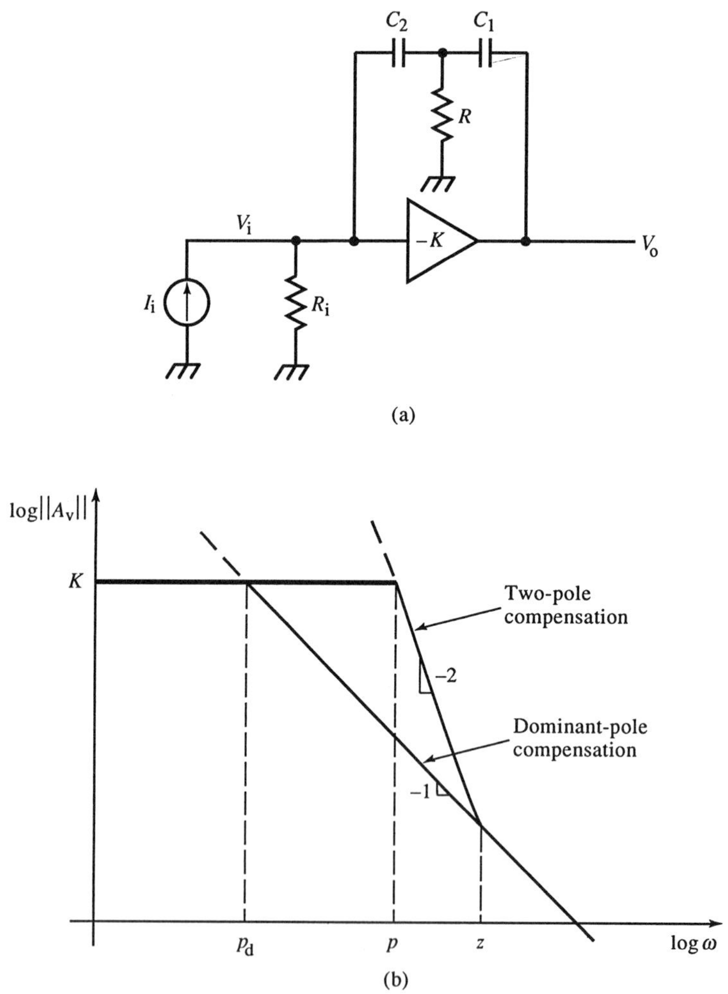

FIG. 6.25 Two-pole compensator (a) and Bode plot (b). Extra gain is available from p_d to z over the dominant single-pole amplifier.

Two-pole compensation is applied to the second stage. The compensation capacitors are typically connected between the output and input of the second stage.

For the ideal op-amp, dc gain is infinite, and the compensation poles reside at the origin. From Fig. 6.25a, as frequency increases, the reactance of the capacitors decreases relative to R until $X_C \ll R$. Then the equivalent circuit consists of the two capacitors in series, shunting the op-amp. The series capacitance is the same value as for the dominant-pole response shown in (b). The zero of the circuit in (a) is located at the frequency for which R becomes negligible relative to X_C.

The circuit of Fig. 6.25a represents an internal op-amp second stage or other amplifier compensation stage. Its loop gain is

$$GH = K \cdot \frac{s^2 R R_i C_1 C_2}{s^2 R R_i C_1 C_2 + s(RC_1 + RC_2 + R_i C_2) + 1} \tag{6.74}$$

where $G = -K$. The closed-loop voltage gain includes

$$\alpha_i = R_i \cdot \frac{sR(C_1 + C_2) + 1}{s^2 R R_i C_1 C_2 + s(RC_1 + RC_2 + R_i C_2) + 1} \tag{6.75}$$

and is

$$\frac{V_o}{V_i} = -K \cdot \frac{sR(C_1 + C_2) + 1}{s^2 R R_i C_1 C_2 (K+1) + s(RC_1 + RC_2 + R_i C_2) + 1} \tag{6.76}$$

where $V_i = R_i I_i$. For an ideal op-amp, $K \to \infty$, and the voltage gain is

$$\left.\frac{V_o}{V_i}\right|_{K \to \infty} = -\frac{sR(C_1 + C_2) + 1}{s^2 R R_i C_1 C_2} \tag{6.77}$$

As K increases, the quadratic term in (6.76) dominates, shifting the poles to the origin. Another useful simplification of (6.76) is to let $R_i \to \infty$. This is the case of a transimpedance amplifier with input I_i. Its transresistance is

$$\left.\frac{V_o}{I_i}\right|_{R_i \to \infty} = -K \frac{sR(C_1 + C_2) + 1}{sC_2[s(K+1)RC_1 + 1]} \tag{6.78}$$

The pole dependent on R_i moves to the origin. Without a finite R_i, the poles cannot be equal. For an op-amp transimpedance amplifier, both R_i and $K \to \infty$. In this case,

$$\left.\frac{V_o}{I_i}\right|_{R_i, K \to \infty} = -\frac{sR(C_1 + C_2) + 1}{s^2 R C_1 C_2} \tag{6.79}$$

This amplifier acts as a dual integrator with a finite zero. For the special case of $C_1 = C_2 = C$:

$$\left.\frac{V_o}{I_i}\right|_{R_i, K \to \infty, C_1 = C_2 = C} = -\frac{s2RC + 1}{s^2 RC^2} \tag{6.80}$$

As $R \to \infty$, (6.76) becomes

$$\left.\frac{V_o}{V_i}\right|_{R\to\infty} = -K\frac{1}{sR_i(C_1 \parallel C_2)(K+1)+1} \tag{6.81}$$

With R open, the circuit defaults to dominant-pole compensation. (The factor $C_1 \parallel C_2$ is the series combination of C_1 and C_2; $\parallel$ is a mathematical operator, not a topological descriptor.) In the case of an ideal op-amp, as $K \to \infty$, the pole approaches the origin and the gain is

$$\left.\frac{V_o}{V_i}\right|_{R,K\to\infty} = -\frac{1}{sR_i(C_1 \parallel C_2)} \tag{6.82}$$

This is a dominant single-pole amplifier response with an amplifier shunt capacitance of C_1 and C_2 in series. For comparison, the plots of (6.77) and (6.82) are shown in Fig. 6.26.

These expressions for closed-loop gain do not of themselves satisfy the requirements for two-pole compensation. The following conditions must also hold:

The poles must be equal (or close), or $p_1 = p_2 = p$. (6.83a)

The poles must be less than the zero, or $z/p > 1$ (6.83b)

The condition of (6.83a) is satisfied for real poles ($\zeta = 1$) when the coefficients of the quadratic pole factor of (6.76) have the relation

$$a = \left(\frac{b}{2}\right)^2$$

Since the $K+1$ factor is in a only, its variation produces the loci of poles for

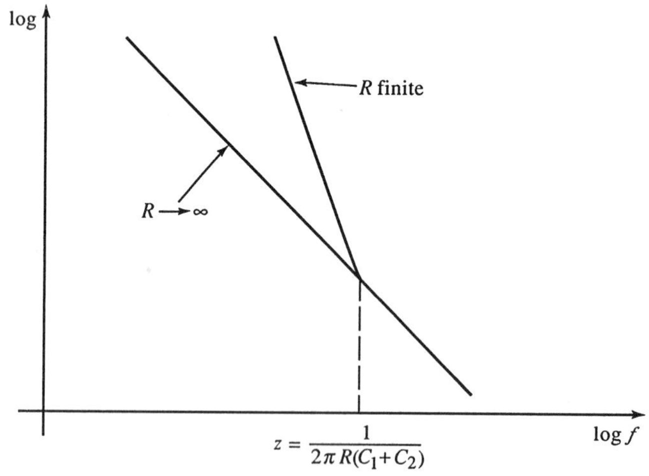

FIG. 6.26 Two-pole roll-off from the origin provides additional gain below z for the amplifier of Fig. 6.25. When R is removed, the default circuit is dominant single pole.

a constant b. (See Fig. 5.15b.) The poles are equal when their value is $-b/2a$, and the corresponding gain is found by setting the discriminant, b^2-4a, to zero and solving for $K+1$:

$$K+1|_{p_1=p_2}=\frac{R^2(C_1+C_2)^2+2R_iR(C_1+C_2)C_2+R_i^2C_2^2}{4R_iRC_1C_2} \tag{6.84}$$

K is set by either the amplifier open-loop gain or by forming an inner-loop, controlled-gain amplifier around which the two-pole compensator is placed. This is often too difficult to design because the forward path gain is determined by an integrated amplifier with an unspecified (or wide tolerance) gain. If the gain is stabilized by use of feedback, the H block interferes with the two-pole compensator. Since it is not usually desirable to set the compensator by adjusting K, we solve (6.84) for one of the compensator elements, R:

$$R=\frac{R_iC_2}{(C_1+C_2)^2}[(2K+1)C_1-C_2\pm 2\sqrt{C_1(K+1)(KC_1-C_2)}] \tag{6.85}$$

where $p_1=p_2$ and, of course, R is positive and real, requiring that $KC_1>C_2$.

Equation (6.85) is rather involved and can be simplified to

$$R\cong R_i\left[\frac{4K(C_1\parallel C_2)}{C_1+C_2}\right],\qquad K\gg 1,\qquad KC_1\gg C_2,\qquad p_1=p_2 \tag{6.86}$$

The second constraint on two-pole compensator realization is (6.83b), or $z>p$. From (6.76),

$$z=\frac{1}{R(C_1+C_2)}=\frac{1}{\tau_z} \tag{6.87}$$

and the positive value of the two poles is

$$p=\frac{b}{2a}=\frac{R(C_1+C_2)+R_iC_2}{2R_iRC_1C_2(K+1)}=\frac{1}{\tau_p} \tag{6.88}$$

Since the poles are equal, $a=(b/2)^2$, and both poles are located on the real axis at

$$\frac{b}{2a}=\frac{2}{b}$$

Then $z>p$ becomes

$$\frac{1}{\tau_z}>\frac{2}{b}=\frac{1}{\tau_p}$$

But from (6.76), $b=\tau_z+R_iC_2$. Substituting, we have

$$\frac{1}{\tau_z}>\frac{2}{\tau_z+R_iC_2}=\frac{1}{\tau_p} \tag{6.89a}$$

or

$$R_iC_2>\tau_z=R(C_1+C_2) \tag{6.89b}$$

If we solve for R_iC_2 in terms of (z/p) from (6.89a), we get the equality

$$R_iC_2 = \tau_z\left(2 \cdot \frac{z}{p} - 1\right) \tag{6.90}$$

By solving $z > p$, using $b/2a$ instead of $2/b$, we get

$$R_iC_2 = \frac{z}{p} \cdot \frac{\tau_z}{2(K+1)[C_1/(C_1+C_2)] - z/p}, \qquad \frac{z}{p} < 2(K+1)\left(\frac{C_1}{C_1+C_2}\right) \tag{6.91}$$

The constraint on z/p is weak for large K but suggests that C_1 be made larger than C_2 for maximum pole-zero separation. A special case of (6.91) is

$$\frac{z}{p} \cong 2(K+1)\left(\frac{C_1}{C_1+C_2}\right), \qquad R_iC_2 \gg \tau_z \tag{6.92}$$

When R_iC_2 dominates b, the pole-zero separation is pushed to the limits of (6.91). In this case, with large K,

$$R \cong R_i\left[\frac{C_2}{4KC_1}\right], \qquad R_iC_2 \gg \tau_z, \qquad K \gg 1, \qquad p_1 = p_2 \tag{6.93}$$

Finally, from (6.89), the constraint on the capacitors is

$$\frac{C_1}{C_2} < \frac{R_i}{R} - 1 \tag{6.94}$$

With these formulas, we can design two-pole compensators with real and equal poles and gain values typical of either op-amps or low-gain amplifiers.

Example 6.9 Two-Pole Compensation

An amplifier (not an op-amp) has the following circuit values:

$$K = 100, \qquad R_i = 33\ \text{k}\Omega, \qquad C_1 = 10\ \text{pF}, \qquad C_2 = 100\ \text{pF}$$

The circuit is shown in Fig. E6.9 along with the Bode plot from circuit simulation. For this amplifier,

$$R = 33\ \text{k}\Omega\left[\frac{100\ \text{pF}}{4(10\ \text{pF})(100)}\right] = 825\ \Omega$$

and the conditions of (6.93) are satisfied. All element values are determined, and the natural frequency of the pole factor, which is the break frequency of the two poles, is found either from (6.89a) or directly from a:

$$f_n = \frac{1}{2\pi\sqrt{a}} = \frac{1}{2\pi\sqrt{R_iRC_1C_2(K+1)}} = 96.5\ \text{kHz}$$

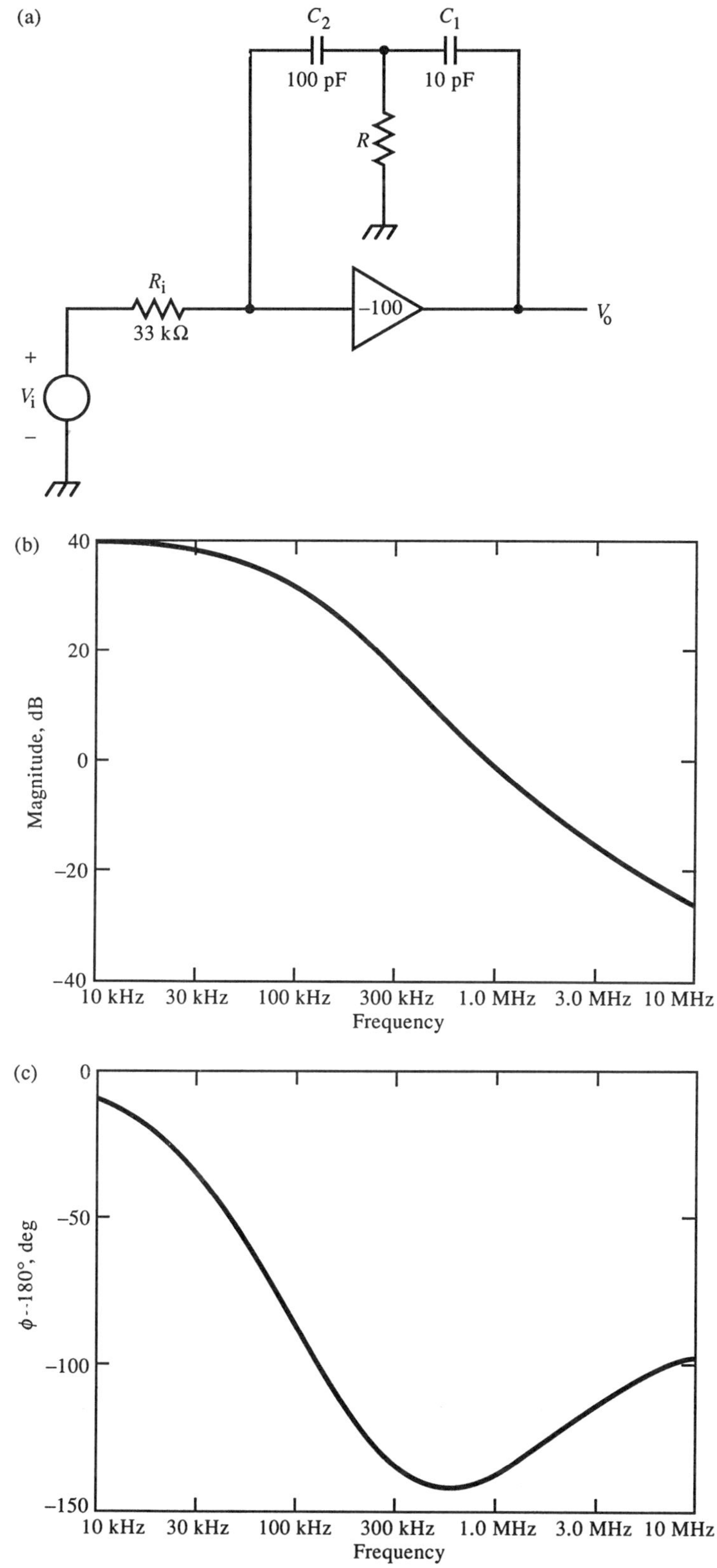

FIG. E6.9 (a) Circuit. (b) Voltage gain. (c) Phase plots of two-pole compensation

As a check, when the poles of a quadratic factor are equal, $\zeta = 1$. From the pole-factor of (6.76), $\zeta = b/2\sqrt{a} = 1.03$. The zero is located at $1/2\pi R(C_1 + C_2) = 1.75$ MHz. From the SPICE simulation, the phase is $-90°$ at 100 kHz, where the poles should be. As a check, the magnitude will be down -6 dB (for two poles) at the break frequency. At 34 dB (down from a dc gain of 40 dB), it is 91 kHz. The maximum closed-loop phase lag occurs at 631 kHz and is $-142°$. The nonmonotonic phase plot, which dips down and comes back up due to the zero, is characteristic of two-pole-compensated amplifiers. The magnitude plot rolls off with a -2 slope at the pole frequency to the zero frequency at about 1.75 MHz. (Because the amplifier is inverting, the Bode plot phase is offset by $-180°$.)

The previous development assumed real poles. A more general set of design formulas takes the given design constraints as independent variables and yields two-pole compensator element values. We start with R_i, K, ω_n, and ζ of the pole factor (now no longer necessarily unity), the pole–zero separation,

$$\frac{z}{\|p\|} = \frac{z}{\omega_n} = \gamma \tag{6.95}$$

and the location of the zero, $z = 1/\tau_z$. Beginning with

$$z = \gamma\omega_n \Rightarrow \frac{1}{\omega_n} = \gamma\tau_z \Rightarrow a = \gamma^2\tau_z^2 \Rightarrow (K+1)RR_iC_1C_2 = \gamma^2R^2(C_1 + C_2)^2$$

and solving for R, we obtain

$$R = R_i \cdot \frac{(K+1)(C_1 \parallel C_2)}{\gamma^2(C_1 + C_2)} \tag{6.96}$$

Now ζ is brought in as

$$\zeta = \frac{b}{2\sqrt{a}} = \frac{b}{2} \cdot \omega_n = \frac{\tau_z + R_iC_2}{2} \cdot \omega_n \tag{6.97}$$

The compensator element C_2 results from solving (6.97) and is

$$C_2 = \frac{(2\zeta\gamma - 1)\tau_z}{R_i}, \qquad \zeta > \frac{1}{2\gamma} \tag{6.98}$$

C_2 of (6.98) is expressed entirely in given parameters and is thereby determined. Next, (6.87) is solved for C_1,

$$C_1 = \frac{\tau_z}{R} - C_2 \tag{6.99}$$

and substituted into (6.96). This results in an expression for R in given parameters and C_2, which is known from (6.98):

$$R = \frac{\tau_z}{C_2}\left[1 - \left(\frac{\gamma^2}{K+1}\right)\left(\frac{\tau_z}{R_i C_2}\right)\right], \qquad R_i C_2 > \left(\frac{\gamma^2}{K+1}\right)\tau_z \tag{6.100}$$

Finally, since R is now known, it is substituted into (6.99) to yield

$$C_1 = C_2\left[\frac{1}{\{1-[\gamma^2/(K+1)][\tau_z/R_i C_2]\}} - 1\right] \tag{6.101}$$

Example 6.10 Two-Pole Compensator Design

The op-amp circuit of Fig. E6.10 has a gain of $K = 10$ k, $R_i = 10\ \text{k}\Omega$ and is two-pole compensated to have a zero at 500 kHz and roll off a decade lower, at 50 kHz. Furthermore, an MFED pole response is desired, where $\zeta \cong 0.866$. Component tolerances are 5%.

The required parameters are

$$\tau_z = \frac{1}{2\pi f_z} = \frac{1}{2\pi(500\ \text{kHz})} = 318\ \text{ns}$$

$$\gamma = \frac{z}{\omega_n} = \frac{500\ \text{kHz}}{50\ \text{kHz}} = 10$$

First we calculate C_2 from (6.98); it is 519 pF. The closest 5% part is

$$C_2 = 520\ \text{pF}$$

Next we calculate R from (6.100), using the calculated value for C_2 (instead of the 5% value) to keep the calculations accurate. (This is important when we get to C_1 because the difference of two large numbers is taken.) Then $R = 613\ \Omega$. The closest value is

$$R = 620\ \Omega$$

Finally, C_1 is calculated from (6.101), or (6.99) if care is taken to retain numerical consistency. It is 0.32 pF. This is a very small discrete capacitor value and suggests that it might be difficult to realize this circuit reliably in manufacture because this value is on the order of parasitic capacitances. The circuit-board layout between the output node and R must minimize stray capacitance. Perhaps the best way to implement C_1 is with a small trimmer capacitor of about 1 pF maximum value. If such a small C_1 is not feasible, then the given parameters must be adjusted to result in larger capacitance. C_1 increases if R increases due to a decrease in C_2. And C_2 decreases when ζ, γ, or τ_z decrease or R_i increases.

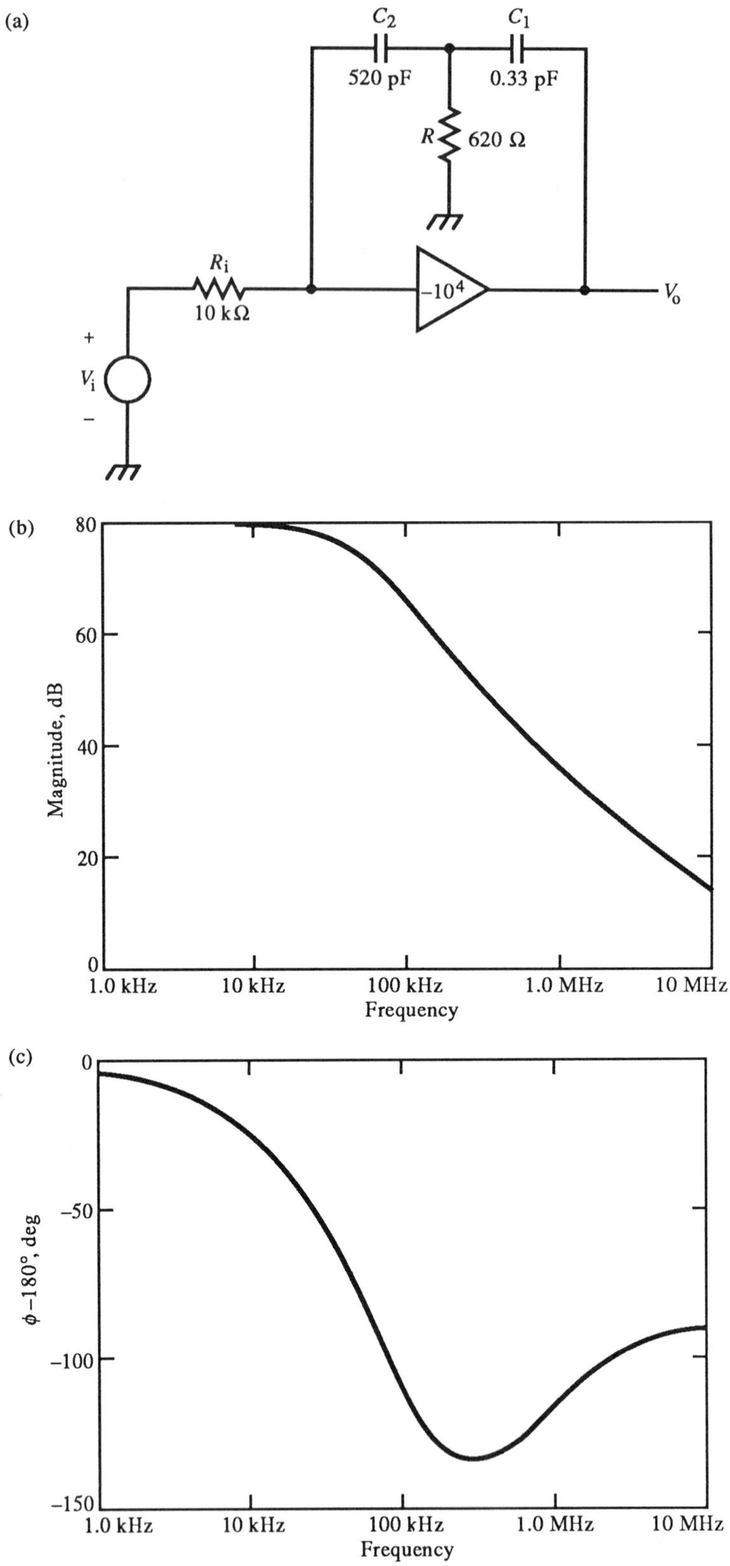

FIG. E6.10 (a) Circuit. (b) Gain, and (c) phase plots of circuit response.

To check our results, we turn from synthesis to analysis and calculate a and b of the pole factor:

$$a = RR_iC_1C_2(K+1) = 1.06\times 10^{-11}\ \mathrm{s}^2 \rightarrow f_n = 48.8\ \mathrm{kHz} \cong 50\ \mathrm{kHz}$$

$$b = R(C_1+C_2)+R_iC_2 = 318\ \mathrm{ns}+5.19\ \mu\mathrm{s} = 5.51\ \mu\mathrm{s} \Rightarrow \zeta = 0.85 \cong 0.87$$

Both f_n and ζ are within the 5% tolerance of the components. Finally, we check our results against the constraints in (6.98) and (6.100):

$$\zeta = 0.87 > \frac{1}{2\gamma} = \frac{1}{20} = 0.05 \qquad \text{(checks)}$$

$$C_2 = 0.33\ \mathrm{pF} > \frac{\gamma^2\tau_z}{R_i(K+1)} = 0.32\ \mathrm{pF} \qquad \text{(checks)}$$

The lower limit of C_2 is approached because C_1 and C_2 are so widely separated.

The final check of this example is made from the SPICE frequency-response simulation.

The amplifier of Fig. 6.25a has no dc feedback and acts like an integrator at dc. Unless it is within a larger feedback loop, the output drifts out of its linear range due to offset errors. For stand-alone applications, R_f must be included for dc stabilization (Fig. 6.27). (The resulting compensator has the topology of a *bridge-T* filter.) The voltage gain for $K \rightarrow \infty$ is

$$\left.\frac{V_o}{V_i}\right|_{K\rightarrow\infty} = -\frac{R_f}{R_i}\frac{sR(C_1+C_2)+1}{s^2R_fRC_1C_2+sR(C_1+C_2)+1} \qquad (6.102)$$

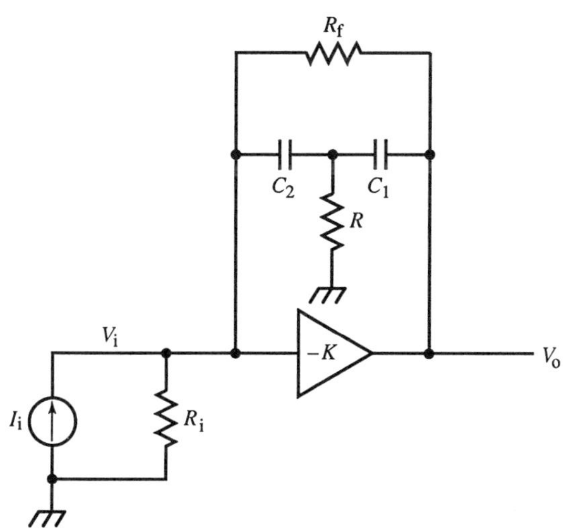

FIG. 6.27 A two-pole compensator with additional R_f for finite dc gain, forming a bridge-T feedback network. The addition of R_f constrains the benefits of two-pole response.

As $R_f \to \infty$, (6.102) approaches (6.77), as it must. This gain differs from (6.76) in that $b = \tau_z$ and does not have the extra degree of freedom that (6.76) does with the $R_i C_2$ term. Consequently, ζ and γ are not independent but are related by

$$\zeta = \frac{b}{2}\omega_n = \frac{\tau_z \omega_n}{2} \Rightarrow \zeta = \frac{1}{2\gamma} \Rightarrow \gamma = \frac{1}{2\zeta} \tag{6.103}$$

Proceeding similarly to the derivation of (6.100), we get

$$R = \frac{\tau_z}{C_2}\left(1 - \frac{\tau_z}{\gamma^2 R_f C_2}\right), \qquad C_2 > \frac{\tau_z}{\gamma^2 R_f} \tag{6.104}$$

C_2 is chosen to satisfy the constraint of (6.104) that $R > 0$. This choice depends on R_f and interacts with it. The pole locus of (6.76) was varied by $(K+1)$ since it was in a but not b. For this compensator, variation with constant b is due to R_f instead. To achieve $\gamma > 1$, as required for a two-pole compensator, the poles must be complex and have a pole angle greater than 60° (Fig. 6.28). Although the pole–zero separation is zero, 60° establishes the minimum M_m as 1.15 (or 1.25 dB) and minimum M_p as 16%. For a useful compensator with one octave of pole–zero separation, $\gamma = 2$. Then $\zeta = 0.25$ ($\phi = 76°$), $M_m = 2.97$ (or 6.3 dB), and $M_p = 44\%$.

Because of the unavoidable underdamped response that accompanies adequate pole–zero separation, the bridge-T two-pole compensator is more limited in its application. Consequently, two-pole compensation is best suited to inner stages of feedback amplifiers instead of being applied as the sole feedback of an amplifier.

Overall, two-pole compensation is not used to increase amplifier stability but to increase mid-frequency performance. It tends to decrease stability and must be applied carefully, making sure that no poles exist below the two-pole break frequency. Otherwise, a three-pole roll-off occurs that, with any significant loop gain, can result in instability.

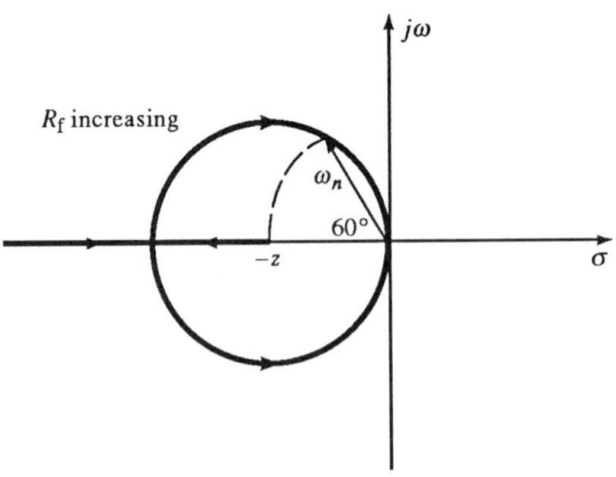

FIG. 6.28 Locus of poles with R_f of Fig. 6.27. At least a 60° pole angle is required to place the frequency of the poles below the zero.

6.10 Output Load Isolation

In some feedback amplifier applications, the load impedance is highly reactive, and the amplifier has a significant output resistance R_o. This combination can add a load-dependent output pole to the loop. A method for isolating capacitive loads is shown in Fig. 6.29, where C_L is the load, and R_o is the amplifier open-loop output resistance.

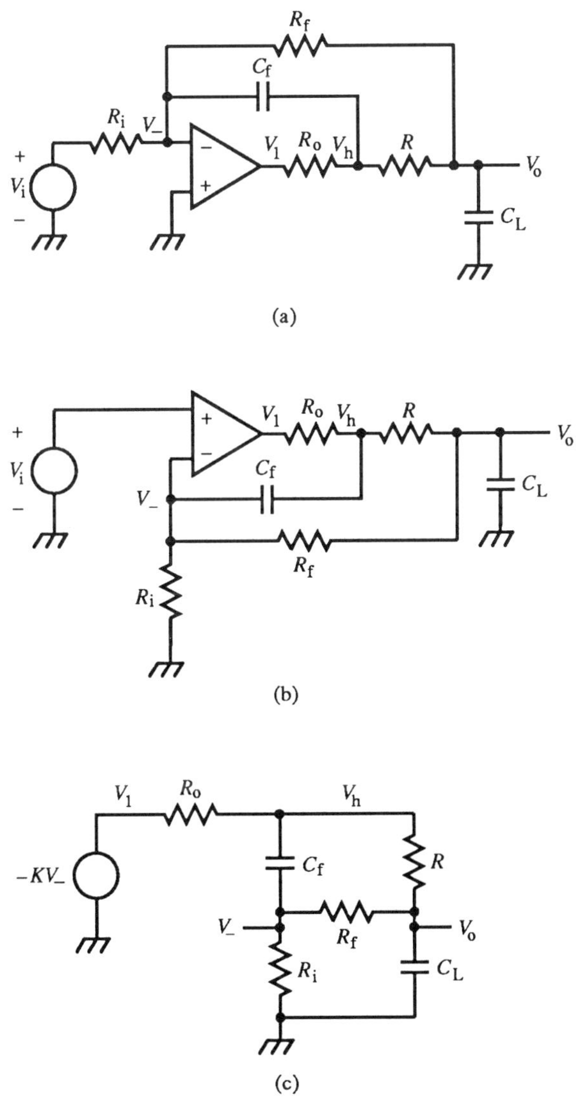

FIG. 6.29 Stabilization of amplifier driving a reactive load by isolation of high frequency feedback from load with R, for (a) inverting and (b) noninverting op-amps. The passive output network has the topology of a bridge (c).

This compensation scheme has two feedback paths, an accurate low-frequency path and a load-isolated high-frequency path. The feedback compensation capacitor C_f is isolated from the load by output decoupling resistor R. The low-frequency feedback through R_f is taken at the output to eliminate dc error due to R_o and R.

With no load isolation or compensation,

$$R = C_f = 0$$

The load introduces a pole in the loop at $-1/(R_f \parallel R_o)C_L$. If C_f is then added to compensate for this pole, the loop gain becomes

$$GH = -\left(\frac{R_f \parallel R_o}{R_f \parallel (-R_m)}\right)\left(\frac{R_i}{R_f + R_i}\right)\frac{\{s[R_f \parallel (-R_m)]C_f + 1\}(sR_fC_f + 1)}{[s(R_f \parallel R_o)(C_f + C_L) + 1][s(R_f \parallel R_i)C_f + 1]} \tag{6.105}$$

where $R_m = R_o/(-K)$. When $C_L = 0$ and $R_i \gg R_o$, the poles are well separated. As C_L increases, pole separation decreases as the higher pole moves down in frequency, reducing stability. C_f introduces a feedback zero and pole as a phase-lead network (Fig. 6.30). Its zero can be placed to cancel the amplifier output pole by setting

$$C_f = \frac{R_o}{R_f} C_L \tag{6.106}$$

From Fig. 6.30, as C_f increases, all poles and zeros shift toward the origin. For $C_L \gg C_f$, the load pole is stationary, and the pole and zero in H move together and away from the load pole.

When R is added and the topology is redrawn (Fig. 6.29c), the output network forms a bridge. The exact solution for this circuit, a nontrivial exercise, is found by applying KCL at the nodes of voltages: V_-, V_1, V_h, and V_o. This

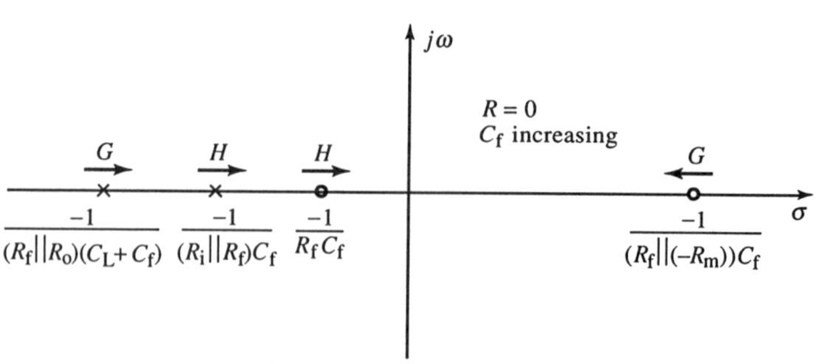

FIG. 6.30 Movement of critical frequencies with increasing C_f and no R, all toward the origin.

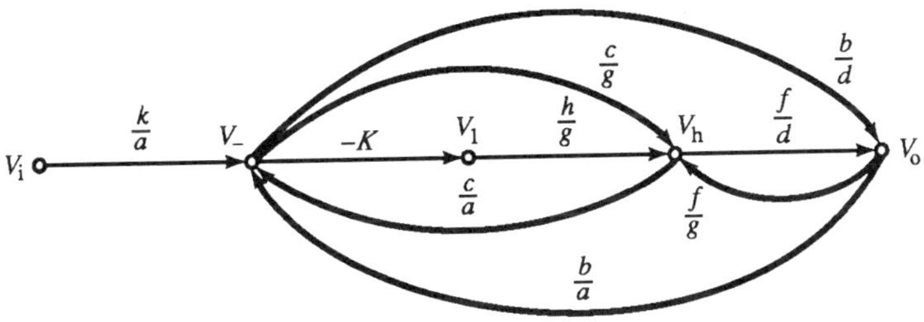

FIG. 6.31 Flow graph of Fig. 6.29a. Paths c/g and b/d contribute little and can be ignored.

results in the flow graph of Fig. 6.31 for the inverting amplifier, where

$$a = \frac{1}{(R_i \parallel R_f \parallel 1/sC_f)}, \quad b = \frac{1}{R_f}, \quad c = sC_f, \quad d = \frac{1}{(Z_L \parallel R_f \parallel R)}$$
$$f = \frac{1}{R}, \quad g = \frac{1}{(R_o \parallel 1/sC_f \parallel R)}, \quad h = \frac{1}{R_o}, \quad k = \frac{1}{R_i} \tag{6.107}$$

The amplifier gain is $-K$ and $V_1 = -KV_-$. Some simplifying assumptions can be made that reduce the complexity of the flow graph (such as removing b/d, c/g, and/or f/g), but the remaining circuit analysis is still unwieldy. We need a more functionally oriented approach.

The low- and high-frequency (lf and hf) feedback paths have been approximated in Figs. 6.32a, c, along with their Bode plots. The lf path in (a) has transmittance

$$\left.\frac{V_-}{V_1}\right|_{\mathrm{lf}} = \left(\frac{R_i}{R_f + R_i}\right)\left(\frac{1}{s(R_o + R)C_L + 1}\right)\left(\frac{1}{s(R_f \parallel R_i)C_f + 1}\right),$$
$$R_o \ll \frac{1}{sC_f}, \quad R_o + R \ll R_f + R_i \tag{6.108}$$

where the simplifying assumptions are that C_f and $R_f + R_i$ do not load the smaller output resistances R_o and R. The hf path is

$$\left.\frac{V_-}{V_1}\right|_{\mathrm{hf}} = \left(\frac{sRC_L + 1}{s(R_o + R)C_L + 1}\right)\left(\frac{s(R_f \parallel R_i)C_f}{s(R_f \parallel R_i)C_f + 1}\right),$$
$$R_o, R \ll \frac{1}{sC_f}, \quad \frac{1}{sC_L} \ll R_f \tag{6.109}$$

The hf-path approximations are similar to those of the lf path. At high frequencies, (6.109) approaches

$$\frac{R}{R_o + R}$$

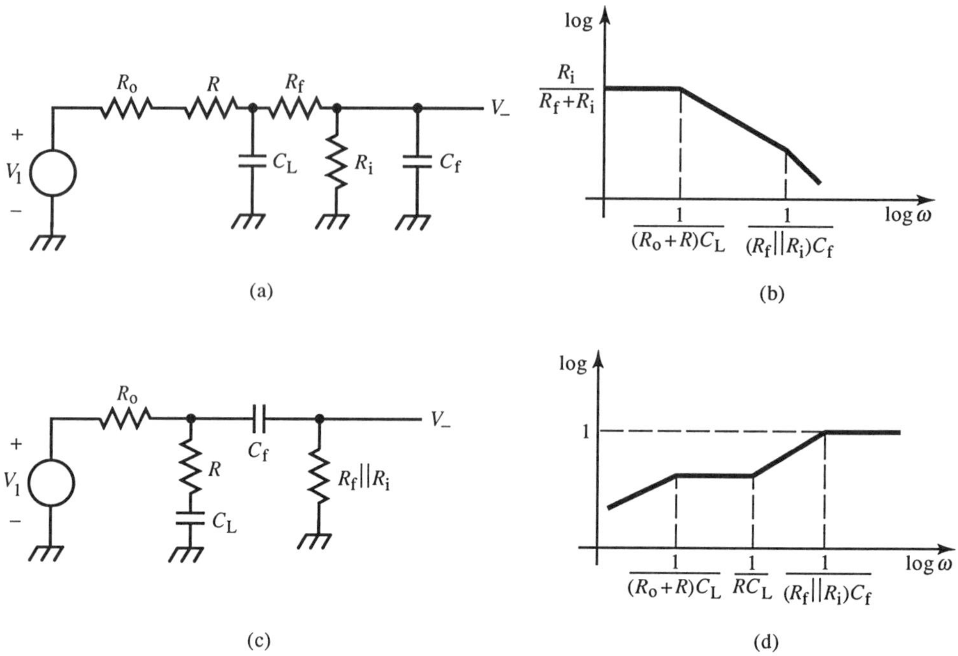

FIG. 6.32 Low-frequency (a) and high-frequency (c) paths of Fig. 6.29a. Bode plot for (a) is (b), and for (c) is (d).

The composite feedback transmittance is the sum of the two paths, or

$$\frac{V_-}{V_1} = \left(\frac{R_i}{R_f + R_i}\right) \frac{s^2 R C_L R_f C_f + s R_f C_f + 1}{[s(R_o + R)C_L + 1][s(R_f \parallel R_i)C_f + 1]} \tag{6.110}$$

Without R, we have one less LHP zero, as in (6.105). The feedback path is an all-pass network when the coefficients of the pole and zero terms are equated. This results in

$$R = \left(\frac{R_i}{R_f}\right) R_o, \qquad C_f = \left(\frac{R_o + R}{R_f}\right)\left(\frac{R_f + R_i}{R_f}\right) C_L \tag{6.111}$$

and the load capacitance pole is removed from the loop gain. The closed-loop response, however, is still affected by C_L. The constraints of (6.108) and (6.109) require that we check $1/R_o C_f$ after applying (6.111), to make sure that it is well above f_T.

Example 6.11 Load Capacitance Compensation

A fast op-amp with $K = 10^5$ and poles at 100 Hz and 4 MHz is used in the inverting configuration to drive a 10 nF load with a voltage gain of −3 (see Fig. 6.29a). $R_f = 30\ \text{k}\Omega$ and $R_i = 10\ \text{k}\Omega$. The open-loop output resistance is 10 Ω. The feedback capacitor C_f and decoupling resistor R are calculated from (6.111):

$$R = 3.3\ \Omega \quad \text{and} \quad C_f = 5.9\ \text{pF}$$

For the uncompensated amplifier, the step response shows obvious ringing (Fig. E6.11a) and peaking in the frequency response and group delay plots (Fig. E6.11b).

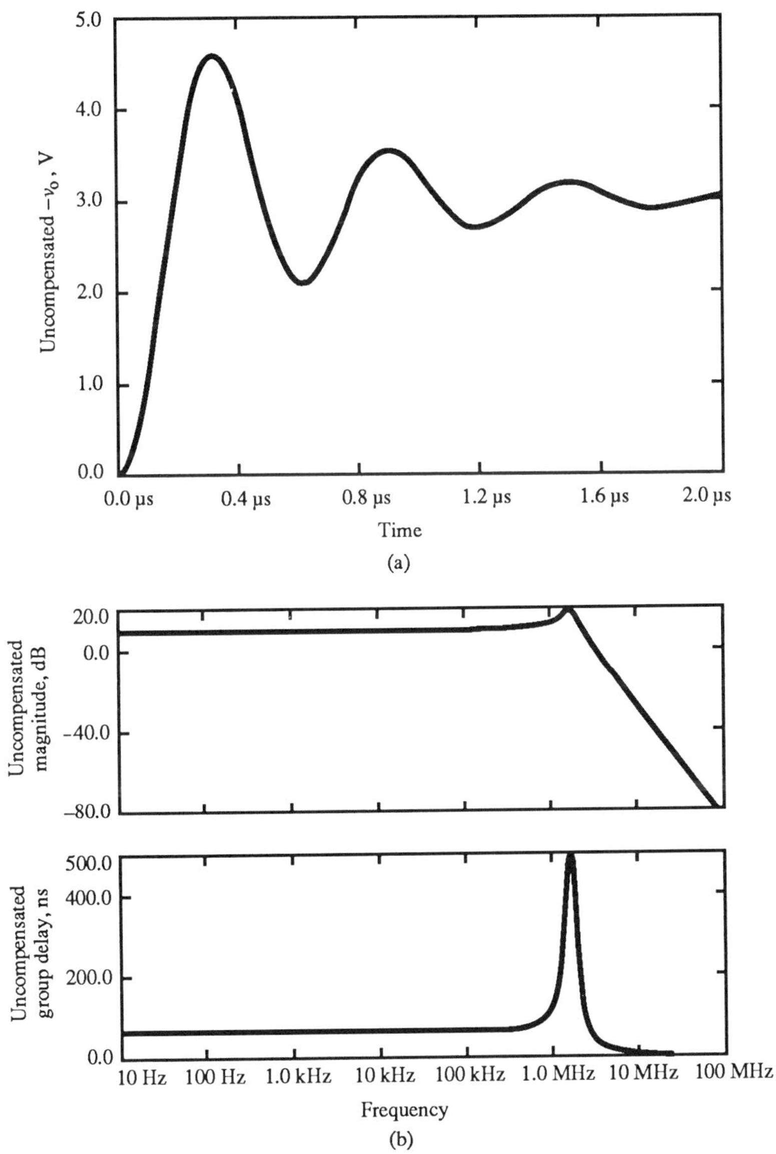

FIG. E6.11 Load-compensated amplifier. (Figure continues.)

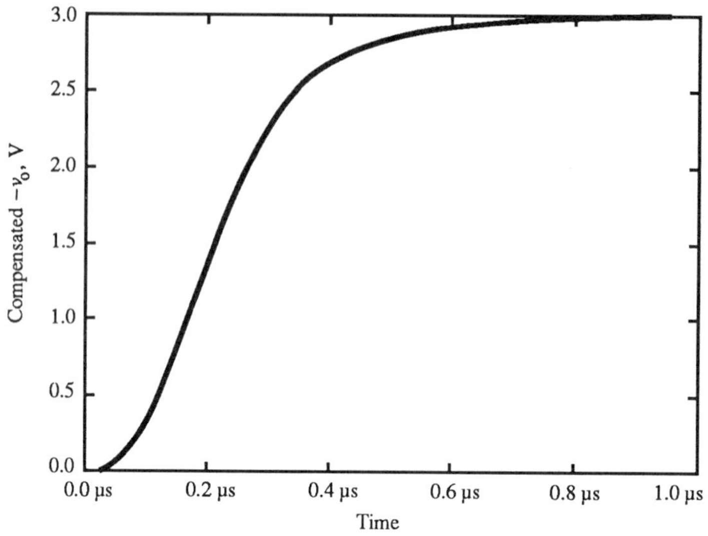

(c)

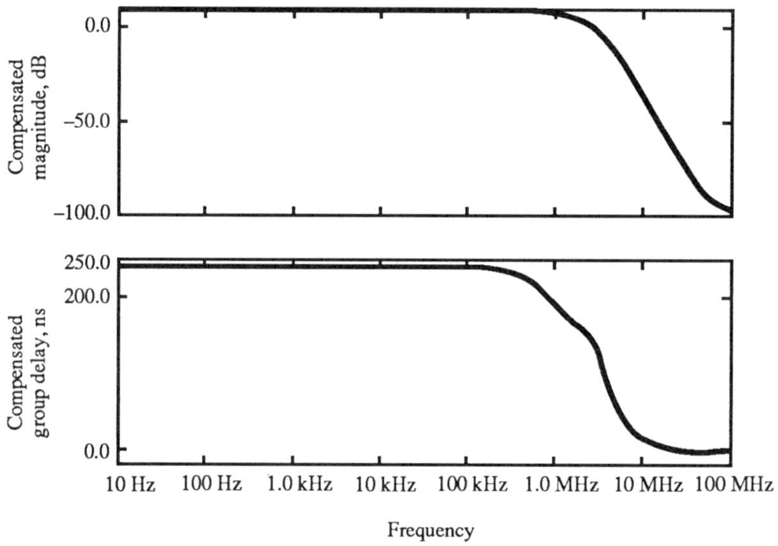

(d)

FIG. E6.11 (continued)

Compensation is now applied, and the response improves. No ringing is evident in the step response (Fig. E6.11c) with no peaking in the group delay or frequency response (Fig. E6.11d). The SPICE listing (Fig. E6.11e) of the compensated amplifier shows how the op-amp is modeled, using RC integrators and buffers to create the poles.

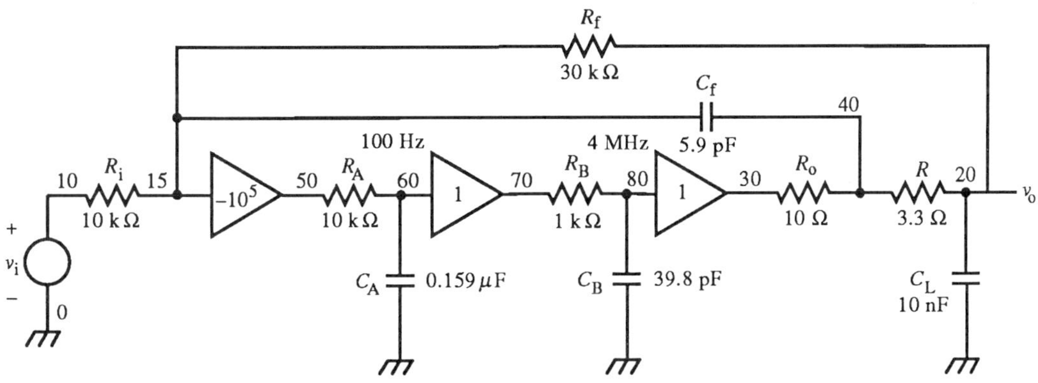

(e)

```
E6.11 Load-compensated amplifier

.OPT NOMOD OPTS NOPAGE
.AC DEC 30 10 100MEG
.TRAN 2n 2u

VI 10 0 AC 1V PULSE (0 1V)
RI 10 15 10k

; amplifier with poles at 100Hz & 4MHz

EA 50 0  15 0  -1E5
RA 50 60 10k
CA 60 0 0.159uF
EB 70 0  60 0 1
RB 70 80 1k
CB 80 0 39.8pF
EC 30 0 80 0 1
RO 30 40 10

R  40 20 3.3
CL 20 0  10nF
CF 40 15 5.9pF
RF 20 15 30k

.PROBE
.END
```

FIG. E6.11 (continued)

Capacitive loads can also be isolated by placing a shunt RL in series with the amplifier output. At dc, the transmittance to the load is unity. At high-frequencies, L appears open, leaving R as isolation. For excessive inductive loading, the load is often characterized by a series RL. If it is fixed, a series RC in parallel with it can form a constant-impedance network.

6.11 Complex Pole Compensation

Previous compensation techniques involved real poles and zeros. Open-loop complex poles appear as resonances that can sometimes be damped by identifying the circuit elements involved in the undesired resonant mode. This identification is not always successful, especially when the circuit has many possible parasitic reactances.

From root-locus criteria, pole angle can be reduced by the addition of a real pole at a lower frequency (Fig. 6.33a). As the pole increases in frequency due to dc loop gain K, the complex pole radius (ω_n) decreases, but so does the pole angle ϕ. This decrease in ϕ is slight in a narrow range of K, making this a marginally useful technique.

A real zero, placed at a higher real frequency than that of the complex poles, draws them out to a larger pole radius and lower pole angle (Fig. 6.33b) with increasing gain.

Complex poles can be compensated directly by complex pole-zero cancellation. Complex zeros are realized by a series RLC network at the output of a transconductance amplifier. Typically, the network is placed in parallel with

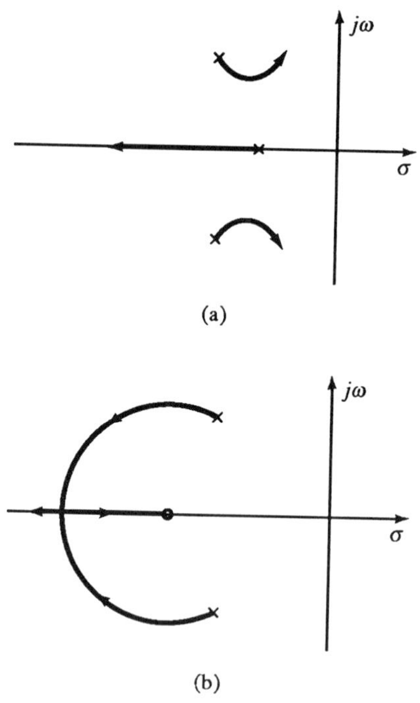

FIG. 6.33 Stabilization of complex pole pair with (a) another pole and (b) a zero. The pole provides only a marginal decrease in pole angle, whereas the zero both decreases pole angle and increases pole radius.

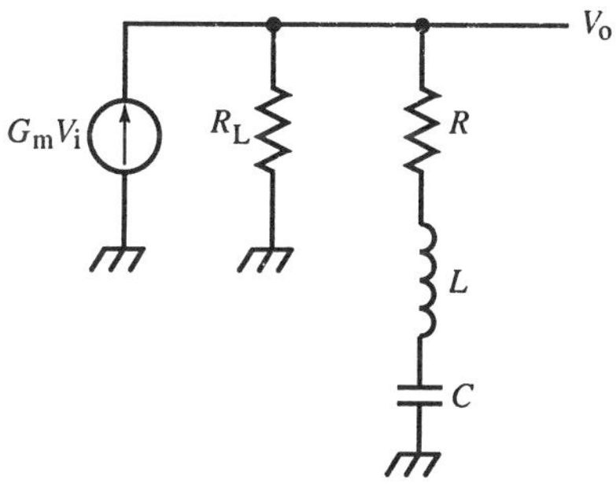

(a)

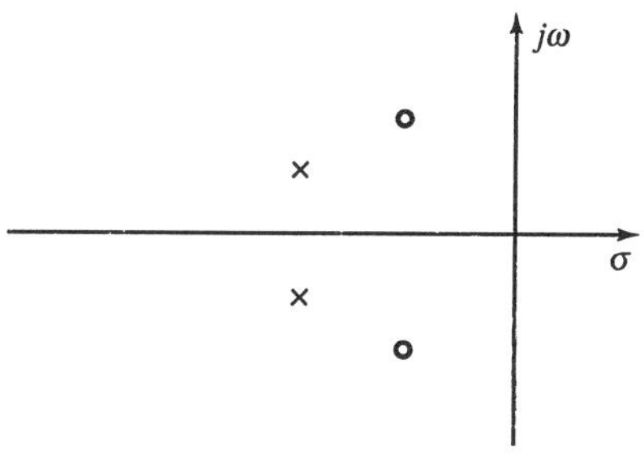

(b)

FIG. 6.34 Complex pole-zero cancellation using an RLC circuit (a); the compensator introduces a pair of complex zeros for cancellation and a complex pole pair at a reduced pole angle (b).

a transistor load resistor (Fig. 6.34a). The transfer function of this circuit is

$$\frac{V_o}{I_o} = Z_o = R_L \cdot \frac{s^2LC + sRC + 1}{s^2LC + s(R + R_L)C + 1} \tag{6.112}$$

The pole-zero placement is shown in Fig. 6.34b. In addition to the desired zeros, there is another pair of poles with a larger linear coefficient (due to R_L). The situation is similar to phase-lead compensation; the addition of zeros by passive linear circuit networks is that the added poles are at a decreased pole angle from the poles cancelled by the zeros with no loss of pole radius.

Empirical compensation of hidden complex pole-pairs begins by observing the ring frequency f_r (which is the damped frequency f_d) and the time constant of its decay, τ_r, from a step response. The value of τ_r can be calculated from the peak overshoot M_p (see Section 5.8). The relationship between τ_r, f_r, and M_p is

$$\tau_r = \frac{1}{2\alpha} = \frac{1}{4f_r \cdot \ln(1/M_p)} \tag{6.113}$$

Compensator element values for an MFED response are calculated by using these empirical parameters:

$$R = \frac{2R_L}{\sqrt{12(\pi\tau_r f_r)^2 + 3} - 2} \tag{6.114a}$$

$$C = \frac{2\tau_r}{R[4(\pi\tau_r f_r)^2 + 1]} \tag{6.114b}$$

$$L = \frac{R\tau_r}{2} \tag{6.114c}$$

It is usually easier to measure the peak overshoot M_p than to estimate τ_r. By using the formula for M_p from (5.131), we find the compensator values for a pole angle of $\cos^{-1}\zeta$ to be

$$R = \frac{R_L \cdot \ln(1/M_p)}{\zeta\sqrt{\pi^2 + [\ln(1/M_p)]^2} - \ln(1/M_p)} \tag{6.115a}$$

$$C = \frac{\zeta\sqrt{\pi^2 + [\ln(1/M_p)]^2} - \ln(1/M_p)}{\pi^2 R_L f_r} \tag{6.115b}$$

$$L = \frac{(\pi^2 + [\ln(1/M_p)]^2) R_L}{4\pi^2 f_r \cdot \zeta\sqrt{\pi^2 + [\ln(1/M_p)]^2} - \ln(1/M_p)} \tag{6.115c}$$

Example 6.12 Compensation by Complex Pole–Zero Cancellation

An amplifier system has an excessively underdamped response. A MFA response is desired. A transadmittance stage in the amplifier is free to be compensated and has a load resistance of 1 kΩ. The response to a step shows a ring frequency of 30 MHz and a peak overshoot of $M_p = 0.25$.

By substituting these values into (6.115) and noting that an MFA response has a pole angle of 45° and $\zeta = \sqrt{2}/2$, we obtain the compensator elements:

$$R = 1.33\ \text{k}\Omega, \qquad C = 3.52\ \text{pF}, \qquad L = 4.10\ \mu\text{H}$$

6.12 Compensation by the Direct (Truxal's) Method

The direct approach to calculation of G and H for a given closed-loop response M is to begin by specifying what M should be. Neglecting α_i and α_o, if any, since they are cascaded with the feedback loop, we describe M as

$$M = \frac{G}{1+GH}$$

Solving for the loop gain, we get

$$GH = \frac{HM}{1-HM} = \frac{N_H N_M}{D_H D_M - N_H N_M} \qquad (6.116)$$

where N and D are numerators and denominators of H and M. For a given G, we can solve directly for H:

$$H = \frac{G-M}{GM} = \frac{1}{M} - \frac{1}{G} = \frac{N_G D_M - N_M D_G}{N_G N_M} \qquad (6.117)$$

Not only M must be chosen to satisfy the system requirements, but also the resulting H must be physically realizable. For a high-order system, M must be high-order for a realizable H. The familiar criteria of amplifier performance are consequently more difficult to express in M. Therefore, this method is of limited use. If the amount of calculation were the limitation, a computer solution would be feasible, but creative design judgment is required in selecting M.

6.13 Power Supply Bypassing

The elimination or compensation of parasitic capacitance and inductance is not a dynamic response compensation method in itself, but it is related. Before a feedback amplifier loop can be compensated, the response of the individual stages of amplification must be acceptable. Oscillating stages must first be stabilized. Parasitic elements arise from both circuit components and layout. (See Sections 9.3 and 9.4 for details.)

Figure 6.14 shows some of the more common parasitic elements for op-amp circuits. The connections of the amplifier to the power supplies involves conductors (wire or circuit-board traces) with a corresponding inductance. Circuit-board trace inductance is difficult to estimate accurately but is roughly 10 nH/cm for a rectangular trace that is much longer than its cross-sectional dimensions [1]. Capacitive bypassing of trace inductance shortens the loop length and decreases the characteristic impedance Z_n to a low value. Oscillation can commonly occur due to the series LC resonance formed by the stray power-supply inductance and stray capacitance from the local supply node to an amplifier input. As L increases, Z_n becomes sizeable with a small C. Consequently, the series resistance required to damp the resonance also must be large. If we add bypassing, L is decreased and C greatly increased. Both effects reduce Z_n so that a smaller input-node resistance damps the series resonance.

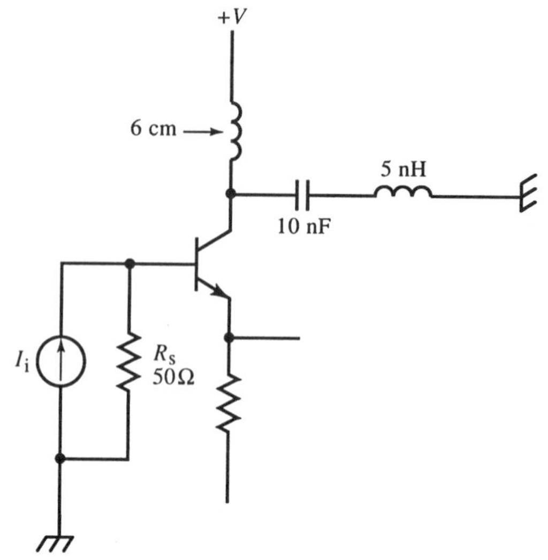

FIG. E6.13

Example 6.13 Damping Oscillation Through Bypassing

The circuit of Fig. E6.13 has a source resistance of 50 Ω shunting the relatively high-resistance base input. $C_{bc} = 3$ pF. The collector supply connection is about 6 cm long.

The inductance is about 60 nH and characteristic impedance is

$$Z_n = \sqrt{\frac{60 \text{ nH}}{3 \text{ pF}}} = 141 \ \Omega$$

R_s is significantly less and oscillation is likely. If a 10 nF bypass capacitor is connected to the collector, a capacitor with 5 nH of parasitic series inductance, then

$$Z_n = \sqrt{\frac{5 \text{ nH}}{10 \text{ nF}}} = 0.7 \ \Omega$$

and the series LC resonance is well damped by R_s.

6.14 Closure

The dynamic response compensation techniques presented here involve design judgment and do not guarantee optimal response. Other techniques will appear later, especially for bandwidth extension. The techniques here provide a foundation in compensator circuits and how to design with them.

Reference

[1] A. E. Ruehli, "Inductance Calculations in a Complex Integrated Circuit Environment," *IBM J. Res. Develop.*, Sept. 1972. 470ff.

CHAPTER 7

Frequency-Related Impedance Transformations

7.1 Active Device Behavior above Bandwidth

Active devices such as transistors or op-amps have bandwidth limitations that can result in instability or undesirable dynamic response in feedback circuits. As gain falls off with frequency, feedback amplifier port resistances change with gain and effectively become reactive. These bandwidth-related reactances can resonate with other circuit elements.

Many linear circuits might not appear to involve "high frequencies" or "wide bandwidths," but these concepts are relative to the bandwidth limitations of the active devices. The frequency range over which bandwidth-limited instability occurs lies between the bandwidth f_{bw} and the unity-gain frequency f_T. The two frequency ranges of interest are

- low-frequency (lf) region, $0 \leq f < f_{bw}$ (below bandwidth),
- high-frequency (hf) region, $f_{bw} \leq f \leq f_T$.

In this chapter, we extend the reduction theorem to the complex-frequency domain. We first find $\beta(s)$ for BJTs and apply it using the β transform. This leads to some interesting impedance gyrations that can create hf resonances. To distinguish between the low-frequency (quasistatic) β of Chapters 2–4 and a frequency-dependent β, we denote

$$\text{low-frequency } \beta \equiv \beta_o$$

For BJTs, current-gain bandwidth is denoted by f_β and unity-current-gain frequency by f_T. They are related by

$$f_T = \beta_o f_\beta \tag{7.1}$$

We take some liberty with the symbol f_T and let it more generally denote the unity-gain frequency (or *gain-bandwidth product*) of any active device. The interpretation of f_T depends on the kind of device: For BJTs, it is the frequency at which β is unity; for voltage-gain amplifiers, it is a unity-voltage-gain frequency.

For a BJT with $f_T = 300$ MHz and $\beta_o = 100$, high-frequency behavior occurs between $f_\beta = f_T/\beta_o$ and f_T, or in the range from 3 MHz to 300 MHz. For power BJTs, f_β can be as low as several hundred kilohertz. The open-loop bandwidth of typical op-amps is less than 100 Hz, and unity-gain frequency is 1 MHz. This range of rather low frequencies is the op-amp hf region.

7.2 Derivation of Bipolar-Junction Transistor High-Frequency Model

We now return to the hybrid-π BJT model of Fig. 2.2b shown without C_μ in Fig. 7.1a and with "ohmic" base and emitter resistances in (b). In actual transistors, C_μ is distributed across r'_b, but here it is shown connected entirely to the internal base node b'. This model is valid for both the lf and hf regions. We develop the hf model equations using Fig. 7.1a.

The idea of the hf model is that as the BJT input-signal frequency increases above $1/r_\pi C_\pi$, a decreasing proportion of I_b flows through r_π as the reactance of C_π decreases. Since

$$Z_\pi = r_\pi \parallel \frac{1}{sC_\pi} = \frac{r_\pi}{sr_\pi C_\pi + 1} \tag{7.2}$$

decreases with frequency, V_{be} also decreases, resulting in decreased collector current. Consequently, β also decreases with frequency. The break frequency

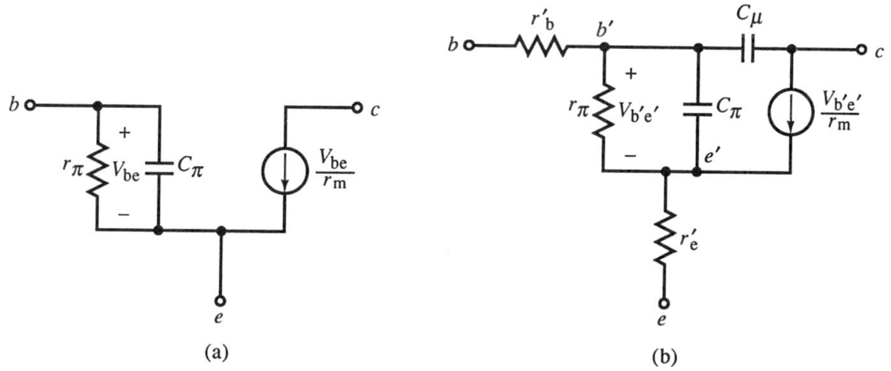

FIG. 7.1 Simplified hybrid-π BJT model (a) used to derive the hf model and (b) a more complete hybrid-π model.

of β is $\omega_\beta = 1/\tau_\beta$, where

$$\tau_\beta = r_\pi C_\pi$$

The frequency-dependent form of β is I_c/I_b or

$$\beta(s) = \beta_o\left(\frac{1/sC_\pi}{1/sC_\pi + r_\pi}\right) = \frac{\beta_o}{s\tau_\beta + 1} \tag{7.3}$$

and

$$\beta(s) + 1 = (\beta_o + 1)\,\frac{s\alpha_o\tau_T + 1}{s\tau_\beta + 1} \tag{7.4}$$

where

$$\tau_\beta = \beta_o \tau_T \tag{7.5}$$

The Bode plot of $\beta(s)+1$ is shown in Fig. 7.2. In the lf region, the transistor model does not require reactive elements (as we assumed in Chapters 2–4). In the hf region, β rolls off with frequency, and above f_T the device has essentially lost its gain (though power gain under the right circuit conditions takes place up to the unity-power-gain frequency f_{MAX}). Other significant factors not accounted for in this model (such as base transit time) cause its error to increase as f_T is approached. The model predicts less phase shift in β than actually occurs due to other transistor delays, yet it is accurate enough to be quite useful.

A simplified model, valid only for the hf region, can be derived from (7.3) by letting $\beta_o \to \infty$. Then

$$\lim_{\beta_o \to \infty} \beta(s) = \beta_{hf} = \frac{1}{s\tau_T} \tag{7.6}$$

The expressions for β from (7.3) and (7.6) can now be used in circuit analysis.

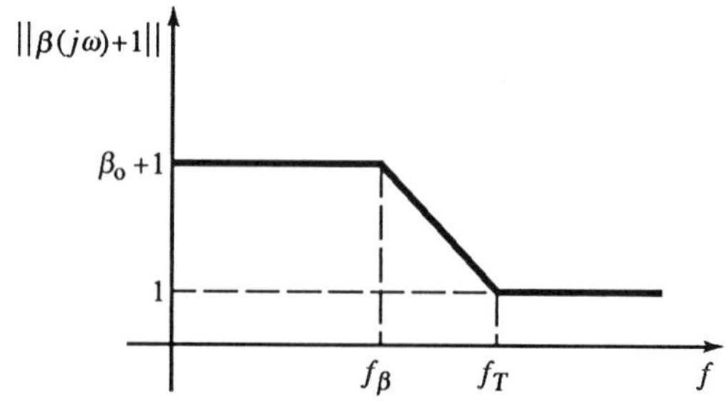

FIG. 7.2 Bode plot of $\|\beta(s)+1\|$.

7.3 Impedance Transformations in the High-Frequency Region

The β transform, as applied to nonreactive BJT circuits in Chapter 2, can be generalized using $\beta(s)$ in reactive circuits. The impedance at the base node due to impedance Z_E at the emitter node is

$$Z_b = [\beta(s)+1]Z_E \tag{7.7}$$

and, from the emitter, the impedance in the base circuit is

$$Z_e = \frac{Z_b}{\beta(s)+1} \tag{7.8}$$

The corresponding circuits are shown in Fig. 7.3, from which we can derive Z_b and verify (7.7). Applying KCL to the emitter node, we obtain

$$\frac{V_o}{Z_E} + \frac{V_o - V_i}{Z_\pi} = \frac{V_{be}}{r_m} \tag{7.9}$$

($V_{be} = V_i - V_o$.) Solving for Z_b gives

$$Z_b = Z_\pi + Z_E\left(1 + \frac{\beta_o}{sr_\pi C_\pi + 1}\right) \tag{7.10}$$

Above f_β, Z_π becomes negligible, and

$$Z_b \cong Z_E\left(1 + \frac{\beta_o}{sr_\pi C_\pi + 1}\right) = Z_E + \frac{1}{sr_\pi C_\pi / Z_E\beta_o + 1/Z_E\beta_o} \tag{7.11}$$

Because $f_T = \beta_o f_\beta$ from (7.1), Z_b can be rewritten as a continued fraction,

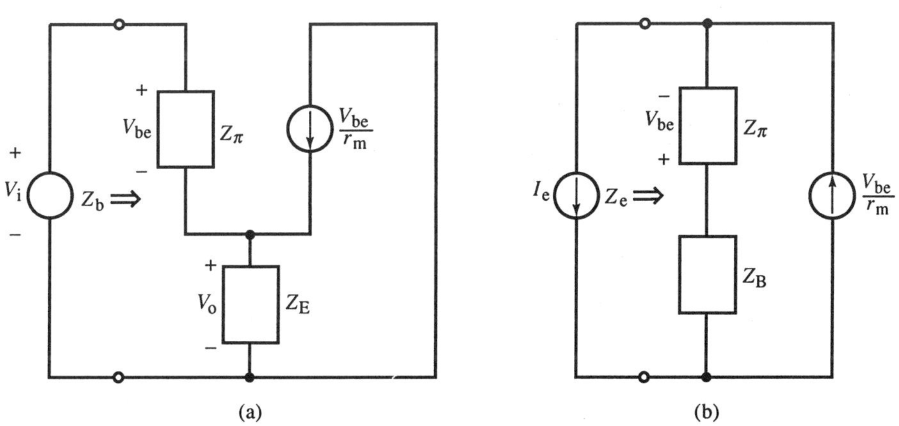

FIG. 7.3 Emitter-follower equivalent circuits for (a) Z_b and (b) Z_e.

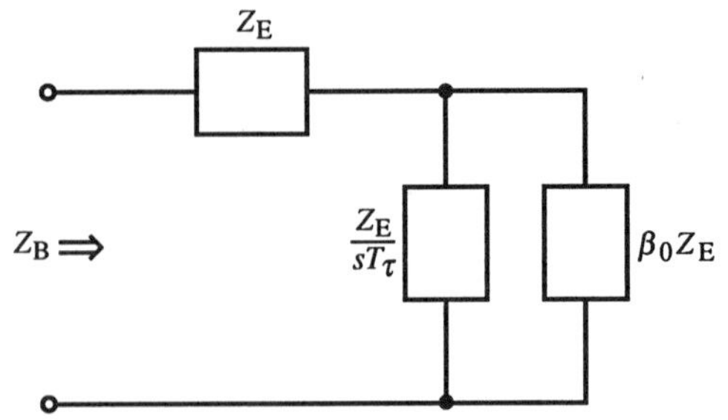

FIG. 7.4 Base node equivalent circuit with emitter impedance Z_E.

which makes the topology explicit in equation form:

$$Z_b \cong Z_E + \cfrac{1}{\cfrac{1}{Z_E/s\tau_T} + \cfrac{1}{\beta_o Z_E}}, \qquad \tau_T = r_m C_\pi \tag{7.12}$$

The corresponding circuit topology is shown in Fig. 7.4. Here $\beta_o Z_E$ is the lf contribution to Z_b, $Z_E/s\tau_T$ is the hf contribution, and the series Z_E is common to both. Below f_β, $Z_E/s\tau_T$ approaches an open circuit so that Z_b is consistent with the lf model. In the hf region, $Z_E/s\tau_T$ dominates Z_b; dividing Z_E by s gyrates the impedance of Z_E by $-90°$ so that

$$R \to C$$
$$C \to -R$$
$$L \to R$$

For the three cases of Z_E (R, L, and C), the transformed impedances are shown in Fig. 7.5. Since this analysis is linear, combinations of the three elements in Z_E can be individually transformed to produce the transformed Z_E.

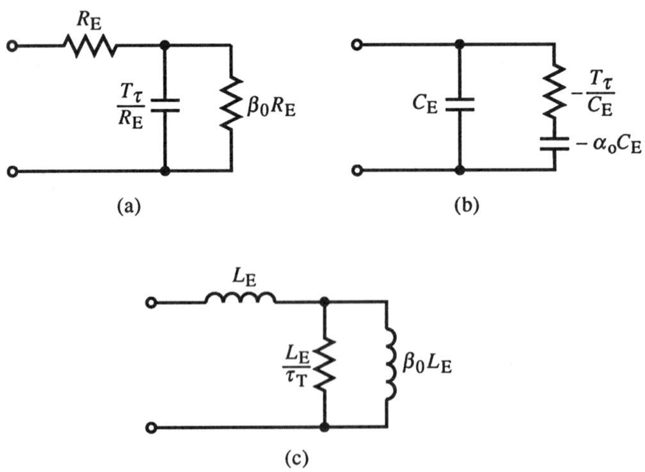

FIG. 7.5 Z_b for Z_E equal to (a) R_E, (b) C_E, and (c) L_E.

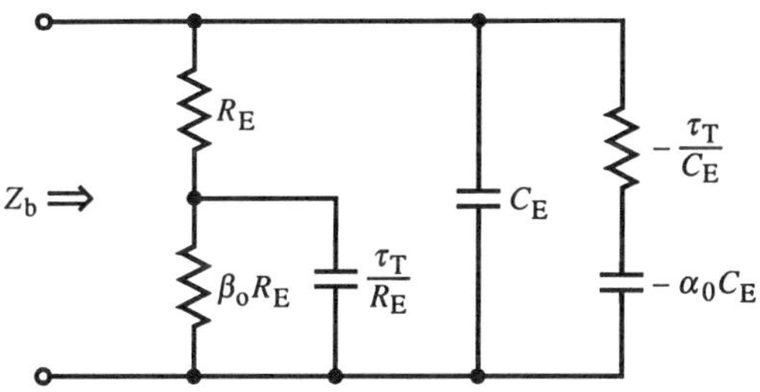

FIG. E7.1

Example 7.1 Shunt *RC*-Loaded CC Amplifier

A CC has a shunt *RC* load for which $R_E = 470\ \Omega$ and $C_E = 10$ pF. The BJT has a $\beta_o = 150$ and $f_T = 300$ MHz at $I_E = 10$ mA (typical of a 2N3904). What is Z_b?

The combination of Figs. 7.5a, b is shown in Fig. E7.1. To find the element values, we calculate $\tau_T = 1/2\pi f_T = 531$ ps and $\alpha_o = 0.993 \cong 1$. Then $\tau_T/R_E = 1.13$ pF and $-\tau_T/C_E = -53.1\ \Omega$. Furthermore, $\beta_o R_E = 70.5\ \text{k}\Omega$ and $-\alpha_o C_E \cong -10$ pF. A hf equivalent circuit omits $\beta_o R_E$. Whether Z_π is negligible depends on the other elements in the circuit. If base reactance creates a resonance with the emitter impedance near f_β, then Z_π is probably significant. For this circuit, $r_e \cong 2.6\ \Omega$ and $C_\pi = \tau_T/r_e = 204$ pF.

A similar circuit derivation for Z_e, based on Fig. 7.3b, results in

$$Z_e = \frac{Z_\pi + Z_B}{1 + \beta_o/(s r_\pi C_\pi + 1)} \tag{7.13}$$

Again approximating $Z_\pi \cong 0$, we obtain the continued fraction

$$Z_e \cong \frac{1}{\dfrac{1}{Z_B} + \dfrac{1}{s\tau_T Z_B + Z_B/\beta_o}} \tag{7.14}$$

The toplogy is shown in Fig. 7.6 and is the dual of Fig. 7.4. Below f_β, $s\tau_T Z_B$ approaches a short circuit and becomes the lf model. The hf contribution of $s\tau_T Z_B$ gyrates Z_e by +90° so that

$$R \rightarrow L$$

$$L \rightarrow -R$$

$$C \rightarrow R$$

The three cases of Z_B are worked out in Fig. 7.7.

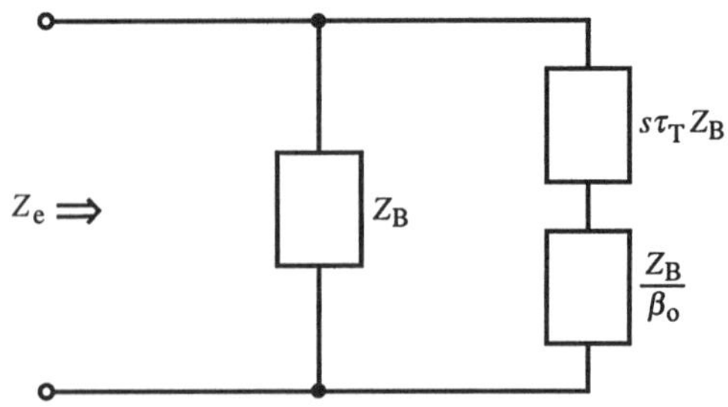

FIG. 7.6 Emitter node equivalent circuit with base impedance Z_B.

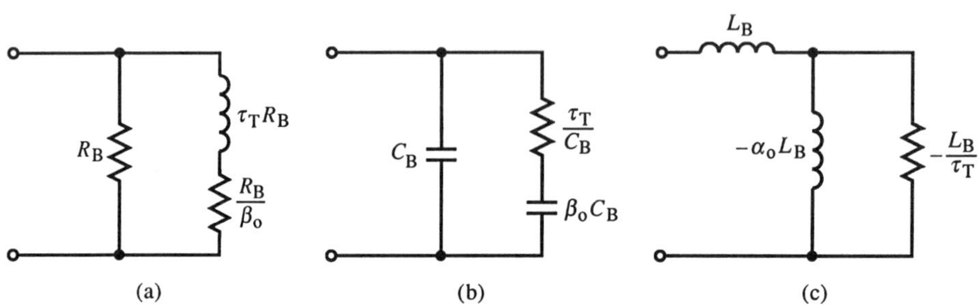

FIG. 7.7 Z_e for Z_B equal to (a) R_B, (b) C_B, and (c) L_B.

These expressions for Z_b and Z_e are valid from dc to f_T. A simpler model, applicable only in the hf region, is derived by using (7.6). The topologies of Figs. 7.4 and 7.6 reduce to those shown in Figs. 7.8 and 7.9. To derive these hf models, we assume

$$\beta_o \to \infty, \qquad Z_\pi = 0 \tag{7.15}$$

Then

$$Z_b(\text{hf}) = Z_E\left(1 + \frac{1}{s\tau_T}\right) \tag{7.16}$$

$$Z_e(\text{hf}) = \frac{Z_B}{1 + 1/s\tau_T} \tag{7.17}$$

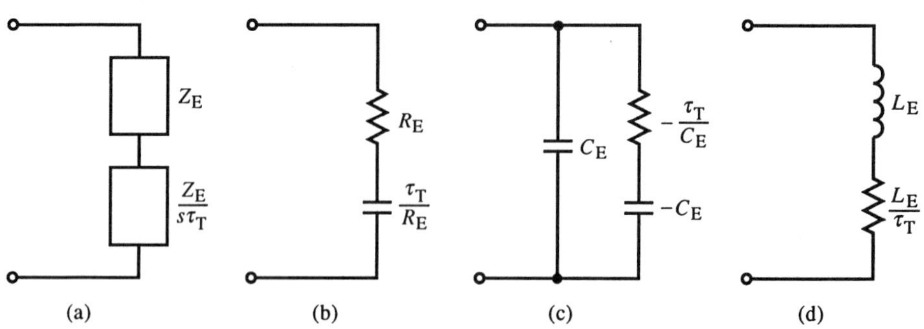

FIG. 7.8 Z_b(hf) for Z_E equal to (a) Z_E, (b) R_E, (c) C_E, and (d) L_E.

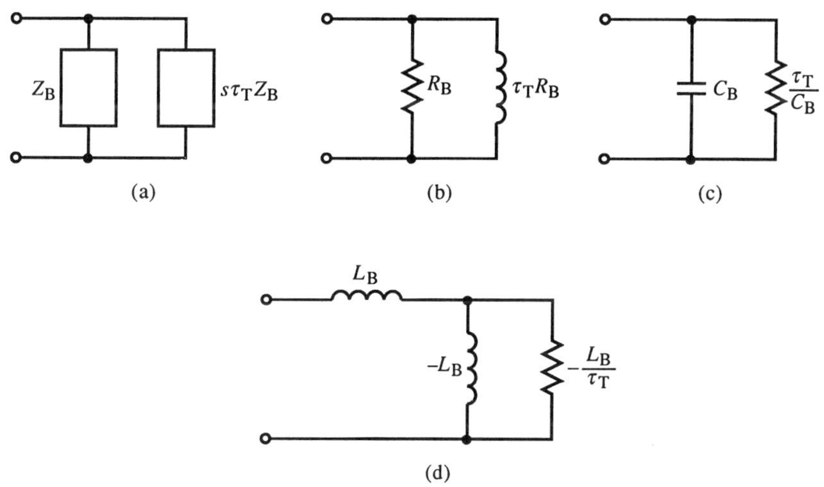

FIG. 7.9 $Z_e(\text{hf})$ for Z_B equal to (a) Z_B, (b) R_B, (c) C_B, and (d) L_B.

The hf models are based on removal of the break frequency of $\beta+1$ (in Fig. 7.2) at f_β so that $\beta(s)$ rolls off from infinity at the origin. The expression for $\beta+1$ becomes

$$\beta_{\text{hf}}+1=\frac{1}{s\tau_T}+1=\frac{s\tau_T+1}{s\tau_T} \tag{7.18}$$

This expression has a pole at the origin which breaks at f_T.

7.4 Reactance Chart Representation of β-Gyrated Circuits

The two representations of hf circuits used thus far, equations and circuit diagrams, are supplemented by a graphic representation using the reactance chart. The reactance plots of some of the impedances of Figs. 7.8 and 7.9 are shown in Figs. 7.10 and 7.11, respectively. The break frequencies of these plots

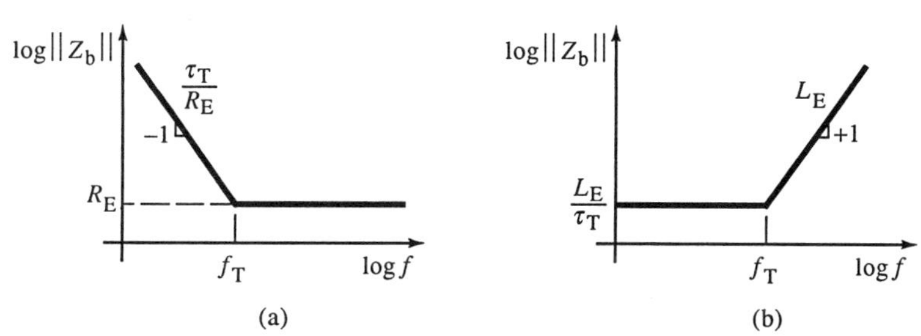

FIG. 7.10 Reactance plots of Z_b for Z_E equal to (a) R_E and (b) L_E.

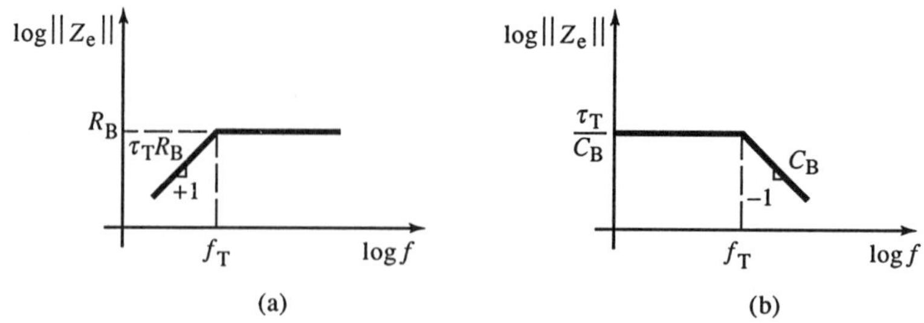

FIG. 7.11 Reactance plots of Z_e for Z_B equal to (a) R_B and (b) C_B.

are f_T, and the nonzero slopes are ± 1 in the hf region. Above f_T the impedance gyration stops since $\beta(s)+1$ stops rolling off with frequency.

Two of the six cases shown in Figs. 7.8 and 7.9 involve negative resistances. Since the reactance chart has a log–log scale, and the logarithm of negative numbers is undefined, it might appear that reactance-chart representation of β-gyrated impedances has limited use. Fortunately, this is not so. The equation for Z_b when $Z_E = 1/sC_E$ can be reformulated as

$$Z_b(C_E) = \frac{s\tau_T + 1}{s^2 \tau_E C_E} \tag{7.19}$$

For the other case, of $Z_e(L_B)$,

$$Z_e(L_B) = \frac{s^2 \tau_T L_B}{s\tau_T + 1} \tag{7.20}$$

These equations can be plotted on a reactance chart and are shown in Fig. 7.12. Below f_T, the plots are of $\tau_T C_E$ with a slope of -2 and $\tau_T L_B$ with a slope of $+2$. Above f_T, the plots are of C_E and L_B.

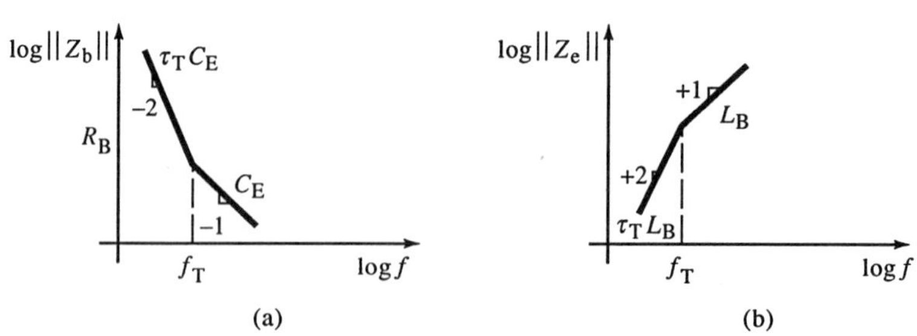

FIG. 7.12 Reactance plots of (a) $Z_b(C_E)$ and (b) $Z_e(L_B)$.

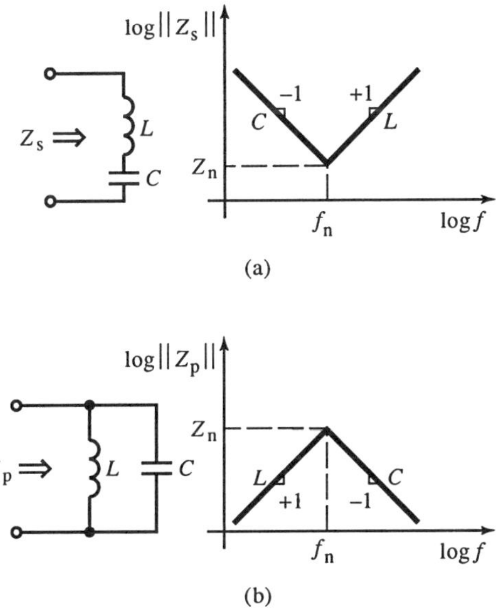

FIG. 7.13 Reactance plots of (a) series and (b) parallel *LC* circuits.

7.5 Reactance Chart Stability Criteria for Resonances

The two common resonant circuits are the series and parallel *LC* circuits shown in Fig. 7.13. We define the intersection of the *L* and *C* plots as the *resonant point*, at which

$$f_n = \frac{1}{2\pi\sqrt{LC}} \tag{7.21}$$

and the characteristic impedance of the resonance is

$$Z_n = \sqrt{\frac{L}{C}} \tag{7.22}$$

Z_n is the reactance of each resonating element at the resonant point.

For all of these reactance plots, asymptotic approximations are used. Exact plots require that a vertical asymptote at f_n be approached on each side by a curve tending to $\pm\infty$. For a series resonance, $Z_s = 0$, which is at $-\infty$ on the reactance chart. For a parallel resonance, $Z_p \to \infty$, which is at $+\infty$ on the reactance chart.

A characteristic feature of resonance is a ± 2 change of slope on the chart: $+2$ for a parallel and -2 for a series resonance. In the case of *LC* circuits, this is a change from ± 1 to ∓ 1. For the cases in Fig. 7.12, however, a change from ± 2 to zero or zero to ± 2 also cause resonances and the potential for oscillation if not sufficiently damped.

The amount of damping of an *RLC* resonance is characterized by the damping ratio ζ. For parallel resonances, this is

$$\text{parallel resonance} \qquad \zeta = \frac{Z_n}{2R} \tag{7.23}$$

and for series resonances, it is

$$\text{series resonance} \qquad \zeta = \frac{R}{2Z_n} \tag{7.24}$$

For critical damping, $\zeta = 1$. Then $R_p = Z_n/2$ and $R_s = 2Z_n$. In both cases, critical damping is achieved by a resistance equal to the combined reactances of the L and C at resonance.

An estimate can therefore be made on the reactance plot as to how well a resonance is damped. For a parallel resonance, the parallel resistance must be below the *LC* resonant point to be well damped, for series resonance, it must be above r.

7.6 Emitter-Follower Reactance Plot Stability Analysis

CC stages are commonly used to drive capacitive loads such as transmission lines. The high current gain of the CC configuration allows it to supply the high transient currents required to quickly charge the capacitive load. When capacitive loading is combined with base resistance, a hf resonance can occur.

Figure 7.14 shows a CC amplifier with its hf and general equivalent circuits of the emitter node. The resulting reactance plot for the hf circuit is shown in Fig. 7.15. Ordinarily, we would combine the parallel Rs and Cs before plotting. Here, sections a and b of the hf circuit are plotted separately because the elements within them are interdependent. Section a of Z_E is resistive up to f_T and is capacitive with value C_B above f_T. Section b is inductive up to f_T, above which it is resistive with value R_B. On the reactance chart, R_B and C_B have been chosen so that $R_B C_B = \tau_\beta$; they intersect at f_β. C_E is much greater than C_B, and it intersects the plot for section b at resonant point r. In Fig. 7.15, this is a parallel resonance; a resistance less than the impedance at r, Z_n, is required to damp it. The plot of the impedance from section a is resistive and less than Z_n at f_n; it damps the resonance. Since this resistance decreases as C_B increases, then increasing base capacitance tends to stabilize a capacitively loaded emitter-follower.

From the reactance plot of Fig. 7.15, we can see what effect changes in the values of circuit elements have. For section a, increasing C_B causes the resistive segment of the a plot to move downward and thus provides more damping at r. At the same time, the diagonal line representing C_B moves to the left. The break frequency does not move but remains constant at f_T, so

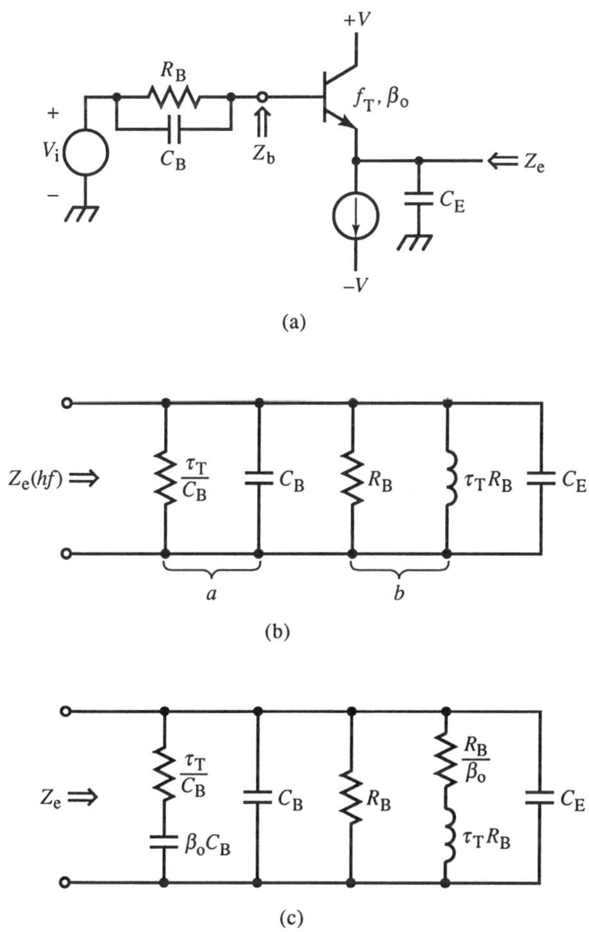

FIG. 7.14 Common-collector (a), hf model (b), and general emitter node equivalent circuit (c).

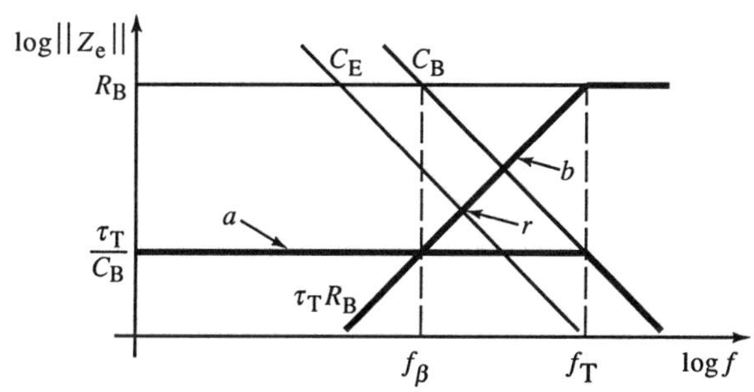

FIG. 7.15 Reactance plot of Fig. 7.14 with $R_B C_B = \tau_T$.

curve *a* moves downward as C_B increases. Similarly, an increase in R_B increases the inductance below f_T in curve *b* while break frequency f_T remains fixed. That is, $R_B \tau_T$ always intersects R_B at f_T. Increasing R_B moves curve *b* upward.

From reactance chart analysis, we can observe that a decrease in C_E or an increase in R_B or C_B tends to stabilize the circuit because Z_n increases relative to the damping resistance. C_E has a range in which instability can occur. As C_E increases, its plot moves to the left, and *r* moves with it and downward until it intersects curve *a* at f_β. Then $1/sC_E = \tau_T/C_B$ at f_β, and *r* is eliminated because impedance gyration does not occur below f_β. This is also true above f_T. If C_E decreases until it crosses R_B at f_T, *r* vanishes. In addition, transistor gain above f_T may be insufficient to sustain oscillation anyway. Since the reactance plots are asymptotic approximations, hf effects extend somewhat above and below the hf region.

This analysis assumes that $C_B \ll C_E$ and $\tau_T/C_B \ll R_B$. More generally, as C_B increases, two effects occur: its β-gyrated resistance, τ_T/C_B, decreases (increasing damping), and C_B also adds to C_E, decreasing Z_n. Adequate damping can occur only when C_E dominates. If C_E becomes negligible relative to C_B, then the resonance cannot be damped better than $\zeta = 0.5$. (See Section 7.7.) For maximum circuit speed, a minimum C_B is desirable to minimize the base input time constant.

7.7. Emitter-Follower High-Frequency Equivalent Circuit Resonance Analysis

An explanation of the CC stage based on the hf equivalent circuit topology (Fig. 7.14b) is that R_B is gyrated at the emitter to become inductive with value $\tau_T R_B$. C_E forms a parallel resonant circuit with $\tau_T R_B$. As R_B increases or C_E decreases, Z_n increases, and the resonance can be damped by higher values of shunt resistance. Although R_B itself provides damping, it is not small enough to be adequate. From (7.23), ignoring C_B,

$$\zeta \cong \frac{1}{2}\sqrt{\frac{\tau_T}{R_B C_E}} \tag{7.25}$$

In the hf region, $R_B C_E > \tau_T$ so that $\zeta < \frac{1}{2}$; the resonance is underdamped. ζ could be increased by decreasing R_B or C_E. (Also, a slower transistor, with increased τ_T, increases ζ.) As $R_B C_E$ approaches τ_β, however, Z_π becomes significant and adds further damping. Also, as $R_B C_E$ approaches τ_T, Z_n approaches R_B and is damped by it.

The addition of C_B damps the resonance because C_B is gyrated at the emitter to a resistance of τ_T/C_B. This resistance can be set as low as needed by increasing C_B. The damping ratio is then

$$\zeta = \frac{Z_n}{2R} \cong \frac{\sqrt{\tau_T R_B/(C_B + C_E)}}{2(\tau_T/C_B) \parallel R_B} = \frac{\tau_T + R_B C_B}{2\sqrt{\tau_T R_B (C_B + C_E)}} \tag{7.26}$$

An acceptable value of ζ can now be obtained by adjusting C_B. The value of C_B can be found directly for a given ζ by solving (7.26) for C_B:

$$C_B = \left(\frac{\tau_T}{R_B}\right)\left[(2\zeta^2 - 1) + 2\zeta\sqrt{\zeta^2 + \frac{C_E}{\tau_T/R_B} - 1}\right] \tag{7.27}$$

(Only one root of C_B is possible for $C_B > 0$.) This simplifies, for $R_B C_B \gg \tau_T$, to

$$C_B \cong \frac{2\zeta^2 \tau_T}{R_B}\left(1 + \sqrt{1 + \frac{R_B C_E}{\tau_T}}\right), \qquad R_B C_B \gg \tau_T \tag{7.28}$$

Furthermore, if $C_E \gg C_B$, then

$$C_B \cong 2\zeta\sqrt{\frac{\tau_T C_E}{R_B}}, \qquad R_B C_B \gg \tau_T, \qquad C_E \gg C_B \tag{7.29}$$

An alternative compensation technique is to adjust R_B for a desired ζ. Increasing R_B increases Z_n so that τ_T/C_B more effectively damps the resonance. When $R_B = \tau_\beta/C_B$, no value of C_E can cause a resonance with $\tau_T R_B$.

Example 7.2 CC Stabilization Using Shunt Base *RC*

The CC of Fig. E7.2, using a 2N3904 BJT, is to be stabilized with C_B, if necessary, so that $\zeta \geq \sqrt{2}/2$ for an MFA response.

The transistor parameters are $\tau_T = 531$ ps and $\alpha_o \cong 1$. The hf resonance is at

$$f_n = \frac{1}{2\pi\sqrt{(531\text{ ps})(1\text{ k}\Omega)(10\text{ pF})}} = 69.1\text{ MHz}$$

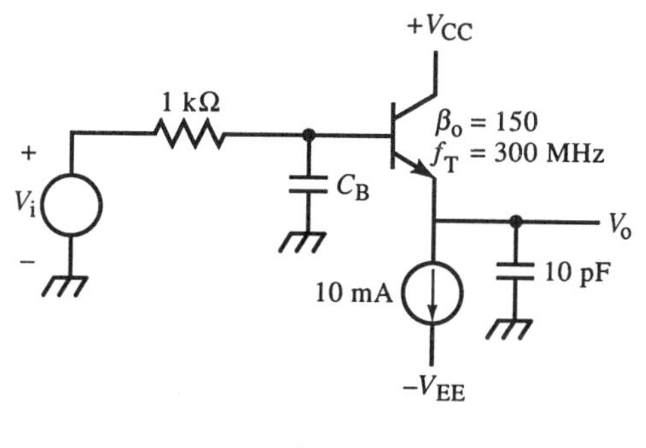

FIG. E7.2

and

$$Z_n = \sqrt{(531\ \text{ps})(1\ \text{k}\Omega)/(10\ \text{pF})} = 230\ \Omega$$

The resonant frequency lies within the hf range of 2 MHz to 300 MHz. Without C_B, from (7.26), $\zeta = 0.115$ ($M_p \cong 0.7$), a very underdamped resonance. The SPICE simulation uses a BJT model with

.MODEL BJT1 NPN (BF=150 TF=531 ps)

and results in $M_p = 0.65$ and $\zeta = 0.136$. This greater damping is partly attributable to R_B/β_o and Z_π.

To achieve the desired ζ, substitute into (7.27) and solve for C_B to get 3.2 pF. Then $R_B C_B = 32\ \text{ns} \gg \tau_T$. If the approximate formulas for C_B are used, (7.28) yields 2.9 pF, and (7.29) yields 3.3 pF. The conditions for (7.29) are met fairly well, and the approximations are valid for the accuracy required for parts selection.

Example 7.3 CC Amplifier Series *R* Compensation

To eliminate hf resonance from the CC of Example 7.2, R_B could be increased instead of increasing C_B. Let $C_B = C_\mu = 5$ pF, and the desired damping is critical ($\zeta = 1$). When (7.28) is solved for R_B, it is

$$R_B = \frac{4\zeta^2 \tau_T (C_E + C_B)}{C_B^2} \tag{E1}$$

For the example, R_B must be greater than 1.27 kΩ; an additional 270 Ω is needed.

This example illustrates why the addition of a small (10 Ω to 1 kΩ) series base resistor sometimes damps an oscillating BJT. The addition of R_B might, however, damp a series *LC* resonance with the collector circuit parasitic inductance instead. With hf resonance analysis, one possible cause of oscillation can be assessed from circuit element parameters.

7.8 Emitter-Follower High-Frequency Compensation

The hf BJT model is adequate for analyzing hf resonances but not for determining optimal compensation schemes since they involve both lf and hf regions.

The approximation of ζ in Section 7.7 ignored Z_π, β_o/C_B, and R_B/β_o. These additional elements cause further damping so that the actual ζ is greater than (7.26) predicts. If we use the general model instead, the emitter impedance of Fig. 7.14c, when derived without C_E, is

$$Z_e = \frac{R_B}{\beta_o + 1} \cdot \frac{s\tau_\beta + 1}{(s\alpha_o\tau_T + 1)(sR_BC_B + 1)} \tag{7.30}$$

This result is obtained by substituting Z_B into (7.14) or by dividing Z_B by $\beta(s)+1$ from (7.4). If $R_BC_B = \tau_\beta$, then the zero is cancelled, and Z_e simplifies to

$$Z_e\big|_{R_BC_B=\tau_T} = \frac{R_B}{\beta_o + 1} \cdot \frac{1}{(s\alpha_o\tau_T + 1)} \tag{7.31}$$

Under this condition, no C_E can cause a resonance with a resistive Z_B.

From the equivalent circuit of Fig. 7.14c, the series RL and RC branches have the form of Example 6.1. This is a constant-resistance network with resistance R_B/β_o when the elements have the relationships

$$\frac{R_B}{\beta_o} = \frac{\tau_T}{C_B} = \sqrt{\frac{R_B\tau_T}{\beta_o C_B}} \tag{7.32}$$

or $R_BC_B = \tau_\beta$. In view of $R_B/\beta_o \parallel R_B = R_B/(\beta_o + 1)$,

$$\begin{aligned} Z_e &= \left(\frac{R_B}{\beta_o + 1}\right) \Bigg\| \frac{1}{sC_B} = \frac{[R_B/(\beta_o+1)](1/sC_B)}{R_B/(\beta_o+1) + 1/sC_B} \\ &= \frac{R_B}{\beta_o + 1} \cdot \frac{1}{s(R_BC_B/(\beta_o + 1)) + 1} \end{aligned} \tag{7.33}$$

The pole time constant can be expressed as

$$\frac{R_BC_B}{\beta_o + 1} = \frac{\tau_\beta}{\beta_o + 1} = \alpha_o\tau_T$$

and (7.33) is equivalent to (7.31).

The condition $R_BC_B = \tau_\beta$ achieves a resistive base or emitter impedance out to near f_T. This eliminates hf resonances but does not always result in maximum circuit speed (minimum risetime or maximum bandwidth). It is feasible for CC amplifiers when C_E varies greatly or is unknown. The zero at f_β can introduce phase-lead out to $1/R_BC_B$ if hf peaking is needed. For maximum circuit speed, R_BC_B is set so that the hf resonance is damped to the desired extent. In this case, $R_BC_B < \tau_\beta$. The hf model equations in Section 7.7 give easy estimates for element values when C_E and τ_T are within known bounds.

Example 7.4 Power Amplifier CC Output Stage

Figure E7.4a shows a typical amplifier bipolar CC output stage. As a symmetrical class AB amplifier, only the top or bottom half is active for most of the output voltage range (except around zero). For analysis, assume that only the upper transistor is conducting. Figure E7.4b shows the simplified equivalent circuit, with $Z_L = 1/sC_E$. The emitter-node hf circuit is shown in Fig. E7.4c with reactance plots in (d) for negligible R_E. R_E causes a +1 slope change in the C_E plot at R_E. Here it is assumed that this zero is above the resonant frequency. The intersection of $\tau_T R_B$ and C_E is the resonant point *r*.

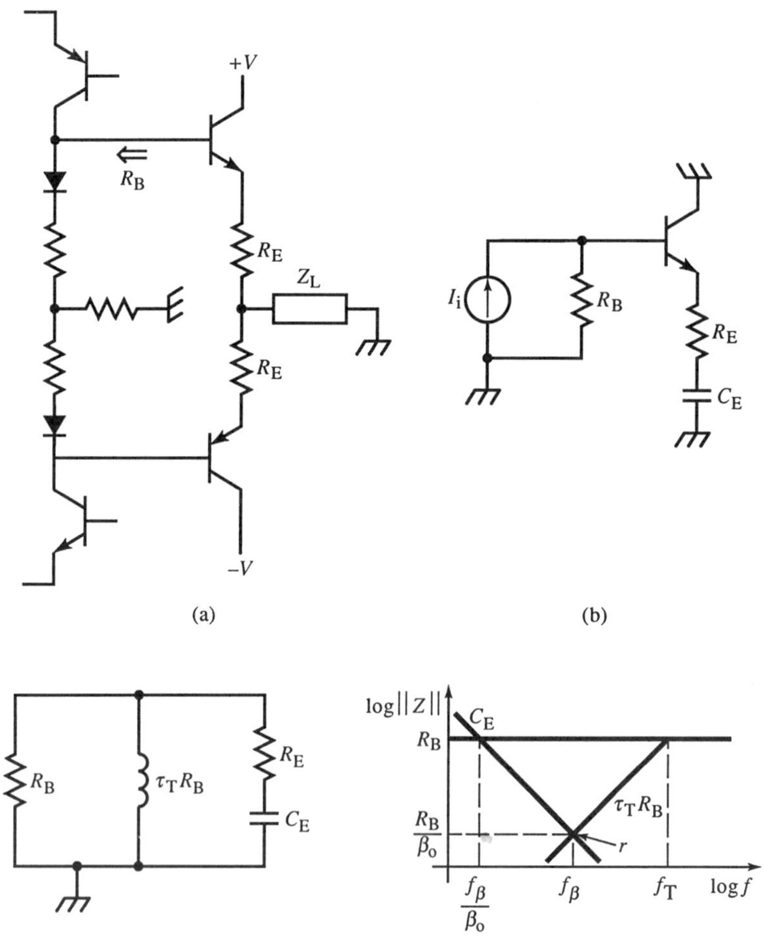

FIG. E7.4

R_B is set by the preceding driver stage. No hf resonance can occur if it is large enough to cause $f_n < f_\beta$, or $R_B C_E \geq \beta_o \tau_\beta$, as shown in Fig. E7.4d. For unconditional stability,

$$f_n < f_\beta \Rightarrow \sqrt{R_B \tau_T C_E} > \beta_o \tau_T$$

or

$$R_B > \frac{1}{(\omega_\beta / \beta_o) \cdot C_E}$$

Alternatively, a sufficiently small R_B places the resonance outside the hf region. Or the series R_E, if large enough, causes $R_E C_E < \tau_n$ and damps the series resonance. (Both series and parallel resonant modes are damped under the same conditions. Any zeros of the series resonance are poles of the parallel resonance.) Depending on r_e of the BJT, Z_π may provide adequate series damping.

7.9 Emitter-Follower Resonance Analysis from the Base Circuit

The CC stage analyses of Sections 7.6 and 7.7 were performed at the emitter node. The analysis from the base node illustrates the plotting of negative impedances. Figure 7.16 shows the hf equivalent base-circuit impedance and

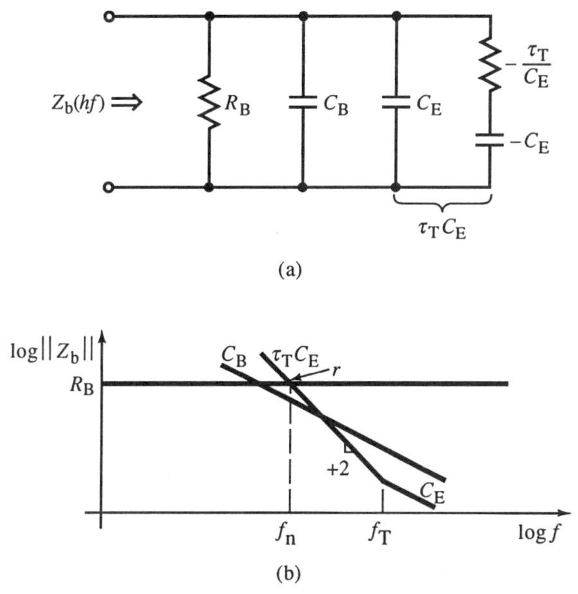

FIG. 7.16 Z_b(hf) equivalent circuit of Fig. 7.14 (a) and reactance plot (b).

reactance plots of R_B, C_B, and the section marked $\tau_T C_E$. The plot of this circuit section is the same kind as in Fig. 7.12a and has a value of $\tau_T C_E$ in the hf region. Above f_T, the curve becomes capacitive with the value of C_E. At resonance point r, the slope of $\tau_T C_E$ intersects R_B with a slope change from -2 to zero indicating a resonance. Unlike the emitter-node analysis, this is a series resonance (because the change in slope across r is positive).

C_B damps the resonance by intersecting R_B at a lower frequency than f_n. Then Z_b rolls off at this intersection along C_B at a -1 slope. When C_B intersects $\tau_T C_E$, the slope changes from -1 to -2, a change of -1. Therefore, no slope change greater than ± 1 occurs, and the resonance at r is damped. If R_B or C_B increases, the break frequency $1/R_B C_B$ decreases, and damping is increased. This result is consistent with the analysis from the emitter node.

From Fig. 7.16a and (7.19),

$$Z_b(\text{hf}) = \left(\frac{s\tau_T + 1}{s^2 \tau_T}\right) \Big\| \ (R_B \,\|\, C_B)$$

$$= R_B \cdot \frac{s\tau_T + 1}{s^2[\tau_T R_B(C_B + C_E)] + s(\tau_T + R_B C_B) + 1} \tag{7.34}$$

From this,

$$\zeta = \frac{b}{2\sqrt{a}} = \frac{\tau_T + R_B C_B}{2\sqrt{\tau_T R_B(C_B + C_E)}} \tag{7.35}$$

This is the same as (7.26), as it must be.

7.10 Emitter-Follower Compensation with a Base Series *RC*

The $\tau_T C_E$ section of Fig. 7.16a can be all-pass compensated by the addition of a series RC branch with positive corresponding element values. The two branches null each other, leaving an open circuit. This leaves C_E shunting R_B and C_B, a nonresonant circuit. This compensation technique is shown in Fig. 7.17. The compensation conditions are

$$C = \alpha_o C_E, \qquad R = \frac{\tau_T}{C_E} \tag{7.36}$$

The reactance chart in Fig. 7.17c shows how this works. At f_T, both the RC curve and the $\tau_T C_E$ curves break. The RC curve rolls off until f_T, where it becomes flat with value R. This curve dominates Z_b below f_T. At f_T, the $\tau_T C_E$ curve breaks to C_E and dominates as the reactance of C_E falls below that of R. The curve for C remains unbroken when $C = C_E$. The result is that

$$Z_b = R_B \,\Big\|\, \frac{1}{sC_E}$$

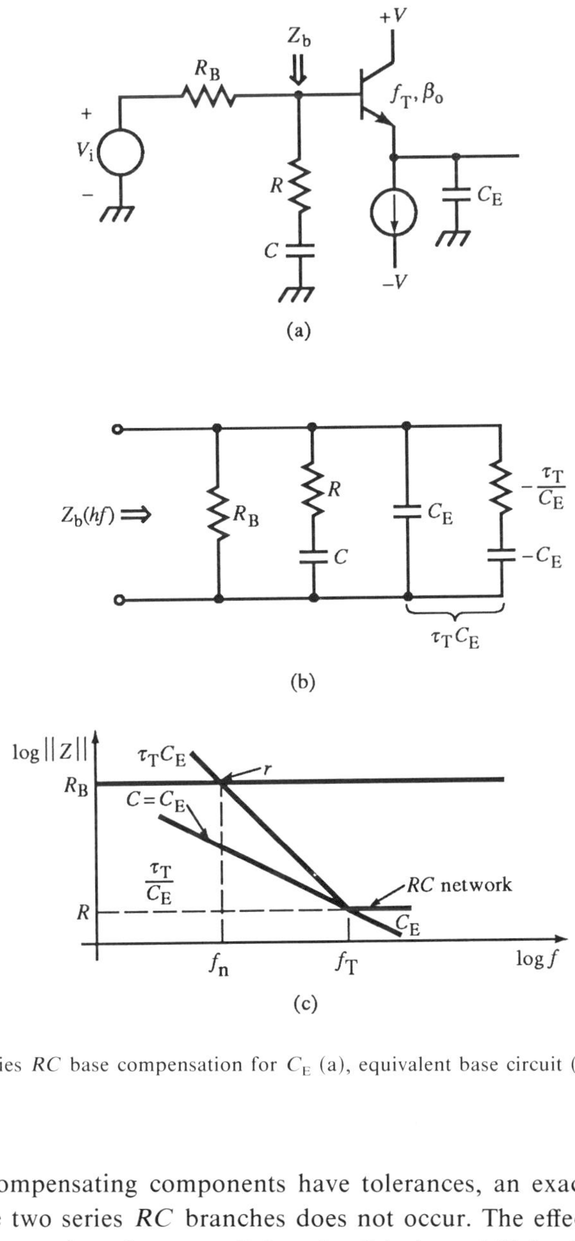

FIG. 7.17 Series *RC* base compensation for C_E (a), equivalent base circuit (b), and reactance plot (c).

Since compensating components have tolerances, an exact cancellation between the two series *RC* branches does not occur. The effect can be seen on the reactance chart, however. Below f_T, C is the stabilizing influence since it causes Z_b to roll off without a resonant change in slope. The resonance at r is damped when C intersects R_B at a frequency below f_n. The condition for this is

$$R_B C > \sqrt{\tau_T C_E R_B} \tag{7.37}$$

or

$$R_B\left(\frac{C^2}{C_E}\right) > \tau_T \tag{7.37a}$$

For unconditional damping (independent of C_E), then $R_B C \le \tau_\beta$. When R and C are set to cancel $\tau_T C_E$, their values depend only on τ_T and C_E. This leaves R_B free to be set to insure adequate damping under variation of τ_T and C_E.

7.11 Bipolar-Junction Transistor Amplifier with Base Inductance

The dual of the circuit with emitter capacitance and base resistance is one with base inductance and emitter resistance, shown in Figs. 7.18a,b with reactance chart in (c). From the base, the emitter resistance gyrates −90° and produces capacitance τ_T / R_E, which series resonates with L_B. As shown, R_E is not large enough to damp it. The circuit can be stabilized in the following ways:

1. Decrease R_E until $f_n < f_\beta$. For this,

$$\sqrt{\frac{L_B \tau_B}{R_E}} > \beta_o \tau_T \rightarrow R_E < \frac{L_B}{\beta_o^2 \tau_T} = L_B \cdot \frac{\omega_\beta}{\beta_o} \tag{7.38}$$

This extreme measure eleminates any hf resonance but slows the circuit more than necessary.

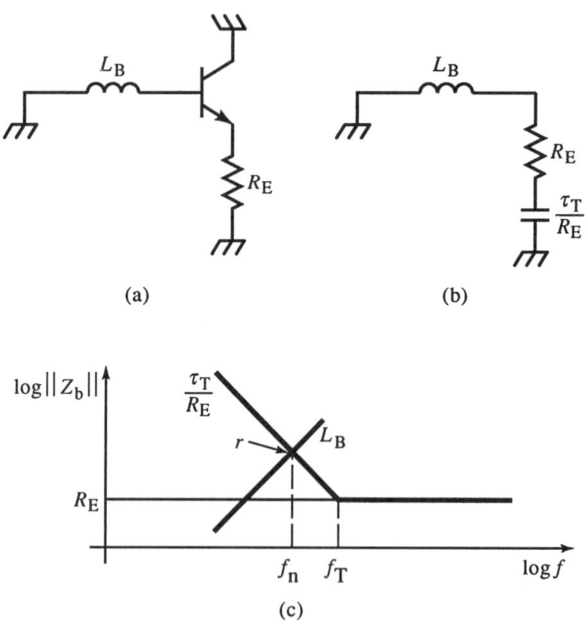

FIG. 7.18 Resonance due to base inductance: ac circuit (a), equivalent circuit with Z_b (b), and reactance plot of Z_b (c).

2. Decrease L_B until $f_n > f_T$. Then $L_B / R_E < \tau_T$ or $L_B < R_E \tau_T$. This approach also eliminates hf resonance by moving the resonance outside the high end of the hf region. If L_B is parasitic, this might not be possible.

3. Increase L_B until $f_n < f_\beta$. Then, from method 1,

$$L_B > \frac{R_E}{\omega_\beta / \beta_o} \tag{7.39}$$

4. Either increase or decrease R_E until hf resonance is eliminated or sufficient damping occurs.

5. Add a series L in the emitter. It is gyrated $-90°$ at the base and provides series damping resistance.

This last approach is the dual of shunt C_B damping.

In the CB stage of cascode amplifiers, base inductance combines with emitter capacitance (Fig. 7.19). C_E is the output capacitance, mainly C_μ of the CE transistor. At f_n, L_B crosses $\tau_T C_E$ with a -3 slope change. By using the hf model, we can derive Z_b:

$$Z_b = \left(\frac{s\tau_T + 1}{s^2 \tau_T C_E}\right) \Big\| \, sL_B = \frac{sL_B(s\tau_T + 1)}{s^3 \tau_T L_B C_E + s\tau_T + 1} \tag{7.40}$$

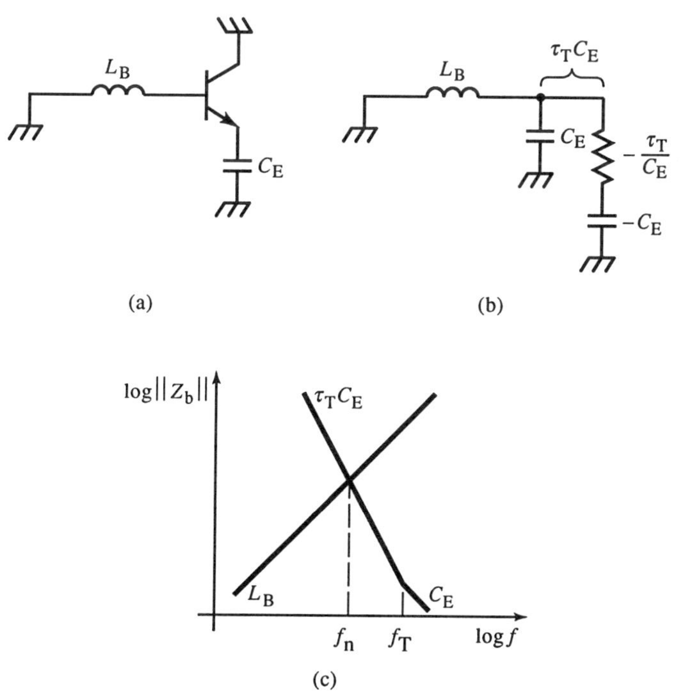

FIG. 7.19 Cascode CB stage ac circuit (a), base node equivalent circuit (b), and reactance plot of Z_b (c).

This circuit can be stabilized several ways:

1. Decrease either L_B or C_E until L_B crosses $\tau_T C_E$ above f_T. Then, $f_n > f_T$.
2. Increase either L_B or C_E until $f_n < f_\beta$; then damp $L_B C_E$ with a series R_E.
3. Shunt L_B with $C_B > C_E$ and damp $L_B C_B$ with a series R_B.
4. Add a series emitter resistance $R_E > L_B/\tau_T$; then $R_E + \tau_T C_E$ crosses L_B above f_T.

Combinations of these techniques are also possible. Method 1 or 3 achieves maximum speed.

7.12 The Effect of r'_b on Stability

We now return to the BJT model of Fig. 7.1b and examine the effect of r'_b on stability. A more accurate model would distribute C_μ across r'_b. This distributed-parameter RC is approximately modeled with segmented lumped-parameter C_μ and r'_b. The base resistance can be divided into several resistances with portions of C_μ connected between them. But the two extreme cases are to connect C_μ to either the internal node (as shown) or to the (external) base terminal node. Actual transistor performance lies within a range bounded by the behavior of these extremes.

For circuits in which $R_B \ll r'_b$, then r'_b appears as $r'_b \tau_T$ from the emitter and can resonate with external emitter capacitance C_E. Since r'_b is inaccessible, a stabilizing C cannot be shunted across the b′ node. If C_μ is internal, it helps damp $r'_b \tau_T$, whereas an external C_μ does not. If the internal C_μ is insufficient, an additional base resistance R_B will increase the total base resistance so that

$$(r'_b + R_B)C_\mu \geq \tau_\beta$$

If adding R_B is infeasible and if r'_e is not enough damping, then R_E can be added in series with the emitter. This could, however, slow the response. A shunt RC, having a time constant of τ_T, added in series with the emitter forms an all-pass network with the base impedance and preserves speed (Fig. 7.20).

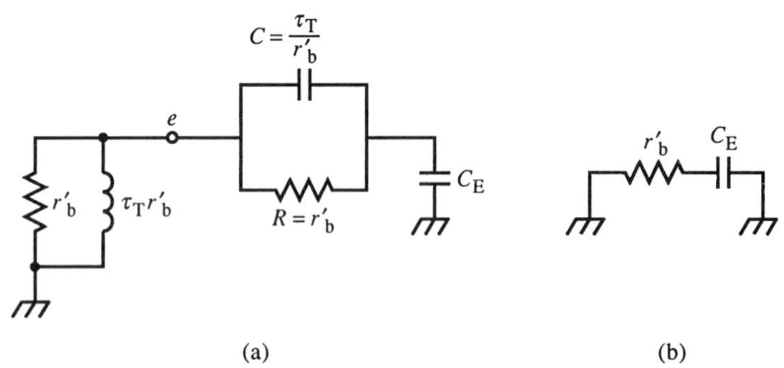

FIG. 7.20 Parallel RC compensation network in series with emitter: (a) equivalent circuit and (b) resulting impedance.

7.13 Field-Effect Transistor High-Frequency Analysis

As stated in the introduction, the BJT hf model can be generalized to apply to other kinds of active devices. If the BJT to FET terminal mapping, (4.6), is used, then the major difference between the BJT and FET models is Z_π. The FET model is configured as a source-follower in Fig. 7.21. The expression for Z_g is derived from the corresponding BJT expression for Z_b. By applying $r_m = r_\pi/\beta_o$, we obtain

$$Z_b = Z_\pi + Z_E\left[1+\frac{1}{s(r_\pi/\beta_o)C_\pi + 1/\beta_o}\right] \tag{7.41}$$

Let $\beta_o \to \infty$. Then for a fixed r_m, $Z_\pi \to 1/sC_\pi$ and

$$Z_b \cong \frac{1}{sC_\pi} + Z_E\left(1+\frac{1}{sr_m C_\pi}\right) \tag{7.42}$$

By analogy, $C_\pi = C_{GS}$ and $Z_E = Z_S$. Substituting yields

$$Z_g = \frac{1}{sC_{GS}} + Z_S\left(1+\frac{1}{sr_m C_{GS}}\right) \tag{7.43}$$

and also, for the source impedance,

$$Z_s = \frac{Z_G + 1/sC_{GS}}{1+1/sr_m C_{GS}} \tag{7.44}$$

C_{GD} is analogous to C_μ, and the time constant $r_m C_{GS}$ is the τ_T of a FET. The BJT equivalent circuits can be applied to FETs with these BJT to FET correspondences.

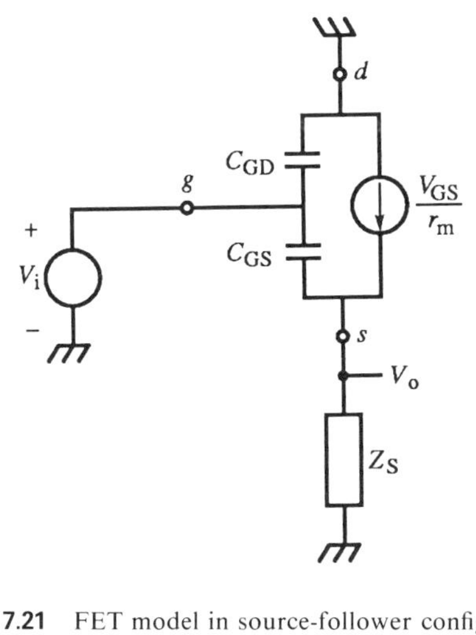

FIG. 7.21 FET model in source-follower configuration.

7.14 Output Impedance of a Feedback Amplifier

Finally, we extend hf modeling to amplifiers having a dominant single-pole response. For an amplifier with an open-loop voltage gain of

$$G = \frac{K}{s\tau_{bw} + 1} \tag{7.45}$$

where $1/\tau_{bw}$ is the small-signal open-loop bandwidth. In the lf region, the open-loop output resistance r_{out} is reduced by the feedback by $1 + GH$. The resulting closed-loop output impedance is

$$Z_{out}(cl) = \frac{r_{out}}{1 + GH} = \frac{r_{out}}{1 + KH} \cdot \frac{s\tau_{bw} + 1}{s(\tau_{bw}/(1 + KH)) + 1} \tag{7.46}$$

where KH is constant. This equation can be expressed as

$$Z_{out}(cl) = \cfrac{1}{\cfrac{1}{r_{out}} + \cfrac{1}{s\tau_{bw}(r_{out}/KH) + (r_{out}/KH)}} \tag{7.47}$$

In continued-fraction form, the corresponding topology of $Z_{out}(cl)$ is explicit (Fig. 7.22a). The lf closed-loop resistance, $r_{out}/(1 + KH)$, gyrates +90° at the

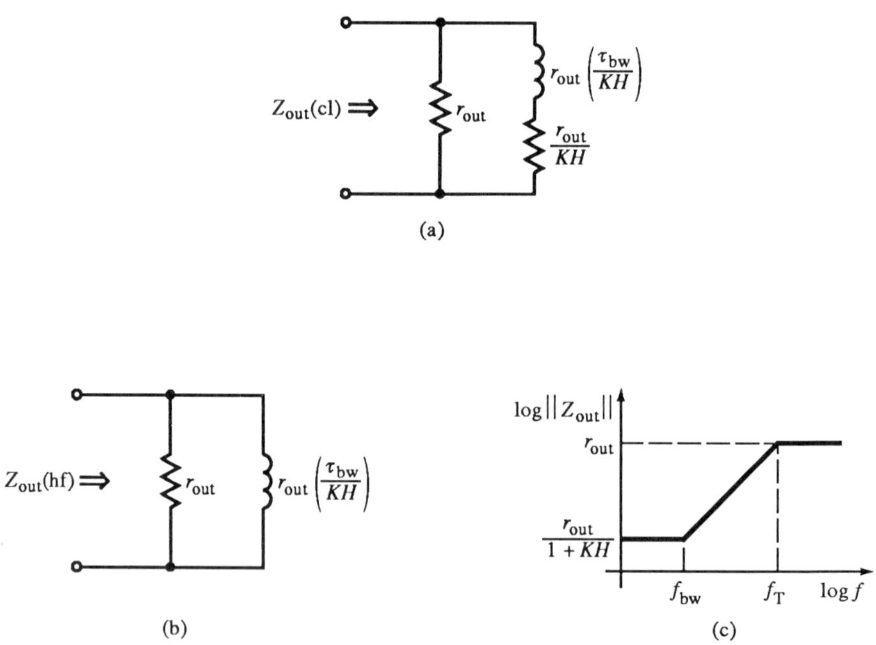

FIG. 7.22 Dominant single-pole feedback amplifier output impedance: (a) equivalent circuit, (b) hf equivalent circuit, and (c) reactance plot of $Z_{out}(cl)$.

open-loop bandwidth to appear inductive out to the unity-gain frequency,

$$f_{bw}(1+KH)$$

Above this unity-gain frequency, $Z_{out}(\text{cl})$ reverts to r_{out}. By analogy, f_{bw} corresponds to the BJT f_β, the unity-gain frequency to f_T, and KH to β_o. The simplified hf equivalent output is derived by letting $KH \to \infty$ as $f \to 0$, with resulting output impedance corresponding to $Z_e(R_B)$. When r_{out} is generalized to Z_{out}, the corresponding BJT models readily apply. The hf equivalent circuit of $Z_{out}(\text{cl})$, as with the BJT model, is only valid above f_{bw}.

Example 7.5 Feedback Amplifier Output Resonance

The amplifier of Fig. E7.5 has a dc gain of -200 and a pole at 1 Ms^{-1} ($\tau_{bw} = 1\ \mu\text{s}$). It has an output resistance of 1 kΩ and a load capacitance of 1 nF.

The loop gain $KH = 100$ and $f_T \cong 15.9$ MHz. Then the gyrated resistance is (10 ns)(1 kΩ) = 10 μH and $Z_n = 100\ \Omega$. For a parallel resonance, $\zeta = Z_n/2r_{out} = 100\ \Omega/2\ \text{k}\Omega = 0.05$. Then $M_p = 0.85$. The simulated circuit $M_p \cong 0.71$, indicating that the hf model estimate of ζ is low. The amplifier simulations for loop gains of 10, 100, and 1000 are tabulated as follows:

KH	1000	100	10
M_p(SPICE)	0.87	0.71	0.25
L	1 μH	10 μH	100 μH
Z_n	31.6 Ω	100 Ω	316 Ω
r_{out}/KH	1 Ω	10 Ω	100 Ω
ζ	0.0158	0.050	$0.158 = Z_n/2r_{out} = (r_{out}/KH)/2Z_n$
M_p	0.95	0.85	0.60
ζ	0.0316	0.10	$0.316 = 1/\sqrt{KH}$
M_p	0.91	0.73	0.35

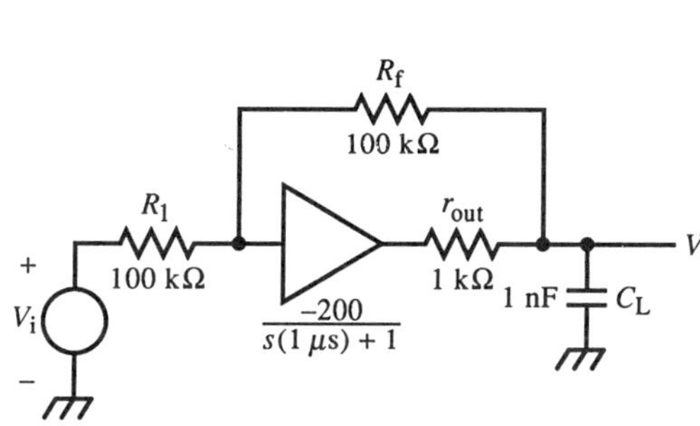

FIG. E7.5

This table reveals that the predicted ζ using either pure parallel or series resonance is always low and that the error increases as KH decreases. If r_{out}/KH and the load capacitance C_L are included in the model (as in Fig. 7.22a), then the expression for $Z_{out}(cl)$ is

$$Z_{out}(cl) = \left(\frac{r_{out}}{1+KH}\right)\frac{s\tau_{bw}+1}{\{s[\tau_{bw}/(1+KH)]+1\}(sr_{out}C_L+1)} \tag{E1}$$

and

$$\zeta = \frac{1}{\sqrt{KH}} \tag{E2}$$

With this more accurate ζ (the lower entry in the table), the agreement with simulation results is much better in the corresponding M_p values. The resonant $Z_{out}(cl)$ of (E1) has the same form as (7.30), from which analogies can be made.

Even with the more exact ζ, the error grows with decreasing KH. This is due to the growing error in the asymptotic approximations of the impedance plot. For $KH = 10$, the error in M_p is quite apparent (40%); but for $KH = 100$, it is much reduced (3%). In this example also, the resonant frequency is near the center of the hf range, thus reducing error due to linear approximation. Near either f_{bw} or f_T, this error becomes large; the approximate calculations of ζ should be used only as a worst-case lower bound.

7.15 Closure

By deriving the complex-frequency expression for β from the BJT hybrid-π model, we can write the β-dependent impedance transformations at base and emitter nodes in a general form. When these impedances are represented graphically on a reactance chart, the effects of circuit element variations on circuit behavior, especially hf resonances, becomes evident. The frequency-dependent β transform is also applicable to FETs and single-pole feedback amplifiers. They show the same impedance gyrations as the BJT, so the BJT results can be easily extended to them also.

References

John A. Archer *et al.*, "Use of transistor-simulated inductance as an interstage element in broadband amplifiers," *IEEE J. Solid-State Circuits* SC-3, Mar. 1968, pp. 12–21.

Arpad Barna, "On the transient response of emitter followers," *IEEE J. Solid-State Circuits* SC-8, Jun. 1973, pp. 233–235.

Alberto Bilotti, "Common-collector impedance transformation sensitivity," *IEEE Trans. Circuit Theory* CT-14, Sept. 1967, pp. 364–366.

Michael Chessman and Nathan Sokol, "Prevent emitter-follower oscillation," *Electronic Design* 13, 21 Jun. 1976, pp. 110–113.

Glenn B. DeBella, "Stability of capacitively-loaded emitter followers—a simplified approach," *Hewlett-Packard Journal* 17, Apr. 1966, pp. 15–16.

Russell G. Gough, "High-frequency transistor modeling for circuit simulation," *IEEE J. Solid-State Circuits* SC-17, Aug. 1982, pp. 666–670.

Joseph L. Kozikowski, "Analysis and design of emitter followers at high frequencies," *IEEE Trans. Circuit Theory* CT-11, Mar. 1964, pp. 129–136.

J. Lindmayer and W. North, "The inductive effect in transistors," *Solid-State Electronics* 8, 1965, pp. 409–415.

Richard I. Ollins and Steven J. Ratner, "Computer-aided design and optimization of a broad-band high-frequency monolithic amplifier," *IEEE J. Solid-State Circuits* SC-7, Dec. 1972, pp. 487–492.

"Amplifier Frequency and Transient Response" course notes, Carl Battjes, Tektronix Education Program, Tektronix, Inc., Beaverton, Oregon.

CHAPTER 8

Wideband Amplification

Some amplifiers are performance-limited mainly by speed. Oscilloscope vertical amplifiers, pulse and function generator output amplifiers, and video and nuclear signal-processing amplifiers are often speed-limited. Fast amplifiers are usually open-loop limited-gain stages, such as those analyzed at low frequency in Chapters 2–4, with one or two transistors per stage. New techniques have increased the speed of op-amp circuits, but the fastest amplifiers consist of limited-gain stages. For the fastest speed, these amplifiers are integrated to reduce parasitic reactances. We first examine the strategy of amplifier design before analyzing various bandwidth extension techniques.

8.1 Multiple-Stage Response Characteristics

The speed of an amplifier can be expressed by its response to a step input. For a single-pole amplifier with pole $p = 1/\tau$, the response to a unit step input can be characterized by its risetime. A single-pole amplifier has a transfer function of the form

$$A(s) = K \cdot \frac{1}{s/p+1} \tag{8.1}$$

with pole at $-p$. The pole is also at the bandwidth, so

$$\text{single-pole } \omega_{\text{bw}} = p$$

From (5.137), the single-pole risetime is

$$t_r = \tau \ln(9) \cong 2.2\tau = \frac{2.2}{p} \cong \frac{0.35}{f_{\text{bw}}} \tag{8.2}$$

The unit step response is

$$v_{\text{step}}(t) = K(1 - e^{-pt}) \tag{8.3}$$

Equations (8.1) and (8.2) assume a linear amplifier (or that small-signal analysis is valid). Large-signal amplifier behavior occurs when a signal quantity reaches the limit of its linear range. What often results is signal slewing, in which signal change is rate-limited and characterized by its *slew rate.* The maximum slew rate of a signal is determined by its maximum instantaneous slope and amplitude. The *full-power bandwidth* f_{BW} of an amplifier is related to the maximum slew rate of a sinusoid that spans the dynamic range of the amplifier. Differentiating the sinusoid and solving for the maximum value, we get

$$\text{maximum slew rate} = \frac{d}{dt}(V_{\text{m}} \sin \omega_{\text{BW}} t)\bigg|_{\text{max}} = \omega_{\text{BW}} V_{\text{m}} \tag{8.4}$$

The sinusoid changes over its full range in the slewing time:

$$t_{\text{slew}} = \frac{2V_{\text{m}}}{\omega_{\text{BW}} V_{\text{m}}} \cong \frac{0.32}{f_{\text{BW}}} \tag{8.5}$$

When $f_{\text{bw}} = f_{\text{BW}}$, t_{slew} is nearly the same as the risetime in (8.2). A more general comparison of large- and small-signal risetime is derived by finding the time it takes to slew from 10% to 90% of its final value. This time is

$$\text{slewing } t_{\text{r}} = \frac{(0.8)2V_{\text{m}}}{\omega_{\text{BW}} V_{\text{m}}} \cong \frac{0.26}{f_{\text{BW}}} \tag{8.6}$$

When an amplifier operates with some slewing, the bandwidth lies between f_{BW} and f_{bw} where always $f_{\text{BW}} < f_{\text{bw}}$.

Another quantity that characterizes speed is time delay t_{d}, defined as the time that the response takes to a unit step input to reach half of its final value. It is found by setting v_{step} of (8.3) to 0.5 and solving for t:

$$t_{\text{d}} = \tau \ln(2) = \frac{\ln 2}{p} \cong \frac{0.69}{p} \cong \frac{0.11}{f_{\text{bw}}} \tag{8.7}$$

Delay time is useful for measuring the propagation delay of linear logic circuits, such as ECL logic gates.

Fast amplifiers usually consist of several cascaded gain stages. An amplifier with n single-pole stages with poles at $-p$ has a transfer function of the form

$$A(s) = K \cdot \frac{1}{(s/p + 1)^n} \tag{8.8}$$

The unit step response of $A(s)$ is

$$n\text{-stage } v_{\text{step}} = \mathscr{L}^{-1}\left\{A(s) \cdot \frac{1}{s}\right\} = Kp^n\left(1 - e^{-pt} \sum_{k=0}^{n-1} \frac{t^k}{k!}\right) \tag{8.9}$$

Calculation of the risetime from (8.9) is not easy. A simpler alternative is to derive expressions for the bandwidth, as we derived (5.153). The bandwidth of a single-pole amplifier is the pole frequency, or

$$\text{single-pole } \omega_{\mathrm{bw}} = p \tag{8.10}$$

The magnitude at bandwidth of a single-pole stage with dc gain of K is found by setting ω to p:

$$\|A(jp)\| = K \cdot \left.\frac{1}{\sqrt{(\omega/p)^2+1}}\right|_{\omega=p} = K \cdot \frac{1}{\sqrt{2}} \tag{8.11}$$

An n-stage amplifier with single-pole stages has a transfer function of the form

$$n\text{-stage } A(s) = K \cdot \frac{1}{(s/p+1)^n} \tag{8.12}$$

When the magnitude of $A(j\omega)$ has rolled off to that of a single-pole stage, as in (8.11), this is the n-stage bandwidth:

$$\|A(j\omega)\| = K \cdot \frac{1}{[(\omega/p)^2+1]^{n/2}} = K \cdot \frac{1}{\sqrt{2}} \tag{8.13}$$

Solving (8.13) for $\omega = \omega_{\mathrm{bw}}$, we express it in terms of p by defining the *bandwidth reduction factor* Ξ as

$$\Xi(n) = \frac{\omega_{\mathrm{bw}}}{p} = \sqrt{2^{1/n}-1} \tag{8.14}$$

Fast amplifiers usually have stages with quadratic pole factors in their transfer functions. The same kind of derivation of $\Xi(n)$ assumes an n-stage amplifier transfer function of the form

$$A(s) = K \cdot \frac{1}{(s^2\tau_{\mathrm{n}}^2 + 2\zeta\tau_{\mathrm{n}}s + 1)^n} \tag{8.15}$$

Setting the magnitude of $A(s)$ to that of a single-pole amplifier at bandwidth results in the expression

$$\|A(j\omega)\| = K \cdot \left(\frac{1}{(1-\tau_{\mathrm{n}}^2\omega^2)^2 + (2\zeta\tau_{\mathrm{n}}\omega)^2}\right)^{n/2} = K \cdot \frac{1}{\sqrt{2}} \tag{8.16}$$

Since the pole factor of (8.15) does not contain p, the bandwidth is related to the single-pole stage by ω_{n}, as in (5.153) and Section 5.12. Solving for ω in terms of ω_{n}, we express the result as

$$\Xi(n) = \frac{\omega_{\mathrm{bw}}}{\omega_{\mathrm{n}}} = (1 - 2\zeta^2 + \sqrt{4\zeta^4 + 4\zeta^2 + 2^{1/n}})^{1/2} \tag{8.17}$$

For n critically damped stages,

$$\Xi(n) = \sqrt{2^{1/2n}-1}; \qquad \zeta = 1, \qquad \text{critically damped stages} \tag{8.18}$$

A one-stage critically damped amplifier has a Ξ of 0.64, and for four stages it is about 0.3. For MFED stages ($\zeta = \sqrt{3}/2$), with $n = 1$, $\Xi \cong 0.786$; with $n = 4$, $\Xi \cong 0.4$. When the number of stages increases to 10, $\Xi \cong 0.25$. For MFED response, n stages have the approximate Ξ of

$$\Xi_{\text{MFED}} \cong \frac{0.786}{\sqrt{n}} \tag{8.19}$$

With these developments, we return to consider the risetime of multistage amplifiers. The transfer function magnitude for a general amplifier of n poles is of the form

$$\|A(j\omega)\| = \frac{K}{[1 + \omega^2 \sum_i (1/p_i^2) + \omega^4 \sum_i \sum_j (1/p_i^2 p_j^2) + \cdots]^{1/2}} \cong \frac{K}{(1 + \omega^2 \sum_i \tau_i^2)^{1/2}} \tag{8.20}$$

where $1/p_i = \tau_i$. The higher-order terms in ω are negligible for widely separated poles at much higher frequencies. The sum of time constants in the ω^2 term can be regarded as an equivalent single-pole time constant of

$$\tau \cong \sqrt{\sum_i \tau_i^2} \tag{8.21}$$

From (8.2), an approximate risetime is therefore

$$t_r \cong 2.2\tau = 2.2\sqrt{\sum_i \tau_i^2} = \sqrt{\sum_i (2.2\tau_i)^2} = \sqrt{\sum_i t_{ri}^2} \tag{8.22}$$

In other words, the approximate risetime of a multipole amplifier is the square root of the sum of the squares of the risetimes of the individual stages. For n identical stages, risetime degrades by approximately $\sqrt{n}$ that of a single stage.

Example 8.1 Instrument System Risetime

A 100 MHz oscilloscope has a probe with a 2 ns risetime. The total risetime is found by first calculating the risetime of the oscilloscope. According to (8.2),

$$t_r \cong \frac{0.35}{10^8 \text{ Hz}} = 3.5 \text{ ns}$$

Then the total risetime is approximately

$$t_r \cong \sqrt{(3.5 \text{ ns})^2 + (2 \text{ ns})^2} = 4 \text{ ns}$$

An accompanying approximation to bandwidth can also be made since τ of (8.21) is an equivalent single-pole time constant. From (8.10) we conclude that

$$\omega_{bw} \cong \frac{1}{\tau} = \sqrt{\sum_i \frac{1}{p_i^2}} \tag{8.23}$$

For n repeated poles, bandwidth, like risetime, degrades approximately by $\sqrt{n}$.

8.2 Amplifier Stage Gain Optimization

As the number of amplifier stages increases, the bandwidth is reduced. For a fast-amplifier design strategy, therefore, the number of stages should be minimized. However, most amplifier designs also require a given overall gain. Reducing the stage count demands increased gain per stage. Amplifier stages have a gain-bandwidth product f_T, affected mainly by the active device. An increase of stage gain decreases stage bandwidth. An optimum stage gain $A_1(s)$ that maximizes amplifier bandwidth ω_{bw} for a given amplifier gain $A(s)$ is derived by first noting that

$$\omega_{bw} = \Xi \cdot \omega_1 \tag{8.24}$$

where ω_1 is the single-stage bandwidth. Then the gain–bandwidth product of the amplifier is expressed in relation to its stages as

$$A^{1/n} \cdot \omega_{bw} = A_1(\Xi \cdot \omega_1) \tag{8.25}$$

assuming the n stages have the same gain,

$$A_1 = A^{1/n} \tag{8.26}$$

Solving (8.25) for ω_{bw} yields

$$\omega_{bw} = A_1 \cdot A^{-1/n} \cdot \Xi(n) \cdot \omega_1 \tag{8.27}$$

The optimum number of stages is found by differentiating (8.27) with respect to n to maximize bandwidth:

$$\frac{d}{dn}\omega_{bw} = A_1 \cdot \omega_1 \cdot \frac{d}{dn}[\Xi(n) \cdot A^{-1/n}] \tag{8.28}$$

For single-pole stages, Ξ is substituted from (8.14). Ξ can be expressed differently by noting that

$$2^{1/n} = e^{\ln(2)/n} = \sum_{k=0}^{\infty} \frac{(\ln 2/n)^k}{k!} \cong 1 + \frac{\ln 2}{n} \tag{8.29}$$

Then

$$\Xi(n) \cong \left(\frac{\ln 2}{n}\right)^{1/2} = \sqrt{\ln 2}\, n^{-1/2} \tag{8.30}$$

Substituting for Ξ in (8.28), the right side becomes

$$A_1 \cdot \omega_1 \cdot \frac{d}{dn}(A^{-1/n} \cdot \sqrt{\ln 2}\, n^{-1/2})$$
$$= A_1 \cdot \omega_1 \sqrt{\ln 2}\left(A^{-1/n}\left(-\frac{1}{2}\right) n^{-3/2} + A^{-1/n} \cdot n^{-1/2} \cdot \frac{1}{n^2} \cdot \ln A\right) \quad (8.31)$$

To find the optimum number of stages n_{opt}, set the derivative to zero and solve. Then

$$n_{\text{opt}} = 2 \ln A \quad (8.32)$$

The optimum stage gain is

$$\text{optimum } A_1 = A^{1/n_{\text{opt}}} = A^{1/2 \ln A} = e^{1/2} = \sqrt{e} \simeq 1.65 \quad (8.33)$$

This is not a large voltage or current gain. In practice, the optimum gain is somewhat larger than this value, usually around 2 to 3, due to bandwidth loss from interstage coupling.

Multistage amplifier frequency-response magnitude approaches a gaussian (probability function) as the number of stages increase. This gaussian response is quickly approached in practice by a few stages. It is derived by first rewriting (8.13) in terms of ω_{bw} from (8.14). Since $p = \omega_{\text{bw}}/\Xi$, then (8.13) becomes

$$\|A(j\omega)\| = \frac{K}{[(\omega/\omega_{\text{bw}})^2(2^{1/n} - 1) + 1]^{n/2}} \quad (8.34)$$

Next, as $n \to \infty$ in (8.29),

$$2^{1/n} - 1 \to \frac{\ln 2}{n}, \qquad n \to \infty \quad (8.35)$$

Substituting this into (8.34), we get

$$\|A(j\omega)\| \cong \frac{K}{[1 + (\omega/\omega_1)^2(\ln 2/n)]^{n/2}} = \frac{K}{[1 + (2/n)(\omega/\omega_1)^2(\ln 2/2)]^{n/2}} \quad (8.36)$$

$\|A(j\omega)\|$ is of exponential form since

$$e^x = \lim_{n \to \infty}\left(1 - \frac{x}{n}\right)^{-n}$$

So as $n \to \infty$,

$$\|A(j\omega)\| = K \cdot e^{-(\omega/\omega_1)^2(\ln 2/2)}, \qquad n \to \infty \quad (8.37)$$

This e^{x^2} form of $\|A\|$ is the gaussian response function.

The maximum achievable bandwidth of an amplifier with a gaussian response is derived based on the unity-power-gain frequency f_{MAX}. If $\|A\|$ of (8.37) is a power gain, then it can be expressed in decibel scaling as

$$\|A(j\omega)\|_{\text{dB}} = 10 \log K - 10\left(\frac{\ln 2}{2}\right)(\log e)\left(\frac{f}{f_1}\right)^2 = \|A(0)\|_{\text{dB}} - c\left(\frac{f}{f_1}\right)^2 \quad (8.38)$$

where c reduces to

$$c = 5 \log 2 \cong 1.51$$

The maximum bandwidth is achieved when $\|A\|$ passes through f_{MAX} at unity gain (0 dB). (From (8.38),

$$\|A(0)\|_{\text{dB}} - c\left(\frac{f_{\text{MAX}}}{f_1}\right)^2 = 0 \tag{8.39}$$

Solving for c and substituting it into (8.38), we get

$$\|A\|_{\text{dB}} = \|A(0)\|_{\text{dB}}\left(1 - \frac{f^2}{f_{\text{MAX}}^2}\right) \tag{8.40}$$

The bandwidth is the frequency at which $\|A\|_{\text{dB}}$ has rolled off by −3 dB:

$$\|A\|_{\text{dB}} - \|A(0)\|_{\text{dB}} = -3 = \|A(0)\|_{\text{dB}}\left(-\frac{f^2}{f_{\text{MAX}}^2}\right) \tag{8.41}$$

The solution of (8.41) for f is the bandwidth f_{bw} and is

$$f_{\text{bw}} = f_{\text{MAX}}\sqrt{\frac{3}{\|A(0)\|_{\text{dB}}}} \tag{8.42}$$

Example 8.2 Oscilloscope Vertical Amplifier

A wideband oscilloscope has a vertical deflection sensitivity of 2 V/cm at the CRT and a deflection plate termination resistance of 350 Ω. The input sensitivity is 50 mV/div into 100 Ω from the source when terminated by the 50 Ω scope input. The power gain is

$$\|A(0)\|_{\text{dB}} = 10 \log\left(\frac{(2\text{ V/div})^2/350\ \Omega}{(50\text{ mV/div})^2/100\ \Omega}\right) \cong 26.6\text{ dB}$$

The f_{MAX} is 2 GHz. The maximum f_{bw} is, from (8.42),

$$f_{\text{bw}} = 2\text{ GHz}\sqrt{\frac{3\text{ dB}}{26.6\text{ dB}}} = 672\text{ MHz}$$

The actual bandwidth of the amplifier (without the CRT) is 550 MHz. The use of maximum bandwidth as a performance index can be taken as the ratio of actual to maximum theoretical bandwidth, or

$$\frac{550\text{ MHz}}{672\text{ MHz}} \cong 82\%$$

That is, 82% of the maximum achievable bandwidth is realized in the vertical amplifier, within 18% of the theoretical limit.

8.3 Pole Determination by Circuit Inspection

With increased circuit complexity, the number of reactive elements increases and makes derivation of the transfer function more difficult. At some complexity threshold (which varies among engineers), the urge to simulate the circuit by computer overwhelms the desire to achieve an intuitive understanding of it. Even for complexity that requires simulation, it is necessary to know what to simulate. Until circuit design is computerized, the choice of numeric values of circuit elements must be based on estimation techniques and qualitative reasoning.

Most circuits can be decomposed into *modules* with well-defined interfaces. Intrastage behavior is relatively free of interaction with other stages. Interaction among modules can be considered apart from interaction within modules. The dynamic behavior of each circuit module can thus be determined individually, reducing the complexity of analysis.

A technique described by Cochrun and Grabel [6] and streamlined by Rosenstark [7] makes estimation of pole locations in active *RC* circuits simpler than solving the circuit for the transfer function. The degree of the characteristic equation (the transfer function denominator set to zero) equals the number of poles and the number of reactive circuit elements. Each capacitor in an *RC* circuit is associated with a pole. The characteristic equation, in a normalized transfer function can be written as

$$D(s) = a_n s^n + a_{n-1} s^{n-1} + \cdots + a_1 s + 1 = 0 \tag{8.43}$$

The technique enables determination of the a_n from inspection of the circuit. The coefficients, in terms of circuit elements, are found as follows. The equation for a_1 is

$$a_1 = \sum_{i=1}^{n} R_i(\text{open}) \cdot C_i = \sum \tau_i(\text{open}) \tag{8.44}$$

As a procedure, (8.44) is

0. a_1 procedure
1. Order the C by numbering them.
2. For each C, beginning with C_1, find the equivalent resistance across its terminals with all other C open. This is R_i(open). Multiply R_i(open) by C for τ_i(open).
3. Sum the τ_i(open) to obtain a_1.

This procedure is expedited by writing the time constants in the first column of a table beginning with τ_1(open) in the top row and τ_n(open) in the bottom row.

Next, for a multicapacitor circuit, a_2 is needed and is

$$a_2 = \sum_{i=1}^{n-1} \sum_{j=i+1}^{n} R_i(\text{open}) \cdot C_i \cdot R_j(C_i \text{ shorted}) \cdot C_j$$

$$= \sum_{i=1}^{n-1} \sum_{j=i+1}^{n} \tau_i(\text{open}) \cdot \tau_j(C_i \text{ shorted}) \tag{8.45}$$

The procedure for a_2 continues the table by filling in the second column and then using the τ_i(open) from the first column. All C not shorted are open when R is being found.

0. a_2 procedure
1. For each C_i, do the following:
   a. Short C_i. For each C after C_i (in the order they were numbered in the a_1 procedure), find the terminal R for C (C_j). Multiply this $R_j(C_i$ shorted) by C_j for $\tau_j(C_i$ shorted).
   b. Multiply the $\tau_j(C_i$ shorted) by τ_i(open).
2. Add the time-constant products from step 1b to obtain a_2.

Each entry in the first column of the table from the a_1 procedure will have $n-i$ entries in the second column for each C_j.

For a_3, three summations are made, extending the a_2 procedure. For the third column, two capacitors are shorted at a time (indices i and j), and k is indexed:

$$a_3 = \sum_{i=1}^{n-2} \sum_{j=i+1}^{n-1} \sum_{j=i+1}^{n} R_i(\text{open}) \cdot C_i \cdot R_j(C_i \text{ shorted}) \cdot C_j \cdot R_k(C_i, C_j \text{ shorted}) \cdot C_k \tag{8.46}$$

A way to keep the indexing straight is to base the entire procedure around the time-constant table. This *Rosenstark table* for three capacitances is shown in Fig. 8.1. The column numbering is for the i index. The procedure amounts

	a_1	a_2	a_3
C_1	τ_1 (open)	τ_2 (C_1 shorted)	τ_3 (C_1,C_2 shorted)
		τ_3 (C_1 shorted)	
C_2	τ_2 (open)	τ_3 (C_2 shorted)	
C_3	τ_3 (open)		

FIG. 8.1 Rosenstark table for a circuit with three capacitances. The time-constants are based on open-circuit resistances, except as indicated.

to filling in the table and then, for a_1, summing the τ_i in column 1; for a_2, summing the products of τ in column 2 with τ from column 1; for a_3, summing the products of τ in column 3 with τ from columns 2 and 1. The summation always involves the τ of a column multiplied by the τ of the columns to the left. When a τ is found, the capacitors indexed in the columns to the left are shorted.

Active devices can change the resistance at a capacitor, for example, due to the Miller effect, and must be taken into account when finding equivalent resistances.

Example 8.3 Op-Amp Circuit Poles from the Cochrun–Grabel Method

The op-amp circuit of Fig. 6.4b can be analyzed using the Cochrun-Grabel method. The Rosenstark table is

	a_1	a_2
C_i	$(R_i \parallel R_s)C_i$	$(R_c+R_f)C_c$
C_c	$(R_c+R_f)C_c$	

The ordering of capacitors, as shown in the table is C_i, C_c. This ordering is arbitrary. The a_1 column is filled in, beginning at the C_i row, by finding the open-circuit resistance across the terminals of C_i. Since an active device, the op-amp, is involved, we must first determine its effect on resistance. The V_- input is a virtual ground for the ideal op-amp. Knowing this, the C_i terminals have across them $R_i \| R_s$ because V_i has zero resistance. The time constant for the first entry, C_i row, a_1 column, is complete. For the C_c row, a_1 column, we examine C_f. R_c is in series with the C_c terminal and goes to ground. From the other terminal, we determine that the V_o node has a resistance of R_f to R_c. Thus, the total resistance across C_c is R_c+R_f. The second entry is complete.

We now begin with the a_2 column. We short C_i (from the first column) and determine resistance across C_c. Again this is R_c+R_f. The table is complete. The a_n are now found from the table. The sum of the first column is

$$a_1=(R_i \parallel R_s)C_i+(R_c+R_f)C_c$$

Then a_2 is found from the second column by multiplying its entries by the first column and adding them. Since there is only one, no addition is needed here, and

$$a_2=(R_c+R_f)C_c \cdot (R_i \parallel R_s)C_i$$

The characteristic equation is

$$a_2 s^2 + a_1 s + 1 = 0$$

This quadratic equation is easily solved by noting that the a_1 terms are the a_2 factors, and the poles are therefore

$$p_1 = \frac{-1}{(R_i \parallel R_s)C_i}, \qquad p_2 = \frac{-1}{(R_c + R_f)C_c}$$

This result agrees with that of the reactance chart method shown in Fig. 6.4c.

Example 8.4 BJT Amplifier Poles from the Cochrun–Grabel Method

The poles of the BJT amplifier of Fig. E8.4a are found by the Cochrun-Grabel method using a Rosenstark table (Fig. E8.4b) to obtain the characteristic equation. The third-degree equation can be solved by computer or approximated by a lower-degree equation that retains an approximation of the dominant poles.

We begin with R of C_π in the first column. With all the other C open, the resistance is found by the steps in Figs. E8.4c–e. In (c), the equivalent circuit is Nortonized to (d). Then the substitution theorem is applied and two branches combined to produce (e). The resistance is

$$R_\pi = r_\pi \parallel \frac{(R_E + R_B + r_b')}{(1 + R_E/r_m)} \tag{E1}$$

This has the form of Miller's theorem, in which $K = R_E/r_m$. Then $R_\pi = 54.8\ \Omega$, and the resulting time constant is 28.9 ns.

For C_μ, the equivalent circuit is shown in Fig. E8.4f. Applying the β transform to R_E gives Fig. E8.4g. The resistance at the b' node to ground is

$$R_{b'} = (r_b' + R_B) \parallel (\beta + 1)(r_e + R_E) \tag{E2}$$

The collector current source is controlled by $V_{b'e}$. From the base loop in Fig. E8.4g,

$$V_{b'e} = V_{b'}\left(\frac{r_\pi}{r_\pi + (\beta+1)R_E}\right) = V_{b'}\left(\frac{r_e}{r_e + R_E}\right) \tag{E3}$$

At the collector,

$$V_c = -R_L \cdot \frac{V_{b'e}}{r_m} = -\alpha \cdot \frac{R_L}{r_e + R_E} \cdot V_{b'} = -KV_{b'} \qquad \text{(E4)}$$

Current injected into the b' node causes the voltage across $b'c$ to change by $1 + V_c/V_{b'} = 1 + K$. This Miller effect causes the voltage across the injecting current source to be larger by $1 + K$ times. This makes the effective resistance $1 + K$ times larger also.

We next change the Norton circuit of the collector loop to a Thévenin source as shown in Fig. E8.4h using (E4). The voltage across $R_{b'}$ is now $(1 + K)V_{b'}$, making $R_{b'}$ appear $(1 + K)$ times larger. This Miller resistance is in series with R_L. Thus the resistance we seek is

$$R_\mu = R_{b'}(1 + K) + R_L$$

$$= (r_b' + R_B) \parallel (\beta + 1)(r_e + R_E)\left[1 + \alpha \cdot \frac{r_e}{r_e + R_E}\right] + R_L \qquad \text{(E5)}$$

Substituting circuit values gives $R_\mu = 6.57\ \text{k}\Omega$ and $\tau = 19.7$ ns.

The time constant for a_1 due to C_L is $R_L C_L = 22.0$ ns. For the a_2 column, the first capacitance C_π is shorted, and R across $b'c$ is again determined. With C_π shorted, $V_{b'e} = 0$, and the transistor current source is nulled. This simplifies the resistance to

$$R_\mu(C_\pi \text{ shorted}) = (r_b' + R_B) \parallel R_E + R_L \qquad \text{(E6)}$$

which is 1.18 kΩ. The time constant is 6.30 ns. Next, C_π remains shorted as we find the resistance across C_L. It is R_L, and $\tau = 22.0$ ns. The last entry in the a_2 column is found by shorting C_μ and finding the resistance across C_L. The collector current source is now across the $b'e$ branch, and the substitution theorem reduces it to r_m. Since $r_m \parallel r_\pi = r_e$,

$$R_L(C_\mu \text{ shorted}) = R_L \parallel (r_e + R_E) \parallel (r_b' + R_B) \qquad \text{(E7)}$$

This resistance is 160 Ω and $\tau = 0.479$ ns.

The final entry, for a_3, is the resistance across C_L with C_π and C_μ shorted. It is

$$R_L(C_\pi, C_\mu \text{ shorted}) = (r_b' + R_B) \parallel R_E \parallel R_L \qquad \text{(E8)}$$

This value is 155 kΩ and $\tau = 3.41$ ns. The table is complete.

We now find the a_n as follows:

$$a_1 = (28.9 + 19.7 + 22.0)\ \text{ns} = 70.6\ \text{ns}$$

$$a_2 = [(6.30)(28.9) + (22.0)(28.9) + (0.479)(19.7)]\ \text{ns}^2 = 827\ \text{ns}^2$$

$$a_3 = (3.41)(6.30)(28.9)\ \text{ns}^3 = 621\ \text{ns}^3$$

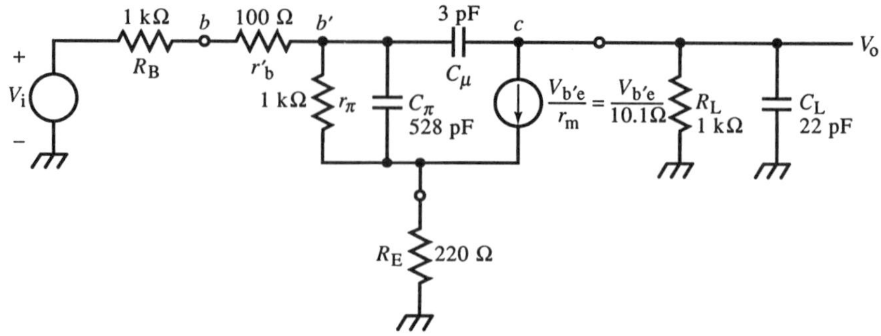

(a)

	a_1	a_2	a_3	ns units
C_π	28.9	6.30	3.41	
		22.0	a_3 = 621 ns^3	
C_μ	19.7	0.479		
C_L	22.0	a_2 = 827 ns^2		
	a_1 = 70.6 ns			

(b)

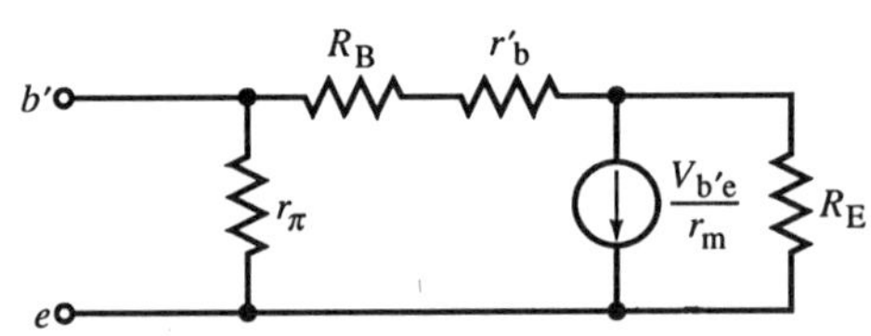

(c)

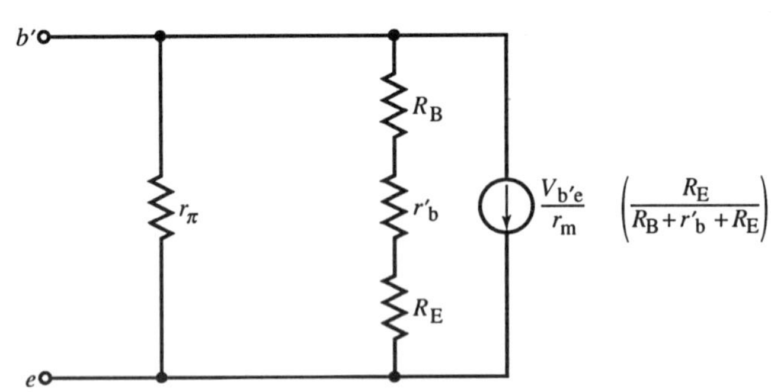

(d)

FIG. E8.4

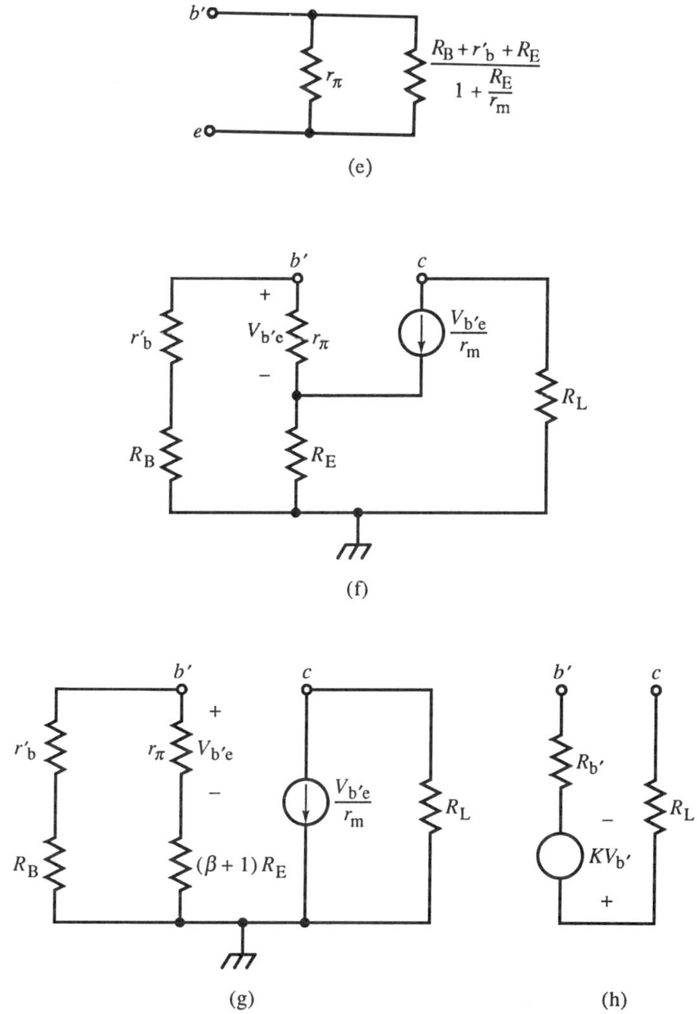

FIG. E8.4 (continued)

The characteristic equation is therefore

$$(621 \text{ ns}^3)s^3 + (827 \text{ ns}^2)s^2 + (70.6 \text{ ns})s + 1 = 0$$

This equation was solved by computer with real roots at

$$-2.83 \text{ MHz}, \qquad 11.6 \text{ MHz}, \qquad 198 \text{ MHz}$$

A SPICE simulation shows a damped response due to the dominant real pole with magnitude roll-off of −3 dB at 2.7 MHz. The two slowest poles combine to yield an approximate bandwidth of 2.75 MHz.

If a computer is not used to solve the characteristic equation for the poles, some approximations can be made by ignoring higher-degree terms. By dropping the s^3 term, we find two poles at 2.85 MHz and 10.7 MHz. By dropping the quadratic term also, we find the single pole at 2.25 MHz, a 16% error. This error is acceptable for many pole estimates and leads to a simplified version of the Cochrun-Grabel approach; instead of building a table, we build only the first column. That is, we sum the open-circuit time constants for each capacitor and invert it for the radian pole frequency. The hard work is in finding R_π and R_μ, but we have done that already, and (E1) and (E5) can be used for BJT analysis (and with the BJT-to-FET transform, for FETs) generally.

The f_T specified by transistor manufacturers is defined as the frequency at which β is unity with the collector ac-shorted to the emitter. Then C_μ shunts C_π. This implies that

$$\text{manufacturer's } f_T = \frac{1}{2\pi\tau_T}, \qquad \tau_T = r_\pi(C_\pi + C_\mu) \cong r_\pi C_\pi \tag{8.47}$$

The manufacturer's f_T for the transistor in this example is 300 MHz and $\beta_o = 99$. In (7.1) and (7.2), f_T is defined with $C_\mu = 0$ in (8.47) to make the resulting theory simpler. This should cause no problem if the manufacturer's f_T and C_μ (given as C_{ob}) values are used to compute f_β and f_T as defined here. In most cases, f_T is close enough already. In this example, the error is 0.6%.

The Cochrun-Grabel method only produces poles. One technique for determining zeros [7] begins by first writing the nodal equations of the circuit. A flow graph is especially helpful here. Those transmittances that lead from input to output are examined for evidence of zeros.

Pole estimation is often applied to interstage coupling, to the pole formed by the load resistance of a CE or CB stage and the following CE stage input (Fig. 8.2a). The input impedance of the loading BJT stage Z_i has branches through C_μ and Z_π.

Consider first the branch involving C_μ. The circuit is idealized in Fig. 8.2b to eliminate the effect of the Z_π path. The BJT is represented by a transconductance amplifier. Its output current, representing collector current, is shown flowing into the amplifier with value $G_m V_i = V_i / R_m$. The input current to the amplifier flows into C_μ and is

$$I_i = \frac{V_i - V_o}{1/sC_\mu} = sC_\mu(V_i - V_o) = \frac{V_o}{R_L} + \frac{V_i}{R_m} \tag{8.48}$$

Solving this KCL equation for V_o, substituting it into the first expression for

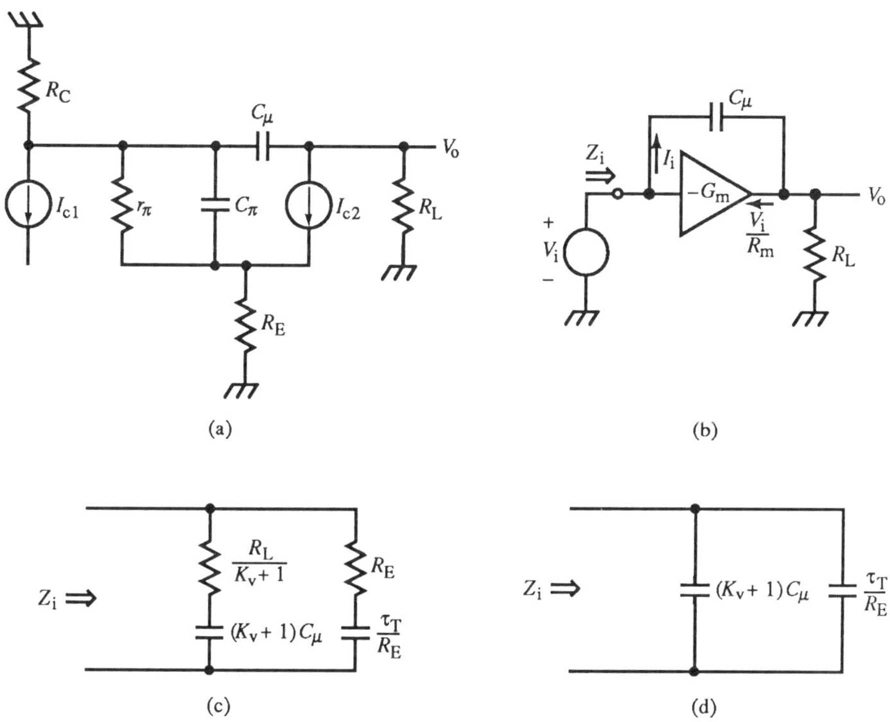

FIG. 8.2 CE amplifier with hybrid-π BJT model (a), an idealized BJT model as a transconductance amplifier (b), equivalent input impedance of a CE (c), and simplified equivalent input circuit (d).

I_i, and solving for V_i/I_i, we get

$$Z_{i\mu} = \frac{1}{s[1+(R_L/R_m)]C_\mu} + \frac{R_L}{[1+(R_L/R_m)]} = \frac{1}{s(K_v+1)C_\mu} + \frac{R_L}{K_v+1} \tag{8.49}$$

Besides the Miller capacitance, R_L is reduced by K_v+1. For large voltage gain, this branch presents a nearly capacitive impedance.

The impedance through Z_π is

$$Z_{i\pi} = (\beta_o+1)(r_e+R_E)\cdot\frac{s\alpha_o\tau_T(R_E/(r_e+R_E))+1}{s\tau_\beta+1} \tag{8.50}$$

For $R_E \gg r_e$ and $\alpha_o \cong 1$, $Z_{i\pi}$ is the hf $Z_b(R_E)$ of Fig. 7.8b. If the series-peaking $\omega_n \ll \omega_T$, we can ignore the zero in (8.50). The result is a shunt RC with R_i shunting $\tau_\beta/R_i \cong \tau_T/R_E$. R_i is large and usually presents negligible shunting. Consequently, $Z_{i\pi}$ reduces to τ_T/R_E. It shunts $Z_{i\mu}$ so that

$$Z_i = Z_{i\mu} \parallel Z_{i\pi} \tag{8.51}$$

When the branches are combined, the resultant Z_i of Fig. 8.2c results. This simplifies, under these assumptions, to Fig. 8.2d. Unless the transistor is very fast (low τ_t) or R_E is small (not much larger than r_e or less), the only significant capacitance is the Miller capacitance. Therefore,

$$Z_i \cong \frac{1}{s(K_v+1)C_\mu}, \qquad R_E \gg r_e, \qquad \beta_o \gg 1, \qquad f \ll f_T, \qquad \frac{R_L}{K_v+1} \cong 0 \quad (8.52)$$

Another assumption of (8.52) is that the capacitance loading R_L from the stage following it is negligible. If not, shunt capacitance across R_L further reduces the impedance in series with C_μ, making (8.52) a better approximation.

A more exacting estimate is based on Z_i of the network of Fig. 8.2c. Let the elements be designated more generally as R_1 in series with C_1 shunting R_2 in series with C_2. Then,

$$Z_i = \frac{(sR_1C_1+1)(sR_2C_2+1)}{[s(C_1+C_2)][s(R_1+R_2)(C_1 \parallel C_2)+1]} \quad (8.53)$$

When the particular element values of Fig. 8.2c are substituted, the zeros are at frequencies of $1/\tau_T$ and $1/R_LC_\mu$; the poles are at the origin, and $1/(R_1+R_2)(C_1 \parallel C_2)$ (where $\parallel$ is a math operator not a topological designator). The capacitance C_1+C_2 dominates Z_i until $1/R_LC_\mu$. The second pole causes Z_i to appear capacitive out to f_T.

8.4 Inductive Peaking

Interstage coupling often degrades bandwidth due to parasitic reactances. For example, collector output capacitance shunts the input base capacitance of the next stage in Fig. 8.3. The load resistor R is shunted by $C = C_{1out} + C_{2in}$. An unwanted pole is created at $1/RC$.

The addition of an inductor can extend the bandwidth by creating a series or parallel resonant circuit with a peak in the frequency or transient responses; hence the technique name of *inductive peaking*. Figure 8.4 shows *series peaking* circuits. The transfer function is not changed by exchanging R and C. In both cases, L is in series with C. The transfer function for series peaking has a

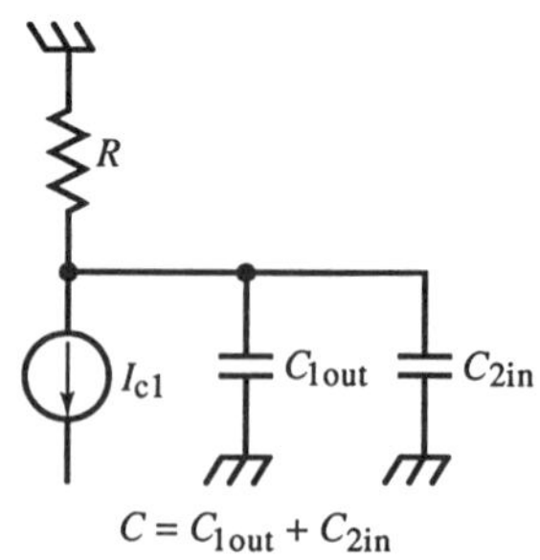

FIG. 8.3 Output node of a typical CE or CB stage, with its output capacitance C_{1out} and that of the next stage C_{2in}.

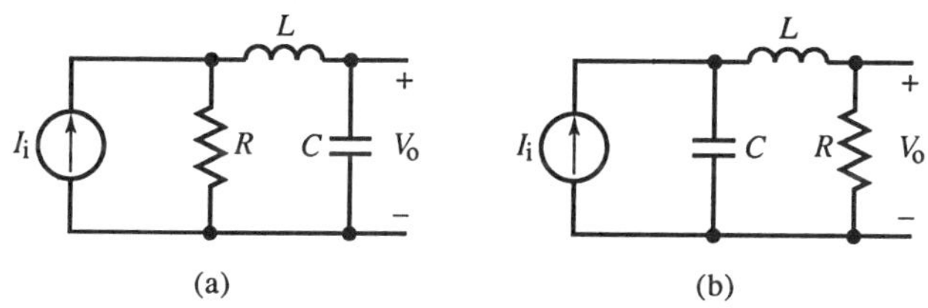

FIG. 8.4 Series inductive peaking has the same effect whether R and C are at the input or the output side of L.

quadratic pole factor:

$$\frac{V_o}{I_i} = R \cdot \frac{1}{s^2LC + sRC + 1} \tag{8.54}$$

This has a familiar quadratic-pole response, as described in Section 5.12. The basic parameters are

$$\omega_n = \frac{1}{\sqrt{a}} = \frac{1}{\sqrt{LC}} = \frac{2\zeta}{RC}; \qquad \zeta = \frac{b}{2\sqrt{a}} = \frac{R}{2Z_n}, \qquad Z_n = \sqrt{\frac{L}{C}} \tag{8.55}$$

Usually, R is chosen to set the gain, and C is parasitic. This leaves L as the design parameter. For a desired ζ,

$$L = \frac{R^2C}{4\zeta^2} \tag{8.56}$$

Since L is in a (the s^2 coefficient) only, the poles move with increasing L as shown in Fig. 8.5a. It is worth noting that if R were varied instead, the pole locus would be as shown in Fig. 8.5b, and for C, as in (c). When the poles become complex as L increases, pole radius ω_n shrinks, and ζ decreases. The most desirable pole locations for a wideband amplifier are in a range slightly off the real axis, near the critically damped pole location $-2/RC$. Here, ω_n is maximum and ζ in a range that gives a desired response.

For variation in R, ω_n remains constant as ζ changes proportionally. Variation in C causes the most trouble. While ζ varies with the square-root

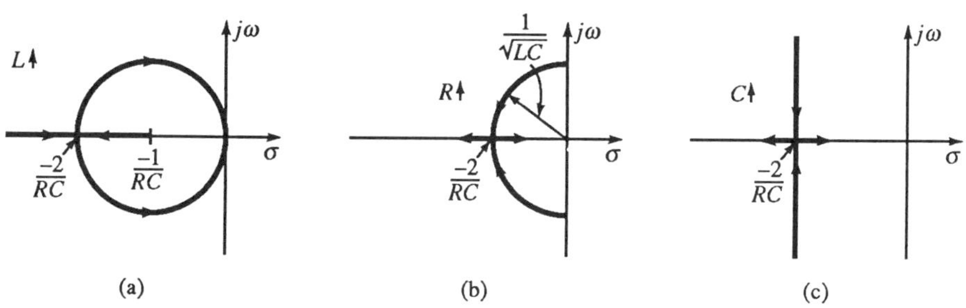

FIG. 8.5 Series-peaking root loci for increasing L (a), R (b), and C (c).

of C, ω_n varies inversely. Consequently, since C is usually parasitic, control over its range of values is least, and though response peaking (in time or frequency) is not so much affected by ΔC, the risetime and bandwidth are. Causes for C, such as transistor process parameters and circuit-board layout, are significant in control of pole radius.

It is interesting to determine how much improvement in bandwidth series peaking can offer. To measure this, we compare bandwidth improvement with the uncompensated RC circuit of Fig. 8.3 by expanding the meaning of the bandwidth reduction factor Ξ to include bandwidth extension. Both definitions of Ξ in Section 8.2 are useful here and are separately denoted as

$$\Xi_n \equiv \frac{\omega_{bw}}{\omega_n}, \qquad \Xi_p \equiv \frac{\omega_{bw}}{p} = \frac{\omega_{bw}}{(1/RC)} \tag{8.57}$$

Ξ_n compares bandwidth with respect to pole radius; Ξ_p compares it to the uncompensated RC circuit.

In the time domain, we have used $t_r\omega_n$ to express relative risetime. Comparison against the risetime of the RC circuit also offers an improvement measurement. The risetime improvement factor is

$$\frac{t_r}{t_r(RC)} \cong \frac{t_r}{2.2RC} \tag{8.58}$$

All of these performance indicators are combined with those already derived in Section 5.12 in a series-peaking summary table:

ζ	L	ω_n	$\frac{\omega_{bw}}{\omega_n}$	$\frac{\omega_{bw}}{p}$	$M_{p'}(\%)$	$t_r\omega_n$	$\frac{t_r}{2.2RC}$	
1.00	$\frac{R^2C}{4}$	$2/RC$	0.644	1.288	0	3.36	0.765	crit. damping
0.866	$\frac{R^2C}{3}$	$1.73/RC$	0.786	1.361	0.433	2.73	0.717	MFED
0.707	$\frac{R^2C}{2}$	$1.41/RC$	1.000	1.414	4.32	2.15	0.692	MFA
0.500	R^2C	$1/RC$	1.272	1.272	16.3	1.64	0.746	$\phi = 60°$

For $\zeta = 0.5$, the poles are at a 60° angle and $\alpha = 1/RC$, the same as the single-pole case. From Fig. 8.5, at critical damping, both poles are at $-2/RC$ and have twice the pole radius of a single-pole RC circuit. As L varies,

$$\left(\frac{\omega_{bw}}{p}\right) = \left(\frac{\omega_{bw}}{\omega_n}\right)(2\zeta) \Rightarrow \Xi_p = 2\zeta\Xi_n \tag{8.59}$$

By adding L for series peaking, we improve the bandwidth by 36% and the risetime by 28% ($\zeta = 1$). This is a significant improvement caused by the addition of one component, but greater improvement is possible.

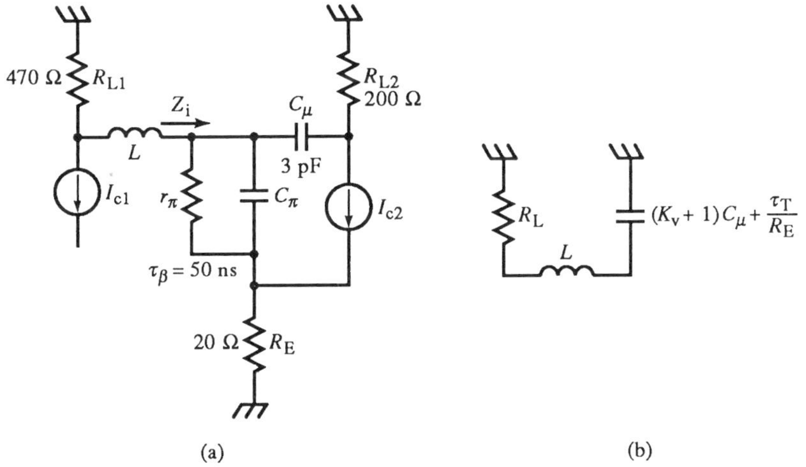

FIG. E8.5

Example 8.5 Series Peaking

An amplifier stage (Fig. E8.5a) output is loaded by the input impedance of a CE stage. The transistors both have $\tau_T = 500$ ps and $\beta_o \cong 100$. The output capacitance of the BJT across I_{c1} is negligible. A MFED response is desired.

The first step is to find the input impedance Z_i of the loading stage. The emitter branch hf capacitance is 500 ps/20 Ω or 25 pF. The voltage gain is about

$$\frac{R_L}{R_E} = 10$$

and the collector capacitance is 11(3 pF) = 33 pF. Because R_E is so small, τ_T / R_E is significant. The series collector resistance <20 Ω. The capacitance in series with L is thus

$$C_i \cong 25 \text{ pF} + 33 \text{ pF} = 58 \text{ pF}$$

The uncompensated bandwidth is

$$f_{bw}(\text{uncomp}) = \frac{1}{2\pi RC} \cong \frac{1}{2\pi(470\ \Omega)(58\ \text{pF})} = 5.8\ \text{MHz}$$

and risetime is

$$t_r \cong 2.2RC = 60 \text{ ns} \qquad \text{or} \qquad t_r \cong \frac{0.35}{f_{bw}} = 60 \text{ ns}$$

From the table, for MFED response,

$$L=\frac{R^2C}{3}\cong\frac{(470\ \Omega)^2(58\ \text{pF})}{3}=4.3\ \mu\text{H}$$

The value of $R=R_L$ assumes negligible series resistance in Z_i. Each path in Z_i has about 20 Ω. From (8.53), Z_i becomes resistive at $1/R_LC_\mu$ or 265 MHz. This is about 50 times larger than f_{bw} (uncomp), and the assumption that Z_i is purely capacitive over the frequency range of interest is valid.

The compensated bandwidth (from the table) is 1.36 times higher, or 7.9 MHz, and the risetime is 43 ns. The series resonance is at $f_n=$ 10 MHz.

An alternative to series peaking is *shunt peaking* (Fig. 8.6). The addition of L in series with R places it in parallel with C and creates a parallel resonance. For a step of input current, most of it charges C at first because current does not change instantaneously in an inductor. Consequently, C charges faster and response speed increases.

The transfer function of the shunt peaking circuit of Fig. 8.6 is

$$\frac{V_o}{I_i}=\frac{s(L/R)+1}{s^2LC+sRC+1} \tag{8.60}$$

The addition of a zero over series peaking improves response speed but also peaks the response more. To compare shunt with series peaking, we need formulas for the performance parameters of a two-pole, one-zero circuit. We now digress to derive them generally. Then we apply them to shunt peaking.

The two-pole, one-zero transfer function can be generally expressed in terms of $\tau_n=1/\omega_n$ and Q as

$$\frac{V_o}{I_i}=\frac{sQ\tau_n+1}{s^2\tau_n^2+(\tau_n/Q)s+1} \tag{8.61}$$

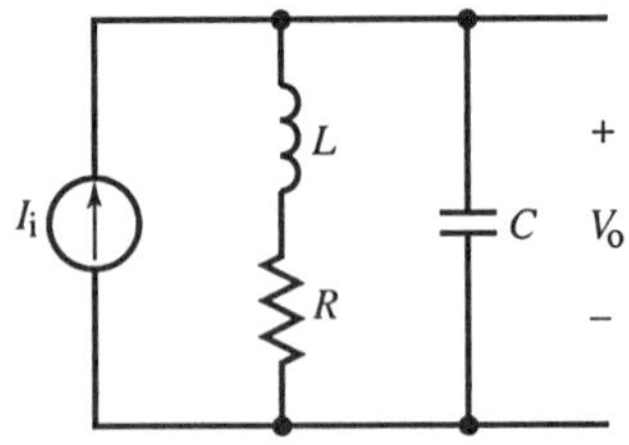

FIG. 8.6 Shunt inductive peaking. L and C form a parallel resonance whereas the load resistor R damps it.

In narrow-band amplifier terminology,

$$Q \equiv \frac{1}{2\zeta} \tag{8.62}$$

This quantity describes the amount of peaking and occurs frequently in resonant circuit equations. The time constant of the zero of (8.61) is

$$\tau_z = Q\tau_n \Rightarrow Q = \left(\frac{\tau_z}{\tau_n}\right) \tag{8.63}$$

Bandwidth is found in the usual way by setting the magnitude of (8.61) to $1/\sqrt{2}$. The general result is

$$\omega_{bw} = \omega_n \sqrt{1 - \frac{1}{2Q^2} + 2Q^2 + \sqrt{\left(1 - \frac{1}{2Q^2} + 2Q^2\right)^2 + 1}} \tag{8.64}$$

The overshoot M_p from [9] is expressed in ζ as

$$M_p = \frac{1}{2\zeta} \cdot \exp\left[\frac{-\zeta}{\sqrt{1-\zeta^2}} \cdot \left(\underset{\substack{\uparrow \\ 270^\circ}}{\frac{3\pi}{2}} - \cos^{-1}\zeta - \tan^{-1}\left\{\frac{\sqrt{1-\zeta^2}}{\zeta}\right\} + \sin^{-1}(-\zeta)\right)\right] \tag{8.65}$$

The unit step response for $\zeta > 1$ is

$$v_{step}(t) = 1 - \frac{1}{2\zeta\sqrt{1-\zeta^2}} \cdot e^{-\zeta\omega_n t} \sin\left(\omega_n t\sqrt{1-\zeta^2} + \cos^{-1}\zeta + \tan^{-1}\left\{\frac{\sqrt{1-\zeta^2}}{\zeta}\right\}\right) \tag{8.66}$$

The 10% and 90% times are found by numerical computer solution for shunt peaking as $t_r\omega_n$. This is a convenient representation since $\omega_n t$ acts as the independent variable in (8.66).

For $\zeta = 1$, the poles are repeated, and the step response is

$$v_{step}\big|_{\zeta=1} = \left[\left(\frac{p}{z} - 1\right)pt - 1\right] e^{-pt} + 1 \tag{8.67}$$

This function also has no closed-form solution and is numerically solved by computer.

We now apply these general results to shunt peaking and (8.60), where the zero is

$$z = \frac{1}{\tau_z} = \frac{1}{L/R} = 2\zeta\omega_n \tag{8.68}$$

and the repeated poles have a frequency of

$$2p = \frac{2}{RC} = \omega_n \tag{8.69}$$

It then follows that, when $\zeta = 1$,

$$\frac{p}{z} = \frac{1}{2} \tag{8.70}$$

and (8.66) reduces to

$$\text{shunt peaking } v_{\text{step}} = (-\tfrac{1}{2}\omega_n t - 1)\, e^{-\omega_n t} + 1 \tag{8.71}$$

The values of $t_r\omega_n$ are numerically computed from this equation.

A table similar to that for series peaking can now be constructed for shunt peaking:

ζ	Q	$\frac{\omega_{bw}}{\omega_n}$	$\frac{\omega_{bw}}{p}$	$M_{p'}$ (%)	$t_r\omega_n$	$\frac{t_r}{2.2RC}$	
1.00	0.500	0.786	1.572	0	3.071	0.699	crit. damping
0.866	0.577	1.086	1.881	0.620	2.319	0.609	$\phi = 30°$
0.707	0.707	1.554	2.198	6.70	1.559	0.502	$\phi = 45°$
0.500	1.000	2.279	2.279	29.8	0.940	0.428	$\phi = 60°$

Shunt peaking is faster than series peaking for the same pole parameters but is less damped in response. It achieves an 88% increase in bandwidth over the *RC* circuit and 39% decrease in risetime for a MFED pole response. These comparisons can be misleading, however. A pole angle of, say, 30° is *not* an MFED response because of the zero. The values of ζ for MFED and MFA responses must be derived as in Section 5.10. For a MFED response, $\zeta \cong 0.881$, and for MFA, $\zeta \cong 0.777$. These values are somewhat higher than those without the zero.

Greater speed improvement can be achieved by using a *T coil.* This is a transformer (Fig. 8.7a) with controlled coupling and a common connection at *c*. An equivalent circuit is shown in (b) where the polarity of coupling determines the polarity of the mutual inductance $-M$ to terminal *c*. With the coupling as shown,

$$L = L_{ab} = (L_1 + M) + (L_2 + M) = L_1 + L_2 + 2M \tag{8.72}$$

The addition of $-M$ in the equivalent circuit produces the correct self-

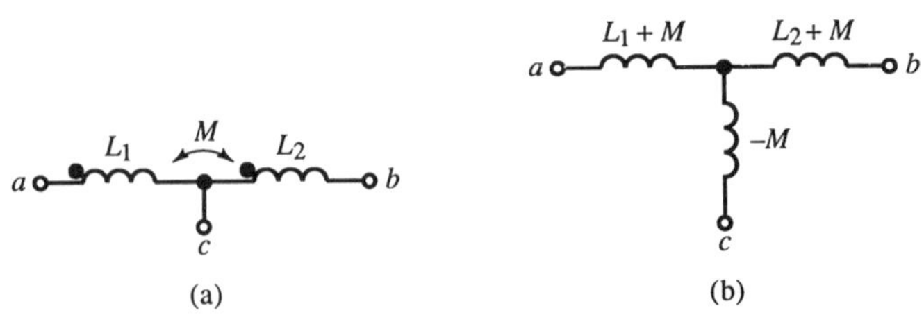

FIG. 8.7 Coupled inductors forming a T coil (a), with mutual inductance M and equivalent circuit (b).

inductances:

$$L_{ac} = (L_1 + M) - M = L_1, \qquad L_{bc} = (L_2 + M) - M = L_2 \tag{8.73}$$

If terminals *a* and *b* are shorted, the inductance from *a* and *b* to *c* is

$$L_{ab,c} = (L_1 + M) \parallel (L_2 + M) - M \tag{8.74}$$

If we make the mutual inductance signed, the coupling coefficient is always positive:

$$k = \left| \frac{M}{\sqrt{L_1 L_2}} \right| \tag{8.75}$$

The use of the T coil for bandwidth extension has resulted in the general form of the *bridged T-coil circuit* (Fig. 8.8). The coil is terminated in R but the load is connected to the center-tap. A general load,

$$Z_L = sL_s + R_s + \frac{1}{sC_L} \tag{8.76}$$

(or series RLC) is similar to the input impedance of BJT stages. At dc, the input impedance of the T coil is R. Because of the bridging capacitor C_B, at high frequencies it is also R. For a given Z_L and by proper choice of L_1, L_2, M, and C_B, $Z_{in} = R$ and is independent of frequency. When the circuit of Fig. 8.8 is solved, the design equations that result are

$$L_1 = \frac{C_L}{4}\left(1 + \frac{1}{4\zeta^2}\right)(R + R_s)^2 - RR_s C_L - L_s \tag{8.77}$$

$$L_2 = \frac{C_L}{4}\left(1 + \frac{1}{4\zeta^2}\right)(R + R_s)^2 - L_s \tag{8.78}$$

$$M = \frac{C_L}{4}\left(R^2 - R_s^2 - \frac{1}{4\zeta^2}(R + R_s)^2\right) + L_s \tag{8.79}$$

$$C_B = \frac{C_L}{16\zeta^2}\left(1 + \frac{R_s}{R}\right)^2 \tag{8.80}$$

In addition to these design equations, the equivalent inductor element values of Fig. 8.8b are

$$L_1 + M = \frac{RC_L}{2}(R - R_s) - L_s \tag{8.81}$$

$$L_2 + M = \frac{RC_L}{2}(R + R_s) - L_s \tag{8.82}$$

$$L = R^2 C_L - 2L_s \tag{8.83}$$

The transfer function has two poles at

$$p_{1,2} = -\frac{4\zeta^2}{RC_L} \pm j\frac{4\zeta}{RC_L}\sqrt{\zeta^2 - 1} \tag{8.84}$$

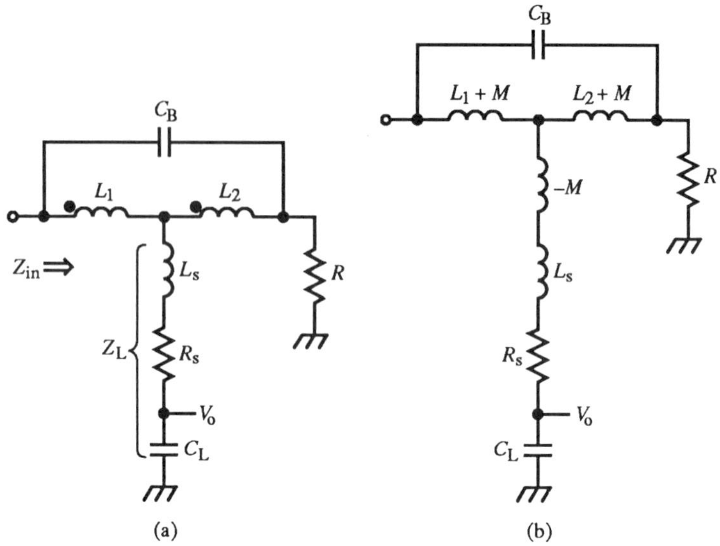

FIG. 8.8 Bridged T coil with *RLC* load (a), and equivalent circuit (b).

The form of transfer function is the same as for series peaking but with twice the speed improvement! For MFED response, $\omega_{bw}/p = 2.72$, nearly three times better than the original *RC* circuit. The greatest improvement is 2.83 for a MFA response.

For $Z_L = 1/sC_L$, the transfer function for the load is

$$\frac{V_o}{I_i} = R \cdot \frac{1}{\frac{1}{4}\left(\frac{k-1}{k+1}\right) R^2 C_L^2 s^2 + \frac{1}{2} \cdot RC_L s + 1}$$

$$C_B = \frac{1}{4}\left(\frac{k-1}{k+1}\right) C_L, \qquad L = R^2 C_L, \qquad L_1 = L_2 = \frac{L}{2(k+1)} \tag{8.85}$$

As k increases, the pole angle decreases for complex poles. With perfect coupling, $k = 1$, and the s^2 term in (8.85) is zero, leaving a single-pole response but with twice the bandwidth of a simple *RC* circuit. This is a simpler, lower-performance T-coil circuit with no bridging capacitance and with $L_1 = 4L_2$. MFED response is achieved when $k = \frac{1}{2}$, a relatively loose coupling, not hard to implement. Then $C_B = C_L/12$, $L_1 = 3L_2$, and $L = R^2 C_L$. A balanced T coil ($L_1 = L_2$) that meets the conditions of (8.85) has no coupling between the coils, $C_B = C_L/4$, and the pole angle is 60°. $L_1 > L_2$ is necessary to meet these conditions with a capacitive load.

Even greater bandwidth improvement is possible by taking advantage of the constant-resistance input of the T-coil circuit. Series peaking can be cascaded in front of the T coil, as shown in Fig. 8.9. The input and output capacitances of Fig. 8.3 are separated and become part of different peaking

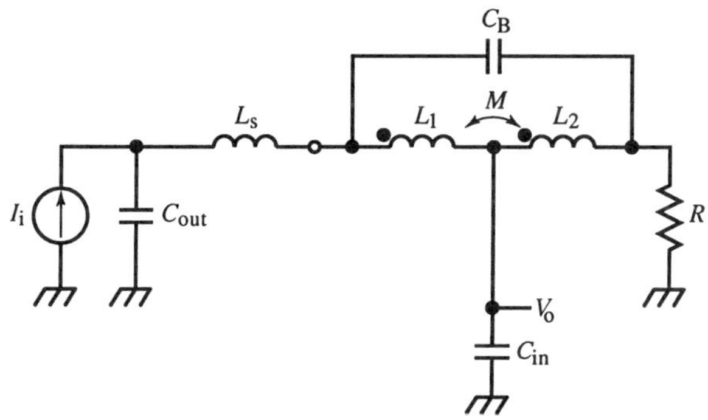

FIG. 8.9 T-coil peaking preceded by series peaking, with capacitive load.

circuits. Since the interstage coupling satisfies the requirement for series peaking, the bandwidth improvement of each circuit remains unchanged. Thus the total improvement is the product of the individual improvement factors.

Amplifiers with bandwidths under 100 MHz usually have tapped solenoidal coils wound on a plastic bobbin. The magnetic path is through air and plastic. For higher frequencies, a common T coil is made of a bifilar-wound loop of magnetic wire. The two wires are twisted together and then formed into a loop. Or, circuit-board traces can be spiraled on opposite sides of the board and connected at the center via plated-through holes. Two traces can be run next to each other to form coupled inductors. For very high-speed circuits, IC bonding wires have even been used to form T coils.

An inductive peaking circuit used in ICs, sometimes referred to as "emitter peaking," realizes a shunt inductance by the high-frequency gyration of a BJT base resistor. The emitter appears inductive. (See Chapter 7.) In Fig. 8.10, the adjustment of R_B adjusts the emitter inductance $R_B\tau_T$.

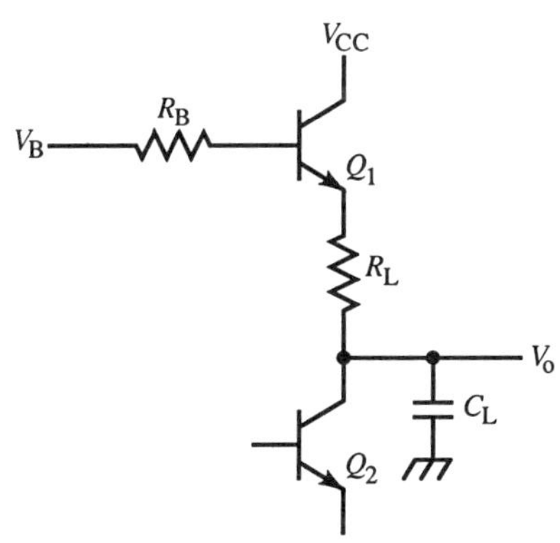

FIG. 8.10 Emitter peaking: Q_1 provides an inductance in series with the load resistor in its hf region, creating a shunt peaking compensator. No actual inductor is required, making the technique attractive for IC design.

Example 8.6 T-Coil Compensation

Example 8.5 is T-coil compensated. The T-coil formulas of (8.77)-(8.80) are applied directly. The loading is $C_L = 58$ pF, and $R_s = L_s = 0$. Also, $R = R_L = 470\ \Omega$. For MFED response, $\zeta = 0.866$. The results are

$$L_1 = L_2 = 4.3\ \mu\text{H}, \qquad M = 2.1\ \mu\text{H}, \qquad C_B = 4.8\ \text{pF}$$

From (8.75), the coupling of the inductors is $k = 0.5$. This is loose coupling and is easily accomplished. The bandwidth is now 15.8 MHz and risetime approximately 22 ns.

8.5 Source-Follower Compensation

CS stages are often used at the input to instrument amplifiers to provide minimal resistive loading. But unlike their BJT counterpart (the CC), FET C_{GS} is typically much smaller than C_π of BJTs. Consequently, the Z_π term of (7.10) cannot be ignored, as in (7.11) for the BJT. The transformation for FETs is made in (7.43). When the effect of C_{GD} is included, it is

$$Z_g = \left[\frac{1}{sC_{GS}} + Z_S\left(\frac{s\tau_T + 1}{s\tau_T}\right)\right] \Big\| \frac{1}{sC_{GD}} \tag{8.86}$$

where $\tau_T = r_m C_{GS}$. The equivalent circuit is shown in Fig. 8.11a and is transformed using the substitution theorem from (b) to (c). For a capacitive load, V_i drives a capacitive divider that causes an input voltage step to immediately rise to a fraction of the step amplitude

$$\frac{C_{GS}}{C_{GS} + C_S}$$

and then continues exponentially, due to r_m, to the input step value. The transfer function of Fig. 8.11 is

$$\frac{V_o}{V_i} = \frac{sr_m C_{GS} + 1}{sr_m(C_{GS} + C_S) + 1} = \frac{s\tau_T + 1}{s(\tau_T + r_m C_S) + 1} \tag{8.87}$$

Since the pole is less than the zero, the response is that of a phase-lag circuit (Fig. 5.15a), and the initial response step is p/z with a time constant of $r_m(C_{GS} + C_S)$. The response can be compensated by adding a phase-lead circuit at another stage in the amplifier.

A second anomaly of the CS is its input impedance. From (8.86), the gyrating factor can be expressed in topological form as

$$Z_S\left(\frac{s\tau_T + 1}{s\tau_T}\right) = Z_S \cdot \frac{1}{1 - 1/(s\tau_T + 1)} \tag{8.88}$$

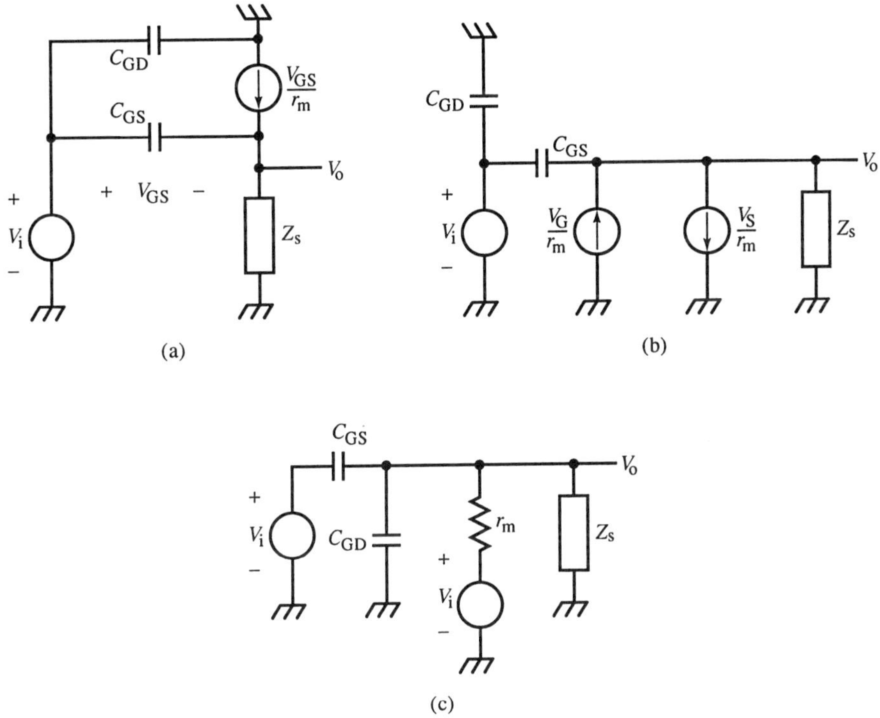

FIG. 8.11 A source-follower (CS) equivalent circuit (a), reduced to (b), then to (c) by the substitution theorem and Norton to Thévenin conversion of the current source. The V_i source driving r_m is separate from the input V_i source but has the same output. With no input impedance, r_m could be made to shunt C_{GS} instead.

Substituting $Z_L = 1/sC_S$, Z_g reduces to

$$Z_g = \frac{1}{sC_{GD}} \,\Big\|\, \left(\frac{1}{sC_{GS}} + \frac{1}{sC_S} \,\Big\|\, \left(-\frac{\tau_T}{C_S} - \frac{1}{sC_S}\right)\right) \tag{8.89}$$

A more useful form, with the same topology as that of Section 7.3, is derived by expressing (8.89) as

$$\begin{aligned} Z_g &= \frac{1}{sC_{GD}} \,\Big\|\, \left(\frac{s(C_{GS}+C_S)\tau_T + C_{GS}}{s^2 C_{GS} C_S \tau_T}\right) \\ &= \frac{1}{sC_{GD}} \,\Big\|\, \left(\frac{1}{s(C_{GS} \parallel C_S)}\right) \,\Big\|\, \left(-r_m \cdot \frac{C_{GS}+C_S}{C_{GS} \parallel C_S} - \frac{1}{s(C_{GS} \parallel C_S)}\right) \end{aligned} \tag{8.90}$$

This topology is shown in Fig. 8.12 and is compensated by the method of Section 7.10, that of shunting the gate with a series RC, which produces a purely capacitive input. The values of the compensating elements are derived from the expression for Z_g from (8.90).

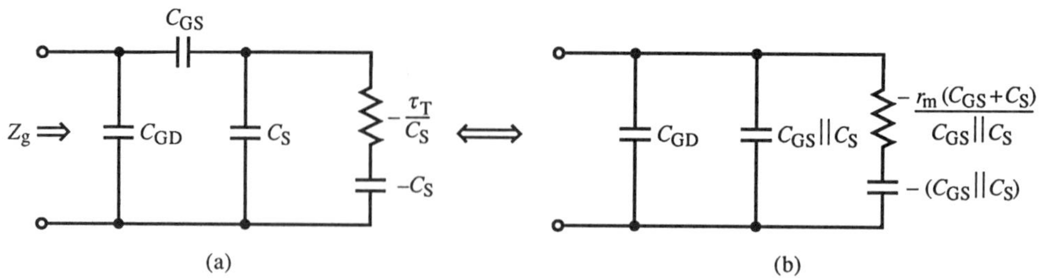

FIG. 8.12 Equivalent hf gate impedance (a) and its equivalent circuit (b), similar to BJT hf base impedance. For FETs, C_{GS} is too large to ignore, and the values of elements in (b) are consequently different.

A source-follower is usually used to prevent loading of a high-impedance source. If the source impedance is resistive, then it forms an uncompensated voltage divider with Z_g. This can be compensated by introducing a shunt RC in series with the input (Fig. 8.13). It forms a compensated divider with Z_g, resulting in a resistive input.

The last CS problem we consider is distortion due to large-signal effects. When a large-amplitude square-wave is applied to the compensated CS, C_{GS} and r_m both change significantly between levels. If C_1 of Fig. 8.13 is adjusted for compensation of the positive transition, then for the negative transition, r_m and C_{GS} increase causing τ_T to increase. An increase of C_{GS} increases the step fraction, causing negative overshoot, or undershoot.

The transfer function is also affected. Both pole and zero decrease, but with significant C_S, the pole decreases less. Consequently, p/z decreases, causing the compensator to overcorrect and produce undershoot. FETs with large C_{GS} have reduced undershoot, and if they also have a large τ_T, undershoot error diminishes more quickly. Similarly, FETs with large pinch-off voltages have less r_m variation with V_{GS}.

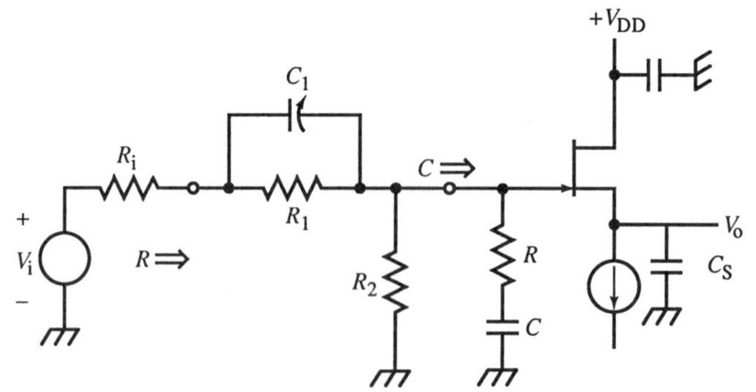

FIG. 8.13 A hf-compensated FET input buffer amplifier. The series RC at the gate makes the input appear capacitive. C_1, R_1, and R_2 form a compensated divider with the input capacitance.

Example 8.7 CD Input Buffer Compensation

A source-follower has a transconductance of 50 mS ($r_m = 200\ \Omega$), $C_{GS} = 6$ pF, and $C_{GD} = 2$ pF. Manufacturer's data sheets give capacitances as

$$C_{iss} = C_{GS} + C_{GD}, \qquad C_{rss} = C_{GD}$$

With a load capacitance of $C_S = 10$ pF, a series RC shunting the gate compensates the FET input. To calculate its values, τ_T is needed:

$$\tau_T = r_m C_{GS} = 1.2 \text{ ns}$$

Then, from Fig. 8.12,

$$R = r_m\left(\frac{C_{GS} + C_S}{C_{GS} \parallel C_S}\right) = 853\ \Omega \Rightarrow 820\ \Omega,$$

$$C = C_{GS} \parallel C_S = 3.75 \text{ pF} \Rightarrow 3.9 \text{ pF}$$

The input now is a shunt RC, where $C_g = C + C_{GD} = 5.9$ pF and $R_2 = 1\ \text{M}\Omega$. To compensate this pole, a shunt RC is placed in series with the input (as in Fig. 8.13). For applications in which the input comes from a probe or passive attenuator, the compensating RC is in the probe body so that the probe itself contains the top part of the voltage divider. For a 10 MΩ input,

$$R_1 = 9\ \text{M}\Omega, \qquad C_1 = \frac{R_2 C_g}{R_1} = \frac{C_g}{9} = 0.66 \text{ pF}$$

If no probe precedes the input, then to avoid dc gain error due to the divider, its attenuation must be near unity, and R_1 must be small relative to 1 MΩ. For large R_1, C_1 is an extremely small capacitance. The R_1 resistor probably has more parasitic shunt capacitance than C_1. As a consequence, practical values of capacitors make it infeasible to try to compensate the input divider. That is why, for example, oscilloscope vertical inputs are marked with labelings such as 1 MΩ, 22 pF.

The circuit of Fig. 8.14 is an alternative approach to CD compensation. It has two signal paths: the main path through the FET and a compensation path through the CB BJT. The FET path transfer function is given by (8.87) whereas for the compensation path, C forms a divider with $Z_\pi/(\beta + 1)$. The BJT-path voltage gain is

$$\text{BJT}\,\frac{V_o}{V_i} = \alpha_o \cdot \frac{s r_{mFET} C}{[s r_{mFET}(C_{GS} + C_S) + 1][s r_e(C_\pi + C) + 1]} \tag{8.91}$$

The paths add to produce the total transfer function. Since both paths share

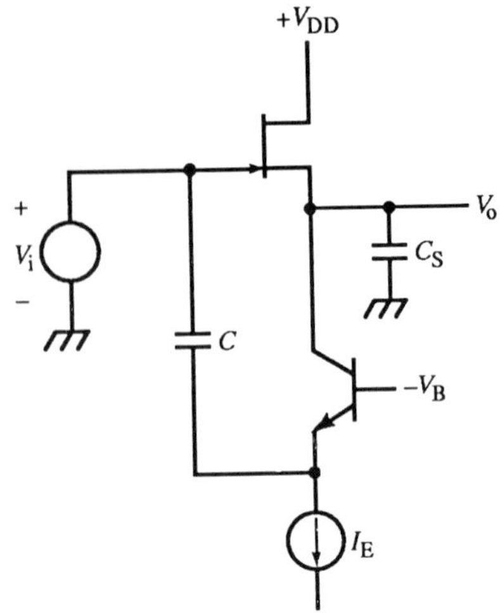

FIG. 8.14 Alternative CS compensator with a CB current source, compensated when $C \cong C_S$.

the pole with time constant,

$$r_{mFET}(C_{GS} + C_S) = \tau_{TFET} + r_{mFET}C_S \tag{8.92}$$

then if the BJT time constant

$$r_e(C_\pi + C) = \tau_{TBJT} + r_eC \tag{8.93}$$

is much smaller than (8.92), its pole can be ignored and the transfer functions of the paths added:

$$\left.\frac{V_o}{V_i}\right|_{\tau_T + r_eC \to 0} \cong \frac{s(\tau_{TFET} + \alpha_o r_{mFET}C) + 1}{s(\tau_{TFET} + r_{mFET}C_S) + 1} \tag{8.94}$$

For flat response, the time constants are equated and

$$\alpha_oC = C_S \Rightarrow C \cong C_S, \qquad \alpha_o \cong 1 \tag{8.95}$$

For step inputs with fast edges, the voltage differentiation of C can cause currents that exceed I_E and drive the BJT into cutoff.

8.6 Emitter Compensation

An impedance in series with the emitter (or source) of CB and CE (or CG and CS) amplifiers creates series feedback (see Section 3.12) and can improve speed. Compensation networks can be connected to the emitter node that correct for speed limitations at the collector. Figure 8.15 shows a CB stage with capacitive output loading. A series RC is placed in parallel with R_E to

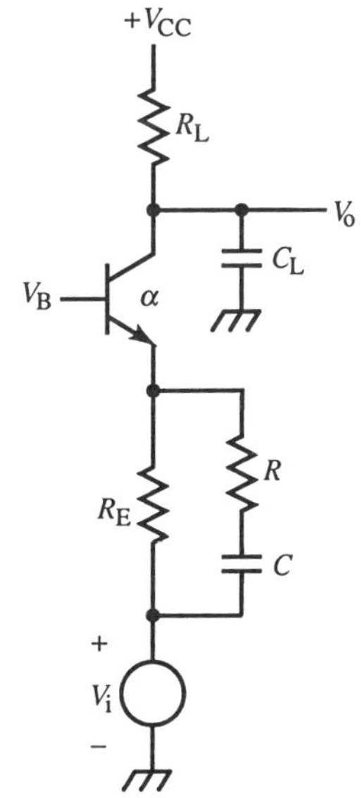

FIG. 8.15 A lf-compensated CB. The series RC across R_E creates a zero in the transfer function that cancels the output pole.

provide correction. In the lf region (or for $\tau_T \to \infty$), the transfer function is

$$\frac{V_o}{V_i} = \alpha \cdot \frac{R_L}{r_e + R_E} \cdot \frac{s(R_E + R)C + 1}{(sR_LC_L + 1)[s(R + r_e \parallel R_E)C + 1]} \tag{8.96}$$

For compensated response, the zero cancels the collector pole at frequency $1/R_LC_L$, leaving a much higher-frequency pole. For a flat frequency response, the compensating elements must have the values

$$R = R_L - R_E, \qquad C = C_L \tag{8.97}$$

and for bandwidth extension,

$$(R + r_e \parallel R_E)C \ll R_LC_L \tag{8.98}$$

For $r_e \ll R$ and R_E, then $R \ll R_L$.

Compensation of CE (or CS) amplifiers is similar. In Fig. 8.16, a similar network is connected to the emitter resulting in negative (8.96). In both cases, we have ignored transistor reactances, and the resulting equations are useful for amplifiers for which the output pole is well below f_β.

Analysis involving the hf region uses the hybrid-π BJT model (and its extension to FETs). The CB stage has the advantage over the CE of no Miller effect. However, Z_π forms an uncompensated voltage divider with R_E, requiring an additional shunt C_E around R_E for compensation. At the output, C_μ contributes to C_L.

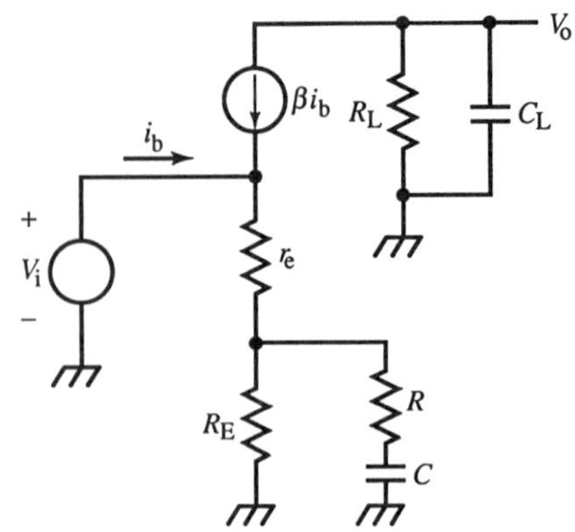

FIG. 8.16 Equivalent circuit of a CE with emitter-network compensator, similar to the CB in Fig. 8.15.

The CE suffers from the Miller effect, the cause of its dominant pole. Also, the input impedance is a hf-gyrated emitter impedance. Networks in the emitter circuit that compensate for output poles have the side-effect of creating input anomalies. To design a fast amplifier with CE input, the Miller effect must be minimized and input impedance controlled. The Miller effect is essentially eliminated by following the CE with a CB (making it a cascode) or operating the CE as a shunt-feedback amplifier. We shall first study the input side of the CE by assuming a cascode configuration. This simplifies analysis and allows us to ignore C_μ for a while and regard the CE as a transadmittance amplifier.

The transresistance approach (now generalized to the transimpedance approach) applies in the hf region, using the hf BJT model of Fig. 7.8. CE output current is $\beta(s)I_b$. The transadmittance (Fig. 8.16) is

$$\frac{I_o}{V_i}=\frac{\beta(s)I_b}{V_i}=\frac{\beta(s)}{Z_b}=\frac{\beta(s)}{[\beta(s)+1]Z_E}=\alpha(s)\,Y_E=\frac{1}{s\tau_T+1}\cdot Y_E \qquad (8.99)$$

Because of $\alpha(s)$, the emitter network admittance Y_E must have a $(s\tau_T+1)$ factor to cancel hf effects. A simple compensation is to let Z_E be a shunt RC:

$$Z_E=\frac{R_E}{sR_EC_E+1} \qquad (8.100)$$

Then,

$$\frac{I_o}{V_i}=\left(\frac{1}{R_E}\right)\frac{sR_EC_E+1}{s\tau_T+1} \qquad (8.101)$$

The response is flat when

$$R_EC_E=\tau_T \qquad (8.102)$$

In the hf region, the input impedance of the CE is

$$Z_b = [\beta(s)+1]Z_E = \left(\frac{s\tau_T+1}{s\tau_T}\right)\left(\frac{R_E}{sR_EC_E+1}\right) \tag{8.103}$$

When (8.102) holds,

$$Z_b = \frac{1}{s(\tau_T/R_E)} = \frac{1}{sC_b} \tag{8.104}$$

The input is a capacitance of value $C_b = \tau_T/R_E$.

In the base circuit, C_b forms a pole with the base node resistance. A more complete hf model includes r_b' and C_μ at the internal (b') base node (Fig. 8.17a). A general expression for the transadmittance is

$$\frac{I_o}{V_i} \cong \frac{I_o}{I_e}\cdot\frac{I_e}{V_e}\cdot\frac{V_e}{V_i} = \frac{\beta}{(\beta+1)Z_E+R_S}\cdot\frac{1}{s[R_S \parallel (\beta+1)Z_E]C_\mu+1} \tag{8.105}$$

where β is $\beta_{hf} = 1/s\tau_T$ and

$$R_S = R_B + r_b'$$

When emitter compensation is added, that is, (8.100) and (8.102) are applied,

$$Z_E = \frac{R_E}{s\tau_T+1} \tag{8.106}$$

and the transadmittance reduces to

$$\frac{I_o}{V_i} = \frac{1}{R_E}\cdot\frac{1}{sR_S[(\tau_T/R_E)+C_\mu]+1} \tag{8.107}$$

In the lf region, for $I_i = V_i/R_B$,

$$K_i = \left.\frac{I_o}{I_i}\right|_{lf} = \alpha_o \cdot \frac{R_B}{R_S/(\beta_o+1)+r_e+R_E} \tag{8.108}$$

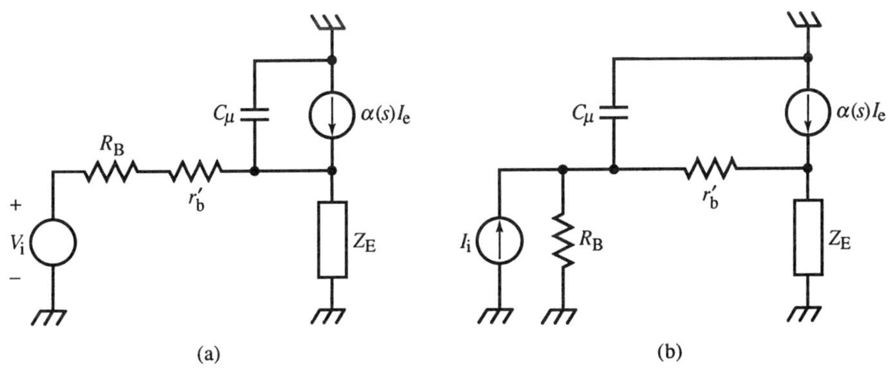

FIG. 8.17 CE transresistance amplifier with internal C_μ (a); CE current amplifier with external C_μ (b).

The hf model has $\alpha_o = 1$ and $r_e = 0$. The lf current gain is then

$$K_i \cong \frac{R_B}{R_E} \tag{8.109}$$

High-speed amplifiers are usually analyzed in terms of current gain because the input variable to a stage is usually a current. For example, the cascode CB has a current input. The input to a CE is usually the collector of another transistor, modeled as a current source. The Thévenin voltage source input of Fig. 8.17a is changed to a Norton equivalent input by setting

$$V_i = I_i \cdot R_B$$

Then (8.107) is used to express current gain as

$$\frac{I_o}{I_i} = \frac{I_o}{V_i/R_B} \cong \frac{K_i}{sR_S[(\tau_T/R_E) + C_\mu] + 1} \tag{8.110}$$

Both (8.105) and (8.110) are approximate because the path to the output through C_μ is ignored. It introduces a RHP zero at $1/R_E C_\mu$. This frequency is usually much higher than the others and can be ignored. The passive path through C_μ causes an output response to occur sooner than the inverted response of the active path. This passive path current is the cause of *preshoot* in the output step response. Instead of rising, the step first dips negative (Fig. 8.18).

The effect of r'_b and C_μ is to degrade speed. Without them, the time constant of the pole in (8.110) is at

$$R_B \frac{\tau_T}{R_E} = K_i \cdot \tau_T \tag{8.111}$$

Given K_i by design choice, there is an optimum value of R_B that minimizes the pole time constant, which can be written in terms of K_i. By multiplying out the pole time constant in (8.110) and setting its derivative to zero, we get

$$\frac{d}{dR_B}\left(K_i \cdot \tau_T + R_B C_\mu + \frac{r'_b K_i \tau_T}{R_B} + r'_b C_\mu \right) = 0 \tag{8.112}$$

The optimum R_B is

$$\text{optimum } R_B = \sqrt{\frac{K_i \cdot \tau_T r'_b}{C_\mu}} \tag{8.113}$$

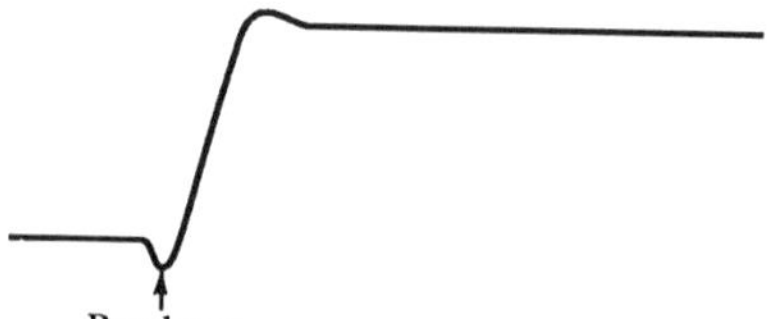

FIG. 8.18 Waveform with preshoot, caused by a RHP zero.

For this value of R_B, the current-gain pole has a time constant of

$$K_i \cdot \tau_T\left(1+2\sqrt{\frac{r_b' C_\mu}{K_i \tau_T}}\right) \tag{8.114}$$

A CE with a lf voltage gain of K_v has an effective C_μ of K_v+1 due to the Miller effect. The optimum R_B and pole time constant are modified by multiplying C_μ by K_v+1.

If C_μ is located at the external base node (on the outside of r_b'), as in Fig. 8.17b, the current gain is

$$\frac{I_o}{I_i} \cong \frac{K_i}{s^2(K_i \tau_T R_B C_\mu) + s[R_B(\tau_T/R_E + C_\mu) + r_b' \tau_T/R_E] + 1} \tag{8.115}$$

$$\uparrow$$

$$[R_B C_\mu + K_i \tau_T(1 + r_b'/R_B)]$$

Again, the RHP zero has been ignored. The CE input now consists of two cascaded RC integrators. The minimum ζ is

$$\min \zeta = \sqrt{1+\frac{r_b'}{R_B}} = \frac{\tau_n}{K_i \tau_T} \tag{8.116}$$

Pole separation is typically not significant enough to approximate the response by a dominant single pole since fast amplifiers have values of R_B not very different from r_b'. Consequently, the r_b' term in (8.115) cannot be ignored. The value of R_B at maximum pole radius occurs when the poles are repeated and $\zeta=1$. This results in a fourth-degree equation in R_B. We can approximate the optimum R_B by assuming independence of the time constants of the RC integrators. Then fastest response occurs when their time constants are equal, or

$$R_B C_\mu = r_b'\left(\frac{K_i \tau_T}{R_B}\right) \tag{8.117}$$

Solving for R_B gives

$$\text{optimum } R_B \cong \sqrt{\frac{K_i \cdot \tau_T r_b'}{C_\mu}} \tag{8.118}$$

Interestingly, (8.118) is the same as (8.113). Whether C_μ is largely internal or external to the base does not strongly affect the optimum R_B value.

8.7 Cascode Compensation of the Common Base Stage

We now turn attention to the cascode CB. Above f_β of the CB transistor, base resistance is gyrated at the emitter to an inductance which resonates with the output capacitance of the CE. This shunt RLC forms a parallel resonance with

$$\zeta = \frac{1}{2R}\sqrt{\frac{L}{C}} = \frac{1}{2}\sqrt{\frac{\tau_T}{R_B C_o}} \tag{8.119}$$

where R_B is the CB base resistance and C_o the CE output capacitance shunting the input to the CB. For a BJT with $f_T = 300$ MHz, $C_o = 3$ pF and $R_B = 100\ \Omega$, $\zeta = 0.665$ or a pole angle of 48°. With a transistor twice as fast, the pole angle is about 60°. This resonance can cause oscillation because C_o is due largely to C_μ of the CE and is connected to the base. The CE provides the gain that causes oscillation.

This resonance can be damped by adding resistor R_S in series with the emitter of the CB, isolating it from C_o. The series damping required for MFED response is 77 Ω. Typically, r'_e is 1 Ω, far less than the resistance required. A series resistance damps the resonance but also creates an uncompensated voltage divider with the CB gyrated base impedance. In addition, it causes voltage gain at the collector of the CE and the Miller effect. The CB transfer function is

$$\frac{I_o}{I_i} = \frac{1}{s^2\{\tau_T R_B C_o[1 + (R_S/R_B)]\} + s(\tau_T + R_S C_o) + 1} \tag{8.120}$$

The pole radius ω_n is reduced by

$$\frac{R_B}{R_S + R_B}$$

Compensation in the CB emitter is shown in Fig. 8.19a as a shunt RC. The hf equivalent emitter circuit (Fig. 8.19b) has a shunt RL due to the gyrated

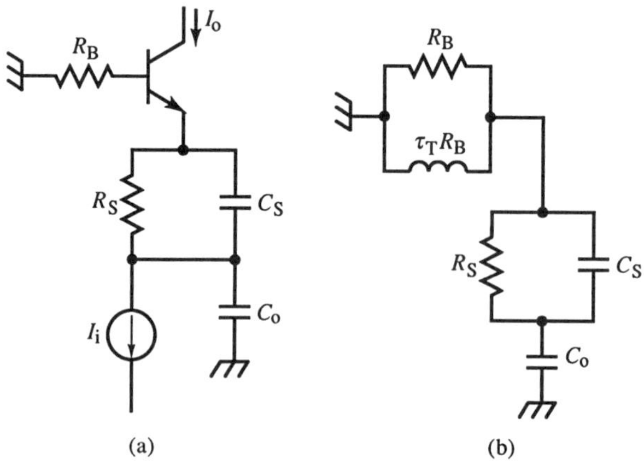

FIG. 8.19 Cascode CB compensation of Z_e with shunt RC (a) and hf equivalent circuit (b), in which an all-pass network is formed, resulting in a resistive emitter equivalent input impedance.

R_B. We encountered this network before (Fig. 7.20). When

$$R_S C_S = \tau_T \tag{8.121}$$

the hf model yields

$$\frac{I_o}{I_i} = \alpha(s) I_e = \frac{1}{s^2(\tau_T R_B C_o) + s(\tau_T + R_S C_o) + 1} \tag{8.122}$$

C_o forms a current divider with the emitter branch; I_e is calculated from the divider formula. This compensation maintains the pole radius the same as the uncompensated CB. For design, the value of C_S is determined by (8.121). R_S is expressed in ζ from (8.120) as

$$R_S = -\frac{\tau_T}{C_o} + 2\zeta\sqrt{\frac{\tau_T R_B}{C_o}} \tag{8.123}$$

For $R_S = R_B$, the network of Fig. 8.19b forms an all-pass constant resistance of R_B. The poles are then located at $-1/R_B C_o$ and $-1/\tau_T$.

An estimation of C_o is required to use (8.123). Current in C_μ of the CE is input current to the base. The resulting collector current is larger by the lf current gain. In effect, current in C_μ results in a total current of $K_i + 1$. Thus, the effective capacitance of C_μ is $(K_i + 1)C_\mu$. (This result suggests a form of Miller's theorem for current amplifiers.)

A CB compensation scheme proposed by John Addis is shown in Fig. 8.20. A series RC is added at the base. Base bias current is ideally supplied from a current source or from a large resistor connected to the collector supply. The emitter impedance is

$$Z_E = \frac{R_B + 1/sC_B}{\beta(s) + 1} = \frac{\tau_T(sR_B C_B + 1)}{(s\tau_T + 1)(C_B)} \tag{8.124}$$

Z_E shunts C_o. The current gain of the CB is

$$\frac{I_o}{I_i} = \frac{1}{s^2(\tau_T R_B C_o) + s[\tau_T(1 + C_o/C_B)] + 1} \tag{8.125}$$

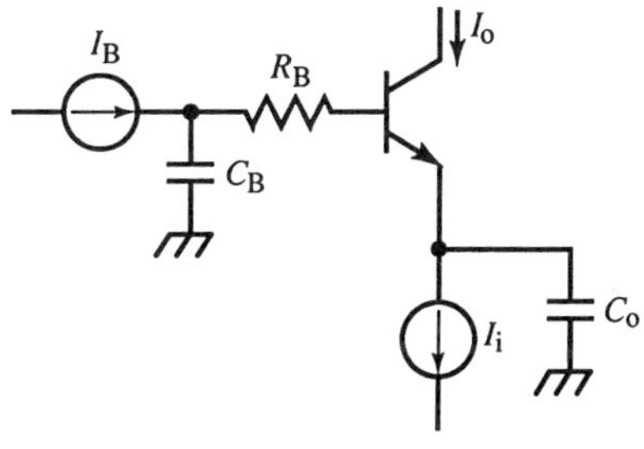

FIG. 8.20 Alternative CB compensation using base series RC, with no reduction in pole radius.

The pole radius is not reduced by this technique, but control of ζ is more limited:

$$\zeta = \frac{1}{2}\sqrt{\frac{\tau_T}{R_B C_o}} + \frac{1}{2}\sqrt{\frac{\tau_T C_o}{R_B C_B^2}} = \zeta_{\text{uncomp}} + \frac{1}{2}\sqrt{\frac{\tau_T C_o}{R_B C_B^2}} \tag{8.126}$$

The appearance of C_o in both terms indicates a minimum ζ dependent on C_o. Because of the unavoidable base spreading resistance, r'_b, R_B is partly constrained in value by the BJT. The design value of C_B is found by solving (8.126):

$$C_B = \frac{C_o}{2\zeta/\sqrt{\tau_T/R_B C_o} - 1} = \frac{C_o}{4\zeta/\zeta_{\text{uncomp}} - 1} \tag{8.127}$$

The lower bound on ζ is that

$$\zeta > \frac{\zeta_{\text{uncomp}}}{4} \tag{8.128}$$

We have yet one other pole in the cascode requiring compensation. At the output, the CB transistor output capacitance forms a pole with the load resistance. This pole can be compensated by peaking the CE. The CE is already compensated by (8.102); output compensation thus requires a more complicated emitter network. The strategy is to cancel the output pole with a zero. Poles of the emitter network impedance are zeros of the cascode transfer function. The network of Fig. 8.21 meets the requirements. Its impedance is

$$Z_E = R_E \cdot \frac{sRC + 1}{s^2(R_E C_E RC) + s(R_E C_E + R_E C + RC) + 1} \tag{8.129}$$

Input impedance compensation requires one of the poles of Z_E to be at τ_T. For pole-zero cancellation of the output pole, with time constant τ_L, the other pole of Z_E must be at $-1/\tau_L$. Therefore, the denominator of (8.129) is constrained to be of the form

$$(s\tau_T + 1)(s\tau_L + 1) = s^2\tau_T\tau_L + s(\tau_T + \tau_L) + 1 \tag{8.130}$$

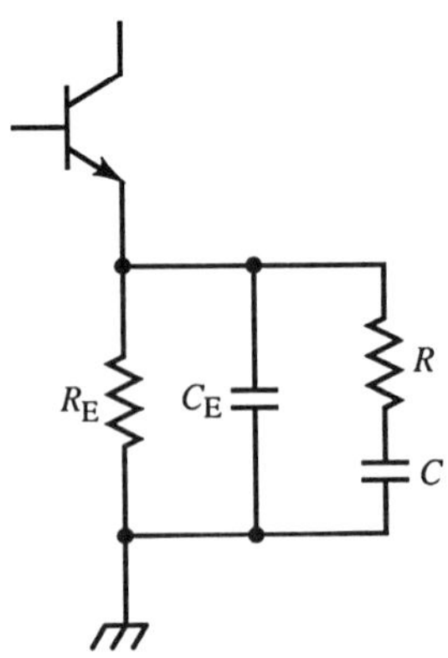

FIG. 8.21 Cascode output pole compensation at the CE emitter.

The zero must lie between τ_T and τ_L. The network acts as a phase-lead compensator, shifting the load pole to the higher frequency at $1/RC$. The cascode transimpedance with Z_E compensation is

$$\frac{V_o}{I_i} = -R_L \cdot K_i \cdot \frac{1}{s^2(K_i\tau_T\tau_L) + s(K_i\tau_T + RC) + 1} \tag{8.131}$$

For design, R_L, K_i, τ_T, and τ_L are given. We can choose RC based on the desired damping ratio without affecting the pole radius:

$$RC = 2\zeta\sqrt{K_i\tau_T\tau_L} - K_i\tau_T \tag{8.132}$$

With RC determined and R_E constrained by K_i, the coefficients of (8.129) and (8.130) are equated and yield

$$C_E = \frac{\tau_T\tau_L}{R_E\, RC} \tag{8.133}$$

$$C = \frac{\tau_T + \tau_L - RC - \tau_T\tau_L/RC}{R_E} \tag{8.134}$$

Since RC is known from (8.132), R is easily found once C is known. Note that with R and C added, the value of C_E is different from that given by (8.102).

Example 8.8 Cascode Dynamic Response Compensation

The cascode amplifier of Fig. E8.8 is to have approximately MFA response and a transresistance of 1 kΩ with maximum bandwidth. The transistors have $f_T = 600$ MHz, $C_\mu = 2$ pF, and $r'_b = 50\ \Omega$. The input current source terminates in a 100 Ω base resistor R_B. The output has 5 pF of load capacitance.

To analyze and compensate this amplifier, we begin with some transistor calculations:

$$\tau_T = \frac{1}{2\pi f_T} = 265 \text{ ps}$$

$$C_{o2} = C_{\mu 2} + C_L = 2\text{ pF} + 5\text{ pF} = 7\text{ pF}$$

Since R_B is given, the optimum R_E or K_i can be determined from (8.118) by solving for R_E:

$$\text{optimum } R_E = \left(\frac{\tau_T}{R_B C_\mu}\right) r'_b = 66.3\ \Omega \Rightarrow 68\ \Omega$$

Then $K_i = R_B/R_E = 1.47$. With R_E, R_L can be found:

$$R_m = \frac{V_o}{V_i} = \frac{V_o}{I_o} \cdot \frac{I_o}{I_i} = R_L \cdot K_i \Rightarrow R_L = \frac{R_m}{K_i} = 680\ \Omega \Rightarrow 680\ \Omega$$

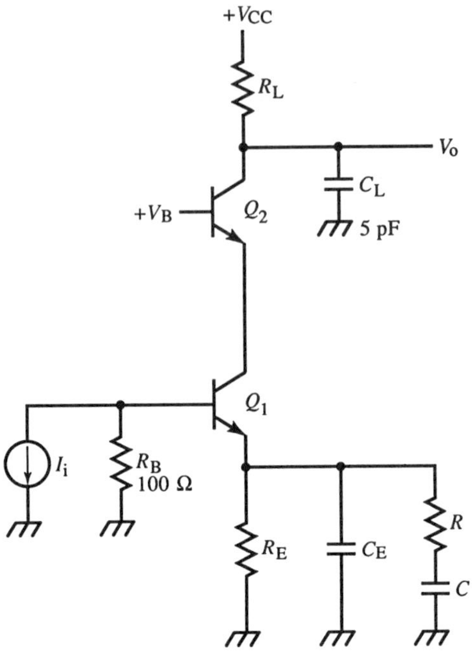

FIG. E8.8

The output pole is at

$$f_L = \frac{1}{2\pi\tau_L} = \frac{1}{2\pi R_L C_{o2}} = \frac{1}{2\pi(4.76\text{ nsec})} = 33.4\text{ MHz}$$

To maximize bandwidth, compensation of τ_L is required. We proceed to compensate for both τ_L and the input impedance of the CE. For MFA response (and assuming a single pole-pair), $\zeta = \sqrt{2}/2 \cong 0.707$. Also,

$$K_i\tau_T = 390\text{ ps}$$

Applying (8.132), we find the series RC compensator in the emitter circuit to have a time constant of

$$RC = 2\zeta\sqrt{K_i\tau_T\tau_L} - K_i\tau_T = (1.414)\sqrt{(390\text{ ps})(4.76\text{ ns})} - 390\text{ ps} = 1.54\text{ ns}$$

Continuing with (8.133), we obtain

$$C_E = \frac{\tau_T\tau_L}{R_E\,RC} = 12.0\text{ pF} \Rightarrow 12\text{ pF}$$

and from (8.134),

$$C = \frac{\tau_T + \tau_L - RC - \tau_T\tau_L/RC}{R_E} = 39.2\text{ pF} \Rightarrow 39\text{ pF}$$

With C known, R can be found from previous calculation:

$$R = \frac{1.54 \text{ ns}}{39 \text{ pF}} = 39.3\ \Omega \Rightarrow 39\ \Omega$$

From series compensation, the new pole is at $1/2\pi RC = 104$ MHz.

The CB base spreading resistance of 50 Ω may require compensation. At the emitter it forms a shunt RLC circuit with $C_{\mu 1}$ with

$$\tau_n = \sqrt{\tau_T r_b' C_{\mu 1}} = 163 \text{ ps} \Rightarrow 978 \text{ MHz}$$

and

$$\zeta = \frac{1}{2}\sqrt{\frac{\tau_T}{r_b' C_{\mu 1}}} = 0.814 \Rightarrow 36°$$

Since this resonance has a high ζ relative to MFA and has f_n at nearly 1 GHz, this pole is not likely to affect the response much and is not compensated.

With R_E now known, we can calculate the CE input time constant. The uncompensated time constant, with only R_E in the emitter circuit, is approximated as

$$\text{uncomp } \tau_i \cong (R_B + r_b' + R_E)\left(C_\mu + \frac{\tau_T}{R_E}\right) = (218\ \Omega)(5.9 \text{ pF}) = 1.29 \text{ ns}$$

This corresponds to a frequency of 124 MHz. The compensated τ_i is

$$\text{comp } \tau_i \cong (R_B + r_b')\left(C_\mu + \frac{\tau_T}{R_E}\right) = 885 \text{ ps} \Rightarrow 180 \text{ MHz}$$

The series R_E and (τ_T/R_E) is replaced by τ_T/R_E alone, and the speed increases. The uncompensated bandwidth is calculated by single-pole approximation of the time constant, from (8.21):

$$\text{uncomp } \tau = \sqrt{\tau_i^2 + \tau_{CB}^2 + \tau_L^2} = \sqrt{(1.29 \text{ ns})^2 + (163 \text{ ps})^2 + (4.76 \text{ ns})^2}$$
$$= 4.93 \text{ ns} \Rightarrow 32.3 \text{ MHz}$$

With compensation,

$$\text{comp } \tau = \sqrt{(885 \text{ ps})^2 + (163 \text{ ps})^2 + (1.54 \text{ ns})^2} = 1.78 \text{ ns} \Rightarrow 89 \text{ MHz}$$

Shunt or series inductive peaking at the output could increase the bandwidth above 100 MHz.

The choice of $\zeta = \sqrt{2}/2$ leads to MFA response for only one pole-pair. However, in this circuit, the CB pole and CE input pole also influences pole angle. The combination is not exactly MFA but is slightly overpeaked from MFA due to the additional poles.

The value of C_μ is not given in manufacturer's data sheets. Since C_μ depends on V_{BC}, its value cannot be specified except at a given voltage. The typical value is at $V_{BC} = 0$ V and is $C_{jc}(0)$ or C_{j0}. Then C_μ is the junction capacitance C_{jc}:

$$C_{jc} = \frac{C_{j0}}{[1 - (V_{BC}/\phi_C)]^m} \tag{8.135}$$

where ϕ_C is the barrier potential and m depends on the junction grading. Typically,

$$\phi_C = 0.75 \text{ V}, \qquad m = 0.5$$

For linearly graded junctions, $m = 0.33$. The SPICE parameters corresponding to these quantities are

$$C_{j0} \Rightarrow \text{CJC}, \qquad \phi_C \Rightarrow \text{VJC(PC)}, \qquad m \Rightarrow \text{MJC(MC)}$$

For a normal-mode NPN, V_{BC} is negative, and the subtraction in (8.135) is the addition of a positive voltage ratio. With the typical values, at 5 V reverse bias a junction has one-third of the capacitance it has at zero volts.

8.8 Compensation Network Synthesis

The emitter compensation network of Fig. 8.21 was chosen because its impedance provided the poles and zero required for compensation. In general, compensation requirements are known in terms of poles and zeros whereas the topology and equations for element values are unknown. The compensation of hf-gyrated impedances was simplified (in Chapter 7) by deriving the equivalent circuits and noting that all-pass networks could be formed with them. Because of the need for compensation networks, we now examine a few common synthesis techniques.

One compensation technique is to make a reactive network with impedance Z resistive and thus independent of frequency. Another compensating impedance Z_c is added in series or parallel with Z (Fig. 8.22).

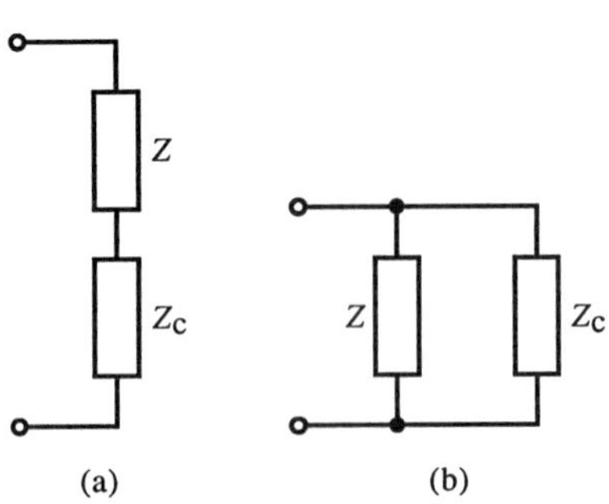

FIG. 8.22 All-pass (resistive input) compensation of network of impedance Z, with compensating network of impedance Z_c in series (a) or parallel (b).

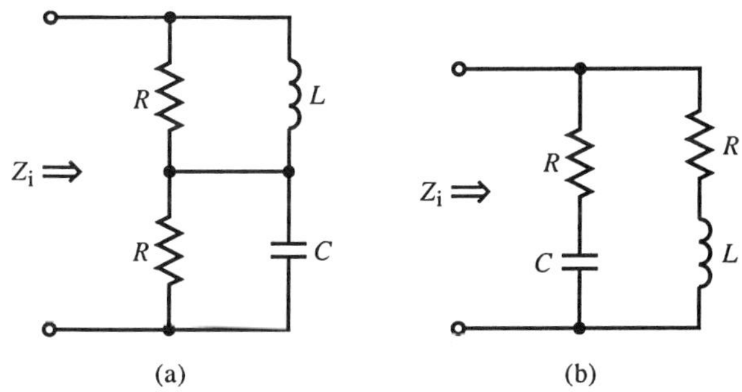

FIG. 8.23 Examples of resistive input networks when $LC = R^2$.

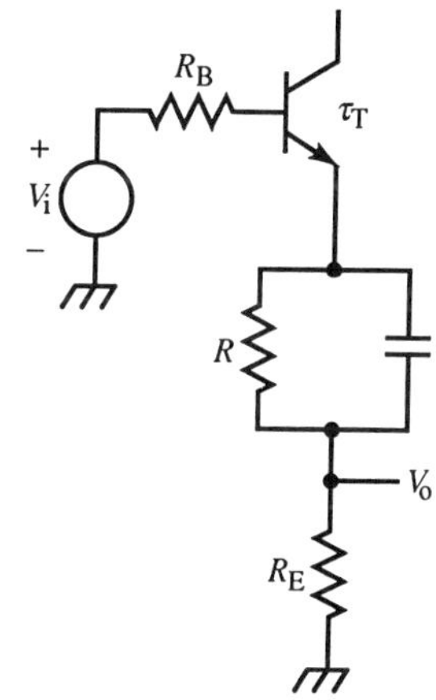

FIG. 8.24 Output impedance compensation of a CC with base resistance.

For the networks in Fig. 8.23, $Z_i = R$ when

$$LC = R^2 \tag{8.136}$$

A shunt RL, such as an emitter-gyrated base resistance, is compensated by adding in series a shunt RC (Fig. 8.24). Then, from (8.136),

$$R = R_B, \qquad C = \frac{\tau_T}{R_B} \tag{8.137}$$

This creates a resistive voltage divider at the emitter. If the load is capacitive, C is made larger to compensate the divider. Then Z of the network from the emitter must still be equivalent to a shunt RC satisfying (8.136).

More generally, a shunt RC can similarly be compensated by adding a shunt RL in series with it. Or a series LC can be compensated by the two resistors shunting each of them (Fig. 8.23a). In these cases, $Z_i = R$ when (8.136) is satisfied.

A series RC, such as a base-gyrated emitter resistance, can be compensated by shunting it with a series RL, as in Fig. 8.23b. The input of a CC with significant C_μ at the internal base node and resistance in the collector supply return line forms a series RC that can compensate the series RL of the base. Since base R and L are both parasitic (r_b' contributes to R and lead inductance to L), they can be made to appear resistive at b'. Adjustment of the collector and base resistance and series base inductance makes it is possible to satisfy (8.136). For $r_b' = 100\ \Omega$, $L_b = 10$ nH, and $C_\mu = 3$ pF, the impedance at b' toward b is resistive and is 100 Ω when the collector resistance is 100 Ω and 20 nH is added to the base circuit. This added inductance might be from the inductive peaking of the previous stage.

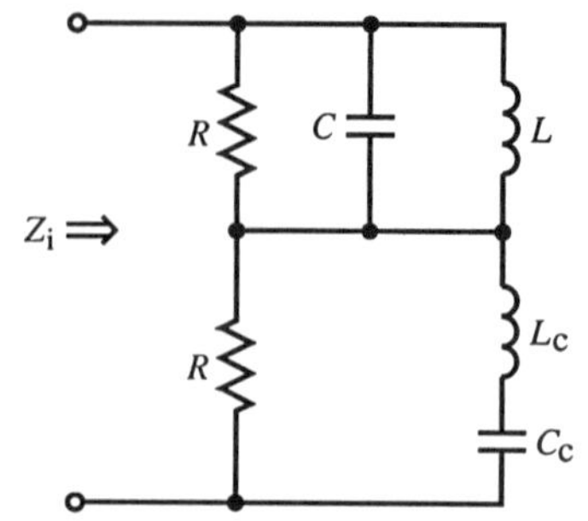

FIG. 8.25 A more complicated all-pass network.

More complicated networks are also possible, such as in Fig. 8.25. Here, the conditions for a resistive input of $Z_i = R$ are

$$LC = L_c C_c = R^2 \tag{8.138}$$

This network is a compensated shunt *RLC*, or a shunt *RC* in which the capacitor has parasitic inductance. This is typical of electrolytic capacitors, which have resonant frequencies around 1 MHz, or for higher frequencies, any capacitors with leads. A monolithic multilayer ceramic capacitor has about 5 nH of inductance with leads of a length needed for insertion into circuit-board holes. Leadless "chip" capacitors are sometimes required for good high-frequency bypass or decoupling of the power supply terminals of active devices.

Two other common networks are shown in Figs. 8.26, the *bridge-T*, and 8.27, the *lattice* or *bridge*. A special case of the bridge-T is applied in T-coil compensators. The input is $Z_i = R$ when

$$Z_a Z_b = R^2 \Rightarrow \frac{V_o}{V_i} = \frac{1}{Z_b/R + 1} \tag{8.139}$$

Z_a and Z_b must be dual reactances (or *reciprocal impedances*); if Z_a is capacitive, Z_b must be inductive.

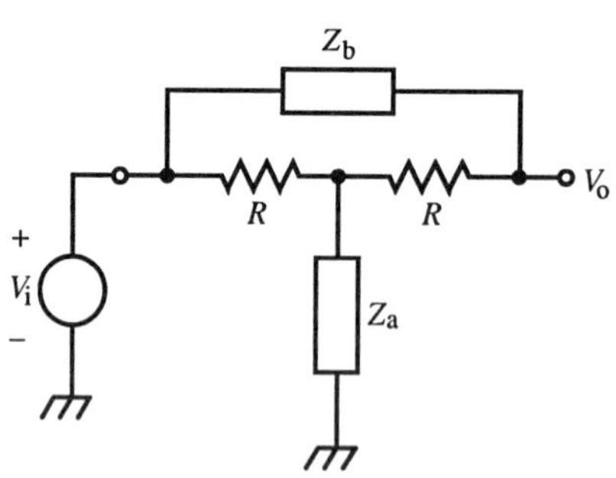

FIG. 8.26 Bridge-T network. The input impedance is resistive when $Z_a Z_b = R^2$.

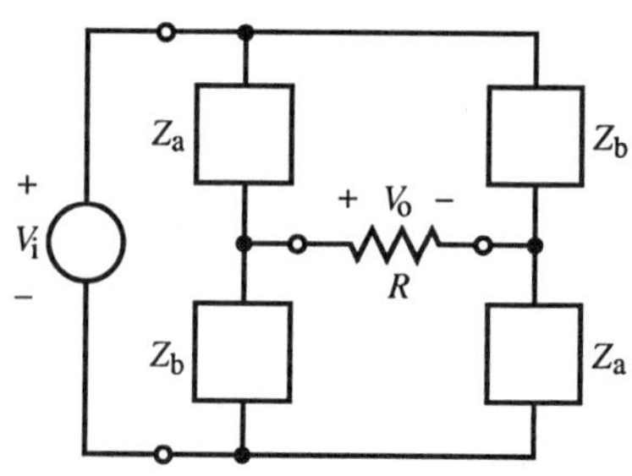

FIG. 8.27 Lattice or bridge network.

The lattice network has the same resistive-input conditions but a different transfer function:

$$Z_a Z_b = R^2 \Rightarrow \frac{V_o}{V_i} = \frac{1 - Z_a/R}{1 + Z_a/R} \tag{8.140}$$

Two general methods are easy to apply to the synthesis of passive networks with a given $Z(s)$:

1. Partial-fraction synthesis, for factored poles or zeros
2. Continued-fraction synthesis: network topology explicit

Partial-fraction synthesis is based on partial-fraction expansion of $Z(s)$. This requires factoring the denominator. If the numerator is easier to factor, expand $Y(s) = 1/Z(s)$ instead. Various network topologies can result, however, and other design considerations could constrain the choice of topology.

Example 8.9 Partial-Fraction Network Synthesis

A network with the following $Z(s)$ is described:

$$Z = R \cdot \frac{s\tau_3 + 1}{(s\tau_1 + 1)(s\tau_2 + 1)} \tag{E1}$$

Z can be written as

$$Z = \frac{A}{s\tau_1 + 1} + \frac{B}{s\tau_2 + 1} \tag{E2}$$

When (E1) is partial-fraction expanded, A and B are

$$A = R\left(\frac{\tau_1 - \tau_3}{\tau_1 - \tau_2}\right), \qquad B = R\left(\frac{\tau_2 - \tau_3}{\tau_2 - \tau_1}\right)$$

A and B are resistances. Equation (E2) can be written as

$$Z = \frac{1}{s(\tau_1/A) + 1/A} + \frac{1}{s(\tau_2/B) + 1/B} = 1/s(\tau_1/A) \left\| \frac{1}{A} + 1/s(\tau_2/B) \right\| \frac{1}{B}$$

Z has the form of two shunt RCs in series.

Continued-fraction synthesis produces a continued-fraction form of Z. A desirable feature of continued-fraction impedances is that the topology is explicit in the form of the expression. The general procedure is to invert rational expressions that are less than unity and to divide by synthetic division.

Example 8.10 Continued-Fraction Network Synthesis

An impedance of the form of Example 8.9 is

$$Z = R \cdot \frac{sc+1}{as^2+bs+1} \tag{E1}$$

R is first multiplied to the numerator in s and the fraction inverted:

$$Z = \frac{1}{\left(\dfrac{as^2+bs+1}{sRc+R}\right)} \tag{E2}$$

The fraction is now greater than unity and can be divided to become

$$Z = \frac{1}{s\left(\dfrac{a}{Rc}\right) + \left(\dfrac{bc-a}{Rc^2}\right) + \dfrac{1}{\dfrac{sRc+R}{\left(\dfrac{c^2-bc+a}{c^2}\right)}}} \tag{E3}$$

The remainder is divided by $sRc+R$ and then inverted. Division is carried out once again, and the final continued fraction results:

$$Z = \frac{1}{s\left(\dfrac{a}{Rc}\right) + \left(\dfrac{bc-a}{Rc^2}\right) + \dfrac{1}{sR\left(\dfrac{c^3}{c^2-bc+a}\right) + R\left(\dfrac{c^2}{c^2-bc+a}\right)}} \tag{E4}$$

The terms in the denominator of Z are admittances. The capacitance a/Rc is in parallel with resistance

$$R\left(\frac{c^2}{bc-a}\right)$$

and with the series RL, where the resistance is

$$R\left(\frac{c^2}{c^2-bc+a}\right) = R_S$$

and the inductance is cR_S.

Continued fractions represent shunt topologies, and partial fractions represent series topologies. In continued-fraction expansion, divisions are executed the usual way, beginning with the highest power in s. If, instead, division begins with the lowest power, the divisor grows in powers of s. The remainder is then a power of s higher than that of the dividend. This approach does not produce the circuit topology of the desired network but can be useful in approximating a network by truncating the quotient. No s^2 term represents a circuit element, but it can be transformed into an equivalent network with negative element values. (See Section 7.3.)

Example 8.11 Differentiator

The circuit of Fig. E8.11a is a differentiator with resistive input and output, suitable for transmission-line coupling. The circuit is analyzed by transforming it to Fig. E8.11b, using the T-coil theory of Section 8.4. This topology is recognized as an approximate bridge-T when $R_L = R$ and the coupled inductors approach being an ideal transformer. Then $L_1 = L_2 = 0$, and $Z_a = sM$, where M is the mutual inductance of L_1 and L_2, and $Z_b = 1/sC$. Applying (8.139), we obtain

$$\frac{V_o}{V_i} = \frac{1}{Z_b/R + 1} = \frac{sRC}{sRC + 1} \tag{E1}$$

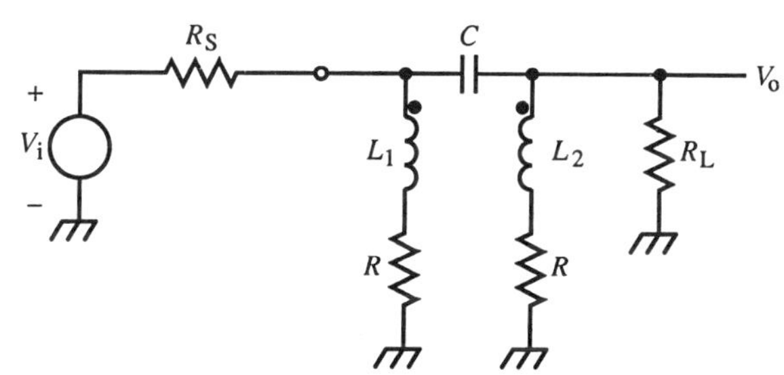

(a)

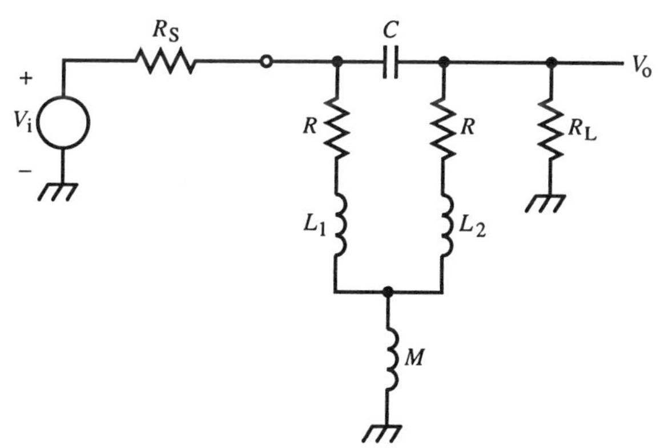

(b)

FIG. E8.11

and $Z_i = R$. The circuit differentiates to a frequency of $1/RC$. A 50 Ω transmission line can drive the differentiator and be terminated properly when $R = 50\ \Omega$. For wideband differentiation to 100 MHz, $RC = 1/2\pi(100\ \text{MHz}) = 1.59$ ns. Then $C = 31.8$ pF and M must be the pulse transformer magnetizing inductance:

$$M = \frac{R^2}{C} = 79.6\ \text{nH}$$

The output-terminating resistance can be the characteristic impedance of another transmission line. In other words, the differentiator can be inserted into a transmission line without causing discontinuity. For example, high-speed differentiation of a ramp can be performed by driving a 50 Ω coaxial cable into a 50 Ω test section wherein the differentiator has been built. This section then terminates in the 50 Ω input of an oscilloscope vertical amplifier.

8.9 Differential-Amplifier Compensation

The two-transistor diff-amp in Fig. 8.28 has a voltage divider formed by R and Z_e of the other transistor. This divider can be compensated by shunting R with a compensating C. The approach is the same for both Π (Fig. 8.28a) and T-section (b) emitter networks, since they are equivalent. We want to compensate

$$Z_e = \frac{Z_\pi + Z_B}{\beta(s) + 1} = \frac{r_e}{s\alpha_o\tau_T + 1} + \left(\frac{Z_B}{\beta_o + 1}\right)\left(\frac{s\tau_\beta + 1}{s\alpha_o\tau_T + 1}\right) \tag{8.141}$$

Below f_T, the first term is approximately r_e. In the hf region, Z_B is gyrated. The situation is similar to that of the CB of the cascode (in Section 8.7). For $Z_B = R_B$, the resistive network of Fig. 8.23a is formed. To present a resistance to the emitter of the other BJT, the compensator time-constant is

$$RC = \frac{\tau_T R_B}{R} \Rightarrow C = \frac{\tau_T R_B}{R^2} \tag{8.142}$$

In practice, significant stray capacitance is often present at the emitter (or current source) node(s). The circuit is then represented by Fig. 8.19, in which C_o is the stray capacitance. The results of Section 8.7 for the cascode can be applied. From (8.119), increasing C_o decreases ζ, causing the circuit to be less damped.

Transistor compensation techniques apply to differential as well as single-ended amplifiers. For balanced differential amplifiers, the shared networks

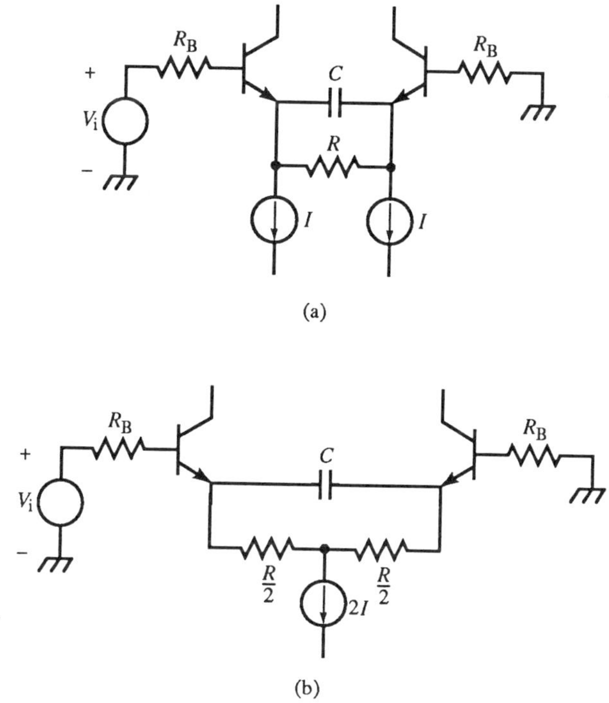

FIG. 8.28 BJT diff-amp compensation in emitter circuit with C, with Π (a) or T (b) biasing network. The two circuits are equivalent.

between sides experience twice the drive of a single-ended network. When gain is involved, their effective impedances are halved.

8.10 Shunt-Feedback Amplifier Design

In Sections 4.15–4.17, we analyzed the frequency-independent shunt-feedback amplifier and derived the closed-loop transresistance, (4.90). A shunt-feedback topology was also considered in Section 6.8 with capacitive Z_f. The general topology, shown in Fig. 6.22, has a transimpedance given by (6.47).

A wideband realization of a shunt-feedback amplifier is the BJT amplifier represented in Fig. 8.29 with frequency-dependent β and general impedances. This amplifier is equivalent to the general topology in Fig. 6.22a but is explicit in $\beta(s)$ so that we can analyze its behavior in the hf region. Assuming a general $\beta(s)$ at first, we find the transimpedance by solving the flow graph of Fig. 8.29b. (This flow graph is a generalized form of Fig. 4.26b.)

$$\frac{V_o}{I_i} = Z_L \cdot \frac{Z_i - \beta Z_f}{Z_f + Z_i + (\beta + 1) Z_L} \tag{8.143}$$

The first term in the numerator represents the passive noninverting path to

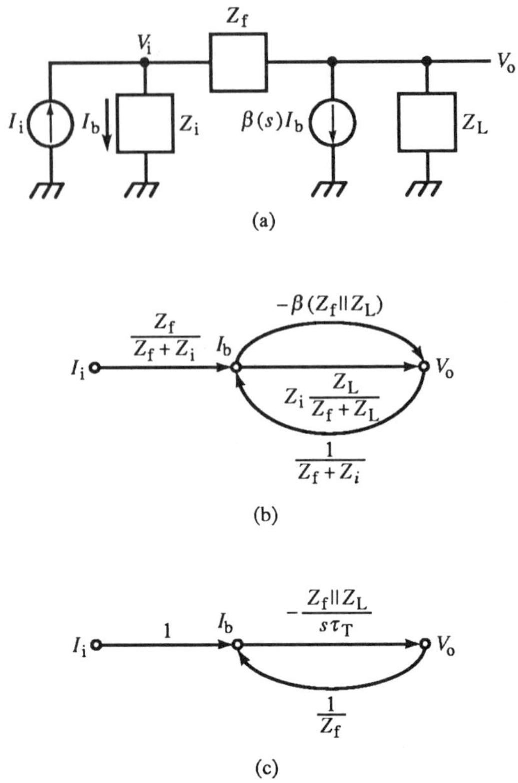

FIG. 8.29 A general equivalent circuit of a shunt-feedback amplifier in $\beta(s)$ (a), its flow graph (b), and its simplification using the BJT hf model, where $\beta \to \infty$ and $Z_\pi = 0$, (c).

the output, and the second term represents the active path through the BJT. The input impedance is derived after (4.100):

$$Z_{in} = \frac{Z_i \parallel Z_f}{1 + GH} = Z_i \cdot \frac{Z_f + Z_L}{Z_f + Z_i + (\beta + 1)Z_L} \tag{8.144}$$

Similarly, following (4.102),

$$Z_{out} = Z_L \left\| \left(\frac{Z_f + Z_i}{\beta + 1} \right) \right. = Z_L \cdot \frac{Z_f + Z_i}{Z_f + Z_i + (\beta + 1)Z_L} \tag{8.145}$$

The hf BJT model has $Z_\pi = 0$. If Z_π is not Z_i, then it is a shunt contributor to it, and when set to zero causes Z_i to be zero. The hf approximation of $\beta(s)$ is $1/s\tau_T$. Making these hf approximations to Fig. 8.29b, we solve the hf flow graph of Fig. 8.29c for the hf transimpedance, which is

$$\left. \frac{V_o}{I_i} \right|_{hf} = -\frac{Z_f}{s\tau_T(1 + Z_f/Z_L) + 1} = -Z_L \parallel \alpha_{hf} Z_f \tag{8.146}$$

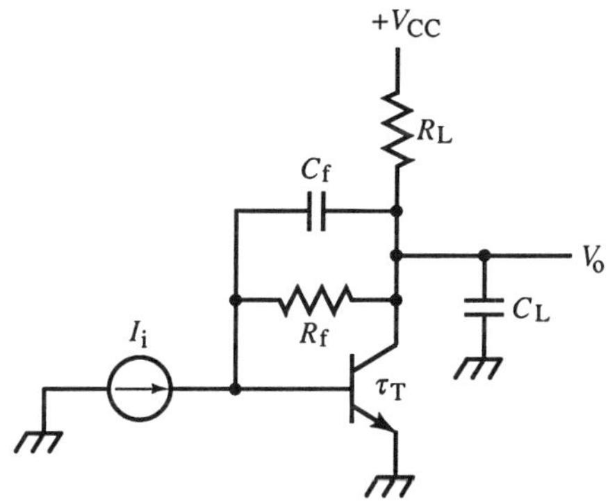

FIG. 8.30 Typical single-BJT shunt-feedback stage with load and feedback capacitances.

Z_{in} is trivially zero, and the hf output impedance is

$$Z_{out}(\text{hf}) = \frac{s\tau_T Z_f}{s\tau_T(1+Z_f/Z_L)+1} = -\frac{(V_o/I_i)|_{hf}}{\beta_{hf}} \qquad (8.147)$$

These general results are applied to a less general single-BJT shunt-feedback amplifier, shown in Fig. 8.30, where

$$Z_f = R_f \Big\| \frac{1}{sC_f} = \frac{R_f}{sR_fC_f+1}; \qquad Z_L = R_L \Big\| \frac{1}{sC_L} = \frac{R_L}{sR_LC_L+1} \qquad (8.148)$$

Substituting into (8.146) and simplifying gives

$$\frac{V_o}{I_i} = -\frac{R_f}{s^2\tau_T R_f(C_f+C_L)+s[\tau_T(1+R_f/R_L)+R_fC_f]+1} \qquad (8.149)$$

Before analyzing this transimpedance, we can easily obtain from (8.147) the expression for output impedance:

$$Z_{out} = \frac{s\tau_T R_f}{s^2\tau_T R_f(C_f+C_L)+s[\tau_T(1+R_f/R_L)+R_fC_f]+1}$$
$$= \left(R_f \| R_L \| \frac{\tau_T}{C_f}\right) \Big\| \frac{1}{s(C_f+C_L)} \Big\| s\tau_T R_f \qquad (8.150)$$

This impedance is represented topologically in Fig. 8.31 as a shunt *RLC*.

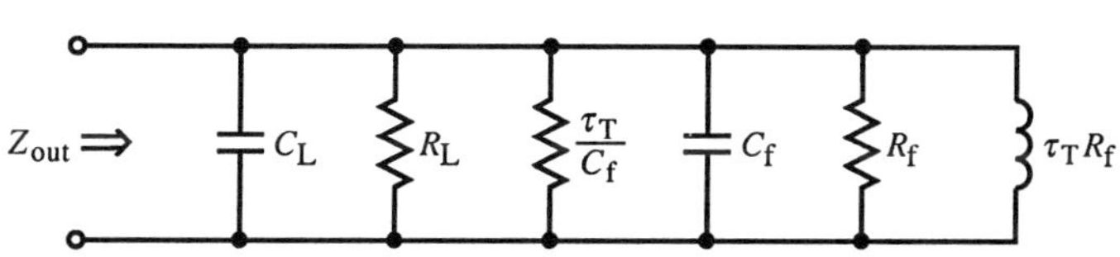

FIG. 8.31 Equivalent output impedance of shunt-feedback amplifier.

From (8.149), the complex pole-pair damping ratio is

$$\zeta = \frac{b}{2\sqrt{a}} = \frac{\tau_T(1+R_f/R_L)+R_fC_f}{2\sqrt{\tau_T R_f(C_f+C_L)}} \tag{8.151}$$

The desired response is set by choosing ζ and solving for the element that is free to be varied. For a given transistor, τ_T is fixed, and the required gain for the stage is also determined by the amplifier design strategy. This sets R_f. Biasing constrains R_L, and C_f is partly determined by C_μ of the BJT. This leaves C_L; its minimum is determined by the capacitive loading of the next stage. The best design insight is gained from the loci of poles when various elements are allowed to vary parametrically. The loci are determined by extending the technique of Section 5.11.

The pole locus is described by geometric equations in the real and imaginary s coordinates, σ and ω. The two basic equations are (5.156) and (5.158). Starting first with R_f as parameter, substitute into (5.156a) from (8.149):

$$\sigma = -\frac{b}{2a} = -\frac{\tau_T(1+R_f/R_L)+R_fC_f}{2\tau_T R_f(C_f+C_L)} \tag{8.152}$$

Solving for R_f, we obtain

$$R_f = \frac{-\tau_T R_L}{\tau_T + R_L C_f + 2\tau_T R_L(C_f+C_L)\sigma} \tag{8.153}$$

Equation (5.158) leads to

$$\omega^2+\sigma^2 = \frac{1}{a} = \frac{1}{\tau_T R_f(C_f+C_L)} = -\frac{\tau_T + R_L C_f + 2\tau_T R_L(C_f+C_L)\sigma}{\tau_T^2 R_L(C_f+C_L)} \tag{8.154}$$

Simplifying the right side and collecting σ terms on the left side yields

$$\omega^2 + \left(\sigma^2 + \frac{2\sigma}{\tau_T}\right) = -\frac{\tau_T + R_L C_f}{\tau_T^2 R_L(C_f+C_L)} \tag{8.155}$$

Completing the square in σ by adding $1/\tau_T^2$ to both sides, we obtain

$$\omega^2 + \left(\sigma + \frac{1}{\tau_T}\right)^2 = -\frac{\tau_T + R_L C_f}{\tau_T^2 R_L(C_f+C_L)} + \frac{1}{\tau_T^2} = \left(\frac{1}{\tau_T}\sqrt{\frac{R_L C_L - \tau_T}{R_L(C_f+C_L)}}\right)^2 = \omega_r^2 \tag{8.156}$$

This equation describes a circular locus centered at $\sigma_0 = -1/\tau_T = -\omega_T$ with a radius of ω_r. Unlike previous loci, the circle does not contain the origin but is offset to the left, as shown in Fig. 8.32. In practice, usually,

$$R_L C_L \gg \tau_T$$

and ω_r simplifies to

$$\omega_r \cong \frac{1}{\tau_T}\sqrt{\frac{C_L}{C_f+C_L}}, \qquad R_L C_L \gg \tau_T \tag{8.157}$$

At $R_f = 0$, the poles are at $-\infty$ and $-\omega_T$. As R_f increases, they move together

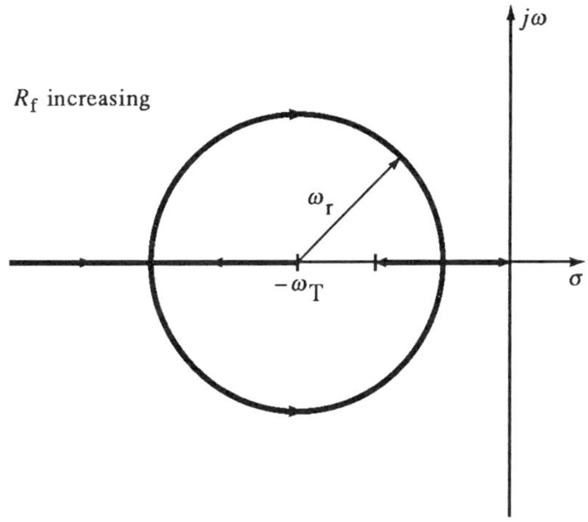

FIG. 8.32 Root locus of shunt-feedback stage poles when R_f varies. The circle does not contain the origin.

and become complex, following the circular locus with decreasing σ. At the σ axis they split; as $R_f \to \infty$, one goes to zero and the other to

$$-\frac{\tau_T / R_L + C_f}{\tau_T(C_f + C_L)} \tag{8.158}$$

The locus equation for parameter τ_T is derived similarly to that for R_f. It is also circular, centered at $-1/R_fC_f$ with

$$\omega_r = \frac{1}{R_fC_f}\sqrt{1 - \frac{R_fC_f}{(R_f \parallel R_L)(C_f + C_L)}}, \qquad R_fC_f < R_LC_L \tag{8.159}$$

The condition of (8.159) is required to keep ω_r real. When R_L is replaced by a current source, (8.159) simplifies to

$$\omega_r = \frac{1}{R_fC_f}\sqrt{\frac{C_L}{C_f + C_L}}, \qquad R_L \to \infty \tag{8.160}$$

At $\tau_T = 0$, the poles are at $-1/R_fC_f$ and $-\infty$. As τ_T increases, the poles move together (Fig. 8.33). The low-frequency pole actually increases in frequency with a slower transistor. The poles form a circular locus with σ decreasing until they separate along the σ axis. At $\tau_T \to \infty$, one pole is at the origin and the other is at

$$-\frac{1}{(R_f \parallel R_L)(C_f + C_L)} \tag{8.161}$$

The locus of C_L is derived similarly and has the same form as before. The center of the circle is in the RHP at

$$\sigma_0 = \left(\frac{C_L}{C_f}\right)\frac{1}{R_fC_f - \tau_T(1 + R_f/R_L)(C_L/C_f)} \tag{8.162}$$

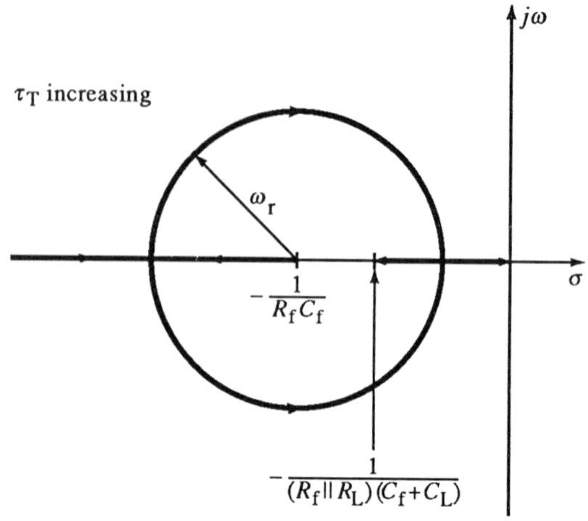

FIG. 8.33 Root locus of shunt-feedback stage when τ_T varies.

with

$$\omega_r = \frac{1}{R_f C_f - \tau_T(1 + R_f/R_L)(C_L/C_f)} \cdot \sqrt{\frac{R_f C_f - \tau_T(1 + R_f/R_L)(C_L/C_f) + \tau_T(C_L/C_f)^2}{\tau_T}} \tag{8.163}$$

Often it is the case that

$$C_L \ll C_f, \qquad R_f C_f \gg \tau_T\left(1 + \frac{R_f}{R_L}\right) \tag{8.164}$$

and Z_f dominates the response. The locus for C_L simplifies to

$$\sigma_0 \cong \left(\frac{C_L}{C_f}\right)\frac{1}{R_f C_f}, \qquad \omega_r \cong \frac{1}{\sqrt{\tau_T R_f C_f}} \tag{8.165}$$

For LHP poles, $\omega_r > \sigma_0$, and (8.165) satisfies this condition for typical values.

The locus for C_f would also be useful but cannot be put in a form similar to the previous parameters. When r_b' is taken into account, the transimpedance gains a pole, and the denominator is cubic. The effect of r_b' is to slightly undamp the amplifier. Under the conditions of (8.164),

$$\frac{b}{2a} = \frac{1}{2\tau_T} = \text{constant}$$

Variation in R_f or C_f moves the poles along the vertical locus. An increase in C_f or R_f reduces pole angle, though pole radius is also reduced somewhat. When $C_L = 0$ and $R_L \to \infty$, the poles are located at $-1/R_f C_f$ and $-\omega_T$.

The astute observer will recognize that Z_{out} (Fig. 8.31) is similar to what would be expected of emitter impedance due to a gyrated shunt RC in the

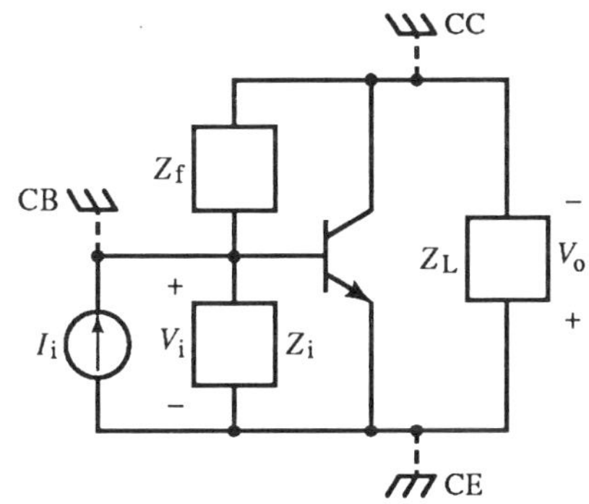

FIG. 8.34 A general configuration-independent BJT amplifier. Which of the three configurations is determined by which node is considered ground.

base. The coincidence is not accidental. Bruce Hofer has observed that the topology of the shunt-feedback amplifier and emitter-follower are identical. Figure 8.34 shows a general BJT circuit with impedances shunting each BJT terminal pair. Which of the three configurations (CE, CB, or CC) is represented depends on where ground is placed, as shown. Since ground is an arbitrary 0 V reference node, the port impedance of V_o is the same for CC and CE. For the CC, Z_f is a shunt base impedance; for the shunt-feedback CE, the output port is the same, except in relation to ground. Since port impedances are independent of grounding conventions, Z_{out} is the same.

This observation also applies to inverting and noninverting op-amp configurations; their topology is identical. In the case of the noninverting op-amp, the input signal is added (in series) with the amplifier.

8.11 Shunt-Feedback Cascode and Darlington Amplifiers

The unavoidable presence of C_μ in the shunt-feedback amplifier has led to a minimization of parasitic feedback capacitance by use of a cascode amplifier as the forward path G (Fig. 8.35a). This involves the additional factor α_2 of the CB. The flow graph (Fig. 8.35b) reduces to a transimpedance of

$$\frac{V_o}{I_i} = -R_f/\{s^3[\tau_{T1}\tau_{T2}R_f(C_f + C_L)] + s^2[\tau_{T1}\tau_{T2}(1 + R_f/R_L) + \tau_{T1}R_f(C_f + C_L)] + s[\tau_{T1}(1 + R_f/R_L) + R_fC_f] + 1\} \tag{8.166}$$

An additional pole due to $\alpha_2(s)$ results in a cubic denominator. For

$$R_fC_f \gg \tau_{T1}, \tau_{T2} \tag{8.167}$$

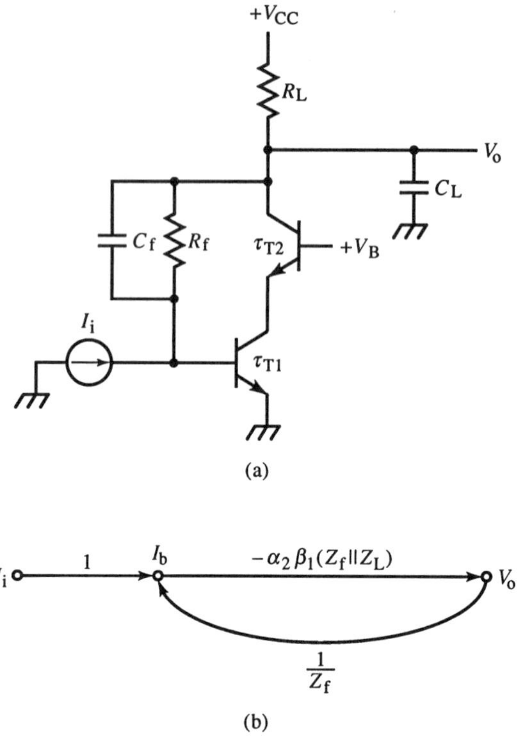

FIG. 8.35 Shunt-feedback cascode amplifier (a), and hf flow graph (b).

the denominator of (8.166) can be factored approximately to yield

$$\frac{V_o}{V_i} \cong -\frac{R_f}{(sR_fC_f+1)(s^2[\tau_{T1}\tau_{T2}(1+C_L/C_f)]+s[\tau_{T1}(1+C_L/C_f)]+1)} \tag{8.168}$$

In the complex pole factor, τ_{T2} is present only in a and not b. This results in a constant ω_n locus for τ_{T2} and a vertical (constant α) pole locus (Fig. 5.13, case 3) for the other parameters.

A transresistance amplifier R_m with load capacitance but with no R_f is shown in Fig. 8.36. The voltage gain for a general load impedance Z_L is

$$\frac{V_o}{V_i} = -\left(\frac{Z_L}{R_m}\right)\frac{-sR_mC_f+1}{sZ_LC_f+1} \tag{8.169}$$

For $Z_L = 1/sC_L$,

$$\frac{V_o}{V_i} = -\frac{-sR_mC_f+1}{sR_m(C_f+C_L)} \tag{8.170}$$

This result has a familiar RHP zero. Equation (8.170) can be visualized as an uncompensated voltage divider in which the upper impedance is a shunt RC consisting of C_f and $-R_m(1+V_o/V_i)$, and the lower impedance is due to C_L.

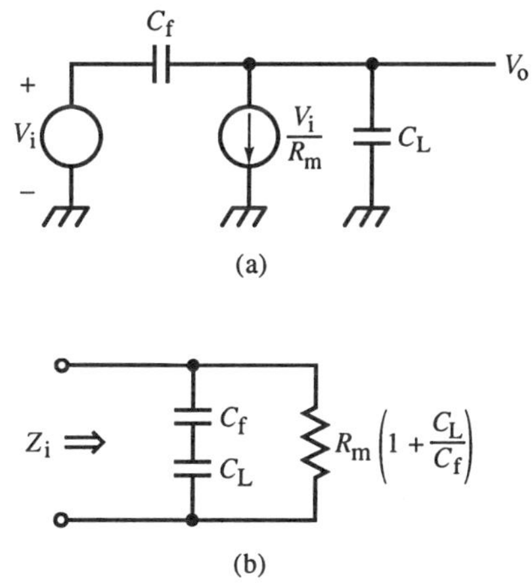

FIG. 8.36 Transconductance amplifier with load capacitance and only capacitive feedback (a); its input impedance (b) has a resistive component.

The input impedance is

$$Z_{\text{in}} = \frac{V_i}{I_i} = \frac{1}{sC_f(1 - V_o/V_i)} \tag{8.171}$$

and has the form of a Miller capacitance. After substitution of (8.170) and simplification, we obtain

$$Z_{\text{in}} = \left[\frac{1}{s(C_f \parallel C_L)}\right] \parallel R_m\left(1 + \frac{C_L}{C_f}\right) \tag{8.172}$$

The equivalent input network is shown in Fig. 8.36b. This is a surprising result because the input has a shunt resistance, but the actual circuit has only a capacitive connection to the input node. This circuit models CE amplifiers with significant C_μ and load capacitance.

Example 8.12 CE with Load and Shunt-Feedback Capacitances

In Fig. E8.12, the BJT amplifier has an input impedance determined by (8.172). Assume that the transistor has $\alpha = 1$ and $r_e \cong 0$. Then the transresistance of the BJT is approximately R_E, or 1 kΩ. The base input impedance is infinite and can be ignored. The analysis applies to the lf region and is not due to hf effects.

An intuitive explanation begins by noting that the path through C_f directly presents a capacitance of the series combination of C_f and C_L, or 0.9 pF, to the input. Second, if a 1 V step is applied to the input, it

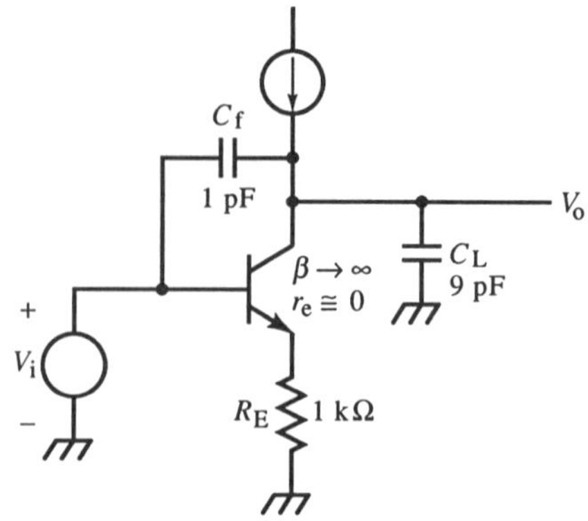

FIG. E8.12

generates 1 mA of collector current. This current divides between C_f and C_L since they form a capacitive current divider. For $C_f = 1$ pF and $C_L = 9$ pF, 0.9 mA flows through C_L whereas 0.1 mA flows out of the input. This component of current corresponds to the resistive path in Fig. 8.36b. The resistance is

$$\frac{1\text{ V}}{0.1\text{ mA}} = 10\text{ k}\Omega$$

Now let us check this result using (8.172). The series combination of capacitances follows immediately. The resistance should be

$$R_m\left(1 + \frac{C_L}{C_f}\right) = 1\text{ k}\Omega\left(1 + \frac{9\text{ pF}}{1\text{ pF}}\right) = 10\text{ k}\Omega$$

and the two solutions agree.

The amplifier of Fig. 8.36 is now modified to conform to a BJT shunt-feedback circuit (Fig. 8.37) for the hf region. For a general output impedance Z_L, the voltage gain is

$$\frac{V_o}{V_i} = -\frac{s\beta C_i}{s(\beta + 1)C_f + 1/Z_L} \tag{8.173}$$

Substituting $\beta_{hf} = 1/s\tau_T$ and $Z_L = 1/sC_L$, we get

$$\frac{V_o}{V_i} = -\left(\frac{C_i}{C_f}\right)\frac{1}{s\tau_T(1 + C_L/C_f) + 1} \tag{8.174}$$

To avoid C_μ, the amplifier can be made a cascode. Then Fig. 8.37 is modified by multiplying α_2 of the CB to the current source. This leads to a complex

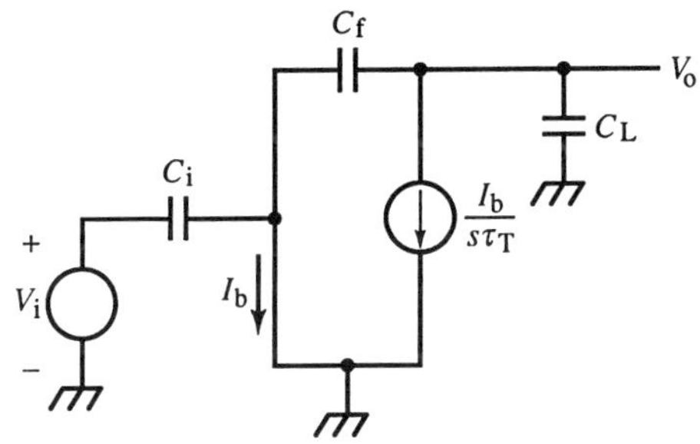

FIG. 8.37 A hf model of a shunt-feedback BJT stage with only capacitive feedback and capacitive input coupling.

pole-pair in the voltage gain:

$$\frac{V_o}{V_i} = -\left(\frac{C_i}{C_f}\right)\frac{1}{s^2\tau_{T1}\tau_{T2}(1 + C_L/C_f) + s\tau_{T1}(1 + C_L/C_f) + 1} \tag{8.175}$$

The response can be designed in the usual way for complex pole-pairs. The damping ratio is

$$\zeta = \frac{1}{2}\sqrt{\left(1 + \frac{C_L}{C_f}\right)\left(\frac{\tau_{T1}}{\tau_{T2}}\right)} \tag{8.176}$$

Comparing this with the ζ for a single-BJT shunt-feedback amplifier, (8.151), we find that C_f has the opposite effect of damping the response. In this circuit, increasing C_f undamps it.

Finally, the idea of isolating Z_f from stray capacitance can be extended to the Darlington configuration. Bruce Hofer has analyzed this circuit as shown in Fig. 8.38. The transimpedance has a cubic denominator and a zero at ω_{T1}. If we again assume a dominant time constant of R_fC_f, the approximate

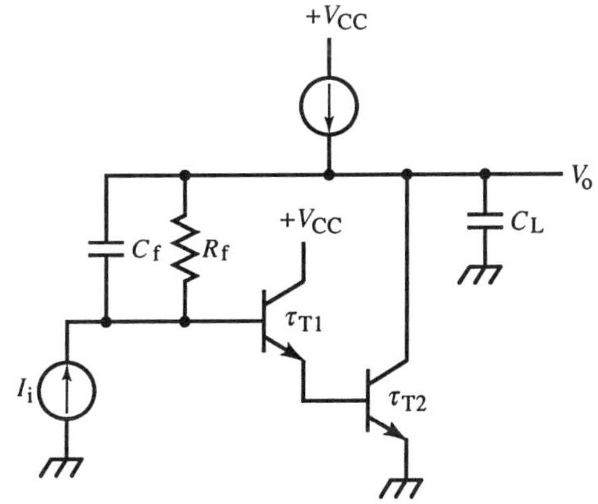

FIG. 8.38 Darlington shunt-feedback amplifier.

transimpedance is

$$\frac{V_o}{I_i} \cong \frac{R_f(s\tau_{T1}+1)}{(sR_fC_f+1)\{s^2\tau_{T1}\tau_{T2}[1+(C_{\mu 2}+C_L)/C_f]+s\tau_{T1}[1+(C_{\mu 2}/C_f)]+1\}} \tag{8.177}$$

The complex pole-pair has the same radius as the shunt-feedback cascode but has a different expression for ζ. Since C_L appears only in a, its root locus has a constant pole radius whereas the other parameters have a constant-α locus. Therefore, C_L is the component of choice for adjustment of response.

8.12 Closure

Several elemental amplifiers have been analyzed and design formulas have been derived. Fast amplifiers consist of several of these stages in cascade. Various combinations are used in fast amplifier designs, such as the diff-amp in cascade with a differential shunt-feedback amplifier (Fig. 8.39). We shall continue development of the fast-amplifier repertoire later, adding more precise, yet fast, amplifier stages.

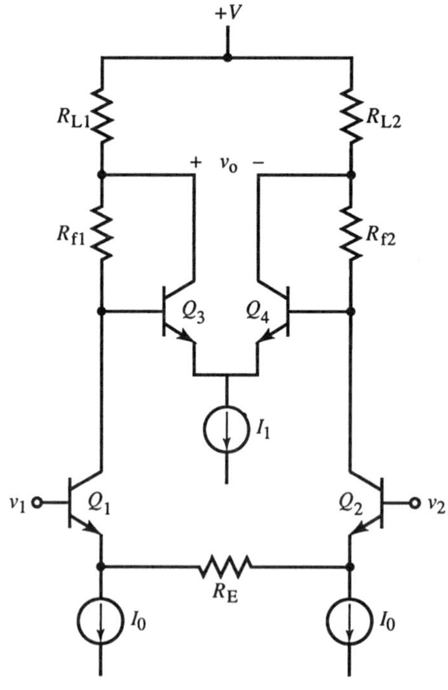

FIG. 8.39 CE diff-amp driving a differential shunt-feedback amplifier.

References

[1] M. S. Ghausi, *Principles and Design of Linear Active Circuits*, McGraw-Hill, 1965.

[2] Carl Battjes, Bruce Hofer, and John Addis, *Amplifier Frequency and Transient Response (AFTR)* course notes, Tektronix, Inc., Beaverton, Oregon.

[3] E. M. Cherry and D. E. Hooper, *Amplifying Devices and Low-Pass Amplifier Design*, Wiley, 1968.

[4] Arpad Barna, *High-Speed Pulse Circuits*, Wiley-Interscience, 1970.

[5] Ernst H. Nordholt, *Design of High-Performance Negative-Feedback Amplifiers*, Elsevier, 1983.

[6] Basil L. Cochrun and Arvin Grabel, "A Method for the Determination of the Transfer Function of Electronic Circuits," *IEEE Trans. Circuit Theory*, Vol. CT-20, No. 1, Jan. 1973. pp. 16–20.

[7] Sol Rosenstark, *Feedback Amplifier Principles*, Macmillan, 1986. pp. 67–77.

[8] A. M. Davis, "Analyze active-network responses without complex manipulations," *EDN*, 20 Feb. 1979. pp. 109–112.

[9] Roberto Saucedo and Earl Schiring, *Introduction to Continuous and Digital Control Systems*, Macmillan, 1968. pp. 273, 275.

CHAPTER 9

Precision Amplification

In the 1960s, speed limited oscilloscope performance as designers sought to extend bandwidth "from dc to daylight." But speed is not the only performance criterion. In audio, bandwidth that is much beyond human hearing degrades performance due to increased noise; the important measure of performance is *fidelity*, the precise reproduction of the input signal. In this looser sense, *precision* means ideal analog signal processing.

9.1 Causes of Degradation in Precision

For amplification, any effect beyond scaling of the input quantity adds error to the scaling function and degrades precision. The output quantity for amplification can be expressed as

$$X_{\text{out}} = \sum_{i=0}^{\infty} a_i X_{\text{in}}^i + X_{\text{noise}} \tag{9.1}$$

where, ideally,

$$X_{\text{out}} = kX_{\text{in}} = (a_1 + \varepsilon)X_{\text{in}} \tag{9.2}$$

and k is the exact scaling coefficient. Basic causes of error are scaling inaccuracy ε and nonlinear terms of X_{in} in (9.1), called *distortion*. Any contribution to X_{out} that is not caused by X_{in} is noise. Noise generated by the circuit itself is *intrinsic*; noise from other electrical activity (electromagnetic interference, or EMI) that interferes with circuit activity is *extrinsic*. A special case of error is the constant term a_0, called *offset error*, due to bias element inaccuracy, static thermal effects, or even noise.

Heat causes noise and distortion. Thermal effects due to the ambient temperature of the circuit environment (or *thermal drift*) cause offset error and affect dynamic circuit parameters. Changes in power dissipation with signal variation cause self-heating of elements whose dynamic parameters change with temperature and cause dynamic thermal effects, or *thermals*.

Bandwidth limitations also degrade precision by failing to scale all frequency components of X_{in} equally (due to magnitude roll-off) and by shifting them relative to each other in time (nonlinear group delay). This causes $X_{out}(t)$ to have a different waveshape from $X_{in}(t)$, and functional accuracy is degraded. A fundamental trade-off occurs between accuracy and bandwidth due to the finite gain–bandwidth product (f_T) of active devices. Greater accuracy requires more settling time, resulting in a lower effective bandwidth. This also applies to feedback amplifiers, in which accuracy is related to loop gain. As loop gain decreases with frequency, loop accuracy degrades. Therefore, larger bandwidth is sometimes necessary to achieve low-frequency accuracy.

9.2 Intrinsic Noise

Intrinsic noise is generated by the components of a circuit. We now examine three mechanisms that generate noise.

Noise is characterized in the frequency domain by its *spectral density*, or power spectrum. This is the Fourier transform of its autocorrelation function,

$$R_{XX}(\tau) = \int_{-\infty}^{+\infty} x(t) \cdot x(t+\tau)\, dt \tag{9.3}$$

For random functions, the greater the time separation of two points on the waveform, the more likely they are to be independent. When $\tau = 0$, the points coincide and $R_{XX}(0)$ is maximum. For continuous $x(t)$, the closer two points are chosen in time, the more likely they are related in amplitude and tend to reinforce, resulting in larger R_{XX}.

The rms noise voltage or current x_n is related to its noise spectral density $\tilde{x}_n$ by

$$x_n^2 = \int_{-\infty}^{+\infty} \tilde{x}^2(f)\, df \tag{9.4}$$

where the expected value of x_n (or the average x_n) is zero. For a limited frequency band of $\Delta f = f_h - f_l$, the limits of integration in (9.4) are the band limits.

Thermal noise arises due to the random motion of particles. Thermal energy is kinetic, and particles move in random paths as they collide. From statistical thermodynamics, the average kinetic energy in any one direction per particle is proportional to kT, the same kT as in the diode v–i relation, where

$$k = \text{Boltzmann's constant} = 1.38 \times 10^{-23}\ \text{J/K}$$

and T is absolute temperature in degrees Kelvin. At $T = 300$ K (27°C), $kT = 4.14 \times 10^{-21}$ J. This kind of noise occurs in electrical circuits because the thermal vibration of ions in a crystalline lattice causes them to collide with free electrons and exchange energy. This is manifested at the macro level as resistance. With no electric field applied, the lattice is at thermal equilibrium, and the average current is zero. But the instantaneous voltage fluctuates about zero and produces an rms voltage across the resistance of

$$\text{rms thermal } v_n = \sqrt{4kTR(\Delta f)} \tag{9.5}$$

where Δf is the frequency band in which the noise occurs. A resistance model that accounts for thermal noise is shown in Fig. 9.1a. Thermal noise is broadband and has a flat spectral density.

With an applied field, average current is due to the average motion of the electrons. Since current is the aggregate motion of many charged particles, it also fluctuates randomly, as does pressure in gaseous systems. This noise due to current is *shot noise.* Fluctuation in the instantaneous current has an rms value of

$$\text{rms shot } i_n = \sqrt{2qI(\Delta f)} \tag{9.6}$$

where $q = 1.60 \times 10^{-19}$ C (the charge of an electron), and I is the average current. Shot noise has a flat spectral density up to optical frequencies. A shot-noise current source is modeled in Fig. 9.1b.

Both thermal and shot noise vary by the square-root of the bandwidth. Manufacturers usually specify noise by its spectral density, or noise/$\sqrt{\text{frequency}}$, in units of V/$\sqrt{\text{Hz}}$ or A/$\sqrt{\text{Hz}}$. The bandwidth Δf is that of an ideal bandpass filter, not an actual circuit. The *noise equivalent bandwidth* Δf is consequently different from any actual bandwidth. For a single-pole bandwidth of f_{bw}, the noise equivalent bandwidth is

$$\int_0^\infty \frac{df}{1 + (f/f_{bw})^2} = \left(\frac{\pi}{2}\right) f_{bw} \cong 1.57 f_{bw}$$

where the integrand is the square of the single-pole transfer function.

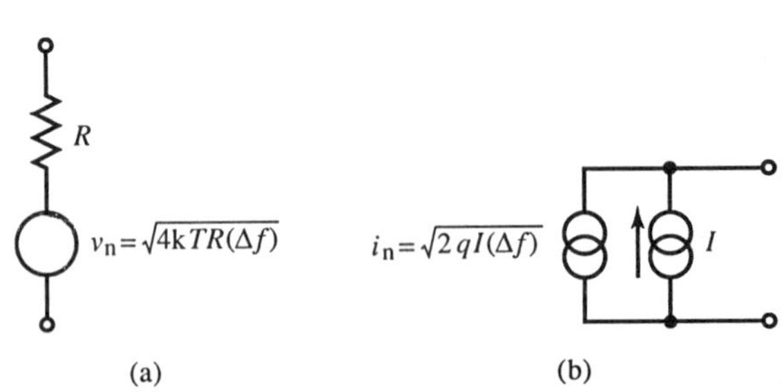

FIG. 9.1 Thermal noise model of R (a); shot noise model of I (b).

The total noise due to multiple sources is combined according to the rms summation formula

$$x_{\mathrm{rms}} = \sqrt{\sum_i x_{\mathrm{rms}}^2(i)} \tag{9.7}$$

where the $x_{\mathrm{rms}}(i)$ are independent (or more generally, uncorrelated) rms noises.

A BJT noise model has three noise sources (Fig. 9.2), with spectral noise densities:

$$\tilde{v}_{\mathrm{nb}} = \text{thermal noise of base resistance, } r_{\mathrm{b}}' = \sqrt{4kTr_{\mathrm{b}}'} \tag{9.8}$$

$$\tilde{i}_{\mathrm{nb}} = \text{shot noise due to base current, } I_{\mathrm{B}} = \sqrt{2qI_{\mathrm{B}}} \tag{9.9}$$

$$\tilde{i}_{\mathrm{nc}} = \text{shot noise due to collector current, } I_{\mathrm{C}} = \sqrt{2qI_{\mathrm{C}}} \tag{9.10}$$

The BJT model can be solved for output noise voltage due to the BJT alone, with a shorted input (so that $R_{\mathrm{S}} = 0$) and open output (so that $R_{\mathrm{L}} \to \infty$). We ignore C_μ. The BJT output noise voltage density $\tilde{v}_{\mathrm{no}}$ has three terms corresponding to the three noise sources just listed:

$$\tilde{v}_{\mathrm{no}}^2 = (\tilde{i}_{\mathrm{nc}} r_{\mathrm{o}})^2 + \underbrace{\left\{\frac{\mu^2}{[\omega r_{\mathrm{b}\pi} C_\pi]^2 + 1}\right\}}_{\substack{\uparrow \\ \text{noise voltage-} \\ \text{gain}^2}} \left[\left(\tilde{v}_{\mathrm{nb}}\left(\frac{r_\pi}{r_\pi + r_{\mathrm{b}}'}\right)\right)^2 + (\tilde{i}_{\mathrm{nb}} r_{\mathrm{b}\pi})^2\right] \tag{9.11}$$

where $r_{\mathrm{b}\pi} = r_\pi \parallel r_{\mathrm{b}}'$ and $\mu = r_{\mathrm{o}}/r_{\mathrm{m}}$. The noise-gain break frequency is at $1/r_{\mathrm{b}\pi}C_\pi$; the noise equivalent bandwidth is at $\omega \cong 1.57/r_{\mathrm{b}\pi}C_\pi$ or $f = 1/4r_{\mathrm{b}}'C_\pi$. The thermal noise voltage of r_{b}' is attenuated by the divider formed by r_{b}' and r_π in the base circuit.

In the BJT circuit, any external resistance in the input (base-emitter) loop, when referred to the base, is r_{S}. It includes r_{b}' and replaces it in (9.11).

Also, for finite load resistance R_{L}, collector resistance $r_{\mathrm{o}} \parallel R_{\mathrm{L}}$ replaces r_{o} in (9.11). To avoid calculation of $\tilde{i}_{\mathrm{nc}} r_{\mathrm{o}}$, the μ transform (Section 4.1) can be applied to refer this noise to the emitter:

$$\frac{\tilde{i}_{\mathrm{nc}} r_{\mathrm{o}}}{\mu + 1} = \tilde{i}_{\mathrm{nc}}(r_{\mathrm{m}} \parallel r_{\mathrm{o}}) \cong \tilde{i}_{\mathrm{nc}} r_{\mathrm{m}}\Big|_{\mu \gg 1} = \tilde{i}_{\mathrm{nc}} \cdot \frac{kT}{qI_{\mathrm{C}}}, \qquad \mu \gg 1 \tag{9.12}$$

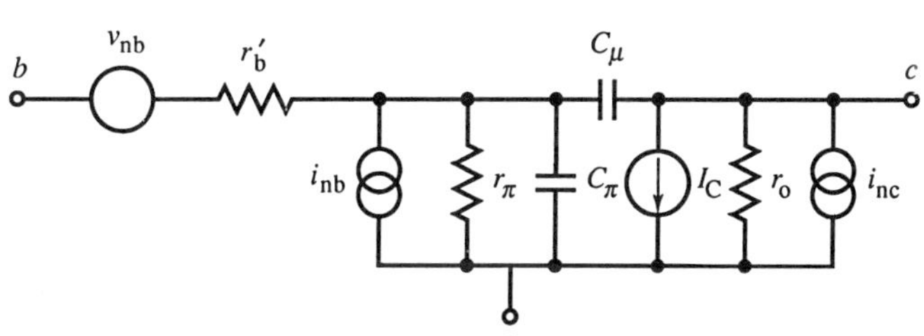

FIG. 9.2 BJT hybrid-π noise model.

Combining (9.10) and (9.12), we obtain the emitter shot noise voltage density:

$$\text{emitter shot } \tilde{v}_n = \sqrt{2qI_C}\left(\frac{kT}{qI_C}\right) = kT\sqrt{\frac{2}{qI_C}} \tag{9.13}$$

This noise voltage decreases with increasing emitter current until thermal noise in r_b' dominates. From (9.8) and (9.11), this noise is

$$\text{input thermal } \tilde{v}_n = \sqrt{4kTr_S}\left(\frac{r_\pi}{r_\pi + r_S}\right) \tag{9.14}$$

The voltage-divider factor can usually be omitted because it is desirable to keep r_S small to reduce thermal noise.

Finally, the third noise term is the shot noise of r_S. From (9.10) and (9.11), this is

$$\text{input shot } \tilde{v}_n = \sqrt{2qI_B}r_S \tag{9.15}$$

The total noise voltage density is the rms sum of (9.13)–(9.15).

Since $I_C = \beta_o I_B$, the noise terms of (9.13) and (9.15) both vary with I_C. An optimum I_C can be found by differentiating the total noise voltage density squared and solving for the current when set to zero. It is

$$\text{minimum noise } I_C = \sqrt{\beta_o}\left(\frac{kT}{qr_S}\right) \tag{9.16}$$

The higher r_m of FETs produces more noise than BJTs, especially at low frequencies (below 10 Hz), except when source resistance is high. Then base current causes dominant shot noise in the BJT source resistance.

Amplifiers have the noise model shown in Fig. 9.3. With a BJT input stage, v_n is due to collector shot noise in r_m and thermal noise in r_b'. The shot noise in r_S, which is external to the amplifier, is accounted for by $i_n \cdot r_S$. An optimum r_S contributes equal amounts of shot and thermal noise, or

$$\sqrt{4kTr_S} = \sqrt{2qI_I}r_S \Rightarrow \text{optimum } r_S = 2\left(\frac{kT}{qI_I}\right) \tag{9.17}$$

At a circuit temperature of 300 K (27°C), the thermal noise voltage density is

$$\text{thermal } \tilde{v}_n \text{ at 300 K} = (129\ \text{pV}/\sqrt{\text{Hz}\,\Omega})\sqrt{R} \tag{9.18}$$

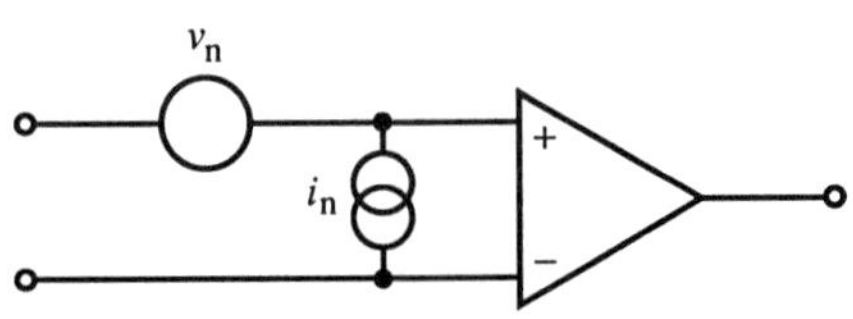

FIG. 9.3 Differential amplifier noise model

and the shot noise current density is

$$\text{shot } \tilde{i}_n \text{ at } 300\ \text{K} = (566 \times 10^{-12}\ \sqrt{\text{A}}/\sqrt{\text{Hz}})\sqrt{I} \tag{9.19}$$

Example 9.1 Op-amp Input Noise

An op-amp has an equivalent input noise voltage density of 20 nV/$\sqrt{\text{Hz}}$ and noise current density of 0.1 pA/$\sqrt{\text{Hz}}$. In the noninverting configuration, no resistors are required for a ×1 buffer; the total noise voltage is 20 nV/$\sqrt{\text{Hz}}$. A ×(−1) inverting op-amp configuration with 100 kΩ input and feedback resistors has a total equivalent input noise voltage at 300 K of

$$\sqrt{(20\ \text{nV}/\sqrt{\text{Hz}})^2 + 2(40.7\ \text{nV}/\sqrt{\text{Hz}})^2 + (5.00\ \text{nV}/\sqrt{\text{Hz}})^2} = 61.1\ \text{nV}/\sqrt{\text{Hz}}$$

or about 3 times (10 dB) as much noise. The input noise gain is that of the noninverting configuration, as it also is for offset voltage, because the noise is in series with the op-amp input terminals.

The op-amp bias current also generates shot noise in the resistors, which must be included if significant. For 1 nA of bias current, the shot noise current is 17.9 fA/$\sqrt{\text{Hz}}$ through 100 kΩ, producing a shot-noise voltage of 1.8 nV/$\sqrt{\text{Hz}}$. This is negligible compared with thermal noise voltage. If the bias current splits between the input and feedback resistors, the shot noise remains unchanged.

Figure 9.4 shows amplifier noise voltage plotted against r_S on a log-log plot. The amplifier noise sets a base independent of external resistance. Then as r_S increases, the thermal noise becomes significant at the thermal noise

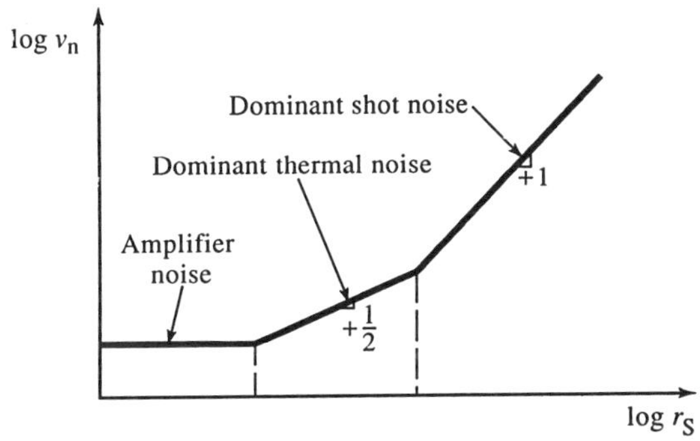

FIG. 9.4 Noise voltage versus source resistance.

corner, where the curve slopes upward with a slope of one-half. Then at the shot noise corner, shot-noise voltage dominates and is proportional to r_S; the slope is unity. The shot noise corner shifts with input current and can lie below the thermal noise corner.

Thermal and shot noise are wideband, or *white* (just as white light is wideband), and are associated with resistance. Another kind of noise, associated with semiconductor surface leakage and conductors in general, is $1/f$ noise, also called *flicker* noise, *excess* noise, or *pink* noise. It rolls off to a break frequency at the circuit white-noise level. This break frequency is typically 1–10 Hz for BJT noise voltage and 10 Hz to 1 kHz for FET noise current. FET $1/f$-noise break frequencies are typically 50 to 100 times higher, and that for CMOS is around 100 times higher.

The $1/f$ noise rolls off at a slope of $-\frac{1}{2}$ and breaks at a frequency where it intersects white noise. This break or corner frequency, f_f, specifies both voltage and current $1/f$ noise. For either voltage or current white-noise density $\tilde{x}_{nw}$, $1/f$ noise density $\tilde{x}_{nf}(f)$, is

$$\tilde{x}_{nf} = \tilde{x}_{nw}\sqrt{\frac{f_f}{f}} \tag{9.20}$$

From (9.4), the band-limited noise is

$$x_n^2 = \int_{f_l}^{f_h} \tilde{x}^2\, df = \tilde{x}_{nw}^2 \cdot f_f \cdot \ln\left(\frac{f_h}{f_l}\right) \tag{9.21}$$

The total noise is the rms sum of white and flicker noise, or

$$\tilde{x}_n = \sqrt{\tilde{x}_{nf}^2 + \tilde{x}_{nw}^2} = \tilde{x}_{nw}\sqrt{\frac{f_f}{f} + 1} \tag{9.22}$$

Manufacturer's data-sheet noise specifications are based on (9.22). Finally, x_n, the noise quantity, is found by substituting (9.22) into (9.4) and integrating. This results in

$$x_n = \tilde{x}_{nw}\sqrt{f_f \ln\left(\frac{f_h}{f_l}\right) + (f_h - f_l)} \tag{9.23}$$

For dc amplifiers, $f_l = 0$, but this yields infinite noise in (9.23). A practical lower limit is the thermal drift frequency, usually a fraction of a hertz. A typical f_l is 10 mHz. Below this, low-frequency noise is indistinguishable from drift.

Now that we have formulas for calculating rms noise, we sometimes are interested in what value the peak noise can achieve. The ratio of peak to rms values is the *crest factor*,

$$\text{crest factor} = \frac{x_m}{x_{rms}} \tag{9.24}$$

For gaussian noise, the probability that $|x|$ exceeds a given crest factor k_c is

$$p(|x| > k_c x_{rms}) = \text{erfc}\left(\frac{k_c}{\sqrt{2}}\right) = \sqrt{\frac{2}{\pi}} \int_{k_c}^{\infty} e^{-(|x|/\sqrt{2}\ x_{rms})^2} d\left(\frac{|x|}{x_{rms}}\right) \qquad (9.25)$$

For $k_c = 1$, $p = 32\%$; $k_c = 2$, $p = 4.6\%$; and for $k_c = 6.6$, $p = 0.1\%$.

A peculiar kind of noise is *burst* or *popcorn noise.* It is caused by process-dependent wafer surface effects and is manifested as random rectangular pulses or shifts in dc level, typically below f_f, that add (algebraically) to the other noises.

9.3 Extrinsic Noise: Radiation and Crosstalk

Extrinsic noise is due to other electrical activity in the environment of the affected circuit. This noise is generally referred to as *electromagnetic interference* (EMI) and can be caused by electromagnetic radiation (far-field), crosstalk (near-field), or conduction. Crosstalk can be either magnetic or electric in origin because in the near field it is not coupled as an electromagnetic wave.

Radiated EMI is reduced by shielding the circuit by enclosing it with conductive material. The shield presents an impedance discontinuity to an incident wave since its characteristic impedance is much below that of a wave in free space,

$$Z_w = \frac{E}{H} = 377\ \Omega \text{ in free space} \qquad (9.26)$$

A 1 MHz wave (in air) impinging on a copper sheet has a transmitted electric field that is 2 millionths (−114 dB) that of the incident wave; the transmitted magnetic field strength is twice the incident field. The characteristic impedance of copper is 0.37 mΩ $\angle 45°$, or nearly zero.

The wave-induced current in the shield flows at the surface and falls off exponentially with penetration depth into the shield material. This depth is characterized by a length constant (similar to a time constant) called the penetration, or *skin depth* δ:

$$\delta = \sqrt{\frac{2}{\omega\mu\sigma}} \qquad (9.27)$$

where μ is permeability and σ is conductivity of the shield material. The surface resistance of the shield material is the same as the dc resistance of the material with a thickness equal to the skin depth. The surface resistance is $1/\sigma\delta$. This resistance is equal to the surface reactance. The surface impedance is analogous to the characteristic impedance of a transmission line and is

$$Z_s \cong \sqrt{\frac{j\omega\mu}{\sigma}} = \sqrt{\frac{\omega\mu}{\sigma}} \angle 45°, \qquad \sigma \gg \omega\varepsilon \qquad (9.28)$$

where ε is the permittivity (dielectric constant). For a good conductor such as copper, steel, or aluminum, the condition of (9.28) is easily satisfied. The attenuation due to reflection is then

$$R = 20\log\left(\frac{Z_W}{4Z_S}\right) = 20\log\left(\frac{Z_W}{4}\sqrt{\frac{\sigma}{\omega\mu}}\right), \qquad Z_W \gg Z_S \tag{9.29}$$

Reflection decreases with frequency and permeability but increases with conductivity.

For poor conductors, a significant amount of the wave is transmitted through the shield and is attenuated more by absorption than by reflection. Absorption is ohmic loss in the shield due to wave-induced eddy currents flowing in a resistive material. Absorption increases with shield thickness, frequency, permeability, and conductivity.

Shields cannot be completely closed surfaces because wires, circuit boards, and adjustment tools must pass through them. These openings are also entrances for interfering waves and act as waveguide apertures. The relevant criterion is that openings have maximum lengths (in any dimension) that are much less than the wavelength of interfering radiation. Seams along case openings and metal slots and holes act as slot antennas. For a maximum slot dimension of

$$d < \frac{\lambda}{2} \tag{9.30}$$

where λ is the wavelength, the slot acts as a dipole antenna and passes frequencies above the cutoff frequency,

$$f_c = \frac{c}{2d}, \qquad c = \text{speed of light} = 3.0\times10^8\ \text{m/s} = 30\ \text{cm/ns} \tag{9.31}$$

For attenuation of waves above frequency

$$f = \frac{c}{\lambda} \tag{9.32}$$

d must be much smaller than λ; $d \ll \lambda/100$ for 60 dB of attenuation at the highest frequency of interest. At the slot, the electric field is maximum at the center because eddy currents in the shield must go farthest around the slot from the center of its maximum length. This creates the largest voltage drop from center to center across the slot. A 0.5 m slot has a shielding effectiveness of only 5.65 dB at 300 MHz. EMI gaskets and EMI-tight enclosure construction techniques provide conductance continuity across slots, thereby maintaining shielding effectiveness.

The near field is the space less than $\lambda/2\pi$ from the radiation source. Wave impedance Z_W depends on the impedance of the source. Electric fields have high wave impedance (E is large in (9.26)), and common metal shields are conductive enough to reflect them effectively. Magnetic fields have low wave impedance. Consequently, the impedance mismatch with the shield is not as great, and reflection as a shielding mechanism is not adequate. Low-frequency magnetic shielding is largely absorptive and requires high-permeability shield material to divert the field.

Shield reflection of electric fields in the near field decreases linearly on a log–log plot with frequency and distance from the source, whereas magnetic field shielding effectiveness increases linearly with frequency and source distance. At low frequencies, electric fields are well shielded by high-conductivity shields, but magnetic fields are attenuated less as frequency decreases. Therefore, low-frequency magnetic fields, such as those from power-supply transformers, commonly cause the most trouble in shielding.

Near-field interference is often due to coupling between signal conductors in a cable or on a circuit board, as shown in Fig. 9.5. In (a), the mutual currents flow through parasitic capacitance between the two lines and cause the same polarity of voltage at each end of the line. The amount of capacitance increases with line length and decreases with spacing.

For inductive coupling, voltage across the secondary loop resistance is due to mutual inductance M between the loops. M depends on the amount of shared area of the loops and their proximity. Coupling between parallel

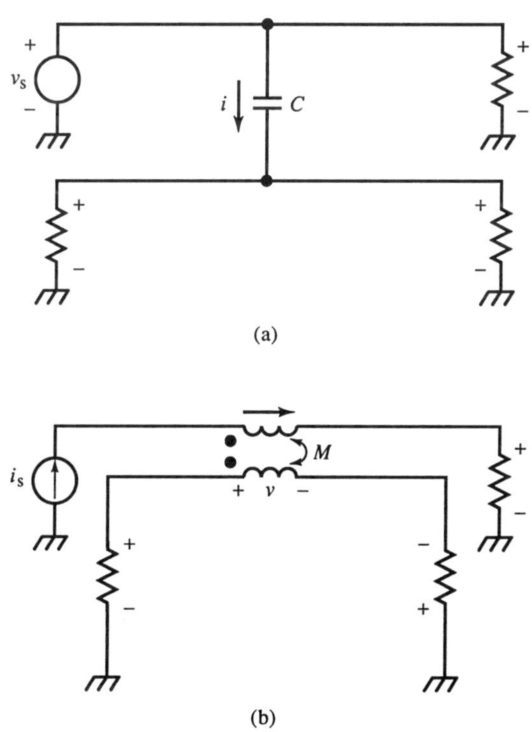

FIG. 9.5 Capacitive (a) and inductive (b) crosstalk between two conductors. The voltage is inverted at the two ends of the receiving loop for inductive crosstalk.

lines causes the source end of the second loop to have the same polarity of induced voltage as the primary loop, while at the load end it is inverted from that of the primary (Fig. 9.5b).

The magnitude of coupled noise depends on the rate of change of source quantities. For inductive coupling, 1 mA/ns induces 1 mV/nH. For capacitive coupling, 1 V/ns causes 1 mA/pF of current.

To predict the amount of crosstalk, estimation of M and C are required. Analytic solutions for crosstalk in Fig. 9.5 are unwieldy. We seek an intuitive ability to estimate and that requires simplified approximations. The length of the lines must be less than $\lambda/4$. At 100 MHz this is 75 cm. The spacing of the two coupled wires is expressed in distance between wires w and number of wire diameters d as w/d, and the separation of the signal lines for each loop from a ground plane is h. When separate return lines are used instead of ground plane, h is half the separation of signal and return lines (assumed the same for both loops). Except for very close spacing, where w/d is close to unity, increasing wire diameter does not appreciably increase C and affects M even less.

As length increases but remains under $\lambda/10$, both C and M increase linearly. For $w/d = 10$ and $h/d = 2$, $C/\text{length} \cong 1$ pF/m. For $h/w = 1$, $M/\text{length} \cong 15$ nH/m. C increases with h approximately linearly for $w/d > 10$ and $h < 100$. M is more sensitive to h than C because increasing h increases loop area. M increases sublinearly with h/w on a log–log plot. More significantly, both C and M decrease quadratically with w/d, or at -40 dB/dec of separation. (A 10 times change in w produces a 100 times change in M or C.) Also, for length less than $\lambda/10$, the amount of signal coupled by C or M increases linearly with frequency.

The impedance of free space, 377 Ω, sets the boundary between which kind of crosstalk dominates. High-impedance (>377 Ω) sources have dominant electric fields and dominant capacitive crosstalk. For low-impedance (<377 Ω) circuits, inductive crosstalk dominates.

Visual or geometric estimation of crosstalk when wiring a circuit or designing a circuit-board layout based on the preceding rules of thumb is often both adequate and at the practical limit of what can be reliably estimated. The following are general wiring and layout guidelines:

1. Maximize spacing between signal lines.

2. Minimize areas between signal lines and their return (ground) paths, running the two as close together as possible. When this is too difficult to do on a single layer, run a ground line alongside either or both of the source and receiving lines. This reduces the relative coupling about five times. Or, add a ground plane layer to the board. For long lines, use twisted pair cable. In flat cable, ground every other conductor. On an existing board, reduction of coupling can be experimentally verified by gluing a ground line of magnet wire between coupling traces.

3. Run signal lines perpendicular on opposite sides of the board. This reduces inductive coupling to intersecting areas formed by line pairs overlapping on opposite sides. Capacitive coupling area is reduced to the crosspoints of lines on opposite sides.

4. For high-speed circuits, confine signals to transmission lines or controlled-impedance environments.

Transmission lines can be as simple as twisted pairs or wires or coaxial cable. Twisted pair cable has a characteristic impedance of

$$Z_n \cong \frac{120\ \Omega}{\sqrt{\varepsilon_r}} \ln\left(\frac{h}{d}\right) \tag{9.33}$$

where h is the distance between conductor centers and d the conductor diameter. The length per twist, or *pitch*, does not affect Z_n, only the propagation delay time, because the line is longer with smaller pitch. For typical wire insulation, some values of dielectric constant are give in the following table.

dielectric material	ε_r
air	1.0
teflon	2.1
polyethylene	2.3
polystyrene	2.5
polyvinyl chloride (PVC)	3.5
epoxy resin	3.6
epoxy glass	4.7
Mylar	5.0
polyurethane	7.0

Typically, Z_n is between 50 and 100 Ω.

Also, wires can be run over a ground plane, with

$$\text{wire-over-ground } Z_n \cong \frac{60\ \Omega}{\sqrt{\varepsilon_r}} \ln\left(\frac{4h}{d}\right) \cong \frac{138\ \Omega}{\sqrt{\varepsilon_r}} \log\left(\frac{4h}{d}\right) \tag{9.34}$$

where ε_r is the relative permeability (dielectric constant) of the medium between wire and ground plane (usually circuit board or air), h is the distance between wire and ground plane, and d is wire diameter. For epoxy glass boards, ε_r typically is 4.7 (G-10 material), and for air it is 1. An order of magnitude increase in h/d results in an increase in Z_n of 138 Ω. A typical wire-over-ground line ($h/d \cong 1.3$) is about 100 Ω and has 3.54 nH/cm and 0.315 pF/cm.

On circuit boards, parallel-plate or *microstrip* transmission lines can be made of the board itself (Fig. 9.6a). With a ground-plane width much greater than the signal-line width w and for $h < \lambda/4$,

$$\text{microstrip } Z_n \cong \frac{87\ \Omega}{\sqrt{\varepsilon_r + 1.41}} \ln\left(\frac{(5.98)h}{0.8w + d}\right) \tag{9.35}$$

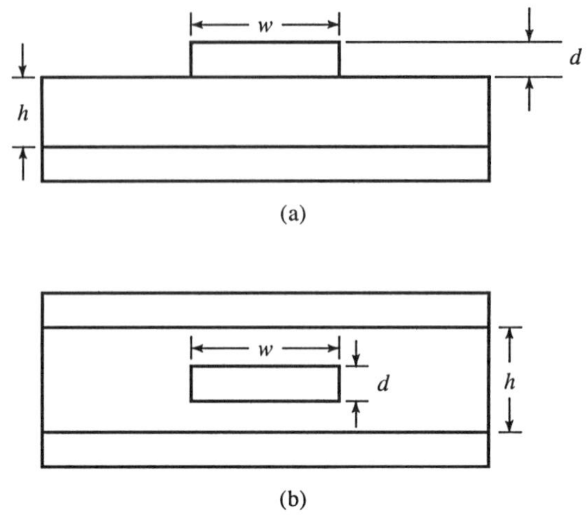

FIG. 9.6 Circuit-board transmission lines: microstrip (a) and stripline (b).

For 1-oz copper board traces, $d \cong 38.1\ \mu\text{m}$; 2-oz board trace thickness is 76.2 μm. A 50 Ω line on a 1/16-in.-thick board is 2.62 mm wide (or about 0.1 in.) using 1-oz copper.

Propagation delay time is

$$t_{pd} \cong 3.34\sqrt{(0.475)\varepsilon_r + 0.67}\ \text{ns/m} \tag{9.36}$$

Z_n of microstrip lines is typically about half that of wire-over-ground lines, or 50–100 Ω.

A symmetrical form of microstrip line, or *stripline*, can be made on a multilayer circuit board as shown in Fig. 9.6b, where the signal conductor is embedded between two ground planes. For stripline,

$$\text{stripline } Z_n \cong \frac{60\ \Omega}{\sqrt{\varepsilon_r}} \ln\left(\frac{4h}{(0.67\pi)w(0.8 + d/w)}\right) \tag{9.37}$$

and

$$t_{pd} \cong 3.34\sqrt{\varepsilon_r}\ \text{ns/m} \tag{9.38}$$

where $w/(h-d) < 0.35$ and $d/h < 0.25$. Striplines typically have the lowest Z_n, about half that of microstrip line.

A degenerate case of a microstrip line is parallel, flat conductors of width w and thickness d, separated by circuit board material of thickness h. For $w \gg h \gg d$,

$$Z_n \cong \frac{377\ \Omega}{\sqrt{\varepsilon_r}}\left(\frac{h}{w}\right) \tag{9.39}$$

Finally, if none of these lines can be implemented, traces run side-by-side with thickness d, width w, and edge-to-edge spacing h, where $w \gg d$, have

$$Z_n = \frac{120\ \Omega}{\sqrt{\varepsilon_r}} \ln\left(\frac{\pi h}{w+d}\right) \tag{9.40}$$

Flat (or ribbon) cables have parallel conductors that can be used as transmission lines. Typical flat-cable wire spacing is 0.050 in. with #28AWG stranded wire. One manufacturer gives the following specifications for such a cable: $Z_n = 105\ \Omega$, 41.3 pF/m, 558 nH/m, and $t_{pd} = 4.49$ ns/m. Flat cable with ground plane, 0.050 in. spacing, and #28AWG stranded wire has $Z_n = 65\ \Omega$, 82.0 pF/m, 558 nH/m, and $t_{pd} = 5.58$ ns/m.

9.4 Extrinsic Noise: Conductive Interference

The third cause of EMI is conductive interference. When two circuit loops share a common path, usually a ground path (Fig. 9.7a), any impedance in that path develops a voltage common to both loops. The most important

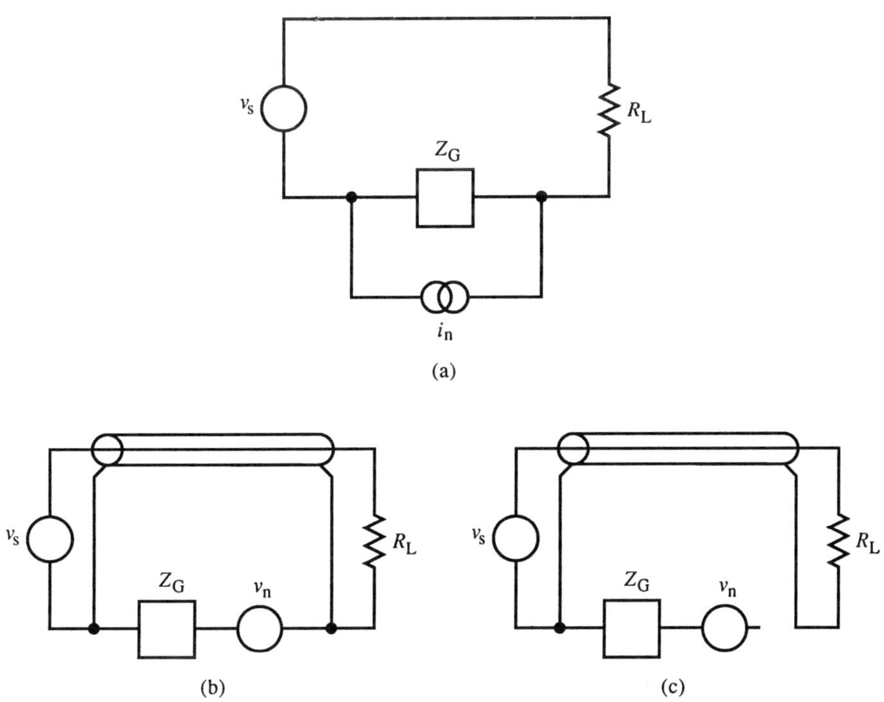

FIG. 9.7 Grounding a shielded conductor. Ground impedance Z_G causes noise voltage drop (a); noise current travels in shield (b); ground loop is eliminated (c).

general guideline for eliminating these *ground loops* is to consider the *complete* path of signals. The return path from the load back to the source is usually where noise gets into a signal loop. This common ground return node is often distributed throughout the subsystem. If it were electrically an ideal node, its impedance Z_G would be zero. To be ideal, it must also have zero length to have zero loop area and thus no magnetic crosstalk among loops. A ground plane approaches a zero-length node.

One of the simplest and most general techniques for preventing ground loops is the *single-point ground.* Separate the return lines for each circuit loop and run them back to their respective source grounds. Then, to connect the source grounds electrically, run separate lines to a single point where they connect, usually at the power supply ground. This technique minimizes external signal currents in a given loop ground return by isolating the signal currents to their own loops.

Signal return path isolation is combined with magnetic crosstalk isolation by the use of a shielded cable. Any of the transmission lines described previously can function as a shielded cable for conductive and crosstalk noise, though a constant Z_n along the cable is not required. A shielded cable provides a separate return path and minimizes loop area. For a coaxial cable, the theoretical loop area is zero, and the outer conductor (the shield) provides the return path, as in Fig. 9.7b. Here, external ground loop noise has been Thévenized. The low shield impedance forms a divider with $Z_G = R_G$ and noise current flows mainly in the shield. Because of the large mutual inductance between signal and return conductor in a shielded cable, ac signal current returns mainly in the shield. For coaxial cable, $M = L_S$, the shield inductance. Above the *shield cutoff frequency,*

$$\omega_c = \frac{R_S + R_G}{L_S} \tag{9.41}$$

signal current is in the shield; R_S is the shield resistance and L_S the shield inductance. The fraction of shield signal current falls off below ω_c to R_G/L_S, where it is then a constant, $R_G/(R_G + R_S)$. These results were obtained by solving the circuit of Fig. 9.7b and using asymptotic approximations to the frequency response of the shield signal current fraction. Cutoff frequency is typically a few kilohertz.

If we ground the shield at only one end, as in Fig. 9.7c, no external noise currents can flow in the shield. Also, no magnetically induced currents can flow in the shield because it does not form a closed loop with the external ground. This is a kind of single-point ground; the source ground terminal is connected to the external ground. For applications in which grounding must occur at both ends, a small resistor (1–10 Ω) can be placed from the load return side to the external ground. This forces most of the signal current into the shield return while achieving a relatively low-resistance path to the main ground line or plane.

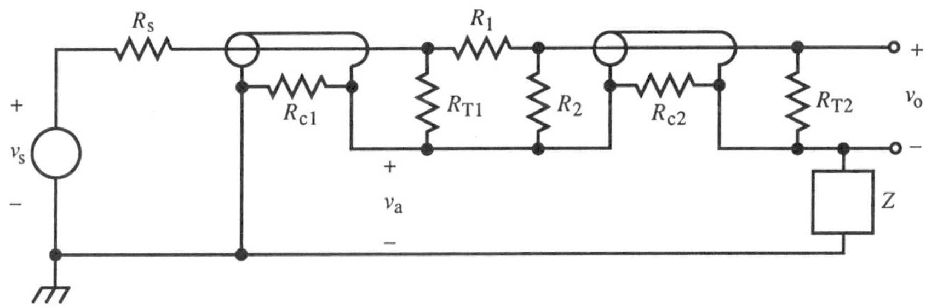

FIG. E9.2

Example 9.2 Shielded Cable Grounding with Attenuator

Shield grounding at both ends is required in equipment for safety. Instruments often require input attenuators that are connected external to the input. The circuit of Fig. E9.2 shows an attenuator, formed by resistors R_1 and R_2 inserted between two cables. "Barrel" attenuators with BNC connectors at each end are commonly used with 50 Ω cables that are terminated in their characteristic impedance by resistors R_{T1} and R_{T2}. The voltage v_a at the attenuator ground is the attenuated source voltage, and

$$\frac{v_a}{v_s} = \frac{R_{c1} \| (R_{c2} + Z)}{R_s + [R_{T1} \| (R_1 + R_2)] + R_{c1} \| (R_{c2} + Z)} \quad \text{(E1)}$$

The output voltage is taken across the load cable terminator R_{T2}. $R_2 + R_{T2}$ shunt R_{c2} and form the top side of a divider with Z. The voltage across the top side of this divider is v_o, and attenuation is

$$\frac{v_o}{v_a} = \frac{(R_2 + R_{T2}) \| R_{c2}}{(R_2 + R_{T2}) \| R_{c2} + Z} \cong \frac{R_{c2}}{R_{c2} + Z}, \qquad R_2 + R_{T2} \gg R_{c2} \quad \text{(E2)}$$

Now, if the load cable shield is grounded, $Z = 0$ in (E2), and v_a contributes to v_o unattenuated. With shield grounding only at the source end, Z is infinite, and attenuator error from shield resistance is zero. In effect, Z bootstraps v_o, and even small values of Z cause large improvements in attenuator accuracy.

A single ground at the source not only eliminates noise currents in the shield but preserves attenuator accuracy. Unfortunately, the chassis or earth ground of the instrument sensing v_o also must connect to the signal ground at its source to minimize its internal noise and provide a low-impedance safety ground fault path.

This problem was solved cleverly in the Hewlett-Packard model 3571A by letting Z be an inductor with a saturable core. A large power-line current through the inductor due to a grounding error would saturate it, reducing its impedance to near zero. Although attenuator error is present at dc, shield resistance is subject to the skin effect and increases with frequency, causing greatest error in the audio frequency range. In this range the inductor has enough impedance to reduce the error significantly (by 30 dB in the HP instrument). At higher frequencies, the mutual inductance of the cables decreases signal flow through Z, in effect making Z large.

All these configurations are susceptible to capacitive coupling into the shield. However, at the low impedances of ground lines ($\ll 377\ \Omega$), magnetic coupling dominates.

Shielded cables are also used in high external field environments to actually shield inner conductor(s). The previously described EMI reduction techniques used a shielded cable to eliminate ground loops and inductive crosstalk, not field radiation. Shielded cable, as a radiation shield, is used to provide a continuous shielded surface between a shielded source and shielded load. Within this closed shield is the signal loop. The return path is a separate conductor within the shield of a two-conductor shielded cable. The advantages of connecting the cable shield to only the source shield apply here, but electrical isolation from the shield enclosing the load leaves an opening possibly accessible to radiation.

External fields cause noise currents in and voltages on the shield. These quantities can couple noise into the signal lines within. For sensitive applications, a second shield is placed around the first to provide additional shielding. This shield can be connected to the rest of the shielding while the inner shield is connected to the source ground inside.

Shielding effectiveness depends on the construction of the enclosing outer conductor, or sheath. A solid sheath is far better than braided cable but is mechanically less flexible. The tighter the braid, the better the shielding. At 1 MHz, the shielding effectiveness of a solid sheath is about 200 dB better than a double-braided cable, which in turn is about 35 dB more effective than single-braided cable. The effectiveness of a solid sheath increases with frequency whereas braided cable remains constant to around 1 GHz and then falls off.

Shields are connected to the enclosed circuit ground at a single point at the source. If the shield is allowed to float, large dv/dt signals can capacitively couple into it. Additional stray capacitance couples from the shield into other signal nodes. In some cases, these noise paths through the shield are amplifier feedback paths that cause instability.

If the signal and return conductors are not magnetically coupled enough to isolate the signal path, their mutual inductance can be increased by placing the two conductors through a ferrite bead or by winding several turns of both conductors together (bifilar wound) around a ferrite toroid. The high permeability of the ferrite ($\cong$2000) forces the signal current to return on the other conductor. The signal current going to the load induces a voltage across the return line, via the flux linkage of the ferrite magnetic path, that causes an equal and opposite current flow in it. Differential-mode currents are passed, and common-mode currents are rejected.

The common-mode rejection of these ferrite transformers offers a second useful function, that of filtering. Common-mode noise currents flow in both conductors in the same direction. The high inductance due to the ferrite core forms a high-frequency filter with the load impedance for common-mode signals. The inductance for the signal currents is the transformer leakage inductance, usually 1% or less than that for common-mode currents.

Differential-mode transformers are the basis of EMI power-line filters. These filters are commonly used to keep power-line noise out of a system and keep system noise from getting onto the power line. A one-stage filter is shown in Fig. 9.8, with an X (for *across*) capacitor across the load for differential-mode filtering and two Y capacitors on the line side for common-mode line filtering. Another X capacitor on the line side improves differential-mode load-noise rejection. Attenuation (or *insertion loss*) increases with frequency to about 50 dB of rejection at 1 to 5 MHz. The break frequency increases with current rating, so low-frequency rejection is less for high-current filters.

The design objective of an EMI line filter is to pass frequencies at 50 to 60 Hz and reject frequencies at which noise is likely to be from 10 kHz to 100 MHz. The filter is modeled as a two-port network with source and load impedances. From the maximum power transfer theorem, the filter load port impedance should be equal to the load impedance at line frequencies and be much different at noise frequencies. This applies as well to the line port. In other words, at noise frequencies the EMI filter is a mismatching network.

The power line has a low impedance, so a high impedance input at noise frequencies is needed: a series inductor input. If the load impedance is high, load port mismatch requires a low-impedance output, which is achieved with

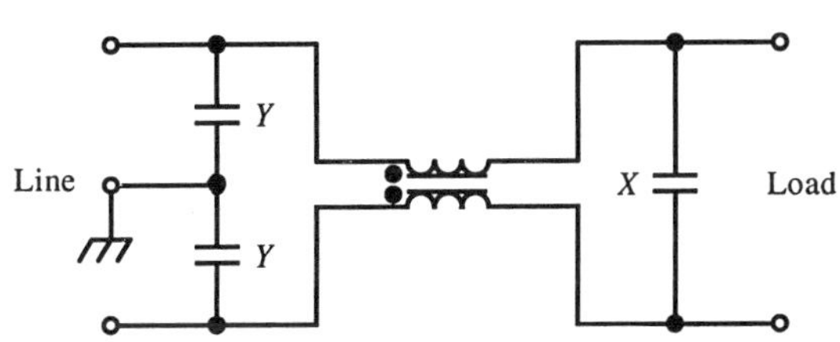

FIG. 9.8 Typical EMI power-line filter. The X capacitor filters differential-mode load noise; Y capacitors filter common-mode line noise.

a shunt capacitor. If load impedance is low, an additional series inductor is added. The X capacitor is large (0.5 μF) and shunts high-current differential-mode pulses from rectifiers and switching logic.

The mean power-line Z_n at 1 MHz is 50 Ω. For 80% of the lines, Z_n is between 10 and 300 Ω. Below 1 MHz, line impedance falls off at about −20 dB/dec to a very low value at line frequency. Differential-mode power-line noise cannot propagate far because of wiring and transformer inductance and does not radiate far because the opposing currents cancel. The dominant power-line noise is common-mode. The Y capacitors at the line input to the filter are intended to shunt common-mode noise to ground. Y capacitor size is limited for safety reasons to limit safety ground leakage current at line frequency. The transformer filters common-mode noise current from the line that the Y capacitors do not shunt.

Since ferrite beads or toroids provide provide the magnetic coupling path in filter transformers, the limits of magnetic materials applies to them. As frequency increases, signal losses in the magnetic circuit increase. As current increases to the point of magnetic saturation, permeability (and mutual inductance) decreases.

We now turn our attention to noise phenomena involving power distribution. Although the analytical models for circuits that we have been using assume that the power supply sources are ideal voltage sources, a power-supply regulator has a finite output impedance. Worse yet, the wiring required to distribute this power to the circuits is inductive and resistive. This impedance is part of the circuit to which power is delivered. The power distribution wiring is also a means for ground loops and crosstalk.

Inductive crosstalk due to ac power-supply currents can be reduced by reducing the area between the supply and ground lines. This is most easily accomplished by running them alongside each other on the same side of the circuit board or opposite each other on two sides. Circuit-board layout practice is to run lines perpendicular to each other on opposite sides of the board to ease interconnection and reduce side-to-side line crosstalk. Therefore, the first approach is usually preferred. Another approach, used commonly in digital circuits, is to run supply lines in parallel on one side, with regular spacing, connected together by a perpendicular line at the edge of the board. Then a similar pattern for ground is put on the other side and offset relative to the power lines.

A better, but more expensive, approach is to use commercially available laminated bus bars. These are flat bars with tabs at regular spacings for power and ground. They have a low, controlled impedance and are dominated by the distributed capacitance of the laminations. They not only reduce magnetic crosstalk but eliminate parasitic line inductance. If multilayer circuit boards are feasible, a similar distribution system can be realized by making the inner two layers ground and supply planes. This has the advantages of low-impedance conduction planes, a small, enclosed area between the supply and ground

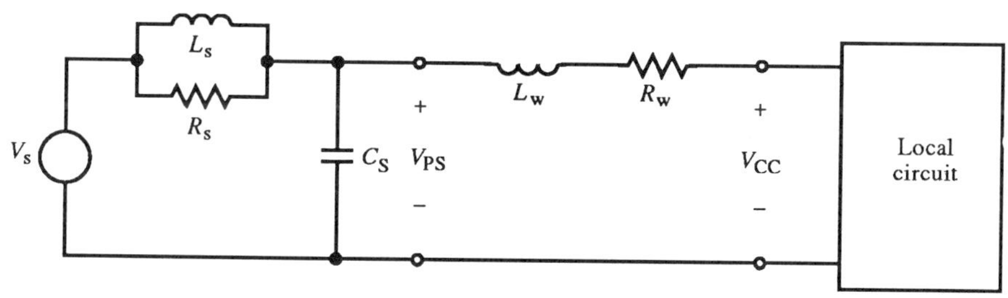

FIG. 9.9 A power supply distribution model.

currents, and the inner planes act as shields between the outer two signal layers of the board.

An ideal voltage source is approached locally at each circuit by reducing the Thévenin equivalent impedance of the supply terminals at the location of the circuit. This equivalent circuit is shown in Fig. 9.9. The power supply regulator has a characteristic shunt *RL* output impedance (see Example 7.5) shunted by an output capacitor C_S to reduce high-frequency impedance. The distribution-wiring impedance is in series with the supply impedance. We desired a low resistive equivalent impedance at the V_{CC} terminals.

Local circuit activity causes current changes on the supply lines that result in voltage changes at V_{CC}. A low-impedance path for ΔI is provided by placing a *bypass* capacitor locally across the supply. The capacitor must be a high-frequency type, typically a ceramic monolithic multilayer and usually not an electrolytic capacitor. A typical 4.7 μF aluminum electrolytic capacitor has a series resistance of about 1 Ω and a series resonance around 1 MHz. Most leaded (through-hole) ceramic capacitors have about 10 nH of inductance. Ceramic chip capacitors have about 5 nH.

The capacitor must be large enough to present a low-impedance source over the frequency range of ac current. At lower frequencies, high-frequency capacitors have a capacitive reactance that is too high, due to a practical limit on their size. The problem is solved by also shunting the supply with a large low-frequency capacitor. The shunt combination results in a wideband low ac impedance.

Both the inductance of the low-frequency capacitor and the noninductance of the high-frequency capacitor can cause trouble. The low-frequency capacitor inductance can resonate with the high-frequency capacitor, and the high-frequency capacitor can resonate with the line inductance. These resonant modes must be adequately damped to prevent low-level ringing on the supply line (Fig. 9.10). The amplitude of this damped sinusoid is less than if no bypassing were installed, but the circuit may not be able to reject it adequately. Amplifier power-supply rejection decreases with frequency. The characteristic impedance of the resonance determines the ring amplitude of the undamped supply in Fig. 9.10a. For a step of current from the circuit of *i*, the voltage

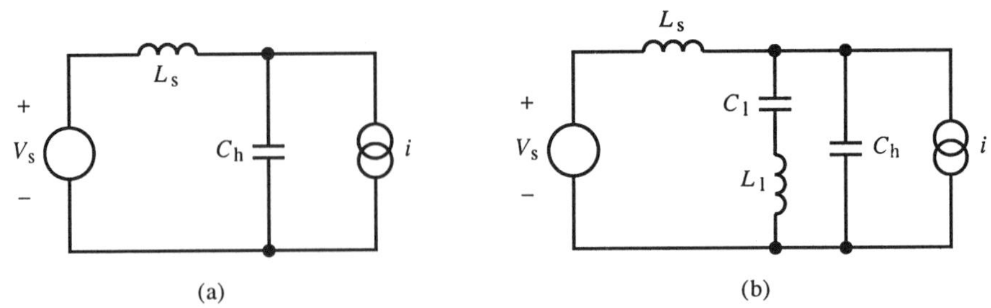

FIG. 9.10 Resonances from bypassing. C_h resonates (a) with supply inductance and (b) with parasitic inductance of C_l.

amplitude is

$$Z_n i = \sqrt{\frac{L_s}{C_h}}\, i$$

This resonance can be damped by inserting resistance in series with the supply line or by increasing C_h. In Fig. 9.10b, an electrolytic capacitor has been added. A second resonant mode is introduced by the parasitic inductance L_l, resonating with C_h. The series resistance of C_l (not shown) is sometimes large enough to damp this resonance; otherwise, an external resistor is added in series with it. The capacitor C_l alone, without C_h, could produce a voltage spike across L_l on the supply with an amplitude greater than if no bypassing were present.

Since several circuits are powered from the same supply, each having different supply performance requirements, the technique of *decoupling* is sometimes used as a kind of EMI filter. A noisy circuit, one with large, fast current changes, can be isolated from the supply by the decoupling circuit shown in Fig. 9.11. An inductor or resistor in series with the line to the circuit

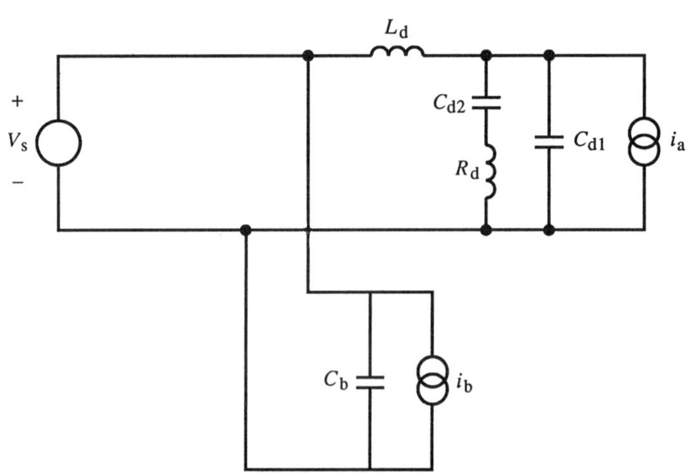

FIG. 9.11 A power-distribution decoupling model.

forms a low-pass filter with the capacitor shunting the circuit. It also filters the circuit noise, keeping it out of the supply. The decoupled circuit could be an output power driver whereas the other circuits are low-level amplifiers. An alternative is to decouple the low-level circuits from a noisy supply. This may be necessary if the series impedance of a decoupling network is too large to supply adequate current to a noisy, high-current circuit. Laudie Doubrava has shown that bypassing and decoupling provide distinct functions; bypassing provides a low-impedance supply source for a local circuit, and decoupling minimizes supply-coupled interactions among circuits.

Multiple signal paths occur in amplifiers and must be individually isolated. In Fig. 9.12a, a general amplifier with ground-referenced input v_i and load R_L is connected to a voltage source, $+V$, with supply line and ground resistances. This circuit has two errors. First, single-point grounding has not been followed. Consequently, the voltage drop across R_G due to amplifier current i_a and load current i_L adds to v_i as noise. Second, the noise voltage, $(i_a + i_L)(R_S + R_G)$, appears across the supply terminals of the amplifier because the terminals are not bypassed.

Figure 9.12b shows the corrected circuit. The reference terminal of v_i is returned to the input reference terminal, the negative supply terminal. Sometimes this cannot be done, and a differential amplifier, which has a separate negative input terminal from the supply, is required to avoid ground noise at the input. The diff-amp inverting input terminal would then be connected to the negative terminal of v_i. The second error is corrected by bypassing the amplifier supply terminals. This has two effects. It keeps signal currents of both the amplifier and the load out of R_G by shunting them around the external

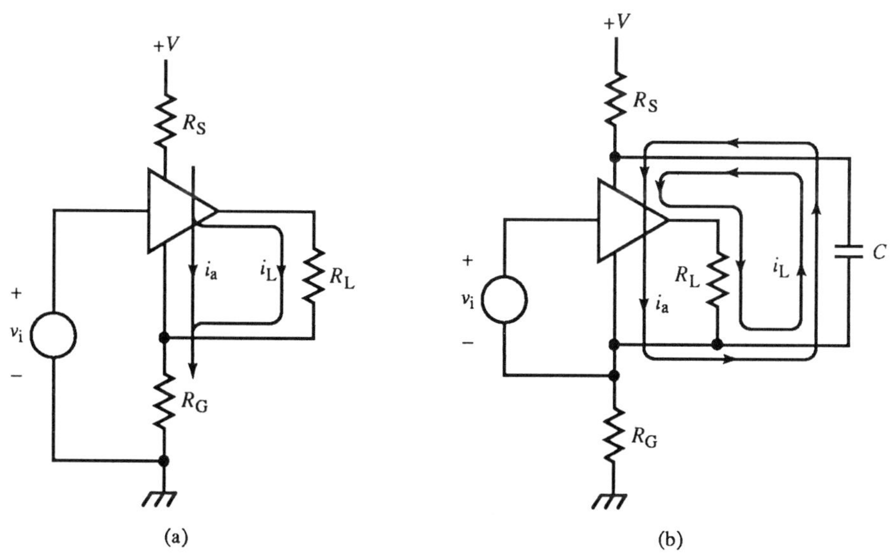

FIG. 9.12 Bypassing currents for an amplifier.

power distribution system and directly to the positive supply terminal. In doing so, it minimizes signal voltage changes across the supply terminals. When the negative terminal of v_i cannot be connected as in Fig. 9.12b, the amplifier negative-supply terminal can also be bypassed to the negative terminal of v_i. This shorts the noise voltage across R_G.

This general illustration of amplifier grounding can be applied to more specific cases. In all cases, it is important first to identify the (complete) current paths involved. Then it is possible to determine the effect of these currents due to parasitic impedances in the supply paths and how to reroute current and sensing loops using isolation, bypassing, and decoupling. In some cases, no solution is possible with the existing circuitry, and a different kind of amplifier is required for isolation of signal paths.

9.5 Differential Amplifiers

The first improvement over an amplifier with a common-ground input is a differential input amplifier, or *diff-amp.* Op-amps are an instance of diff-maps, but their gain is too large to be of use without feedback. We usually want an amplifier with a fixed gain and nonloading differential input. These finite-gain diff-amps can be made from op-amps. A one-op-amp diff-amp is shown in Fig. 9.13, where

$$v_i = v_{i+} - v_{i-} \tag{9.42}$$

and

$$v_o = A_{v+} v_{i+} + A_{v-} v_{i-} \tag{9.43}$$

or

$$v_o = \left(\frac{R_2}{R_1 + R_2}\right)\left(\frac{R_4 + R_3}{R_3}\right) v_{i+} + \left(-\frac{R_4}{R_3}\right) v_{i-} \tag{9.44}$$

The noninverting gain path has a voltage-divider preceding the op-amp noninverting gain (in the first term). The second term represents the inverting path.

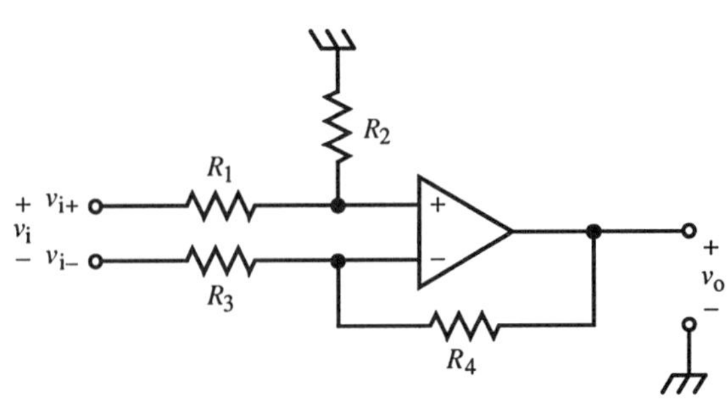

FIG. 9.13 One-op-amp differential amplifier.

For a differential amplifier, $A_{v+} = A_{v-}$. Equating gains and simplifying yields

$$\frac{R_2}{R_1} = \frac{R_4}{R_3} = A_v = \frac{v_O}{v_{i+} - v_{i-}} \tag{9.45}$$

Since its inputs are resistances, this diff-amp can operate with common-mode input voltages far larger than the op-amp linear input dynamic range. Its main disadvantage is the finite input resistance of the noninverting input. In precision applications, the resistance of the ground return and input source resistance are common causes of gain error.

Example 9.3 Voltage Supply Current Sensing

The one-op-amp diff-amp input common-mode range allows it to be used to sense the current of a high-voltage supply (Fig. E9.3). A single-supply op-amp operated from a 5 V converter is used to measure the current out of the 12 V battery that supplies the converter. The battery voltage must also be acquired by an A/D converter requiring 2 V input for a battery voltage of 12 V.

The 0.2 Ω sense resistor is insignificantly shunted by the two 100 kΩ input resistors to the op-amp. The battery voltage input to the A/D requires a divider attenuation of 6. The noninverting input to the op-amp can be used to supply this voltage. This requires that the bottom side of the divider be 20 kΩ.

For differential current sensing, the op-amp feedback resistor must also be 20 kΩ, according to (9.45). Consequently, the scale of V_O to battery current is 40 mV/A and 2 V corresponds to 50 A.

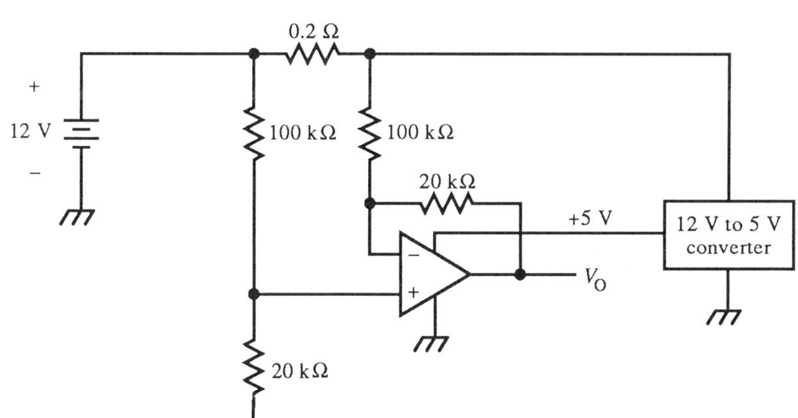

FIG. E9.3

The trade off for a wide common-mode sensing range is reduced gain, and even attenuation, from the one-op-amp diff-amp.

The speed of the one-op-amp diff-amp can be improved by adding a second op-amp within its loop, with a closed-loop gain of A'_v (Fig. 9.14a). The second op-amp increases the loop gain. For op-amps with the same open-loop frequency characteristics, the Bode plots of each and their combination are shown in Fig. 9.14b. Although the combined gain-bandwidth is no better, more loop gain is available at higher frequencies with the additional amplifier. Variations on this theme are shown in Fig. 9.15. [1] An additional amplifier can also be placed in the feedback path.

A conceptually simple two-op-amp diff-amp is shown in Fig. 9.16. Amplifier B is an inverting amplifier that sums inputs from v_{i-} and amplifier A, which inverts v_{i+}. The output is

$$v_o = \left(\frac{R_5}{R_3}\right)\left(\frac{R_2}{R_1}\right) v_{i+} - \left(\frac{R_5}{R_4}\right) v_{i-} \tag{9.46}$$

The condition for differential inputs is

$$\frac{R_2}{R_3} = \frac{R_1}{R_4} \tag{9.47}$$

This condition is realized in two ways:

$$\begin{cases} R_2 = kR_3 \\ R_1 = kR_4 \end{cases} \quad \text{or} \quad \begin{cases} R_1 = kR_2 \\ R_3 = kR_4 \end{cases} \tag{9.48}$$

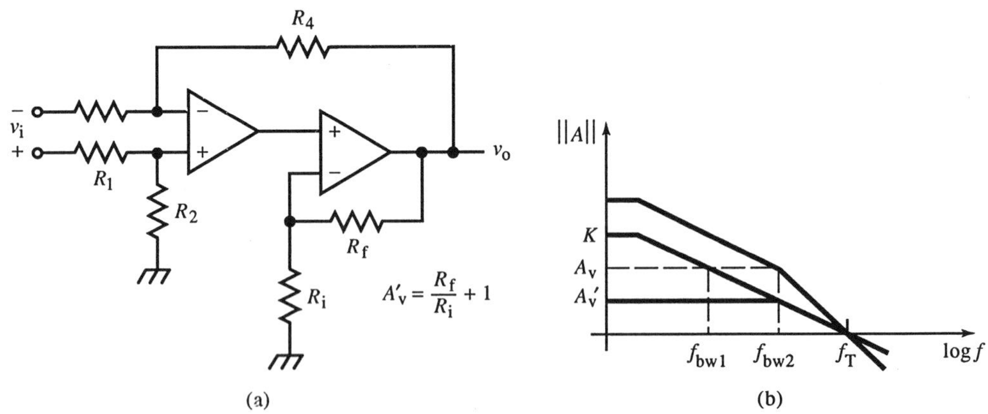

FIG. 9.14 Faster diff-amp, using another closed-loop amplifier in the forward path (a). Loop gain is increased at high frequencies (b).

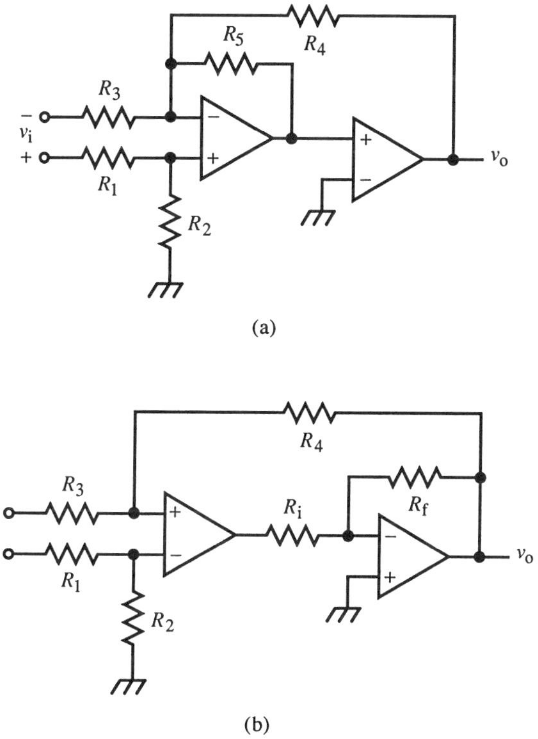

FIG. 9.15 One-op-amp diff-amps with an additional amplifier in the forward path.

The differential gain, where $v_i = v_{i+} - v_{i-}$, is

$$\frac{v_o}{v_i} = \left(\frac{R_5}{R_4}\right) \tag{9.49}$$

This amplifier has no advantage over the basic one-op-amp diff-amp because it has similar input impedance and involves two op-amp input error sources instead of one. But its noninverting form (Fig. 9.17) has the advantage of high (ideally infinite) input impedance. The v_{i-} input is amplified by an op-amp

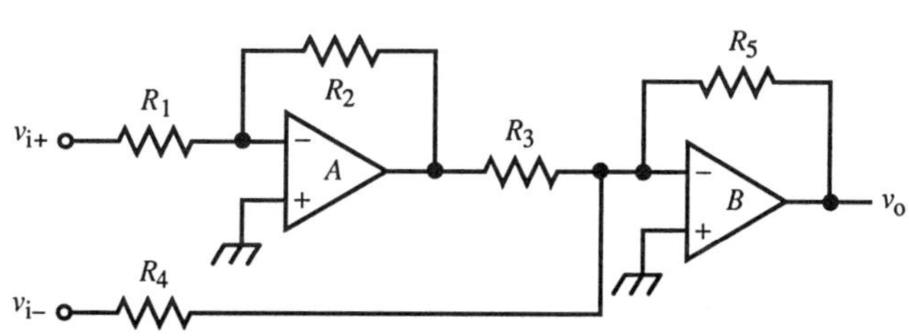

FIG. 9.16 Two-op-amp diff-amp with inverting op-amp summing.

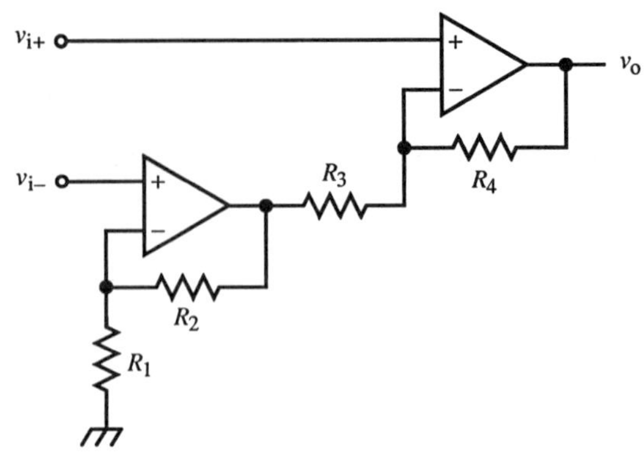

FIG. 9.17 Two-op-amp diff-amp with noninverting op-amp summing. Both inputs are high-impedance.

in the noninverting configuration and then is inverted by the output amplifier. The output voltage is

$$v_o = \left(\frac{R_4}{R_3} + 1\right) v_{i+} - \left(\frac{R_4}{R_3}\right)\left(\frac{R_2}{R_1} + 1\right) v_{i-} \tag{9.50}$$

and the differential-input condition is

$$\frac{R_4}{R_3} = \frac{R_1}{R_2} \tag{9.51}$$

This results in a voltage gain of

$$\frac{v_o}{v_i} = \frac{R_4}{R_3} + 1 = \frac{R_1}{R_2} + 1 \tag{9.52}$$

A useful variation on this diff-amp is to add R_G (Fig. 9.18). R_G complicates

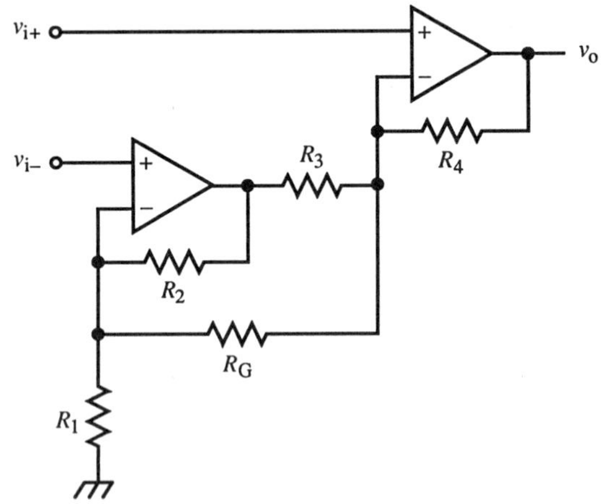

FIG. 9.18 Two-op-amp diff-amp of Fig. 9.17 with the addition of R_G, which can set the diff-amp gain.

the analysis somewhat, but for ideal op-amps, it is across the (virtual) input voltage and affects only the differential gain. By straightforward analysis,

$$v_o = \left[\left(\frac{R_4}{R_3 \parallel R_G} + 1\right) + \left(\frac{R_4}{R_3} \cdot \frac{R_2}{R_G}\right)\right] v_{i+} - \left[\frac{R_4}{R_G} + \frac{R_4}{R_3}\left(\frac{R_2}{R_1 \parallel R_G} + 1\right)\right] v_{i-} \quad (9.53)$$

Solving for the differential input condition, we find

$$\frac{R_4}{R_3} = \frac{R_1}{R_2} \quad (9.54)$$

and the gain is

$$\frac{v_o}{v_i} = \frac{R_1 + R_4}{R_G} + \frac{R_4}{R_3} + 1 \quad (9.55)$$

A *programmable-gain amplifier* (PGA) is an amplifier for which the gain can be set parametrically to given values. A PGA can be realized with this topology with a simple gain-setting strategy. Set the minimum gain and then program the additional gain with R_G. With R_G open, R_3 and R_4 set the minimum gain. With finite R_G and R_4 already determined, a fixed (nonprogrammable) R_1 allows R_G to change the gain in an additive manner.

The gain of (9.55) can be found in a simple way by superposition. When R_G is removed, the gain is that of the amplifier in Fig. 9.17, or (9.50). Since v_i appears across R_G, the current through it (i_G) flows through R_2 and R_4, causing the additional gain of

$$\left.\frac{v_o}{v_i}\right|_{R_G} = \underbrace{\frac{-R_2}{R_G}\left(-\frac{R_4}{R_3}\right)}_{\substack{i_G \text{ flows} \\ \text{through } R_2 \\ \text{out of } R_G}} + \underbrace{\frac{R_4}{R_G}}_{\substack{i_G \text{ flows} \\ \text{through } R_4 \\ \text{into } R_G}} \quad (9.56)$$

Then by superposition of gains,

$$\frac{v_o}{v_i} = \left.\frac{v_o}{v_i}\right|_{R_G \to \infty} + \left.\frac{v_o}{v_i}\right|_{R_G} \quad (9.57)$$

which is equal to the gain of (9.55).

The two-op-amp diff-amp of Fig. 9.19 has the second op-amp in the feedback loop, with v_{i+} inserted there. This odd topology can be analyzed by finding v_1. It is the voltage divider attenuation times the output of op-amp B, or

$$v_1 = \left(\frac{R_4}{R_3 + R_4}\right)\left[\left(\frac{R_2}{R_1} + 1\right) v_{i+} - \left(\frac{R_2}{R_1}\right) v_o\right] = v_{i-} \quad (9.58)$$

Since op-amp A input error is nulled, $v_1 = v_{i-}$. Solving for v_o, we get

$$v_o = \left(\frac{R_1}{R_2}\right)\left[\left(\frac{R_2}{R_1} + 1\right) v_{i+} - \left(\frac{R_3}{R_4} + 1\right) v_{i-}\right] \quad (9.59)$$

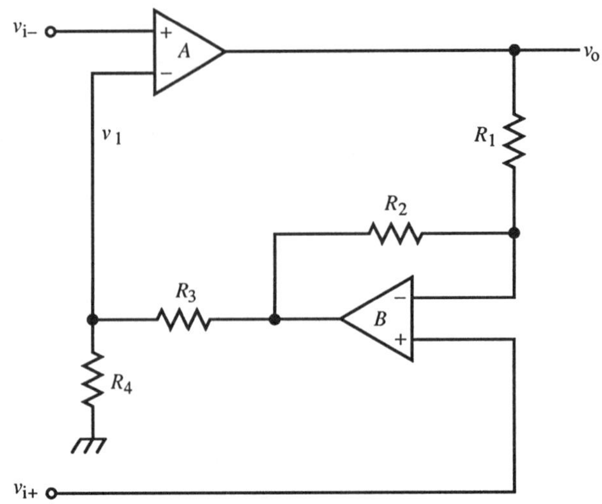

FIG. 9.19 Two-op-amp diff-amp with op-amp B and v_{i+} input in feedback path.

The differential-input condition is

$$\frac{R_2}{R_1} = \frac{R_3}{R_4} \tag{9.60}$$

The differential gain is

$$\frac{v_o}{v_i} = \frac{R_1}{R_2} + 1 \tag{9.61}$$

The amplifiers of Figs. 9.16–9.19 input v_{i+} and v_{i-} into different amplifiers. As a result, common-mode rejection (CMR) does not tend to be as good as for the single op-amp diff-amp. This is especially true as a function of frequency since the signals paths from the two inputs are asymmetric; one input goes through two amplifiers, the other through only one.

9.6 Instrumentation Amplifiers

An amplifier with programmable gain, high input resistance, and high CMR is an *instrumentation amplifier* (IA). The gain can be set to a given value, or programmed, by setting one resistor R_G. The most common instance is the three-op-amp diff-amp (Fig. 9.20). It has two stages. The input stage is a differential amplifier (at both input and output), and the second stage is the one-op-amp diff-amp. The differential input voltage v_i appears across R_G, creating a current flow of v_i/R_G in both feedback resistors R_f. The gain of the first stage is thus $2R_f/R_G + 1$. Since the amplifier is symmetric, the center-point of R_G is a virtual "ground" or null point, and each side behaves as though it

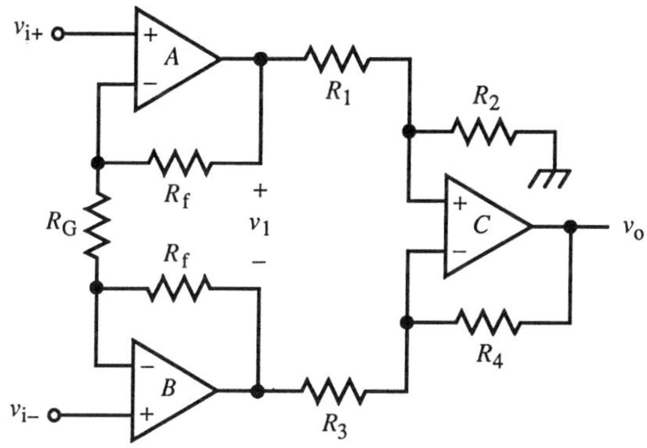

FIG. 9.20 Standard instrumentation amplifier topology; the three-op-amp diff-amp. Gain is set by R_G.

were connected to $R_G/2$ to the null point at the average input voltage. For the second stage to be differential, (9.45) applies, and the total amplifier gain is

$$\frac{v_o}{v_i} = \left(\frac{R_2}{R_1}\right)\left(2\frac{R_f}{R_G} + 1\right) \tag{9.62}$$

The gain of this IA is easily programmed by setting R_G, and it is amenable for use as a PGA by switching in different values of R_G.

This topology is commonly used as an input amplifier for remotely located transducers. Figure 9.21 shows a typical application, in which the transducer voltage v_i is connected to the IA inputs via two conductors with resistances

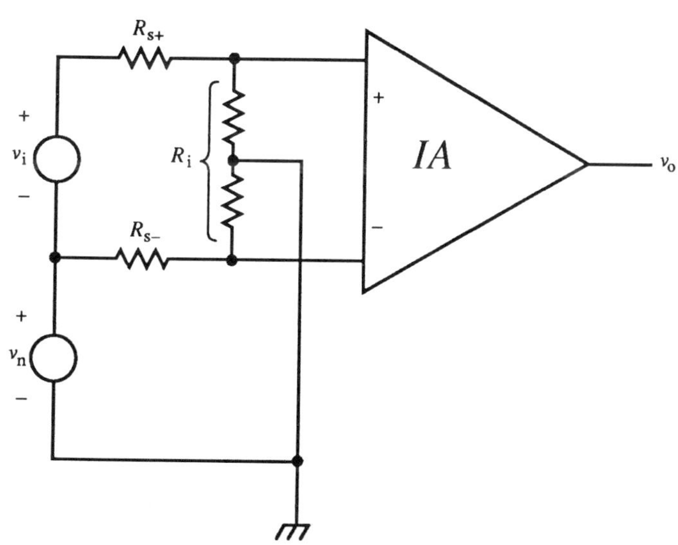

FIG. 9.21 IA input noise rejection with R_i. CMR requires balanced R_i.

R_{s+} and R_{s-} of generally different values. Common-mode noise v_n appears at the input, attenuated by different amounts due to R_i. A diff-amp with ideal CMR cannot reject differing amounts of noise at the two inputs. An IA has an ideally open input so that R_i is infinite and v_n unattenuated from the two conductors. In effect, v_n is retained as purely common-mode.

Since CMR is a major parameter of the IA, topologies with improved differential inputs have been developed. The use of two op-amps in the differential input stage involves two sets of differential inputs that must be matched for high CMR. A simpler realization of IA inputs uses a single differential transistor stage (Fig. 9.22). The differential input voltage is buffered by the BJT diff-amp and appears across R_G.

The linearity of this input stage is limited by the variation in dynamic emitter resistances with signal current. Furthermore, as gain is changed by changing R_G, the bandwidth changes since the amplifier has a fixed gain-bandwidth product. Both of these errors can be minimized by use of feedback (Fig. 9.23). The BJT collector voltages are buffered and fed back to the input BJT emitters through feedback resistors. (This amplifier is similar to that of Fig. 3.14.) These feedback buffers are more linear than emitter-followers and more linearly transfer v_i across R_G. The output v_o is applied to the input of the second stage.

Both stage and loop gains are affected by R_G. The feedback loop gain varies inversely with R_G since it is part of the feedback path attenuator and is within the loop. The stage gain v_o/v_i is $2R_f/R_G$ and also varies inversely with R_G. These gains track with changes in R_G. When R_G is decreased, both

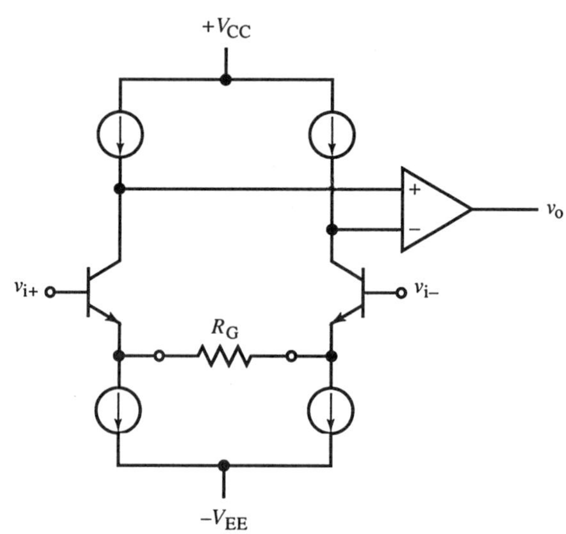

FIG. 9.22 IA input stage.

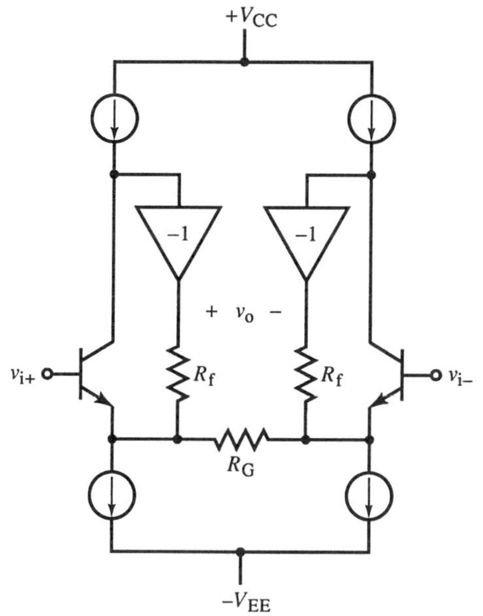

FIG. 9.23 IA input stage with linearizing and speed-enhancing local feedback.

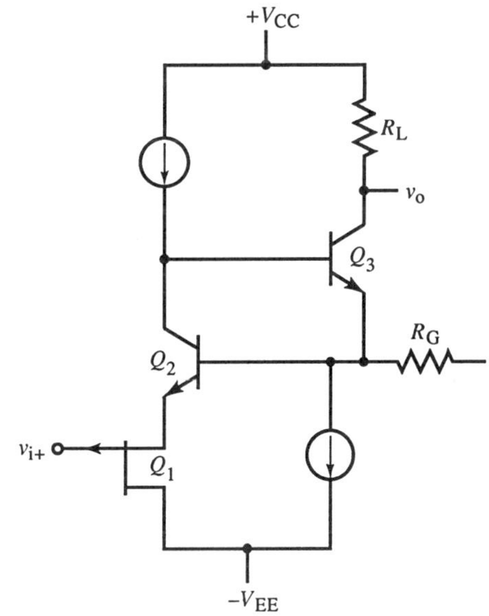

FIG. 9.24 Realization of the feedback buffer loops of Fig. 9.23.

open-loop and closed-loop gains increase the same amount. Consequently, the closed-loop dynamic response remains unchanged with gain changes.

In some IAs, the input transistors are FETs. The r_m variations of FETs is greater than r_e variations of BJTs, leading to less linearity and greater advantage in a feedback input stage. A clever variation on input-stage buffering is found in the PMI AMP-05. One side of the input stage is shown in Fig. 9.24. Input FET Q_1 is in series with BJT Q_2 forming a complementary differential pair. The output of Q_2 is buffered by Q_3 with no feedback resistance, fixing the closed-loop gain at unity. The stage output is taken from the load resistor of Q_3.

Guarding is a technique for improving dynamic CMR by bootstrapping the input cable shield of remote sources. The shield acts as a guard by not allowing input signal voltage variation across the cable capacitance. The situation can be generalized from Fig. 9.21 by letting R_i be differential input capacitance C_i. Any capacitive difference between the two sides results in asymmetrical *RC* filter circuits and different dynamic responses. Different responses cause a degradation in CMR with frequency. By minimizing C_i by guarding, we improve CMR. Guarding is used in the buffer amplifier of Fig. 9.25.

Guarding also bootstraps cable shunt resistance for high input-resistance (or low signal-current) applications, in which leakage currents are critical. Circuit-board surfaces provide stray leakage paths for current. The inputs to

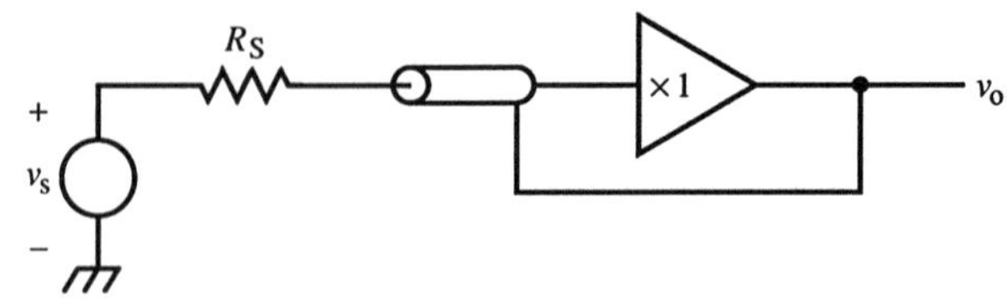

FIG. 9.25 A guarding technique. The buffer drives the shield, effectively reducing shield capacitance.

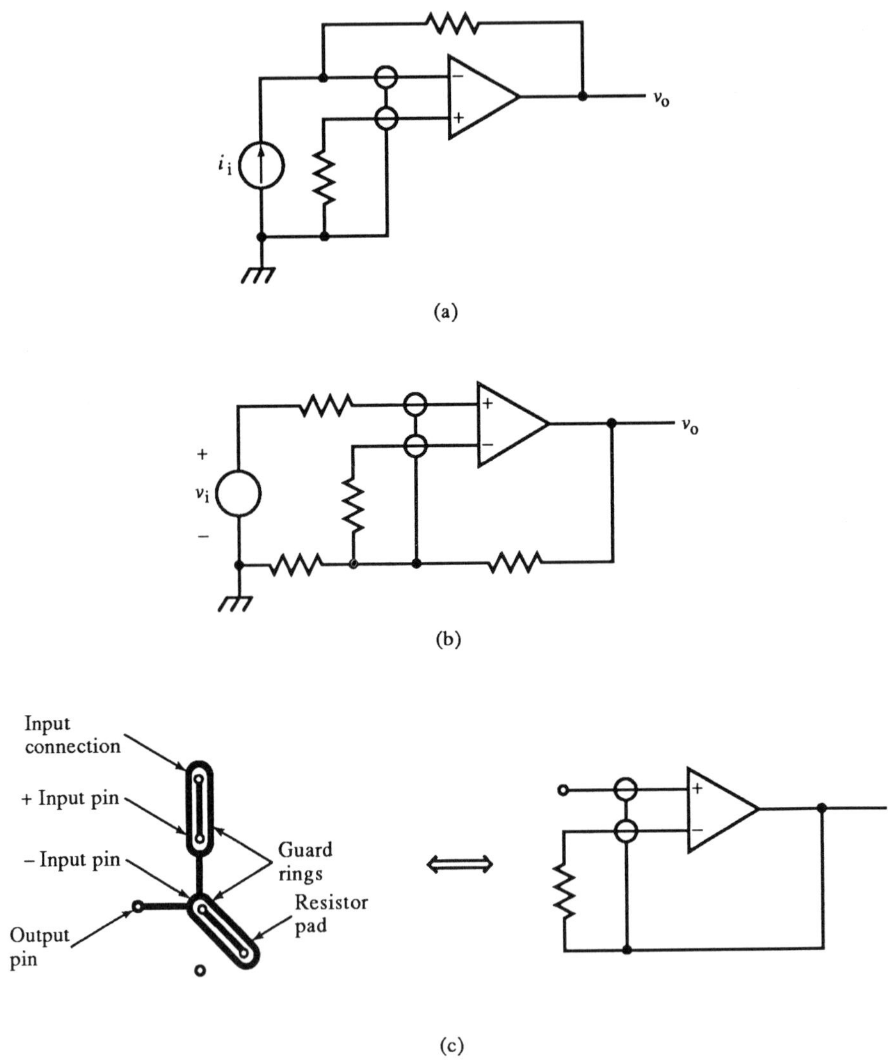

FIG. 9.26 Guarding applied to inverting op-amp (a) and noninverting op-amp (b). The circuit-board layout for guarding against leakage (c), showing part of 8-pin DIP socket and foil pattern for unity-gain buffer.

the op-amp circuits of Fig. 9.26 are guarded by enclosing the op-amp IC input pins, as shown in 9.26c for a noninverting ×1 buffer. The board surface is bootstrapped at the same voltage as the input pins by the output, which easily absorbs leakage into the guard rings.

Leakage through the board, or bulk leakage, is reduced by about ten times by placing guard rings on both sides of the board. For critical applications, the through-hole connection on the board is replaced by a polyethylene standoff with feedthrough so that no conductor contact of critical nodes is made with the board.

Guarding is commonly used with instrumentation amplifiers. Figure 9.27 shows three techniques. In (a), the first-stage output common-mode voltage is supplied by a balanced divider. A split-gain resistor also has the common-mode voltage at its center-tap, as in (b), where it is buffered. These two guard circuits reduce cable capacitance and leakage due to common-mode voltages. In (c), differential guarding bootstraps both common and differential-mode voltage variations.

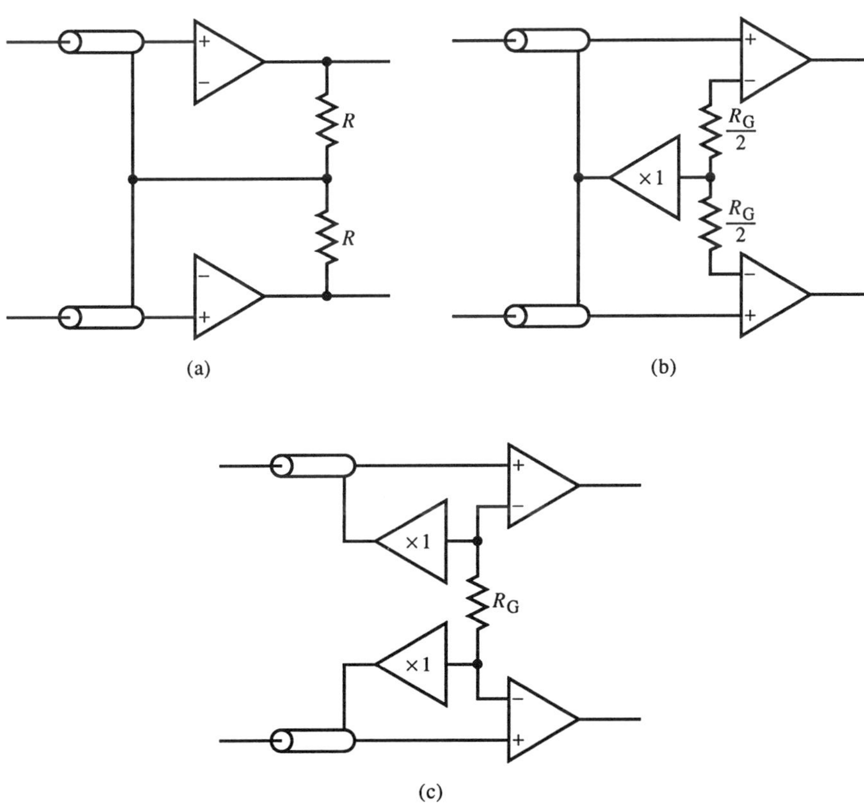

FIG. 9.27 IA guarding techniques. Shield is driven by common-mode output of first stage (a), by buffer amplifier from common-mode input off R_G (b), and by differential-mode input across R_G through buffers.

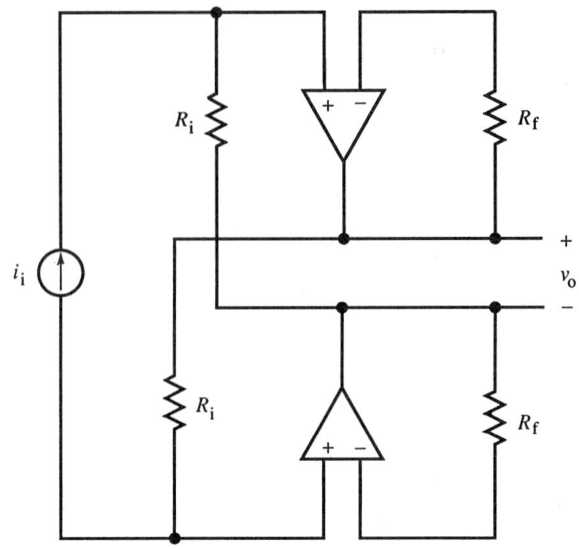

FIG. 9.28 Current-input IA topology with zero voltage compliance.

The high input impedance of the IA applies well to precision voltage amplification. The IA input stage can also be made into a transresistance amplifier for current amplification with the circuit of Fig. 9.28, a differential inverting op-amp topology. The transresistance is $2R_i$. The advantage of this circuit over that of an IA with a resistor across its input is that it has no input voltage drop ($R_G = 0$) and is well suited for current sources with limited compliance.

9.7 Low-Level Amplification and Component Characteristics

Very small voltage or current signals are *low-level* signals, small enough so that circuit-board and cable leakage, thermal gradients, and thermocouple effects are significant. At these levels, random noise and EMI considerations are important. Most low-level signal-processing circuits are limited in speed as a tradeoff for low-frequency precision.

High precision requires that circuits use components with low temperature coefficients (TCs) to minimize drift of the bias or quiescent operating point. For example, a change in op-amp bias current can cause a change in offset voltage and output voltage error. This error can be nulled by adjusting the offset, but to remain nulled, the bias current must not change.

High precision also requires minimization of low-level quasistatic thermal noise. Two sources of this kind of noise are *thermal gradients* and *thermoelectric effects*. A thermal gradient across the two sides of a differential amplifier causes junction temperatures to be asymmetric, resulting in offset and gain error. These gradients arise from convection currents of air across the circuit board or thermal conduction along an IC substrate. They are minimized for a discrete circuit by thermally shorting balanced components with a thermally conductive path between them. By mounting two transistors on the same metal heat sink, their static temperatures track because of the low thermal conductivity of metal. The metal acts as the thermal equivalent of a single electrical node since it is the same temperature everywhere (or *isothermal*). The thermal-electrical analogy is

$$\text{temperature } T \Leftrightarrow \text{voltage}$$

$$\text{thermal power } P_\theta \Leftrightarrow \text{current}$$

$$\text{thermal resistance } R_\theta \Leftrightarrow \text{resistance}$$

The thermal analog of Ohm's law is

$$T = P_\theta R_\theta \tag{9.63}$$

Analogs also exist between "thermal mass" (mass × specific heat) and capacitance, leading to dynamic thermal effects, to be studied later. A heat sink has $R_\theta \cong 0$, thus minimizing ΔT between differential transistor pairs. In some cases, a box is built around sensitive circuitry to act as a thermal convection shield.

For ICs, these techniques cannot be applied. Instead, balanced components are placed across symmetric thermal gradients. This requires thermal as well as electrical consideration of the IC layout. Alternatively, circuit elements are constructed so that balanced pairs receive the same thermal stimulus. For example, a two-transistor differential pair can be constructed of two pairs of transistors in a square pattern, with parallel devices situated diagonally (Fig. 9.29). For isotherms along the x axis, A–C and B–D are at the same temperature; isotherms along the y axis heat C–B and A–D the same. In each case, thermal symmetry is preserved, and the electrical effects are cancelled.

The effect of diagonal isotherms is minimized if the gradient is linear. A gradient with isotherms of slope −1 heats A and B to the same temperature, whereas each of C and D is hotter and colder than A–B. If the average

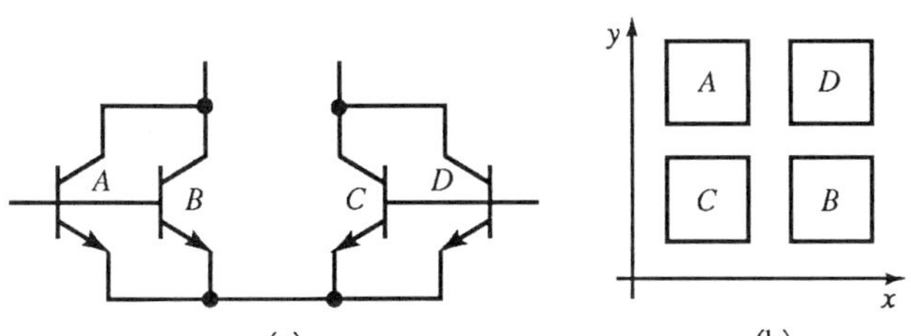

FIG. 9.29 A BJT diff-amp IC layout for minimizing thermal effects.

temperature of C and D is the temperature of $A-B$, thermal effects are cancelled to the extent that the thermal-to-electrical transfer function is linear. The pn junction relation shows that it is not for saturation current, $I_S(T)$. Therefore, close layout is desired to minimize temperature differences among transistors.

Thermoelectric voltage generation occurs when dissimilar metals are joined, as in thermocouples. Thermocouple voltages are generated at solder joints and connectors. IC leads are often made of the alloy kovar. A copper-kovar joint generates 40 μV/°C, whereas a copper-solder joint (with tin-lead solder) generates 1-3 μV/°C. A type K (chromel-alumel) thermocouple has a room-temperature TC of 39 μV/°C, for comparison. These thermocouple joints are compensated by symmetry, by having the same number of them on each side of a differential input, and by having them track with temperature.

Resistors affect performance according to their type:

resistors	carbon film	metal film	bobbin wirewound
accuracy	0.5%	0.1%	0.01%
stability	1%/kh	0.5%/kh	0.5%/kh
TC	500 ppm/°C	25-150 ppm/°C	20 ppm/°C
R range	1 Ω-150 MΩ	10 Ω-10 MΩ	0.05 Ω-6 MΩ

Resistor stability is the fractional amount the resistance changes over time, usually in units of 1000 hours. Wirewound resistors are constructed in various ways. Winding resistance wire on a bobbin results in the highest performance, as given in the table.

Variable resistors (potentiometers and rheostats) are also affected significantly by temperature and time.

pots	carbon	cermet	wirewound
stability	±10%/kh	±3%/kh	±2%/kh
TC, ppm/°C	400-800	100	50
R range	100 Ω-5 MΩ	10 Ω-2 MΩ	10 Ω-50 kΩ

Contact resistance variation (CRV) as the wiper is moved along the resistive element is the main cause of adjustment noise in a variable resistor. CRV is typically 1% maximum, but the application determines the extent of the effect. As a potentiometer with no wiper current, CRV does not affect circuit operation. For rheostats, in which all the current flows in the wiper, the effect is maximum. Generally, wirewound pots have the lowest CRV, followed by conductive plastic, molded carbon, and cermet.

Resistors have parasitic capacitance and inductance. The shunt terminal capacitance of a $\frac{1}{4}$ W carbon resistor is about 0.5 pF. Series inductance is more significant, especially in wirewound resistors. Noninductive winds of resistance wire (usually manganin) can reduce inductance. This is significant when power resistors are used to sense current in magnetic deflection amplifiers, current-mode switching power supplies, and motor drives.

Capacitors also affect circuit precision. Selection of the optimum capacitor type for a given application is a small specialty in itself, greatly aided by familiarity with manufacturer's specifications. A concise chart giving typical capacitor characteristics is appealing, but variations among values on existing charts suggest a more general approach. Some general facts about capacitors are helpful in design, with emphasis on plastic film capacitors.

Capacitor plates are either metal foil or a metallized deposition on the dielectric film itself. Foil capacitors are larger because the foil is thicker but have lower series resistance. They also cost less but can have higher TCs.

Figure 9.30 shows a general model of a capacitor. Capacitor C has series inductance L and series resistance R_s, called *equivalent series resistance* (ESR) on data sheets. R_s is a limiting parameter in power applications, involving large ripple currents, and is characterized directly; a 1 mF, 25 V aluminum electrolytic has a typical series resistance of 50 mΩ.

In small-signal applications, R_s is characterized by its *dissipation factor* (DF) the ratio of R_s/X_c, where X_c is capacitive reactance. Alternative characterizations abound. The arctangent of the ratio is the *loss angle*, and its reciprocal is the *quality factor*. The *power factor* is the sine of the loss angle. Dissipation factor is a function of both temperature and frequency.

Current leakage, both through the dielectric and across the body of the capacitor, is modeled by the parallel resistance R_p. The dielectric resistance is expressed by the geometric resistance formula, $R = \rho l/A$, where ρ is resistivity, l the length, and A the area. The length is the dielectric thickness, and

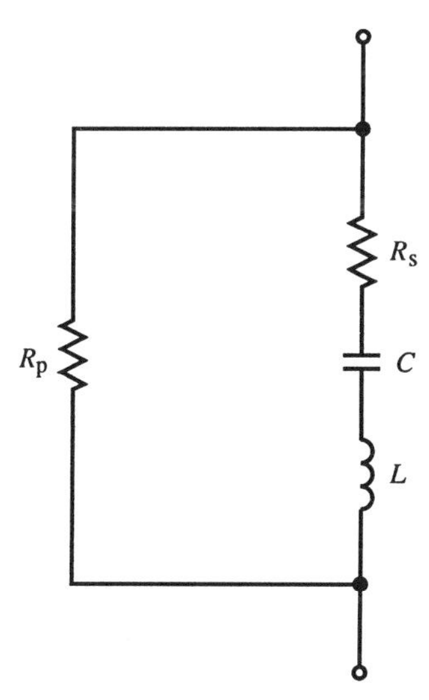

FIG. 9.30 Capacitor model with ESR R_s, leakage resistance R_p, and parasitic series inductance L.

A is plate area. Since the geometric capacitance formula is $C = \varepsilon l/A$, R_p is proportional to C and is usually given in units of ohm-farads (or megaohm-microfarads). Insulation resistance decreases with temperature for all film capacitors. Foil capacitors have about five times higher insulation resistance then metallized types. For low-leakage applications such as slow ramp generators or long-interval timers, the outer surface of the capacitor must be clean. A capacitor with fingerprints shows a noticeable increase of leakage, by orders of magnitude. Moisture also degrades insulation and creates current paths.

For bypassing, a minimum L is desired. Electrolytic capacitors have the largest L, with series-resonant frequencies typically between 100 kHz and 5 MHz. Other capacitors are usually not limited in their frequency by L but by dissipation factor. The best general high-frequency capacitors are ceramic.

Dielectric materials tend to become polarized by an electric field that remains after the field is removed. If a capacitor is shorted for a while so that its terminal voltage is zero and then the short is removed, the capacitor exhibits a nonzero voltage! This phenomenon is due to the partial polarization of the dielectric material into an electret, the electric equivalent of a permanent magnet. It is modeled in Fig. 9.31 by several series RC elements in parallel. This characteristic is quantified as *dielectric absorption*, measured as the fraction of the applied voltage that the capacitor exhibits after being shorted for a fixed time, usually 5 seconds. This effect is good for making transducers but causes anomalies in circuit response. For high-performance integrators or sample-and-hold circuits, dielectric absorption must be minimized.

Capacitors are made by winding alternate layers of foil and dielectric film, or layers of metallized film, together with opposing plates offset to opposite sides of the roll. In "noninductive" capacitors, the extra foil coming out of each end is smashed against the end and soldered to a lead. The plates are electrically paralleled; this reduces both L and R_s. In an "inductive" construction, the foils are brought together at the ends of the wrap. One lead is placed

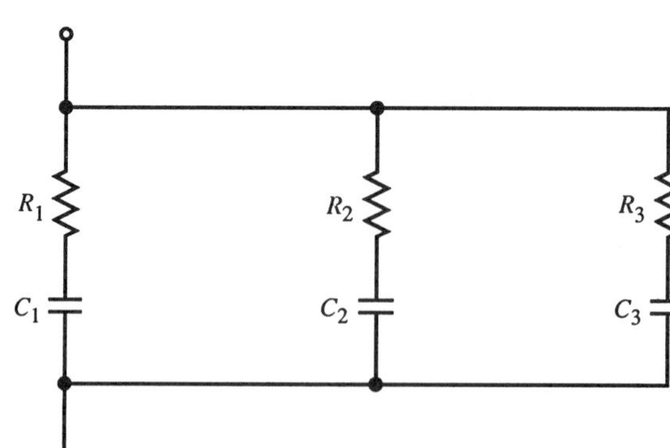

FIG. 9.31 Capacitor model of dielectric absorption.

in the center and the other at the perimeter. Both leads run the width of the capacitor. Since the length of the foil or film wind is much greater than its width, L and R_s are greater. But this method of construction is simpler and is the only way to build metallized film capacitors.

The plastic films used as dielectrics in capacitors number about a half dozen. The common name for polyester is Mylar. For general use, where high performance is not required, polyester is preferred because it is low-cost and has good volumetric efficiency (charge density). Capacitor size depends mainly on dielectric constant and dielectric strength, the maximum electric field before breakdown occurs. Polyester has a dielectric constant of 3.2. Its TC is bad at temperature extremes. It monotonically increases from $-5\%/°C$ at $-55°C$, with an inflection point at 25°C where the TC levels off. At its maximum temperature of 125°C, the TC is a huge $+14\%/°C$. But from 0° to 50°C, the TC is about 1%/°C, and then only at the extremes. The typical DF of polyester is 0.5%, dielectric absorption is 0.2%, and foil insulation resistance is the worst of the films at $10^5\ \Omega F$ at 25°C.

Polycarbonate capacitors are slightly larger than polyester capacitors with a similar shape of TC versus temperature, increasing with temperature and flat with zero TC at 25°C. The TC extremes are $-1.5\%/°C$ at $-55°C$ and $+1.5\%/°C$ at 125°C. Its dissipation factor is about half that of polyester, and foil insulation resistance is twice as high. Polyester and polycarbonate insulation resistance decreases superlinearly with temperature; the better capacitors decrease linearly. Dielectric absorption is typically 0.08%. Both decrease superlinearly with temperature. It is a medium-grade capacitor and a likely choice when polyester is not quite good enough. Its biggest weakness is its moisture sensitivity; it is the worst of the film capacitors. Although they are not abundant, polysulfone capacitors are similar to polycarbonates but have less moisture sensitivity and DF, have a flatter TC, and can operate up to 150°C.

The most commonly used high-performance film capacitors are polypropylene and polystyrene. Polypropylenes are comparable in price to polyesters, are only slightly larger, and have a maximum temperature of 105°C. Polystyrene is the better of the two, electrically, but costs more, is three times the size of polyester, and operates to only 85°C. For polypropylene, typical values are $DF \cong 0.02\%$ over temperature, foil insulation resistance $\cong 8 \times 10^5\ \Omega F$ at 25°C, dielectric absorption $= 0.02\%$ with negligible moisture sensitivity. The TC is a linear -250 ppm/°C. Polystyrene is similar but has half the DF and maintains insulation resistance at higher temperatures much better. Its TC is also linear but varies with material. The standard TC is -120 ppm/°C ± 50 ppm/°C. Polypropylene capacitors are a good choice for high-frequency power applications due to their low DF.

The common name for polytetrafluoroethylene (PTFE) is Teflon. It is the highest performance dielectric and highest in cost. PTFE is comparable to polystyrene in most properties and is twice the size of polyester. It has twice the foil insulation resistance of polystyrene, the best available. It has a linear negative TC of -200 ppm/°C and is, along with polystyrene, the most stable in capacitance over time, at 0.1%/year. The worst is metallized polyester, at 0.5%/year, with the others around 0.2%/year. DF is flat with temperature. It is the highest in operating temperature: 200°C.

One approach to zero TC is to make a hybrid dielectric of materials with opposite TCs that cancel. One such hybrid capacitor uses films of polyester and polypropylene with a TC of zero ±100 ppm/°C. It retains some of the worst properties of polyester, however: DF = 0.5% and dielectric absorption = 0.15%. Two new dielectrics, polyphenylene sulfide and polyvinylidene fluoride, can be used to make the most stable capacitors. Polyphenylene sulfide has a DF comparable to that of polypropylene.

In summary, and with some simplification, there are three performance classes of film capacitors: polyester is low-performance, polycarbonate is medium-performance, and polypropylene, polystyrene, and PTFE (in order of improvement) are high-performance. Except for frequency characteristics, they are better than ceramic capacitors except for "zero" TC NPO ceramics. Mica capacitors give high performance at a high price and accuracy, with a range of TC down to ±70 ppm/°C. The ultimate capacitor, except for size and cost, has a vacuum dielectric. Glass and air approach it in stability and low TC but have extremely low charge density.

Another important kind of capacitor is that made from glass epoxy circuit-board material. These capacitors are sometimes intentional but are usually parasitic. The most common board materials, G-10 and fire-retardant FR-4, have a dielectric constant of about 4.8 and a DF (at 1 MHz) of 0.02%, comparable to that of polypropylene. The volume resistivity of G-10 is 5×10^8 MΩ-cm, five times that of FR-4. The surface resistivity of G-10 is 4×10^8 MΩ, whereas for FR-4 it is an order of magnitude less. Both materials can be used up to 130°C. Low-cost low-performance phenolic boards have a dielectric constant of 4.1, a 1 MHz DF of 0.03%, volume resistivity of 5×10^6 MΩ-cm, surface resistivity of 5×10^4 MΩ, and a maximum temperature of 125°C.

Boards that have not been cleaned after assembly contain solder flux that can be a cause of excessive leakage. When even clean board leakage is excessive and guarding techniques inadequate, critical high-resistance nodes can be constructed with Teflon standoffs or mechanically stable parts connected in midair.

9.8 Isolation Amplifiers

Some applications require extreme isolation of the amplifier input from power supply and output. These include patient monitoring, floating measurements involving high-voltage circuits, digital voltmeters, and in environments with severe ground loops.

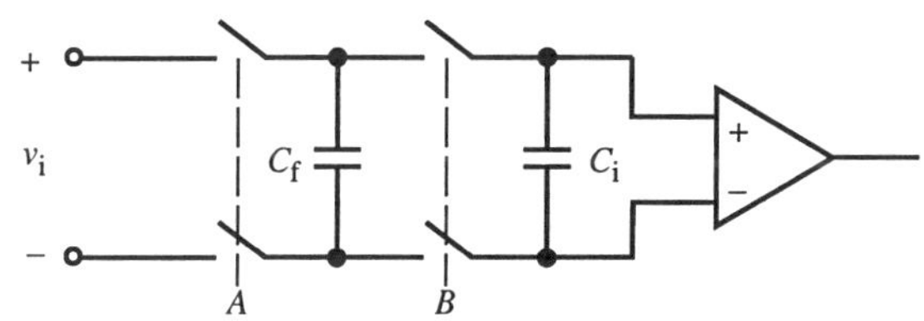

FIG. 9.32 Flying-capacitor input isolation technique.

A long-used isolation technique is *flying capacitor* isolation (Fig. 9.32). Switches activated by A are closed, charging capacitor C_f to the input voltage. The A switches are then opened, and the B switches are closed, charging the amplifier stray input capacitance C_i. For negligible loss of voltage, $C_f \gg C_i$. By using reed relays for the switches, we can achieve large maximum voltage ratings and low switch resistance. Limitations are finite switch-cycle life, contact bounce, contact noise, relatively large size and power requirements, and slow switching speed. A solid-state version overcomes these problems but suffers from lower voltage ratings; the LTC1043 "switched-capacitor building block" contains two flying-capacitor circuits and an oscillator to drive them, realized in monolithic silicon.

A more general approach is shown in Fig. 9.33. An input amplifier drives a kind of signal transmitter that drives a signal coupler. This coupler has no dc conductive (or *galvanic*) path between input and output sides and preserves an isolation barrier. Energy transmission is done by some means other than electrical conduction, such as optical coupling using optoisolators. Magnetic coupling is common using a transformer. Since dc amplification is desired, the signal transmitter in this case is a square-wave amplitude modulator and the receiver on the output side, a demodulator and filter. The form of transmission is *pulse amplitude modulation* (PAM).

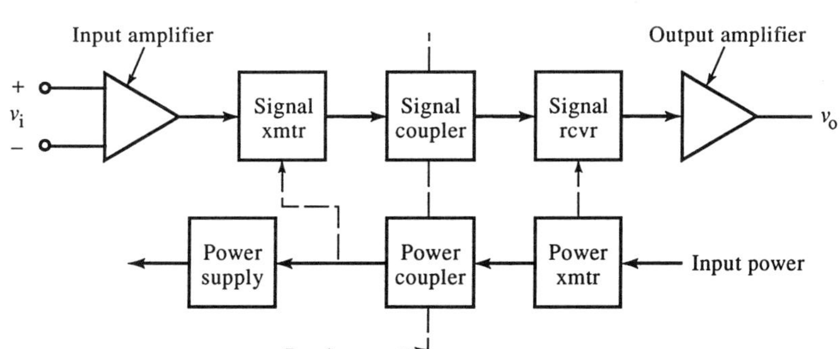

FIG. 9.33 Block diagram of an isolation amplifier. Input and output conductive paths are broken by couplers.

Because the circuits on the input side require power, a separate isolated power coupler is required with a power oscillator driving it and a power supply on the input side. If the power port is also isolated from both input and output sides and supplies isolated power to both, then the system has three isolated ports, the most general case of isolation. But a two-port isolation amplifier, with power and output ports unisolated, is often adequate.

The dotted paths in Fig. 9.33 indicate that the switched power waveform can be supplied to the modulator and demodulator for synchronous demodulation and a doubling of the signal-to-noise ratio.

Other forms of modulation can be used. Instead of PAM, voltage can be encoded in frequency with a voltage-to-frequency converter (VFC). Pulses coupled across the barrier are converted back to a voltage by a FVC, or their frequency is measured digitally. With pulse-width modulation (PWM), the voltage is encoded as a duty-ratio so that the coupled pulse frequency is not related to voltage accuracy. For any modulation technique, the signal bandwidth is limited by the carrier (or modulated) frequency. Optical coupling is dc, but LEDs and phototransistors are nonlinear, and their scaling is difficult to control. Consequently, matched optical paths are used. One is the input amplifier feedback loop, and the other is on the output side. Other dc-responding devices, such as Hall-effect devices, have similar difficulties and have not become commonplace for such coupling.

A common industrial instrumentation interface is the *20 mA current loop.* Signals are encoded as analog currents with 4 mA as zero scale (zs) and 20 mA as full scale (fs). Several channels can be remotely connected to the inputs of isolation amplifiers using a cable containing a twisted pair of wires for each channel. At the input end of the cable, transducers with 20 mA current-loop interfaces send back sensor signals with little voltage variation, causing minimal capacitive coupling between twisted pairs in a cable. The transducers are required to operate on a maximum of 4 mA. Then the two-wire interface supplies power as a voltage, and power supply current is measured as the transducer output signal.

When transformers are used to isolate circuits—not only in isolation amplifiers but in systems with EMI in general—parasitic capacitance between primary and secondary windings provides an electrical path for noise or ac leakage current and degrades isolation. Interwinding capacitance is minimized by placing the windings on opposite sides of a toroidal core or separating them on opposite ends of an E core, using a split bobbin. Capacitive imbalances arise from winding distribution, in which interwinding capacitance on one end of a winding is larger than on the other end. These differential imbalances cause differential outputs due to common-mode inputs. Interwinding capacitance can be minimized by placing a nonshorting turn of insulated copper sheet between concentric transformer windings as an electrical or *Faraday shield.* This shield is connected to the side generating the shield current, providing a return to the generating source.

9.9 Autocalibration

Precise signal processing is achieved by making the components precise. Since this is limited by technological capability, compensation methods are required to improve performance. Several techniques are based on a general idea, called *autocalibration*. Instead of trying to null circuit errors with more circuitry, circuit behavior is characterized using known inputs. The deviation of the resulting outputs from the ideal are measured and can be used to correct the unknown signal output.

A general autocalibration scheme is shown in Fig. 9.34a for an analog-to-digital converter (ADC). An analog multiplexer switches zero-scale (zs) and full-scale (fs) reference voltages into the ADC, and the output is sent to a microcomputer (μC) that is controlling the multiplexer sequencing. If the ADC is linear, a linear model is assumed for its transfer function, shown in (b). With two measurements based on known inputs, the slope and offset are computed. Then measurements of V_x are corrected according to these parameters. Autocalibration thus requires two modes, calibration and measurement.

A general autocalibration equation is based on two arbitrary reference voltages V_{R1} and V_{R2} and their measured values $\hat{V}_{R1}$ and $\hat{V}_{R2}$. Then measurement values $\hat{V}_X$ of the unknown input V_X are calculated from

$$V_X = \left(\frac{V_{R2} - V_{R1}}{\hat{V}_{R2} - \hat{V}_{R1}}\right)(\hat{V}_X - \hat{V}_{R1}) \tag{9.64}$$

Integrating ADCs often have an *autozero* feature that corrects for offset error in this way. Without a microcomputer, a clock oscillator drives switches that first ground (or null) the ADC input and store the error on a capacitor. This offset error is then switched in during measurement so that it subtracts from the input signal.

Autozeroing is used in the autobalanced CMOS comparator (Fig. 9.35). A CMOS inverter is connected in series with capacitor C and switches. When the A switches are closed on the first half-cycle of a clock, the inverter input and output are forced to be equal at threshold, V_{TH}. The input side of C is

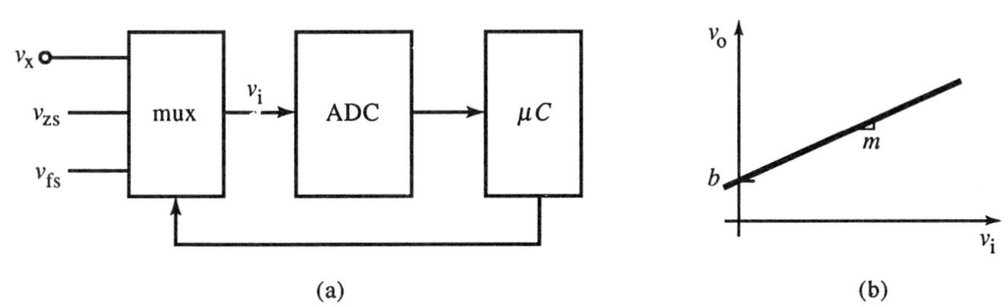

FIG. 9.34 Autocalibration technique for A/D input to microcomputer μC. The μC performs calibration by sequencing zero and full-scale inputs to the ADC, and calculating slope m and output offset b.

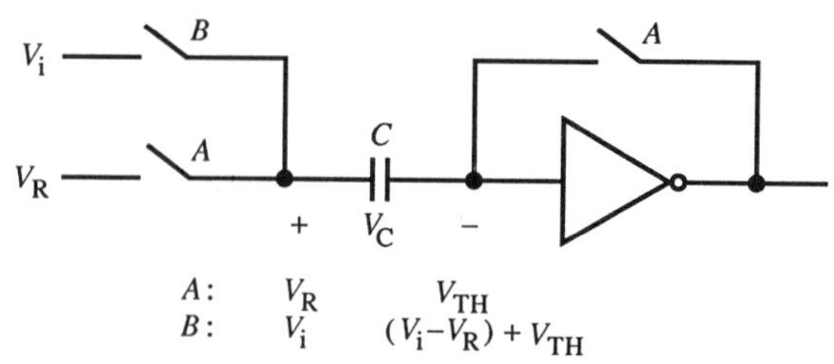

FIG. 9.35 Switched-capacitor comparator with CMOS inverter. The inverter threshold offset is common-mode rejected by differencing inputs across capacitor *C*.

at the comparator reference voltage V_R. On the second half-cycle of the clock, the *A* switches are opened, and *B* is closed, connecting the input voltage. The inverter input voltage is then

$$V_{in} = V_i + V_C = V_i - (V_R - V_{TH}) = V_i - V_R + V_{TH} \tag{9.65}$$

Because the inverter input is offset by V_{TH}, the difference between input and reference is also offset by V_{TH}, nulling the offset. The inverter output consequently responds to $V_i - V_R$ and performs the comparator function. A large number of these comparators have been used to implement parallel ADCs and other mixed analog and digital systems with digital IC processes. A similar input scheme is used in the LTC1040 dual micropower comparator.

Amplifier input offset voltage error can be reduced with autozeroing by nulling the amplifier inputs and charging a capacitor with the input error voltage. Then the signal is switched in, and offset error is nulled. A fault in this approach is that by switching between ground and signal, the switching waveform becomes part of the output. Also, the input signal is interrupted. Such amplifiers are called *chopping* or *chopper-stabilized* amplifiers. Multiple-path amplifier topologies have been devised to deal with these problems (see Chapter 10).

The chopping or switching of the input signal makes autocalibration techniques discontinuous, or discrete, in time. Systems with continuous amplitude and discrete time functions are called *sampled-data systems* and exhibit a range of behavior investigated in Chapter 12. The bandwidth of these systems is limited by the chopping or *sampling rate*; the input signal bandwidth must be much less than the sampling frequency. In practice, autocalibrating or switched-capacitor circuits have a limited bandwidth (10 Hz to 100 kHz) but excellent input offset voltages (1–10 μV). They are well suited as thermocouple and vacuum-system ion gage input circuits.

9.10 Distortion

Besides noise, another major cause of degradation in precision is distortion. Its two main causes are device nonlinearity and thermal effects.

Active devices are inherently nonlinear. Nonlinearity can be minimized by setting the operating point in the most linear region. Circuit techniques that minimize nonlinear behavior can be used, such as the minimization of BJT base-width modulation for the CE in a cascode stage. These attempts at linearization are often limited by other constraints. We now examine distortion due to nonlinearity and develop a simple way of estimating the amount of it.

Bruce Hofer has developed a technique for estimating *harmonic distortion* in amplifiers. He begins with the truncated series expansion of an amplifier output:

$$v_O = \sum_{k=0}^{3} a_k v_I^k = V_{os} + a_1 v_I + a_2 v_I^2 + a_3 v_I^3 \tag{9.66}$$

The technique is based on measurement of the amplifier gain at the center and positive and negative peaks of an input sinusoid. The incremental gain is

$$A = \frac{dv_O}{dv_I} = a_1 + 2a_2 v_I + 3a_3 v_I^2 \tag{9.67}$$

We now solve for a_k at $v_I = 0$, V, and $-V$. Substituting the three values, we obtain

$$A(0) = a_1 \tag{9.68a}$$

$$A(V) = a_1 + 2a_2 V + 3a_3 V^2 \tag{9.68b}$$

$$A(-V) = a_1 - 2a_2 V + 3a_3 V^2 \tag{9.68c}$$

Equations (9.68) are solved for a_k by adding (b) and (c) and then subtracting (c) from (b). The solutions are

$$a_2 = \frac{A(V) - A(-V)}{4V} \tag{9.69}$$

$$a_3 = \frac{A(V) + A(-V) - 2A(0)}{6V^2} \tag{9.70}$$

For a sinusoidal input of

$$v_I = V \sin \omega t \tag{9.71}$$

the output is

$$v_O = V_{os} + a_1 V \sin \omega t + a_2 V^2 \sin^2 \omega t + a_3 V^3 \sin^3 \omega t \tag{9.72}$$

Applying trigonometric identities,

$$\sin^2 \omega t = \tfrac{1}{2} - \tfrac{1}{2} \cos 2\omega t, \qquad \sin^3 \omega t = \tfrac{3}{4} \sin \omega t - \tfrac{1}{4} \sin 3\omega t \tag{9.73}$$

to (9.72), we have

$$v_O = \underbrace{(V_{os} + \tfrac{1}{2} a_2 V^2)}_{\text{dc term}} + (a_1 + \tfrac{3}{4} a_3 V) V \sin \omega t - \tfrac{1}{2} a_2 V^2 \cos 2\omega t - \tfrac{1}{4} a_3 V^3 \sin 3\omega t \tag{9.74}$$

The distortion due to the nth harmonic is defined as

$$nHD \equiv \frac{\|X_n\|}{\|X_1\|} = \frac{\text{amplitude of } n\text{th harmonic}}{\text{amplitude of fundamental}} \tag{9.75}$$

The second-harmonic distortion is

$$2HD = \left|\frac{\frac{1}{2}a_2 V^2}{a_1 V + \frac{3}{4}a_3 V^2}\right| \cong \frac{V}{2}\left|\frac{a_2}{a_1}\right| = \left|\frac{A(V) - A(-V)}{8A(0)}\right|, \qquad \tfrac{3}{4}|a_3| V \ll |a_1| \tag{9.76}$$

The condition of (9.76) is that the distortion be small. Similarly, for third-harmonic distortion,

$$3HD = \left|\frac{\frac{1}{4}a_3 V^3}{a_1 V + \frac{3}{4}a_3 V^2}\right| \cong \frac{V^2}{4}\left|\frac{a_3}{a_1}\right| = \left|\frac{A(V) + A(-V) - 2A(0)}{24A(0)}\right|, \qquad \tfrac{3}{4}|a_3| V \ll |a_1| \tag{9.77}$$

The total harmonic distortion can be estimated as the rms sum of the nHD. To estimate distortion for an amplitude other than V, say, $V' = kV$, we substitute into (9.67), resulting in

$$a_2' = k a_2 \qquad \text{and} \qquad a_3' = k^2 a_3$$

If the amplitude is doubled, 2HD doubles and 3HD quadruples.

Example 9.4 Estimation of CE Harmonic Distribution

The CE amplifier stage in Fig. E9.4 is analyzed for second and third harmonic distortion for an input sinusoid with $V = 10$ mV amplitude. The BJT is biased with 1 mA of emitter current.

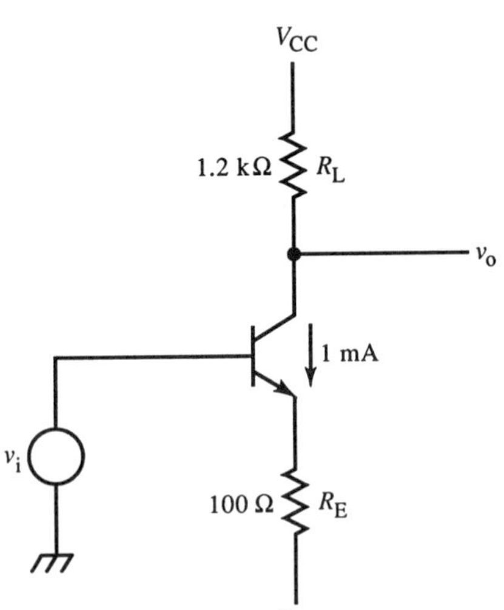

FIG. E9.4

The distortion is due to variation in r_e due to i_e. We first find the gain at the three input levels: -10 mV, 0 V, and $+10$ mV. The gain in general is

$$A = -\frac{R_L}{r_e + R_E}$$

Then

$$A(0) = -\frac{1.2\ \text{k}\Omega}{100\ \Omega + 26\ \Omega} = -9.52$$

At $V_I = 10$ mV, $\Delta I_E \cong 10\ \text{mV}/126\ \Omega \cong 80\ \mu\text{A}$. Then $r_e \cong 33\ \Omega$. The value of $A(V)$ is

$$A(10\ \text{mV}) = -\frac{1.2\ \text{k}\Omega}{100\ \Omega + 24\ \Omega} = -9.68$$

Similarly,

$$A(-10\ \text{mV}) = -\frac{1.2\ \text{k}\Omega}{100\ \Omega + 28\ \Omega} = -9.38$$

With the calculated gains, we now solve for the distortion values:

$$2HD \cong \left| \frac{(-9.68) - (-9.38)}{8(-9.52)} \right| = 0.39\%$$

$$3HD \cong \left| \frac{(-9.68) + (-9.38) - 2(-9.52)}{24(-9.52)} \right| = 0.086\%$$

If the input amplitude is increased to 25 mV,

$$2HD \cong (0.39\%)\left(\frac{25\ \text{mV}}{10\ \text{mV}}\right) = 0.99\%$$

and

$$3HD \cong (0.086\%)\left(\frac{25\ \text{mV}}{10\ \text{mV}}\right)^2 = 0.54\%$$

Example 9.5 Estimation of CC Harmonic Distortion

The CC of Fig. E9.5 has a sinusoidal input of 0.25 V in amplitude and is biased at an emitter current of 1 mA. The calculations of gain are similar to those of Example 9.4, with voltage gain,

$$A = \frac{R_E}{R_E + r_e}$$

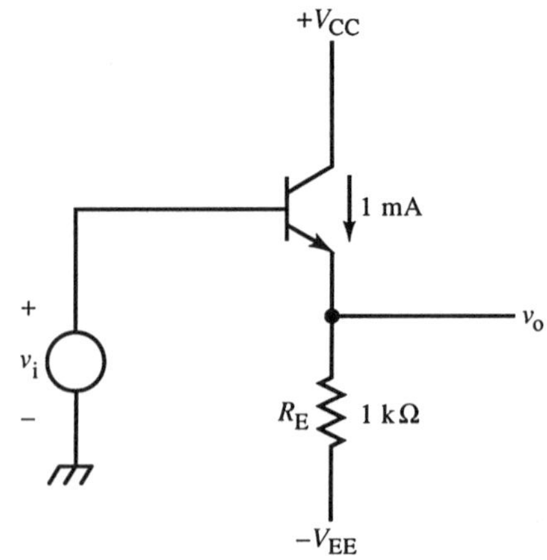

FIG. E9.5

The three gains are

$$A(0) = \frac{1\ \text{k}\Omega}{1\ \text{k}\Omega + 26\ \Omega} = 0.975$$

At $V_I = 0.25$ V, the emitter current changes by approximately

$$\frac{0.25\ \text{V}}{1.026\ \text{k}\Omega} = 0.244\ \text{mA}$$

Then $I_E \cong 1\ \text{mA} + 0.244\ \text{mA} = 1.24\ \text{mA}$, $r_e \cong 21\ \Omega$, and

$$A(0.25\ \text{V}) = \frac{1\ \text{k}\Omega}{1\ \text{k}\Omega + 21\ \Omega} = 0.980$$

And similarly,

$$A(-0.25\ \text{V}) = \frac{1\ \text{k}\Omega}{1\ \text{k}\Omega + 34\ \Omega} = 0.967$$

The distortion is

$$2HD \cong \left| \frac{(0.980) - (0.967)}{8(0.975)} \right| = 0.17\%$$

$$3HD \cong \left| \frac{(0.980) + (0.967) - 2(0.975)}{24(0.975)} \right| = 0.013\%$$

When analyzed by this technique, the two-transistor diff-amp is found to have no 2HD and dominant 3HD. In general, odd transfer functions, in which $v_O(v_I) = -v_O(-v_I)$, have only harmonics that are odd-integer multiples of the fundamental, or *odd harmonics*, whereas even functions, in which $v_O(v_I) = v_O(-v_I)$ have only even harmonics.

Harmonic distortion consists of frequencies that are harmonically related (that is, integer multiples of) the fundamental. Nonlinearity also produces interactions between frequency components of the input signal when it is not a sinusoid. This form of distortion is *intermodulation distortion* (IM). It produces sum and difference frequencies as a modulator or multiplier does.

In practical systems, the dominant linear term produces a scaled input, and the nonlinear terms, the distortion. Intermodulation distortion can be demonstrated by a system with only a quadratic term, so it produces only distortion terms in the output:

$$v_O = v_I^2 \tag{9.78}$$

Instead of a single sinusoid, v_I is set to be the sum of two sinusoids at different frequencies:

$$v_I = V_1 \sin \omega_1 t + V_2 \sin \omega_2 t \tag{9.79}$$

Substituting v_I into (9.78) gives the output:

$$v_O = \underbrace{\frac{V_1^2 + V_2^2}{2}}_{\text{dc offset error}} \underbrace{- \frac{V_1^2}{2} \cos 2\omega_1 t - \frac{V_2^2}{2} \cos 2\omega_2 t}_{\text{2HD}} \underbrace{+ V_1 V_2[\cos(\omega_1 - \omega_2)t - \cos(\omega_1 + \omega_1)t]}_{\text{IM}} \tag{9.80}$$

Since (9.78) is purely nonlinear, all terms of (9.80) are distortion terms. Besides second-harmonic distortion of each input frequency, sum and difference frequencies are also produced. These are the IM terms.

9.11 Transconductance Linearity of Bipolar-Junction Transistor Diff-Amp

The BJT differential amplifier stage, or emitter-coupled pair (Fig. 9.36), is of interest because of its frequent use in precision circuits. As the external emitter

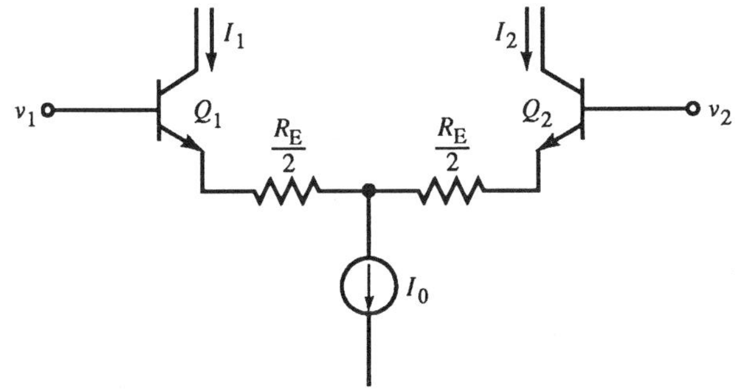

FIG. 9.36 BJT diff-amp with emitter resistance and current-source I_0.

resistance R_E is increased, the input dynamic range is extended, and the effect of r_e decreases because of the much larger series R_E. Thus Δr_e is less significant and linearity increases. To gain greater insight into the BJT diff-amp, we derive its differential transconductance. Let the output current be

$$i_O = i_2 - i_1 \tag{9.81}$$

and let the emitter bias current be

$$I_0 = i_1 + i_2 \tag{9.82}$$

The differential input is

$$v_I = v_2 - v_1 \tag{9.83}$$

Assume $\alpha = 1$ and matched junctions $I_{S1} = I_{S2}$. Then KVL is applied around the input loop:

$$\begin{aligned} v_I &= V_T \ln\left(\frac{i_2}{I_S}\right) + i_2\left(\frac{R_E}{2}\right) - i_1\left(\frac{R_E}{2}\right) - V_T \ln\left(\frac{i_1}{I_S}\right) \\ &= V_T \ln\left(\frac{i_2}{i_1}\right) + i_O\left(\frac{R_E}{2}\right) \end{aligned} \tag{9.84}$$

The first term can be expressed as

$$V_T \ln\left(\frac{1 + i_O/I_O}{1 - i_O/I_O}\right) = 2V_T \tanh^{-1}\left(\frac{i_O}{I_0}\right) \tag{9.85}$$

The incremental transconductance is

$$G_m = \frac{di_O}{dv_I} \tag{9.86}$$

G_m is found by implicitly differentiating (9.84) and solving:

$$G_m = \frac{1}{\dfrac{2V_T}{I_0[1 - (i_O/I_0)^2]} + \dfrac{R_E}{2}} \tag{9.87}$$

Since G_m depends on i_O, it is not linear. When $v_I = 0$, the amplifier is balanced, and $i_O = 0$. Then G_m is maximum and is

$$G_m(0) = \frac{1}{2V_T/I_0 + R_E/2} = \frac{1}{r_e + R_E/2} \tag{9.88}$$

This result is consistent with the transresistance method.

Instead of calculating distortion, we define the error in G_m as

$$\varepsilon = -\frac{G_m - G_m(0)}{G_m(0)} = \frac{(i_O/I_0)^2}{(R_E I_0/4V_T)[1 - (i_O/I_0)^2] + 1} \tag{9.89}$$

With $i_2 = 0.75$ mA and $i_1 = 0.25$ mA, then $i_O/I_0 = 0.5$. Assuming $R_E = 104\ \Omega$, $\varepsilon = 14.3\%$. With $R_E = 0\ \Omega$, ε is simply $(i_O/I_0)^2$ or 25%.

When $R_E = 0$, (9.84) can be solved for i_O. First,

$$\frac{i_2}{i_1} = e^{(v_I/v_T)} \tag{9.90}$$

With (9.82), i_1 can be expressed as

$$i_1 = \frac{I_0}{e^{(v_I/V_T)} + 1} \tag{9.91}$$

By definition,

$$\tanh x \equiv \frac{e^x - e^{-x}}{e^x + e^{-x}} \tag{9.92}$$

and

$$\tfrac{1}{2}[1 - \tanh x] = \frac{1}{e^{2x} + 1} \tag{9.93}$$

By change of variable, let $x \to x/2$. Then,

$$\tfrac{1}{2}\left[1 - \tanh\frac{x}{2}\right] = \frac{1}{e^x + 1} \tag{9.94}$$

Using this relationship with (9.91), we get

$$i_1 = \tfrac{1}{2} I_0 \left(1 - \tanh\frac{v_I}{2V_T}\right) \tag{9.95}$$

The differential output current is

$$i_O = i_2 - i_1 = I_0 - 2i_1 = I_0 - I_0\left(1 - \tanh\frac{v_I}{2V_T}\right)$$

or

$$i_O = I_0 \tanh\frac{v_I}{2V_T} \tag{9.96}$$

The hyperbolic tangent, an odd function, is linear around the origin and flattens out to ± 1 for large inputs. At room temperature, $2V_T \cong 52$ mV. The deviation of i_O/I_0 from a line tangent at the origin for a few values of v_I demonstrates the useful input range of v_I.

v_I, mV	$v_I/2V_T$	i_O/I_0	i_O error (%)
10.4	0.20	0.197	−1.31
17.1	0.33	0.319	−3.48
25.9	0.50	0.462	−7.58
38.8	0.75	0.635	−15.3
51.7	1.00	0.762	−23.8
77.6	1.50	0.905	−39.7
103.5	2.00	0.964	−51.8
258.7	5.00	1.000	−80.0

About 100 mV of input is the practical limit of the dynamic range. Above this, severe compression of i_O results. With zero R_E, the transconductance is

$$G_m\big|_{R_E=0} = \frac{I_0}{2V_T}\left[1 - \left(\frac{i_O}{I_0}\right)^2\right] \tag{9.97}$$

The effect of R_E is to linearize the hyperbolic tangent curve around the origin and add a linear term to (9.97).

Various schemes have been devised to improve the linearity of the BJT diff-amp. The *cross-quad* circuit of Fig. 9.37 is one of them. The input voltage is applied across four b–e junctions, two per side, and the common emitter resistance R_E. The voltage drop around the input loop is

$$v_2 - v_1 = v_B + v_C + v_{RE} - v_D - v_A = V_T \ln\left(\frac{i_B i_C}{I_S^2}\right) + i_{RE}R_E - V_T\left(\frac{i_D i_A}{I_S^2}\right) \tag{9.98}$$

where, for instance, the base-emitter junction drop of transistor A is v_A. Equation (9.98) can be further simplified for matched transistors to

$$v_2 - v_1 = V_T \ln\left(\frac{i_B i_C}{i_D i_A}\right) + (i_A - i_B)R_E \tag{9.99}$$

For $\alpha \cong 1$, $i_B = i_D$ and $i_C = i_A$. This reduces the log term of (9.99) to zero, leaving only the linear term. The large-signal transconductance is

$$G_M = \frac{i_O}{v_I} \cong \frac{i_B - i_A}{v_2 - v_1} = -\frac{1}{R_E} \tag{9.100}$$

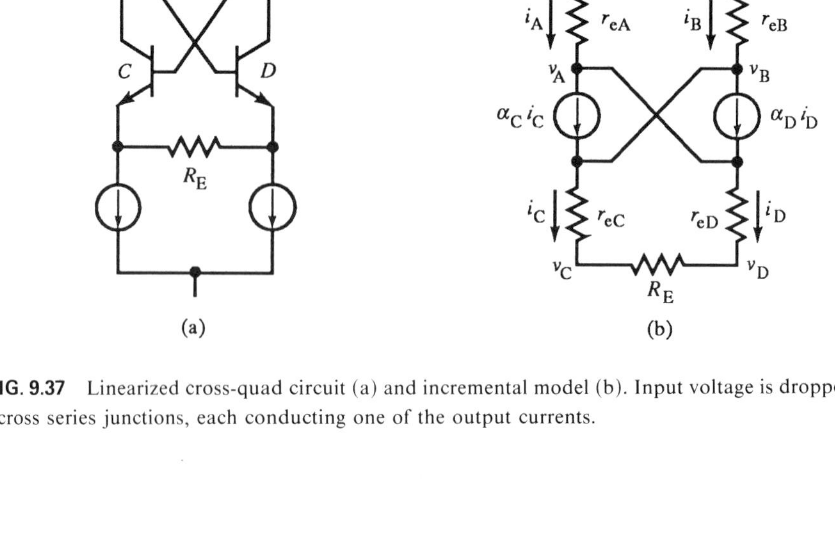

FIG. 9.37 Linearized cross-quad circuit (a) and incremental model (b). Input voltage is dropped across series junctions, each conducting one of the output currents.

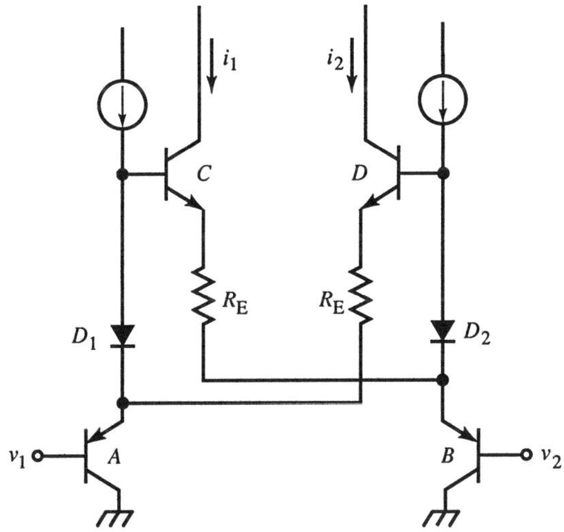

FIG. 9.38 Complementary diff-amp scheme using the cross-quad idea.

The input loop is linearized by summing voltages across junctions that are conducting currents from both sides. A variation on this idea, using complementary pairs of BJTs and diodes for biasing, is shown in Fig. 9.38. The diodes bias the NPN pair to approximately the same currents as the PNP pair at zero input voltage.

Incremental analysis of the cross-quad circuit around $v_I = 0$ can assume equal dc current in the four BJTs so that r_e is the same for all of them. Assuming matched transistors with equal β gives the inverted small-signal transconductance:

$$\frac{v_I}{i_O} = -\frac{1}{\alpha}\left[\left(\frac{2}{\beta-1}\right)r_e + \left(\frac{\beta+1}{\beta-1}\right)\frac{R_E}{2}\right] \cong -\left[\frac{2r_e}{\beta} + \frac{R_E}{2}\right], \quad \beta \gg 1 \qquad (9.101)$$

Without the cross-BJTs, $2r_e$ would not be divided by β. In effect, the nonlinearity is reduced by β. Note that the cross-quad amplifier is inverting because input voltage v_2 mainly drives BJT C, which generates i_1.

9.12 Bipolar-Junction Transistor and Field-Effect Transistor Diff-Amp Temperature Characteristics

Ambient temperature variations can cause changes in offset voltage of a diff-amp. This *offset voltage drift* is derived by applying KVL to the input loop of a BJT diff-amp. The input offset voltage,

$$V_{os} = V_2 - V_1 = V_T \ln\left(\frac{I_{E2}}{I_{S2}}\right) - V_T \ln\left(\frac{I_{E1}}{I_{S1}}\right) = V_T \ln\left(\frac{I_{E2}}{I_{E1}}\right) - V_T \ln\left(\frac{I_{S1}}{I_{S2}}\right) \qquad (9.102)$$

Since the thermal voltage $V_T = kT/q$ is in both terms of (9.102), they are both temperature dependent. The first term has drift due to emitter-current mismatch; the second is due to mismatched transistor I_S. BJT matching, especially monolithic matching, minimizes the second term. The first term requires that the dc currents of the differential pair be equal. The drift TC is

$$\frac{dV_{os}}{dT} = \frac{k}{q} \ln\left(\frac{I_{E2}}{I_{E1}}\right) \cong (198\ \mu\text{V}/^\circ\text{C}) \log\left(\frac{I_{E2}}{I_{E1}}\right) \tag{9.103}$$

A 1 μV/°C TC requires a 1% match of emitter currents. For perfectly matched BJTs, zero TC occurs at zero offset voltage. Slight mismatches cause the second term of (9.102) to be significant, leading to a modified (9.103) and nonzero offset when the emitter currents are equal. In addition, ohmic emitter resistance variations and surface leakage introduce offset error.

Another useful quantity is the drift TC per offset voltage, for matched BJTs:

$$\frac{dV_{os}/dT}{dV_{os}} \cong 3.3 \frac{\mu\text{V}/^\circ\text{C}}{m\text{V}}, \qquad T = 300\ \text{K} \tag{9.104}$$

A 1 mV offset produces a drift of 3.3 μV/°C.

Two adjustments are required to precisely balance a practical diff-amp because offset voltage drift is not nulled at zero offset voltage. Offset voltages are also caused by load resistor and r_o mismatch. The emitter currents are adjusted first for zero TC. This adjustment introduces its own output offset voltage, which is nulled by adjusting the load resistor balance.

For JFET diff-amps, the drift mechanisms are carrier mobility μ and gate-channel junction barrier voltage ϕ. With a constant applied V_{GS}, the terminal V_{GS} is the effective $V_{GS} + \phi$. Then $dV_{GS} = d\phi$. The V_{GS} TC is thus,

$$\frac{dV_{GS}}{dT} = \frac{\partial V_{GS}}{\partial \phi} \cdot \frac{d\phi}{dT} + \frac{\partial V_{GS}}{\partial I_D} \cdot \frac{\partial I_D}{\partial \mu} \cdot \frac{d\mu}{dT} \tag{9.105}$$

We need the TCs of ϕ and μ. The TC of ϕ is that of a pn junction, or about -2 mV/°C at room temperature. For an n-channel JFET, $d\phi/dt > 0$. Mobility varies with doping concentration and type, and only an average approximation can be given. It is about -0.5%/°C. The drain current of a JFET in the current saturation region is

$$i_D = I_{DSS}\left(1 - \frac{v_{GS}}{V_P}\right)^2 \tag{9.106}$$

where I_{DSS} is the drain current when $V_{GS} = 0$ and V_P is the pinch-off voltage. $V_P < 0$ for an n-channel JFET. The saturation region is where

$$V_{DS} \geq V_{GS} - V_P \tag{9.107}$$

I_D is a function of device geometry and electrical parameters. It is directly dependent on channel mobility, so that

$$\frac{dI_D}{I_D} = \frac{d\mu}{\mu} \tag{9.108}$$

Equation (9.105) can now be simplified after noting that

$$\frac{dI_D}{dV_{GS}} = g_m = -\frac{2I_{DSS}}{V_P}\left(1 - \frac{V_{GS}}{V_P}\right) \tag{9.109}$$

Substituting into (9.105) gives

$$\frac{dV_{GS}}{dT} \cong (2\text{ mV/°C}) + (-0.5\%/\text{°C})\left(\frac{I_D}{g_m}\right) \tag{9.110}$$

For zero voltage offset TC,

$$\left.\frac{I_D}{g_m}\right|_{TC(V_{GS})=0} \cong 0.4\text{ V} \tag{9.111}$$

Divide I_D from (9.106) by g_m of (9.109), resulting in

$$\frac{I_D}{g_m} = \tfrac{1}{2}(V_{GS} - V_P) \tag{9.112}$$

Set this equal to (9.111) and solve for the zero TC V_{GS}:

$$V_{GSZ} = V_{GS}\big|_{TC=0} \cong V_P + 0.8\text{ V} \tag{9.113}$$

In other words, the zero-drift V_{GS} is about 0.8 V above pinch-off. The typical range of V_{GSZ} is −2 V to −4 V. The corresponding zero TC I_D is

$$I_{DZ} = I_D\big|_{V_{GS}TC=0} = I_{DSS}\left(-\frac{0.8\text{ V}}{V_P}\right)^2 \tag{9.114}$$

I_{DZ} is typically several hundred microamps. Equation (9.110) can be written in terms of V_{GSZ} by substituting for V_P from (9.113):

$$\frac{dV_{GS}}{dT} \cong (-0.25\%/\text{°C})(V_{GS} - V_{GSZ}) \tag{9.115}$$

This formula is equivalent to (9.103) for a FET. The TC per volt of offset from V_{GSZ} is 0.25%/°C or 2.5 μV/°C per millivolt of offset. The drift TC per offset error is comparable to that of a BJT.

To find the effect of I_D on TC(V_{GS}), substitute I_{DSS} of (9.114) into (9.106), retaining the sign within the square:

$$I_D = I_{DZ}\left(\frac{V_P - V_{GS}}{-0.8\text{ V}}\right)^2 \tag{9.116}$$

Then replace V_P from (9.113), solve for $V_{GS} - V_{GSZ}$ and substitute it into (9.115). The result is

$$\frac{dV_{GS}}{dT} = (-2\text{ mV/°C})\left(1 - \sqrt{\frac{I_D}{I_{DZ}}}\right) \tag{9.117}$$

The sign can be checked by noting that when $I_D > I_{DZ}$, the TC is positive. The $TC(V_{GS})$ is opposite in sign to $TC(I_D)$. Below I_{DZ}, $TC(I_D)$ is positive because it is dominated by $TC(\phi)$. As $|\phi|$ decreases with increasing temperature, the effective V_{GS} increases, and more current flows. If I_D is forced to be constant, then V_{GS} must decrease. Thus, below V_{GSZ}, $TC(V_{GS}) < 0$. Above I_{DZ}, the $TC(I_D) < 0$ and is dominated by $TC(\mu) < 0$. Here, $V_{GS} > V_{GSZ}$ and $TC(V_{GS}) > 0$. For constant I_D, V_{GS} must increase with increasing temperature.

These results apply to a single JFET. For a matched differential pair with currents I_{D1} and I_{D2},

$$V_{os} = V_{GS2} - V_{GS1} \tag{9.118}$$

and

$$\frac{dV_{os}}{dT} = \frac{dV_{GS2}}{dT} - \frac{dV_{GS1}}{dT} = (-2\ \text{mV/°C})\left(\sqrt{\frac{I_{D1}}{I_{DZ}}} - \sqrt{\frac{I_{D2}}{I_{DZ}}}\right) \tag{9.119}$$

The diff-amp is compensated when each drain current is I_{DZ}, and

$$I_{D1} + I_{D2} = 2I_{DZ} \tag{9.120}$$

The radicals can be written in the form

$$\frac{I_{D1}}{I_{DZ}} = \frac{2I_{D1}}{2I_{DZ}} = \frac{(I_{D1} + I_{D2}) - (I_{D2} - I_{D1})}{2I_{DZ}} = 1 - \frac{I_O}{2I_{DZ}} \tag{9.121a}$$

$$\frac{I_{D2}}{I_{DZ}} = \frac{(I_{D1} + I_{D2}) + (I_{D2} - I_{D1})}{2I_{DZ}} = 1 + \frac{I_O}{2I_{DZ}} \tag{9.121b}$$

Equation (9.119) can now be expressed as

$$\frac{dV_{os}}{dT} = (2\ \text{mV/°C})\left(\sqrt{1 + \frac{I_O}{2I_{DZ}}} - \sqrt{1 - \frac{I_O}{2I_{DZ}}}\right) \tag{9.122}$$

where $I_O = I_{D2} - I_{D1}$ is the output current offset. This error is usually small relative to I_{DZ}, so the binomial approximations,

$$(1 \pm x)^{-1/2} \cong 1 \pm \frac{x}{2} - \frac{x^2}{8}, \qquad x \ll 1 \tag{9.123}$$

can be applied. Then the TC reduces to

$$\frac{dV_{os}}{dT} = (2\ \text{mV/°C})\left(\frac{I_{D2} - I_{D1}}{2I_{DZ}}\right) \tag{9.124}$$

Comparing this result to (9.103) for the BJT diff-amp, we find that a TC of 1 μV/°C occurs when the currents are mismatched by 0.1%. For the same TC, current matching must be an order of magnitude better for FETs than BJTs, which is why the input offset specification of FET-input op-amps is generally worse than their BJT counterparts.

The preceding derivations were for JFETs, but the results, including (9.106), hold for MOSFETs too. The difference is in I_{DSS}:

$$I_{DSS}(\text{JFET}) = \frac{\mu C}{2L^2}, \qquad I_{DSS}(\text{MOSFET}) = \frac{\mu \varepsilon W}{2LT} = \frac{\mu}{2L^2} \cdot \frac{\varepsilon L W}{T} = \frac{\mu}{2L^2} \cdot C_G \tag{9.125}$$

where μ is mobility, L the channel length, W the channel width, T the gate oxide thickness, and ε the gate-silicon dielectric constant. As shown in (9.125), even the form of I_{DSS} is the same. The difference is in how the gate and channel capacitances relate to the different device structures. Also, the MOSFET analog of pinch-off voltage V_P is threshold voltage V_T. The difference here is largely semantic.

9.13 Thermal Distortion

As the voltages and currents of electronic components vary, their power dissipation varies also. For components with parameters that are significantly affected by temperature, this self-heating can be regarded as a temperature signal. The electrical response is a kind of dynamic thermal distortion or noise.

Semiconductor devices are strongly sensitive to temperature variation. A pn junction has a voltage TC of typically -2 mV/°C. As a diode is heated, its v–i curve moves toward the vertical (current) axis, as seen on a curve tracer. The β of BJTs is sensitive to temperature but is hard to derive theoretically. It is typically about $+1\%$/°C for silicon BJTs. The most significant parameter for gain variation with temperature is r_e, which is directly proportional to the thermal voltage,

$$V_T = \frac{kT}{q}$$

At an ambient temperature of $T = 300$ K, the fractional TC(V_T) is

$$\text{TC\%}(V_T) = \frac{dV_T/dT}{V_T} = \frac{1}{T} = \frac{1}{300\ \text{K}} \cong 0.33\%/\text{K} = 0.33\%/°\text{C}$$

As a BJT heats on a curve tracer, not only does the v–i curve decrease in voltage, its slope at a given current becomes less steep; that is, r_e increases. These three thermal effects must be considered in circuit design.

A well-designed amplifier is not very β-sensitive; α appears in front of most gain expressions and depends on β as

$$\frac{d\alpha}{dT} = \left(\frac{1}{\beta + 1}\right)^2 \frac{d\beta}{dT} \tag{9.126}$$

For $\beta = 100$, TC(α) $\cong 1$ ppm and can be ignored.

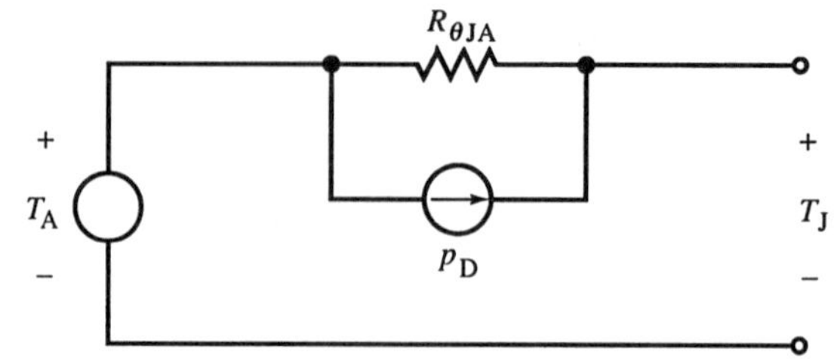

FIG. 9.39 Thermal-model electrical analog. Temperature is analogous to voltage and power dissipation to current.

Good amplifier design also minimizes the dependence of gain on r_e. The remaining effect is the TC of v_{BE}. BJT b–e junctions are highly doped and have a TC of about −2.2 mV/°C; diodes are lightly doped to increase breakdown voltage and have a TC closer to −1.8 mV/°C, similar to the b–c junction of BJTs when operated in the inverse mode. Thermal effects are therefore mainly the results of a b–e bias shift with temperature change. This shift appears as a dynamic b–e signal and is modeled as a voltage source in series with a fixed V_{BE}.

Junction temperature T_J can be derived from the thermal model shown in Fig. 9.39. Ambient temperature T_A is modeled by a voltage source. In series with it is the thermal resistance from junction to ambient. Dissipated in this resistance is power p_D, modeled as a current. The junction temperature is

$$T_J = p_D R_{\theta JA} + T_A \tag{9.127}$$

The maximum power that a transistor can dissipate is limited by the maximum junction temperature. For silicon, this is about 200 °C. The thermal model of (9.127) leads to the power derating curve of Fig. 9.40. Above $T_A = 25$ °C, maximum power decreases linearly up to the maximum junction temperature. The slope of the derating curve is the thermal conductance in

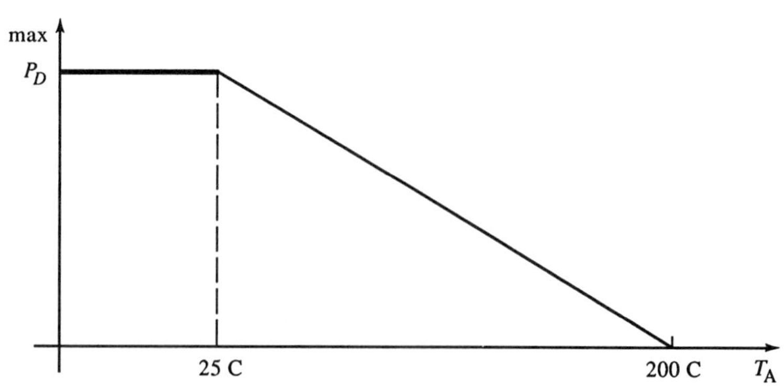

FIG. 9.40 Typical transistor power derating curve.

W/°C. A 16-pin dual-in-line (DIP) IC package has a thermal resistance of about 100°C/W and a TO-92 transistor package about 300°C/W.

Minimization of dynamic thermal effects requires minimization of either $R_{\theta JA}$ or ΔP_D. Semiconductor packaging often limits the minimum $R_{\theta JA}$, though the addition of heat sinks can significantly reduce it. We shall seek ways of minimizing Δp_D instead. Direct reduction in p_D comes from reducing P_D, the static power dissipation. Signal swings around this quiescent power are scaled down accordingly. Unfortunately, since power is often related to signal quantities, they also scale down with thermal noise. The approach we take here is to find the operating point at which signal excursion causes minimal change in power dissipation.

The simple CE circuit of Fig. 9.41a illustrates the basic idea of thermal compensation. BJT power dissipation, by Watt's law, is

$$p_D = v_{CE} i_E \cong v_{CE}\left(\frac{V_{CC} - v_{CE}}{R_L}\right) = \left(\frac{V_{CC}}{R_L}\right) v_{CE} - \frac{v_{CE}^2}{R_L} \tag{9.128}$$

The curve described by (9.128) is parabolic (Fig. 9.41b). Maximum p_D occurs at $V_{CC}/2$ and is

$$\max p_D = \frac{V_{CC}^2}{4R_L}, \qquad v_{CE} = \frac{V_{CC}}{2} \tag{9.129}$$

The slope of dp_D/dv_{CE} is minimum at maximum power. If we bias the transistor at maximum power, signal excursions from quiescence cause minimal changes in power.

This amplifier is highly β and r_e dependent but is not sensitive to Δv_{BE}. When a step of base current is applied, i_C increases causing v_{CE} to decrease, and the power steps down the curve from the peak. The transistor cools. As it does, the junction temperature decreases with time, with a cooling rate determined by the mass heated between the junction and ambient and the

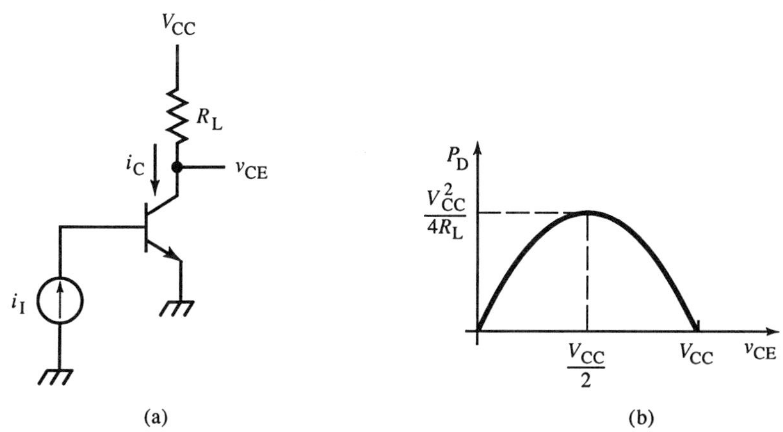

FIG. 9.41 Variation of power dissipation in a CE BJT over v_{CE}.

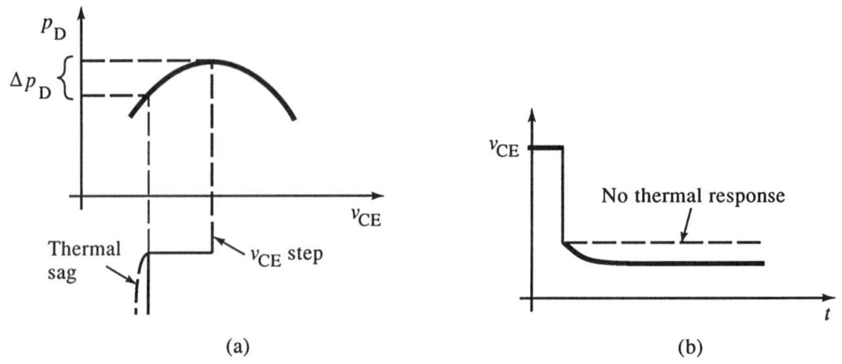

FIG. 9.42 The effect on step response (b) of dynamic change in power dissipation (a).

specific heat of the mass. As T_J decreases, v_{BE} increases, and V_T and r_e decrease, causing the gain to increase. This further decreases v_{CE}, so it contributes a thermal "sag" (Fig. 9.42) due to increasing gain. But the dominant effect is due to β-dependency. With cooling, β decreases, causing i_C to decrease more than it increases from Δr_e. The two effects tend to cancel, but β dominates here. This circuit is not typical of good design, but sometimes β and r_e thermal effects are unavoidable.

A better amplifier circuit is the CE of Fig. 9.43. It is relatively β and r_e insensitive but has a voltage-source input and amplifies Δv_{BE}. Assuming $\alpha = 1$

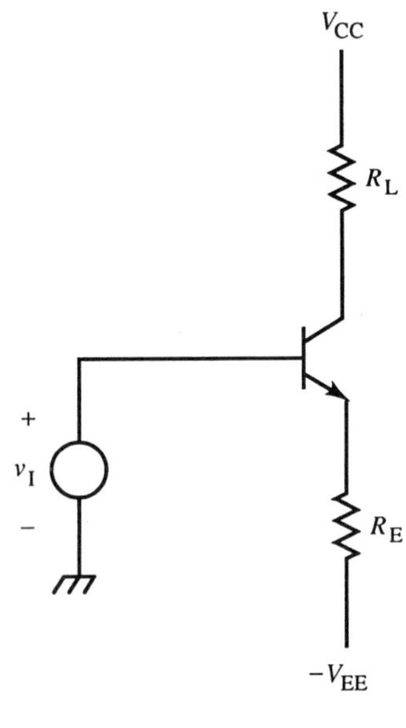

FIG. 9.43 CE circuit with emitter resistance.

and $R_E \gg r_e$, we get the BJT power dissipation:

$$p_D \cong v_{CE} \cdot i_C = \left[V_{CC} - \left(\frac{v_I - V_{EE}}{R_E} \right) (R_L + R_E) - V_{EE} \right] \left(\frac{v_I - V_{EE}}{R_E} \right) \quad (9.130)$$

Differentiating p_D and solving for v_I gives

$$v_I \Big|_{\max p_D} = \left(\frac{V_{CC} - V_{EE}}{2} \right) \left(\frac{R_L + R_E}{R_E} \right) + V_{EE} \quad (9.131)$$

Usually, v_I is determined by dc input level V_I. The supply voltage $-V_{EE}$ can be adjusted instead. Solving (9.131) for V_{EE} gives

$$V_{EE} \Big|_{\max p_D} = \frac{(2v_I - V_{CC})R_E - V_{CC}R_L}{R_E - R_L} > 0 \quad (9.132)$$

When the supply voltages are given, V_{EE} can be made a Thévenin equivalent voltage by the addition of an emitter resistor to ground.

Example 9.6 CS Buffer Thermal Compensation

The FET ×1 buffer amplifier of Fig. E9.6 (also Fig. 2.15a) uses a matched FET as the current source. Its source resistor is chosen to set V_{GS} to V_{GSZ} so that $I_D = I_{DZ}$, and the current has zero TC. (See Section 9.12.) The current from this source is I_0. Our goal is to choose V_{CC} so that thermal balance is achieved when $R_L = 1\ \text{k}\Omega$ and $I_0 = 5$ mA.

In this case, two devices are involved. We want the power difference between the FETs to change minimally since a power change in either of them relative to the other introduces thermal distortion. The power

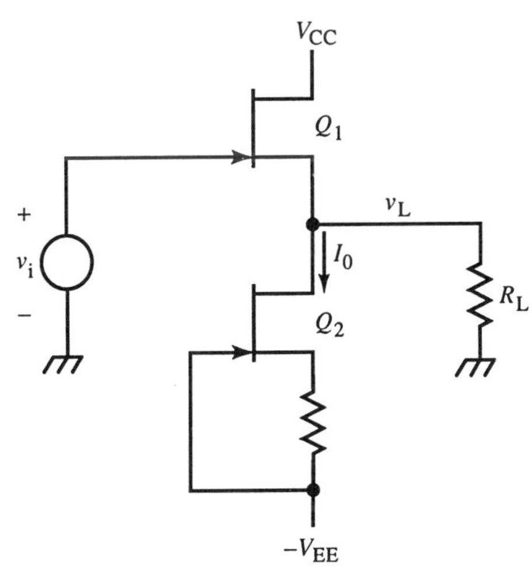

FIG. E9.6

difference between them is

$$\Delta p_{\rm D} = p_1 - p_2 = \left[(V_{\rm CC} - v_{\rm L})\left(I_0 + \frac{v_{\rm L}}{R_{\rm L}}\right)\right] - [(v_{\rm L} + V_{\rm EE})I_0]$$

$$= -\frac{v_{\rm L}^2}{R_{\rm L}} + \left(\frac{V_{\rm CC}}{R_{\rm L}} - 2I_0\right) v_{\rm L} + (V_{\rm CC} - V_{\rm EE})I_0 \qquad \text{(E1)}$$

The maximum power difference occurs at

$$v_{\rm L}\big|_{\max \Delta p_{\rm D}} = \frac{V_{\rm CC}}{2} - I_0 R_{\rm L} \Rightarrow V_{\rm CC}\big|_{\max p_{\rm D}} = 2(v_{\rm L} + I_0 R_{\rm L}) \qquad \text{(E2)}$$

where $v_{\rm L}$ is the quiescent value. Substituting into (E2), we get

$$V_{\rm CC} = 10\ \text{V}$$

If the available supply is not 10 V, a Thévenin equivalent supply can be constructed from a larger supply. If a series resistor $R_{\rm C}$ is inserted in the drain of Q_1, then

$$V_{\rm CC} = 2\left[v_{\rm L}\left(1 + \frac{R_{\rm C}}{R_{\rm L}}\right) + I_0(R_{\rm C} + R_{\rm L})\right] \qquad \text{(E3)}$$

For $V_{\rm CC} = 12$ V, $R_{\rm C} = 200\ \Omega$. To avoid the Miller effect, the drain of Q_1 is bypassed with a capacitor to load ground.

For applications in which dc offset must be minimized, the addition of the zero-TC bias resistor in the source of the lower FET must be matched by a resistor of equal value between the source of the upper FET and the output. This resistor drops the same voltage as the $V_{\rm GS}$ of the lower FET, thus compensating for the same $V_{\rm GS}$ in the top FET when the same current flows through it. The output loading also must be minimal so that I_0 flows through both FETs.

A change in temperature at the b–e junction of a BJT is a thermally generated noise. From the power formula for a BJT, the change in power can be calculated, given the input signal change. The resulting power change $\Delta p_{\rm D}$ results in a change in $v_{\rm BE}$. This $\Delta v_{\rm BE}$ is in series with the base and adds to $v_{\rm I}$ (Fig. 9.43). Knowing the gain, we can find the effect on the output of $\Delta v_{\rm BE}$ once we know

$$\frac{dv_{\rm BE}}{dp_{\rm D}} = \frac{dv_{\rm BE}}{dT_{\rm J}} \cdot \frac{dT_{\rm J}}{dp_{\rm D}} \cong (-2\ \text{mV/°C})R_{\theta\text{JA}} \qquad (9.133)$$

For a 16-pin DIP package,

$$\frac{dv_{\rm BE}}{dT} \cong (-2\ \text{mV/°C})(100\text{°C/W}) = -200\ \text{mV/W} \qquad (9.134)$$

For dynamic analysis, $R_{\theta JA}$ is a thermal impedance, but here only the resistive component is considered. The reactive component, a thermal capacitance, approximates the response of the thermal effect. The actual response is the solution of Laplace's equation and is only approximated by a first-order capacitive model. In practice, compensation networks use several time constants to approximate the thermal response. The complete BJT model, with thermal system included, is shown in Fig. 9.44.

Differential amplifiers reject thermal noise as a common-mode voltage when electrically and thermally balanced. With transistors on each side operating at maximum p_D at quiescence and operating on the same power curve, a positive step input to the diff-amp of Fig. 9.45a causes the action shown in (b) and (c). Both transistors move away from their maximum power point. Q_2 conducts more current and its v_{CE} decreases, whereas Q_1 conducts less with increased v_{CE}. Each moves down the power curve the same amount, symmetrically. The effect is that the same thermal Δv_{BE} occurs for both, as a small common-mode input.

We now derive an approximate formula for the step error, beginning with the expressions for p_D for each side. Each transistor is conducting I_C at V_{CE}. An input step v_i perturbs the amplifier from quiescence. Q_1 now conducts $i_{C1} = I_C - i_c$ at $v_{CE1} = V_{CE} + v_{ce}$. For Q_2, $i_{C2} = I_C + i_c$ at $v_{CE2} = V_{CE} - v_{ce}$. The power expressions are

$$p_1 = i_{C1} v_{CE1} = (I_C - i_c)(V_{CE} + v_{ce}) \tag{9.135a}$$

$$p_2 = i_{C2} v_{CE2} = (I_C + i_c)(V_{CE} - v_{ce}) \tag{9.135b}$$

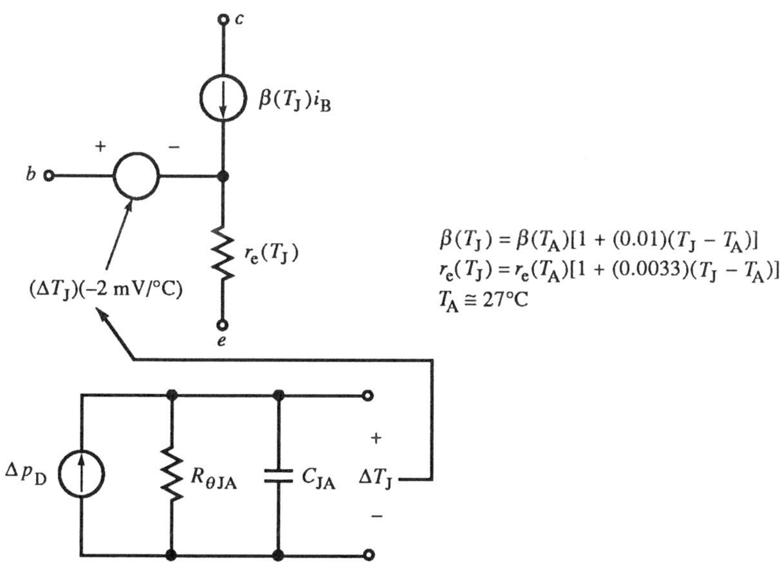

FIG. 9.44 BJT T model including thermal effects on β, r_e, and V_{BE}. A first-order thermal response is assumed.

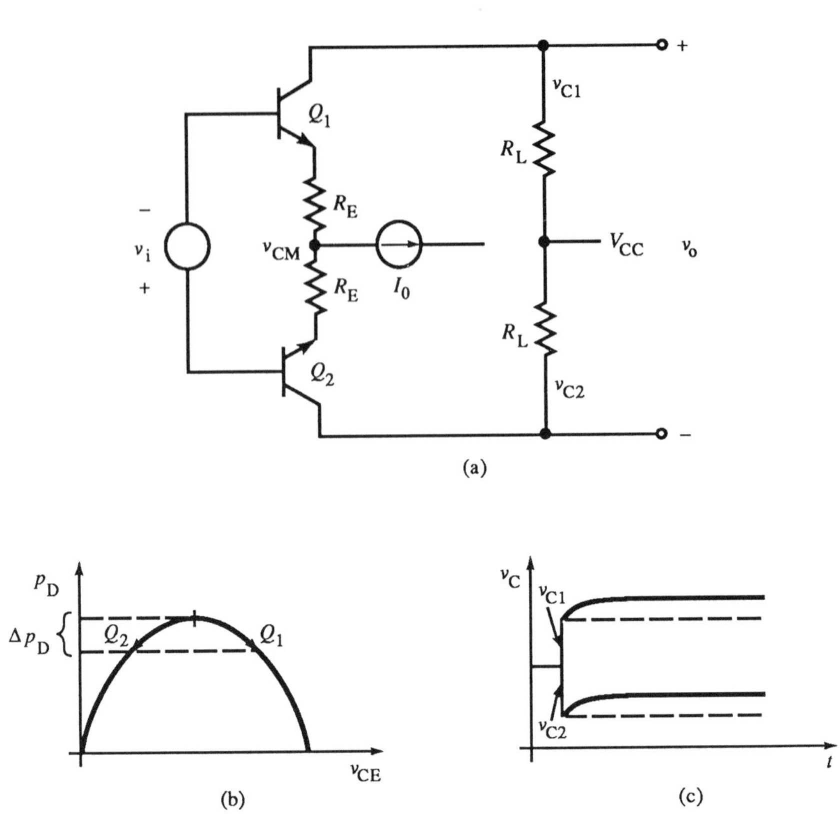

FIG. 9.45 A CE diff-amp (a), power dissipation (b), and step response. The thermal sag in the step response is common-mode rejected in a thermally compensated diff-amp.

Thermal error results from a change in the differential power,

$$\Delta p_D = p_1 - p_2 = -2(V_{CE} i_c - I_C v_{ce}) \tag{9.136}$$

The thermal response of v_{BE} due to Δp_D is

$$\text{thermal } \Delta v_{BE} \cong \frac{dv_{BE}}{dp_D} \cdot \Delta p_D \tag{9.137}$$

We now express the incremental variables i_c and v_{ce} in v_i. Neglecting α and r_e, and applying the transresistance method, we obtain

$$i_c \cong \frac{v_i}{2R_E} \tag{9.138}$$

$$v_{ce} = i_c(R_L + R_E) = \frac{R_L + R_E}{2R_E} \cdot v_i \tag{9.139}$$

where the polarity of gain for v_{ce} is already in (9.135). Substituting these expressions and (9.136) into (9.137), we get the expression for Δv_{BE}. The thermal error, referred to the input, is

$$\frac{\text{thermal } \Delta v_{\text{BE}}}{\text{signal } v_{\text{i}}} \cong -\frac{V_{\text{CE}} - I_{\text{C}}(R_{\text{L}} + R_{\text{E}})}{R_{\text{E}}}\left(\frac{dv_{\text{BE}}}{dp_{\text{D}}}\right) \tag{9.140}$$

When V_{CE} and the drop across $R_{\text{L}} + R_{\text{E}}$ are equal, they are both half of the supply, and the thermal error is zero.

Equation (9.136) is applicable for differential stages in general. For a particular stage, two incremental, single-ended gains, $v_{\text{ce}}/v_{\text{i}}$ and $i_{\text{c}}/v_{\text{i}}$, are required to compute the thermal error. The general formula for fractional input-referred thermal error is

$$\text{thermal error} = -\left(V_{\text{CE}} \cdot \frac{i_{\text{c}}}{v_{\text{i}}} - I_{\text{C}} \cdot \frac{v_{\text{ce}}}{v_{\text{i}}}\right)\left(\frac{dv_{\text{BE}}}{dp_{\text{D}}}\right) \tag{9.141}$$

The general condition for thermal compensation is that thermal error be zero. Then

$$\text{zero thermal error} \rightarrow \frac{v_{\text{ce}}}{i_{\text{c}}} = \frac{V_{\text{CE}}}{I_{\text{C}}} \tag{9.142}$$

Figure 9.46 shows a differential shunt-feedback amplifier. (Refer to Sections 4.15–4.17 for shunt-feedback theory.) Equation (9.141) can be expressed as

$$\text{thermal error} = -\left(V_{\text{CE}} - I_{\text{C}} \cdot \frac{v_{\text{ce}}}{i_{\text{c}}}\right)\left(\frac{i_{\text{c}}}{v_{\text{i}}}\right)\left(\frac{dv_{\text{BE}}}{dp_{\text{D}}}\right) \tag{9.143}$$

The output resistance $v_{\text{ce}}/i_{\text{c}}$ is found by assuming an ideal transconductance amplifier for the BJT. Then

$$\frac{v_{\text{ce}}}{i_{\text{c}}} = \frac{v_{\text{ce}}}{i_{\text{i}}} \cdot \frac{i_{\text{i}}}{i_{\text{c}}} \cong R_{\text{f}}\left(\frac{R_{\text{L}}}{R_{\text{f}} + R_{\text{L}}}\right) = R_{\text{f}} \parallel \text{R}_{\text{L}} \tag{9.144}$$

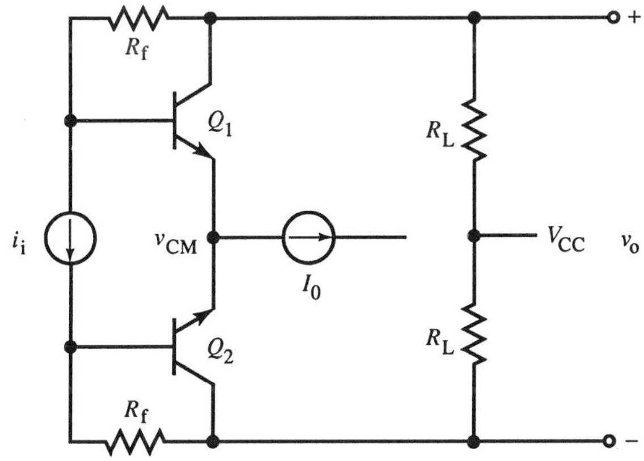

FIG. 9.46 A differential shunt-feedback amplifier.

The first factor is the ideal shunt-feedback transconductance. The second factor is a current divider for i_c. Ideally, $v_i = 0$. Thus, R_f and R_L are in parallel at the collector node. The second gain is

$$\frac{i_c}{v_i} = \frac{1}{R_f} \tag{9.145}$$

A voltage in series with the base causes current that can only flow in R_f. The ideal amplifier responds by sinking the current as i_c. This causes a drop across R_f of v_i, keeping the input a virtual ground. The current is determined by R_f. When these expressions are substituted into (9.143), the thermal error for a differential shunt-feedback amplifier is

$$\text{shunt-feedback thermal error} = -\frac{V_{CE} - I_C(R_f \| R_L)}{R_f}\left(\frac{dv_{BE}}{dp_D}\right) \tag{9.146}$$

When $V_{CE}/I_C = R_f \| R_L$, thermal compensation is achieved.

Finally, the CE of the cascode amplifier (Fig. 9.47a) requires thermal compensation. Its collector is at the fixed voltage of the CB base. The thermal error is found by noting that $v_{ce} = v_i$ and $i_c/v_i = 1/R_E$. Then $v_{ce}/i_c = R_E$. Substituting into (9.143) gives

$$\text{cascode thermal error} = -\frac{V_{CE} - I_C R_E}{R_E}\left(\frac{dv_{BE}}{dp_D}\right) \tag{9.147}$$

Thermal compensation requires $V_{CE}/I_C = R_E$. This is often not possible because V_{CE} is determined partly by V_B. To compensate, add a collector series resistor R_C. Then the CE thermal error is expressed by (9.140). The addition

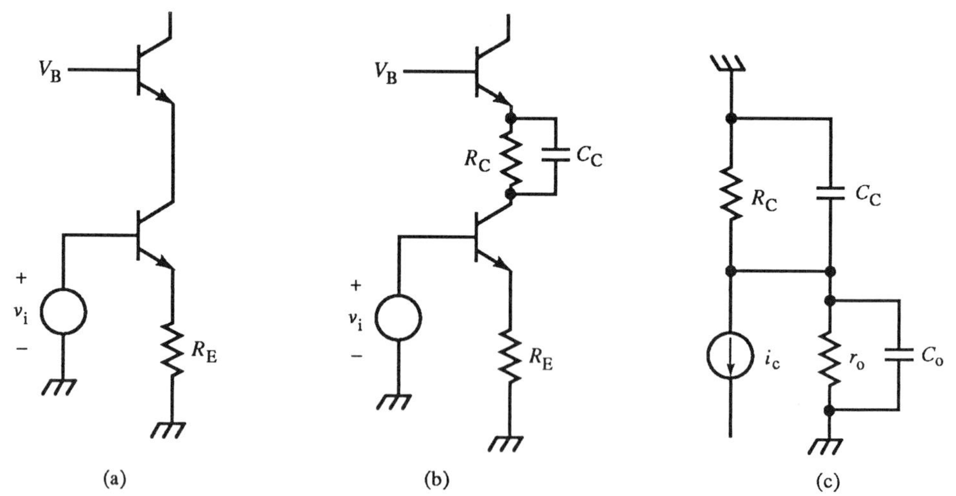

FIG. 9.47 Thermal compensation of the cascode amplifier CE (a) with a series resistor R_C (b), that is dynamically compensated by C_C based on r_o and C_o of CE, (c).

of R_C has dynamic response consequences and can be bypassed with capacitor C_C (see Fig. 9.47b). Since $R_C C_C$ forms a time constant, it is reasonable to wonder by what criterion its value should be chosen. First, C_C must be large enough to have negligible reactance in the hf region of the CB transistor. But this still leaves a wide range for C_C. Bruce Hofer has noted that the output capacitance of the CE forms a shunt *RC* time constant with r_o of the CE as in Fig. 9.47c. By setting

$$R_C C_C = r_o C_o \tag{9.148}$$

we form a compensated current divider for the CE collector current, and i_e of the CB is

$$\text{CB } i_e = \left(\frac{r_o}{r_o + R_C}\right) i_c \tag{9.149}$$

In single-ended and unbalanced differential amplifiers, after Δp_D is minimized, thermal effects remain. At this point, electrical compensation is required. A series *RC* network is commonly used in the emitter circuit to cancel thermal error. Furthermore, the simple single-pole thermal model we have used is only a dominant-pole approximation to most heat-transfer temperature functions. Thus, it is not unusual to find that several series *RC* networks are required to achieve acceptable response.

9.14 Complementary Emitter-Follower Output Amplifier

A common need in circuit design is for a bipolar voltage buffer with current-drive capability. This is usually the output stage of a power amplifier, used to drive transmission lines, cables with large capacitance, transducers, or large power devices. The common requirement is that the driver have large-signal dynamic range relative to its transistors and that the output resistance be constant, usually 50 Ω for the transmission lines used to interconnect electronic laboratory equipment.

One of the simplest drivers is the CC configuration. Its near-unity voltage gain and a current gain of β make it an attractive output buffer stage. Its disadvantage is that it can provide current in only one direction; it is inherently unipolar. To achieve bipolar drive, a CC of the opposite polarity (a *complementary* CC) is connected in parallel (Fig. 9.48). Without the dc voltage sources (depicted as batteries), Q_1 conducts when v_I is positive and Q_2 when negative. Because of $V_{BE} \cong 0.6$ V for silicon, there is a deadband of reduced gain around zero output voltage where neither transistor conducts. The gain around zero is much less than the gain for $|v_I| > V_{BE}$. This gain variation results in a kind of nonlinearity, called *crossover distortion*, than can occur whenever the output drive "crosses over" from one transistor to another. The nonlinearity is due

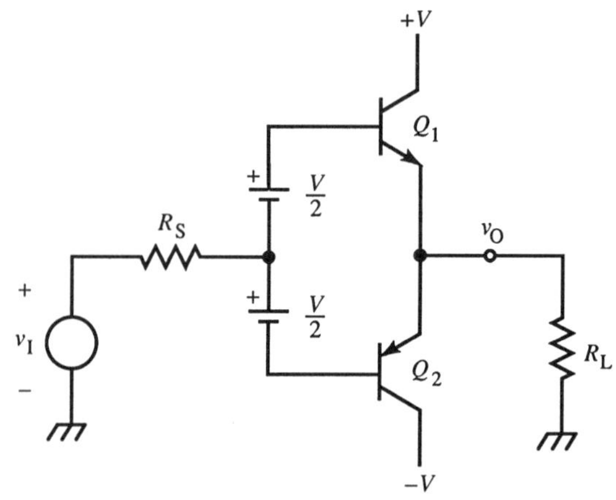

FIG. 9.48 Complementary emitter-follower buffer with input bias of *V*.

to the change in output resistance with output voltage. The voltage gain is

$$A_V = \frac{R_L}{R_L + r_{out}} \tag{9.150}$$

where r_{out} is the incremental (small-signal) output resistance. As v_O varies, r_{out} varies, and so does A_v.

The deadzone caused by V_{BE} can be narrowed and even eliminated by adding the base offset voltages shown in Fig. 9.48. The transfer characteristics of $v_O(v_I)$ are shown in Fig. 9.49 for V of 1.0–1.8 V in 0.2 V steps and $R_L = 1\ k\Omega$. At $V = 1.8$ V, nonlinearity is undiscernible from the graph. This base bias

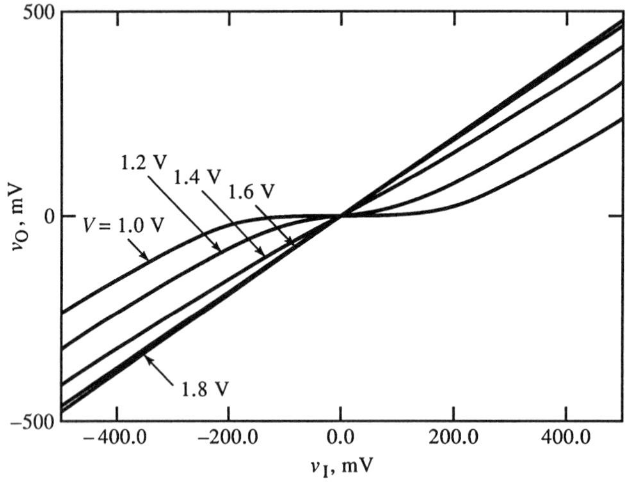

FIG. 9.49 Transfer functions for Fig. 9.48 when R_L is 1 kΩ and V as parameter. The deadzone is apparent for small *V*.

causes conduction of current I_0 through the two transistors. As v_O increases with v_I, some of I_0 is diverted into the load R_L. With further increase, all of I_0 is diverted from Q_2, and it approaches cutoff, leaving Q_1 to drive the load alone. The emitter currents of Q_1 and Q_2 are found by applying KVL to the input loop:

$$V = V_T \ln\left(\frac{i_{E1}}{I_{S1}}\right) + V_T \ln\left(\frac{i_{E2}}{I_{S2}}\right) = V_T \ln\left(\frac{i_{E1} i_{E2}}{I_{S1} I_{S2}}\right) \tag{9.151}$$

Assuming $I_{S1} = I_{S2} = I_S$ and solving for the emitter current product, we get

$$i_{E1} i_{E2} = I_S^2 e^{(V/V_T)} \neq 0 \tag{9.152}$$

For a fixed V/V_T and I_S, neither emitter current goes to zero but only approaches it in the limit. This desirable property keeps both transistors on to reduce Δr_e and r_{out}; however, the ratio of emitter currents has such a wide range that r_{e1} and r_{e2} do also, and Δr_{out} is large.

Another problem with this circuit is thermal instability. In (9.152), the i_E product varies with I_S and V_T. Both are temperature dependent. The dominant effect is the exponential variation of current with temperature due to V_T that can cause excessive I_0. This circuit is thermally unstable because increased I_0 causes increased junction heating leading to further increases in current. This phenomenon in BJTs is called *thermal runaway* and requires stabilization. Two modifications are shown in Fig. 9.50: the addition of emitter resistors R and

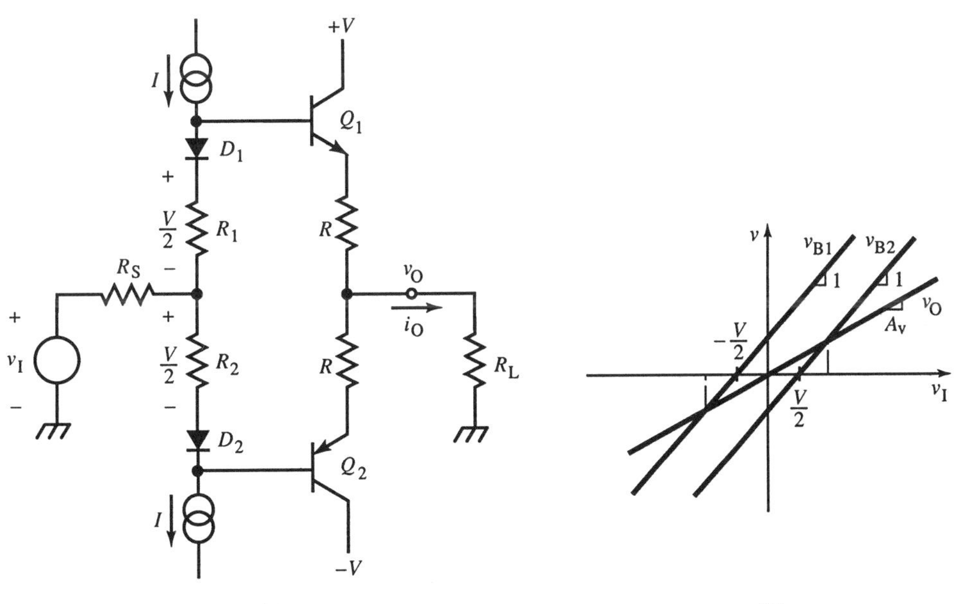

FIG. 9.50 CE buffer with bias circuit compensation for V_{BE} (a). The crossover region is identified by the intersection of the curves in (b).

the replacement of dc voltage sources with a more practical biasing network. The current sources cause voltage drops across the series resistors and diodes that bias the CCs. The diodes provide thermal tracking of the CC b–e junctions and the resistors R_1 and R_2 are set to each drop $V/2$. With ideal junction compensation, diode voltages equal the CC v_{BE}, and $V/2$ appears stably across R in the emitters.

When I_0 is large and Δr_e is negligible relative to R, the cutoff points of the transistors are at the values of v_I where the voltage across R is zero. In Fig. 9.50b, v_O is plotted along with the base voltages v_{B1} and v_{B2}. As v_I increases, v_{B2} increases at the same rate. Since $A_V < 1$, v_O increases at a more gradual rate, and the voltage across R of Q_2 decreases until it is zero; Q_2 is cut off. A similar argument applies to decreasing v_I and Q_1 cutoff. The input crossover voltages are

$$\text{crossover } v_I = \pm\frac{V/2}{1 - A_V} \tag{9.153}$$

and

$$\text{crossover } v_O = \left(\frac{A_V}{1 - A_V}\right)\left(\frac{V}{2}\right) \tag{9.154}$$

The transfer characteristic of Fig. 9.50b is approached at relatively large values of I_0, where excessive power is dissipated due to biasing. At a reduced I_0, r_e variation is significant, and the gain is not constant over the output range. The resulting transfer function is asymptotically approximated as shown in Fig. 9.51. In the crossover region, the gain has been linearly approximated as the gain at the origin; the circuit is symmetrical there, with equal dynamic emitter resistances $r_e(0)$. Then r_{out} consists of two shunt paths, each with a

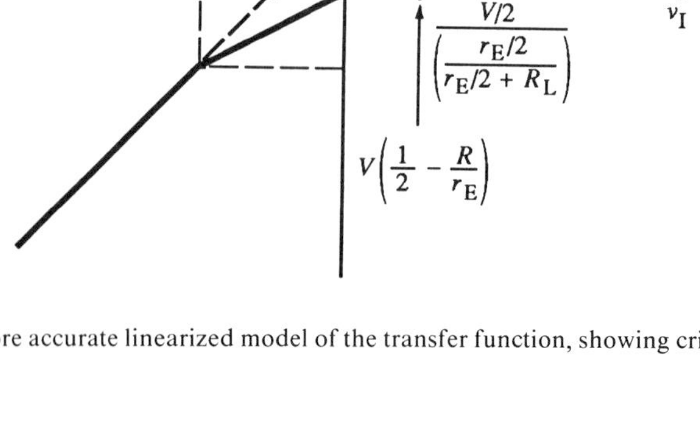

FIG. 9.51 A more accurate linearized model of the transfer function, showing critical parameters.

dynamic resistance of $R+r_e$. Base resistance is referred to the emitter and included in R. Outside the crossover region, r_{out} is $R+r_e$ and approaches R with increasing v_O. The gain is asymptotic with

$$A_v \cong \begin{cases} \dfrac{R_L}{R_L+[R+r_e(0)]/2}, & \text{crossover region} \\ \dfrac{R_L}{R_L+R}, & \text{outside region} \end{cases} \tag{9.155}$$

The effect of Δr_e on incremental gain over the amplifier range is shown in Fig. 9.52a for $V=0$ V. The smaller values of R show a decrease in gain around zero input due to the large increase in r_e. This is due to a low value

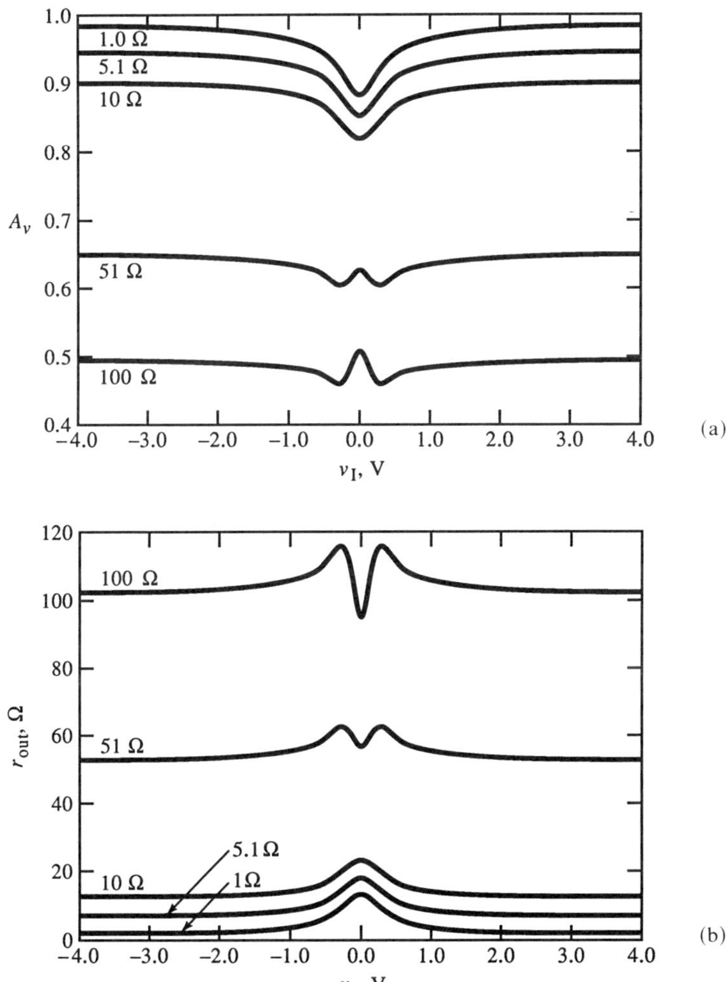

FIG. 9.52 Incremental voltage gain A_v curve (a) and output resistance r_{out} curve (b) for buffer of Fig. 9.50, with $R_L=100\ \Omega$, $V=0$ V, and parameter R.

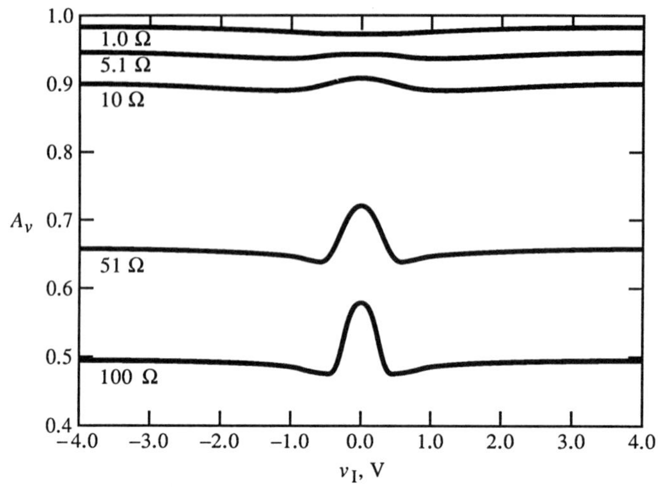

FIG. 9.53 Incremental voltage gain for $V = 100$ mV; otherwise, same conditions as for Fig. 9.52a.

of I_0 resulting from inadequate V. For larger R, the gain begins to dip but then peaks at zero input. In these cases, r_e contributes a smaller fraction of the total r_{out} per side. When both sides conduct equally around zero, their shunt resistance is lowest and the gain peaks. In Fig. 9.53, $V = 100$ mV, and this peak around zero is more pronounced since the shunting effect is significant over the wider crossover range. With greater I_0, r_e does not increase as much before the crossover region is entered.

The zero-bias output resistance is shown in Fig. 9.52b. It determines the gain and varies inversely with the gain, as (9.150) predicts. Figure 9.54 shows

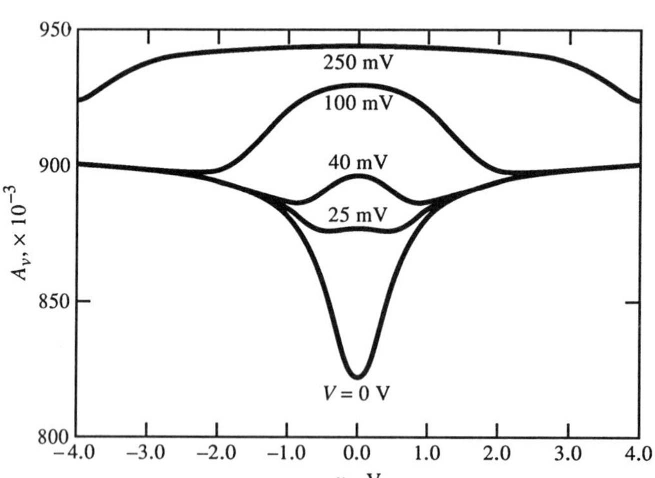

FIG. 9.54 A_v with $R = 10\ \Omega$ and parameter V. The dips in the curves are from increasing r_e near zero input; the peaks are from the two BJTs shunting their outputs.

the gain with $R = 10\,\Omega$ and V as parameter. As V increases, the dip in gain around zero due to increased r_e begins to show the effect of the two sides shunting, the central peak. As shunting grows in dominance, the dip disappears and only the peak remains. The peak then broadens as the crossover range broadens with V.

The endpoints of the crossover region and the transfer curve offsets are symmetrical about the v_O axis and are found by a total-variable analysis. Since

$$i_{E1} = \frac{(v_I + V/2) - v_O}{R}, \qquad i_{E2} = \frac{v_O - (v_I - V/2)}{R} \tag{9.156}$$

the endpoints are found by setting the emitter currents to zero, substituting the crossover-region gain for v_O and solving for v_I. The asymptotic crossover endpoints are

$$\text{crossover extremum of } v_I = \pm\left(\frac{R_L + r_E/2}{r_E/2}\right)\left(\frac{V}{2}\right), \qquad r_E = R + r_e(0) \tag{9.157}$$

These correspond to asymptotic output values of

$$v_{Ox} = \text{crossover extremum of } v_O = \pm\left(\frac{R_L}{r_E}\right)V \tag{9.158}$$

The transfer function outside the crossover region has an input offset of

$$v_{Ix} = A_v \text{ input offset voltage} = \pm\left(\frac{r_e(0) - R}{r_e(0) + R}\right)\left(\frac{V}{2}\right) \tag{9.159}$$

Since

$$r_e(0) = \frac{V_T}{I_0} = \frac{V_T}{((V/2)/R)} = \left(\frac{V_T}{V}\right)2R \tag{9.160}$$

all of the values indicated in Fig. 9.51 can be expressed in R, R_L, V_T, and V. For example, (9.159) is

$$v_{Ix} = \pm\left(\frac{V}{2}\right)\left(\frac{2V_T - V}{2V_T + V}\right) \tag{9.161}$$

From the numerator, $v_{Ix} = 0$ when $V = 2V_T$. This gives the most linear design:

$$\text{optimum } V = 2V_T \tag{9.162}$$

Since V_T varies with temperature, optimum performance over temperature requires that V track V_T.

The effect of $r_e(0)$ on v_O and gain is derived from Fig. 9.51 for several interesting cases in Fig. 9.55. When $r_e(0)$ is very large, as in (a), a deadzone of V exists. As $r_e(0)$ decreases but remains larger than R, the effect of increasing r_e in the crossover region is shown in (b) with $A_v(v_I)$. The dip in A_v is evident in previous figures when $r_e(0)$ is large relative to R. This occurs when V is small. As $r_e(0)$ increases, the optimum value is reached when $V = 2V_T$, shown

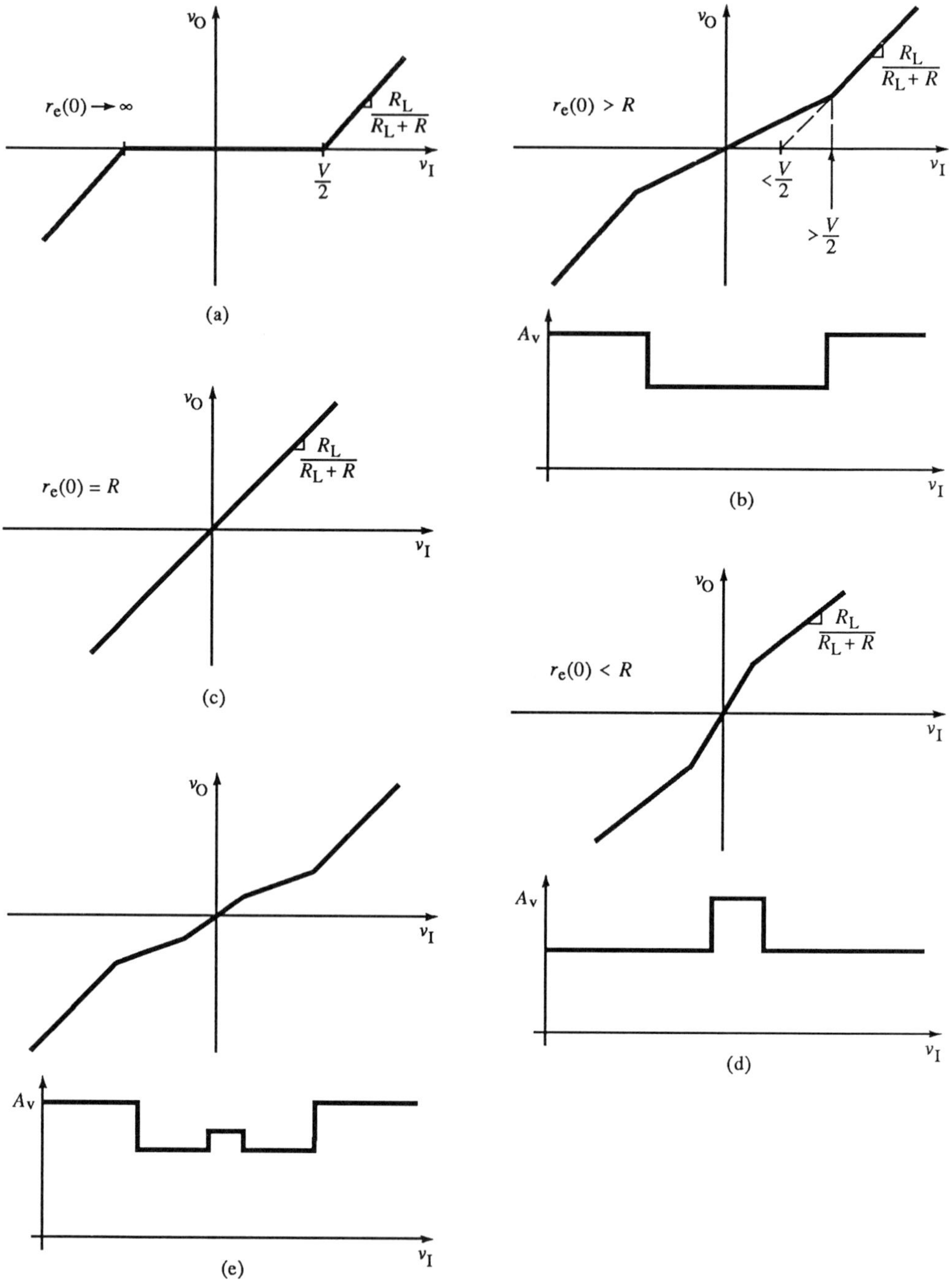

FIG. 9.55 Transfer curves for $r_e(0)$ varying from infinity (a) to less than R (d). A more accurate piecewise-linear model of v_O includes increase in A_v around origin (e).

in (c). As $r_e(0)$ continues to decrease, the crossover gain is larger than that of the outside regions as in (d). The plot of A_v shows the characteristic peak of previous gain plots, due to the shunting effect. When $r_e(0)$ is zero, R dominates and $v_{Ix} = \mp V/2$.

The small peak in a large dip is common in the $A_v(v_I)$ plots. The plots of Fig. 9.55e are a piecewise-linear approximation of this phenomenon. Increasing r_e reduces gain in the outer part of the crossover region. At its center, shunting increases the gain and causes the peak.

For high-current buffers, power dissipation in R can be reduced by diode shunts. As $|v_I|$ increases, the voltage across R increases until the diode conducts most of the current. In the crossover region, R dominates, but when the diodes dominate, the effective R decreases (when $r_d \ll R$) and the gain increases. The circuit of Fig. 9.56 illustrates this. Without the diodes, Fig. 9.57a results; with diodes, (b). In (b), for the larger R, the diodes cause the gain to increase when they dominate conduction. Since their dynamic resistance r_d is much less than R, gain increases appreciably. The diodes thus improve gain accuracy outside the crossover region.

A second circuit modification that reduces Δr_{out} when the input to the stage has a low Thévenin resistance is the addition of R_{IO} (Fig. 9.58). In the crossover region, R_{IO} supplies load and bias current to the output from the buffer input source. Then $r_{out} < R_{IO}$ for the entire v_I range. This dramatic reduction in r_{out} is evident by comparing Fig. 9.57b (no R_{IO}) with Fig. 9.59 (with R_{IO}).

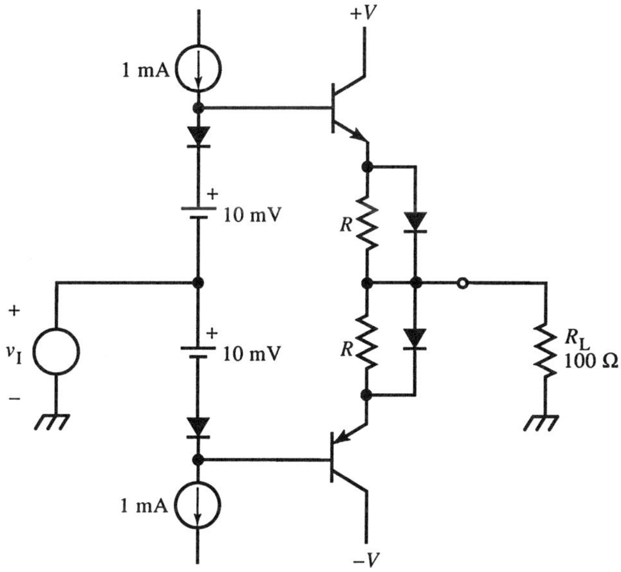

FIG. 9.56 Complementary CE buffer with shunt emitter diodes.

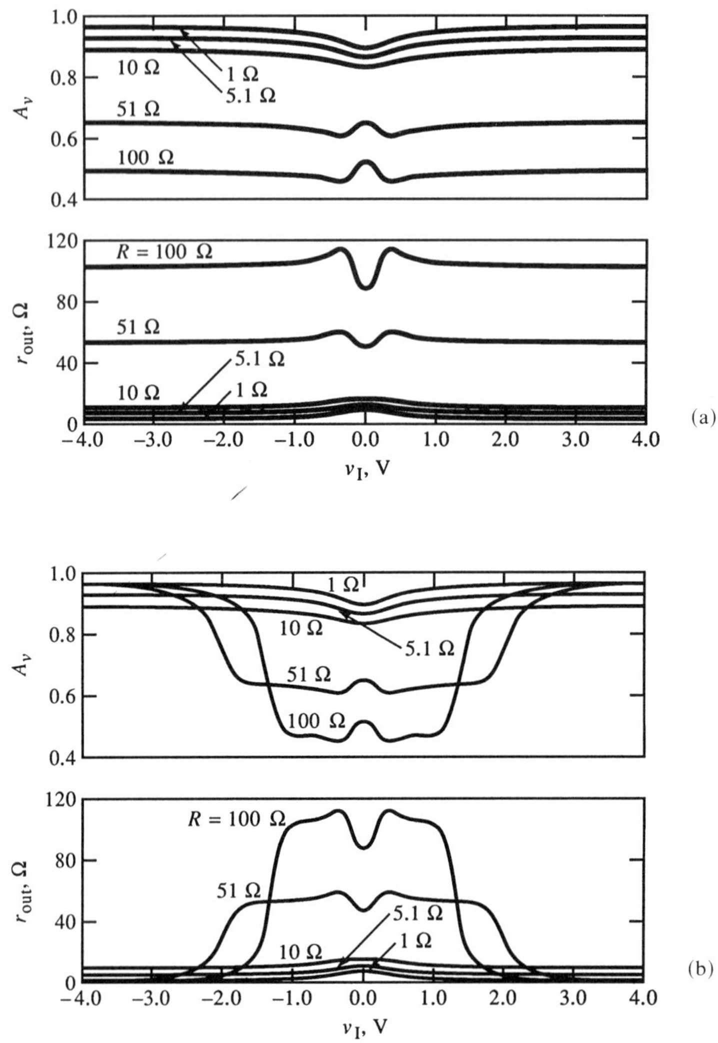

FIG. 9.57 A_v and r_{out} for Fig. 9.56 circuit without diodes (a), and with diodes (b). R is parameter. Diodes appreciably decrease r_{out} when they conduct, increasing A_v for large R.

In some cases (such as Fig. E7.4a with R_B removed) the complementary CC buffer is driven by a high-resistance source. The output resistance is then also very high, reflecting the high base resistance. In this case, the buffer defaults to a current amplifier, and the ratio of gains of the two sides is the ratio of the βs of their transistors. Matched-β transistors are then important for linearity.

Feedback reduces both β-mismatch and crossover distortions by $1+GH$. Since the nonlinear stage is a part of the loop, GH is affected by it. For the minimum gain of the stage, the loop gain is minimum and distortion reduced

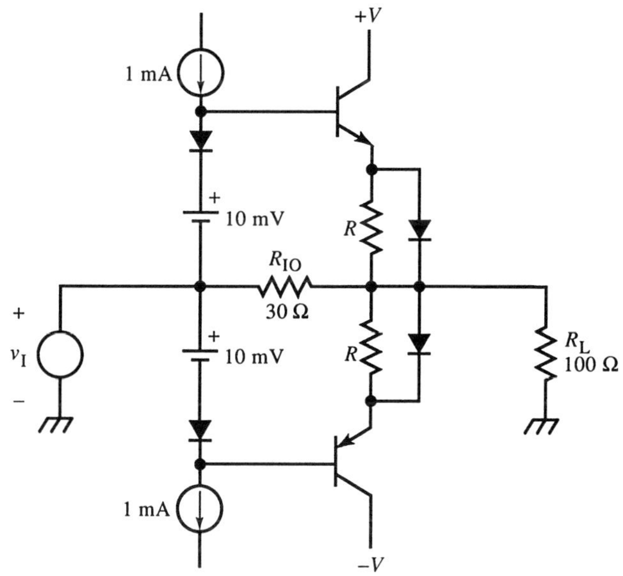

FIG. 9.58 R_{IO} included in circuit of Fig. 9.56.

least. For the pathological case of Fig. 9.55a, the gain is zero in the deadband, and no amount of feedback improves linearity there. Too wide a gain variation due to nonlinearity can also produce feedback instability in high-gain regions of the dynamic range. Reduction of loop gain required for stability then compromises distortion reduction in the low-gain regions. Therefore, a good general strategy is to make the open-loop stages as linear as possible before closing the loop.

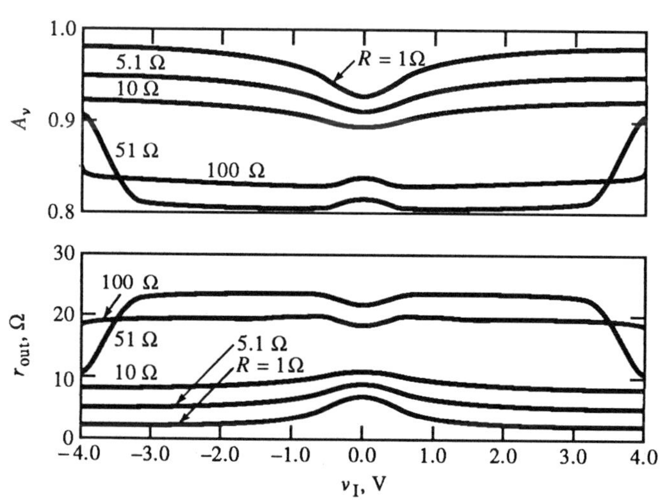

FIG. 9.59 A_v and r_{out} of circuit with R_{IO} (Fig. 9.58). For large R, r_{out} is reduced around zero input.

9.15 Closure

Precision is a general idea with many facets. In this chapter, we have not exhausted all of the techniques for improving precision. In particular, we would like methods of designing precision amplifiers that are also fast. These high-performance amplifiers are explored in the next chapter.

References

[1] Ahmed M. Soliman, "Instrumentation Amplifiers with Improved Bandwidth," *IEEE Circuits and Systems Magazine*, Vol. 3, no. 1, 1981. pp. 7-9.

[2] Bruce Hofer, Harmonic distortion estimation notes.

[3] Chu-Sun Yen, "Distortion in Emitter-Driven Variable-Gain Pairs," *IEEE JSSC*, Vol. SC-14, No. 4, Aug. 1979. pp. 771–773.

[4] Al Ryan and Tim Scranton, "D-C Amplifier Noise Revisited," *Analog Dialogue*, Vol. 18, No. 1, Analog Devices, Inc., 1984. pp. 3–10.

[5] National Semiconductor Corp., *Linear Applications Databook*, Application Note AN-222, 1986. pp. 517ff.

[6] Fairchild FAST Circuits and Application Seminar.

[7] Henry Ott, "Ground—A Path for Current Flow," *EMC Technology*, Jan–Mar. 1983. pp. 44–48.

[8] Laudie Doubrava, "Proper handling of voltage spikes safeguards circuit designs," *EDN*, 5 Mar. 1979. pp. 83-87.

[9] Laudie Doubrava, "Design Techniques for Controlling Point-of-Load High Frequency Performance of Power Supplies."

[10] Laudie Doubrava, "High-Frequency Bypass and Decoupling Design," *Proc. Powercon* 5, May 1978. pp. H1–1 through H1–11.

[11] Ralph Morrison, "Grounding in Instrumentation Systems," *EMC Technology*, Jan.–Mar. 1983. pp. 50–52.

[12] Edward F. Vance, "Cable Grounding for the Control of EMI," *EMC Technology*, Jan.–Mar. 1983. pp. 54–58.

[13] Edward R. Oates, "Good grounding and shielding practices," *Electric Design* 1, 4 Jan. 1977. pp. 110–112.

[14] David A. McCully, "Preservation of Sensor Data Integrity in Noisy Environments," Computer Products, Inc. MC1081, Jan. 1980. Fort Lauderdale, Florida.

[15] Werner Neuman, "Switching Power Supplies, Radio Frequency Interference, & Power Line Interference Filters," Corcom, Inc., Chicago, Illinois.

[16] "Characteristics of Flat Cable," inset, *EDN*, 7 Jan. 1981. pp. 110–111.

[17] Tom Balph, "Interconnection Techniques for Motorola's MECL 10,000 Series Emitter Coupled Logic," Motorola Application Note AN-556, 1972.

[18] *Compliance Engineering 1988*, Glen Dash, ed., Vol. IV, Compliance Engineering, Acton, Massachusetts.

[19] Donald R. J. White and Michel Mardiguian, *EMI Control Methodology and Procedures*, (4th edition), Interference Control Technologies, Don White Consultants, Inc., Gainsville, Virginia, 1985.

[20] John Severinsen, "Designer's Guide to EMI Shielding, Part I," *EDN*, 5 Feb. 1975. pp. 47–51.

[21] William Jarva, "Design EMI shielding more accurately," *Electronic Design* 6, 15 Mar. 1977. pp. 88ff.

[22] David McElvein, "Common Misconceptions In The Use of EMI Gaskets," *IEEE Symp. EMC*, 7–9 Oct. 1980, Baltimore, Maryland; from Chomerics, Inc., Woburn, Massachusetts.

[23] Bernard M. Oliver, "Distortion in Complementary-Pair Class-B Amplifiers," *Hewlett-Packard Journal*, Vol. 22, No. 6, Feb. 1971. pp. 11–16.

CHAPTER 10

High-Performance Amplification

The basic conflict between speed and precision in amplifier design has led to the development of techniques that are both fast and precise; this high-performance amplification is applied as analog preprocessing for analog-to-digital converters (ADC), as postprocessing for digital-to-analog converters (DAC), and for instrumentation such as oscilloscope amplifiers. We first examine novel subsystem-level amplifier topologies involving multiple signal paths, and then single-stage amplifiers. These new topologies are applied to buffer amplifier design, continued from the previous chapter. Finally, the versatility of controlling or *programming* amplifier gain leads to a discussion of multipliers and programmable-gain amplifiers.

10.1 Current-Input and Current-Feedback Amplifiers

The transistor diff-amp amplifies differential input voltages, not currents. Because of the Miller effect and the dominance of stray capacitance over stray inductance, ac voltages are often more easily degraded than ac currents. It is thus desirable to have op-amp topologies that sum currents instead of voltages at their inputs.

A current-differencing amplifier input stage, shown in Fig. 10.1, is commercially implemented in the National Semiconductor LM3900 (which is called a *Norton amplifier*). The noninverting input current I_{i+} flows mainly through the diode and develops a voltage across the b–e junction of Q_1. If the diode

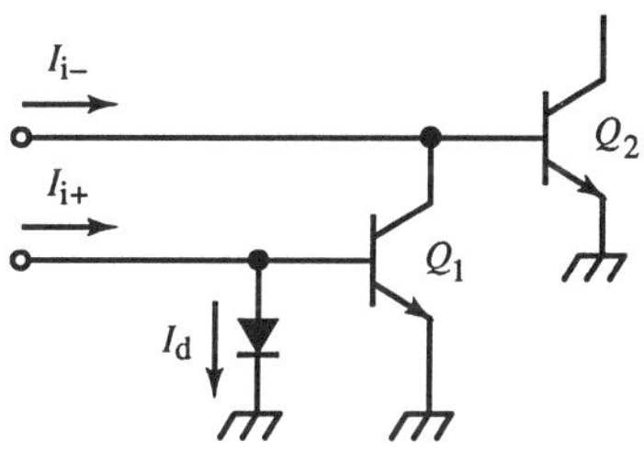

FIG. 10.1 Differential-current amplifier input uses Q_1 as current mirror to subtract current from input of Q_2.

junction is matched with Q_1, then

$$I_d = I_{e1} \tag{10.1}$$

The diode current is

$$I_d = I_{i+} - I_{b1} \tag{10.2}$$

Since $I_{c1} = \alpha I_{e1}$, then

$$I_{c1} = \alpha I_d = \left(\frac{\beta}{\beta + 2}\right) I_{i+} \cong \alpha I_{i+} \tag{10.3}$$

The differential input current is

$$I_i = I_{i+} - I_{i-} \cong -I_{b2} = I_{c1} - I_{i-} \tag{10.4}$$

Current differencing accuracy is limited by β of Q_1 and is compensated by decreasing the area of the diode relative to the Q_1 *b–e* junctions or by a better current-mirror topology.

The feedback amplifier of Fig. 10.2 has a noninverting high-impedance voltage input and an inverting low-impedance current input. This strange combination has some advantages over feedback amplifiers with voltage-differencing inputs. The inverting input is, ideally, a voltage source ($R = 0$). Its voltage follows the noninverting input. This eliminates the possibility of error-voltage summation at this node; summation of feedback current results in the error current I_i. For this reason, this topology is called a *current-feedback op-amp.* The forward-path transmittance is a transimpedance amplifier with input I_i and output V_o. From the op-amp terminals, the topology is identical to that of a conventional voltage-gain, noninverting op-amp. But the dynamic response characteristics, both small- and large-signal, are quite different.

Figure 10.2b shows the amplifier flow graph. The input voltage, via the input buffer amplifier, causes error current I_i to flow through the Thévenin resistance $R_i \parallel R_f$ at the inverting terminal. Then I_i is amplified by the transimpedance Z_m to produce V_o. The feedback current is the current through R_f generated across the voltage difference $V_o - V_i$. Feedback analysis of the

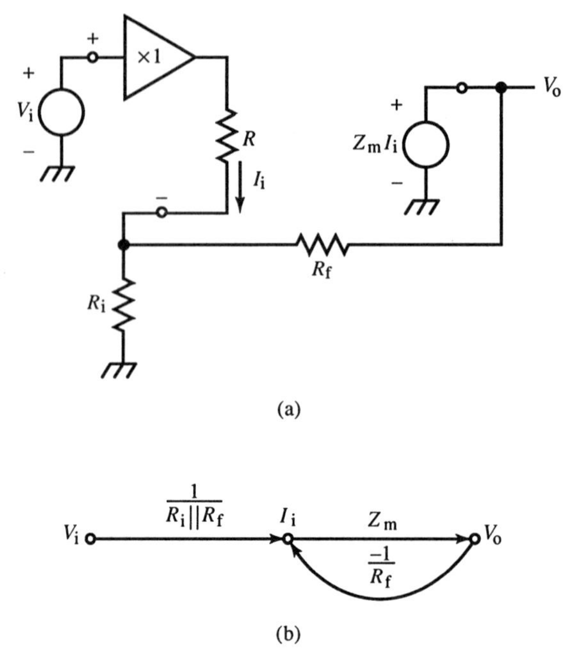

FIG. 10.2 Current-feedback amplifier topology (a) and flow graph (b).

closed-loop voltage gain yields

$$A_v = \alpha_i \cdot \frac{G}{1+GH} = \left(\frac{1}{R_i \parallel R_f}\right)\frac{Z_m}{1+Z_m(1/R_f)} = \left(\frac{R_f}{R_i}+1\right)\frac{Z_m}{Z_m+R_f} \tag{10.5}$$

Note that the first factor of (10.5) is the conventional noninverting op-amp closed-loop gain. For $Z_m \gg R_f$, it is the current-feedback amplifier closed-loop gain also.

The frequency response is derived from (10.5) by assuming, as we did for the voltage-gain op-amp, a single-pole roll-off. Let the transimpedance be

$$G = Z_m = R_{mo} \cdot \frac{1}{s\tau_{bw}+1} \tag{10.6}$$

Substituting into (10.5), we get a closed-loop gain of

$$A_v(s) = \left(\frac{R_f}{R_i}+1\right)\left(\frac{R_{mo}}{R_{mo}+R_f}\right)\frac{1}{s[R_f/(R_{mo}+R_f)]\tau_{bw}+1} \tag{10.7}$$

This result is significant. Unlike its conventional op-amp counterpart, the closed-loop bandwidth of this amplifier does not depend on the closed-loop gain. [Compare with (6.12) from Fig. 6.5. Also note (6.18).] In effect, there is no gain–bandwidth product; the bandwidth is independent of the gain. Both gain and bandwidth can be set independently if the bandwidth is first set with R_f and then the gain with R_i. Ideally, for infinite R_{mo}, the gain is exactly that of the noninverting op-amp formula, and the closed-loop bandwidth is infinite. Bandwidth actually increases slightly with increasing forward-path gain. In practice, τ_{bw} also depends on R_{mo} since the transimpedance amplifier itself has a gain–bandwidth product.

The bandwidth is independent of the gain only when the input buffer is ideal. With output resistance R, the inverting input is not constrained to be V_i, and a voltage-feedback interpretation could be given. But keeping with current feedback, we can modify the transmittances to account for R:

$$\alpha_i = \frac{1}{R + R_i \parallel R_f}, \qquad H = -\left(\frac{R_i}{R_f + R_i}\right)\left(\frac{1}{R + R_i \parallel R_f}\right) \tag{10.8}$$

G remains the same. These transmittances are found by Nortonizing the external feedback circuit to the inverting input and then solving for the current dividers in each direction. The resulting closed-loop gain is

$$A_v = A_{vc}\left(\frac{R_{mo}}{R_{mo} + R_f + RA_{vc}}\right)\frac{1}{s\tau_{bw}(R_f + RA_{vc})/(R_{mo} + R_f + RA_{vc}) + 1} \tag{10.9}$$

where

$$A_{vc} = \frac{R_f + R_i}{R_i} \tag{10.9a}$$

R is effectively increased by the closed-loop gain A_{vc} and also has the effect of adding to R_f. As A_{vc} increases, the time constant of (10.9) approaches the open-loop value of τ_{bw}. In good design, $R_{mo} \gg R_f$; therefore, the effective increase in R_f due to RA_{vc} only slightly decreases bandwidth. Equation (10.9) can be used to determine more precise values of R_f and R_i than (10.7), given R.

Current-feedback amplifiers also have a large-signal advantage over voltage op-amps. To show the contrast, we first examine the typical op-amp (Fig. 10.3) with three stages. (See Section 6.9.) The first stage has a transconductance of $G_m = 1/2r_e$ when balanced, loaded by the high-gain second stage with compensation capacitor C_C. This small on-chip capacitance is multiplied by the Miller effect and effects dominant single-pole compensation. The transconductance stage can supply, at most, I_0 to charge the second-stage input capacitance C_i. For large K_v,

$$C_i = (K_v + 1)C_C \cong K_v C_C, \qquad K_v \gg 1 \tag{10.10}$$

With a step input, the voltage, Δv_C, becomes slew-rate limited as Δv_I increases. The output current of the first stage is

$$i_O = G_m v_I \tag{10.11}$$

A maximum i_O of I_0 causes the output voltage to change at the maximum rate of

$$\text{slew rate of } v_O = \frac{dv_O}{dt} = K_v \cdot \frac{dv_C}{dt} = K_v \cdot \frac{I_c}{C_i} \cong \frac{I_0}{C_C} \tag{10.12}$$

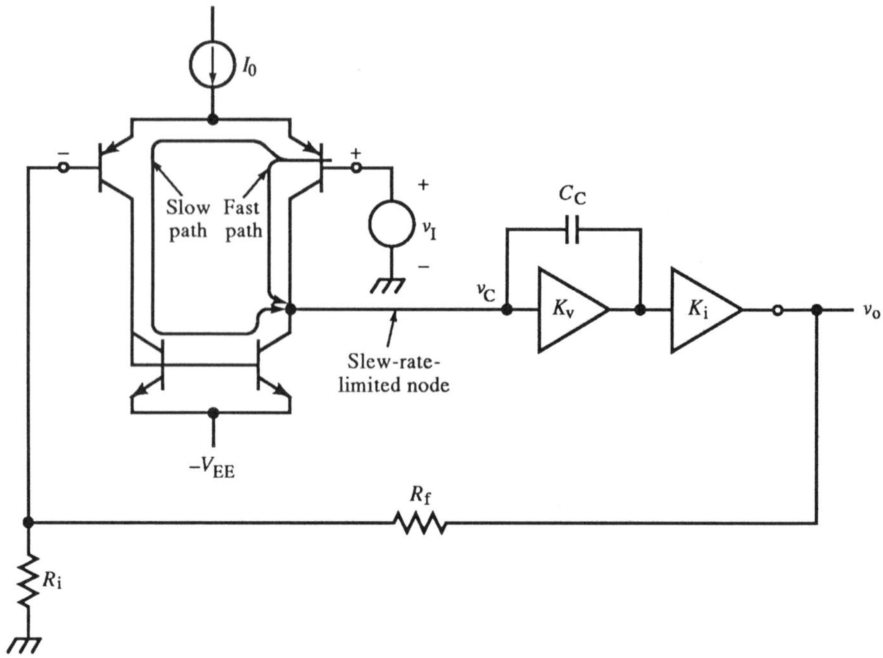

FIG. 10.3 Conventional three-stage op-amp. The input diff-amp has two paths from the noninverting input to the second-stage input.

The unity-gain frequency f_T corresponds to a time constant τ_T determined by the compensation. The small-signal exponential response to an input step has a maximum slope of

$$\max \frac{dv_O}{dt} = \frac{\Delta v_O}{\tau_{bw}} = \frac{\Delta v_I}{\tau_T} \tag{10.13}$$

The maximum small-signal step at the onset of slewing is thus

$$\text{maximum incremental step} = \max \Delta v_I = \frac{I_0}{C_C} \cdot \tau_T \tag{10.14}$$

The small-signal dynamic range of the input stage is limited to

$$\max \Delta v_I = \frac{I_0}{G_m} = I_0(2r_e) = I_0\left(2 \cdot \frac{V_T}{I_0/2}\right) = 4V_T \cong 100\text{ mV} \tag{10.15}$$

Then τ_T is determined by equating (10.14) and (10.15) and solving:

$$\tau_T = \frac{4V_T C_C}{I_0} \Rightarrow f_T = \frac{I_0}{8\pi V_T C_C} \tag{10.16}$$

The small-signal dynamic range at the input can be increased by decreasing G_m. This can be achieved by adding external R_E to the emitter circuit. In addition, the small-signal response of the input stage is complicated by its two signal paths. In the noninverting configuration of Fig. 10.3, v_i travels directly through the input transistor to the second stage. This is the direct path and is faster than the path through the other diff-amp transistor and the current mirror. The C_{be} of the mirror transistors adds an additional pole to this route, leading to a combined response in v_c.

This split-path response is eliminated in the circuit of Fig. 10.4, in which the additional shunt diff-amp at the input provides differential drive to the second stage. It also drives the current mirror with a common-mode dc current, the effect of cancelling inverted signals. The resulting current biases the mirror. No gain is lost due to elimination of the slow path because the second stage is driven differentially. The LF400 JFET-input amplifier uses this technique.

Another "transconductance spoiler" is realized by making the diff-amp transistors with split collectors. The large area collectors are connected to the supply and shunt most of the signal current. In effect, the G_m is reduced by collector current-divider action. An example of the use of this technique is the National Semiconductor LM346 programmable op-amp.

The large-signal behavior of current-feedback amplifiers does not have this slew-rate limitation on input dynamic range. Figure 10.5 shows simplified circuitry of a typical current-feedback amplifier, the Analog Devices AD846. The inverting input connects to the output of a buffered complementary CC driven by the noninverting input. The output CC collector currents are a differential output of the input current difference at v_{I-}. The first stage of the transimpedance amplifier is a differential current mirror—one current mirror

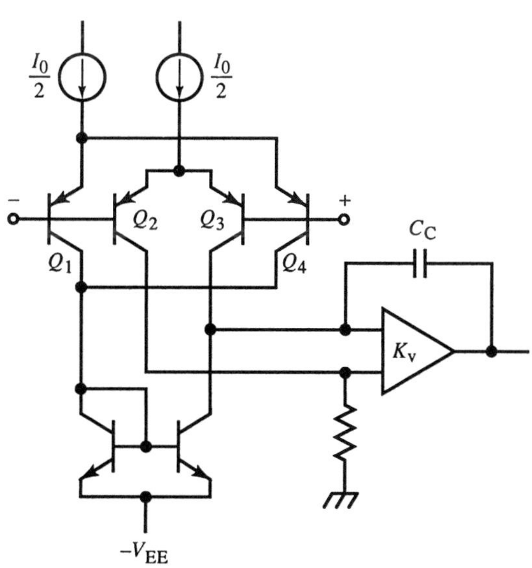

FIG. 10.4 The dynamic slow path of Fig. 10.3 is eliminated by common-mode biasing the current-mirror with a second diff-amp.

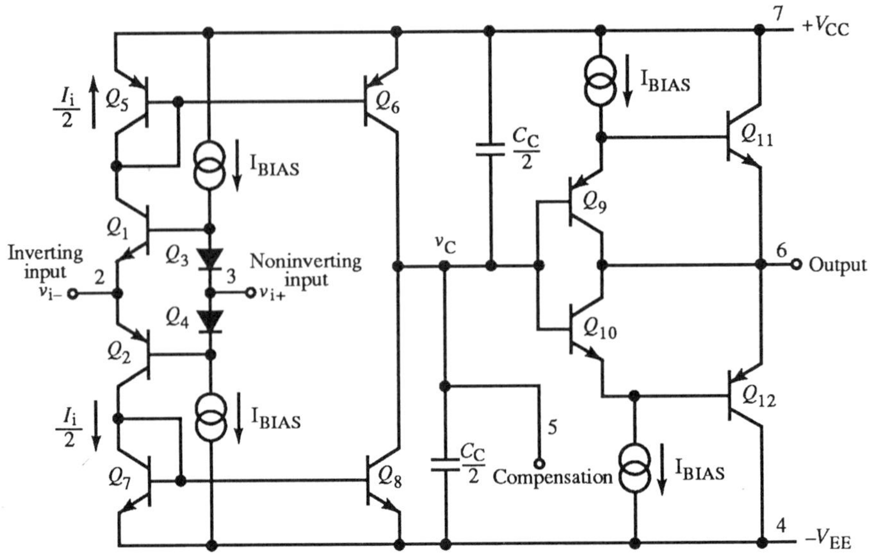

FIG. 10.5 Simplified topology of the AD846 current-feedback amplifier. I_i is the input error current to the transimpedance amplifier.

per side—that drives another complementary CC buffer. The dominant pole is determined at the high-resistance input node of this buffer, with dominant-pole capacitance C_C split between the power-supply "rails."

Slewing does not occur at v_C for a wide range of input voltages because a larger Δv_I produces a larger current into the transimpedance amplifier. This current is generated by Δv_I across the resistance driven by the inverting-input buffer,

$$r_{in-} = R + R_i \parallel R_f \tag{10.17}$$

The output slew rate is

$$\max \frac{dv_O}{dt} = \frac{i_C}{C_C} = \frac{(\Delta v_I / r_{in-})}{C_C} = \frac{\Delta v_i}{r_{in-} C_C} = \frac{\Delta v_O}{K_v r_{in-} C_C} \tag{10.18}$$

For $R = 0$, this reduces to

$$\max \left. \frac{dv_O}{dt} \right|_{R=0} = \frac{\Delta v_O}{R_f C_C} \tag{10.19}$$

and

$$\tau_{bw} = R_f C_C \tag{10.20}$$

In contrast to (10.16), no dependence on large-signal parameters appears in (10.20); the current-feedback amplifier is free of the slew-rate limitations of conventional op-amps.

Although the inverting input of a current-feedback amplifier has low open-loop impedance, since the error quantity is the terminal current I_i, it is nulled by feedback to a low value. Consequently, current-feedback op-amps can be used with the same external circuits as voltage-gain op-amps.

The current-feedback concept also applies to instrumentation amplifiers. Figure 10.6 shows the Precision Monolithics AMP-01 simplified topology, a typical current-feedback IA. This topology is different from the current-feedback op-amp because it has a voltage diff-amp input and a voltage-divider feedback path. The divider is followed by a voltage-to-current (V/I) converter consisting of a BJT diff-amp with emitter resistor R_S. The error current is generated by the cascoded diff-amps and in feedback nomenclature is

$$i_E = i_i - i_B = \frac{v_i}{R_G} - \left(\frac{R_i}{R_f + R_i}\right)\frac{v_o}{R_S} \tag{10.21}$$

The output voltage is

$$v_o = R_m i_E \tag{10.22}$$

where R_m is the transmittance of the transimpedance amplifier. Combining

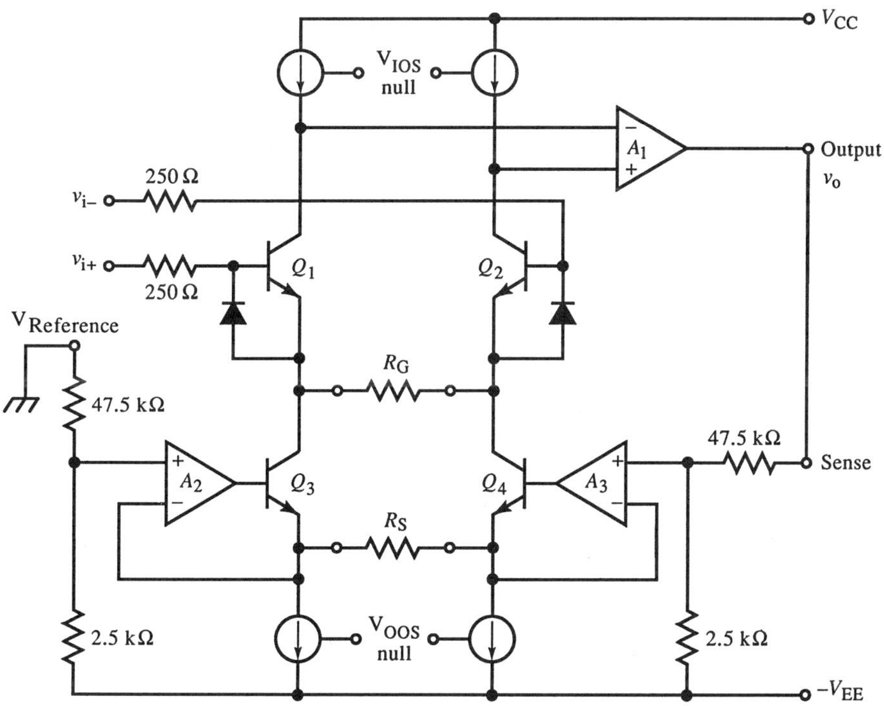

FIG. 10.6 Simplified topology of the AMP-01 current-feedback instrumentation amplifier. Error current is summed in the cascode stage.

(10.21) and (10.22) and solving for the closed-loop voltage gain, we obtain

$$A_v = \left(\frac{R_f + R_i}{R_i}\right)\left(\frac{R_S}{R_G}\right)\frac{1}{1 + [(R_f + R_i)/R_i](R_S/R_m)}$$

$$= A_{vo}\left(\frac{R_S}{R_G}\right)\frac{1}{1 + A_{vo} \cdot (R_S/R_m)} \tag{10.23}$$

Ideally, $R_m \gg R_S$, and the gain becomes

$$A_v\big|_{R_m \to \infty} = \left(\frac{R_f + R_i}{R_i}\right)\left(\frac{R_S}{R_G}\right) \tag{10.24}$$

The error current is generated as emitter current in the input diff-amp transistors. An input increase v_{i+} causes Q_1 to conduct more, resulting in output increase v_o. Through the feedback network, Q_4 is made to conduct more and Q_3 less. The error current for $i_{e1} = v_i/R_G - i_{c3}$, where the first term is i_i. A similar expression for i_{e2} can be written. Current-feedback IAs have the same basic properties as current-feedback op-amps; bandwidth remains relatively independent of gain.

In passing, note that the AMP-01 has two adjustments to correct offset error. (See Section 9.12.) The IA requirement for two adjustments is somewhat different from that of a two-transistor diff-amp. At the output, the total offset voltage is the sum of gain-dependent (input) and gain-independent (output) offsets:

$$V_{OS} = A_v V_{IOS} + V_{OOS} \tag{10.25}$$

A similar equation holds for the offset voltage TCs. The input offset voltage adjustment varies the current ratio of the collector current sources at the output of the first stage. With R_G shorted, the offset voltage of Q_1 and Q_2 is most sensitive to their emitter current ratio. With R_G unshorted, the output offset voltage is adjusted by the emitter current ratio of Q_3 and Q_4 since they are part of the output feedback path. That is, the gain from output to R_S is $1/A_{vc}$, whereas the gain from the collector current sources to the output - the transimpedance amplifier gain - is very large.

10.2 Split-Path, Low-Frequency Feedback, and Feedbeside Amplifiers

In Chapter 4, we investigated the effect of r_o in simple BJT amplifiers and the passive forward path through feedback networks. These paths are largely unintentional and generally degrade performance. Amplifiers designed to benefit from multiple paths are *composite* amplifiers. Various topologies at the subsystem level are possible. The *split-path amplifier* (Fig. 10.7) purposely has

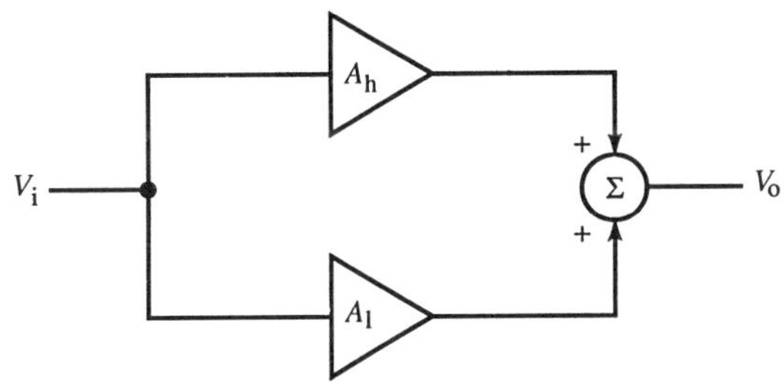

FIG. 10.7 Split-path amplifier.

parallel paths to improve performance. Because fast amplifiers often lack good dc characteristics, the idea is to combine the best of both in one amplifier. A common strategy is to shunt a fast amplifier with a low-speed, precision amplifier. The two paths are combined at the output.

A requirement for flat frequency response is that the fast and slow paths have complementary gains in the crossover frequency range so that their sum is constant. The crossover gain must also be the same as the low- and high-frequency gains. A simple design strategy is to let the slow path have a single-pole response:

$$A_l = K\left(\frac{1}{s\tau_l + 1}\right) \tag{10.26}$$

The fast path is an ac amplifier (no gain at dc) that rolls up in gain as A_l rolls off. At a high frequency of $1/\tau_h$, it also rolls off:

$$A_h = K\left(\frac{s\tau_l}{s\tau_l + 1}\right)\left(\frac{1}{s\tau_h + 1}\right) \tag{10.27}$$

The composite gain is

$$A = A_l + A_h = K\left(\frac{1}{s\tau_h + 1}\right)\left(\frac{s(\tau_l + \tau_h) + 1}{s\tau_l + 1}\right) \tag{10.28}$$

With $\tau_h = 0$, the two paths combine to give a flat response. To approach this ideal, the crossover frequency $1/\tau_l$ must be much less than the fast path bandwidth of $1/\tau_h$.

This constraint to flat response can be eliminated by adding a compensation stage A_c in the slow path (Fig. 10.8). It compensates for τ_h of the fast path and is

$$A_c = \frac{1}{s\tau_h + 1} \tag{10.29}$$

The modified slow-path response is thus

$$A_{lc} = K\left(\frac{1}{s\tau_l + 1}\right)\left(\frac{1}{s\tau_h + 1}\right) \tag{10.30}$$

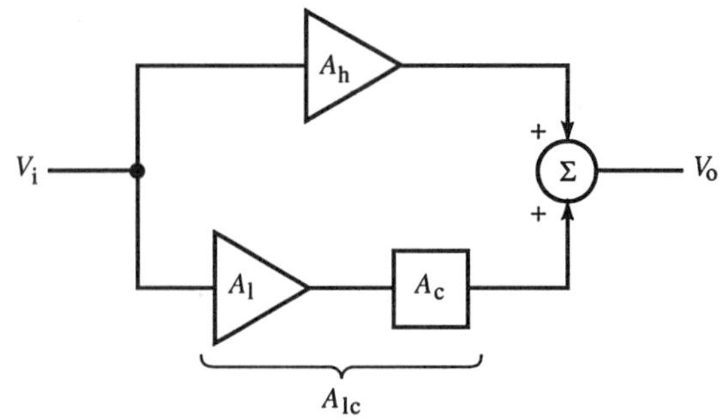

FIG. 10.8 Compensated split-path amplifier, in which frequency response is independent of the crossover frequency.

Now the composite gain is

$$A = A_{lc} + A_h = K\left(\frac{1}{s\tau_h + 1}\right) \tag{10.31}$$

and the crossover frequency $1/\tau_l$ is independent of fast-path bandwidth.

Example 10.1 Split-Path Composite Trigger Amplifier

The amplifier of Fig. E10.1 is typical of the input to an oscilloscope trigger generator. Various sources are selected for v_i, and either lf or hf

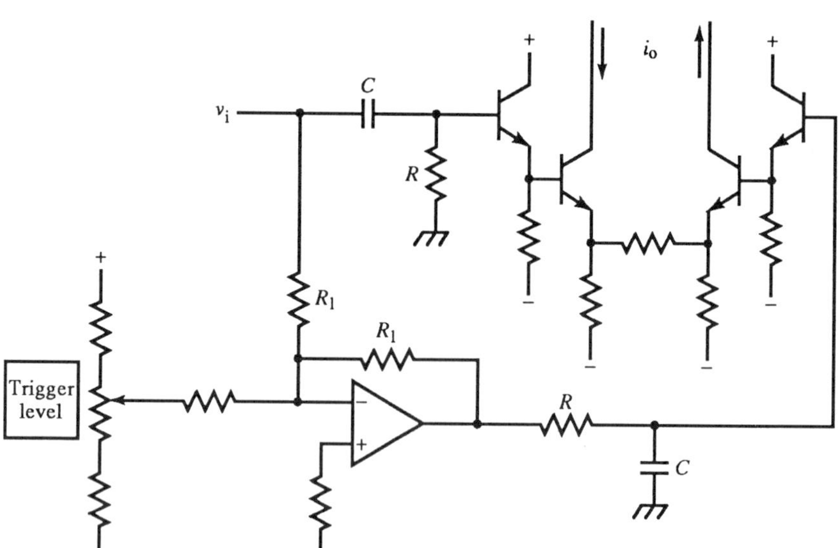

FIG. E10.1

paths can be turned off for "lf reject" or "hf reject" trigger functions. The single-ended v_i is converted to a differential signal for the diff-amp input by the inverting op-amp in the slow path. The trigger level control is summed in the slow path. For good dynamic response, the fast-path RC differentiator and slow-path RC integrator time constants must be equal. The low-pass pole of the slow path could be realized by placing a capacitor around the slow-path op-amp, but its inability to reject high frequencies makes this capacitor an undesirable hf feedthrough. A passive RC integrator following the op-amp is a broadband low-pass filter.

Another composite amplifier topology that is intended to accomplish the same design objective is the *low-frequency feedback* topology (Fig. 10.9). Low-frequency correction is made at the input to the fast amplifier by adding to V_i the error quantity,

$$V_l = A_l(V_i - HV_o) \tag{10.32}$$

The output is

$$V_o = A_h(V_i + V_l) \tag{10.33}$$

Combining (10.32) and (10.33) yields the composite gain:

$$A = A_h \cdot \frac{1 + A_l}{1 + HA_hA_l} = \underset{\substack{\uparrow \\ \text{fast path}}}{\frac{A_h}{1 + HA_hA_l}} + \underset{\substack{\uparrow \\ \text{slow path}}}{\frac{A_hA_l}{1 + HA_hA_l}} \tag{10.34}$$

Both paths benefit from feedback, with a loop gain of HA_hA_l. For large A_l at low frequencies, the loop gain is large, and dc characteristics are improved over those of A_h alone. To frequency-compensate this amplifier, we assume

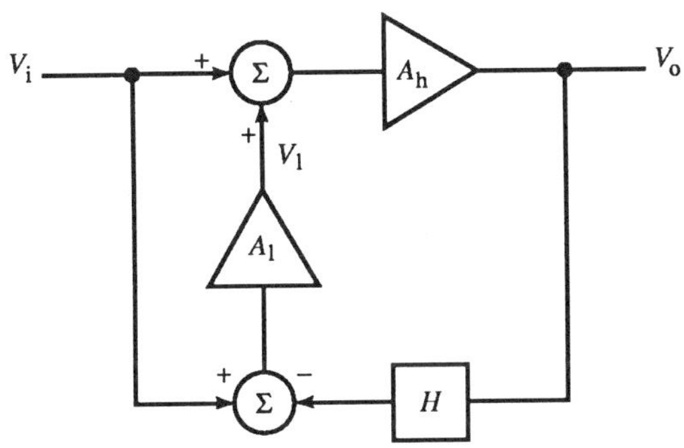

FIG. 10.9 Composite low-frequency feedback amplifier.

that A_l rolls off with frequency, so that

$$\lim_{f \to \infty} A_l = 0 \tag{10.35}$$

Consequently,

$$\lim_{f \to \infty} HA_h A_l = 0 \tag{10.36}$$

and the fast-path response is

$$\lim_{f \to \infty} \frac{A_h}{1 + HA_h A_l} = A_h \tag{10.37}$$

The slow-path response is

$$\lim_{f \to \infty} \frac{A_h A_l}{1 + HA_h A_l} = 0 \tag{10.38}$$

At high frequencies, only the fast path contributes A_h to the gain. If we also assume that A_l is an op-amp with infinite gain at dc, then the fast-path response at dc is

$$\lim_{f \to 0} \frac{A_h}{1 + HA_h A_l} = 0 \tag{10.39}$$

and for the slow path,

$$\lim_{f \to 0} \frac{A_h A_l}{1 + HA_h A_l} = \lim_{f \to 0} \frac{A_h}{1/A_l + HA_h} = \frac{1}{H} \tag{10.40}$$

Then $A = 1/H$ at dc. This is the same gain as for the inverting op-amp configuration. For flat response, the gain at dc must be the same as at high frequencies, or

$$\frac{1}{H} = A_h \Rightarrow H = \frac{1}{A_h} \tag{10.41}$$

This condition is necessary but not sufficient for flat response; the midfrequency range might not be flat. In (10.34), the $(1 + A_l)$ factor in the numerator cancels the denominator only under this condition and ensures wideband flatness. The compensated gain is

$$A|_{H = A_h^{-1}} = A_h \tag{10.42}$$

This topology has the practical disadvantage that a slow amplifier A_l is driven by a fast signal V_i. Unless the feedback compensation is correct, A_l responds to a fast error voltage but is unable to follow it. The effect is a low-frequency response anomaly; the cause, however, is that a slow amplifier is responding to a fast input. In addition, because of loop delay (or phase lag) through A_h and H, the output of H cannot match V_i in phase. At frequencies for which the delay is significant, this error quantity becomes large.

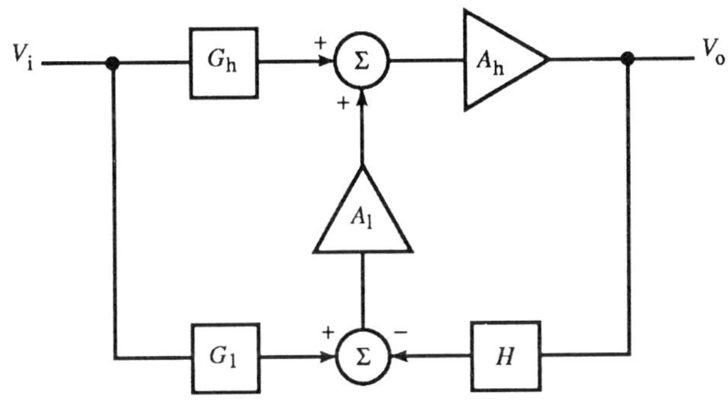

FIG. 10.10 General low-frequency feedback topology.

The general low-frequency feedback topology is shown in Fig. 10.10. Its voltage gain is

$$A = A_h \cdot \frac{G_h + G_l A_l}{1 + H A_h A_l} \tag{10.43}$$

The fast path goes through G_h and A_h and is represented by the first term; the slow path goes through G_l, A_l, and then A_h, which are factors of the second term of (10.43). The feedback loop contains the blocks represented in the denominator.

A special case of Fig. 10.10 has $H = G_h = 1$. It has the same problem with high-frequency input to the slow path as that of Fig. 10.9, but coming from V_o. For it, the composite gain is

$$A = \frac{A_h(1 + G_l A_l)}{1 + A_h A_l} = \frac{A_h}{1 + A_h A_l} + \frac{1 + G_l A_l}{1 + A_h A_l} \tag{10.44}$$

With the compensation criterion of this amplifier, the gain is

$$A\big|_{G_l = A_h} = A_h \tag{10.45}$$

This topology has the advantage that a matching A_h can be used to provide tracking compensation of the fast path.

Example 10.2 Low-Frequency Feedback Composite Amplifier

The amplifier design of Fig. E10.2 is based on the following reasoning; because the dc characteristics of A_h are poor, the ac couples to it through an *RC* differentiator and adds op-amp feedback from A_l at its input. The G_l block of Fig. 10.10 is set according to (10.45) and to make A_2 differential. Will the amplifier provide a flat frequency response?

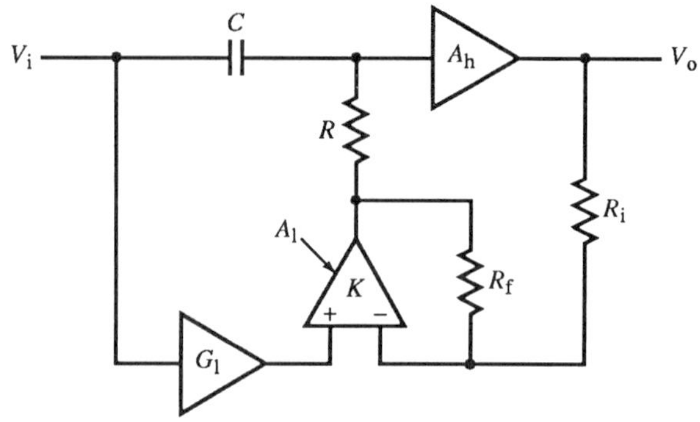

FIG. E10.2

For this amplifier, the blocks of Fig. 10.10 are as follows:

$$G_l = A_h\left(\frac{R_f}{R_f + R_i}\right), \qquad G_h = \frac{sRC}{sRC+1}$$

Then

$$A = A_h\left(\frac{G_h + A_l}{1 + A_l}\right), \qquad A_l = K\left(\frac{R_f}{R_i}\right)$$

Substituting for G_h,

$$A = A_h\left(\frac{A_l}{1+A_l}\right)\frac{sRC\left(\frac{1+A_l}{A_l}\right)+1}{sRC+1}$$

and has a flat response only for infinite K.

Example 10.3 Low-Frequency Feedback Input Buffer

The buffer amplifier of Fig. E10.3 has a high-impedance input because of the FET CD input stage. The FET has voltage offset and drift that are corrected by a slow path that varies the FET current to achieve zero dc offset voltage.

This amplifier is represented by the general low-frequency feedback topology of Fig. 10.10 and (10.43). Since it is a buffer (with a ×1 gain), $A_h \cong 1$. From the circuit diagram, the FET CD is $G_h = 1$, and $G_l = H$. Both G_l and H are RC integrators. A_l is thus driven only by low-

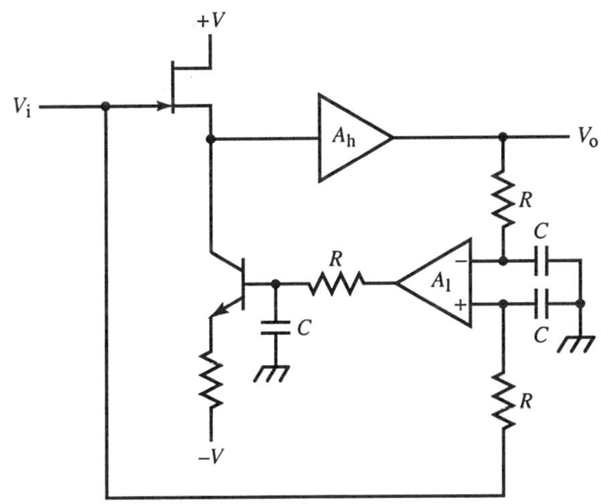

FIG. E10.3

frequency signals. Substituting into (10.43) gives

$$A = A_h \cdot \frac{1 + G_l A_l}{1 + G_l A_l A_h}$$

If $A_h = 1$, the fraction is unity, and the gain is A_h; otherwise, gain error in A_h contributes to compensation error. For large A_l, the fraction approaches $1/A_h$, and A approaches unity. The RC time constant is set well within the bandwidth of A_l. This buffer eliminates the need for matched high-frequency FETs.

The two concepts of split-path and low-frequency feedback are combined in the amplifier of Fig. 10.11. Its gain is

$$A = \frac{A_h + G_i A_l}{1 + G_o A_l} \tag{10.46}$$

Factoring A_h from the numerator, we get the gain A_h under the condition

$$A\big|_{G_i = G_o A_h} = A_h \tag{10.47}$$

If G_o is a low-pass filter, then the input of A_l is always low in frequency. This eliminates high-frequency rectification at the inputs due to op-amp inability to servo quickly enough. It also keeps A_l from responding slowly to fast inputs.

Flat frequency response is not the only criterion of precision. Fast amplifiers typically have poorer linearity and more thermal drift and distortion than low-frequency amplifiers. Nonlinearity is greatest at the extremes of the

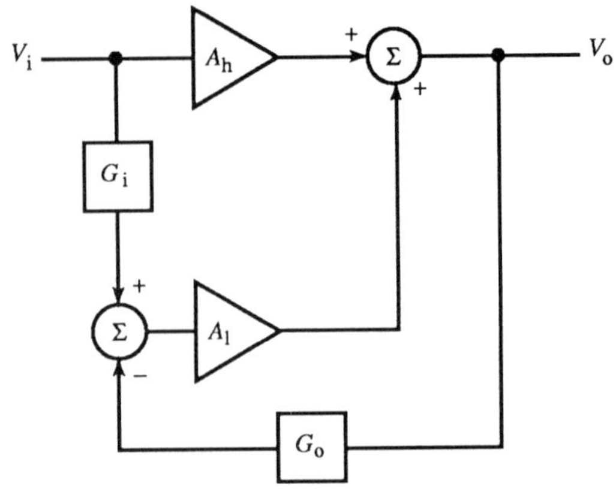

FIG. 10.11 Combined split-path and low-frequency feedback topologies.

dynamic range, as in Fig. 10.12a. Usually, gain decreases at the extremes of V_i. The amplifier has constant gain over its dynamic range at low frequencies (Fig. 10.12b) due to feedback. The step response is shown in (c); the gain of A_h is less than the low-frequency gain of A. As the slow path begins to respond, the gain increases.

To correct for this nonlinearity, A_h can be linearized. Also, temperature effects of A_h can be compensated. But it is better in wideband amplifiers to minimize the complexity of the compensation networks at high-frequency nodes due to their additional stray reactance. A complicated slow-path linearity or thermal correction network avoids this problem.

One such scheme (Fig. 10.13) sums at the input to A_h. The input is sampled by a slow compensation path that corrects for imperfections in A_h. This scheme, called *feedbeside*, is open-loop. Its compensation network parameters are adjusted to correct for low-frequency imperfections in A_h. Its gain is

$$A = A_h \cdot (1 + GA_l) \tag{10.48}$$

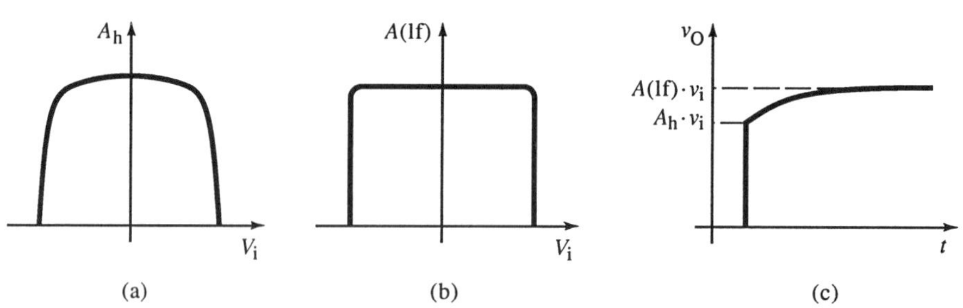

FIG. 10.12 Fast-amplifier gain curve (a) showing typical compressive effect near input range limits. Low-frequency amplifier path (b) has better characteristics. The combined-path transient response (c) at range extremes.

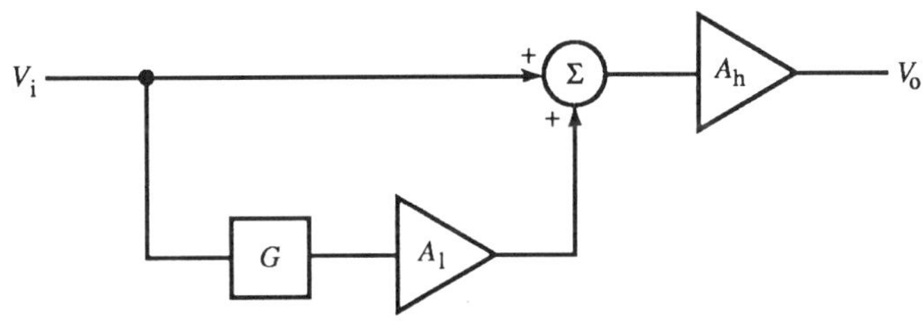

FIG. 10.13 Feedbeside amplifier.

where G is a passive, attenuating compensation network. A_l provides scaling and inversion, if needed, so that GA_l matches the reciprocal of the error in A_h. The feedbeside path can itself be a split-path amplifier, in which each path independently compensates for an anomaly in A_h.

An amplifier involving feedbeside is that of Fig. 10.14a. A fast op-amp A_h requires input offset-voltage correction provided by the slower, more precise

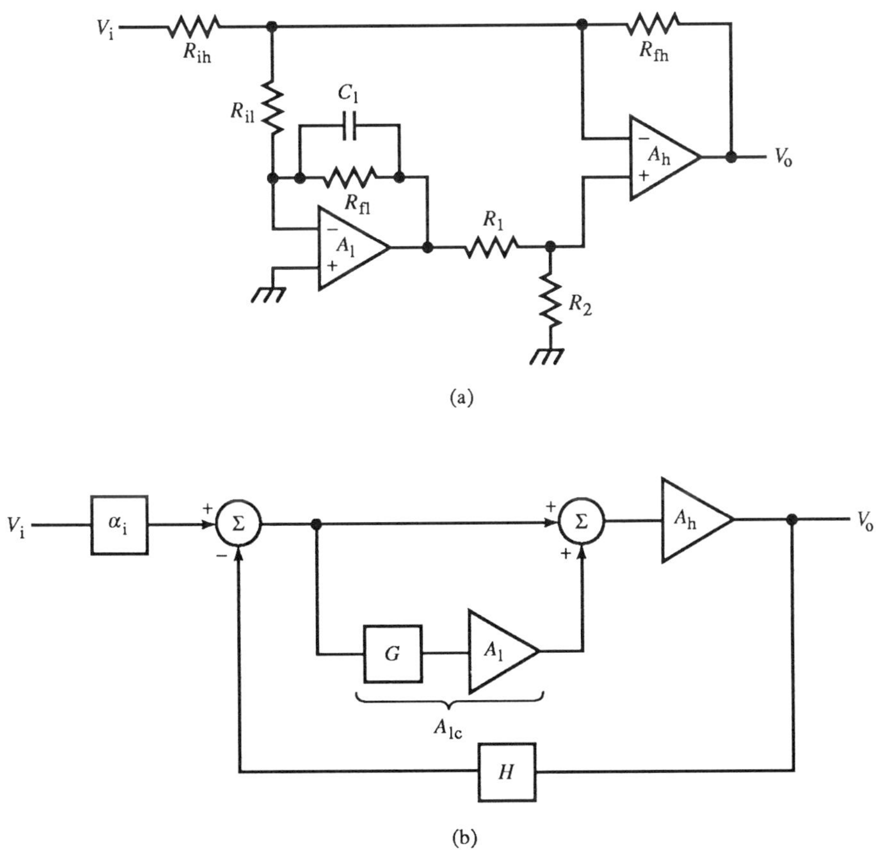

FIG. 10.14 Input offset nulling scheme (a) combines feedbeside within a feedback loop.

A_l. C_l limits the bandwidth of the feedbeside path to near dc; R_1 and R_2 reduce the gain of the slow path.

This amplifier has the block diagram of Fig. 10.14b and is not a pure feedbeside topology since the slow path is within the fast-path feedback loop. The general form of the gain found from (b) is

$$A = \alpha_i \cdot \frac{A_{hc}}{1 + HA_{hc}}, \qquad A_{hc} = A_h(1 + GA_l) \tag{10.49}$$

This topology has the form of classical feedback with feedbeside within the loop, modifying A_h. The feedbeside loop provides added gain in the fast loop at low frequencies. The attenuator consisting of R_1 and R_2 reduces the gain of this low-frequency loop to stabilize it.

The feedbeside topology is used in the Tektronix 7104 1 GHz oscilloscope vertical amplifier to correct wideband vertical stages for thermal effects. It is used in a different way in the LMC669 as an offset voltage compensator for op-amp inputs. This auto-zero IC samples the inverting virtual ground input of an op-amp and controls its noninverting input to null the voltage offset, much like the circuit of Fig. 10.14.

10.3 Feedforward and Linearized Differential Cascode Amplifiers

Feedback compares output with input and drives the forward-path amplifier with the error, thus correcting the output. Instead of applying the error to the input, the *feedforward* scheme adds a compensating error quantity at the output. The feedforward topology (Fig. 10.15) has a forward path through G. Its output is fed back through H and subtracted from the amplifier input

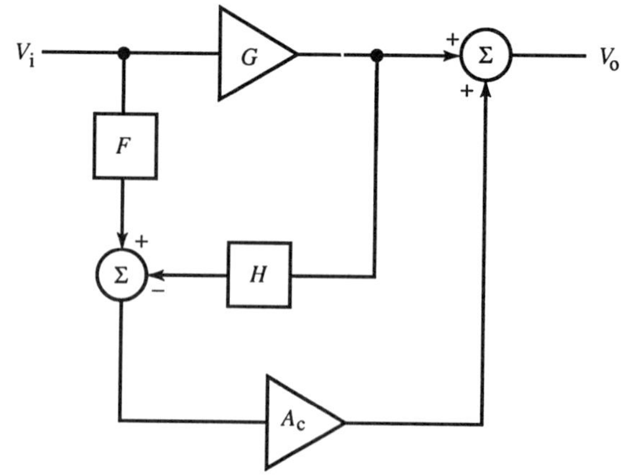

FIG. 10.15 Feedforward amplifier.

through F, resulting in the error of G in amplifying the input. This error is scaled by compensation amplifier A_c to the same magnitude as the output of G. When it is then added to the output of G, the error terms in GV_i are nulled.

The gain of the feedforward topology is

$$A = G + A_c(F - GH) \tag{10.50}$$

To demonstrate error reduction, let

$$G = K + \varepsilon \tag{10.51}$$

where K is the linear gain of G and ε is the nonlinear distortion terms. For feedforward error nulling, the scaling of F, H, and A_c must be correct. F and H are passive attenuators and are assumed linear. The input to A_c should be only error terms; V_i is nulled by scaling F and H so that

$$KH = F \Rightarrow H = \frac{K}{F} \tag{10.52}$$

In addition, since F has scaled down the input, A_c must amplify it to correct for scaling and have a gain of K/F. Since A_c also has distortion, and assuming nonlinear terms scale with amplitude, then let

$$A_c = \frac{K + \varepsilon_c}{F} \tag{10.53}$$

Substituting (10.51)-(10.53) into (10.50), we get the gain:

$$A = K - \frac{\varepsilon\varepsilon_c}{K} \tag{10.54}$$

The feedforward advantage is that distortion is reduced from ε to a magnitude of about ε^2/K. For small ε, this is a large improvement of linearity. In practice, it requires accurate scaling and gain-matching of transmittances, whereas feedback does not. Feedforward amplifiers can be made faster than feedback amplifiers since loop delay is not a limitation. However, the error-path delay must match the main path for correct output summation. Although F and H are passive and can have little delay, the gain-bandwidth demand on A_c is greater than G since it has more gain and requires slightly greater bandwidth due to F and H. But since it is amplifying an already small error, its linearity need not be high.

Example 10.4 Op-Amp Feedforward Amplifier

The feedforward amplifier of Fig. E10.4 uses two inverting op-amps to give a gain of -10. Input summing is done at the inverting input of the error amplifier, and output summing is done with resistors. The output

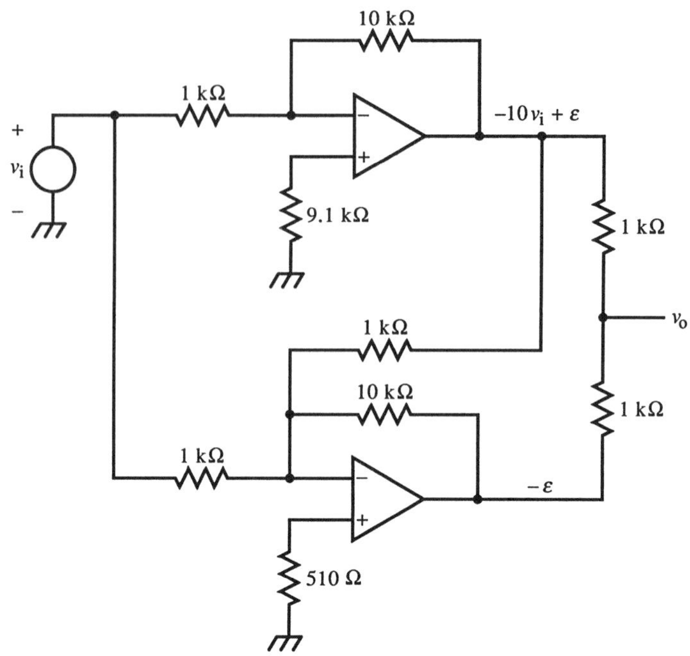

FIG. E10.4

might need to be buffered, and the output summer could be the input of another inverting op-amp. A dual op-amp IC provides matched op-amps. The resistors must be matched or, at least, have 1% tolerances.

The feedforward topology can be used to linearize fast BJT diff-amps. Nonlinearity is caused by variation in transistor parameters r_e, β, r_o, and f_T. The most significant is r_e. As v_i varies, the emitter current,

$$i_e = \frac{v_i}{r_{e1} + r_{e2} + 2R_E + 2R_B/(\beta + 1)} \tag{10.55}$$

also varies with r_{e1}, r_{e2}, and β. Transconductance is largest when v_i is zero (Section 9.11) and decreases with $|v_I|$. This is a *compressive* (versus *expansive*) gain characteristic. The emitter-referred base resistance term in (10.55) also varies with $\beta(i_E)$. Input impedance varies expansively with r_e variation, forming a nonlinear divider with base resistance. Since f_T also varies with emitter current, C_π varies, causing input capacitance to vary too. Since r_e is proportional to V_T and hence absolute temperature, gain varies with temperature. To stabilize, either R_E must be increased or the emitter current source I_0 must track V_T.

All of these causes of distortion can be compensated individually, but this approach is complicated and difficult to achieve. A simpler, more elegant solution is to remove them all together by nulling the combined error with feedforward. In particular, we concentrate on b–e junction nonlinearity. Techniques for correcting β or α follow.

In 1976, Pat Quinn invented the feedforward amplifier of Fig. 10.16. Q_1 and Q_2 comprise the diff-amp and Q_3 and Q_4 the error amplifier, also a diff-amp. The error amplifier output current is summed with the main output current at their collectors. Instead of feeding back from the main collectors, the error is sensed at the emitters and summed with resistors R_1 through R_4. Applying KVL around the main diff-amp input loop, we get

$$v_I = v_2 - v_1 = v_{BE2} - v_{BE1} + (i_{E2} - i_{E1})R_E = \Delta v_{BE} + \Delta i_E \cdot R_E \quad (10.56)$$

By superposition, the input voltages to the error amplifier are

$$v_3 = v_1\left(\frac{R_3}{R_1 + R_3}\right) + v_{E2}\left(\frac{R_1}{R_1 + R_3}\right) \quad (10.57)$$

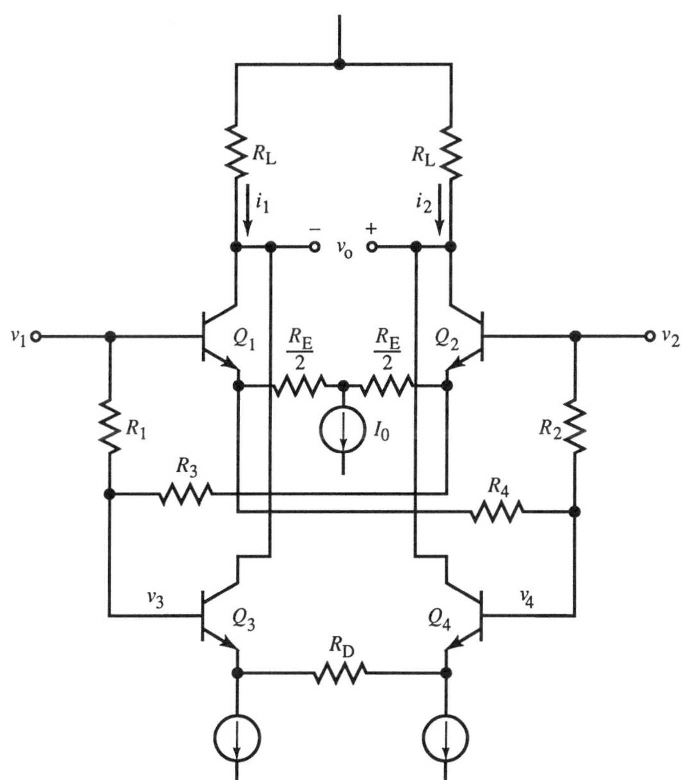

FIG. 10.16 Diff-amp feedforward amplifier with diff-amp error amplifier.

and

$$v_4 = v_2\left(\frac{R_4}{R_2+R_4}\right) + v_{E1}\left(\frac{R_2}{R_2+R_4}\right) \tag{10.58}$$

Feedforward scaling requires

$$R_1 = R_3, \qquad R_2 = R_4 \tag{10.59}$$

Then the dividers on each side have the same attenuation F of $\frac{1}{2}$, and the error amplifier differential input is

$$v_4 - v_3 = \frac{\Delta v_{BE}}{2} \tag{10.60}$$

This input contains only the scaled Δv_{BE} error term of (10.56), as required for feedforward. The error amplifier transconductance must now be scaled to cancel the Δv_{BE} terms of the main output current. The main amplifier output current is

$$i_O = \alpha \cdot \Delta i_E = \alpha\left(\frac{v_I}{R_E} - \frac{\Delta v_{BE}}{R_E}\right) \tag{10.61}$$

where $\alpha_1 = \alpha_2 = \alpha$. The output current of the error amplifier is

$$i_O = \frac{\Delta v_{BE}}{2} \cdot \frac{\alpha}{R_D + r_{e3} + r_{e4} + 2R_B/(\beta+1)} \tag{10.62}$$

where $R_B \cong$ the Thévenin resistance of the dividers. Error amplifier gain scaling requires that

$$R_E = 2\left(R_D + r_{e3} + r_{e4} + \frac{2R_B}{\beta+1}\right) \cong 2R_D, \qquad R_D \text{ dominant} \tag{10.63}$$

With this scaling, Δv_{BE} currents cancel at the output, resulting in an amplifier gain of

$$A = -2\alpha \cdot \frac{R_L}{R_E} \tag{10.64}$$

with b-e nonlinearity removed. The error amplifier has twice the gain of the main amplifier. If the transistors are of the same type, then precision amplification extends to about half the bandwidth of the main path.

The feedforward diff-amp was succeeded by the *cascomp* (Fig. 10.17), another invention of Pat Quinn at about the same time. The main amplifier is a differential cascode. Because the same currents (apart from i_B) flow through the CB as the CE, the nonlinearity of the CE transistors is duplicated at the emitters of the CB. Because the CB base voltages are fixed, the differential emitter voltage is the error signal, conveniently ground-referenced. The resistive dividers are not needed, and the error amplifier is driven directly by the CB Δv_E. The error amplifier can be another diff-amp (Fig. 10.17b).

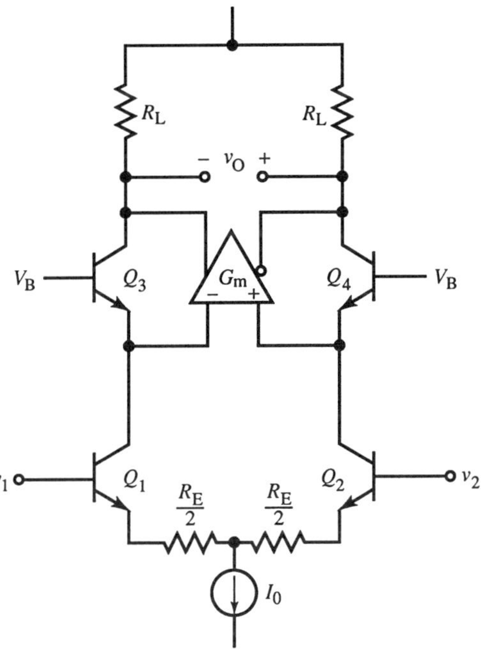

(a)

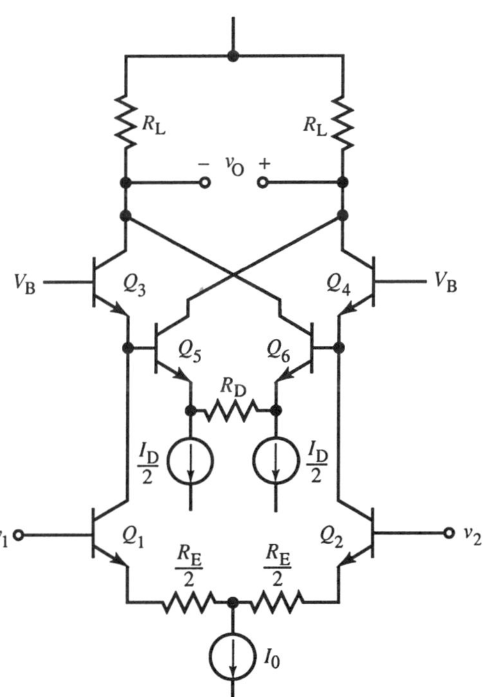

(b)

FIG. 10.17 The general cascomp topology (a), with diff-amp error amplifier (b).

The error is nulled when

$$G_m = \frac{\alpha}{R_D + r_{e5} + r_{e6}} = \frac{\alpha}{R_E} \tag{10.65}$$

For dominant R_D, the condition for cascomp error-path gain is that $R_D = R_E$. The error amplifier corrects for main amplifier error, which is worst at the limits of its dynamic range. It is here that error-amplifier correction is most needed, and its gain is matched at these range limits.

Since the emitter bias current I_D is arbitrary, criteria for its optimization can be sought. Increasing I_D reduces error amplifier Δr_e, increasing its linearity. It also increases error-amp base current I_B, which is part of the CB emitter current. This causes $v_{BE}(CB)$ error but tends to compensate for dc α loss of the CE input transistors. By adding I_B-compensating sources at the error amplifier bases, total amplifier linearity is improved and dynamic range extended. For $I_D/I_0 > 3$, a total current $I_D + I_0$ in an uncompensated cascode reduces its Δr_e to the point of comparable linearity to the cascomp. An optimum ratio of I_D/I_0 is about 2.

The error amplifier must be linear over an input range of $\Delta v_{BE}(CB)$, which is limited by I_0. Its maximum dynamic range is

$$\Delta v_{BE}(CB) = V_T \ln\left(\frac{I_0}{I_B}\right) \tag{10.66}$$

since the minimum emitter current of Q_3 and Q_4 is I_B. This base current reduces $\Delta v_{BE}(CB)$, effectively reducing the gain and dynamic range of the error amplifier. Also, the CE emitter current generates $v_{BE}(CE)$, but its collector current generates $v_{BE}(CB)$. This current mismatch causes a corresponding mismatch between $\Delta v_{BE}(CE)$ and $\Delta v_{BE}(CB)$. This "α error" can be compensated by adding a base resistor to the CB. It drops additional voltage that compensates for the loss of i_B.

The cascomp is thermally compensated by stacking another differential CB onto the output of the cascomp CB transistors. The output CB base voltage is set to provide the same thermal operating point as the CE stage beneath.

10.4 α-Compensated Gain Cells

Amplifier stages with sub-1% nonlinearities must compensate for α error due to finite β and loss of base current. Two multiple-transistor stages with feedback that function as a gain unit, or *cell*, were devised by Ken Schlotzhaur (Fig. 10.18) and Stewart Taylor (Fig. 10.19). These gain cells both use α-compensation techniques.

The Schlotzhaur cell is a diff-amp with CC feedback to Q_2. The feedback loop has little delay, and the cell is fast. Two of these cells are used as ×1 buffer amplifiers. They apply the differential input voltage that appears across

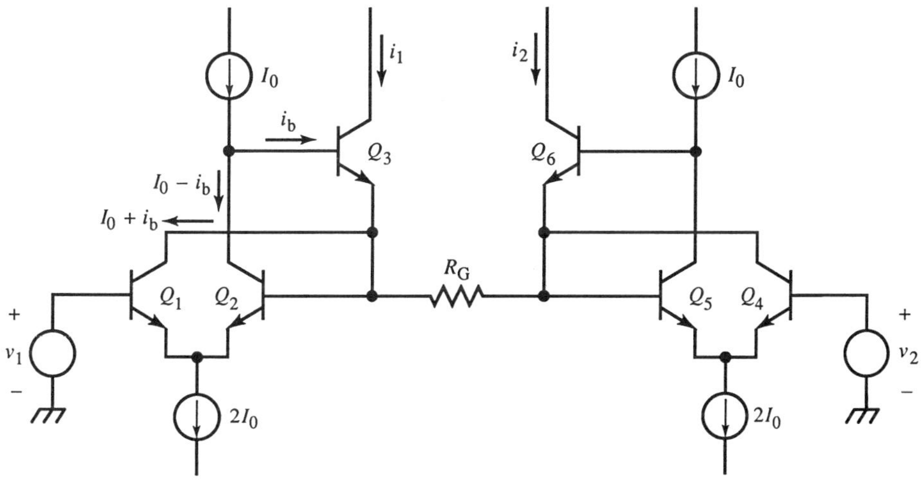

FIG. 10.18 Schlotzhaur gain cell with α compensation.

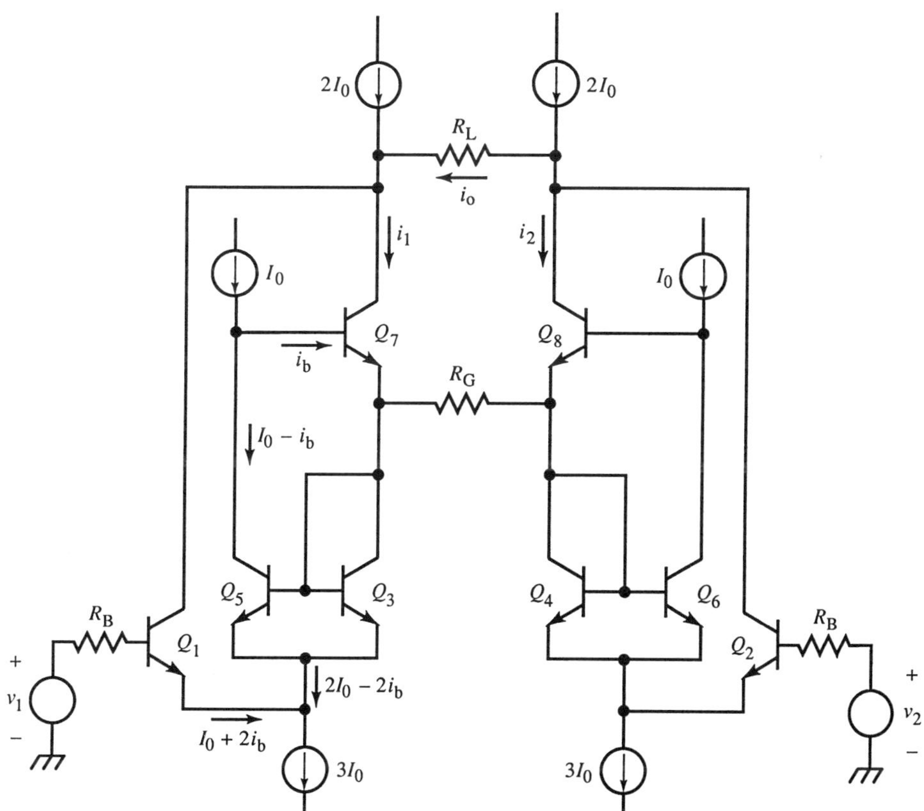

FIG. 10.19 Taylor gain cell with α compensation.

R_G. The differential transconductance is therefore R_G, or $R_G/2$ per side. Since it is a noninverting configuration, the input impedance is large due to feedback.

In the left cell, the α loss in the CC (Q_3) is compensated by connecting the collector of Q_1 to the emitter of the CC instead of the supply. The current through R_G is then the output current, i_1. The base current added to i_1 is removed as Q_1 collector current. Since

$$i_{c1} + i_{c2} = 2I_0 = \text{constant} \tag{10.67}$$

the base current of Q_3 lost from i_{C2} must add to i_{C1}. The resulting current in R_G is i_1. The shunting base impedance of Q_2 is negligible if the loop gain is large.

The error quantity on the left side is

$$E = v_1 - v_{B2} \tag{10.68}$$

Applying feedback analysis, the forward gain to R_G is

$$G = \frac{v_{B2}}{v_1} = \alpha \cdot \frac{(\beta+1)(r_e + R_G/2)}{2r_e} \cdot \left(\frac{R_G/2}{R_G/2 + r_e}\right) - \alpha \cdot \frac{R_G/2}{2r_e} \tag{10.69}$$

assuming symmetrical circuitry for a virtual ground at the midpoint of R_G and equal r_e. The first term is due to i_{C2} and the second to i_{C1}. The load resistance of Q_2 is the input resistance of Q_3. G simplifies to

$$G = \alpha \cdot \frac{\beta(R_G/2)}{2r_e} \tag{10.69a}$$

and $H = 1$, due to the direct connection. The closed-loop voltage gain is thus

$$\frac{v_{B2}}{v_1} = \frac{R_G/2}{2r_e/\alpha\beta + R_G/2} \tag{10.70}$$

When $2r_e/\alpha\beta \ll R_G/2$, the gain approaches one, and the nonlinearity due to r_e is less than that of a diff-amp with $R_G/2$ emitter resistance by β times. The single-ended incremental input resistance is

$$r_{in} = \frac{v_1}{i_{b1}} = \frac{v_1}{i_{b3}/\beta} \cong \beta \cdot \frac{v_{b3}}{i_{b3}} = \beta^2 \cdot \frac{v_{b3}}{i_{c3}} = \beta^2 \cdot \frac{R_G}{2} \tag{10.71}$$

an improvement of about β times over a diff-amp.

The conceptually similar Taylor cell of Fig. 10.19 is a differential amplifier. The input voltage appears across R_G via Q_1 and Q_3 on the left side and Q_2 and Q_4 on the right. It causes an incremental current through R_G that flows through Q_7 and Q_8 to the output. The goal is to correct the α error of Q_7 and Q_8. Analyzing the left side, beginning with Q_5, we get its collector current:

$$i_{C5} = I_0 - i_{B7} \tag{10.72}$$

Q_5 and Q_3 comprise a current mirror, and i_{C5} is replicated as

$$i_{C3} = i_{E7} = I_0 - i_{B7} \tag{10.73}$$

We are assuming that the areas of Q_5 and Q_3 are ratioed for a current-mirror

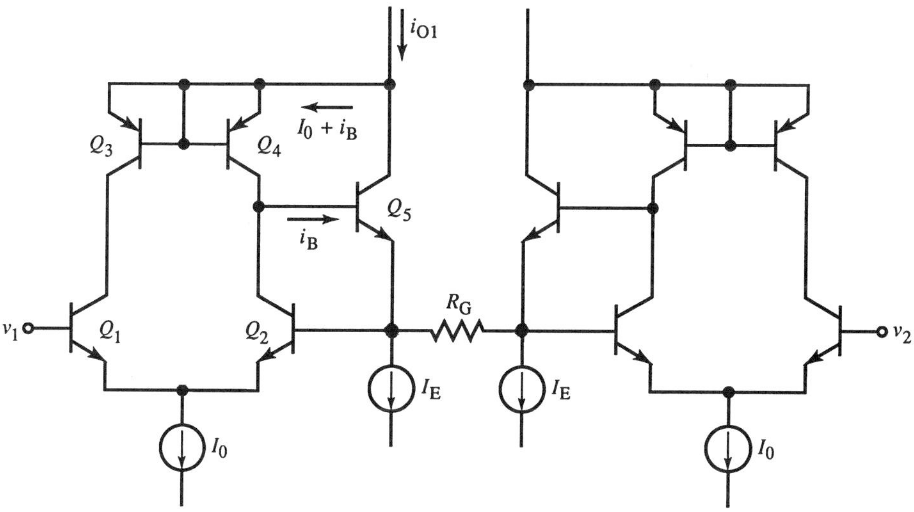

FIG. 10.20 Current-mirror variation on the Schlotzhaur cell.

gain of 1, not $(\beta+2)/\beta$ as for equal areas. Q_7 loses i_{B7} by α loss and

$$i_{C7} = I_0 - 2i_{B7} \tag{10.74}$$

At the emitter node of Q_1, KCL yields

$$i_{E1} = 3I_0 - (i_{E5} + i_{E3}) = 3I_0 - (2I_0 - 2i_{B7}) = I_0 + 2i_{B7} \tag{10.75}$$

At the output node, i_{C1} and i_{C7} add. Applying KCL, we get

$$i_1 + i_{C1} - 2I_0 = i_{C7} + i_{C1} - 2I_0 = \left(\frac{1}{\beta}\right)(I_0 + 2i_{B7}) \cong 0 \tag{10.76}$$

If Q_1 has $\alpha = 1$, then (10.76) is zero, and Q_7 α error is compensated. When Q_1 $\alpha < 1$, α error is reduced by $\beta/2$.

The last cell, in Fig. 10.20, is similar to the Schlotzhaur cell except that a current mirror is used. The total mirror current is I_0, conducted by the diff-amp Q_1-Q_2, plus the base current of the output CC, Q_5. The mirror current is part of the output, and since it includes i_{B5},

$$i_{O1} = I_0 + i_{E5} = I_0 + I_E + \frac{v_1}{R_G/2} \tag{10.77}$$

The cell is also differential, and R_G determines the transconductance.

10.5 f_T Multipliers

The f_T (gain-bandwidth product) of current-feedback amplifiers is not gain-limited but this property is not unique to them. Amplifiers with a fixed f_T can

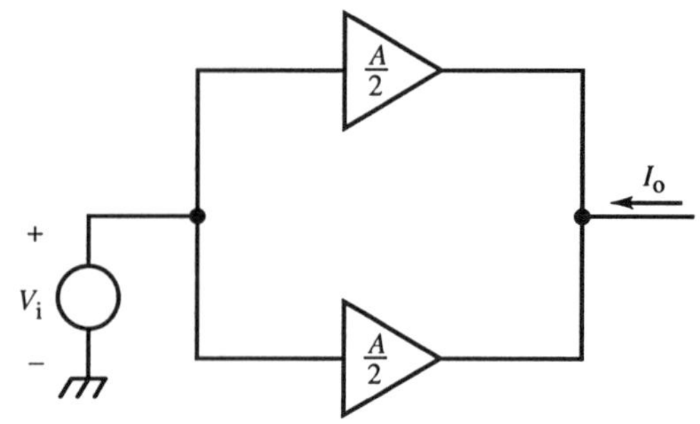

FIG. 10.21 Basic f_T-doubler amplifier topology.

be combined to achieve a greater f_T. A f_T-multiplying stage output can avoid bandwidth reduction by summing current outputs from individual amplifiers and converting them to a voltage in a succeeding stage. Or for very fast amplifiers, a current input creates an input voltage to paralleled transconductance amplifiers across source resistance R_S. The composite amplifier then has a current gain instead of a voltage gain.

As the gain of an individual amplifier is reduced, its bandwidth increases. Figure 10.21 shows two amplifiers connected in parallel. The transfer function for each amplifier individually is

$$A = \frac{A_o}{s(\tau_T/A_o)+1} \tag{10.78}$$

where the bandwidth time constant is

$$\tau_{bw} = \frac{\tau_T}{A_o} \tag{10.79}$$

When the lf gain A_o is halved, the bandwidth doubles. Two of these amplifiers in parallel have an additive output, so the total gain is twice that of (10.78), or

$$\text{composite } A = 2\left(\frac{A_o/2}{s\left(\dfrac{\tau_T}{A_o/2}\right)+1}\right) = \frac{A_o}{s\left(\dfrac{\tau_T}{A_o/2}\right)+1} \tag{10.80}$$

Comparing (10.80) with (10.78), we find that the composite amplifier has the same lf gain but twice the bandwidth. In effect, f_T has doubled.

Such amplifiers, called *f_T doublers*, were invented by Carl Battjes for use in oscilloscope vertical amplifiers. He extended the idea to an arbitrary number of amplifiers; three amplifiers triples f_T. In practice, parasitic elements of most f_T-multiplier topologies cause diminishing returns above two amplifiers. The input loading increases, and the input pole decreases bandwidth faster than extra amplifiers increase it.

The common circuit realization of the f_T doubler is shown in Fig. 10.22. Two diff-amps are connected so that their inputs are in series across V_i, and

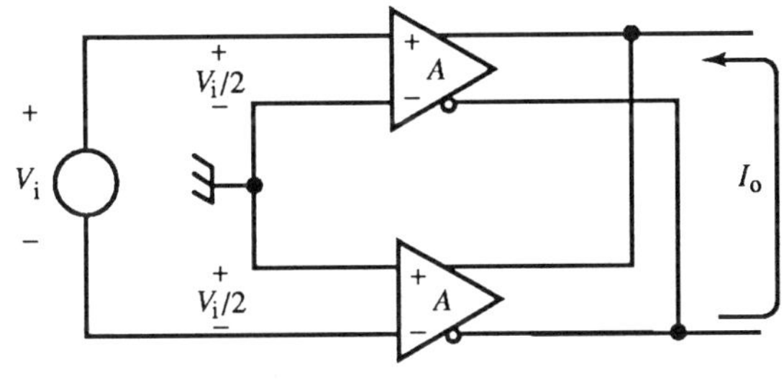

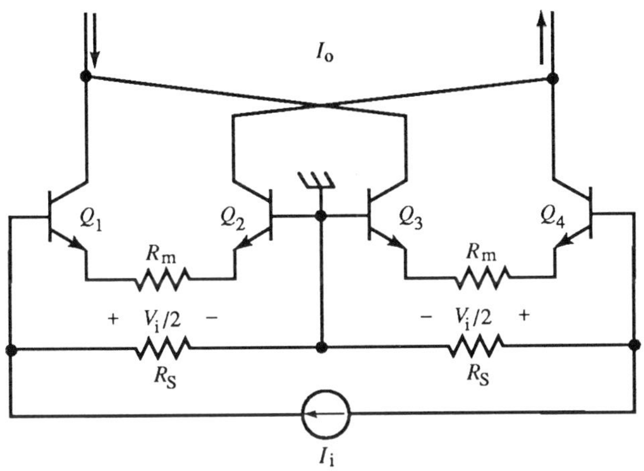

FIG. 10.22 Differential f_T-doubler block diagram (a) and circuit realization with two diff-amps with series voltage inputs and parallel current outputs.

their outputs are in parallel, cross-coupled so that they add. With a gain of A and input of $V_i/2$ for each diff-amp, the f_T-doubler output is AV_i, the same as a single diff-amp with input of V_i and gain of A. But the f_T doubler has twice the gain at the same bandwidth. The difference is illustrated in one equation for f_T multiplication, m:

$$\underset{\substack{\uparrow \\ mf_T}}{m\left[A\left(\frac{V_i}{m}\right)\right]} = \underset{\substack{\uparrow \\ f_T}}{AV_i} \tag{10.81}$$

Another differential f_T doubler is shown in Fig. 10.23. The input to the second diff-amp is taken from across $2R_m$ of the input diff-amp. For $R_m \gg r_e$, the input to the second diff-amp is approximately V_i also.

Single-ended f_T doublers are also possible (Fig. 10.24). The topology is basically that of a Darlington configuration except for the diode. This gain

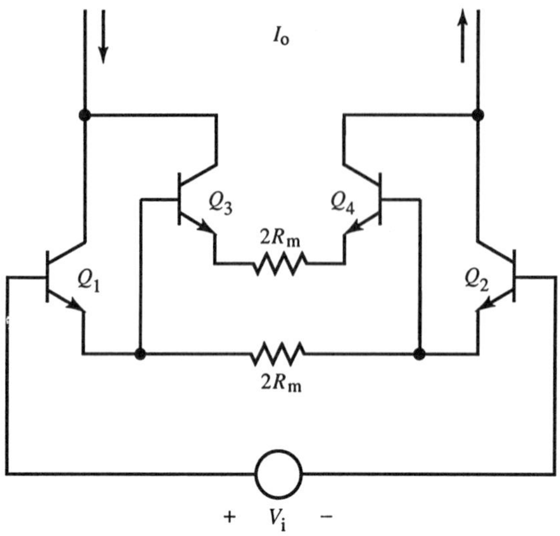

FIG. 10.23 An f_T doubler in which the voltage inputs are in parallel.

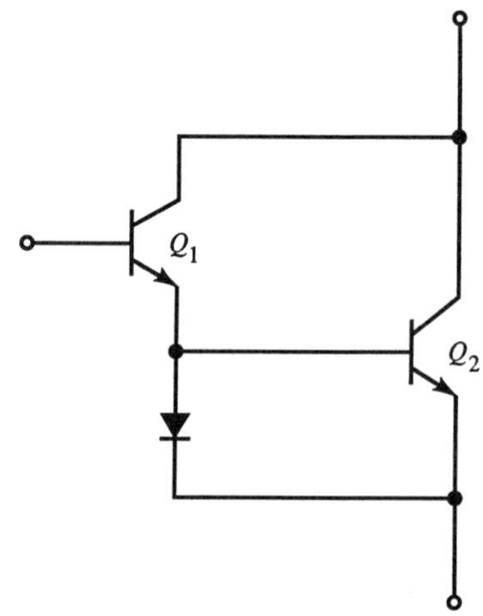

FIG. 10.24 Single-ended f_T doubler.

cell has three terminals and can be substituted for a single BJT. The base current of Q_1 generates emitter current i_{E1}, which flows through the diode, creating the same voltage drop, v_{BE2}, as v_{BE1}. This results in i_{C2}, which sums with i_{C1}, producing $2i_{C1}$. This current output is caused by an input of i_{B1}; therefore, the current gain is $2i_{C1}/i_{B1} = 2\beta$. The current gain is double that of a single BJT with the same f_β. Thus, f_T has doubled.

More precisely, some of the i_{E1} goes into the base of Q_2. With matched junctions, the diode current is i_{C1}, and the current mirror replicates this current in the emitter of Q_2, or $i_{E2} = i_{C1}$. The output current is slightly less than before, or

$$i_O = (1+\alpha)i_{C1} \cong 2i_{C1}, \qquad \beta \gg 1 \tag{10.82}$$

The f_T-doubler current gain is consequently

$$\text{single-ended } f_T \text{ doubler} \quad A_i = (1+\alpha)\beta(s) \tag{10.83}$$

The idea of adding the outputs of multiple amplifiers instead of cascading them leads us to the fastest amplifier topology, the *distributed amplifier*, shown in simplified form in Fig. 10.25. We start with two discrete delay lines. The input source drives the input line. As the signal propagates along the line, it drives the inputs of gain stages. They are typically a single transistor amplifier with large input resistance and a controlled input capacitance that is part of the delay line. Its output feeds a tap on the output line, also with controlled output capacitance. As the individual stages respond, their outputs accumulate as they propagate down the output line toward the load. Input and output signals propagate in synchronism for minimum phase error in the summation of outputs. Each line is terminated properly at each end.

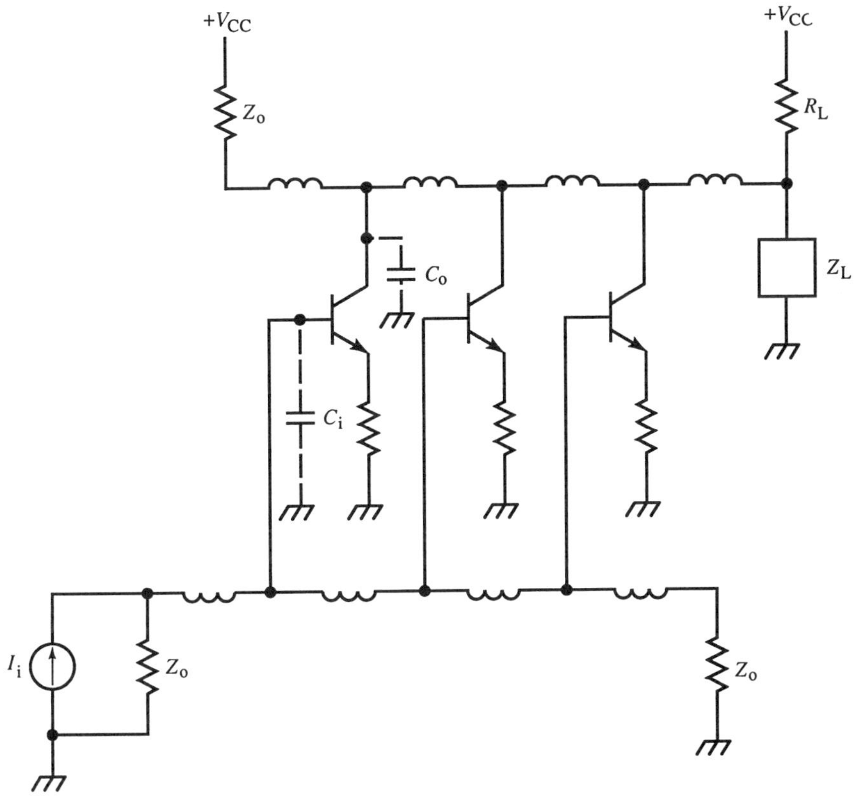

FIG. 10.25 Distributed amplifier.

Although distributed amplifiers have been implemented in various oscilloscope vertical amplifiers, they are a last resort at speed improvement because of the number of stages required for a given gain. For n stages with gain A, the composite gain is nA rather than A^n for cascaded stages. For large gain requirements, the number of stages could be excessive. With increasing IC density, however, a novel form of IC distributed amplifier with silicon-based delay-line structures might be feasible.

10.6 High-Performance Buffer Amplifiers

Amplifiers with a high-impedance input, accurate (usually ×1) voltage gain, and a low-impedance, large-signal output are called buffer amplifiers, or *buffers.* We continue from where Section 9.14 ended with the design of these amplifiers. They are typically used to drive low-impedance loads and have a large-signal dynamic range. Examples of loads are video distribution cables, CRT deflection plates, magnetic deflection yokes, electromechanical devices, and pulse- and function-generator outputs.

The two-stage complementary emitter-follower in Fig. 10.26a – a variation on that of Section 9.14 – is preceded by a complementary CC stage. This gives a higher input impedance. In addition, instead of returning the input-stage emitters to the supplies, they are connected to the output. This has the effect of reducing the output-stage deadzone by supplying output drive from the input stage. Since the deadzone is around 0 V output, the output current requirement within it is not large, and the input stage can supply much of the needed current, putting otherwise wasted signal to use.

The buffer of Fig. 10.26b also supplies input-stage signal to the output and reduces the deadzone. Its complementary CC input stage is inverted from that of Fig. 10.26a, and the collectors connect to the output. In this circuit, as v_I increases, Q_2 increases conduction, supplying emitter current to Q_3. In the deadzone, this current biases Q_3 on, reducing r_{e3} and increasing gain. By symmetry, Q_1 similarly biases Q_4. This buffer does not necessarily require input-bias voltage sources since the b–e junction voltages of Q_1–Q_3 and Q_2–Q_4 tend to cancel. The choice of R in both (a) and (b) circuits of Fig. 10.26 is based on a trade off between input characteristics and deadzone reduction.

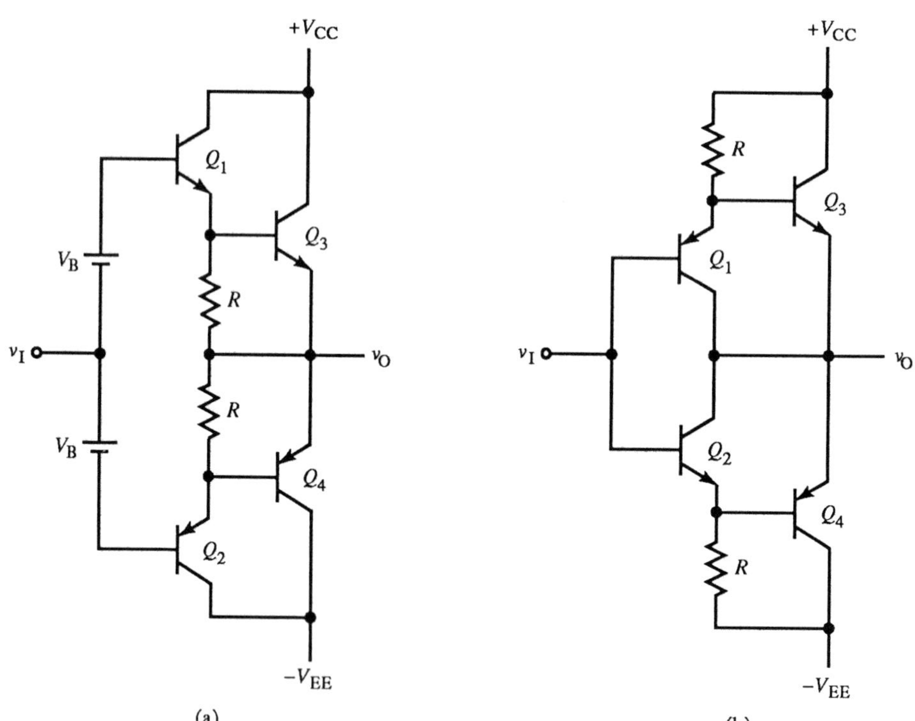

FIG. 10.26 Two-stage buffer with low-level output drive from first stage (a), and with complementary first stage (b).

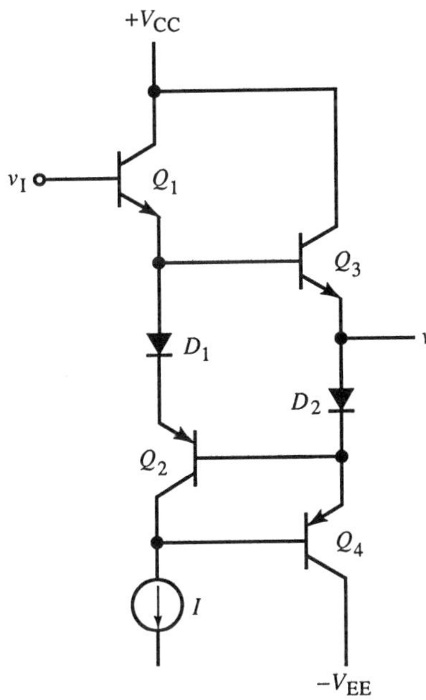

FIG. 10.27 Output buffer stage with current-mirror sink.

The output buffer of Fig. 10.27 has equal diode currents at 0 V output. The current source sinks I through Q_2 and D_1. Q_2 and Q_4 are coupled to exchange base currents and are thereby α compensated. Since

$$v_{D1} + v_{BE2} = v_{BE3} + v_{D2} \tag{10.84}$$

with matched junctions and no load current, $i_{D2} = i_{D1}$. Consequently, the base currents of Q_2 and Q_4 are equal (for matched devices) and cancel. As v_I varies, load current upsets the compensation somewhat since v_{BE3} no longer matches v_{BE2}. However, the deadzone is reduced because Q_3 is biased by an emitter current source of I.

The buffer can sink up to $(\beta+1)I$ current. Applying KCL at the bases of Q_2 and Q_4 and solving, we obtain

$$i_{D2} = (\beta+1)I - \frac{\beta^2+\beta+1}{\beta+1} \cdot i_{D1} \cong (\beta+1)I - \beta i_{D1} \tag{10.85}$$

For large negative inputs, $i_{D1} = 0$, and I is the base current of Q_4. Thus, I can be kept small to reduce the base current of Q_1. This buffer has an inherent voltage offset of $v_{BE1} + v_{BE3}$.

Output current-limiting does not appreciably reduce the dynamic range in the noninverting feedback buffer of Fig. 10.28. Without Q_3 and R_E, the maximum output occurs when Q_2 saturates, and is V_{CC}. When the current-limit circuit is added, some drop occurs across R_E, but it can have a small value because of the gain through Q_3 to the amplifier input loop.

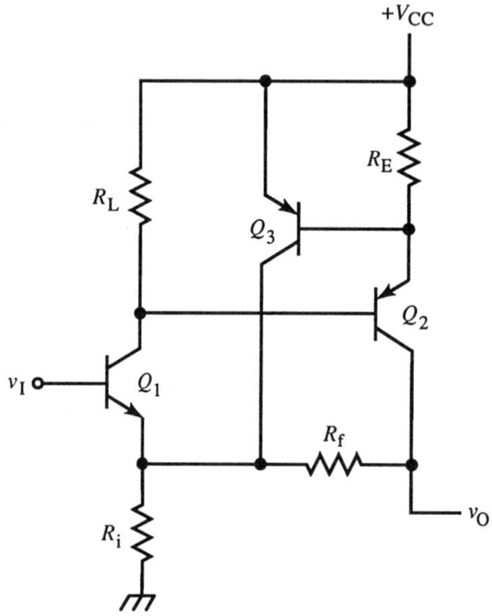

FIG. 10.28 Scheme for current limiting with Q_3.

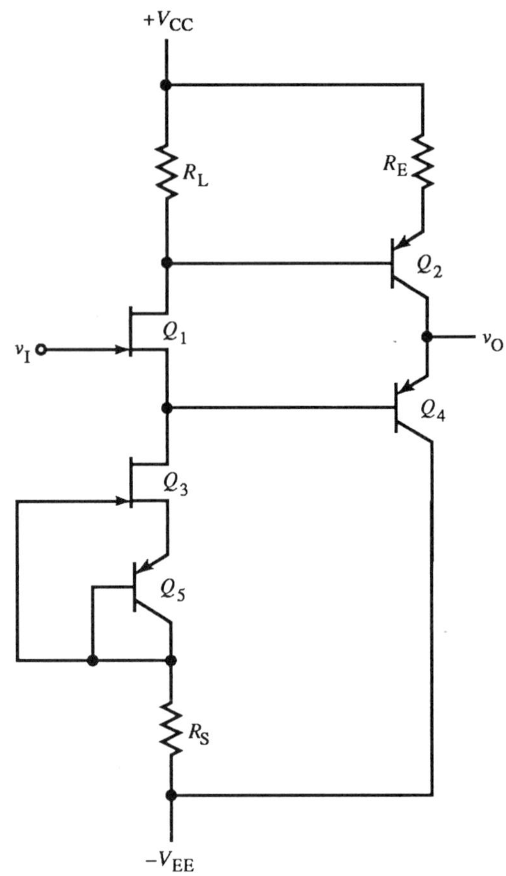

FIG. 10.29 Composite buffer with v_{BE} compensation of Q_4 by Q_5, and active source Q_2.

More precise buffer amplifiers have better input characteristics and less offset voltage. In Fig. 10.29, Q_4 is matched with Q_5. Then with matched FETs,

$$V_{GS1} = -V_{BE5} = -V_{BE4}$$

and offset voltage is cancelled. Since V_{GS3} is forced to be V_{BE5}, any FET drift in Q_3 causes a corresponding drift in Q_1, thus nulling FET TC effects on offset voltage.

The current-sourcing capability of the buffer is improved by the Q_2 CE, though this path is slower than for current-sinking through Q_4. R_S may be needed for thermal compensation.

A similar two-path buffer topology is found in the Linear Technology LT1010 and is simplified in Fig. 10.30. Q_1 and Q_2 are complementary CCs of the fast path. Q_2 also functions as a CE to drive the slow path through a gain of K to Q_3, which provides active current-sinking, similar in function to Q_2 of Fig. 10.29.

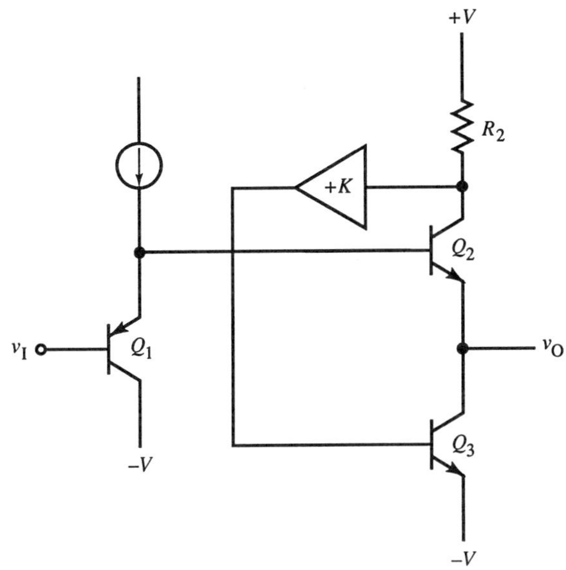

FIG. 10.30 Two-path buffer topology of the LT1010.

The general technique of bootstrapping is applied to the buffer of Fig. 10.31. (This circuit is also used in the LT1010.) Q_3, Q_4, and R_B form the bootstrap circuit. Q_4 is an integrated split-collector BJT with one collector connected to the base. This is equivalent to the circuit in Fig. 10.31b and functions as a simple current mirror. The bootstrap circuit is driven from the input to Q_2 and forms an active parallel path to its base, supplementing the drive of the passive path through R_B. To increase efficiency, the emitter current of Q_3 is supplied to the output.

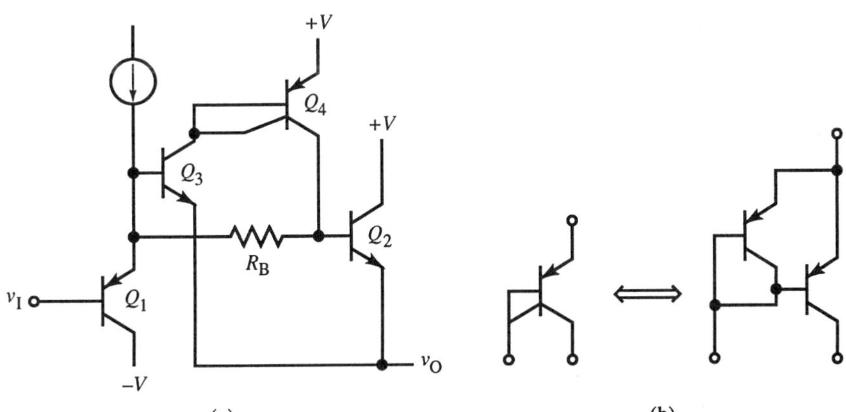

FIG. 10.31 Bootstrapped buffer (a), with integrated split-collector BJT current mirror Q_4 (b).

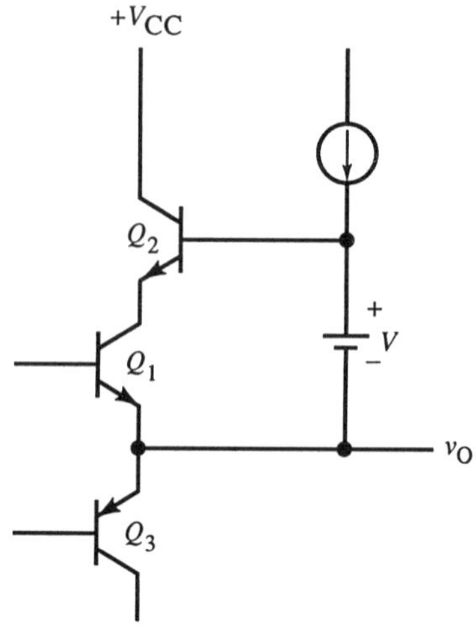

FIG. 10.32 Bootstrapping technique for increasing dynamic range of an output stage.

Finally, a general bootstrapping technique for increasing the dynamic range of an output stage is shown in Fig. 10.32. Q_2 is bootstrapped via V atop Q_1. This minimizes power dissipation in Q_1 but gives it more collector voltage as the output requires it. A similar complementary circuit is applied to Q_3.

10.7 Unipolar Voltage-Translating Amplifiers

Some amplifier stages require control from near circuit ground and supply a unipolar drive at an elevated voltage. Examples are the current-sourcing drivers of H-bridge power switches, oscilloscope horizontal deflection amplifiers, and voltage level-translators. The basic problem to be solved is that of supplying adequate drive to the elevated output device from ground. The basic situation is shown in Fig. 10.33. Q_1 operates near ground and drives Q_2. Since Q_1 can only sink current, it cannot actively drive Q_2 on. R_L supplies Q_2 drive but must be kept large to limit Q_1 current. The capacitance C_o at the base of Q_2 and load capacitance referred to the base node limit amplifier speed. Various schemes have been devised to replace R_L with a high-side driver to reduce power dissipation in Q_1 and increase slew rate.

If we make Q_2 a Darlington stage instead (Fig. 10.34a), less drive current is needed and the slew-rate-limiting base node is further isolated from the load. But the time constant, $R_L C_o$ is not reduced, and with the higher input resistance of the Darlington, the node response time is increased for small signals.

A complementary Darlington (Fig. 10.34b) is used in the LM3900 with some advantages over that of (a). The output voltage range is increased by a

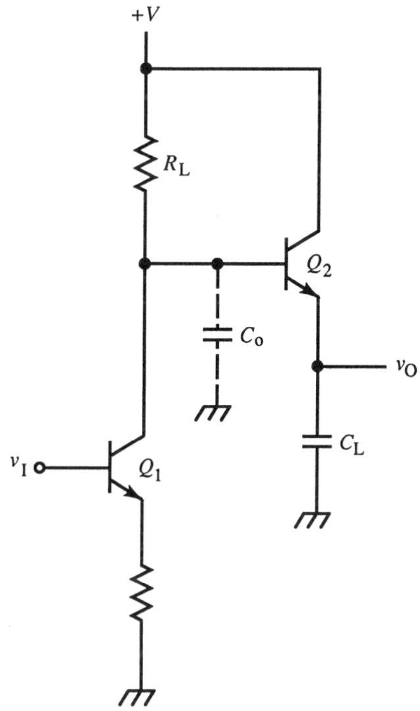

FIG. 10.33 Basic high-side driver. C_o limits slew rate.

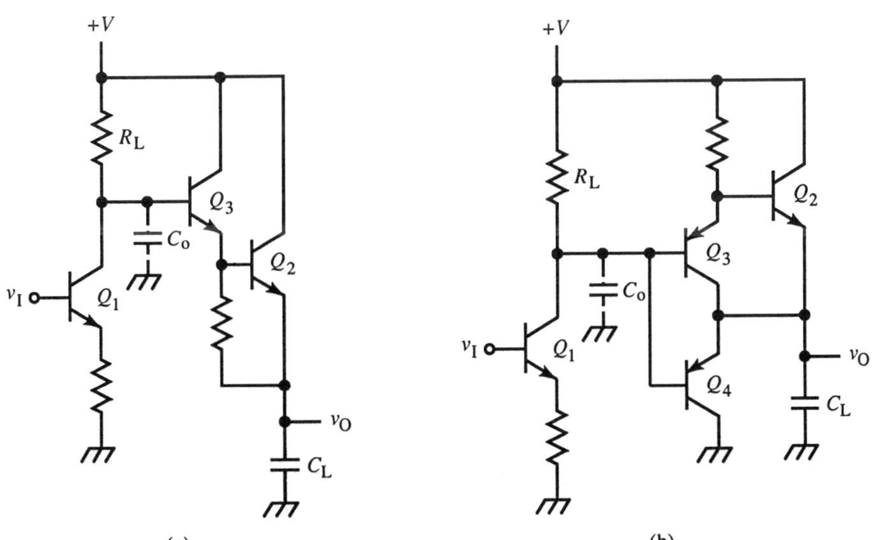

FIG. 10.34 Darlington output (a) reduces base drive current needed to charge C_o; complementary Darlington with active current sink Q_4 (b).

junction drop. Also, the b–c junction of Q_3 conducts in the forward direction for large, quick negative-going outputs and provides a low-impedance path from the output to Q_1. In effect, Q_1 is the output current-sinking driver. (This feature can be easily added to Fig. 10.33 by placing a diode in reverse across the b–e junction of Q_2 but with no additional gain advantage.) When the output voltage offset from the input is small (as in the LM3900), Q_4 can be added to buffer Q_1 for sinking output current. In amplifiers with large V_{BC4}, Q_4 (like Q_1 in Fig. 10.33) can dissipate excessive power. Therefore, high-side drivers are needed that do not sink large currents to ground across large voltage drops.

One approach is to minimize C_o. The input stage can be made a cascode; the Miller effect is eliminated and C_o substantially reduced. A shunt-feedback cascode with high side driver has feedback benefits as well, but feedback itself cannot overcome slew-rate limitations.

In the early 1970s, Carl Battjes devised the floating current mirror scheme of Fig. 10.35. The current mirror consists of Q_3 and Q_4 with area ratio $A_3/A_4 = K$. The emitter current splits in the proportions of

$$i_{E3} = \frac{K}{K+1} \cdot i_1, \qquad i_{E4} = \frac{1}{K+1} \cdot i_1 \tag{10.86}$$

The output sink current is

$$i_3 = i_{E3} + i_{B4} = \frac{K}{K+1} \cdot i_1 + \frac{i_1/(K+1)}{\beta+1} \tag{10.87}$$

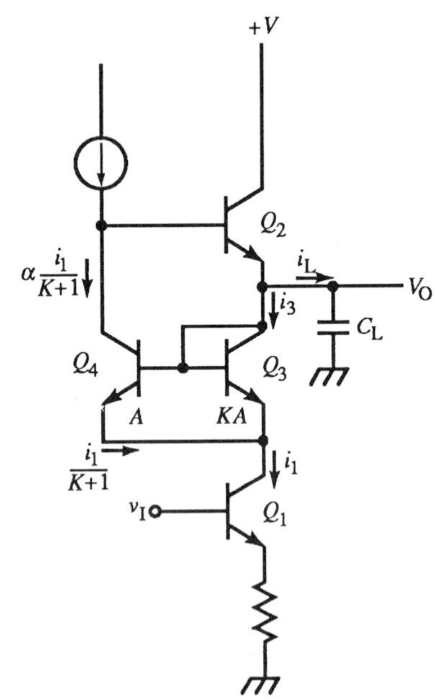

FIG. 10.35 Floating current-mirror driver.

The source current from the emitter of Q_2 is $-(\beta+1)i_{C4}$. Voltage inversion occurs at the collector of Q_4, reducing i_{E2} when i_1 increases. Then

$$i_{E2} = -(\beta+1)\left[\alpha \cdot \frac{i_1}{K+1}\right] = -\frac{\beta}{K+1} \cdot i_1 \tag{10.88}$$

Then the transfer ratio of

$$\frac{i_L}{i_1} = \frac{i_{E2} - i_3}{i_1} = -\frac{\beta + 1/(\beta+1) + K}{K+1} \tag{10.89}$$

The floating current mirror sinks output current and drives Q_2.

Bootstrapping techniques can be applied to the high-side driver from either the input (Fig. 10.36a) or output (b). Both are dynamic bootstraps, as shown, to improve speed. In (a), the cascode input drives Q_2 through C, making it an active current source for fast inputs.

Figure 10.36b is a circuit developed by Art Metz for oscilloscope horizontal amplifiers. The output, through C, bootstraps R_2; as v_O increases, v_1 does also. This keeps the voltage across R_2 and its current from decreasing as v_2 rises, thereby maintaining current drive to Q_2. He further improved the circuit by making Q_1 and Q_2 f_T doublers (as in Fig. 10.24). For Q_2, the gain and bandwidth are higher, and base current drive is reduced, leaving more current

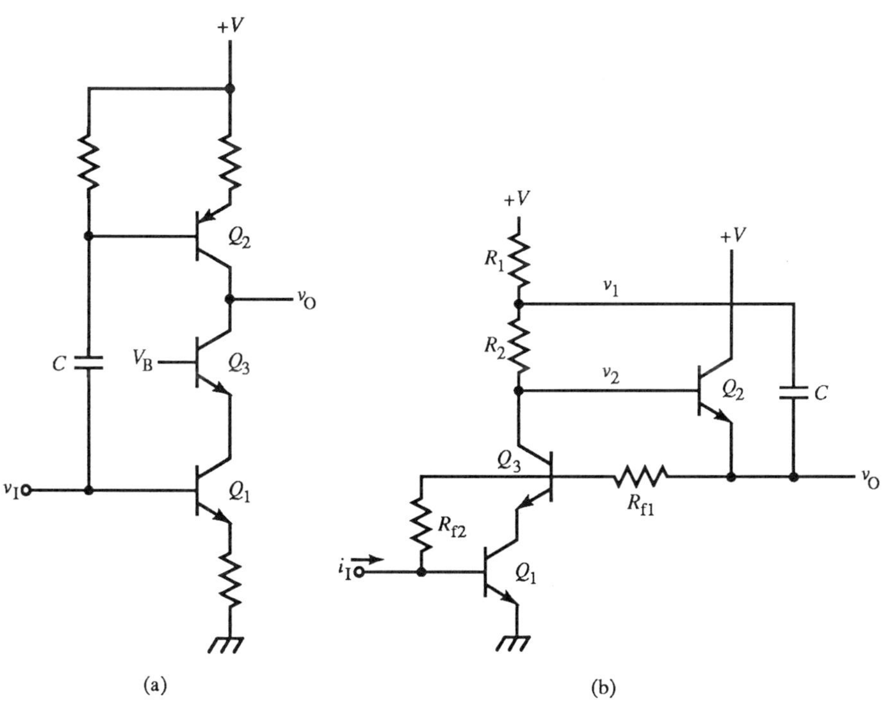

FIG. 10.36 Input bootstrap drive to active source Q_2 (a); output bootstrap of R_2 maintains charging current as output voltage approaches $+V$ (b).

to charge node capacitance. The final embellishment is to tap the shunt-feedback resistor and drive the base of the cascode CB, Q_3, from the tap. The c–b junction of Q_3 is bootstrapped since output and base voltage vary together. This increases the dynamic range and reduces C_o and the breakdown voltage requirements of Q_3. The topology is similar to that of the split cascode in Fig. 4.32.

10.8 Bootstrapped Input Stages

Bootstrapping is commonly applied to the inputs of op-amps to increase their dynamic range and improve performance. A simplified topology of the 741 op-amp is shown in Fig. 10.37. The input differential BJT pair, Q_1–Q_2, is a CC stage driving a complementary differential CB stage, Q_3–Q_4. The CB bases are bootstrapped to follow the inputs and are controlled through the feedback loop through Q_5. As the input pair conduct more current due to a common-mode voltage increase, Q_5 operates as a current mirror and sources more current, causing the CB bases to rise. This bootstrapping action takes the form of noninverting feedback and also bootstraps the input impedance.

The more conventional diff-amp input of Fig. 10.38 drives bootstrapped CB transistors. In the circuit, the common-mode input voltage is taken from the emitter virtual ground node. The voltage translator, shown as a battery, is typically a zener diode or resistor driven by a current source.

The Linear Technology LT1011 comparator has an input like that in Fig. 10.39. A differential two-stage follower is bootstrapped to control the emitter bias currents in the first stage CCs. The variation in bias currents of the output CCs is tracked in the input CCs.

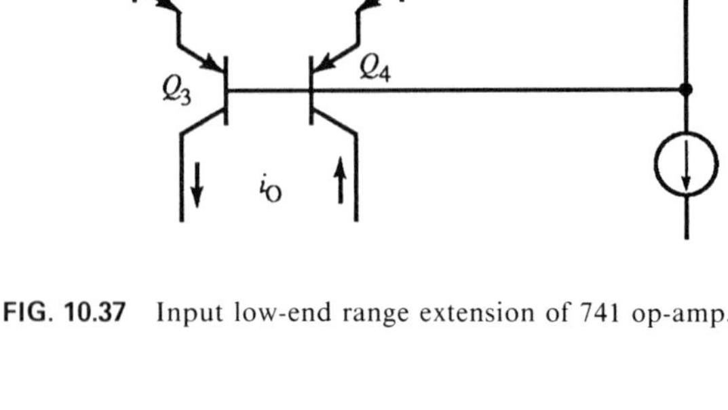

FIG. 10.37 Input low-end range extension of 741 op-amp.

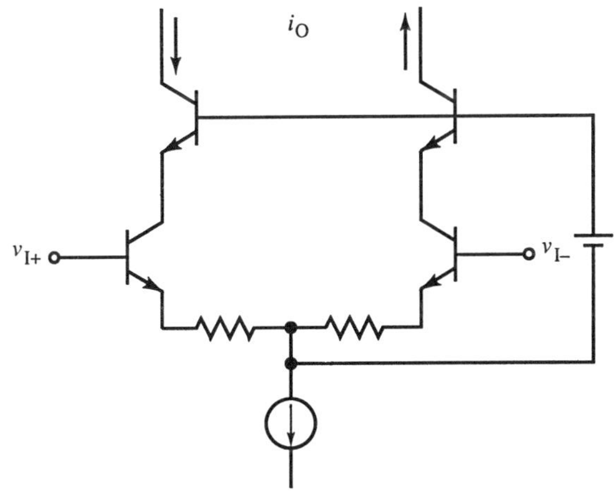

FIG. 10.38 Input high-end range extension: a bootstrapped cascode diff-amp.

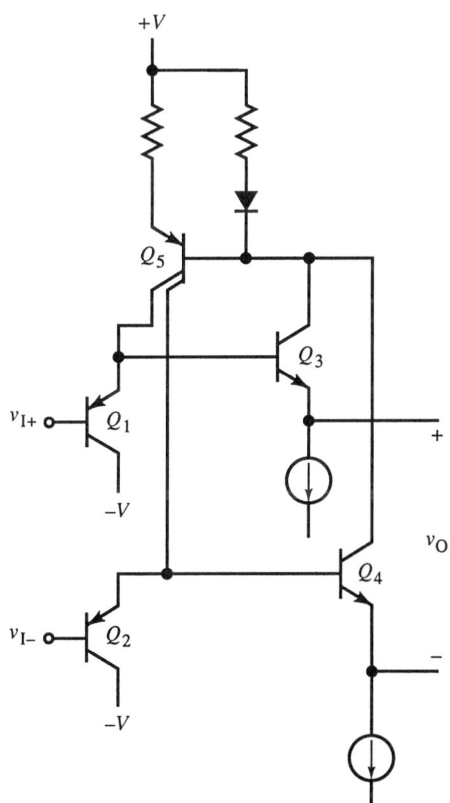

FIG. 10.39 Bootstrapping for offset-voltage compensation by control of emitter currents in the input stage.

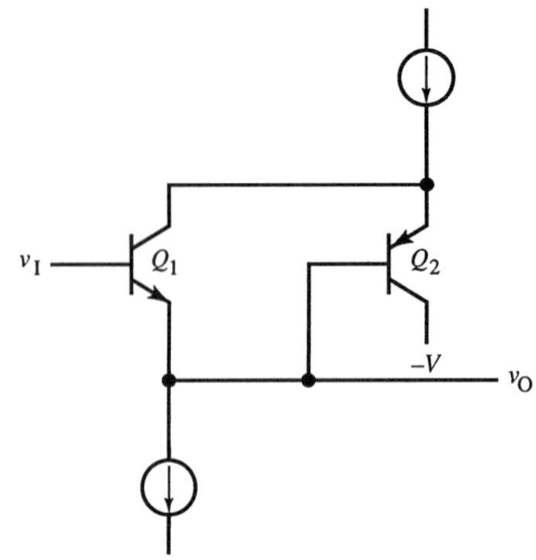

FIG. 10.40 Simple input bootstrapping scheme for collector of Q_1, provided by Q_2.

Finally, the simple input bootstrapping scheme for the collector of Q_1 in Fig. 10.40 is provided by Q_2. Although V_{CE} of Q_1 is limited to V_{BE2}, its collector follows its emitter as the complementary CC output of Q_2.

10.9 Composite-Feedback and Large-Signal Dynamic Compensation

The composite amplifiers of Section 10.2 had a single feedback path. We now consider composite amplifiers with multiple feedback paths. In Fig. 10.41, the paths are nested. The inner loop of G_1H_h is part of the forward path of the outer loop. Here, an op-amp drives a more powerful, reactively loaded buffer amplifier. The high-frequency path through C_f is isolated by the buffer and R_f from the load, whereas dc feedback is taken at the load through R_f. (This topology is an alternative to that of Fig. 6.29.) The block diagram transfer function reduces to

$$\frac{V_o}{V_i} = \alpha_i \cdot \frac{G_1 G_2}{1 + G_1 H_h + G_1 G_2 H_l} \tag{10.90}$$

The noninverting version omits α_i. H_h, H_l, and α_i have poles at $1/\tau = 1/(R_f \parallel R_i)C_f$, and H_h has a zero at the origin since it is an RC differentiator. These blocks are of the form

$$\alpha_i = \frac{\alpha_{io}}{s\tau + 1}, \qquad -H_h = \frac{s\tau}{s\tau + 1}, \qquad -H_l = \frac{H_o}{s\tau + 1} \tag{10.91}$$

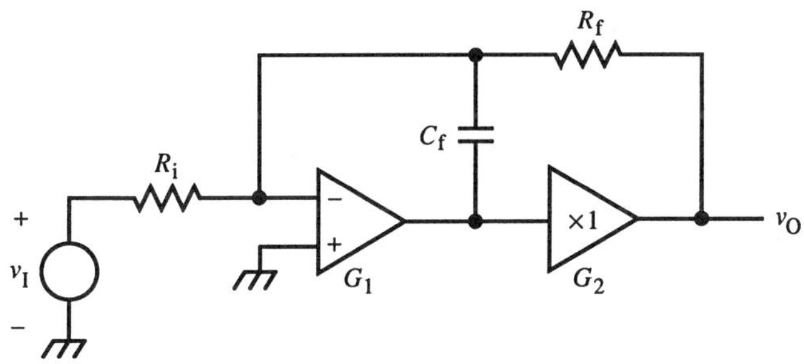

(a)

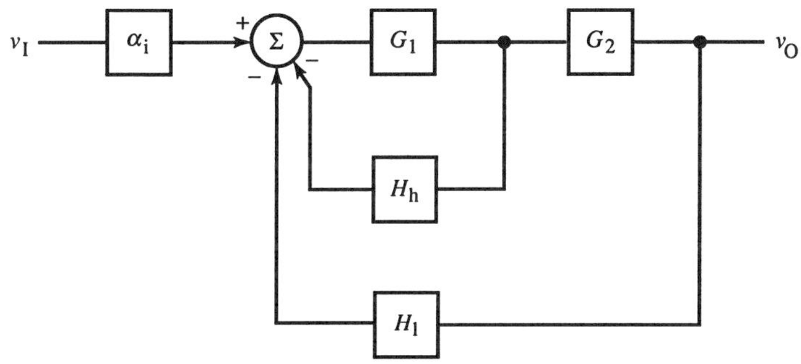

(b)

FIG. 10.41 Multiple-feedback amplifier for isolating the hf path from the load (a), and its block diagram (b).

where

$$\alpha_{io} = \frac{R_f}{R_f + R_i}, \qquad H_o = \frac{R_i}{R_f + R_i} \tag{10.91a}$$

Substituting into (10.90), setting $G_1 = -K$ and $G_2 = 1$, the voltage gain is

$$\frac{V_o}{V_i} = -\alpha_{io}\left(\frac{K}{1 + KH_o}\right)\frac{1}{s\tau\left(\dfrac{1+K}{1+KH_o}\right)+1} \tag{10.92}$$

For infinite K,

$$\frac{V_o}{V_i} = -\frac{R_f}{R_i} \cdot \frac{1}{sR_fC_f + 1} \tag{10.93}$$

The low-frequency gain is that of a standard inverting op-amp with a pole at $1/R_fC_f$. This pole can be placed at a high frequency, away from the poles of the forward path and load.

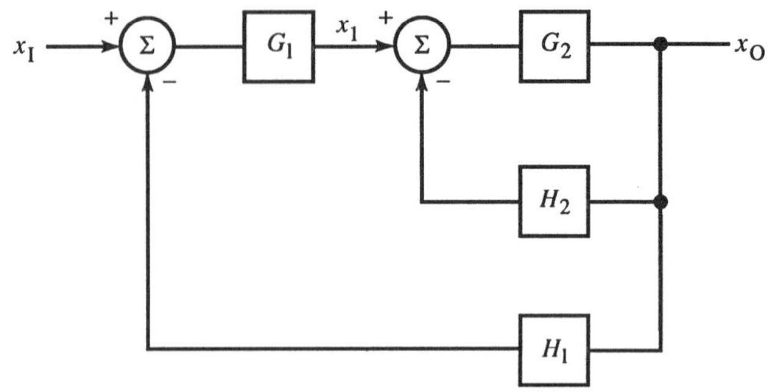

FIG. 10.42 Nested feedback amplifier.

Nested feedback loops are common in power electronics. In both power converters and motor controllers, a current-controlling inner loop is controlled by a voltage or speed-controlled outer loop. In position controllers for motors, a third outer position loop is added to allow speed to be controlled by position error. The general topology is shown in Fig. 10.42. For a position controller, x_1 is speed and G_2H_2 is a speed-control loop. For a motor speed controller, x_1 is current (or scaled torque); and for current-mode switching power supplies, x_I is voltage and x_1 is current. Voltage-mode power converters have the same topology, with x_1 as duty ratio and the inner loop as a pulse-width modulator. This topology can be equally well applied to purely electronic applications and is sometimes called *pseudoderivative feedback* (PDF) control.

The most common form of compensator for this topology is a variant of lead-lag compensation called *proportional integral differential* (PID) control (Fig. 10.43). It is a multipath topology with gain, integral, and derivative paths that add. Each path can be adjusted independently, with transfer function:

$$\text{PID compensator} = K + sA + \frac{B}{s} = \frac{s^2(A/B) + s(K/B) + 1}{sB} \tag{10.94}$$

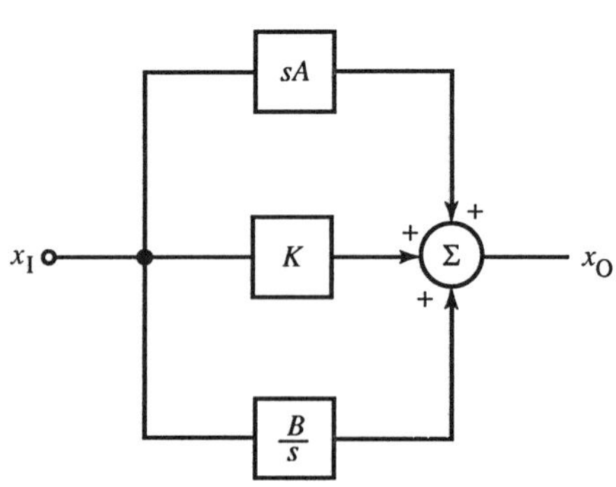

FIG. 10.43 PID control compensator, a multipath topology.

Since A, K, and B are independently adjustable, the coefficients can be set with only scale interaction. The dc gain, $1/B$, scales the coefficients and is set first. Note that this transfer function has more zeros than poles and can deter phase lag more effectively than lead-lag compensation.

The complexity of topologies beyond those given here requires a more abstract and simplifying formulation. State-variable control theory is a good foundation for more complex designs, and as with classical control, some techniques do appeal to design insight. Formulations must be sought that provide physically meaningful insight into circuit operation [see Lorenz (1986) in the reference list].

A design technique for compensating feedback amplifiers with large feedback resistors is shown in Fig. 10.44. The problem caused by the large value of R_f is that C_f must be made too small to be practical for proper compensation. Not uncommonly, fractions of a picofarad are required. Another problem caused by C_f, due to the Miller effect, is large input capacitance. Some sources must have minimum capacitive loading, such as D/A outputs.

A larger C_f can be used in the circuit in Fig. 10.44b, in which a voltage divider is formed to drive C_f. If the divider Thévenin resistance is negligible, then the loop gain is reduced by the divider. The voltage gain across C_f is also reduced from $(A_v + 1)$, so the Miller capacitance is

$$C_{in} = C_f(A_v + 1)\left(\frac{R_2}{R_1 + R_2}\right) \qquad (10.95)$$

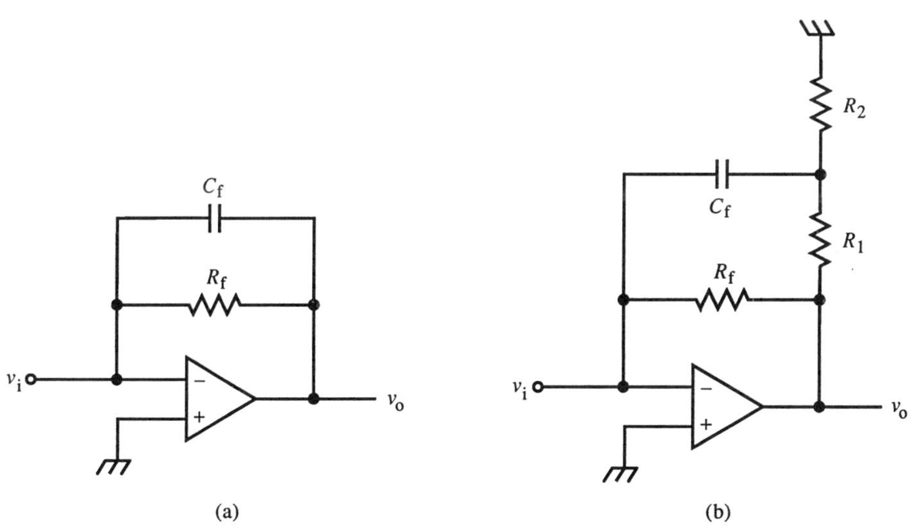

FIG. 10.44 Conventional feedback-compensated op-amp with large R_f and impractically small C_f (a); modified circuit using voltage divider in feedback path and larger C_f (b).

If $R_1 \| R_2$ is not negligible, various poles and zeros appear. A divider in the feedback path can similarly be used to reduce R_f.

Nonlinear dynamic compensation is often necessary in circuits with large dynamic ranges. Transistor parameters change with the signal, causing otherwise well-compensated circuits to show errors in their response. The b–c capacitance of BJTs varies nonlinearly with v_{BC}. The circuits of Fig. 10.45 compensate for ΔC_{bc} with another b–c junction of a similar transistor. In (a), emitter compensation is applied by reverse-biasing the b–c junction of Q_2 to track C_{bc} of Q_2. V_B biases C_{bc} of Q_2 to track Q_1. The tracking is more easily

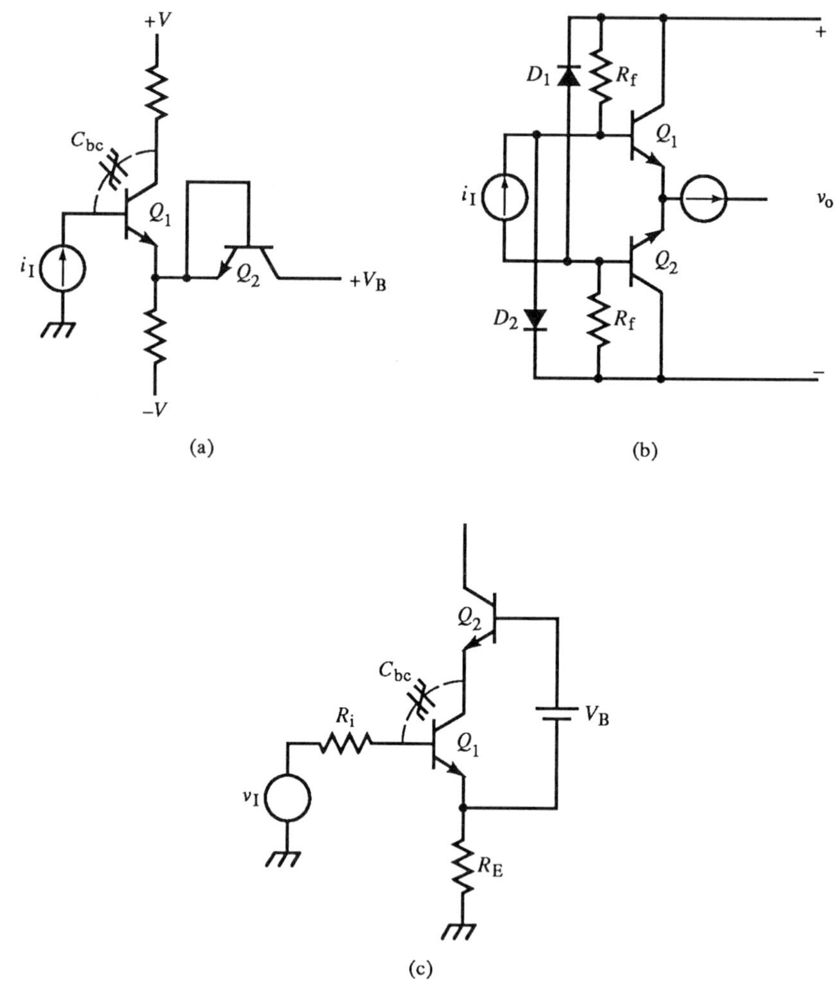

FIG. 10.45 Dynamic large-signal compensation of b–c junction of Q_1 with b–c junction of Q_2, as emitter compensation (a); differential compensation with reverse-biased diodes (b); bootstrapped cascode compensation of C_{bc} (c).

accomplished in a differential amplifier, as in (b), where the opposing sides of the shunt-feedback amplifier vary by $\pm\Delta v_{BC}$ around the same bias point.

In Fig. 10.45c, C_{bc} is bootstrapped from the emitter of Q_1. As v_I varies, v_{E2} follows it, keeping the voltage across C_{bc} constant. However, now that v_{B2} varies, the effect is transferred to C_{bc} of Q_2. But if V_{CB} of Q_2 is large relative to that of Q_1, its C_{bc} is smaller and varies less.

The basic technique of adjusting a semiconductor parameter with a dc quantity for dynamic compensation is used in high-speed IC circuits, in which compensation of internal nodes is only feasible by means of an external adjustment outside the high-frequency signal path.

10.10 The Gilbert Gain Cell and Multiplier

In the mid-1960s, Barrie Gilbert discovered another basic amplifier technique with wide bandwidth and high linearity. Instead of increasing circuit complexity to compensate for circuit error, the technique is based on the accurate logarithmic function of b–e junctions and good junction matching.

A differential current mirror, or *translinear cell*, based on the one in Fig. 10.1, is shown in Fig. 10.46; it consists of two pairs of emitter-coupled transistors. The input pair is connected as diodes, with input current $i_{I2} - i_{I1}$. If the areas are matched, then I_S is the same for all transistors. The differential

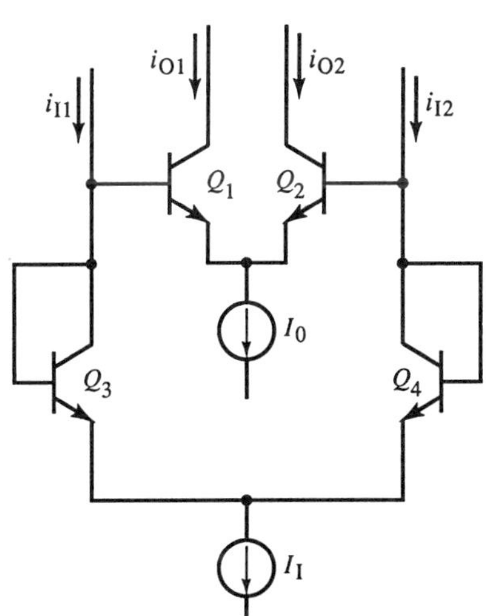

FIG. 10.46 Translinear cell.

voltage across the bases of Q_1 and Q_2 is the same as across Q_3 and Q_4, or

$$\Delta v_{BE} = V_T \ln\left(\frac{i_{O2}}{i_{O1}}\right) = V_T\left(\frac{i_{I2}}{i_{I1}}\right) \Rightarrow \left(\frac{i_{O2}}{i_{O1}}\right) = \left(\frac{i_{I2}}{i_{I1}}\right) \tag{10.96}$$

In other words, the ratio of output currents equals the ratio of input currents, assuming that BJT $\alpha = 1$. Since this result is a current gain, it cannot exceed β but is very linear up to gains near β. Linearity is reduced by junction area mismatch and ohmic base and emitter resistance but is much better than for comparable diff-amps with emitter resistance. Also, since Q_3 and Q_4 have $V_{CE} = V_{BE} \cong 0.8$ V, additional voltage drop across the collector ohmic resistance can cause $V_{C'E}$ to approach zero, causing diode error. The bandwidth is limited by BJT f_T.

The differential current gain is found by expressing the current ratios of (10.96) in the form

$$\frac{i_{O2}}{i_{O1}} = \frac{(i_{O2}+i_{O1})+(i_{O2}-i_{O1})}{(i_{O2}+i_{O1})-(i_{O2}-i_{O1})}$$
$$= \frac{i_{I2}}{i_{I1}} = \frac{(i_{I2}+i_{I1})+(i_{I2}-i_{I1})}{(i_{I2}+i_{I1})-(i_{I2}-i_{I1})} \tag{10.97}$$

Solving (10.97) for current gain yields

$$A_i = \frac{i_O}{i_I} = \frac{i_{O2}-i_{O1}}{i_{I2}-i_{I1}} = \frac{i_{O1}+i_{O2}}{i_{I1}+i_{I2}} = \frac{I_0}{I_I} \tag{10.98}$$

The significance of this result is that ratios of input and output variables can be expressed as ratios of their sums and differences. In general,

$$\frac{a}{b} = \frac{c}{d} \Rightarrow \frac{a-b}{a+b} = \frac{c-d}{c+d} \tag{10.99}$$

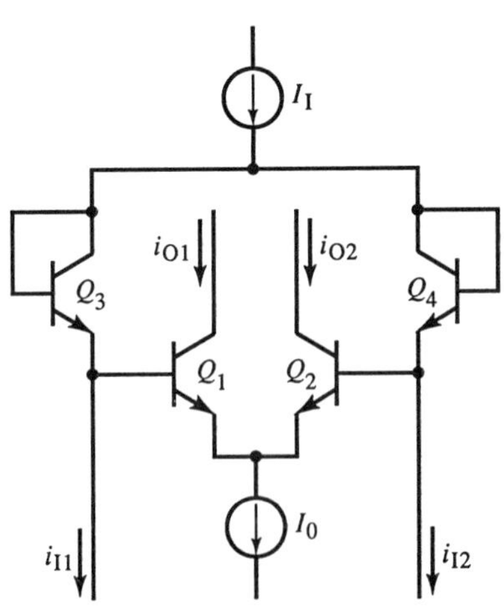

FIG. 10.47 Inverting translinear cell.

A gain-inverting translinear cell is shown in Fig. 10.47. The outer pair Δv_{BE} is negative that of Fig. 10.46, with current ratios

$$\frac{i_{O2}}{i_{O1}} = \frac{i_{I1}}{i_{I2}} \tag{10.100}$$

and

$$A_i = -\frac{I_O}{I_I} \tag{10.101}$$

Complementary pairs of inner and outer junctions are also possible, though I_S-matching is more difficult.

The *Gilbert gain cell* follows from Fig. 10.47 by connecting Q_3 and Q_4 as CB transistors and cross-coupling their collectors with Q_1 and Q_2 for additive outputs (Fig. 10.48). The current gain of the outer pair alone is $\alpha \cong 1$, apart from the additional current gain of the inner pair. The inner-pair gain is that of (10.101), or

$$\frac{i_{C2} - i_{C1}}{i_{I2} - i_{I1}} = -\frac{I_O}{I_I} \tag{10.102}$$

Approximating α as unity again, we find the Gilbert-cell gain:

$$A_i = \frac{i_{O2} - i_{O1}}{i_{I2} - i_{I1}} = \frac{(i_{C1} + i_{I2}) - (i_{C2} + i_{I1})}{i_{I2} - i_{I1}}$$

$$= -\frac{i_{C2} - i_{C1}}{i_{I2} - i_{I1}} + 1 = \frac{I_O + I_I}{I_I} \tag{10.103}$$

Two currents, such as i_{I1} and i_{I2}, can be expressed as fractions of I_I. Let the "modulation index" be the fraction x, so

$$i_{I1} = xI_I \qquad \text{and} \qquad i_{I2} = (1 - x)I_I \tag{10.104}$$

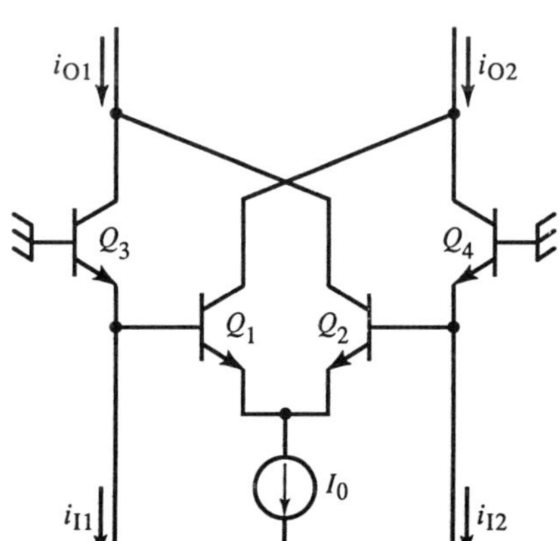

FIG. 10.48 Gilbert gain cell.

Their sum is still I_I. Applying (10.100) to the Gilbert cell, we find that i_O of (10.100) is i_C of the Gilbert cell. Their ratio is

$$\frac{i_{I1}}{i_{I2}} = \frac{x}{1-x} = \frac{i_{C2}}{i_{C1}} \tag{10.105}$$

Thus,

$$i_{C2} = xI_0 \quad \text{and} \quad i_{C1} = (1-x)I_0 \tag{10.106}$$

The Gilbert cell output currents are then

$$i_{O2} = i_{C1} + i_{I2} = (1-x)(I_0 + I_I) \tag{10.107a}$$

$$i_{O1} = i_{C2} + i_{I1} = x(I_0 + I_I) \tag{10.107b}$$

The dynamic range is $0 \le x \le 1$; unlike the cascomp, I_0 and I_I are fully used as signal currents over this range. Gilbert cells are easily cascaded with little bandwidth loss due to interstage coupling.

The differential current mirror and Gilbert cell have important uses beyond current amplification. Equations (10.98) and (10.101) suggest that they also function as multipliers. Since I_0 and I_I can both be varied, the current gain can be changed; I_0 multiplies the gain and I_I divides it. If I_0 is not held constant but allowed to be the signal i_Y, then the output from (10.98) is

$$i_O = \frac{i_I i_Y}{I_I} \tag{10.108}$$

A divider is similarly realized if I_I is a signal instead. Since i_I is differential, it is a bipolar signal; i_Y is unipolar, resulting in a two-quadrant multiplier. Two differential pairs (Fig. 10.49) give full four-quadrant multiplication. The collectors of the pairs are cross-coupled. When $i_{Y1} = i_{Y2}$, the BJT pairs have

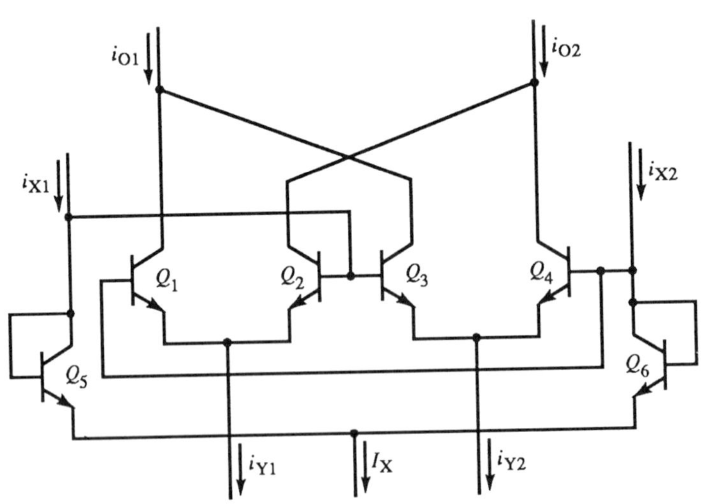

FIG. 10.49 Gilbert four-quadrant transconductance multiplier based on translinear cells.

equal but opposite gains, and their output signal currents cancel. At either range extremum, i_{Y1} or $i_{Y2}=0$, and only one pair amplifies. Within the range, the proportions of collector currents are set by the ratio of i_{Y1} and i_{Y2}.

The four-quadrant multiplier output is derived in terms of the inputs,

$$i_X = i_{X2} - i_{X1} \qquad \text{and} \qquad i_Y = i_{Y2} - i_{Y1} \tag{10.109}$$

where

$$i_{X1} + i_{X2} = I_X \qquad \text{and} \qquad i_{Y1} + i_{Y2} = I_Y \tag{10.110}$$

Define the ratio of input currents to be

$$i_{X2} = xI_X, \qquad i_{X1} = (1-x)I_X \tag{10.111}$$

Then from (10.109),

$$i_X = xI_X - (1-x)I_X = (2x-1)I_X \tag{10.112}$$

The output currents are

$$i_{O1} = i_1 + i_3 = xi_{Y1} + (1-x)i_{Y2} = -xi_Y + i_{Y2} \tag{10.113}$$

and

$$i_{O2} = i_2 + i_4 = (1-x)i_{Y1} + xi_{Y2} = xi_Y + i_{Y1} \tag{10.114}$$

The differential output current is

$$i_O = i_{O2} - i_{O1} = 2xi_Y - i_Y = (2x-1)i_Y = \frac{i_X}{I_X} \cdot i_Y \tag{10.115}$$

The output current is a fraction of the total output current I_Y, or

$$\frac{i_O}{I_Y} = \frac{i_X}{I_X} \cdot \frac{i_Y}{I_Y} \tag{10.116}$$

The result is ideal; i_X and i_Y are multiplied and scaled by I_X and I_Y. Since the fractional output depends on I_X, it too can be varied to provide division without affecting the output scaling. In practice, v_{BE} mismatch causes even harmonic distortion, junction resistance causes odd harmonics and thermal noise, and C_{bc} of Q_1–Q_4 are a feedthrough path for high-frequency signal leakage, causing appreciable error around zero output. The circuit in Fig. 10.49 is the multiplier core. To multiply voltages, diff-amp transconductance amplifiers drive the multiplier and ratio I_X and I_Y by their inputs.

Two-quadrant multipliers are used as *voltage-controlled amplifiers* (VCA), in which only magnitude (not phase) of the signal is controlled. A typical two-quadrant multiplier is shown in Fig. 10.50 using an inverting translinear cell. I_X controls the amplitude of the input v_Y. The fractional output $i_O/I_X = i_Y/I_Y$.

An improved multiplier topology is the two-quadrant *controlled-cascode multiplier* cell (Fig. 10.51). The input v_Y generates i_Y as collector currents of a differential cascode amplifier. The translinear input junctions are now the

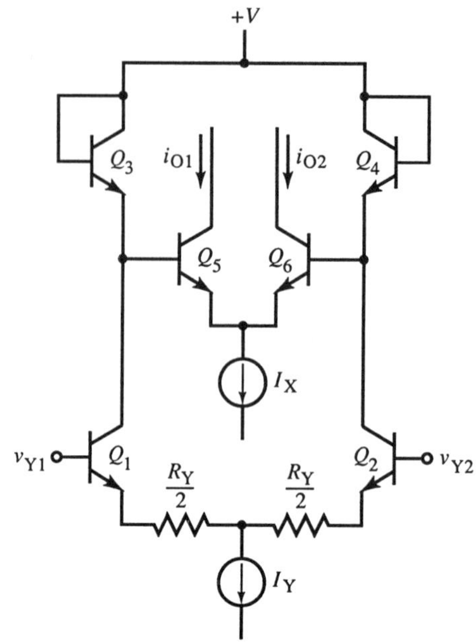

FIG. 10.50 Two-quadrant multiplier, or VCA.

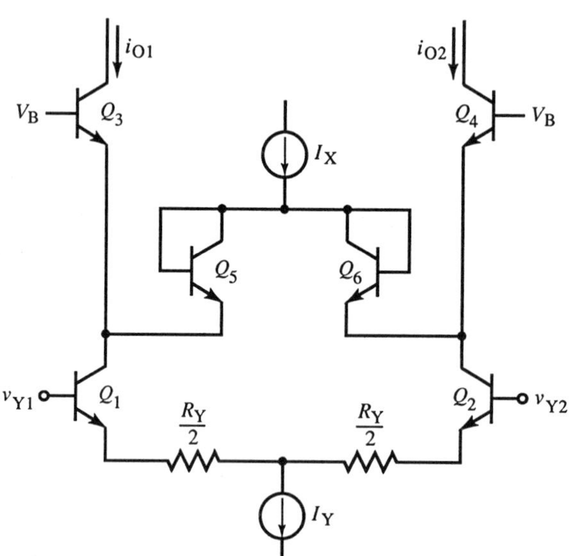

FIG. 10.51 Controlled-cascode two-quadrant multiplier.

b–e junctions of the CB stage instead of diodes. Consequently, i_Y becomes the output current. I_X drives the other translinear pair of junctions, Q_5 and Q_6, and is proportioned by the CB currents so that xI_X subtracts from xI_Y. The outputs are the differences:

$$i_{O1} = i_{Y1} - i_{X1} = (1-x)I_Y - (1-x)I_X = (1-x)(I_Y - I_X)$$
$$i_{O2} = i_{Y2} - i_{X2} = xI_Y - xI_X = x(I_Y - I_X) \tag{10.117}$$

Thus the differential output is

$$i_O = (2x-1)(I_Y - I_X) \tag{10.118}$$

and the fractional output is

$$\frac{i_O}{I_Y} = (2x-1)\left(1 - \frac{I_X}{I_Y}\right) \tag{10.119}$$

As a two-quadrant multiplier, I_X varies from zero to I_Y, whereas the input v_Y varies x over the range from zero to one. The scaling effect for I_X is reversed since an increase in I_X decreases the output.

This topology has less high-frequency feedthrough because of the CB isolation of C_{cb}. In the circuit of Fig. 10.50, the fractional output varies inversely with input-current scaling I_Y. In (10.119), the sensitivity of i_O to I_Y is $I_Y/(I_Y - I_X) > 1$, and scaling is more sensitive to I_Y. This topology also has less nonlinearity from area mismatch and junction resistance. The Analog Devices AD539 is a dual cascode-controlled multiplier of Gilbert's design, with an i_Y gain-independent bandwidth of 60 MHz, I_X bandwidth of 5 MHz, and less than 1% nonlinearity.

A four-quadrant cascode multiplier using Gilbert gain cells was developed by Art Metz. The BJT pair, Q_1–Q_2, in Fig. 10.52 forms a Gilbert cell with Q_5–Q_6. The second Gilbert cell, with Q_3–Q_4, is reversed from the first cell. When $i_X = 0$, both cells have the same gain and their outputs cancel, leaving i_Y unattenuated. At the extremes, only one cell is on and either adds or subtracts from i_Y. The output range is from zero to twice i_Y when $i_{X2} = I_X = I_Y$. The outputs in general are

$$i_{O1} = i_{Y1} + xi_{X1} + (1-x)i_{X2} \tag{10.120a}$$

$$i_{O2} = i_{Y2} + xi_{X2} + (1-x)i_{X1} \tag{10.120b}$$

For the translinear cell,

$$\frac{i_Y}{I_Y} = \frac{i_X}{I_X} = 2x - 1 \tag{10.121}$$

The output is then

$$i_O = i_Y + (2x-1)i_X = i_Y\left(1 + \frac{i_X}{I_Y}\right) \tag{10.122}$$

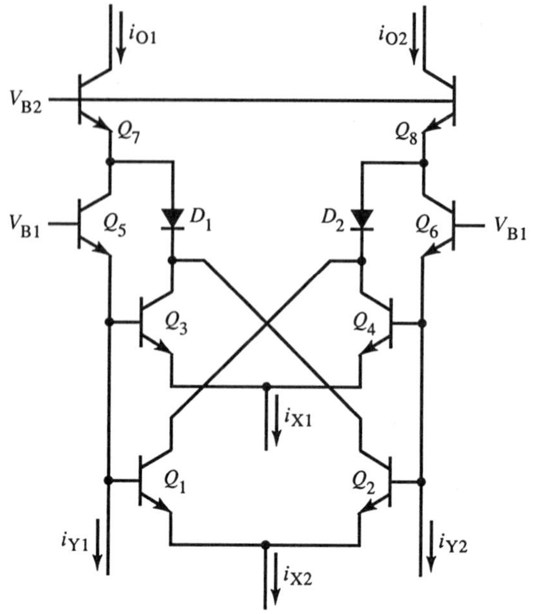

FIG. 10.52 Four-quadrant Gilbert gain-cell multiplier.

The multiplier acts as a controlled-gain amplifier of i_Y, with current gain

$$A_i = \frac{i_O}{i_Y} = 1 + \frac{i_X}{I_Y} \qquad (10.123)$$

Since i_X is bipolar, the gain is unity when i_X is zero.

This multiplier has the same advantages as that of Fig. 10.51: high linearity with junction mismatch and resistance. Both also have low thermal distortion.

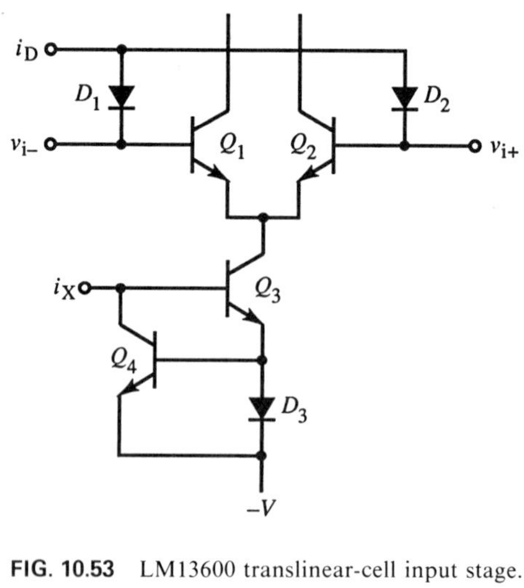

FIG. 10.53 LM13600 translinear-cell input stage.

The diodes D_1 and D_2 in Fig. 10.52 make the operating-point c–b voltage of the Gilbert-cell transistors the same as for the CB. D_1 reduces v_{CE1} and v_{CE3} to compensate for v_{BE5}. If $I_X = I_Y$, the emitter currents are also equal, and thermal balance results. The additional CB of Q_7 and Q_8 provides feedthrough isolation for the Gilbert cells.

A translinear cell is also found in the input of the LM13600 dual transconductance op-amp employing "linearizing diodes" (Fig. 10.53). A method of biasing the translinear cell is shown in Fig. 10.54a. Balanced currents of $I_D/2$

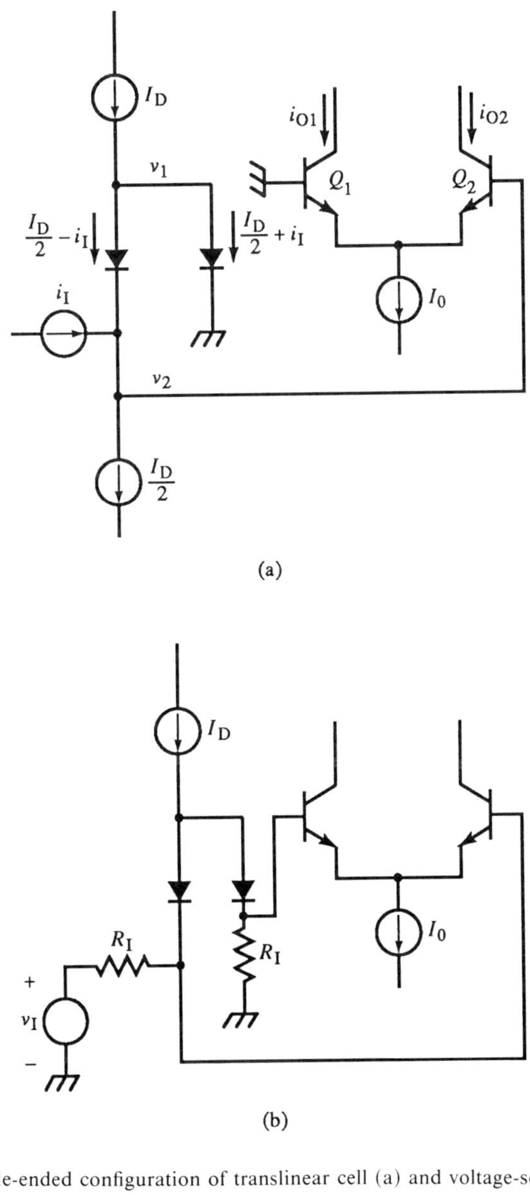

FIG. 10.54 Single-ended configuration of translinear cell (a) and voltage-source version (b).

flow through both diodes, resulting in zero input voltages to the diff-amp. Input current i_I upsets the balance so that the junction voltages around the translinear-cell loop are

$$v_2 - v_1 = V_T \ln\left(\frac{I_D/2 + i_I}{I_D/2 - i_I}\right) = V_T \ln\left(\frac{i_{O2}}{i_{O1}}\right) = V_T \ln\left(\frac{I_0 + i_O}{I_0 - i_O}\right) \qquad (10.124)$$

which simplifies to

$$i_O = i_{O2} - i_{O1} = 2\left(\frac{I_0}{I_D}\right) i_I, \qquad |i_I| < I_D/2 \qquad (10.125)$$

This topology is a single-ended-to-differential translinear cell. Since matching of bias-current sources is harder to achieve than matching resistors, the voltage-input version of Fig. 10.54b is easier for discrete design. When signal amplification must be linear but amplitude control need not be, the linearizing diodes of the translinear cell can be discarded. The linear input is i_X, and VCA control input is v_I.

A class of variable-gain circuits with minimal control bandwidth and linearity requirements are *automatic gain control* (AGC) and *compandor* (compressor-expander) circuits, used in radios to maintain constant loudness. A variety of variable-gain circuits exist such as the one in Fig. 10.55. By increasing I_X, r_e in Q_3 and Q_4 are reduced and gain increased.

FET forms of the Gilbert multiplier have been studied, but since the FET is a square-law rather than an exponential device, the simplicity of high-performance multiplication has not been achieved. The *quarter-square multiplier* is based on square-law devices and the quarter-square formula:

$$\tfrac{1}{4}[(x+y)^2 - (x-y)^2] = xy \qquad (10.126)$$

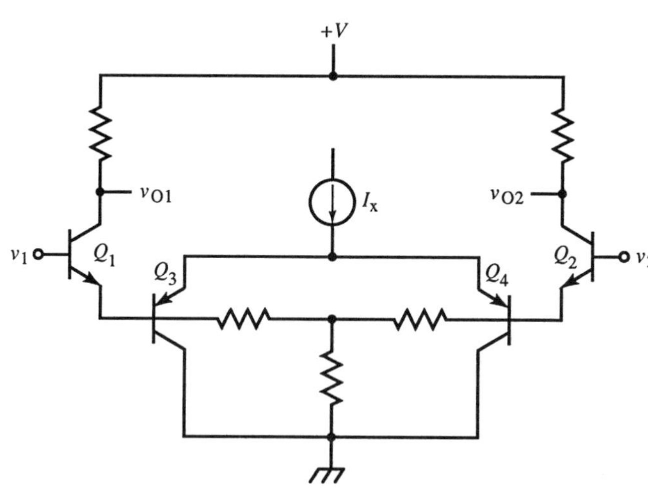

FIG. 10.55 A simple variable-gain amplifier for AGC applications. I_X controls gain by varying r_e of Q_3 and Q_4.

This technique requires three summations, two squarings and scaling and has been implemented with MOSFETs.

10.11 Programmable-Gain Amplifiers

For amplifier applications in which inputs cover a wide range, amplifier range and gain constraints can conflict. For small signals, a high gain is needed to produce a full-scale (fs) output; but for large signals, the same gain drives them out of range. If the scaling accuracy is not critical, a logarithmic amplifier or compander can solve this problem; both have a nonlinear transfer curve, so for small signals the gain is high but decreases with amplitude. The output, of course, is distorted by the log-amp or compander gain variation.

For linear systems, a simple solution is to change the amplifier gain to match the signal amplitude. This minimizes amplifier error and keeps the signal within the amplifier linear range. An input x_i is subject to scaling (gain) and offset errors in the amplifier, so the output, referred to the input, is

$$x_o = (1 + \varepsilon_s)x_i + \varepsilon_o \tag{10.126}$$

where ε_s is scaling error, and ε_o is offset error. The amplifier error is defined as

$$\varepsilon \equiv \frac{x_o - x_i}{x_i} = \frac{\varepsilon_s x_i + \varepsilon_o}{x_i} = \varepsilon_s + \frac{\varepsilon_o}{x_i} \tag{10.127}$$

The error has two terms: ε_s is the "range error," and ε_o/x_i is the "reading error." Accuracy specifications are often given as ε_s because it is a fraction of the fs amount for all x_i. As x_i approaches zero, the fixed offset error becomes an increasingly large fraction of the total error and varies inversely with x_i. The zero-scale (zs) end of the range partly depends on how much error can be tolerated.

Amplifier gain can be changed over a continuous range using a multiplier. But usually it is simpler to switch gain-determining elements for a small number of gain settings. These *gain-switched* or *programmable-gain* amplifiers are commonly found on measurement instruments, such as oscilloscopes, in which a very wide range of input amplitudes is allowed. The vertical sensitivity control on oscilloscopes typically spans a range from 2 mV/div to 50 V/div. (The *div* unit is a division of vertical deflection on the CRT screen graticule; eight divisions is standard for full scale.) Instrument designers have standardized on a 1-2-5 sequence of gain settings within a decade. Since these discrete settings fall short of matching the amplifier fs to the signal, some additional error due to ε_o is accepted.

For a given gain setting, the range is from zs to fs for input quantity x_i, or

$$x_{zs} = \frac{x_{fs}}{a} \leq x_i \leq x_{fs}, \qquad a > 1 \tag{10.128}$$

where x_{fs} is the maximum x_i at fs and minimum x_i at zs is x_{zs}, the fraction $1/a$ of fs. Thus a defines the extent of the range. Now the average error over the range is

$$\bar{\varepsilon} = \frac{1}{x_{fs} - x_{fs}/a} \int_{x_{fs}/a}^{x_{fs}} \left(\varepsilon_s + \frac{\varepsilon_o}{x_i} \right) dx_i = \left(\frac{a \ln a}{(a-1)x_{fs}} \right) \varepsilon_o + \varepsilon_s \tag{10.129}$$

for $a > 1$. For amplifiers with continuously adjustable gain, for all x_i, gain is adjusted so that x_i is x_{fs}. The ratio of offset errors for the discrete to continuous cases is

$$\frac{\varepsilon_{od}}{\varepsilon_{oc}} = \left(\frac{a}{a-1} \right) \ln a \tag{10.130}$$

For a range of a:1, the larger a is, the larger the error ratio, to the disadvantage of discrete-gain settings.

In computer-based data-acquisition systems, a 1-2-5 gain sequence is not necessary since the computer can rescale the measurements. The question then arises as to the advantage of regular gain settings over the 1-2-5 settings. For minimum offset error, many ranges with small a are desired. If we assume that at fs, ε_s can be nulled, then for three gain settings per decade, the 1-2-5 scheme has the following a and normalized offset error:

x_{fs}	x_{zs}	a	$(\bar{\varepsilon} - \varepsilon_s)/\varepsilon_o$
2	1	2.0	1.39
5	2	2.5	1.53
10	5	2.0	1.39

The maximum normalized error is 1.53. For three regularly spaced settings covering a decade,

$$a = 10^{1/3} \cong 2.15$$

and maximum normalized error is 1.43. Therefore, the improvement of regular settings over 1-2-5 settings is $1.43/1.53 = 0.93$, for 7% improvement. This is usually not significant, and the 1-2-5 sequence appears to be well chosen.

The major consideration in gain-switched amplifiers is the imperfection of the switches. Reed relays are closest to ideal electrically but are slow, bulky, and power-intensive. Solid-state analog switches are used in all but the most demanding applications, with their parasitic series, shunt (leakage) resistances, and shunt capacitance.

Diodes, BJTs, JFETs, and MOSFETs function as switches of voltage or current. JFETs (including those of biFET implementation) have constant on-resistance r_{ON} of typically 100 Ω over the input voltage range. MOSFET switches have lower r_{ON}, typically 10–20 Ω, with large Δr_{ON}. Discrete power MOSFETs have r_{ON} values of less than 100 mΩ but with correspondingly large shunt capacitances. MOSFETs have a parasitic diode between the body and drain. Since three-terminal MOSFETs have the body connected to the source, current flows through this diode in the opposite direction to the gate-controlled flow. Two MOSFETs in series, back to back, form a bipolar switch. Because of MOSFET r_{ON} variation, n- and p-channel devices are used together as CMOS switches since their variations tend to cancel. The combined r_{ON} still varies more than for a JFET switch and peaks at the midrange input voltage.

Discrete diodes and BJTs are sometimes used as gain switches, as in the BJT amplifier of Fig. 10.56. The diff-amp emitter resistors are different ($R_1 \neq R_2$), and gain is selected by activating one diff-amp. The diff-amps are switched by switching the emitter-current source to the selected diff-amp. This scheme is extendable by adding more diff-amp pairs and current switches.

Of the two switching modes, voltage and current, parasitic elements in switches mainly determine which mode to use. A switch with high r_{ON} is a poor choice for switching in a voltage divider; r_{ON} adds to the divider resistance, changing the attenuation. Except for voltage range (compliance) limitations on current sources, r_{ON} does not affect current-mode switching. An input source is selected by current-switching into the virtual ground of an inverting op-amp (Fig. 10.57). Current-mode switching also reduces the effects of shunt capacitance by minimizing voltage variation.

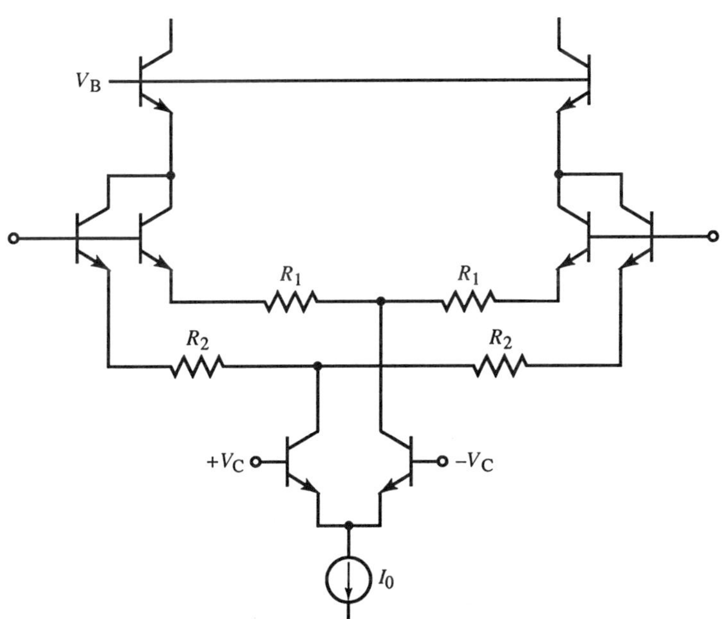

FIG. 10.56 BJT gain-switching via the current source of diff-amps with different gains.

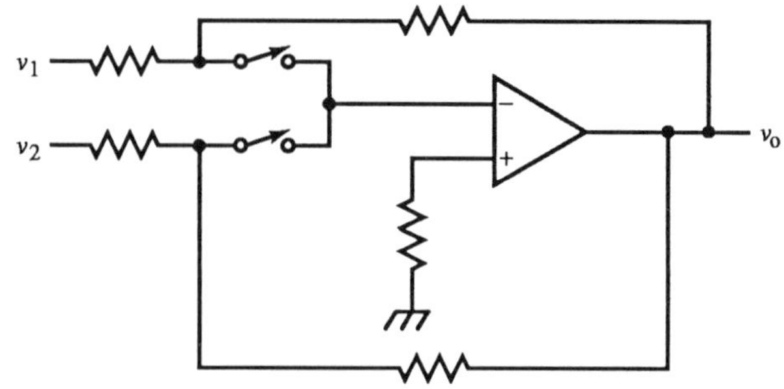

FIG. 10.57 Current switching at virtual ground of inverting op-amp.

Sometimes voltage-mode switching is necessary and can be error-free if the switch load is an open-circuit. That is, switch current (and voltage drop) is zero. A common example is gain-switching in noninverting op-amps (Fig. 10.58). To keep the op-amp loop from momentarily opening during switching, the switches should be make-before-break types.

10.12 Closure

This chapter has introduced several new amplifier concepts based on current amplification, to get around gain-bandwidth limitations, and the use of composite topologies to improve performance. The translinear cell opens a wide range of possibilities for improved analog signal processing. Here, we investigated amplification and multiplication but function generation is another use for it. The study of amplification has been our focus to this point, but with the introduction of gain switching, nonlinear analog circuit functions become a consideration.

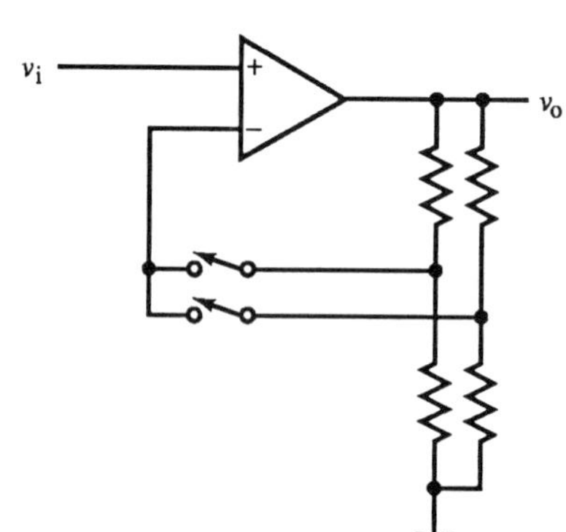

FIG. 10.58 Voltage switching into open circuit in noninverting op-amp.

References

Robert D. Lorenz, "Synthesis of State Variable Controllers for Industrial Servo Drives," *Conf. Appl. Motion Control 1986,* p. 247ff.

Arthur Metz, "Circuit-design/process combo speeds horizontal-amp slewing," *EDN*, 4 Aug. 1983. pp. 173–180.

Jim Williams, "Monolithic power-buffer IC drives difficult loads," *EDN*, 9 Aug. 1984. pp. 153–159.

Carl R. Battjes, "A Wide-Band High-Voltage Monolithic Amplifier," *IEEE JSSC*, Vol. SC-8, No. 6, Dec. 1973. pp. 408–413.

Winthrop Gross, "Use npn and pnp devices effectively in semicustom arrays," *EDN*, 27 Oct. 1988. pp. 297–308.

Shaun Simpkins and Winthrop Gross, "Cascomp Feedforward Error Correction in High Speed Amplifier Design," *IEEE JSSC*, Vol. SC-18, No. 6, Dec. 1983. pp. 762–764.

Pat Quinn, "Feedforward Amplifier," Tektronix patent #4,146,844.

Sergio Franco, "Current-feedback amplifiers benefit high-speed designs," *EDN*, 5 Jan 1989. pp. 161–172.

Alan Hansford, "Use of transimpedance amplifiers minimizes design tradeoffs," *EDN*, 26 Nov. 1987. pp. 205–214.

"A New Approach to Op Amp Design," Comlinear Corp. Application Note 300-1, Mar. 1985.

George R. Wilson, "A Monolithic Junction FET-n-p-n Operational Amplifier," *IEEE JSSC*, Vol. SC-3, No. 4, Dec. 1968. pp. 341–348.

Barrie Gilbert, "A New Wide-Band Amplifier Technique," *IEEE JSSC*, Vol. SC-3, No. 4, Dec. 1968. pp. 353–365.

Barrie Gilbert, "A Precise Four-Quadrant Multiplier with Subnanosecond Response," *IEEE JSSC*, Vol. SC-3, No. 4, Dec. 1968. pp. 365–373.

Willy M. C. Sansen and Robert G. Meyer, "An Integrated Wide-Band Variable-Gain Amplifier with Maximum Dynamic Range," *IEEE JSSC*, Vol. SC-9, No. 4, Aug. 1974. pp. 159–166.

Barrie Gilbert *et al.*, "Build fast VCAs and VCFs with analog multipliers," *EDN*, 18 Oct. 1984. pp. 289–299.

Jesús S. Peña-Finol and J. Alvin Connelly, "A MOS Four-Quadrant Analog Multiplier Using the Quarter-Square Technique," *IEEE JSSC*, Vol. SC-22, No. 6, Dec. 1987. pp. 1064–1073.

CHAPTER 11

Signal-Processing Circuits

Besides amplification and multiplication, various other signal-processing functions are a part of the analog circuit design repertoire. Most of these functions are nonlinear. We shall survey a variety of these signal-processing circuits.

11.1 Voltage References

Stable and accurate voltage sources are needed as references for measurement circuits and power supplies. The Zener diode is a simple voltage-reference device. Although it has been in use a long time, it is still the most stable kind of reference available (other than reference standards such as temperature-controlled batteries or superconducting quantum-effect devices). A simple Zener-based reference is shown in Fig. 11.1a. Zener diodes combine two mechanisms, tunneling and avalanche breakdown. Tunneling has a negative TC, and avalanche has a positive TC. At around 5 V the mechanisms cancel, but the tolerance for 5 V Zeners is not good, making selection necessary for low TC. The TC of Zeners increases reliably with Zener voltage V_Z above about 6 V at about 1 mV/°C per volt, or 0.1%/°C. At a V_Z of 5.6 V, the TC is that of a forward-biased diode, about −2 mV/°C. By placing a diode in series with a 5.6 V Zener, we get a zero TC 6.3 V Zener reference diode. Manufacturers' literature shows that low-TC diodes are around 6.3 V. Low-TC Zeners at higher voltages are also possible by stacking more diodes in series, but tracking makes repeatable manufacture of zero-TC devices more difficult.

Zener diodes are noisy, especially at low currents. Consequently, they are bypassed with a capacitor (Fig. 11.1a). In ICs, high-performance Zeners are

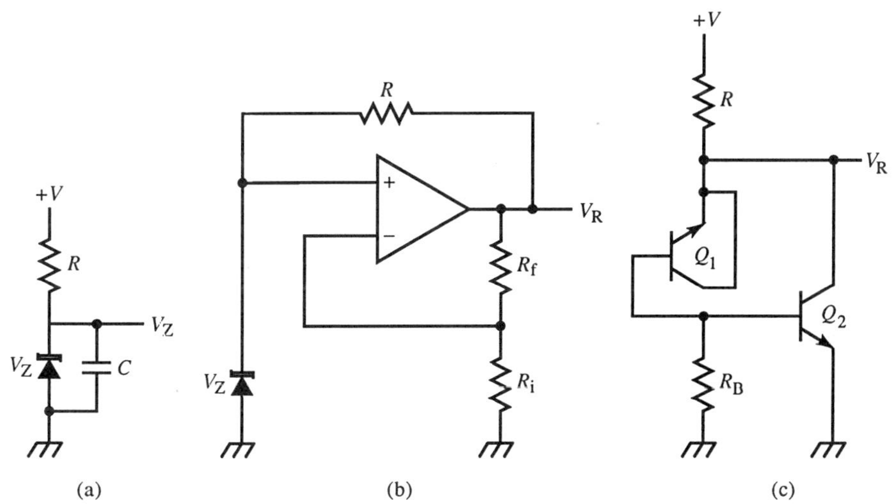

FIG. 11.1 Zener diode voltage references: (a) basic reference, (b) bootstrapped reference with scaling, and (c) BJT Zener with series b–e junction compensation of Q_2.

built below the IC surface as *subsurface Zeners*, which are less noisy because surface effects are eliminated. Lateral ion-implanted Zeners have low-tolerance voltages (typically less than 1%) and are commonly used as references in IC circuits. With a substrate temperature controller on the same chip, monolithic references with 1 ppm/°C are commercially available.

A minimum TC also depends on Zener current I_Z, typically 5–10 mA. The circuit of Fig. 11.1a is subject to I_Z variation with the voltage supply. The resistor supplying I_Z can be bootstrapped (Fig. 11.1b) with an op-amp. The op-amp output,

$$V_R = \left(\frac{R_f}{R_i} + 1\right) V_Z \tag{11.1}$$

supplies a stable Zener current of $(V_R - V_Z)/R$. This circuit requires a starting circuit for the Zener, when power is first applied. A simpler bootstrap circuit (Fig. 11.1c) uses a transistor b–e junction as the Zener diode, for which V_Z is 6–7 V, with a TC of around 2 mV/°C. The Zener Q_1 is in series with the forward-biased b–e junction of Q_2. The combination forms a reference Zener and has a low TC. Q_2 provides shunt regulation to reduce output resistance. Zener diodes have resistances of around 10 Ω, increasing with V_Z to 100 Ω. Thus, their load regulation is unacceptable for high stability and must be buffered. Another disadvantage to Zeners is that a smaller reference voltage is desired for monolithic 5 V regulators and other devices, such as ADCs and DACs, that operate from 5 V.

A newer kind of voltage reference is based on the temperature characteristics of pn junctions themselves. Junction voltage has a negative TC of about

$-2\ \mathrm{mV/°C}$. The differential voltage across two matched junctions (as in a diff-amp) is

$$\Delta V = V_{\mathrm{T}} \ln \frac{I_2}{I_1} = \frac{kT}{q} \cdot \ln \frac{I_2}{I_1} \tag{11.2}$$

When the current ratio is held constant, ΔV has a positive, linear TC of

$$\mathrm{TC}(\Delta V) = \frac{k}{q} \cdot \ln \frac{I_2}{I_1} = \left(86.17 \frac{\mu\mathrm{V}}{°\mathrm{C}}\right) \ln \frac{I_2}{I_1} \tag{11.3}$$

By scaling ΔV and adding it to junction voltage V, we obtain an output of

$$V_{\mathrm{O}} = V + g(\Delta V) = V + g V_{\mathrm{T}} \ln \frac{I_2}{I_1} \tag{11.4}$$

where g is the gain required to amplify ΔV so that its TC is opposite that of V. If the junction areas of Q_1 and Q_2 are not equal, the more general form of ΔV is

$$\Delta V = V_{\mathrm{T}} \ln \frac{J_2}{J_1} = V_{\mathrm{T}} \ln \left[\left(\frac{I_2}{I_1}\right)\left(\frac{A_1}{A_2}\right)\right] \tag{11.5}$$

where J is current density and A is the b–e junction area; $J = I/A$. The TC of V_{O} is

$$\mathrm{TC}(V_{\mathrm{O}}) = \frac{dV_{\mathrm{O}}}{dT} = \frac{dV}{dT} + g \cdot \frac{d\,\Delta V}{dT} \tag{11.6}$$

We can substitute for the TC of V and ΔV, set $\mathrm{TC}(V_{\mathrm{O}})$ to zero, and solve for the gain g. To find $\mathrm{TC}(V)$, we first differentiate V:

$$\left.\frac{dV}{dT}\right|_I = \frac{d}{dT}\left(V_{\mathrm{T}} \ln \frac{I}{I_{\mathrm{S}}}\right) = -\left(\frac{1}{I_{\mathrm{S}}} \cdot \frac{dI_{\mathrm{S}}}{dT}\right) V_{\mathrm{T}} + \frac{V}{T} \cong -2 \frac{\mathrm{mV}}{°\mathrm{C}} \tag{11.7}$$

The expression,

$$\mathrm{TC}\%(I_{\mathrm{S}}) = \left(\frac{1}{I_{\mathrm{S}}} \cdot \frac{dI_{\mathrm{S}}}{dT}\right) \tag{11.8}$$

is the fractional $\mathrm{TC}(I_{\mathrm{S}})$.

From semiconductor physics, saturation current is

$$I_{\mathrm{S}} = qA\left(\frac{D_{\mathrm{h}} p_{\mathrm{no}}}{L_{\mathrm{h}}} + \frac{D_{\mathrm{e}} n_{\mathrm{po}}}{L_{\mathrm{e}}}\right) = kTAn_{\mathrm{i}}^2\left(\frac{\mu_{\mathrm{h}}}{L_{\mathrm{h}} N_{\mathrm{D}}} + \frac{\mu_{\mathrm{e}}}{L_{\mathrm{e}} N_{\mathrm{A}}}\right) \tag{11.9}$$

where D are diffusion coefficients, L the diffusion length constants, p_{no} and n_{po} the equilibrium minority hole and electron concentrations, N the ion doping concentrations, μ the carrier mobilities, and n_{i} the intrinsic carrier concentration, where

$$n_{\mathrm{o}} p_{\mathrm{o}} = n_{\mathrm{i}}^2 \tag{11.10}$$

and n_{o} and p_{o} are the equilibrium electron and hole concentrations. In (11.9),

μ and n_i^2 are temperature dependent; the other constants are fixed by geometry, doping, or materials properties. Delving deeper into solid-state physics, we have

$$n_i^2 \propto T^3 \, e^{-E_{go}/kT} \tag{11.11}$$

where E_{go} is the semiconductor bandgap energy, linearly extrapolated to 0 K. For silicon it is

$$E_{go} = 1.205 \text{ eV} \qquad \text{(Si)} \tag{11.12}$$

Because $V_T = kT/q$, (11.11) can be expressed as

$$n_i^2(T) \propto T^3 \, e^{-V_{go}/V_T} \tag{11.13}$$

I_S also depends on $\mu(T)$. For silicon, $\mu(T) \propto T^{-2.6}$. With this value, it follows from (11.9) that

$$I_S \propto T \cdot T^{-2.6} \cdot T^3 \, e^{-V_{go}/V_T} = T^{1.4} \, e^{-V_{go}/V_T} \tag{11.14}$$

By taking the derivative and dividing it by I_S, we get

$$\text{TC\%}(I_S) = \frac{1}{I_S} \cdot \frac{dI_S}{dT} = \frac{V_{go}/V_T + 1.4}{T} \tag{11.15}$$

We now have an expression for (11.8). At 300 K, this is

$$\text{TC\%}(I_S)\big|_{T=300\text{ K}} = 15.53\%/^\circ\text{C} + 0.47\%/^\circ\text{C} = 16\%/^\circ\text{C} \tag{11.16}$$

The dominant effect on TC%(I_S) with temperature is the first term, involving the bandgap. The mobility (second term) affects it only about 3%.

Returning to (11.7), we can now solve for the value of

$$\left.\frac{dV}{dT}\right|_I = -\left(\frac{1}{I_S} \cdot \frac{dI_S}{dT}\right) V_T + \frac{V}{T} = \frac{V - V_{go} - 1.4 V_T}{T} \tag{11.17}$$

At 300 K and $I_S = 10^{-14}$ A, for $I = 1$ mA, $V = 0.655$ V, and

$$\frac{dV}{dT} = -1.95 \frac{\text{mV}}{^\circ\text{C}} \cong -2.0 \frac{\text{mV}}{^\circ\text{C}}.$$

While we are calculating TCs, the fractional TC of I is

$$\text{TC\%}(I)\big|_V = \frac{1}{I} \cdot \left.\frac{dI}{dT}\right|_V = \frac{1}{I_S} \cdot \frac{dI_S}{dT} - \frac{1}{T} \cdot \frac{V}{V_T} \tag{11.18}$$

At 300 K and $V = 0.655$ V, as before,

$$\text{TC\%}(I)\big|_V = 16\%/^\circ\text{C} - 8.44\%/^\circ\text{C} \cong 8\%/^\circ\text{C} \tag{11.19}$$

We already know that

$$\text{TC\%}(V_T) = \frac{1}{V_T} \cdot \frac{dV_T}{dT} = \frac{1}{T} \tag{11.20}$$

At 300 K, TC%(V_T) = 0.33%/°C. Returning to (11.6), we must yet find

$$\frac{d\,\Delta V}{dT} = \frac{V_T}{T} \cdot \ln \frac{J_2}{J_1} \tag{11.21}$$

Substituting (11.17) and (11.21) into (11.6), we obtain

$$\left.\frac{dV_O}{dT}\right|_I = \frac{V}{T} - \left(\frac{1}{I_S} \cdot \frac{dI_S}{dT}\right) V_T + g \ln\left(\frac{J_2}{J_1}\right)\left(\frac{V_T}{T}\right) \tag{11.22}$$

From (11.17), this can also be expressed in V_{go} as

$$\left.\frac{dV_O}{dT}\right|_I = \frac{V_T(\ln(I/I_S) + g \ln(J_2/J_1) - 1.4) - V_{go}}{T} \tag{11.23}$$

For $TC(V_O) = 0$, the required gain is

$$g = \frac{(1/I_S)(dI_S/dT)T - (V/V_T)}{\ln(J_2/J_1)} = \frac{V_{go}/V_T - (\ln(I/I_S) - 1.4)}{\ln(J_2/J_1)} \tag{11.24}$$

The constraint on achieving a zero TC is that I be held constant.

Example 11.1 Bandgap Reference Design

The CA3086 BJT array has $I_S = 10^{-15}$ A at 300 K. Let I be 1 mA. Then $V = 0.715$ V, and from (11.22),

$$\frac{dV_O}{dT} = 2.38\,\frac{\text{mV}}{°\text{C}} - 4.14\,\frac{\text{mV}}{°\text{C}} + \left(86.17\,\frac{\mu\text{V}}{°\text{C}}\right) \cdot g \ln \frac{J_2}{J_1}$$

The first two terms have a combined TC of −1.76 mV/°C. The gain required to null this TC is, from (11.24),

$$g = \frac{(0.16/\text{K})(300\text{ K}) - (0.715\text{ V})/(25.87\text{ mV})}{\ln(J_2/J_1)} = \frac{20.4}{\ln(J_2/J_1)}$$

Finally, from (11.4),

$$V_O = 0.715\text{ V} + (20.4)V_T = 0.715\text{ V} + 0.527\text{ V} = 1.242\text{ V}$$

The result for V_O of Example 11.1 is curiously close to V_{go}. If we substitute (11.24) into (11.4), we get

$$V_O = V_{go} + 1.4\,V_T = V_g = 1.242\text{ V} \tag{11.25}$$

The output of a bandgap reference is V_g, the ambient-temperature bandgap voltage. A concise expression for V_O, in terms of g, follows from substituting

(11.5) into (11.4):

$$V_O = V_T \ln\left(\frac{I}{I_S}\right) + gV_T \ln\left(\frac{J_2}{J_1}\right) = V_T \ln\left[\left(\frac{I}{I_S}\right)\left(\frac{J_2}{J_1}\right)^g\right] \tag{11.26}$$

Substituting g from (11.24), we can reduce this to (11.25).

The gain formula of (11.24) can be expressed more simply using (11.25) as

$$g = \frac{V_g - V}{\Delta V} \tag{11.27}$$

This simpler formula for g is expressed entirely in dc circuit voltages.

A simple bandgap circuit (Fig. 11.2) is the *Widlar bandgap reference*, after Bob Widlar (pronounced "wide-ler"), who invented the bandgap concept. R_1 sets current I_1 through Q_1. The current, or current density, of Q_1 must be larger than that of Q_2 to create a positive ΔV across R_2. Assuming $\alpha = 1$ for Q_2, we get the gain $V_{R3}/\Delta V = R_3/R_2$. The application of (11.4) involves V_{R3} and V_{BE3}:

$$V_O = V_{BE3} + V_{R3} = V_T \ln\frac{I_{E3}}{I_{S3}} + \left(\frac{R_3}{R_2}\right) V_T \ln\frac{J_2}{J_1} \tag{11.28}$$

and this must equal V_O of (11.25). The junction currents are kept constant by bootstrapping their sources from the stable output, supplied by I_0. Q_3 also shunt regulates the output. In this analysis, α error has been ignored, and biasing constrains $V_{BE3} > V_{BE1}$ to keep from saturating Q_2. Consequently, $I_2 < I_1 < I_3$ for equal areas. The topology places a limit on how large ΔV can be made.

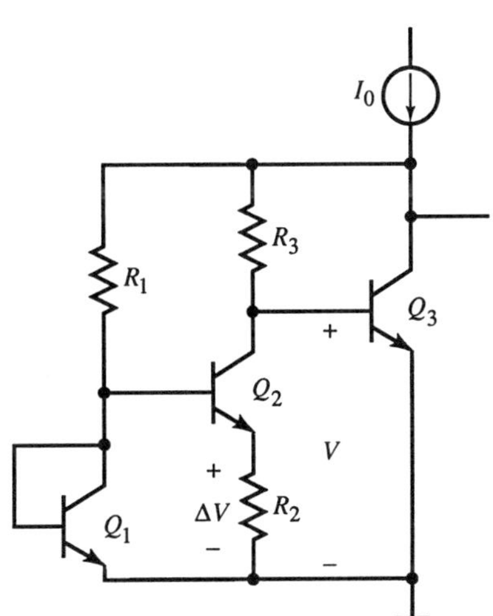

FIG. 11.2 Widlar bandgap reference.

Example 11.2 Widlar Bandgap Reference

Based on the topology of Fig. 11.2 and Example 11.1, $I_{E3} = 1$ mA and $V_{BE3} = 0.715$ V. From (11.25), the output voltage is $V_g = 1.24$ V. Then

$$V_{R3} = V_O - V_{BE3} = 0.527 \text{ V}$$

We must set $I_1 < I_{E3}$ to reverse-bias the b–c junction of Q_2, and for $\Delta V = V_{R2} > 0$, $I_2 < I_1$. For maximum I_1/I_2, let $V_{CB2} = 0$ V, the same as V_{CB1}. At low currents, the drop across r_c' is negligible and saturation of Q_2 avoided. Then

$$V_1 = V_{BE3} \Rightarrow I_1 = I_{E3}$$

By choosing I_1, we choose ΔV as

$$\Delta V = V_T \ln \frac{I_{E3}}{I_2}$$

Let $I_2 = 100\ \mu$A. Then

$$\Delta V = V_T \ln \frac{I_1}{I_2} = 59.6 \text{ mV}$$

and

$$g = \frac{V_g - V}{\Delta V} = 8.85$$

Furthermore,

$$R_3 = \frac{V_O - V}{I_2} = \frac{0.527 \text{ V}}{100\ \mu\text{A}} = 5.27 \text{ k}\Omega$$

$$R_2 = \frac{R_3}{g} = 595\ \Omega$$

and

$$R_1 = \frac{V_O - V_1}{I_1} = \frac{0.527 \text{ V}}{100\ \mu\text{A}} = 5.27 \text{ k}\Omega$$

Total current is 2.1 mA, considerably less than the zero-TC current of typical Zeners. Much lower currents are also feasible.

The Widlar reference has the disadvantage of a determined output voltage V_g that is not optimal for many applications. A bandgap reference with arbitrary output would be better. The differential bandgap circuit of Fig. 11.3 has output V_R that is set by R_f and R_i. Q_2 has a higher current density than Q_1, so ΔV

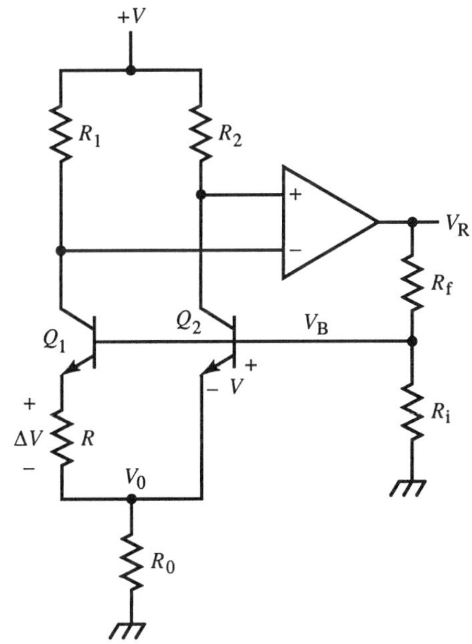

FIG. 11.3 Differential bandgap reference.

is the difference of the V_{BE} voltages:

$$\Delta V = I_1 R = V_{BE2} - V_{BE1} = V - V_T \ln \frac{I_1}{I_S} = V_T \ln \frac{J_2}{J_1} \tag{11.29}$$

The op-amp inputs are kept at the same voltage by feedback so that

$$I_1 R_1 = I_2 R_2 \tag{11.30}$$

If J_2/J_1 is chosen, then ΔV and I_1 are determined by R. By choosing the b–e junction areas, we also determine I_2. Thus, R_1 and R_2 are determined. Then

$$V_0 = (I_1 + I_2) R_0 \tag{11.31}$$

is set by R_0. The base voltage is then

$$V_B = V + V_0 = V_T \ln\left(\frac{I_2}{I_S}\right) + \left(\frac{R_0}{R}\right)\left(1 + \frac{I_2}{I_1}\right) V_T \ln\left(\frac{J_2}{J_1}\right) \tag{11.32}$$

Finally, the output is

$$V_R = V_B \left(1 + \frac{R_f}{R_i}\right) \tag{11.33}$$

For zero TC(V_R), we must have TC(V_B) = 0. For this circuit, (11.32) has the form of (11.4), the basic bandgap-reference equation. V_B is V_O, V_0 corresponds to ΔV, $I_2 = I$, and

$$g = \frac{V_0}{\Delta V} = \left(\frac{R_0}{R}\right)\left(1 + \frac{I_2}{I_1}\right) \tag{11.34}$$

For $TC(V_B) = 0$, g must satisfy (11.27), or

$$V_0 = V_g - V = V_g - V_T \ln\left[\left(\frac{I_1}{I_S}\right)\left(\frac{I_2}{I_1}\right)\right] \qquad (11.35)$$

The $TC(V_0) > 0$ and is linear with absolute temperature.

Example 11.3 Differential Bandgap Reference

The circuit in Fig. E11.3 is based on a five-transistor CA3086 array of matched transistors with equal areas. The goal is to design a 2.50 V reference. For CA3086 BJTs,

$$I_S = 10^{-15}\ \text{A}$$

For this design, let

$$\frac{J_2}{J_1} = 10, \qquad I_1 = 60\ \mu\text{A}, \qquad I_2 = 600\ \mu\text{A}$$

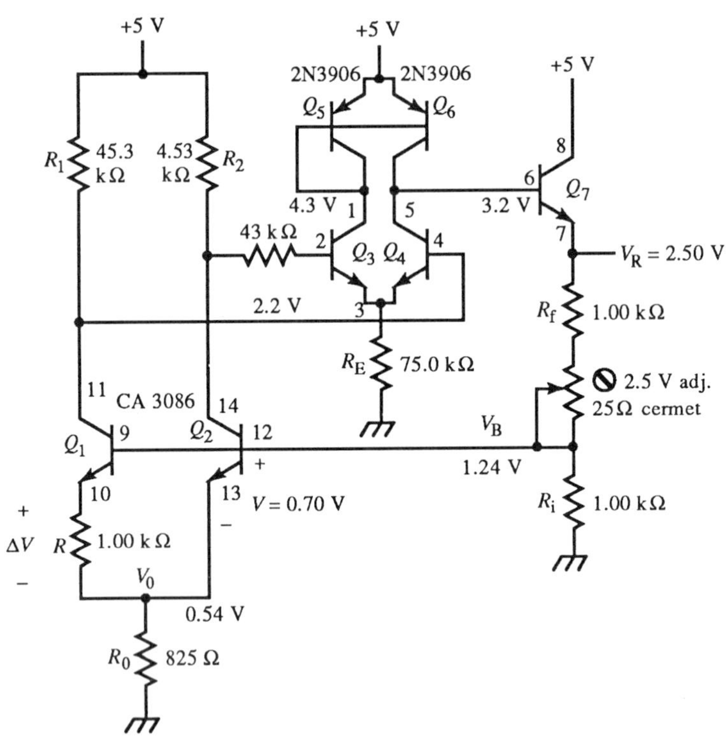

FIG. E11.3

Then

$$R = \frac{\Delta V}{I_1} = \frac{V_T \ln(10)}{60\ \mu A} = \frac{59.56\ mA}{60\ \mu A} = 992\ \Omega \Rightarrow 1.00\ k\Omega, \quad 1\%$$

We now proceed to find R_0 by first calculating V:

$$V = V_{BE2} = V_T \ln\left(\frac{600\ \mu A}{10^{-15}\ A}\right) = 0.702\ V$$

Then, from (11.31) and (11.35),

$$V_0 = R_0(660\ \mu A) = 1.242\ V - 0.702\ V = 0.541\ V$$

Solve for R_0:

$$R_0 = 819\ \Omega \Rightarrow 825\ \Omega, \quad 1\%$$

Now calculate R_1 and R_2 from (11.30). Since this is a ratio, we need another constraint to determine actual values. Note that the V_{CB} of Q_1 and Q_2 are in series with that of Q_3 and Q_4. V_{C3} is a junction drop down from the supply of 5 V, or about 4.3 V. But V_{C4} is less; it is a junction drop up from the output of 2.5 V or 3.3 V. And V_B is $V_g = 1.242$ V. We split the voltage difference between the series b–c junctions so that

$$V_{B3} = V_{B4} = \frac{3.3\ V - 1.24\ V}{2} + 1.24\ V = 2.3\ V$$

Then

$$R_1 = \frac{5\ V - 2.3\ V}{60\ \mu A} = 45\ k\Omega \Rightarrow 45.3\ k\Omega, \quad 1\%$$

and

$$R_2 = \frac{R_1}{10} = 4.5\ k\Omega \Rightarrow 4.53\ k\Omega, \quad 1\%$$

To compensate the diff-amp for bias current, set the source resistance equal. This requires a base resistor for Q_3 of

$$R_{B3} = 45.3\ k\Omega - 4.53\ k\Omega = 40.8\ k\Omega \Rightarrow 43\ k\Omega$$

The diff-amp emitter bias current is set by R_E. If we choose it to be 20 μA, then base current for $\beta \cong 100$ is about 100 nA, a small fraction of 60 μA. The emitter voltage is a junction drop down from V_{B3}, or about 1.5 V. Then

$$R_E = \frac{1.5\ V}{20\ \mu A} = 75\ k\Omega$$

Finally, the feedback divider is

$$\frac{R_f}{R_i}+1=\frac{2.500\ \text{V}}{1.242\ \text{V}}=2.013$$

Choose $R_i = 1.00\ \text{k}\Omega$, 1%, a convenient value. Then $R_f = 1.013\ \text{k}\Omega$. To allow adjustment of the output to correct for parts tolerances, place a trim-pot (screwdriver-adjusted potentiometer) in series with

$$R_f = 1.00\ \text{k}\Omega,\quad 1\%$$

having twice the remaining resistance, to center the pot, or

$$R_{adj} = 2(1.013\ \text{k}\Omega - 1.00\ \text{k}\Omega) = 25.8\ \Omega \Rightarrow 25\ \Omega$$

A cermet pot has a low TC, required for the application, but such a low value may not be available in cermet, and R_f must be reduced to make the trim-pot larger.

All parts values have now been determined, and the circuit can be "prototyped" to verify performance. A prototype was built with the following deviations:

$$R_1 = 49.9\ \text{k}\Omega,\ 1\%, \qquad R_2 = 4.99\ \text{k}\Omega,\ 1\%, \qquad R_0 = 820\ \Omega,\ 5\%$$

The supply measured 5.01 V, V_0 was 0.540 V, and V_B was 1.242 V. These measurements were taken on a warm spring evening in a building without air conditioning; the temperature was approximately 300 K. The CA3086 was heated to about 50°C above ambient temperature with a soldering iron, and V_B became 1.237 V. The circuit was then cooled with circuit cooler (an aerosol); V_B was then 1.248 V. A rough calculation indicates that TC(V_B) is roughly 100 ppm, about the same as the metal-film 1% resistors. In a refined design, this discrete implementation should have all 1% metal-film resistors (no 5% composition resistors). Better yet, it should be an integrated circuit. But for the 30 minutes it took to build, it demonstrated the validity of the derived design equations.

In integrated form, a simple differential bandgap reference can have the topology of Fig. 11.4, in which a nonunity emitter-area ratio is used, where $A_1 > A_2$. Q_3 is a current mirror, and Q_4 and Q_5 form a Darlington buffer to the output.

A third bandgap-reference topology (Fig. 11.5) uses an op-amp with inputs from the emitter (instead of collector) circuit of the bandgap cell, Q_1 and Q_2. The analysis is similar to the previous one. Equations (11.29) and (11.33) are valid. So are (11.30) and (11.31), where V_0 equals the expressions of (11.30).

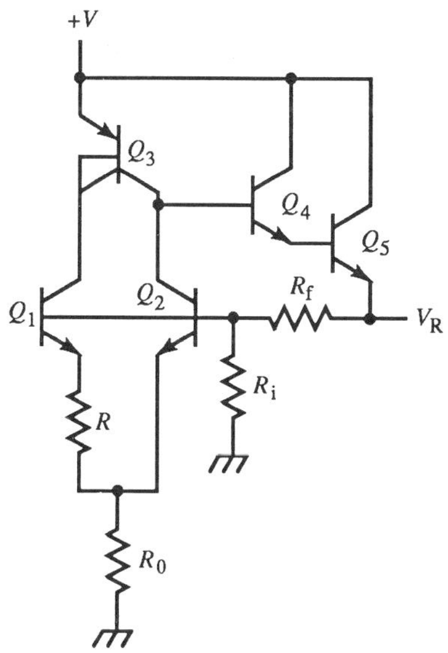

FIG. 11.4 Discrete BJT realization of differential bandgap reference.

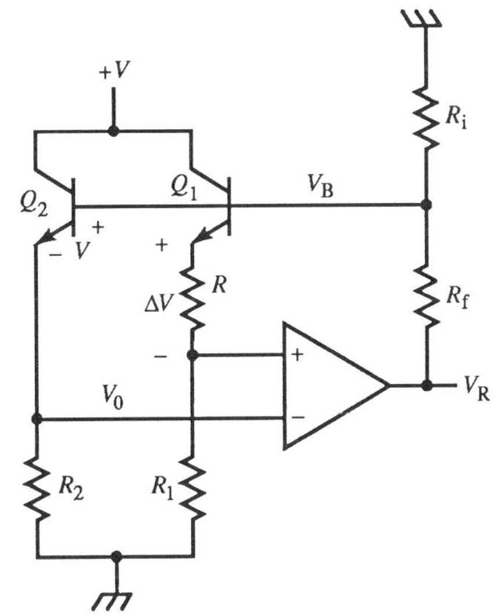

FIG. 11.5 Bandgap voltage reference with emitter inputs to op-amp.

Equation (11.32) is slightly modified:

$$V_B = V + V_0 = V_T \ln\left(\frac{I_2}{I_S}\right) + I_1 R_1 = V_T \ln\left(\frac{I_2}{I_S}\right) + \left(\frac{R_1}{R}\right) V_T \ln\left(\frac{J_2}{J_1}\right) \quad (11.36)$$

By comparison,

$$g = \frac{V_0}{\Delta V} = \frac{R_1}{R} \quad (11.37)$$

and (11.35) also applies.

Besides these three popular bandgap circuits, various other topologies have been used in commercial ICs.

In some designs, a very simple voltage reference is needed that does not require a low TC. A shunt-feedback voltage source is shown in Fig. 11.6a. We want V_O to be insensitive to the temperature and supply voltage V for good *power-supply rejection* (PSR).

This circuit has no closed-form dc solution but can be designed without iteration, given V, V_O, and I_S. Assume that the diode and BJT are matched. The supply current is I_E; choose I_E. Then

$$V_B = V_T \ln \frac{I_E}{I_S} \quad (11.38)$$

By KVL,

$$V_O = V - I_E R_L - V_D = V - I_E R_L - V_T \ln \frac{\alpha I_E}{I_S} \quad (11.39)$$

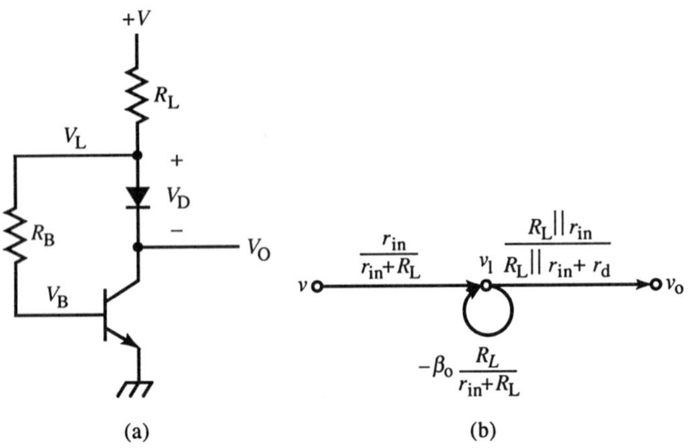

FIG. 11.6 Shunt-feedback voltage reference.

Applying (11.38) reduces the last term to $V_B + V_T \ln \alpha$. Then

$$R_L = \frac{V - (V_O + V_B + V_T \ln \alpha)}{I_E} \tag{11.40}$$

For $\alpha \cong 1$, $V_T \ln \alpha \cong 0$. KVL is applied again to the base circuit:

$$\frac{V - I_E R_L - V_B}{R_B} = \frac{I_E}{\beta + 1} \tag{11.41}$$

Solving for R_B yields

$$R_B = (\beta + 1)\left[\left(\frac{V - V_B}{I_E}\right) - R_L\right] \tag{11.42}$$

The PSR is expressed in small-signal quantities as v_o / v. The flow graph, shown in Fig. 11.6b, reduces to

$$\frac{v_o}{v} = \frac{r_{in}}{r_{in} + (\beta+1)R_L} \cdot \frac{R_L \| r_{in}}{R_L \| r_{in} + r_d} \cong \frac{r_{in}}{r_{in} + (\beta+1)R_L}, \qquad r_d \cong 0 \tag{11.43}$$

where

$$r_{in} = R_B + (\beta + 1) r_e \tag{11.44}$$

PSR is often expressed as the *PSR ratio* (PSRR):

$$\text{PSRR} \equiv 20 \log \frac{v}{v_o} \tag{11.45}$$

The TC of the shunt-feedback reference largely depends on TC(β). The output voltage is

$$V_O = V_B - V_D + I_B R_B = V_T \ln \alpha + \frac{I_E}{\beta + 1} R_B \tag{11.46}$$

Since the junctions nearly cancel, the first term is negligible. In the second term, $I_B(T)$ varies with $\beta(T)$ at about 1%/°C. Less base current is required to sustain V_L with increasing β. Thus, V_O has a negative TC. For small changes in V_O and V, (11.43) is the transfer function; v_o varies inversely with β.

The dynamic output resistance is reduced from r_{in} due to feedback to

$$r_{out} \cong \frac{r_{in}}{1+\beta R_L/(r_{in}+R_L)} = r_{in} \cdot \frac{r_{in}+R_L}{r_{in}+(\beta+1)R_L} \tag{11.47}$$

Example 11.4 Shunt-Feedback Voltage Reference

The reference of Fig. 11.6 has the following design parameters:

$I_S = 10^{-15}$ A, $\beta = 99$, matched junctions, $V = 5$ V, $V_O = 2.5$ V

Let $I_E = 1$ mA. This value is chosen so that any load current is negligible in comparison. Applying (11.38), (11.40), and (11.42) yields

$$V_B = V_T \ln \frac{1\text{ mA}}{10^{-15}\text{ A}} = 0.715\text{ V}$$

$$R_L = \frac{5\text{ V} - [2.5\text{ V} + 0.715\text{ V} + (25.87\text{ mV})(\ln 0.99)]}{1\text{ mA}} = 1.79\text{ k}\Omega \Rightarrow 1.8\text{ k}\Omega$$

$$R_B = (100)\left[\frac{5\text{ V} - 0.715\text{ V}}{1\text{ mA}} - 1.79\text{ k}\Omega\right] = 250\text{ k}\Omega \Rightarrow 240\text{ k}\Omega$$

The PSR is calculated as follows:

$$r_{in} = 240\text{ k}\Omega + (100)(26\ \Omega) = 243\text{ k}\Omega$$

$$\frac{v_o}{v} = \frac{243\text{ k}\Omega}{243\text{ k}\Omega + (100)(1.8\text{ k}\Omega)} = 0.57$$

This does not amount to much power-supply rejection. As PSRR, it is only 4.8 dB. The dynamic output resistance is reduced by about the same fraction:

$$r_{out} = (243\text{ k}\Omega)(0.58) = 140\text{ k}\Omega$$

In this example, the shunt-feedback voltage reference has little advantage over a resistive divider because R_L is not sufficiently large relative to r_{in}. If we choose $I_E = 100\ \mu$A instead, the result is actually better:

$$R_L = 18\text{ k}\Omega,\quad R_B = 2.4\text{ M}\Omega,\quad r_{in} = 50\text{ k}\Omega,\quad \frac{v_o}{v} = 0.027,$$

$$r_{out} = 1.84\text{ k}\Omega$$

The improvement is due to increased loop gain resulting from increased R_L.

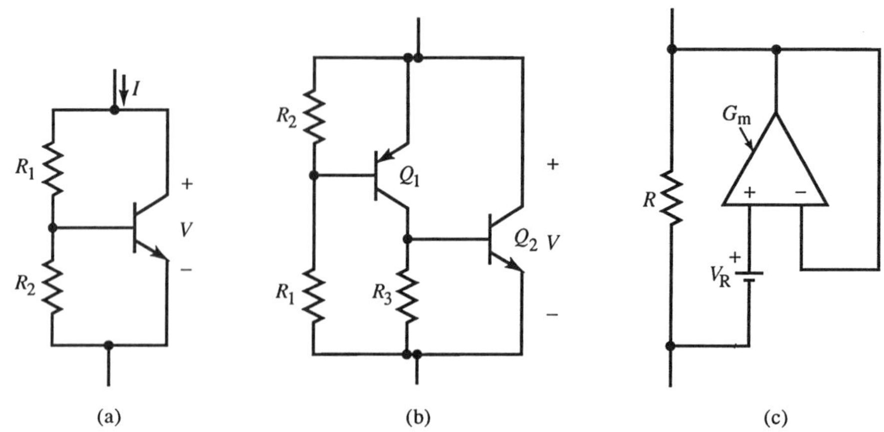

FIG. 11.7 V_{BE} multipliers: (a) basic circuit, (b) two-stage circuit with additional gain, (c) ideal transconductance amplifier realization with voltage reference V_R.

A simple voltage source that is easily floated is the V_{BE} *multiplier* (Fig. 11.7a). When driven by a current source, it acts as a shunt-feedback amplifier with voltage

$$V = R_1 I_B + V_{BE}\left(1+\frac{R_1}{R_2}\right) = \frac{IR_1}{\beta+1} + V_{BE}\left(1+\alpha\cdot\frac{R_1}{R_2}\right) \tag{11.48}$$

where I is the total current. Its main advantage over a current-driven resistor is its dynamic resistance,

$$r_{out} = (R_1 + R_2 \parallel r_\pi) \parallel r_m\left(\frac{R_1 + R_2 \parallel r_\pi}{R_2 \parallel r_\pi}\right) \tag{11.49}$$

where r_m is the BJT transresistance of r_e/α and $r_\pi = (\beta+1)r_e$. The first shunt resistance is the divider resistance, and the second is the equivalent BJT resistance. If the resistive-divider loading is negligible, then

$$r_{out} \cong r_m\left(\frac{V}{V_{BE}}\right) = \frac{1}{\alpha}\cdot\frac{V_T}{I_E}\cdot\frac{V}{V_T \ln(I/I_S)} = \frac{V/I}{\alpha \ln(I/I_S)} \tag{11.50}$$

The numerator is the value of a current-driven resistor; the denominator is the improvement factor due to the BJT.

The voltage source driving this circuit is V_{BE}, which drifts with temperature. From (11.48),

$$\frac{dV}{dT} = \frac{-IR_1}{\beta+1}\cdot \mathrm{TC\%}(\beta) + \left(1+\alpha\frac{R_1}{R_2}\right) V_{BE}\cdot \mathrm{TC\%}(V_{BE}) \tag{11.51}$$

assuming $\mathrm{TC}(\alpha) \cong 0$. The fractional TC of junction voltage with constant I, which is found, in general, from (11.17), is

$$\mathrm{TC\%}(V)\big|_I = \frac{1}{V}\cdot\frac{dV}{dT}\bigg|_I = \frac{1}{V}\cdot\frac{V-V_g}{T} = \frac{1}{T}\left(1-\frac{V_g}{V}\right) \tag{11.52}$$

Then (11.51), with a typical $V_{BE} = 0.7$ V at 27°C, becomes

$$\text{typical} \quad \frac{dV}{dT} = \frac{-IR_1}{\beta + 1} \cdot (1\%/°\text{C}) - \left(1 + \alpha \frac{R_1}{R_2}\right)(0.7\ \text{V})(0.26\%/°\text{C}) \tag{11.53}$$

and the TC(V) < 0. This circuit is commonly used in the base circuit of complementary CC buffers, as an alternative in Fig. 9.50a to $R_1D_1R_2D_2$. It can be designed so that its TC tracks the CC output BJTs.

A V_{BE} multiplier with an additional gain stage is shown in Fig. 11.7b. The circuit in (c) is the general form of the V_{BE} multiplier. The voltage source V_R drives the shunt transconductance amplifier across R. With a current of I, the voltage across it is

$$V = (I + V_R G_m)\left[R \,\Big\|\, \frac{1}{G_m}\right] \tag{11.54}$$

The incremental dynamic resistance is

$$r_{out} = R \,\Big\|\, \frac{1}{G_m} \tag{11.55}$$

Without the amplifier, it is R. For large G_m, it approaches zero.

11.2 Current Sources

The three-terminal current-source IC of Fig. 11.8 has a bandgap cell Q_1-Q_2 and a transconductance amplifier with an output of $14I_0$. Each BJT conducts

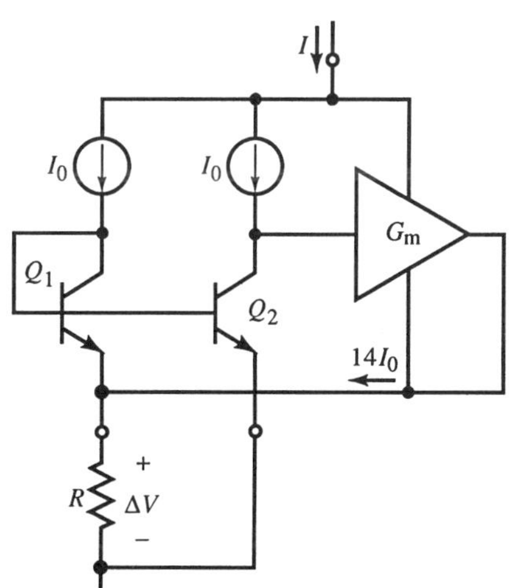

FIG. 11.8 Simplified LM334 bandgap current source.

I_0, and the area of Q_1 is 14 times that of Q_2, or $A_1 = 14A_2$. With the same current,

$$\frac{J_2}{J_1} = 14 \tag{11.56}$$

The terminal current at 25°C is

$$I = 16I_0 = 16\left(\frac{\Delta V}{15R}\right) = \left(\frac{16}{15}\right)\frac{V_T \ln(14)}{R} = \frac{(1.067)(67.77\text{ mV})}{R} = \frac{72.3\text{ mV}}{R} \tag{11.57}$$

This circuit is based on the National Semiconductor LM334. The data book specification indicates that the voltage across R is 67.7 mV. Since I varies with V_T, it has a TC of 0.336%/°C at 25°C. A shunt RD combination in series with this part adds a negative TC. If the shunt R is chosen properly, the TC can be set to zero.

A current source based on the V_{BE}-multiplier concept is shown in Fig. 11.9. The general topology is given in (a), where the amplifier has voltage gain K. Otherwise, the topology is the same as the V_{BE}-multiplier voltage source (Fig. 11.7c). The circuit equations are

$$V = K(V_R - IR) \tag{11.58}$$

or

$$I = \frac{V_R - V/K}{R} \tag{11.59}$$

and at the output,

$$I = \frac{V_L}{R_L} = \frac{1}{R_L}\left(\frac{R_L}{R_L + R}\right) V \tag{11.60}$$

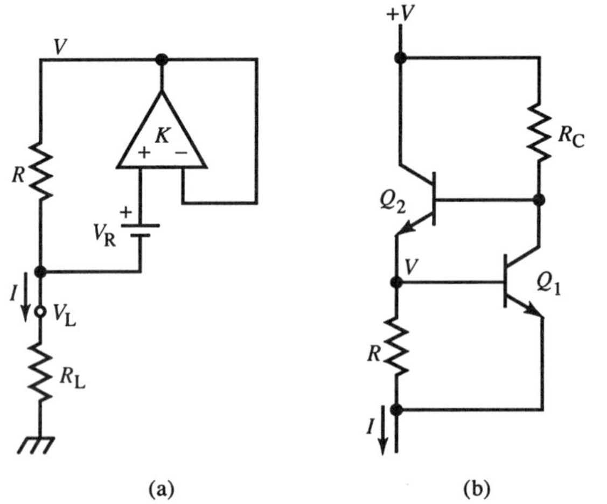

FIG. 11.9 Precision op-amp current source (a) and BJT realization (b).

or

$$V = \left(\frac{R_L + R}{R_L}\right) V_L \tag{11.61}$$

Solving (11.58) and (11.60) for *I*, we have

$$I = \frac{V_R}{R_L/K + R[(K+1)/K]} \tag{11.62}$$

For an op-amp, $K \to \infty$, and

$$I\big|_{K\to\infty} = \frac{V_R}{R} \tag{11.63}$$

I is independent of the load, as desired. The dynamic output resistance, which ideally is infinite, is

$$r_{out} = \frac{dV_L}{dI} = \frac{dV_L}{dV} \cdot \frac{dV}{dI} = \left(\frac{R_L}{R_L + R}\right)(-KR) = (-K)(R \parallel R_L) \tag{11.64}$$

As $K \to \infty$, $r_{out} \to \infty$, as desired. R should be made as small as feasible to maintain high r_{out} for small R_L. R_L begins to affect I significantly as it approaches the value of R.

A BJT realization of the V_{BE}-multiplier current source is that of Fig. 11.9b, in which V_R is V_{BE} of Q_1, and K is the loop gain with V_{BE1} as input. It is

$$K = \frac{v}{v_{be1}} = \alpha \cdot \frac{R_C \parallel [(\beta+1)(r_{e2} + R \parallel r_{\pi 1} + R_L)]}{r_{e1}} \tag{11.65}$$

R should be made small for load insensitivity and R_C large for high K. Since $I \propto V_R = V_{BE1}$ and $TC(V_{BE}) \cong -2$ mV/°C,

$$TC\%(I) = TC\%(V_{BE1}) \tag{11.66}$$

In the BJT current-source, the diff-amp input is the *b-e* junction of Q_1, and I_{E1} also contributes to *I*, or

$$I = I_{E1} + I_{E2} \tag{11.67}$$

If I_{E1} is chosen, then V_{BE1} is determined and R calculated from

$$R = \frac{V_T \ln(I_{E1}/I_S)}{I - I_{E1}} \tag{11.68}$$

Next, R_C must be chosen to satisfy dc constraints. Given R_L and V, and with BJT parameter I_S, then

$$I_{E2} = I - I_{E1} + \frac{I_{E1}}{\beta + 1} = I - \alpha I_{E1} \tag{11.69}$$

and

$$V_{BE2} = V_T \ln \frac{I_{E1}}{I_S} \tag{11.70}$$

Then the current through R_C, corrected for I_{B2}, is

$$I_{RC} = \alpha I_{E1} + \frac{I_{E2}}{\beta + 1} = \frac{[\beta^2/(\beta+1)]I_{E1} + I}{\beta + 1} \tag{11.71}$$

With these calculated values, we can now find R_C:

$$R_C = \frac{V - (V_{BE2} + V_{BE1} + IR_L)}{I_{RC}} \tag{11.72}$$

Example 11.5 V_{BE}-Multiplier Current Source

The current source of Fig. 11.9b has a 1 kΩ nominal load to ground and $V = 5$ V. We want I to be 1 mA. The BJTs have $\beta = 99$ and $I_S = 10^{-15}$ A. Then, if we let

$$I_{E1} = 100\ \mu\text{A}$$

$$V_{BE1} = V_T \ln \frac{100\ \mu\text{A}}{10^{-15}\ \text{A}} = 0.655\ \text{V}$$

Consequently,

$$R = \frac{0.655\ \text{V}}{1\ \text{mA} - 100\ \mu\text{A}} = 728\ \Omega \Rightarrow 750\ \Omega$$

From (11.69), $I_{E2} = 0.901$ mA, and $V_{BE2} = 0.712$ V. $IR_L = 1$ V. Next, $I_{RC} = 10.80\ \mu$A, and finally, from (11.72),

$$R_C = \frac{5\ \text{V} - 2.367\ \text{V}}{10.80\ \mu\text{A}} = 24.4\ \text{k}\Omega \Rightarrow 24\ \text{k}\Omega$$

K is calculated from (11.65) as 80. This makes $r_{out} = -34$ kΩ. The negative value means that an increase of V_L is accompanied by a slight decrease in I. Negative resistances can cause oscillations, and instability should be evaluated (as in Chapter 7). Since $R \cong R_L$, this design could be improved by a choice of smaller I_{E1}, causing R to be smaller. But since I_{E1} is already a tenth I_{E2}, we are at a point of diminishing improvement. An op-amp realization would get around the lower limits on R.

This circuit is quite sensitive to the value of R. It was built using a 750 Ω, 5% resistor; the resulting I was about 7% low. With a trim-pot adjusted to 728 Ω, the error was about 0.2%. Therefore, a 1% value of 732 Ω would be better to use for R.

An op-amp-based current source was invented in 1963 by Brad Howland at MIT. It is the *Howland current source*, shown with a floating voltage source input in Fig. 11.10. This circuit has positive feedback to the noninverting input.

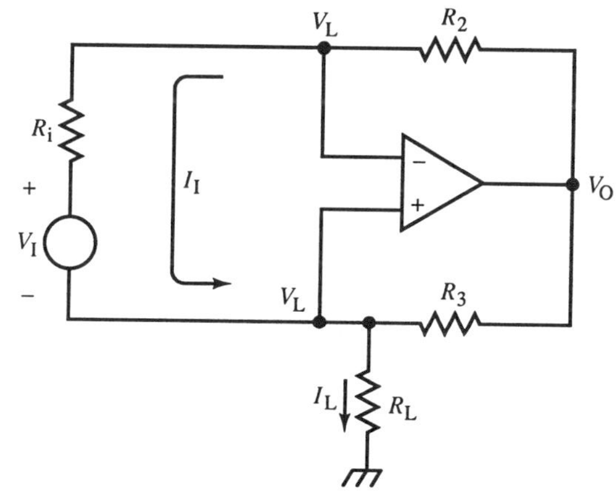

FIG. 11.10 Howland op-amp current source with differential input voltage.

With a sufficiently large load resistance, the circuit becomes unstable. The positive feedback provides a bootstrap effect that keeps the load current I_L constant.

Since the op-amp keeps its inputs at the same voltage, they are both at the load voltage V_L. The same voltage appears at both ends of the input branch through which flows the input current,

$$I_I = \frac{V_I}{R_i} \tag{11.73}$$

This current flows through R_2, causing

$$V_O = I_I R_2 + V_L = \frac{R_2}{R_i} V_I + V_L \tag{11.74}$$

V_O is thus established. It causes a current through R_3; applying KCL at the load node and substituting (11.74) gives

$$I_L = I_I + \frac{V_O - V_L}{R_3} = \frac{V_I}{R_i} + \left(\frac{R_2}{R_i}\right)\left(\frac{1}{R_3}\right) V_I = \left(\frac{R_2 + R_3}{R_i R_3}\right) V_I \tag{11.75}$$

The cancellation of V_L in the numerator of (11.75) represents the bootstrapping. V_O tracks V_L, so I_L is independent of V_L and hence is a current source.

Floating voltage sources are usually inconvenient. A more general Howland circuit (Fig. 11.11) has two voltage inputs, V_1 and V_2, with differential input

$$V_I = V_2 - V_1 \tag{11.76}$$

What is different is that the currents in R_1 and R_3 can be different. The circuit is solved similar to the previous one. V_O from the inverting side is

$$V_O = V_L + I_2 R_2 = V_L + \frac{V_L - V_1}{R_1} \cdot R_2 = \left(-\frac{R_2}{R_1}\right) V_1 + \left(\frac{R_2}{R_1} + 1\right) V_L \tag{11.77}$$

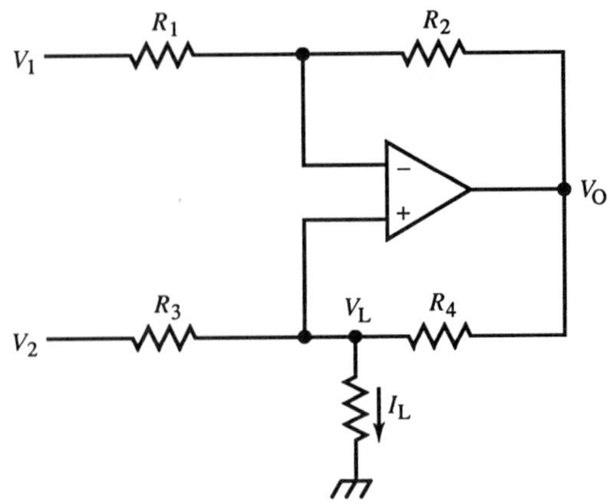

FIG. 11.11 Howland current source.

On the noninverting side, applying KCL and substituting for V_O,

$$I_L = \frac{V_2 - V_L}{R_3} + \frac{V_O - V_L}{R_4} = \left(-\frac{R_2}{R_1 R_4}\right) V_1 + \frac{V_2}{R_3} + \left(\frac{R_2}{R_1 R_4} - \frac{1}{R_3}\right) V_L \tag{11.78}$$

This general expression for I_L is not independent of V_L, as required of a current source. The coefficient of V_L is set to zero under the condition

$$\frac{R_2}{R_1 R_4} = \frac{1}{R_3} \Rightarrow \frac{R_2}{R_4} = \frac{R_1}{R_3} \quad \text{or} \quad \frac{R_2}{R_1} = \frac{R_4}{R_3} \tag{11.79}$$

Under this condition, (11.78) reduces to

$$\text{current source } I_L = \frac{V_2 - V_1}{R_3} = \frac{V_i}{R_3} \tag{11.80}$$

The output resistance is found by regarding the dc quantities of (11.78) as variables, and then differentiating and inverting:

$$r_{out} = \frac{\partial V_L}{\partial I_L} = 1 \bigg/ \left(\frac{R_2}{R_1 R_4} - \frac{1}{R_3}\right) = \frac{R_4}{R_2/R_1 - R_4/R_3} \tag{11.81}$$

Under the condition of (11.79), r_{out} is infinite.

The modified Howland source of Fig. 11.12 has an additional resistor R_4 and a buffer between the load and noninverting input. This increases compliance (load-voltage range) and load-current range because R_5 can be made small while R_4 satisfies the gain requirement of a current source. If $I_{R4} \ll I_L$, the buffer can be omitted and R_4 connected to R_5. For this circuit, we assume finite op-amp gain K and apply superposition to the op-amp inputs:

$$V_- = \left(\frac{R_1}{R_1 + R_2}\right) V_O + \left(\frac{R_2}{R_1 + R_2}\right) V_1 \tag{11.82}$$

$$V_+ = \left(\frac{R_3}{R_3 + R_4}\right) V_L + \left(\frac{R_4}{R_3 + R_4}\right) V_2 \tag{11.83}$$

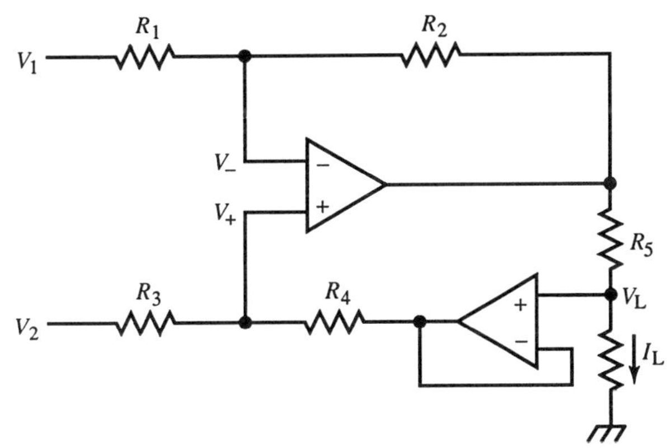

FIG. 11.12 Precision Howland current source.

The op-amp output voltage is

$$V_O = K(V_+ - V_-) = \frac{K}{[(R_1 + R_2)/R_1] + K} \cdot (V_O|_{K\to\infty}) \tag{11.84}$$

When $K \to \infty$, V_O is

$$V_O|_{K\to\infty} = \left(\frac{R_1 + R_2}{R_1}\right)\left[\left(\frac{R_4}{R_3 + R_4}\right) V_2 - \left(\frac{R_2}{R_1 + R_2}\right) V_1 + \left(\frac{R_3}{R_3 + R_4}\right) V_L\right] \tag{11.85}$$

The load current then is

$$I_L = \frac{V_O - V_L}{R_5} \tag{11.86}$$

Substituting (11.85) yields an expression in V_1, V_2, and V_L. When the coefficient of V_L is set to zero, the current-source condition results. Not surprisingly, it is the same as (11.79) because the feedback topology is the same as the previous circuit. Then

$$I_L = \left(\frac{R_2}{R_1}\right)\frac{V_I}{R_5} = \left(\frac{R_4}{R_3}\right)\frac{V_I}{R_5} \tag{11.87}$$

where $V_I = V_2 - V_1$. When the buffer is omitted, I_L is reduced by I_{R4}.

Figure 11.13 shows an inverting current-gain amplifier that uses positive feedback, a variation on the Howland topology. Since the voltages at the op-amp inputs are kept equal, R_1 and R_2 drop the same voltage. It then follows that

$$\frac{I_O}{I_I} = \frac{R_1}{R_2} \tag{11.88}$$

An application of the inverting op-amp current amplifier (Fig. 11.13b) reverses the DAC output-current polarity and scales it for input to the inverting op-amp.

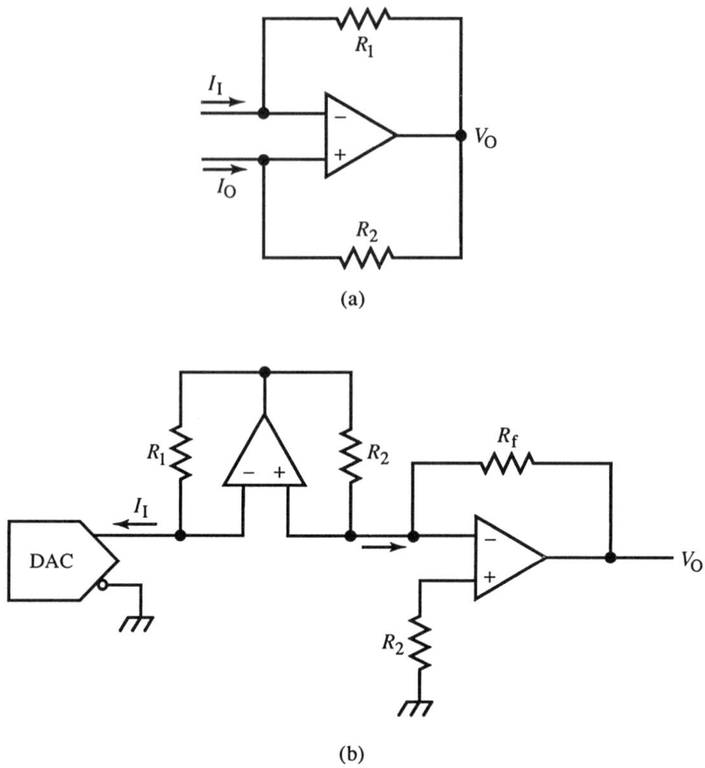

FIG. 11.13 Current inverter (a) and its use in a DAC interface to an op-amp.

The voltage output is negative. Of significance is the DAC output node, which is kept at the same voltage as the virtual ground (inverting input) of the op-amp, meeting the constraint of a limited-compliance DAC.

The noninverting current amplifier of Fig. 11.14 applies to R_2, through the $\times 1$ buffer, the same voltage that is across R_1. The current gain is

$$\frac{I_O}{I_I} = 1 + \frac{R_1}{R_2} \tag{11.89}$$

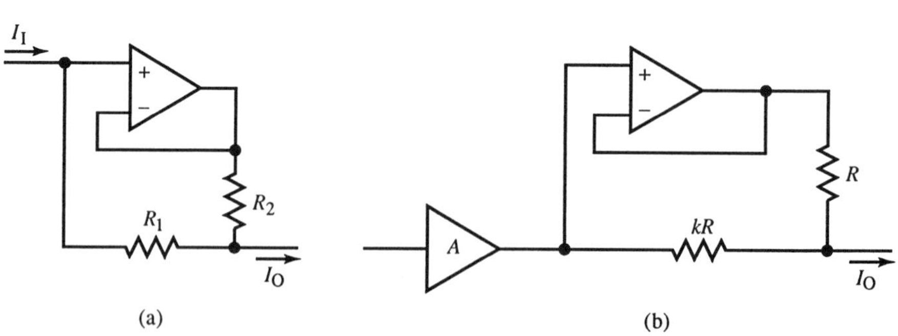

FIG. 11.14 Precision current shunt (a) and its use as an amplifier output-current booster.

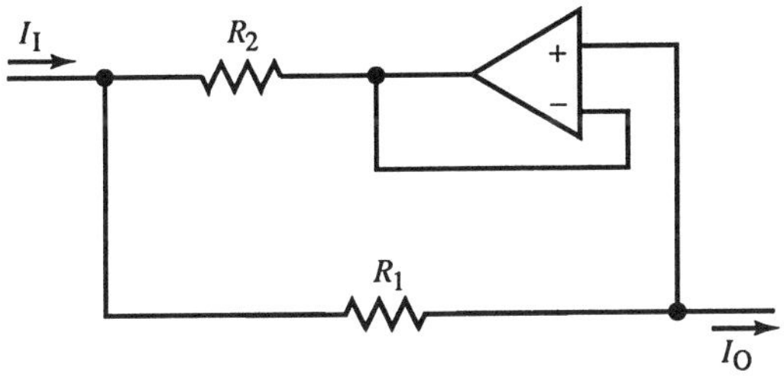

FIG. 11.15 Precision current divider or attenuator.

In (b), this current amplifier is used to boost the output current I_O of an amplifier by k times.

In Fig. 11.15, by reversing the buffer, current gain becomes attenuation, a precision floating current shunt. This circuit is similar to the current source of Fig. 11.9a; V_R is removed, and R is driven by I_I instead. The current gain is the current divider formula,

$$\frac{I_O}{I_I} = \frac{R_2}{R_1 + R_2} \tag{11.90}$$

Example 11.6 Bipolar Simulated Resistance

A circuit with similar topology to the Howland current source (Fig. E11.6) also uses positive feedback. This circuit provides a precision, adjustable, bipolar input resistance and can be used in the one-op-amp diff-amp in place of the grounded resistor (R_2 in Figs. 9.13 and 9.20). This is sometimes necessary due to unavoidable parasitic resistance in the ground return path. Applying KCL twice gives

$$R_i = \frac{V_I}{I_I} = R_1\left(1 + \frac{R_4}{R_3}\right) - R_2 \tag{E1}$$

where R_2 and R_4 include the trim-pot resistances.

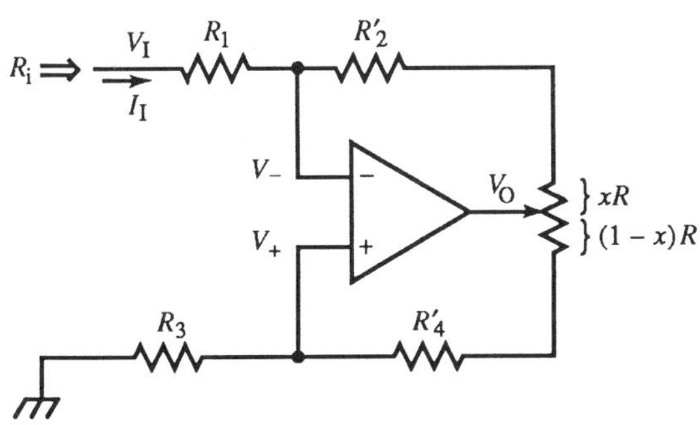

FIG. E11.6

When $R_1 = R_3 = R$, then

$$R_i = R + (R_4 - R_2) \tag{E2}$$

When the trim-pot is centered, $R_2 = R_4$, and $R_i = R$. If we set $R_2' = R_4'$, the trim-pot allows adjustment of R_i around R as center value.

Example 11.7 Inverting Howland Current Source

The goal in this example is to design an op-amp current source based on the bootstrapping action of the Howland source but with an inverted output. In Fig. E11.7 is a proposed circuit. The $\times(-1)$ amplifier can be an inverting op-amp. Then

$$v_2 = -v_1 = -\left[v_I - \left(\frac{R}{R - R_S}\right)(v_L - v_I)\right]$$
$$= \left(\frac{R}{R - R_S}\right)v_L - \left(\frac{2R - R_S}{R - R_S}\right)v_I \tag{E1}$$

The load current is, by KCL,

$$i_L = \frac{v_2 - v_L}{R_S} - \frac{v_L - v_I}{R - R_S} \tag{E2}$$

Substituting for v_2 and reducing, we find that the coefficient of v_L is zero, as required for a current source. Then i_L depends only on v_I:

$$i_L = -\frac{2v_I}{R_S} \tag{E3}$$

This circuit therefore functions as a current source.

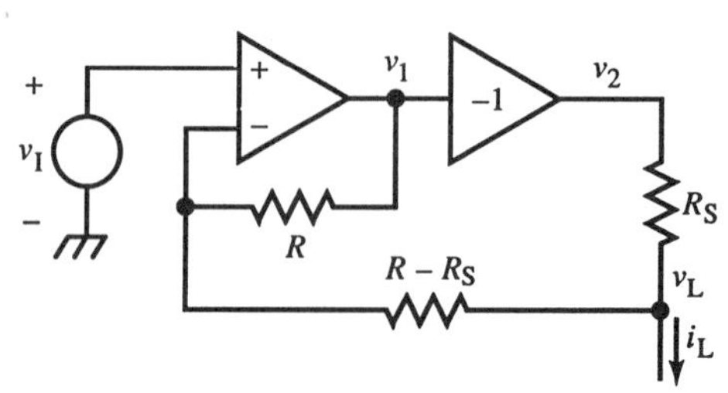

FIG. E11.7

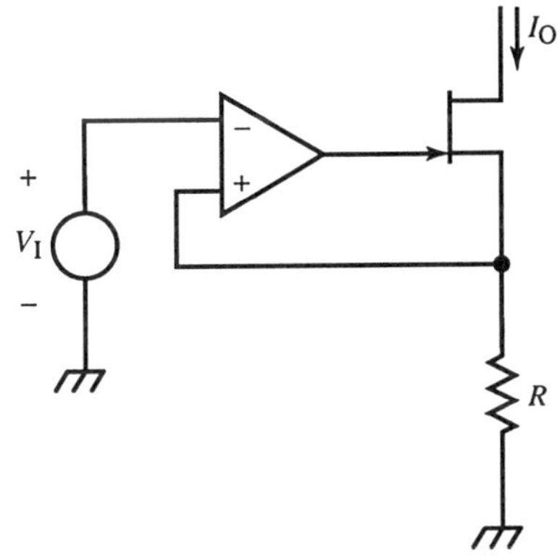

FIG. 11.16 Precision voltage-to-current converter.

A common way to generate a current I_O from a given voltage V_I is to use an op-amp voltage-to-current (V/I) converter (Fig. 11.16). The op-amp keeps V_I across R so that

$$I_O = \frac{V_I}{R} \tag{11.91}$$

The FET can be replaced by a BJT or Darlington, but it avoids error due to α loss. This circuit need not be grounded. Ground can be replaced by $-V_{EE}$ or, for the complementary V/I converter using the opposite polarity of transistor, by $+V_{CC}$.

The current mirrors of Section 2.12 are current-gain amplifiers and can be used as current-driven current sources. For high precision, the mirror of Fig. 2.13b should have a junction in the collector of Q_2 for electrical symmetry between Q_1 and Q_2, especially if R_1 and R_2 are not used (Fig. 11.17a). An application for the complementary form of this current mirror, in (b), is similar to that of Fig. 11.13b except that the output is a bipolar current. The total DAC output current is

$$I_{fs} = I_1 + I_2$$

D_1 keeps Q_3 out of saturation when D_2 conducts.

11.3 Filters

Filters are characterized generally by their transfer functions in the complex-frequency domain. As rational functions of s, numerator and denominator can be factored into first- and second-order factors. Higher-order filter polynomials are products of these lower-order factors. Filter responses can be categorized broadly as low-pass (LP), high-pass (HP), or band-action filters, which are either band-pass (BP) or band-reject (or "notch") filters.

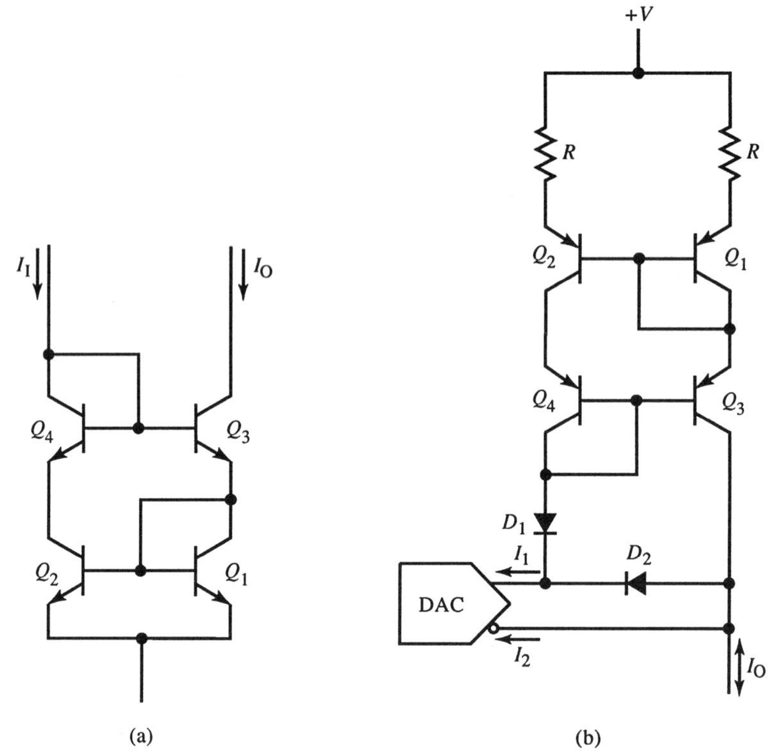

FIG. 11.17 Thermally balanced Wilson current source (a), applied as differential-to-bipolar DAC interface (b).

In Chapter 5, amplifier analysis assumed a low-pass response. In Chapter 10, we saw some use for high-pass filters in composite amplifiers. In radio communications, highly resonant, or tuned, circuits are used as band-pass filters. These circuits have low ζ or high Q ($Q = 1/2\zeta$). Their complex poles are very underdamped and have dominant imaginary components. That is, they lie near the $j\omega$ axis (Fig. 11.18a.) The conjugate pole-pair p_1 and p_2 is:

$$p_{1,2} = -\alpha \pm j\omega_{\mathrm{d}} = -\zeta\omega_{\mathrm{n}} \pm j\omega_{\mathrm{n}}\sqrt{1-\zeta^2} = -\frac{\omega_{\mathrm{n}}}{2Q} \pm j\omega_{\mathrm{n}}\sqrt{1-\frac{1}{(2Q)^2}} \quad (11.92)$$

For $Q \gg 1$, the poles have imaginary component $\pm j\omega_{\mathrm{d}} \cong \pm j\omega_{\mathrm{n}}$. The steady-state sinusoidal (or $j\omega$-axis) response is found (as in Section 5.9) from the zero-vector lengths divided by the pole-vector lengths. Note that $j\omega - p_1$ varies significantly in both magnitude and angle around $j\omega_{\mathrm{n}}$, where peaking of the magnitude response occurs. At $j\omega_{\mathrm{d}} \cong j\omega_{\mathrm{n}}$, $\|j\omega - p_1\|$ is minimum and the band-pass transfer function is maximum. There, $\angle(j\omega - p_1)$ passes through 0°, an indication of resonance. From the geometry of Fig. 11.18a, variations in $j\omega$ around $j\omega_{\mathrm{d}}$ have little effect on the length or orientation of the conjugate pole vector,

$$j\omega - p_2 \cong j\omega_{\mathrm{n}} - (-j\omega_{\mathrm{n}}) = 2j\omega_{\mathrm{n}}, \qquad j\omega \cong j\omega_{\mathrm{n}} \quad (11.93)$$

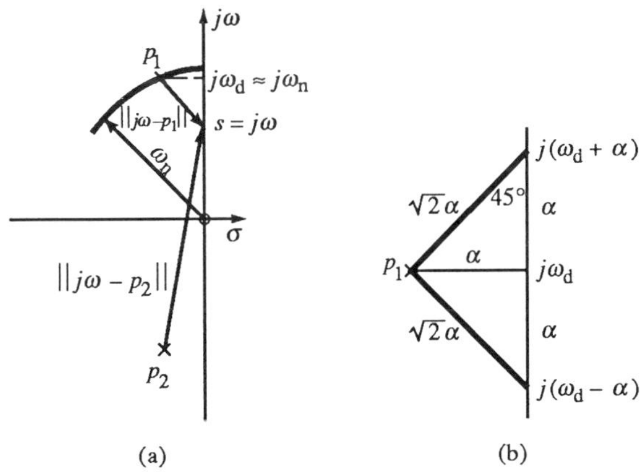

FIG. 11.18 Narrowband approximations: (a) high-Q circuit response near resonant frequency; (b) geometric tuned-circuit bandwidth derivation.

Similarly, the zero at the origin does not change by much either, so the net effect of the conjugate pole and zero is

$$\left.\frac{j\omega}{j\omega - p_2}\right|_{j\omega \cong j\omega_n} \cong \frac{1}{2} \tag{11.94}$$

These *narrowband approximations* assume that the poles are near $\pm j\omega_n$ and that the frequency range for $j\omega$ is around $j\omega_n$. The second-order resonant response is consequently reduced to a first-order approximation of the pass-band response:

$$\left.\frac{s\omega_n}{(s-p_1)(s-p_2)}\right|_{s \cong j\omega_n} \cong \frac{1}{2} \cdot \frac{\omega_n}{j\omega - p_1} \tag{11.95}$$

Application of the narrowband approximations effects a band-pass to low-pass filter transformation. The first-order result is the response centered around $j\omega_n$ instead of the s-plane origin.

A critical parameter of tuned circuits is their bandwidth relative to their resonant frequency. The less damped a resonant circuit is, the narrower its bandwidth and the more selective its response to a particular frequency channel. In Fig. 11.18b, a closer view of the s-plane near p_1 is shown. As previously defined, bandwidth was the frequency at which the magnitude of the response rolled off to $1/\sqrt{2}$ of its low-frequency value. In this case, two frequencies are centered about $j\omega_d$ where roll-off to $1/\sqrt{2}$ occurs. For bandpass response, bandwidth is defined by those frequencies. The magnitude of (11.95) rolls off to bandwidth magnitude when $\|j\omega - p_2\| = \sqrt{2}$ (Fig. 11.18b). At this vector length, the pole angles are 45°, and by geometry the bandwidth frequencies are $j(\omega_d + \alpha)$ and $j(\omega_d - \alpha)$. Under the narrowband assumption, the bandwidth frequencies are

$$j\omega \cong j(\omega_n \pm \alpha) \tag{11.96}$$

The bandwidth is consequently

$$\omega_{bw} = j(\omega_n + \alpha) - j\omega_n = \alpha \tag{11.97}$$

and the width of the passband is

$$\Delta\omega = j(\omega_n + \alpha) - j(\omega_n - \alpha) = 2\alpha = 2\omega_{bw} \tag{11.98}$$

Now from (11.92),

$$\alpha = \frac{\omega_n}{2Q} \Rightarrow Q = \frac{\omega_n}{2\alpha} = \frac{\omega_n}{\Delta\omega} \tag{11.99}$$

In this formula, the significance of expressing ζ as Q is made explicit; Q is the selectivity. The larger Q is, the narrower the band-pass width relative to the center frequency.

A geometric interpretation [see Angelo (1969) in the reference list] in the s-plane also eases locating the frequency of maximum magnitude or gain ω_m for a complex pole-pair (Fig. 11.19). When the pole vectors form a right angle at ϕ, the vertex on the $j\omega$ axis is at $j\omega_m$. Let the vectors be ρ_1 and ρ_2, as shown. Then we seek to maximize the magnitude response

$$\frac{\omega_n^2}{\|\rho_1\|\,\|\rho_2\|} \tag{11.100}$$

It is maximum when $\|\rho_1\|\,\|\rho_2\|$ is minimum. From geometry, the area of the triangle that is formed by ρ_1, ρ_2, and the vertical (dashed) line between the poles, of length $2\omega_d$, is

$$A = \tfrac{1}{2} \cdot \alpha(2\omega_d) = \alpha\omega_d = \tfrac{1}{2}\|\rho_1\|\,\|\rho_2\| \sin\phi \tag{11.101}$$

As the $j\omega$ vertex moves, the area remains constant. The magnitude response is thus

$$\frac{\omega_n^2}{2\alpha\omega_d} \cdot \sin\phi \cong \frac{\omega_n}{2\alpha} \cdot \sin\phi = Q \sin\phi \tag{11.102}$$

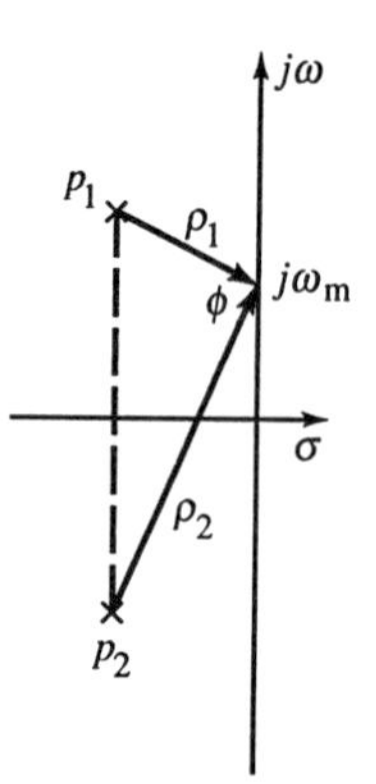

FIG. 11.19 Geometric derivation of ω_m.

When $\phi = 90°$, $\sin \phi$ is maximum as is the response. At ω_m, the peak magnitude is Q. From the Pythagorean theorem,

$$\omega_m^2 = \omega_d^2 - \alpha^2 \quad (11.103)$$

The triangles themselves are not physically significant but are a mnemonic device for reasoning in the s-plane.

Cascaded stages of identical tuned circuits improve selectivity. This scheme of *synchronous tuning* has a bandwidth reduction factor previously calculated as (8.14). The factor is, for n stages,

$$\sqrt{2^{1/n} - 1}$$

In this case, bandwidth reduction is desirable because it improves selectivity.

A shunt RLC is a parallel resonant circuit with an impedance of

$$Z_p = \frac{sL}{s^2LC + s(L/R) + 1} = \frac{s/C}{s^2 + s(1/RC) + 1/LC} \quad (11.104)$$

with parameters

$$\omega_n = 1/\sqrt{LC}, \qquad Q = \frac{RC}{\sqrt{LC}} = \frac{R}{Z_n}, \qquad Z_n = \sqrt{\frac{L}{C}} \quad (11.105)$$

The band-pass width is $\Delta\omega = 1/RC$ and is not affected by L. Thus L can be adjusted to tune the circuit without affecting $\Delta\omega$. These parameters describe a circuit in which Z_p is driven by a current source. For example, it can be a collector or drain load of a tuned amplifier stage.

A more accurate model of an LC parallel-resonant ("tank") circuit, commonly found in radios, includes the series resistance of the inductor R_s. We then have three parallel branches with admittance,

$$Y = sC + \frac{1}{sL + R_s} + \frac{1}{R_p} \quad (11.106)$$

Solving for $Z = 1/Y$ gives

$$Z = (R_s \parallel R_p) \cdot \frac{s(L/R_s) + 1}{s^2(LC) + s[(L/R_p) + R_sC] + 1} \quad (11.107)$$

As usual, $\omega_n = 1/\sqrt{LC}$, but Q is now

$$Q = \frac{1}{\sqrt{L/C}/R_p + R_s/\sqrt{L/C}} \quad (11.108)$$

Q is infinite when R_p is infinite and R_s is zero.

For large Q, (11.105) suggests that L must be small or C large. Parasitic elements associated with components limit the practical range of values of L or C. Also, interstage loading resistance can be too small to allow high Q. In these cases, impedance transformation in the resonant circuit is often the solution. In Fig. 11.20a, a tapped inductance transforms load resistance R_p/n^2

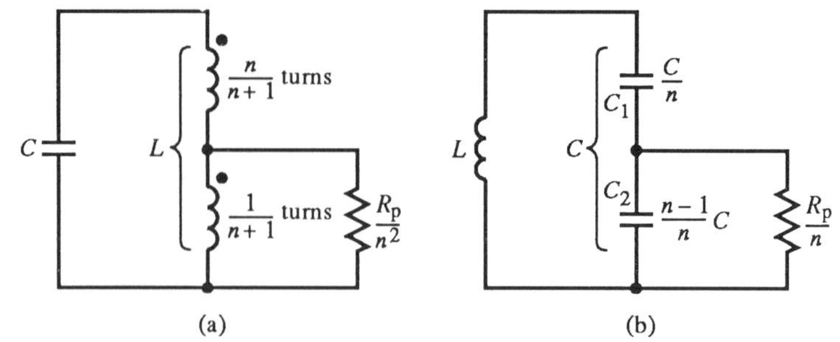

FIG. 11.20 Impedance transformation in parallel resonant circuits with (a) tapped inductor or autotransformer and (b) capacitive divider.

to R_p across the shunt LC. The inductor is an autotransformer with a high coupling coefficient ($k \cong 1$). The mutual inductance causes the LC voltage to be n times that across the resistance, where n is the turns ratio of the top to bottom windings. The current is reduced n times, causing R_p to appear $1/n^2$ times smaller across the shunt LC.

In Fig. 11.20b, a capacitive divider is used in a similar way except that the capacitors do not have a mutually coupled field. The impedance from the inductor terminals is, for $\omega \gg 1/(R_p/n)(C_1+C_2)$:

$$Z = \frac{1}{sC_1} + \frac{1}{sC_2} \left\| \frac{R_p}{n} = \frac{1}{s(C_1 \| C_2)} \right\| \frac{R_p}{n} \cdot \left(\frac{C_1+C_2}{C_1}\right) \tag{11.109}$$

The impedance shunting L is a series C_1C_2 branch shunting R_p. If the equivalent shunt LC resistance is R_p, as assumed, then n must be

$$n = \left(\frac{C_1+C_2}{C_1}\right) \tag{11.110}$$

We now examine two of the most popular op-amp second-order filters, the *Sallen–Key* and *multiple-feedback* LP and BP filters. Figure 11.21a shows a Sallen–Key LP filter, which can be analyzed as a feedback amplifier for voltage gain or by application of KCL and divider formulas. The transfer function is

$$\frac{V_o}{V_i} = \frac{1}{K} \cdot \frac{1}{s^2(R_1R_2C_1C_2) + s\{[(K-1)/K]R_1C_1 + (R_1+R_2)C_2\} + 1} \tag{11.111}$$

If we let the amplifier be a $\times 1$ buffer, (11.111) collapses to

$$\frac{V_o}{V_i} = \frac{1}{s^2(R_1R_2C_1C_2) + s(R_1+R_2)C_2 + 1}, \qquad K = 1 \tag{11.112}$$

An op-amp need not be used for the buffer; in some cases, an emitter-follower is good enough. The resonant frequency is at

$$\omega_n = \frac{1}{\sqrt{R_1R_2C_1C_2}} \tag{11.113a}$$

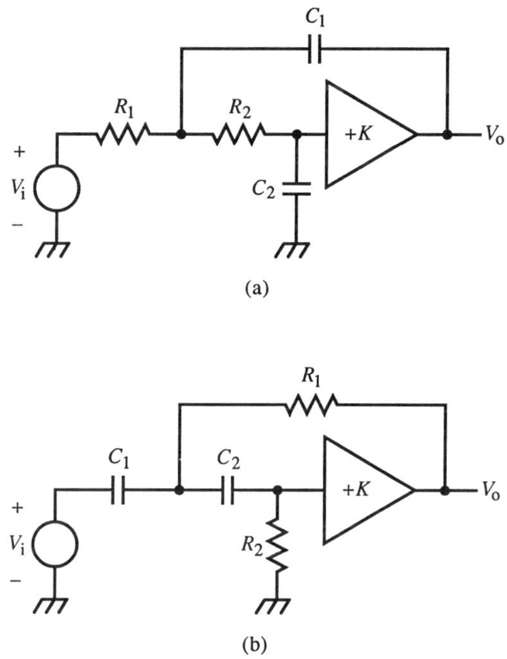

FIG. 11.21 Sallen–Key low-pass filter (a) and band-pass filter (b).

and

$$Q = \sqrt{\frac{(R_1 \parallel R_2)C_1}{(R_1 + R_2)C_2}} \tag{11.113b}$$

For minimum waveform distortion, a Bessel or MFED response requires a Q corresponding to a pole angle of 30° ($\zeta = \sqrt{3}/2$) or

$$Q(\text{MFED}) = \frac{\sqrt{3}}{3} \cong 0.577 \tag{11.114}$$

and, from (11.113), $(R_1 \parallel R_2)C_1/(R_1 + R_2)C_2 = \frac{1}{3}$.

The band-pass filter of Fig. 11.21b has the same topological form as in (a) but with transfer function

$$\frac{V_o}{V_i} = K \cdot \frac{s^2 R_1 R_2 C_1 C_2}{s^2(R_1 R_2 C_1 C_2) + s\{R_1 C_1 + [(K-1)/K]R_2 C_2\} + 1} \tag{11.115}$$

For $K = 1$,

$$\frac{V_o}{V_i} = \frac{s^2 R_1 R_2 C_1 C_2}{s^2 R_1 R_2 C_1 C_2 + sR_1 C_1 + 1}, \qquad K = 1 \tag{11.116}$$

Compared with the LP filter, ω_n is the same and

$$Q = \sqrt{\frac{R_2 C_2}{R_1 C_1}} \tag{11.117}$$

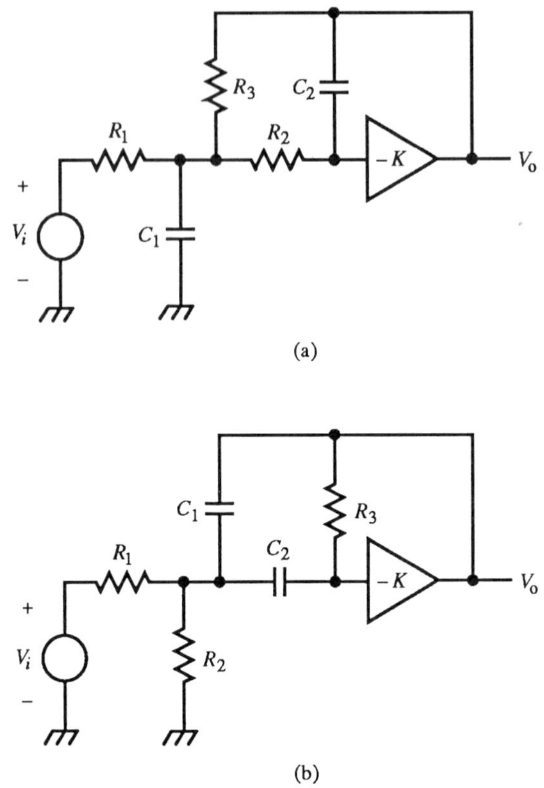

FIG. 11.22 Multiple-feedback low-pass filter (a) and band-pass filter (b).

The multiple-feedback topology has two feedback paths, as in the LP filter of Fig. 11.22a and the BP filter of (b). These circuits cannot achieve high Q values without appreciable attenuation of the input signal, but they provide a simple second-order filter with good phase linearity. For infinite op-amp gain, the LP filter transfer function is

$$\frac{V_o}{V_i} = -\frac{R_3}{R_1} \cdot \frac{1}{s^2 R_2 R_3 C_1 C_2 + s\left\{R_3 C_1 + \left[\left(\frac{R_3}{R_1}+1\right)R_2 + R_3\right]C_2\right\} + 1} \tag{11.118}$$

For the BP filter,

$$\frac{V_o}{V_i} = -\frac{sR_1 C_1^2 (R_1 \parallel R_2) R_3 C_2}{s^2 (R_1 \parallel R_2) R_3 C_1 C_2 + s(R_1 \parallel R_2)(C_1 + C_2) + 1} \tag{11.119}$$

In the multiple-feedback topology, all elements affect both ω_n and Q. In practice, Q is limited to about 20.

The *state-variable* filter topology gets around the design difficulty of interacting filter parameters at the expense of additional circuitry. This approach is that of the analog computer; cascaded integrators output state

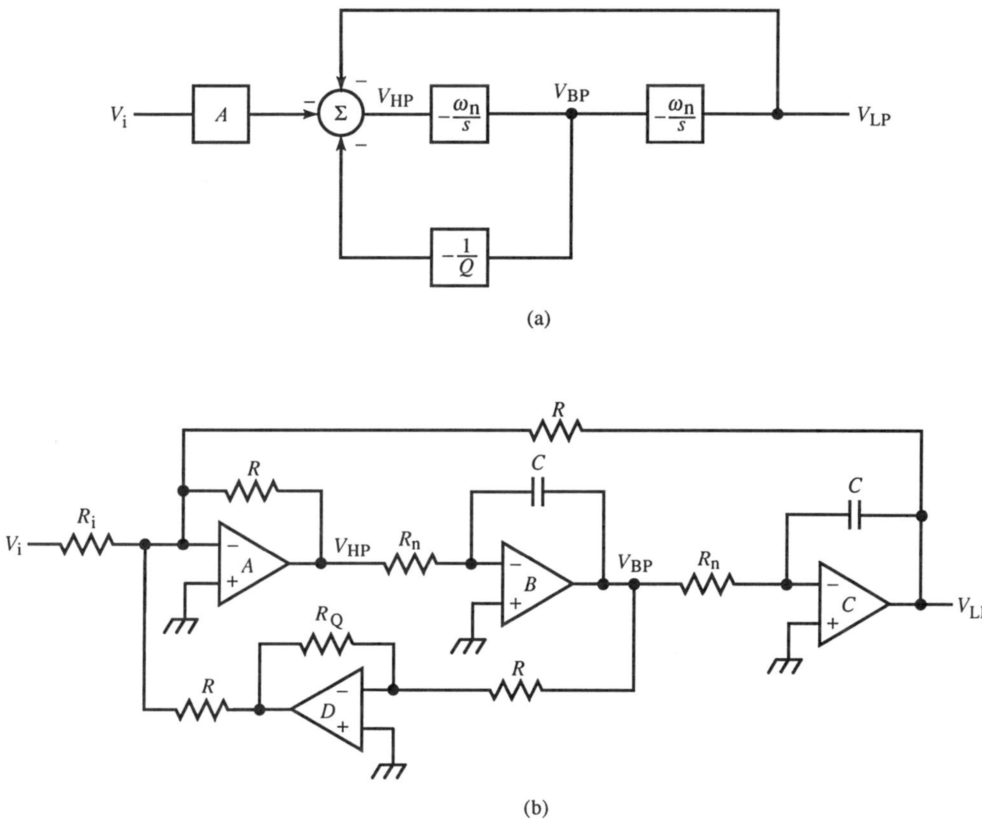

FIG. 11.23 State-variable filter block diagram (a) and op-amp realization (b).

variables that are weighted, combined, and fed back or output. One topology (Fig. 11.23) produces HP, BP, and LP outputs. A quad op-amp IC suffices for gain blocks. Op-amp A is an input summing block, B and C are op-amp integrators, and D is a scaling feedback block. The transfer functions for the three filter types are

$$\frac{V_{LP}}{V_i} = A_{vo} \cdot \frac{1}{(s/\omega_n)^2 + (1/Q\omega_n)s + 1} \qquad (11.120)$$

$$\frac{V_{BP}}{V_i} = A_{vo} \cdot \frac{s/\omega_n}{(s/\omega_n)^2 + (1/Q\omega_n)s + 1} \qquad (11.121)$$

$$\frac{V_{HP}}{V_i} = A_{vo} \cdot \frac{(s/\omega_n)^2}{(s/\omega_n)^2 + (1/Q\omega_n)s + 1} \qquad (11.122)$$

where

$$A_{vo} = -\frac{R}{R_i}, \qquad \omega_n = \frac{1}{R_n C}, \qquad Q = \frac{R}{R_Q} \qquad (11.123)$$

In the characteristic equation, R_Q occurs in the linear term but not the quadratic term, thus leaving it free to adjust Q independent of ω_n. Its adjustment has the locus of a semicircle centered at the origin (case 1 of Fig. 5.13). State-variable filters can achieve a high Q relative to multiple-feedback filters.

A similar filter topology is the *biquad* filter, named for the biquadratic form of the transfer function:

$$\frac{s^2 + ds + e}{s^2 + bs + c} \tag{11.124}$$

It is similar in form to a state-variable filter, but damping is adjusted within the cascaded loop of blocks at A in Fig. 11.24. This topology, like the state-variable filter, has multiple filter outputs: Both BP and LP are available. By weighting and combining outputs from two or three of the op-amps in a fourth op-amp output stage, we obtain the biquadratic filter function.

For the filter of Fig. 11.24,

$$\frac{V_{BP}}{V_i} = -\frac{R_5}{R_i} \cdot \frac{s\left(\dfrac{R_2 C_2}{R_4/R_3}\right)}{s^2\left(\dfrac{R_2 C_2 R_5 C_1}{R_4/R_3}\right) + s\left(\dfrac{R_2 C_2 R_5}{(R_4/R_3) R_1}\right) + 1} \tag{11.125}$$

and

$$\frac{V_{LP}}{V_i} = \frac{R_5}{R_i} \cdot \frac{1}{R_4/R_3} \cdot \frac{1}{s^2\left(\dfrac{R_2 C_2 R_5 C_1}{R_4/R_3}\right) + s\left(\dfrac{R_2 C_2 R_5}{(R_4/R_3) R_1}\right) + 1} \tag{11.126}$$

The filter parameters are then

$$\omega_n = \sqrt{\frac{R_4/R_3}{R_2 R_5 C_1 C_2}}, \qquad Q = R_1 C_1 \omega_n \tag{11.127}$$

The biquad filter has an advantage over previous filter topologies in that the band-pass $\Delta\omega$ is adjustable independent of center frequency ω_n. From the expression for Q in (11.127),

$$\Delta\omega = \frac{\omega_n}{Q} = \frac{1}{R_1 C_1} \tag{11.128}$$

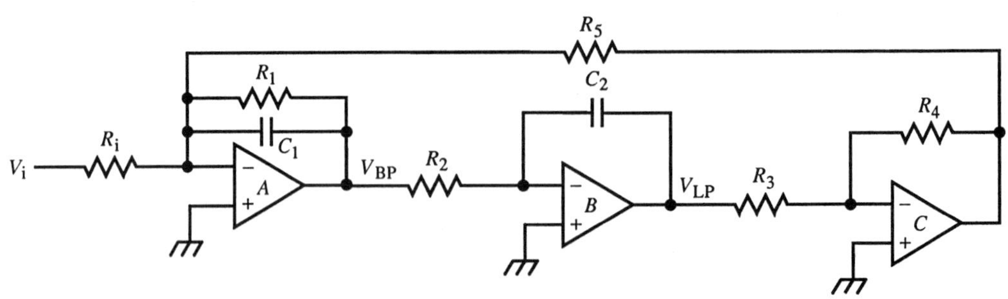

FIG. 11.24 Biquad filter.

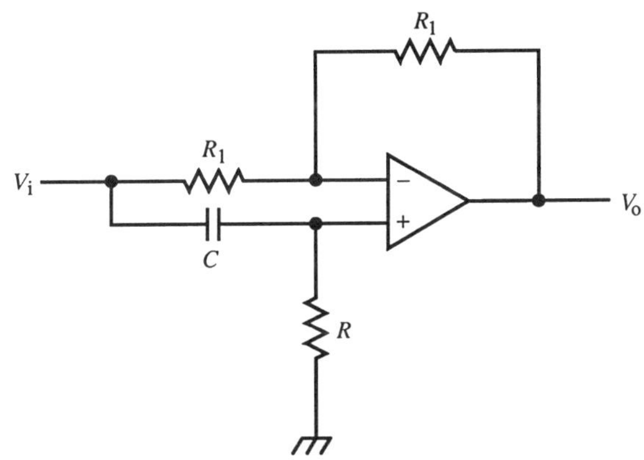

FIG. 11.25 Phase-shifter, an all-pass filter.

Since ω_n is independent of R_1, it independently adjusts $\Delta\omega$. The band-pass function can be simplified to

$$\frac{V_{BP}}{V_i} = -\frac{R_f}{R_i} \cdot \frac{sR_fC}{s^2(R_fC)^2 + s[R_fC(R_f/R_1)] + 1} \tag{11.129}$$

where

$$R_2 = R_5 = R_f, \qquad R_4 = R_3, \qquad C_1 = C_2 = C \tag{11.129a}$$

Gain is independently set by R_i, and $\Delta\omega$ by R_1.

One other filter is shown in Fig. 11.25, an all-pass filter that operates as a phase shifter. Its transfer function is

$$\frac{V_o}{V_i} = \frac{2sRC}{sRC+1} - 1 = \frac{sRC-1}{sRC+1} = -\frac{-sRC+1}{sRC+1} \tag{11.130}$$

The delay time through the filter is

$$t_d = 2RC \tag{11.131}$$

and phase can be adjusted by adjusting R. This can be done electronically using a CMOS DAC or FET.

11.4 Hysteretic Switches (Schmitt Triggers)

Comparators are usually inadequate in providing a single output transition when the input polarity changes. Slowly changing input signals with some noise causes the output to "dither" between the high and low states near the input threshold. This dithering is reduced or eliminated by an input deadzone or deadband, an input range around the threshold where no output change

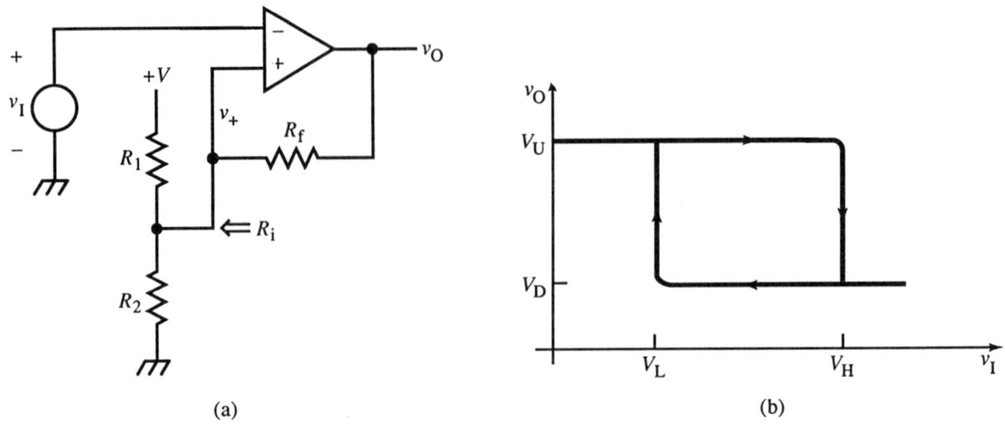

FIG. 11.26 Inverting Schmitt trigger or hysteretic comparator (a) and clockwise hysteresis loop describing circuit behavior (b).

can occur. Furthermore, if the deadzone is state dependent, the effect is called *hysteresis.* Figure 11.26b shows a characteristic square hysteresis loop. The accompanying circuit is an inverting *hysteretic comparator,* or *Schmitt trigger* circuit. The state dependence is achieved by use of positive feedback. In effect, the Schmitt trigger is a bistable memory device.

Assume that the output is in the high state; then $v_O = V_U$. Thévenize the divider R_1, R_2 so that its Thévenin voltage is V_T, the threshold voltage, and its resistance is R_i:

$$V_T = \left(\frac{R_2}{R_1 + R_2}\right) V, \qquad R_i = R_1 \parallel R_2 \tag{11.132}$$

With v_O high, v_+ is set from the feedback divider R_f, R_i. When v_I increases to where the comparator inputs are equal, the output changes to a low state. This input voltage is V_H, the upper hysteresis threshold. By setting $v_+ = V_H$, by superposition, we obtain

$$V_H = \left(\frac{R_i}{R_f + R_i}\right) V_U + \left(\frac{R_f}{R_f + R_i}\right) V_T \tag{11.133}$$

Once v_O is low, then v_I must decrease to V_L, the lower threshold, before v_O becomes high. By letting $v_+ = V_L$ and again applying superposition, we get

$$V_L = \left(\frac{R_i}{R_f + R_i}\right) V_D + \left(\frac{R_f}{R_f + R_i}\right) V_T \tag{11.134}$$

The deadzone size is the width of the hysteresis loop. This *hysteresis window* is

$$\text{inverting } \Delta v_I = V_H - V_L = \left(\frac{R_i}{R_f + R_i}\right)(V_U - V_D) \tag{11.135}$$

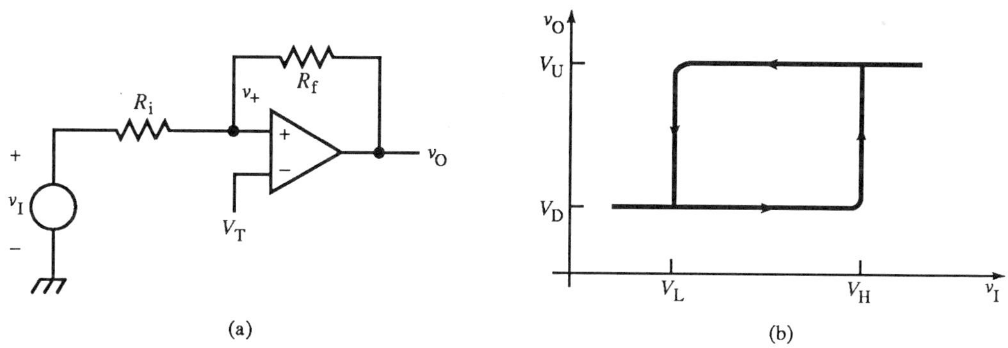

FIG. 11.27 Noninverting Schmitt trigger (a) with counter-clockwise hysteresis loop (b).

A noninverting form of hysteretic comparator (Fig. 11.27) has a similar hysteresis loop, but it is traversed in the counterclockwise direction. The circuit has a positive feedback divider and can be analyzed by superposition. When v_O is either high or low, the threshold for v_+ is fixed at V_T. Hence, we must solve for V_L and V_H after applying superposition. The results are

$$V_L = \left(\frac{R_f}{R_i}+1\right)\left[V_T - \left(\frac{R_i}{R_f+R_i}\right)V_U\right] = -\left(\frac{R_f}{R_i}\right)V_U + \left(\frac{R_f}{R_i}+1\right)V_T \quad (11.136)$$

and

$$V_H = -\left(\frac{R_f}{R_i}\right)V_D + \left(\frac{R_f}{R_i}+1\right)V_T \quad (11.137)$$

with the following input hysteresis window:

$$\text{noninverting } \Delta v_I = V_H - V_L = \left(\frac{R_i}{R_f}\right)(V_U - V_D) \quad (11.138)$$

A two-transistor discrete realization of a hysteresis comparator is the *emitter-coupled* Schmitt trigger of Fig. 11.28. This is a diff-amp with positive feedback from the collector of Q_1 through a voltage translator V_Z to the base of Q_2. The hysteresis loop goes clockwise (Fig. 11.26b). The additional inversion at the collector of Q_2 reverses the direction of the loop at v_O. V_Z provides an additional degree of freedom in setting the thresholds.

This circuit introduces another facet of hysteretic comparators; as the input approaches the threshold, the diff-amp transconductance increases. Away from the threshold, one of the diff-amp transistors is cut off, and diff-amp transconductance is low. As a result, loop gain is low. But when the threshold is approached, the cut-off transistor begins to conduct, and loop gain increases. When it reaches unity, the positive-feedback loop becomes unstable and transitions to the other state. To find the threshold voltages accurately, an iterative solution is required since diff-amp gain changes with input voltage. In other words, a large-signal analysis is necessary using (9.87).

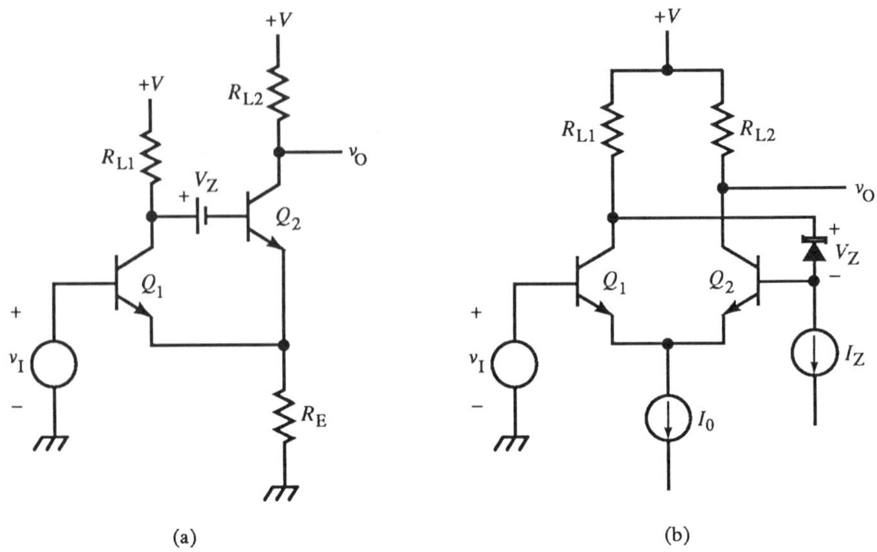

FIG. 11.28 Two-BJT Schmitt trigger (a); redrawn emitter current source and Zener diode voltage translator (b).

Since the hysteretic comparator is a positive-feedback amplifier, it is inherently unstable. But this instability must be controlled in the design; instability is only allowed for changing state. If the loop gain has peaking, then as the input voltage approaches the threshold, unity loop gain is reached first at ω_m, the frequency at which the loop-gain magnitude peaks. Before the output changes, the loop oscillates with frequency ω_m. Therefore, otherwise stable loops require a loop gain without peaking.

11.5 Discrete Logic Circuits

Even analog designs are likely to require some logic functions. In discrete designs, it is often unnecessary to add logic ICs to perform simple logic functions. Figure 11.29 shows a variety of diode and BJT logic circuits using only two diodes or one transistor. In Fig. 11.30, two more circuits are shown, using four transistors, to realize an exclusive-or gate (a) and an and-or-invert (AOI) gate (b). These circuits have no particular merits other than their simplicity.

11.6 Clamps and Limiters

Nonlinear circuits that modify waveforms in some manner involving limits are called *clamps* or *limiters*. Depending on the particular application, they

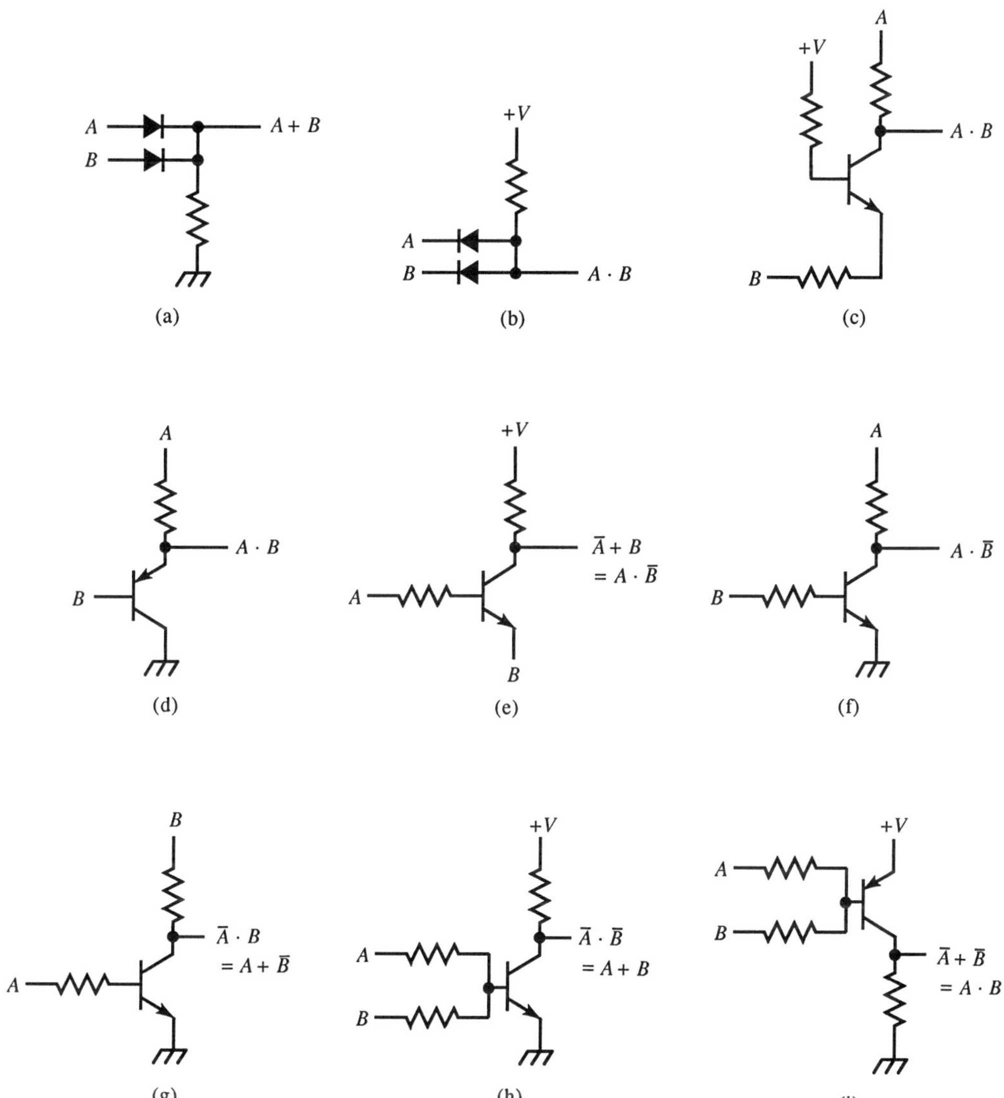

FIG. 11.29 Discrete logic circuits.

might have other names. In Fig. 11.31a, diodes are used to limit the range of v_I by "clipping" the signal outside the range of $\pm V$. This circuit is commonly used as an input protection circuit in MOS ICs and oscilloscope trigger inputs. It is sometimes called a *clipping circuit.* Figure 11.31b shows a type of clamp that establishes an ac signal at a given dc level. An application is as a *baseline restorer* in video signal processing. The negative extremes (sync tips) are established near ground by the clamp diode.

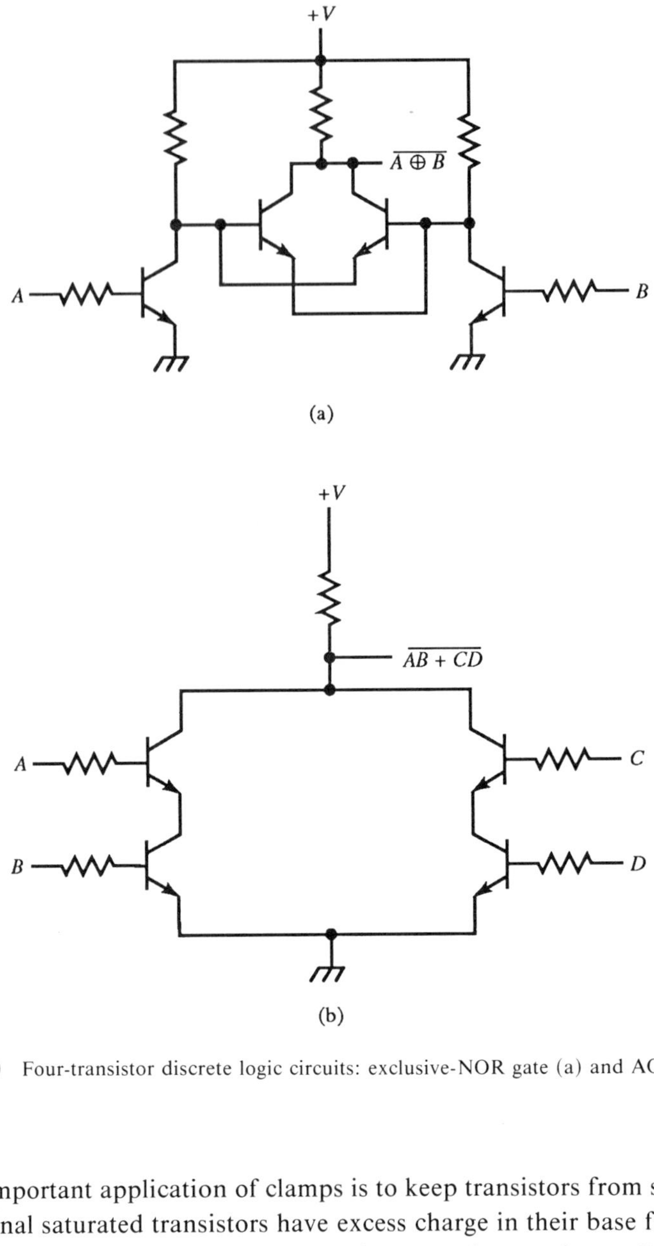

FIG. 11.30 Four-transistor discrete logic circuits: exclusive-NOR gate (a) and AOI gate (b).

An important application of clamps is to keep transistors from saturating. Small-signal saturated transistors have excess charge in their base from being overdriven. This charge must be removed to turn the transistor off, and with limited base-current drive the base storage time increases. This causes a delay in turn-off. Fall time is not significantly affected.

In large-signal (power) transistors, although excess base charge is a storage-time factor, another effect dominates, causing fall times of unclamped transistors to be larger. With large collector currents, a BJT operates in the high-level injection region, where the collector minority-carrier concentration (majority carriers from the base) approaches that of the collector majority concentration.

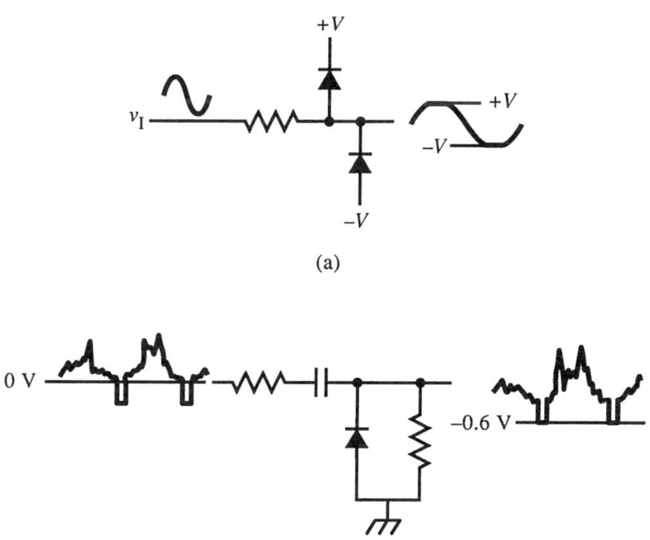

FIG. 11.31 Bipolar diode clamp (a); video sync clamp or baseline restorer (b).

Under strong high-level injection, the collector side of the b–c junction actually inverts in charge polarity due to a dominance of carriers from the base. This *Kirk effect* causes conductivity modulation of the base since the base width effectively increases. This effect occurs at a v_{CE} just above saturation, in the quasisaturation region, and causes excess rounding in the collector family of curves at low v_{CE} (of a few volts). (A related effect, called *crowding*, is due to ohmic v_{BE} drop laterally across the base, which causes less conduction in the center of the base than in the outer ring closer to the base contact.) Conductivity modulation also affects the fall time. As excess charge is swept out of the base, the excess base width begins to decrease, and turn-off commences. By decreasing excess base drive, conductivity modulation also decreases, along with both storage and fall times in power transistors.

The *Baker clamp* prevents saturation by adding two diodes to a BJT (Fig. 11.32a). Because the b–c junction is a diode, it is in series with D_1. Together, they conduct with a voltage drop of two junctions. D_2 shunts them and conducts sooner, with one junction drop. Thus, D_2 clamps the D_1-b–c path and keeps it from conducting.

Quantitatively, by KVL,

$$v_{CE} \cong v_{BE} + v_{D1} - v_{D2} \cong v_{BE} \tag{11.139}$$

For approximately equal diode voltage drops, v_{CE} is clamped at v_{BE} or with $v_{CB} \approx 0$ V. Transistors with appreciable r'_c or large collector current may require an additional diode in series with D_1 to clamp v_{CB} at about a junction drop.

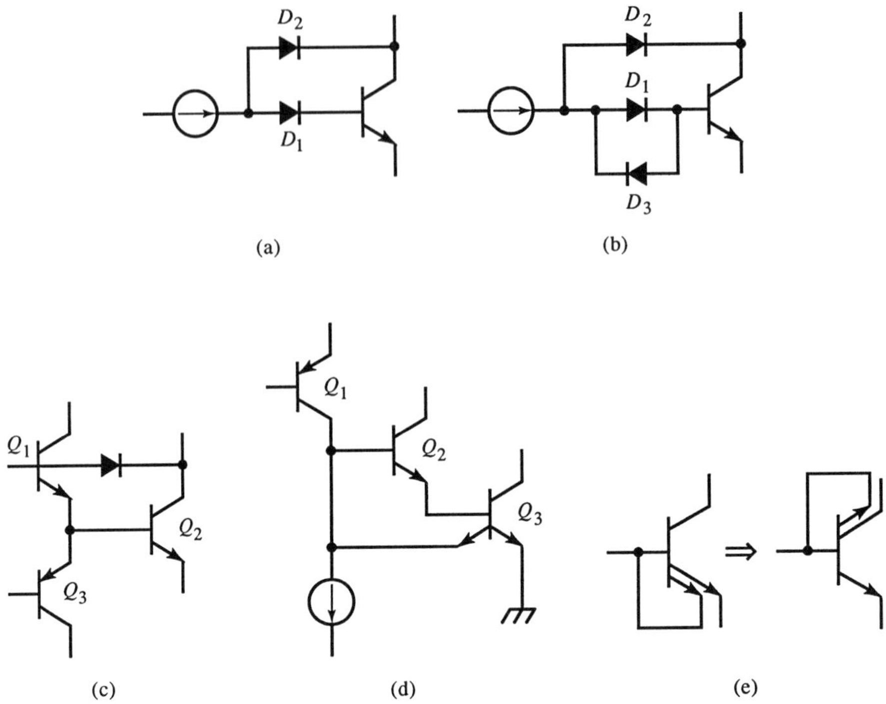

FIG. 11.32 Baker clamp circuits: (a) basic clamp; (b) clamp with base turn-off diode; (c) BJT IC realization; (d) Murphy clamp; (e) second emitter operates as a collector in (d).

The disadvantage of the Baker clamp is the higher on-state v_{CE}, typically a half volt. The *Schottky clamp* has reduced v_{CE}(on). It is a variation of the Baker clamp and is very simple; a Schottky diode, with forward voltage of about 0.4 V, shunts the *b–c* junction. When v_C decreases to 0.4 V below the base voltage, the Schottky diode turns on, clamping v_{CE} at about 0.3 V. In Schottky logic output stages, this clamp prevents hard saturation while allowing lower v_{CE}(on) than a standard Baker clamp.

Since D_1 is in series with the *b–e* junction, no turn-off path exists without a shunt *b–e* resistor or, as in Fig. 11.32b, a diode shunting D_1 and reversed (or antiparallel) relative to D_1. If a complementary CC driver is used to drive the clamped transistor, as in (c), only one added diode is required for the Baker clamp. The *b–e* junction of Q_1 takes the place of D_1 in (b) and the *b–e* junction of Q_3 for D_3. The diodes themselves also must have suitably fast turn-off recovery capability.

The feedback clamp scheme of Fig. 11.32d, used in ICs, is a form of the *Murphy clamp*, shown in (e). A second emitter is added to the transistor to operate as a clamp. In (e), this antisaturation emitter is connected back to the base. Since the *b–c* junction, as a diode, also points outward (like the emitter arrow does), the second emitter can be regarded as a second collector, as shown. It is more heavily doped than the collector and shunts current from it at low v_{BC}. When the collector voltage decreases to near saturation, the second emitter dominates; for the same reverse bias, it has more minority carriers to inject into the base. As a result, the *b–c* junction does not conduct heavily in the forward direction, avoiding hard saturation.

In Fig. 11.32d, Q_3 avoids saturation by its second emitter. This emitter is reverse-biased by v_{BE2}, allowing v_{CB3} to decrease to v_{BE2} before the antisaturation emitter takes effect. In addition, the shunted collector current is taken from the input drive current. The current gain of Q_2 amplifies this limiting action, resulting in a "sharp" limiting response.

Diodes and op-amps are combined in the precision clamps of Fig. 11.33. They use the op-amp input as a comparator when the loop is open and diode nonconducting: in (a), when $v_I < V_L$, and in (b), when $v_I > V_H$. For the input ranges where the diodes conduct, the op-amps operate as ×1 buffers. In effect, these clamps are half-wave rectifiers with programmable limiting voltages.

A common limiter in power supply circuits is the *foldback current limit*, with one realization of it in Fig. 11.34. As load current i_L increases, the voltage drop across R_S increases until the *b–e* junction of Q_1 is turned on. The transistor conducts, diverting base drive from the series regulator Darlington. The current

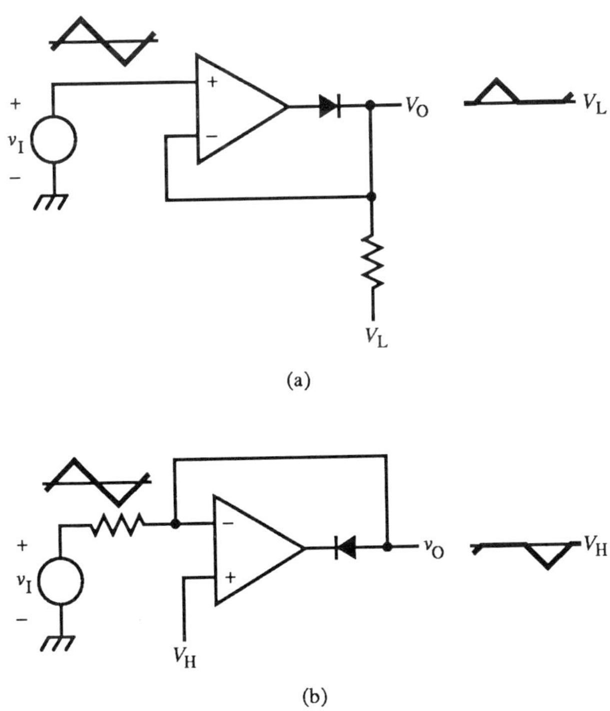

FIG. 11.33 Precision diode clamp: minimum voltage limiter (a); maximum voltage limiter (b).

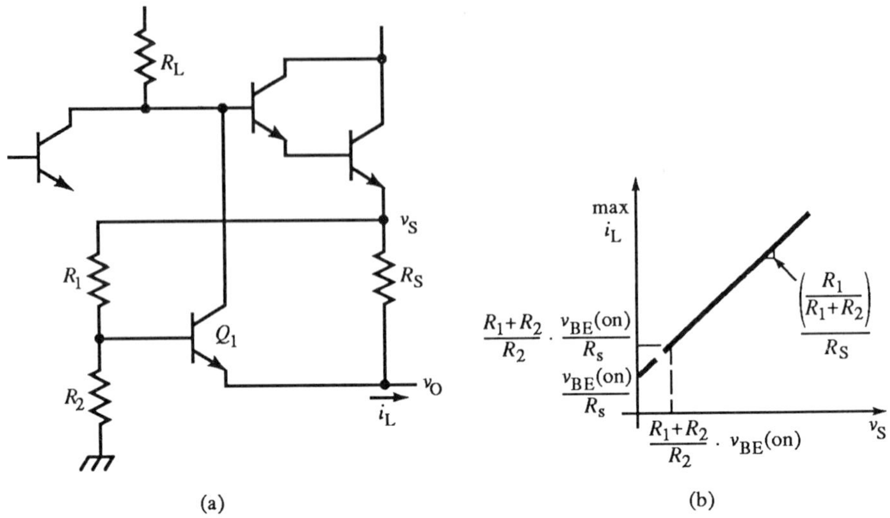

FIG. 11.34 Foldback current limiter (a) and v–i characteristics (b).

is limited to

$$\max i_L = \frac{v_{BE}(on) + v_S[R_1/(R_1+R_2)]}{R_S} \qquad (11.140)$$

where $v_{BE}(on)$ is the v_{BE1} for which the loop gain is barely sufficient to sustain the limiting value of i_L. The maximum current depends on v_S, graphed in Fig. 11.34b. The current limit decreases or "folds back" with reduced voltage so that a short conducts reduced current and thus power dissipation. As the short "clears" (load resistance increases), more current is allowed.

11.7 Multivibrators and Timing Circuits

Multivibrators (MVs) are positive-feedback (or regenerative) switching circuits with analog timing of switching behavior. They can be *bistable*, having two stable states (such as Schmitt trigger circuits); *monostable*, having one stable state; or *astable*, having no stable states. Monostable multivibrators (MMV) are also called "one-shots"; they change output state upon input of a trigger signal. This quasistable state lasts for a timed interval (until the MMV "times out"), at which time it reverts to its stable state. MMVs are used to generate a triggered pulse of a given duration. Astable MVs are digital oscillators, or *clock generator* circuits and sometimes are called "free-running" MVs.

The standard astable MV topology consists of two capacitively coupled CE stages (Fig. 11.35). When v_{C2} goes low, the negative transition is coupled through C_1, cutting off Q_1. Since R_1, C_1 form an RC differentiator, v_{B1} begins

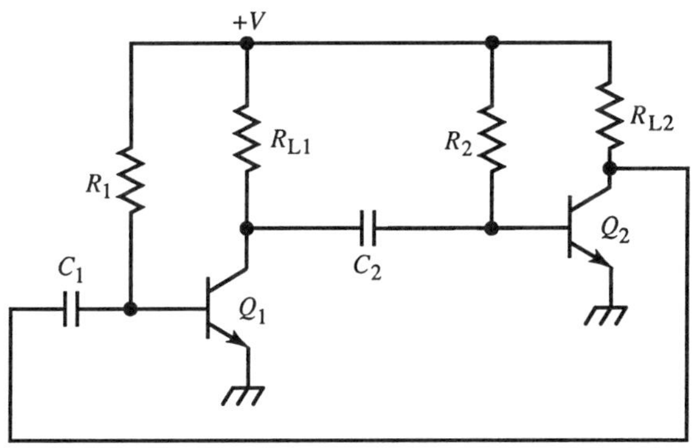

FIG. 11.35 Classic astable multivibrator.

to increase exponentially as C_1 charges through R_1. (We assume that the b–e junction of Q_1 is not in reverse breakdown.) When v_B reaches $v_{BE1}(\mathrm{on})$, Q_1 conducts, causing its collector to transition to near ground. This cuts off Q_2 for the second half-cycle of oscillation in the same way Q_1 was cut off. For identical CE stages, the duty-ratio,

$$D = \frac{t_{on}}{t_{on} + t_{off}} = \frac{t_{on}}{T} \tag{11.141}$$

is 50%. When Q_1 turns on and drives Q_2 off, v_{C2} goes high (to $+V$), and C_1 is charged in the other direction through R_{L2} and the b–e junction of Q_1. This recharging time constant must be shorter (at least 5τ) than R_1C_1 to fully recharge the capacitor. Therefore, R_1, R_2 must be at least five times R_{L1}, R_{L2}.

MV frequency is increased by reducing coupling capacitor C. To analyze the effect of C_{be} on timing, assume that the base timing resistor R is a current source I instead. This assumption is valid for off-times that are much less than the time constant RC when the exponential base voltage can be approximated as linear. Then C and C_{be} form a capacitive voltage divider so that a negative input step of $-V$ causes v_B to step down to

$$v_B(0^+) = -V\left(\frac{C}{C + C_{be}}\right) + v_{BE}(0) \tag{11.142}$$

Without the effects of C_{be}, the timing ramp of v_B has a slope of I/C and times out at

$$t_1 = \frac{C}{I}V \tag{11.143}$$

With C_{be},

$$v_{BE}(t) = \left(\frac{I}{C + C_{be}}\right)t - v_B(0^+), \qquad t > 0 \tag{11.144}$$

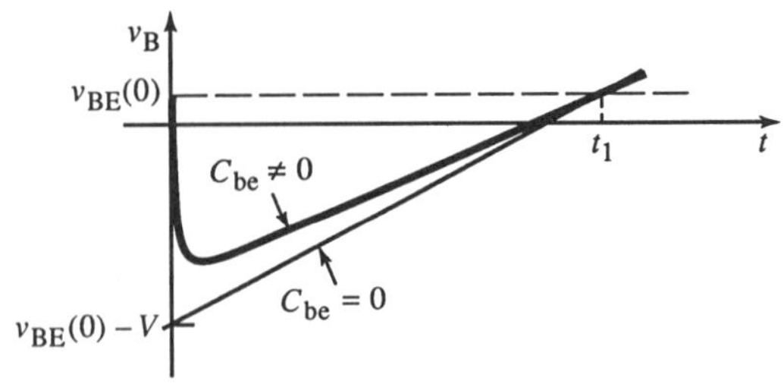

FIG. 11.36 BJT C_{be} does not affect time-out.

The offset of $v_{BE}(0)$ does not affect timing since it is also the voltage threshold for determining t_1 (Fig. 11.36). If we set $v_{BE}(t_1) = v_{BE}(0)$ and solve for t_1, it is the same as in (11.143). Therefore, C_{be} does not affect timing.

The previous analysis assumed zero fall time of the negative input step. To examine this assumption, consider the effect of switching time on timing by approximating the negative transition at the collector with a linear approximation to the waveshape and its average slope m. Let the transition time be t_f. Then, from Fig. 11.37, the slower the transition, the smaller the negative excursion of the timing voltage $-v_f$; thus,

$$t_1 - t_f = \frac{C}{I} v_f \qquad (11.145)$$

and

$$v_f = -\left(\frac{I}{C} + m\right) t_f \qquad (11.146)$$

with $m = -V/t_f$. Substituting into (11.145) and solving for t_1, we obtain

$$t_1 = \frac{C}{I}\left(\frac{-I}{C} + \frac{V}{t_f}\right) t_f + t_f = \frac{C}{I} V \qquad (11.147)$$

Therefore, when $t_f < t_1$, the falltime has no effect on timing.

The final timing analysis determines the effect of collector resistance r_c of the conducting transistor. From Fig. 11.38, although the slope of $v_B(t)$ is

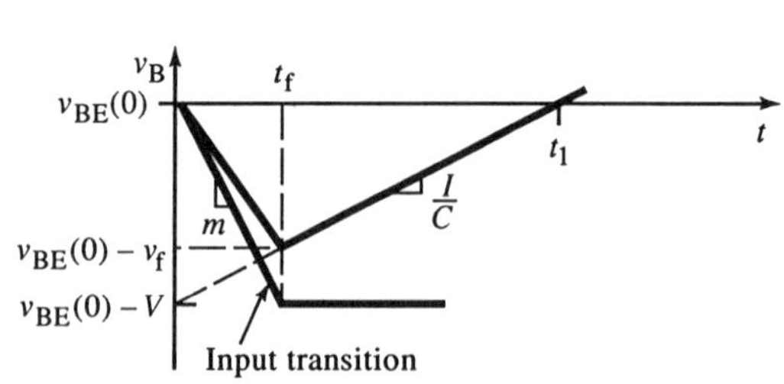

FIG. 11.37 BJT switching time does not affect time-out.

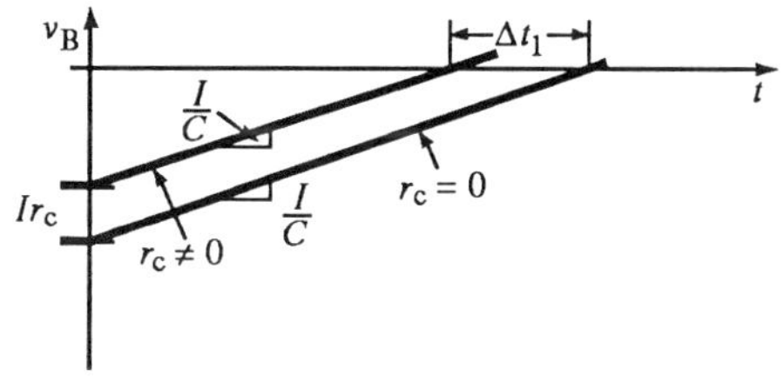

FIG. 11.38 Collector resistance decreases time-out.

independent of r_c, the initial step size is not. With r_c, I causes an opposing step of Ir_c that translates into a time error of

$$\Delta t_1 = \frac{Ir_c}{I/C} = r_c C \tag{11.148}$$

The fractional error is Ir_c/V. Consequently, to minimize r_c, the BJTs must be driven well into saturation by R_1 and R_2.

The classic astable MV timing is independent for each half-cycle. This makes adjustment of half-cycle timing easy, but since the period depends on both half-cycles, each can contribute to period error. Two separate timing circuits are not necessary. The astable MV in Fig. 11.39, based on an open-collector comparator such as the LM393, has only one timing capacitor. This single-supply clock generator uses regenerative feedback through R_f to effect a hysteretic comparator while the timing is done by R and C. When $v_O = V_U$ (high state), the timing waveform at v_- is increasing and crosses V_H at the v_+ input. Then v_O goes low (to V_D), and the timing waveform decreases toward it while $v_+ = V_L$. When it reaches V_L, v_O switches high again.

For single-supply operation, R_1 and R_2 set a voltage around which the input hysteresis limits of V_L and V_H are chosen. When $v_O = V_U$, the timing

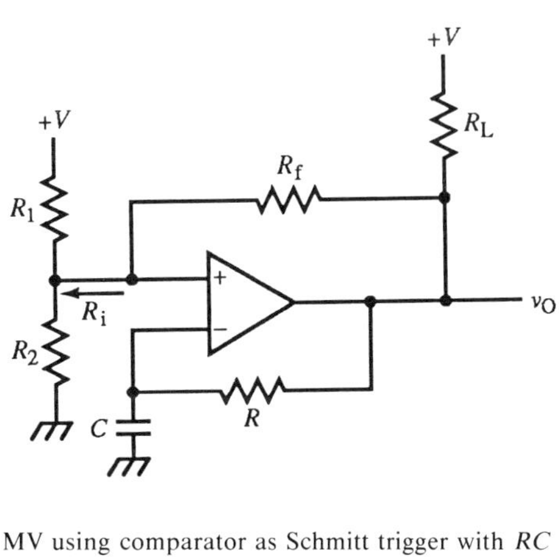

FIG. 11.39 Astable MV using comparator as Schmitt trigger with RC integrator for timing.

resistance is

$$R_U = R + R_L \parallel (R_f + R_i), \qquad R_i = R_1 \parallel R_2 \tag{11.149}$$

V_U and V_H are calculated from the divider formed by R_L, R_f, and R_i. The timing voltage v_- begins at V_L and heads for V_U. The high-state time duration is

$$t_H = R_U C \ln\left(\frac{V_U - V_L}{V_U - V_H}\right) \tag{11.150}$$

During the low output state, $v_O = V_D$ and $v_+ = V_L$. The comparator BJT saturates, and $V_D \cong 0$ V. Then v_- decreases from V_H toward V_D, and

$$t_L = RC \ln\left(\frac{V_H - V_D}{V_L - V_D}\right) \cong RC \ln\left(\frac{V_H}{V_L}\right) \tag{11.151}$$

Finally, the output period is

$$T = t_H + t_L \tag{11.152}$$

One of the most versatile MV circuits is the *timer*, notably the 555 bipolar and 7555 CMOS ICs, with block diagram in Fig. 11.40. This timer is mainly applied as a MMV or clock generator. It consists of two comparators with trigger ($\overline{TR}$) and threshold (TH) inputs. The other inputs are taken from a resistive divider from the supply, V_{CC}. For the bipolar version, $R = 5$ kΩ; the CMOS version $R > 100$ kΩ. The threshold-comparator input threshold is at

$$V_H = \tfrac{2}{3} V_{CC}$$

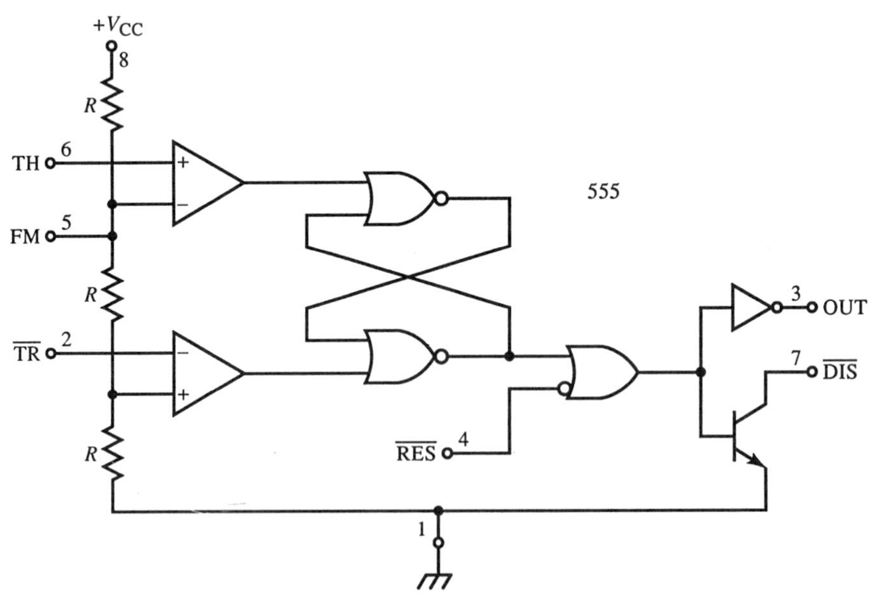

FIG. 11.40 555 timer IC block diagram.

and the trigger-comparator is at

$$V_L = \tfrac{1}{3} V_{CC}$$

The comparator outputs drive a NOR-gate RS flip-flop (FF). The trigger input overrides the threshold input for control of the output. When $\overline{\text{TR}}$ is asserted $[v(\overline{\text{TR}}) < V_L]$, the output is forced high. The reset input $\overline{\text{RES}}$, however, overrides all other inputs. A separate transistor (BJT in 555, MOSFET in 7555) with open collector (or drain) marked $\overline{\text{DIS}}$ (for "discharge") is an alternative output for MV control.

An astable MV circuit using the timer is shown in Fig. 11.41a. Initially, $\overline{\text{TR}}$ is low and the output high. The discharge transistor is off. C charges through R_1 and R_2 until the timing voltage crosses V_H. The output goes low, $\overline{\text{DIS}}$ sinks current, and C discharges to ground through R_2. When the timing voltage crosses V_L, the cycle repeats. The time duration for a high output is

$$t_H = (R_1 + R_2)C \ln\left(\frac{V_{CC} - V_L}{V_{CC} - V_H}\right) = (R_1 + R_2)C \ln 2 \cong (0.693)(R_1 + R_2)C \tag{11.153}$$

and for a low output,

$$t_L = R_2 C \ln\left(\frac{V_H}{V_L}\right) = R_2 C \ln 2 \cong (0.693) R_2 C \tag{11.154}$$

The period is therefore

$$T = t_H + t_L = (R_1 + 2R_2)C \ln 2 \cong (0.693)(R_1 + 2R_2)C \tag{11.155}$$

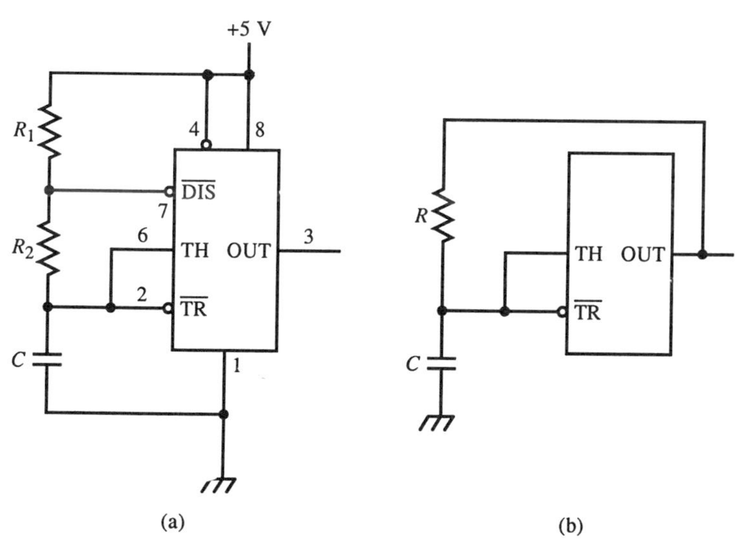

FIG. 11.41 555 timer connected as an astable MV (a); square-wave generator topology using 7555 (b).

with duty-ratio

$$D = \frac{R_1 + R_2}{R_1 + 2R_2} \tag{11.156}$$

Since $t_L < t_H$, a duty-ratio of $D < 0.5$ is not possible with this circuit. An alternative circuit achieves a longer t_L by placing an additional resistor in series with $\overline{\text{DIS}}$. In t_L, R_2 is replaced by $R_2 + R_3$. For $R_1 = R_3$, $D = 0.5$. A CMOS alternative with accurate 50% duty-ratio is shown in Fig. 11.41b. The output is used instead of $\overline{\text{DIS}}$ to control timing, and the timing elements are the same for both half-cycles. The period is

$$T = 2RC \ln 2 = 1.386RC \tag{11.157}$$

A timer-based MMV (Fig. 11.42) is triggered by a negative-going pulse. It sets the output high, and C begins to charge through R. The initial voltage on C is 0 V, and it charges to V_H. The time-out is thus

$$t_H = RC \ln\left(1 - \frac{V_H}{V_{CC}}\right) = RC \ln\left(\frac{1}{3}\right) \cong 1.1RC \tag{11.158}$$

If the trigger pulse duration exceeds t_H, the output is kept high until the trigger goes high (Fig. 11.42b).

An MMV that begins its time-out after the trigger pulse goes high is shown in Fig. 11.43a. The additional PNP transistor keeps C discharged until the trigger releases. This MMV is retriggerable in that the output remains high as long as trigger pulses continue to occur before time-out. Each new pulse resets the timing and retriggers the MMV.

The FM terminal (pin 5) gives control of the comparator thresholds in Fig. 11.43b, where a positive transition on the FM terminal triggers the MMV. The $\overline{\text{TR}}$ input is biased at $V_{CC}/4$. The trigger-comparator divider voltage is raised by the positive step of the trigger to the point at which it exceeds $V_{CC}/2$ and starts the time-out.

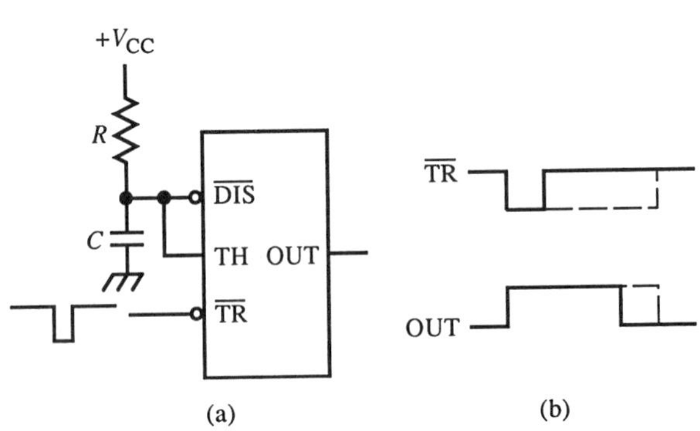

FIG. 11.42 555 timer monostable MV (a) and timing diagram (b). If trigger input stays low longer than time-out, the output remains high.

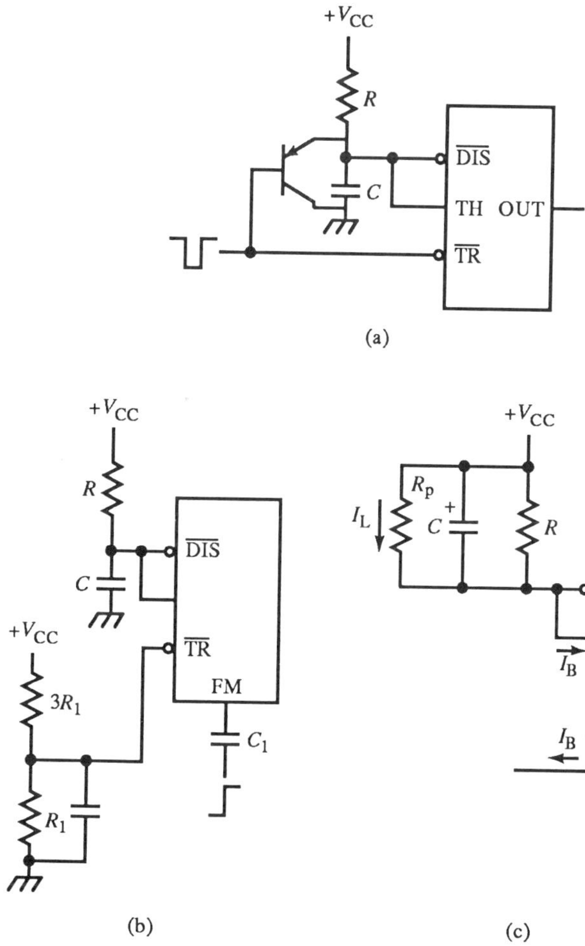

FIG. 11.43 555 timer circuits: (a) retriggerable MMV; (b) positive-edge trigger drives threshold divider; (c) improved topology for leaky timing capacitors and long time-outs.

For long time-outs, either a large R, a large C, or both are needed. For a large R, timer-comparator bias currents can cause timing error. The bipolar timer has an NPN diff-amp input stage in its threshold comparator and a PNP stage for the trigger comparator. The bias currents (Fig. 11.43c) place a limit on the minimum charging current. For large C, electrolytic capacitors are likely to be used. Their limitation is their leakage current. If C is placed as in (c), its leakage current I_L path is modeled by a shunt resistance R_p, shunting the timing resistor R. The maximum limit on R is thus set by R_p. This circuit has less timing variation (*jitter*) because the target voltage of C is 0 V instead of V_{CC}, the same voltage that R_p would discharge C to. When C is ground-based, R_p opposes charging instead of aiding it.

11.8 Capacitance and Resistance Multipliers

Timing circuits with long time-outs often require large capacitors. For accurate timing, these capacitors are plastic. Large-value plastic capacitors are volumetrically large and expensive. The *capacitance multiplier* is a circuit that uses gain to make a small capacitor appear electrically large. One realization is shown in Fig. 11.44. Here, a current source drives the C-multiplier to generate a ramp waveform. The ×1 buffer causes R_i and R_f to form a current divider because it keeps the voltage across R_f the same as across R_i. The charging current is a fraction of the input current, or

$$\frac{i_C}{i_I} = \frac{R_f}{R_f + R_i} \tag{11.159}$$

The equivalent capacitance C_{eq} is based on the relation

$$i_I = C_{eq} \cdot \frac{dv_O}{dt} \Rightarrow \frac{i_I}{C_{eq}} = \frac{dv_O}{dt} = \frac{i_C}{C} \tag{11.160}$$

The last equality follows because $v_O = v_C$. Applying (11.159) gives

$$C_{eq} = \left(\frac{i_I}{i_C}\right) C = \left(\frac{R_i}{R_f} + 1\right) C \tag{11.161}$$

This is a "transcapacitance" multiplier because v_O is not across the same terminals that i_I flows through. The input is C_{eq} in series with $R_f \| R_i$.

A true capacitance multiplier can be based on the Miller effect (Fig. 11.45). The ×1 buffer drives an inverting op-amp with a gain of $-R_f/R_i$. Applying Miller's theorem, we obtain

$$C_{eq} = \left(\frac{R_f}{R_i} + 1\right) C \tag{11.162}$$

Above the bandwidth of the amplifier branch, the input is no longer purely capacitive but also has a shunt RL in series with C_{eq}.

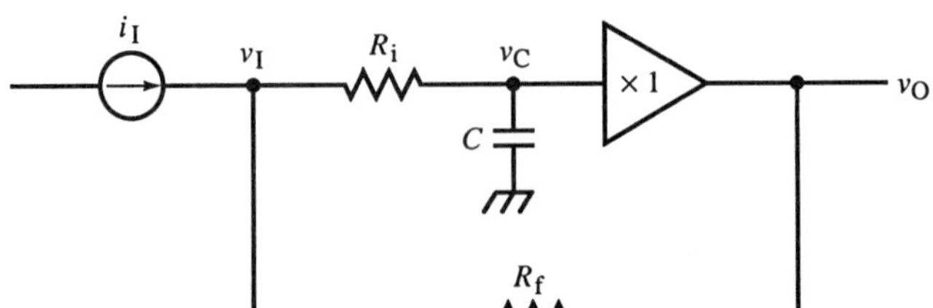

FIG. 11.44 Capacitance multiplier based on current divider.

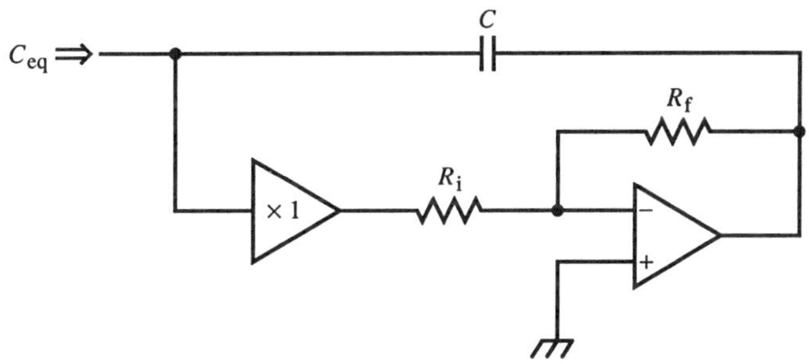

FIG. 11.45 Capacitance multiplier based on Miller effect.

Example 11.8 Timer with Capacitance Multiplier

The threshold-terminal bias current of a 555 timer limits its useful timing range as a MMV. A capacitance multiplier of the kind in Fig. 11.44 is used to extend the time-out t_H by connecting it as in Fig. E11.8a. The TH input is now driven by the op-amp output, eliminating bias-current error from TH but introducing op-amp offset current and voltage error. An equivalent circuit in (b) is derived as follows. The op-amp circuit with R_f and R_i is Thévenized and floated on v_C. The op-amp offset voltage V_{OS} is divided by the resistors so that its Thévenin voltage is

$$V_{OS}\left(\frac{R_i}{R_f + R_i}\right)$$

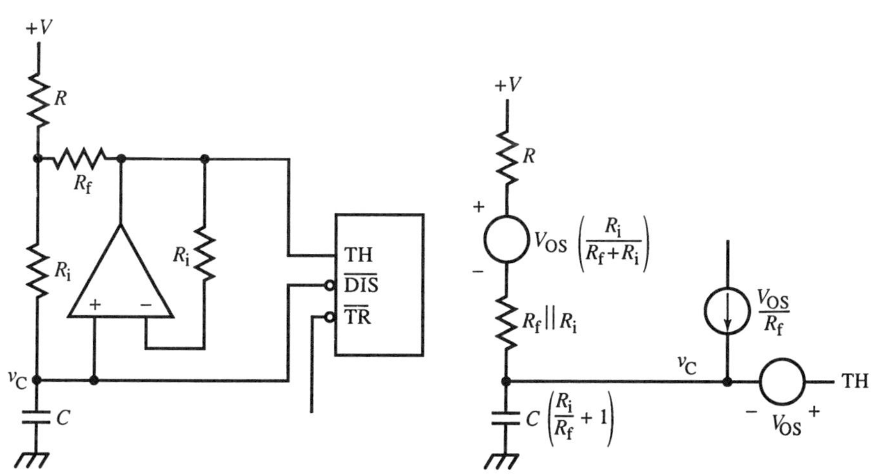

(a) (b)

FIG. E11.8

in series with $R_f \parallel R_i$. Since the op-amp has ×1 gain, V_{OS} also is in series with the TH input. In addition, V_{OS} at the op-amp output contributes V_{OS}/R_f to the timing current. From this model, timing error is calculated.

For significant multiplication of C, $R_i \gg R_f$. The series Thévenin voltage source is then about V_{OS}, and the I_{OS} term in the error current dominates.

Capacitance multiplication is achieved in the preceding circuits by applying bootstrapping to a current divider. The idea can be extended to resistance multiplication. A previous instance is the bootstrapped CC of Fig. 2.6c, described by (2.54) and (2.56). A more deliberate and precise resistance multiplier, in Fig. 11.46, uses an op-amp buffer instead of a CC or CS. Solving the circuit for r_{in} gives

$$r_{in} = R_1\left(1 + \frac{R_3}{R_2}\right) + R_3 \tag{11.163}$$

11.9 Trigger Generators

A trigger generator is a kind of precision synchronizer. In television deflection systems, horizontal and vertical scans or *sweeps* of the CRT are synchronized to the video signal by perturbing a free-running oscillator with the synchronization (or *sync*) signal, forcing it to lock to the sync frequency. The sync pulse also corrects the phase each cycle.

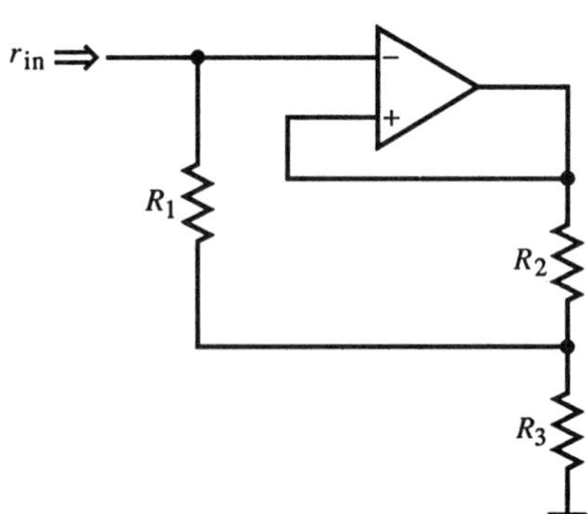

FIG. 11.46 Resistance multiplier based on voltage bootstrapping.

In synchronous digital systems, asynchronous events must be synchronized to the system clock. This is usually done using a flip-flop (or *flop*) clocked by the system clock; the asynchronous pulse is the data input to the flop. The problem with this scheme is that if the data pulse changes state too soon before the active edge of the clock, the flop setup time is insufficient, and the output state is indefinitely indeterminate. A second flop cascaded with the first can reduce the indeterminate time skew at the expense of one clock period of delay. Of course, the first flop output may remain indeterminate for longer than one clock period, but the probability diminishes rapidly with time.

This synchronizing problem is especially acute in oscilloscopes. The vertical signal is fed to an event processor. On a selected slope and at the voltage of the trigger-level control, a comparator generates an *event*, an output transition that is used to start the sweep. When the sweep reaches the right end of the CRT screen, the CRT beam is turned off, and the beam retraces back to the left side, where it settles to the same starting position.

During sweep retrace, a hold-off pulse keeps input signals from firing the sweep until it is settled in its starting position. This hold-off pulse is asynchronous with the trigger events. If an event occurs while hold-off is releasing, time skew occurs in starting the sweep; this *trigger jitter* causes successive traces to be horizontally misaligned; the trace appears fuzzy, and multiple traces can be observed.

The trigger generator of Fig. 11.47 reduces jitter by synchronizing the trigger events, *TR*, and hold-off pulses, *HO*. The circuit consists of two D flip-flops and a delay device, such as a digital delay-line or the propagation delay of some logic devices. The flops are clocked on positive (rising) edges by *TR*. The sweep gate is asserted low (as $\overline{GATE}$) and lets the sweep generator operate. The negative (falling) edge of $\overline{GATE}$ must occur consistently at a fixed delay time relative to the rising edge of *TR*. Assume *HO* is low; *TR* clocks flop A. As its Q output settles, *TR* is delayed and then clocks flop B. If the delay is long enough, flop A output becomes valid and sets up the D input of flop B for its setup time. Then when the delayed *TR* clocks B, the output edge time is determinate. Most of the time, *HO* does not violate flop A D-input setup time, but when it does, flop B is required to synchronize its release with the trigger event.

This trigger generator is adequate for oscilloscopes of up to about 50 MHz bandwidth. For higher performance, the faster trigger generator of Fig. 11.48a

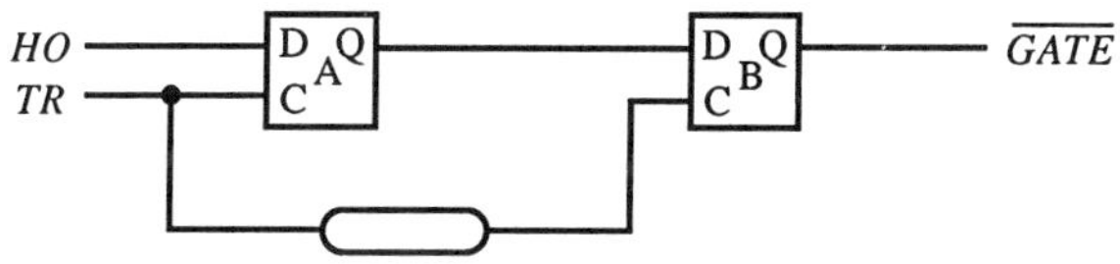

FIG. 11.47 A dual D flip-flop trigger generator, which synchronizes asynchronous input *HO* with clock *TR* to produce jitter-free output $\overline{GATE}$.

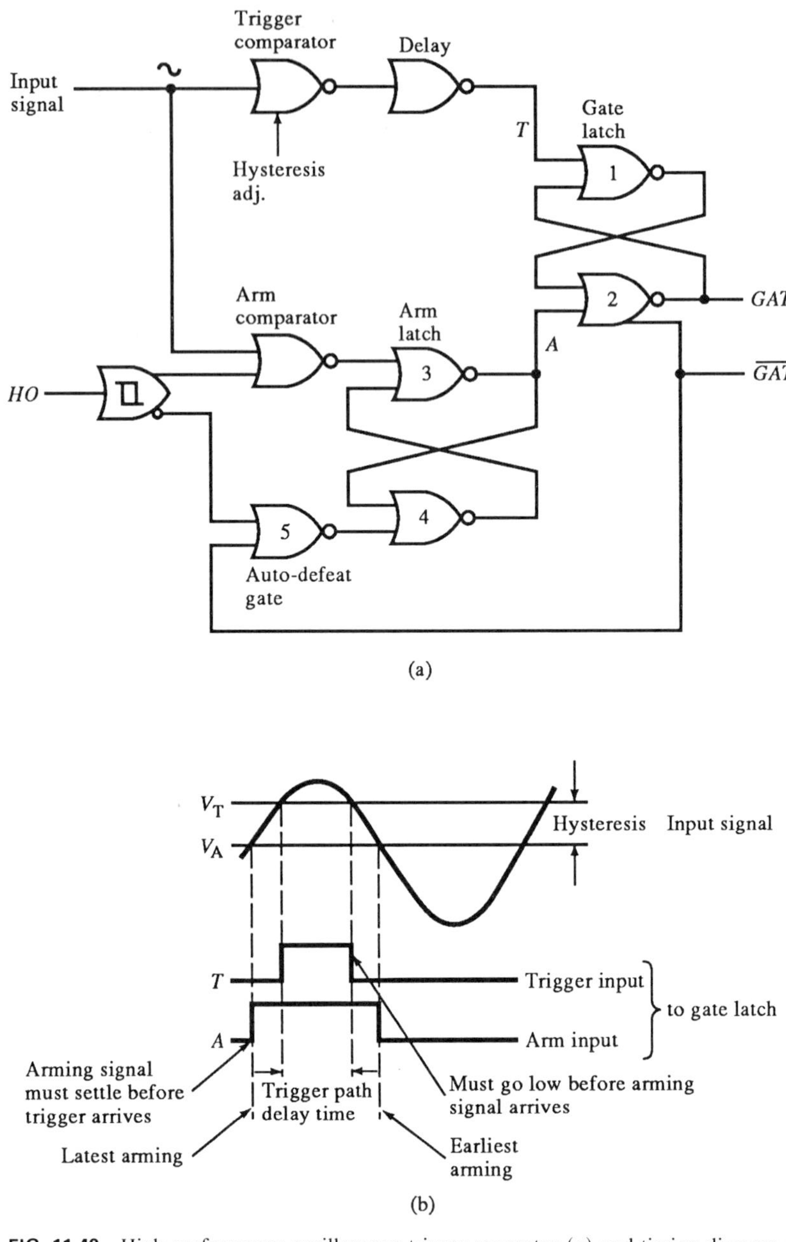

FIG. 11.48 High-performance oscilloscope trigger generator (a) and timing diagram (b).

is implemented as an IC with ECL logic. The comparators are ECL gates. The D flops of Fig. 11.47 are replaced with faster RS flops or latches: a gate latch (corresponding to flop B) and an arm latch (corresponding to flop A). The trigger signal drives both trigger and arm comparators. The trigger event out of the trigger comparator sets the gate latch high (*GATE* is high) if *A* was low. If so, the sweep runs, and its end is detected, causing *HO* to assert. *HO* forces reset of the arm latch; A is forced high. This forces *GATE* low and resets the gate latch. When *HO* goes low, before *GATE* can assert gain, A must be asserted low by setting the arm latch from the arm comparator. Then the gate latch is "armed" and can be set by the trigger comparator.

The detailed sequence of events is shown in Fig. 11.48b. *HO* has become low. When the input signal goes below the arm-comparator threshold V_A, the arm latch is set (A is low) through gate 3. The range of time when arming can occur is shown in (b) as the negative half-cycle of the input signal. It crosses the trigger-comparator threshold V_T, is delayed, and attempts to set the gate latch by asserting T high. If A has been low long enough, *GATE* goes high without jitter.

The hysteresis in Fig. 11.48b is the difference in comparator thresholds. Since the signal has a finite slope, a time delay is generated between the latest possible arming A and the earliest trigger T. The additional gate delay in the trigger path lets A settle to a valid level at the gate latch input.

The relative time delay of the trigger and arm paths is critical to proper synchronization. Otherwise, two trigger anomalies can occur: trigger jitter and double triggering.

Consider first the case in which the trigger-path delay is *too short* relative to the arm path. A low logic level at A must be established through the arm path preceding trigger-path assertion at T (of a high level). With insufficient delay in the trigger path, T could assert before A is settled. This occurs when the input signal, on its positive slope, is crossing V_A just as *HO* releases at the latest possible arming time. Gate 3 of the arm latch will be driven by a quick pulse, barely enough to cause the arm latch to change state after some time. This leaves A indeterminate. When the trigger-path input to gate 2 goes low, A causes an uncertain starting edge at *GATE*. Trigger jitter is the result.

Now consider the case in which the trigger-path delay is *too long* relative to the arm path. When the negative slope of the input signal decreases through V_T, the trigger path propagates a low level at T. Meanwhile, the arm path also propagates, through the arm latch, a low level at A. This arming of the gate latch must be preceded by a low level at T. If the trigger-path delay is excessive, the release of the gate latch at A occurs while T is still high, causing *GATE* to go high. This produces an extra *GATE* or double trigger. Under correct operation, only the positive slope asserts *GATE*. Double triggering occurs when *HO* releases just as the input signal crosses the arm comparator threshold at the earliest possible arming. On the screen, the waveform appears to be triggering on both slopes.

The optimal delay between the paths is somewhere between these two extremes. Delay-time tolerance is provided by the comparator threshold hysteresis and finite slope (or slew rate) of the input signal. Since the trigger path to gate 2 is one gate longer than the arm path, the hysteresis delay on the negative slope of the signal must be at least one gate delay, t_{pd}, to avoid double triggering. An approximate maximum t_{pd} can be calculated by assuming a maximum amplitude sinusoidal input of frequency f_{max}. The gate propagation delay must be

$$t_{pd} < \frac{\text{hysteresis}}{\text{slew rate}} = \frac{V_T - V_A}{2\pi V_{fs} f_{max}} \tag{11.164}$$

A hysteresis adjustment of the comparator threshold on the trigger comparator allows the trigger generator to be adjusted for maximum-frequency fault-free triggering.

Trigger jitter can be observed on an oscilloscope screen (Fig. 11.49). Spurious signals to the left of the main signal are due to late triggering. A signal that is advanced in phase (hence, late) appears shifted to the left. This may seem counterintuitive since later time is to the right on the screen. But it is *GATE* that is late relative to the signal. If the sweep gate had started on time, the waveform would not have advanced as far in phase.

Slew-rate limiting of the input signal in the trigger system causes it to shift in time on the screen as the trigger level is adjusted. The limited signal is time distorted, and its phase error varies with amplitude relative to the vertical signal. A time-domain test of trigger-generator performance is to let the input signal be a pulse of varying width. The minimum width that achieves a stable trigger is an index of generator speed capability.

In Fig. 11.48, the function of gate 5 has not yet been described. The *auto trigger* mode causes the sweep to run at a low rate (typically <50 Hz) to display a trace on the screen when no signal is present (or when the trigger controls are not adjusted properly). An auto-mode retriggerable MMV is driven by *GATE*. If the sweep has not run for a while, the auto-MMV times out and gates the sweep on directly. When a triggered gate occurs, the MMV is reset and the free-run mode turned off. Now, if *HO* were the input to gate 4, at high sweep speeds, it would be difficult to get out of the free-run mode without gate 5.

At high sweep rates, *HO* has a large duty-ratio. That is, it takes much longer to retrace and recover from a sweep than the sweep time. If a slow trigger signal is applied with a period much greater than the sweep time (when *HO* is unasserted), the probability is low that during the sweep the signal

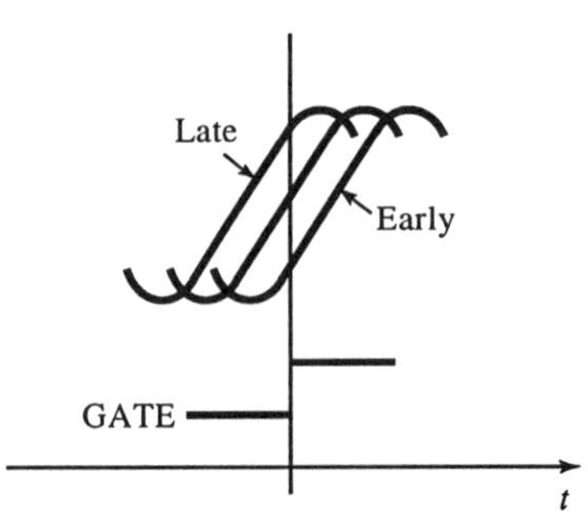

FIG. 11.49 Trigger jitter as displayed on oscilloscope screen.

would cross the hysteresis window. To escape the free-run mode, a *GATE* pulse is needed to reset the auto-MMV. The autodefeat gate (gate 5) is added to provide the pulse.

When *HO* unasserts during free run of the sweep, if the trigger signal is below V_A, the arm latch output A goes low. Since the sweep was run by the auto-gate signal, *GATE* is low, $\overline{GATE}$ is high, and gate 5 thus blocks *HO* from resetting the arm latch. This allows the input signal time to cross V_T and set the gate latch on its next half-cycle. If it does while *HO* is low, a triggered gate is asserted and the auto-MMV reset. More likely, if *HO* is high, gate 5 allows $\overline{HO}$ to reset the arm latch, and *A* goes high, resetting the gate latch low. *GATE* is high for the propagation time around the loop of gates 2-5-4-3-2. This is a few nanoseconds, enough time to reset the auto-MMV.

Another approach to auto triggering, *autolevel*, is to generate a triangle-wave with a counter and DAC and sum it with the trigger-level control output. The triangle-wave scans the level through the input range. When it intersects the input signal, a trigger event is generated, and triggering then occurs. If no signal is present, it autogenerates a trigger when the dc input level is crossed. An alternative approach to automatic triggering, *peak-to-peak auto*, uses positive and negative peak detectors to generate dc voltages at the extrema of the input signal. The trigger-level potentiometer is then placed between these peak voltages so that its control range is always within the signal range.

11.10 Ramp and Sweep Generators

Oscilloscope time-base systems consist of a trigger generator followed by a sweep generator. The sweep generator is a gated ramp or sawtooth generator that drives the horizontal deflection amplifier. Ramp generators are also used in magnetically deflected CRT display systems to generate deflection-coil currents. Pulse-width modulators also require sawtooth ramps.

A *bootstrap ramp generator* (Fig. 11.50) uses the bootstrapping technique to maintain a constant voltage across a timing resistor *R*. As *C* charges, the

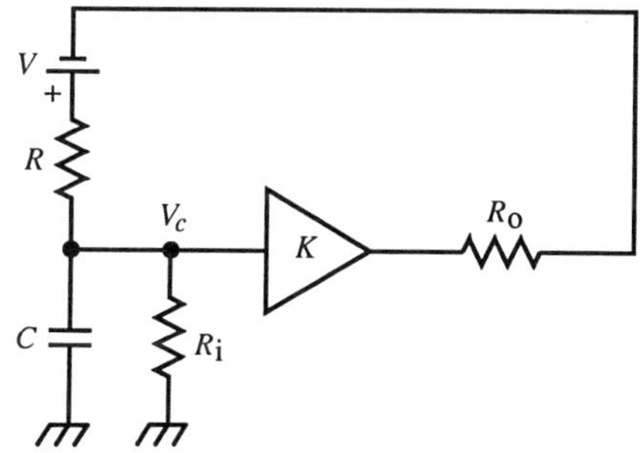

FIG. 11.50 Bootstrap ramp generator.

top end of R follows it, driven by the buffer. The result is a linear ramp output. With floating voltage source V, the ramp slope is

$$\frac{dv}{dt}=\frac{V/R}{C}=\frac{V}{RC} \tag{11.165}$$

This perfect scheme is spoiled by shunt resistance R_i (due to the capacitor and buffer input), buffer output resistance R_o, and buffer gain deviation from unity. Let V be gated on at $t=0$ as

$$V(s)=Vu(t) \tag{11.166}$$

Then solving the circuit in s, we get

$$\frac{V_c(s)}{V(s)}=\frac{R_i}{R+R_o+(1-K)R_i}\cdot\frac{1}{s\left[\left(\frac{R+R_o}{1-K}\right)\,\|\,R_i\right]C+1} \tag{11.167}$$

where K is the buffer voltage gain. The bootstrap effect appears as the increase in effective resistance of $R+R_o$ by $1/(1-K)$. This increases the time constant by the same factor; in effect, the ramp is generated as the initial segment of a long exponential curve. K, R_o, and R_i all contribute to the time-constant deviation from RC.

Another approach to ramp generation is to use an op-amp integrator. This *Miller ramp generator* (Fig. 11.51) has a transfer function of

$$\frac{V_o(s)}{V(s)}=-K\cdot\frac{1}{s(1+K)RC+1} \tag{11.168}$$

For an op-amp, $K\to\infty$, and the transfer function approaches $-1/sRC$, an ideal integrator. With finite gain, the output is an exponential with time constant multiplied by $(1+K)$; the early part of the curve is approximately linear. The fractional deviation from linear response is

$$\text{fractional nonlinearity}\cong\frac{t}{2\tau} \tag{11.169}$$

where τ is the effective time constant. Equation (11.169) was derived by

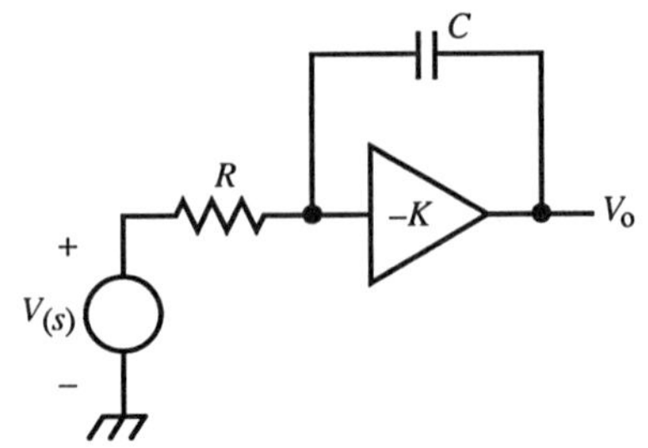

FIG. 11.51 Miller ramp generator.

series-expanding the response exponential to the quadratic term, subtracting the (ideal) linear and constant terms, and dividing by the linear term. Error grows with time as the exponential becomes increasingly sublinear. This error formula applies to both bootstrap and Miller ramp generators since the responses of both are exponentials.

Finite gain also causes a slope error in the Miller integrator. In (11.168), let $V(s)$ be that of (11.166). Then (11.168) corresponds to the normalized time-domain response of

$$\frac{dv_o(t)/V}{dt}$$

Applying the initial value theorem gives

$$\left.\frac{dv_o(t)/V}{dt}\right|_{t=0} = \lim_{s\to\infty} s\left(-K\cdot\frac{1}{s(1+K)RC+1}\right) = -\left(\frac{K}{1+K}\right)\frac{1}{RC} \quad (11.170)$$

The initial slope of the ideal ramp response, $-1/sRC$, is $-1/RC$. Thus,

$$\text{fractional slope error} = \frac{K}{1+K} \quad (11.171)$$

Fast Miller ramp generators are driven by an input current source instead of a voltage source and R. This produces a more linear response and reduces the effect of input impedance. The ideal response of $-1/sC$ is only approximate. More precisely,

$$\frac{V_o(s)}{I_i(s)} = \frac{1}{sC\left(\dfrac{1+K(s)}{K(s)}\right)} = \left(\frac{K(s)}{1+K(s)}\right)\frac{1}{sC} \quad (11.172)$$

The ideal response is multiplied by the s-domain equivalent of the fractional slope error. To approach the ideal, the op-amp must maintain high gain at high frequencies. This is a major limitation, especially since the op-amp output impedance gyrates inductively. For

$$K(s) = \frac{K_o}{s\tau_{bw}+1} \quad (11.173)$$

$$\frac{K(s)}{1+K(s)} = \left(\frac{K_o}{1+K_o}\right)\frac{1}{s[\tau_{bw}/(1+K_o)]+1} = \left(\frac{K_o}{1+K_o}\right)\frac{1}{s\tau_T+1} \quad (11.174)$$

The op-amp adds an additional pole at its unit-gain frequency f_T.

An improved ramp generator is based on the simplicity of a gated current source charging a capacitor, followed by a buffer amplifier. Feedback loops are avoided, and the step response is faster than for the previous two schemes. Bruce Hofer has recognized the topological equivalence of this current-source ramp generator with the Miller generator, compared in Fig. 11.52. The difference is in where the ground is put. In the Miller, the BJT is a CE; in

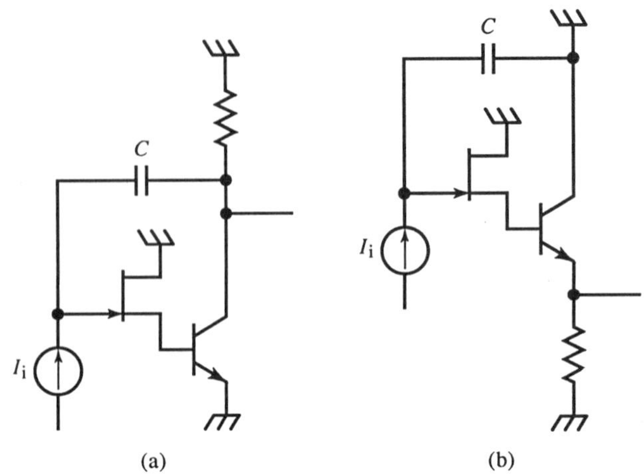

FIG. 11.52 Topological equivalence of Miller generator (a) and current-source ramp generator (b).

the current-source, a CC. In the Miller, any anomalous switching voltage at the FET gate is coupled through C to the output; in the current-source, it is bypassed by C to ground.

11.11 Logarithmic and Exponential Amplifiers

Logarithmic amplifiers are useful for compressing a wide dynamic-range signal, for multiplying, and for function generation. Figure 11.53 shows a simple log-amp based on a BJT b–e junction characteristic, (2.1). The input current i_I is the BJT collector current, and the output voltage

$$v_O = -v_{BE} = -V_T \ln\left(\frac{i_E}{I_S}\right) = -V_T \ln\left(\frac{i_I}{\alpha I_S}\right) \qquad (11.175)$$

This log-amp reflects the temperature sensitivity of the b–e junction and drifts due to both V_T and I_S.

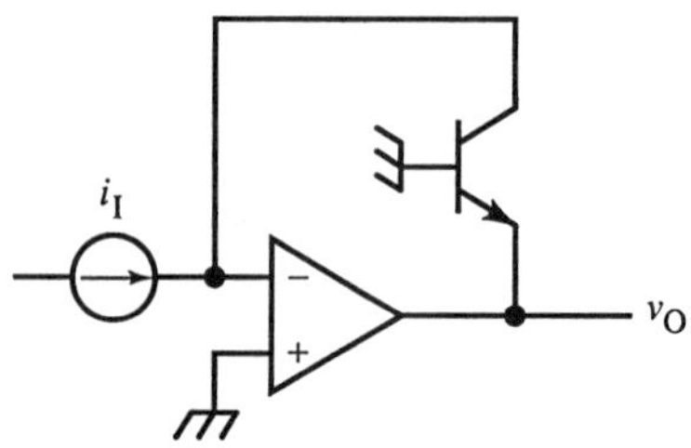

FIG. 11.53 A logarithmic amplifier based on BJT transconductance.

Detailed expressions for i_C and i_E [see Gibbons and Horn (1964) in the reference list] indicate that several error terms vanish when v_{CB} is zero. The logarithmic relation extends to lower currents for i_C (because $v_{CB}=0$ and $v_{BE}\neq 0$) and is accurate over about nine decades. The high end (typically 10 mA) is limited by ohmic voltage drops in series with v_{BE} and at the low end (typically 100 pA) by I_S. From (11.175),

$$v_{BE} = V_T \ln i_C + V_T \ln \alpha I_S \tag{11.176}$$

The slope of v_{BE} versus ln i_C is linear with a slope of

$$\frac{dv_{BE}}{d(\log i_C)} = \frac{dv_{BE}}{d(\ln i_C)} \cdot \frac{d \ln i_C}{d \log i_C} = V_T \cdot \frac{1}{\log e} = 59.56 \frac{\text{mV}}{\text{dec}} \tag{11.177}$$

For each decade of change in i_C, v_{BE} changes by about 60 mV at 300 K over, typically, nine decades of i_C.

The effect of I_S can be eliminated with a matched pair of BJTs, as in the *log-ratio amplifier* (Fig. 11.54). Assuming equal I_S, we have

$$\Delta v_{BE} = v_{B2} = v_{BE2} - v_{BE1} = V_T \ln\left(\frac{I_R}{i_I}\right) \tag{11.178}$$

The output is v_{B2} scaled by the divider, or

$$v_O = \left(\frac{R_2}{R_1}+1\right) v_{B2} = -\left(\frac{R_2}{R_1}+1\right) V_T \ln\left(\frac{i_I}{i_R}\right) \tag{11.179}$$

The second op-amp B functions similar to A in keeping $i_{C2} = I_R$, a reference current. For $i_I < I_R$, $v_O > 0$. Op-amp B controls the emitter currents $i_I + I_R$, while op-amp A controls v_{B2} to keep $i_{C1} = i_I$. R_E is selected so that the output voltage range of B can span the range of $i_I + I_R$.

This circuit eliminates dependence on I_S but is subject to V_T. The 0.33%/°C TC of v_{B2} is sometimes compensated by making R_1 a positive TC (PTC) thermistor. Its TC is found by differentiating v_O with respect to T, setting it

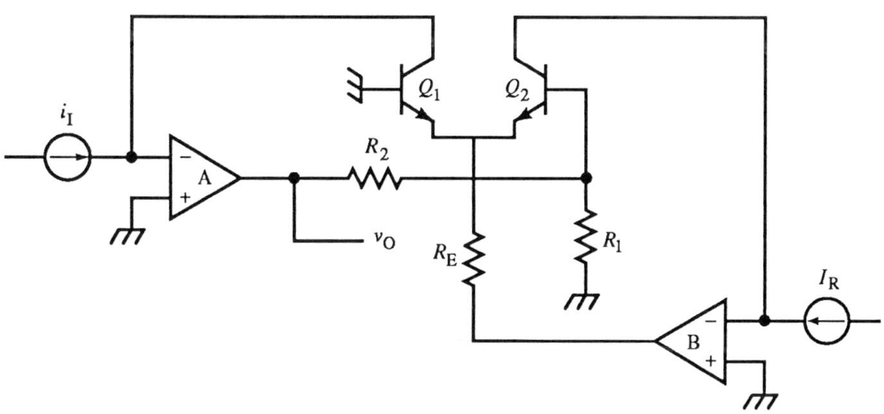

FIG. 11.54 Log-ratio circuit.

to zero, and solving for dR_1/dT. The fractional TC equation is

$$\frac{1}{v_O}\cdot\frac{dv_O}{dT}=\mathrm{TC\%}(v_O)=\frac{1}{T}-\frac{1}{R_1}\cdot\frac{dR_1}{dT}\left(\frac{R_2}{R_1+R_2}\right)$$

$$=\frac{1}{T}-\mathrm{TC\%}(R_1)\left(\frac{R_2}{R_1+R_2}\right) \qquad (11.180)$$

When $\mathrm{TC\%}(v_O)=0$,

$$\mathrm{TC\%}(R_1)=\left(\frac{1}{T}\right)\left(\frac{R_1+R_2}{R_2}\right)\cong(0.33\%/°\mathrm{C})\left(\frac{R_1+R_2}{R_2}\right)\text{ at 300 K} \qquad (11.181)$$

Example 11.9 Log-Amp Design

A log-amp based on Fig. 11.54 is to have an input range of 0.1 μA to 1 mA with minimal temperature drift. The output must be 0 V at an input of 1 mA (fs) and 4 V at 0.1 μA (zs).

The scaling is 4 V per four decades of 1 V/dec of i_I. The divider must be

$$\left(\frac{R_2}{R_1}+1\right)=\frac{1\text{ V/dec}}{60\text{ mV/dec}}=16.8$$

PTC thermistors with 1 kΩ values are available, so let $R_1=1$ kΩ. Then from the divider ratio, $R_2=15.8$ $k\Omega$, 1%. The TC of R_1 must be, from (11.181),

$$\mathrm{TC\%}(R_1)=(0.33\%/°\mathrm{C})\left(\frac{16.8}{15.8}\right)=(0.35\%/°\mathrm{C})$$

For $v_O=0$ V at 1 mA input, $I_R=1$ mA. This can be supplied from a voltage reference of 5 V through a 4.99 kΩ, 1% resistor. When i_I is supplied from v_I through a 100 kΩ, 1% resistor, the input range is 10 mV to 100 V. Finally, for a minimum op-amp B output voltage of about −4.7 V, $R_E=4\text{ V}/2\text{ mA}=2$ kΩ.

Op-amp input capacitance C_i can destabilize the log-amp. A feedback capacitor C_f forms a pole with collector resistance r_c. The loop gain is

$$GH=\frac{K}{s\tau_{bw}+1}\cdot\frac{r_c}{r_M}\cdot\frac{sr_MC_f+1}{sr_c(C_f+C_i)+1} \qquad (11.82)$$

where K and τ_{bw} are op-amp open-loop gain and bandwidth, and r_c is collector resistance:

$$r_c=(\beta+1)r_o \qquad (11.183)$$

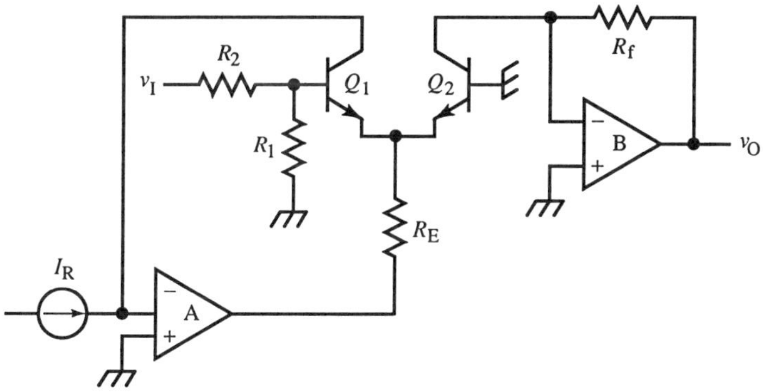

FIG. 11.55 Exponential amplifier.

Also,

$$r_M = \frac{r_e + r'_e}{\alpha} = r_m + \frac{r'_e}{\alpha} \tag{11.184}$$

The zero depends on r'_e, which varies with i_C. For stability, the poles must be separated and the zero placed for lead compensation near unity loop gain.

The inverse function, exponentiation, is achieved by modifying the log-amp (Fig. 11.55). In this exp-amp (or antilog-amp), op-amp A maintains a constant $i_{C1} = I_R$. Then

$$v_{B1} = \left(\frac{R_1}{R_1 + R_2}\right) v_I = v_{BE1} - v_{BE2} = V_T \ln\left(\frac{I_R}{i_{C2}}\right) \tag{11.185}$$

and at the output,

$$v_O = R_f i_{C2} \tag{11.186}$$

Solving for i_{C2} in (11.185) and substituting yields

$$v_O = R_f I_R \exp\left[\left(\frac{R_1}{R_1 + R_2}\right)\left(\frac{v_I}{V_T}\right)\right] \tag{11.187}$$

Temperature compensation is also required due to V_T.

Log-amps, exp-amps, and amplifiers combine to form function-generating circuits. Four op-amps and matched BJTs combine in Fig. 11.56 to form a multiplier or divider based on the relation:

$$xy = \log^{-1}(\log x + \log y) = \exp(\log x + \log y) \tag{11.188}$$

The BJTs are connected with b–e junctions in series so that

$$V_T \ln\left(\frac{i_1}{I_S}\right) + V_T \ln\left(\frac{i_2}{I_S}\right) = V_T \ln\left(\frac{i_3}{I_S}\right) + V_T \ln\left(\frac{i_4}{I_S}\right) \tag{11.189}$$

or

$$i_4 = \frac{i_1 i_2}{i_3} \tag{11.190}$$

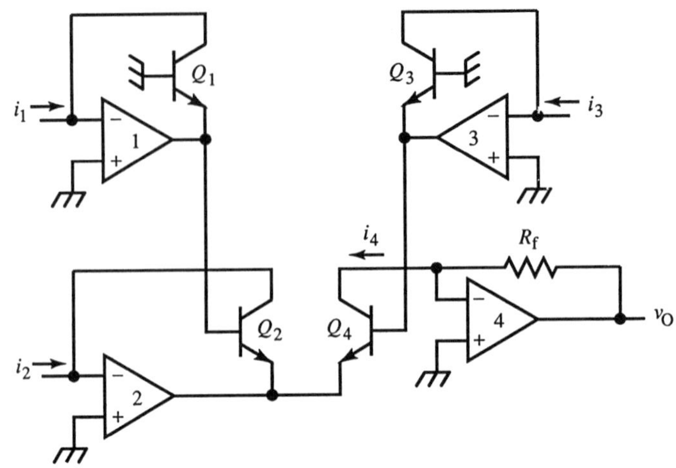

FIG. 11.56 Log-antilog multiplier: $i_4 = i_1 i_2 / i_3$.

In practice, the basic log-amp is stabilized, and voltage input and offset (for bipolar operation) are provided, as in Fig. 11.57.

V_T can be temperature compensated with a thermistor, but a more exacting approach uses two log-ratio amps and a divider. The first log-amp has inputs i_I and I_R, and the second is a temperature compensator; it has inputs of I_R and kI_R, where k is a scale factor. The log output is the quotient of the two log-ratio amp outputs:

$$\frac{V_T \ln(i_I/I_R)}{V_T \ln(kI_R/I_R)} = \frac{\ln(i_I/I_R)}{\ln k} \tag{11.191}$$

A log-ratio amplifier with output

$$v_O = V_T \ln\left(\frac{i_X}{i_Z}\right) \tag{11.192}$$

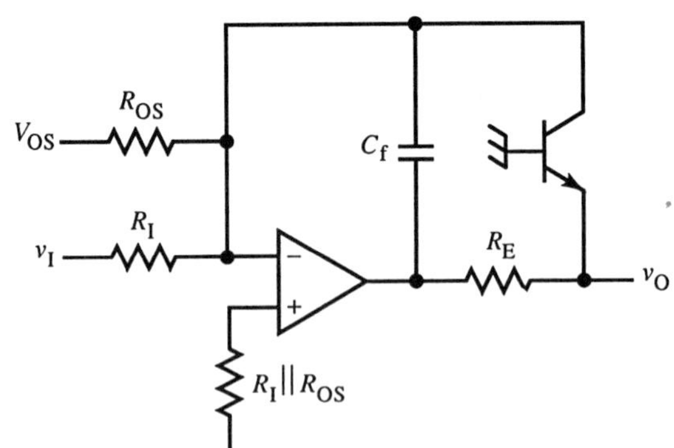

FIG. 11.57 Log-amp with frequency and bias-current compensation. An offset input adjusts scale for bipolar operation.

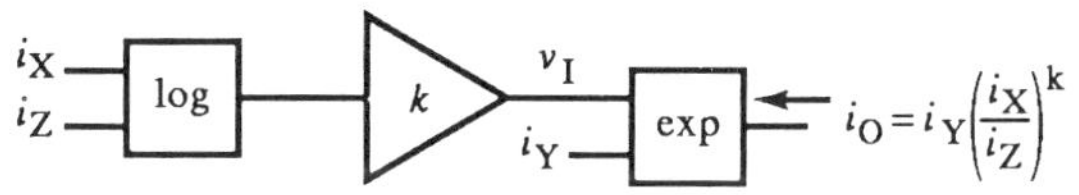

FIG. 11.58 This circuit raises a ratio of inputs to a power and multiplies by a third input, a versatile function module.

and an exp-amp with output

$$i_O = i_Y\, e^{(v_I/V_T)} = i_Y \exp\left(\frac{v_I}{V_T}\right) \tag{11.193}$$

are combined with an amplifier of gain k in Fig. 11.58. The output is

$$i_O = i_Y\left(\frac{i_X}{i_Z}\right)^k \tag{11.194}$$

For $k<1$, the amplifier is replaced with a voltage divider. This function is quite versatile since scaled powers of ratios include squares, cubes, square roots, and truncated power series of transcendental functions. For example,

$$\sin x \cong x - \frac{x^3}{6.79}, \qquad 0 \le x \le \frac{\pi}{2} \quad \pm 1.35\% \text{ error} \tag{11.195}$$

With fractional powers, a better approximation is [see *Nonlinear Circuits Handbook* (1974) in the reference list]:

$$\sin x \cong x - \frac{x^{2.827}}{6.28}, \qquad 0 \le x \le \frac{\pi}{2} \quad \pm 0.25\% \text{ error} \tag{11.196}$$

Also,

$$\tan^{-1} x \cong \frac{\pi}{2} \cdot \frac{x^{1.2125}}{1 + x^{1.2125}} \tag{11.197}$$

The inverse function $f^{-1}(w)$ of a function circuit $f(y)$ can be realized by placing $f(y)$ in the feedback path of an op-amp (Fig. 11.59). The output y is

$$y = K(x - f(y)) \tag{11.198}$$

or

$$x = \frac{y}{K} + f(y) \tag{11.199}$$

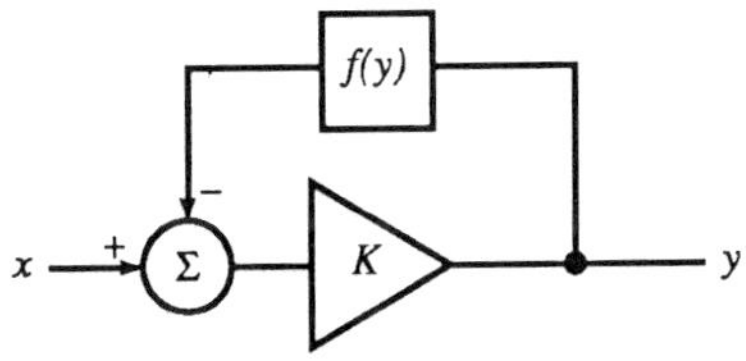

FIG. 11.59 Inversion of $f(y)$ by placing it in the feedback path of an op-amp.

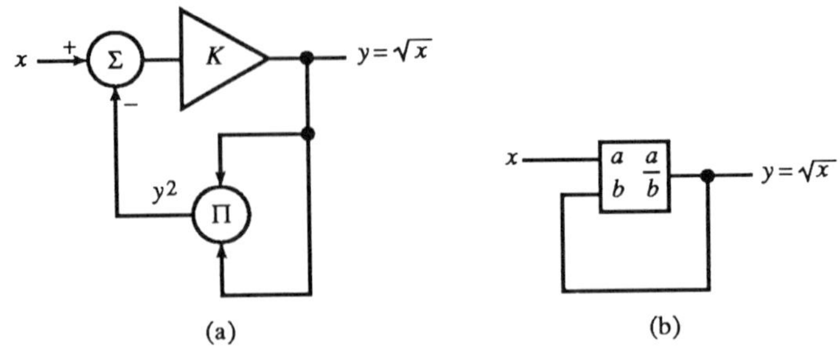

FIG. 11.60 Square-root circuits: (a) inversion of the squaring function using a multiplier; (b) implicit root-taking using a divider.

For infinite K, $x = f(y)$, or $y = f^{-1}(x)$. This is useful in generating functions implicitly. In Fig. 11.60a, a multiplier (indicated by the Π symbol) is in the feedback path of a noninverting op-amp. The circuit outputs the square root of the input:

$$y = K(x - y^2) \Rightarrow \frac{y}{K} = x - y^2 \tag{11.200}$$

For infinite K, $x = y^2$ or $y = \sqrt{x}$. Implicit function generation, as in Fig. 11.60b, is often more accurate for the same complexity of function blocks because the range over which intermediate variables must maintain accuracy is reduced.

This is especially true of the commonly used root sum of squares (rss) function, the *vector magnitude* function or Pythagorean formula:

$$y = \sqrt{\sum_{k=1}^{n} x_k^2} \tag{11.201}$$

The range of x_k^2 can be large but need not be if computed implicitly. An implicit formula is derived from (11.201). Squaring each side, adding yx_n, factoring, and dividing, we get

$$y = \frac{\sum_1^n x_k^2 + yx_n}{y + x_n} = \frac{\sum_1^{n-1} x_k^2 + x_n(y + x_n)}{y + x_n} = \frac{\sum_1^{n-1} x_k^2}{yx_n} + x_n \tag{11.202}$$

In this formulation of y, it is fed back to divide x_k^2, thus reducing the required input dynamic range.

A function describing the output of a bridge circuit with a sensor in one branch is

$$y = \frac{x}{1 + x} \Rightarrow y = x(1 - y) \tag{11.203}$$

More generally,

$$y = \frac{Ax^k}{1 + Ax^k} = (A - y)x^k \tag{11.204}$$

Finally, a two-term power series with $y(0) = 0$ is

$$y = Ax + Bx^2 = Ax + x(Bx + Cy) = \frac{Ax + Bx^2}{1 - Cx} \tag{11.205}$$

11.12 Function Generation

Function generation by log-amp circuits is based on the logarithmic nature of BJT junctions. The translinear cell of Section 10.10 is also a basis for function generation. The basic cell can be generalized by placing an arbitrary number of junctions in series in either the input or output side of the cell. To illustrate by simple example, the current squarer of Fig. 11.61 has two junctions in series on the input side. As we did with translinear-cell analysis, assume that β is infinite. The output current is derived from the circuit equations:

$$v_{BE3} = 2V_T \ln\left(\frac{i_I}{I_S}\right) \tag{11.206a}$$

$$i_O = I_S \exp\left(\frac{v_{BE3}}{V_T}\right) = \frac{i_I^2}{I_S} \tag{11.206b}$$

More generally, m series input junctions results in i_I^m output. The translinear cell is not dependent on I_S as this circuit is. In Fig. 11.62, the improved current squarer adds Q_3, biased at a constant current I_3. Its additional junction drop provides the needed correction for I_S:

$$v_{B4} = V_T \ln\left(\frac{i_O}{I_S}\right) = v_{B3} - v_{BE3} = 2V_T \ln\left(\frac{i_I}{I_S}\right) - V_T \ln\left(\frac{I_3}{I_S}\right)$$

or

$$i_O = \frac{i_I^2}{I_3} \tag{11.207}$$

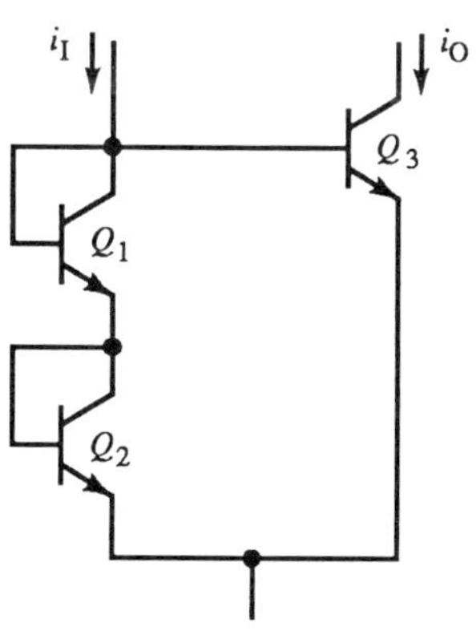

FIG. 11.61 A current squarer based on the translinear concept.

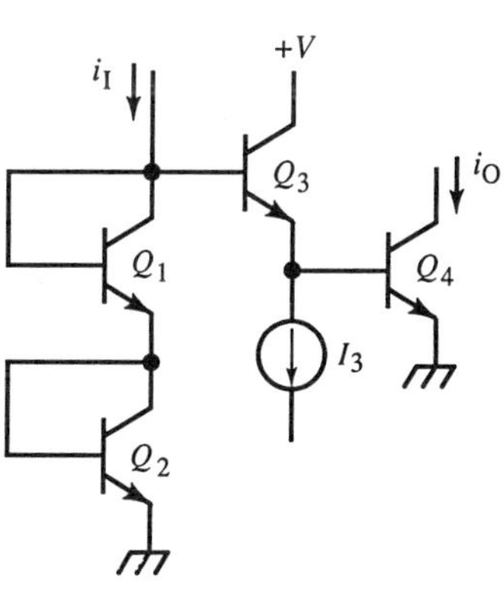

FIG. 11.62 Compensated current squarer.

This circuit concept can be extended to m input diodes and $m-1$ stages like Q_3 with currents I_1 to I_{m-1}. The output current is then

$$i_O = \frac{i_I^m}{I_1 I_2 \cdots I_{m-1}} \tag{11.208}$$

A more general scheme is shown in Fig. 11.63. Here, two stacks of diode junctions drive the BJT. The stack biased by I_R is buffered to keep BJT current separate from I_R. Applying KVL to the loop and eliminating $V_T \ln$ gives

$$\frac{i_O}{I_S} = \left(\frac{i_I}{I_S}\right)^m \left(\frac{I_S}{I_R}\right)^{m-1} \Rightarrow i_O = \frac{i_I^m}{I_R^{m-1}} \tag{11.209}$$

Fractional powers result by exchanging i_I and i_O (Fig. 11.64), with appropriate modifications. The BJT is moved to the former input string, and its first diode is now at the new input. Applying KVL and discarding $V_T \ln$ gives

$$\left(\frac{i_O}{I_S}\right)^m = \left(\frac{i_I}{I_S}\right)\left(\frac{I_R}{I_S}\right)^{m-1} \Rightarrow i_O = i_I^{1/m} \cdot I_R^{(m-1)/m} \tag{11.210}$$

A rss circuit based on a generalized translinear cell (Fig. 11.65) can be extended to three or more inputs. The voltage at the emitter, across the common diode, is

$$v_E = V_T \ln\left(\frac{i_1 + i_2}{I_S}\right) \tag{11.211}$$

Applying KVL and removing $V_T \ln$ gives

$$\begin{aligned} \frac{i_1}{I_S} + \frac{i_2}{I_S} &= \left(\frac{i_X}{I_S}\right)^2 \left(\frac{I_S}{i_1 + i_2}\right) + \left(\frac{i_Y}{I_S}\right)^2 \left(\frac{I_S}{i_1 + i_2}\right) \\ &= \frac{i_X^2 + i_Y^2}{i_1 + i_2} \cdot \frac{1}{I_S} \end{aligned} \tag{11.212}$$

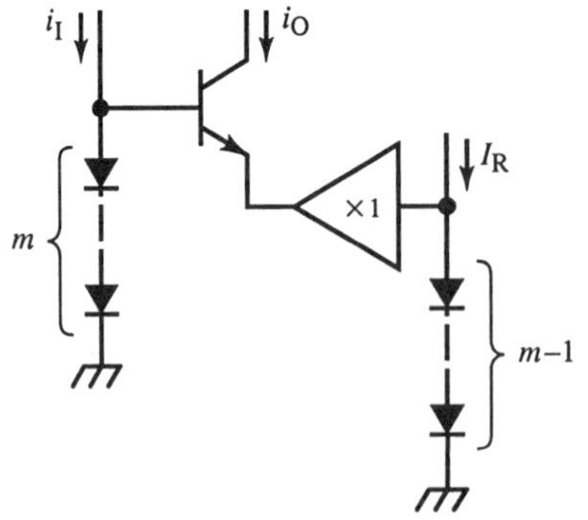

FIG. 11.63 Precision generator of an integer power.

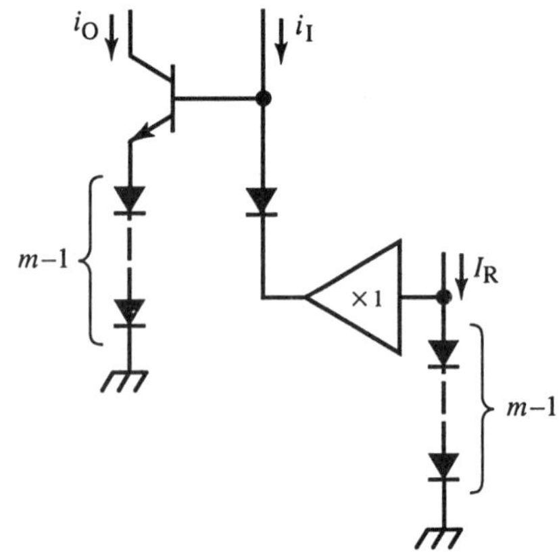

FIG. 11.64 Precision generator of an integer root.

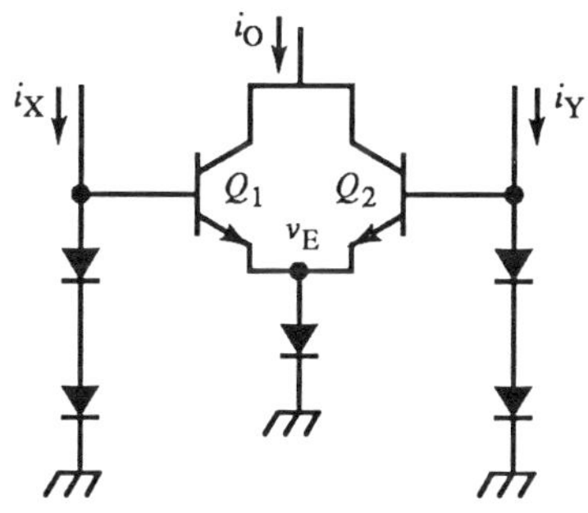

FIG. 11.65 A rss (vector magnitude or polar converter) circuit.

The output current is

$$i_O = i_1 + i_2 = \sqrt{i_X^2 + i_Y^2} \qquad (11.213)$$

A log-antilog approach to a rss circuit (Fig. 11.66) has common output current

$$i_O = i_1 + i_2 \qquad (11.214)$$

and two loops to which KVL is applied to yield

$$\left(\frac{i_X}{I_S}\right)^2 = \left(\frac{i_1}{I_S}\right)\left(\frac{i_O}{I_S}\right) \quad \text{and} \quad \left(\frac{i_Y}{I_S}\right)^2 = \left(\frac{i_2}{I_S}\right)\left(\frac{i_O}{I_S}\right) \qquad (11.215)$$

These equations reduce to

$$i_X^2 = i_1 i_O \quad \text{and} \quad i_Y^2 = i_2 i_O \qquad (11.216)$$

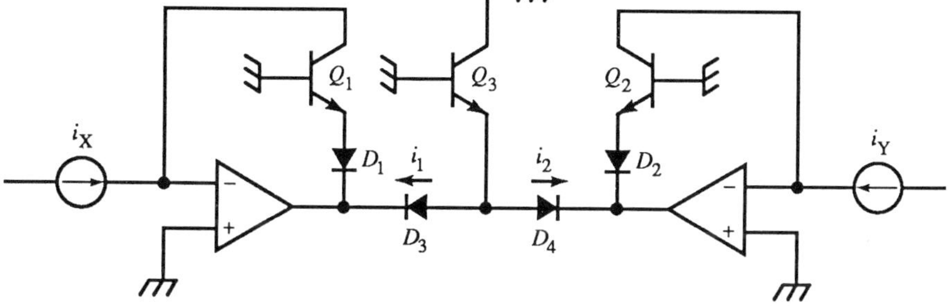

FIG. 11.66 Op-amp realization of rss circuit.

Adding them results in

$$i_O(i_1+i_2)=i_O^2=i_X^2+i_Y^2 \Rightarrow i_O=\sqrt{i_X^2+i_Y^2} \tag{11.217}$$

A rms circuit similar to these rss circuits consists of three cascaded blocks: a squarer, averager (integrator), and square-root block. The rms circuit requires the intermediate averaging function.

11.13 Triangle-Wave Generators

Function generators are a low-cost and versatile signal source. Their oscillator is a kind of multivibrator combined with a dual-slope ramp generator. This *triangle-wave generator* (TWG) outputs both triangle and square waves. TWGs are voltage-controlled oscillators (VCOs); their frequency can be accurately controlled over several decades by an input voltage, designated voltage-controlled frequency (VCF) or voltage-controlled generator (VCG) on commercial function generator (FG) instruments. This makes them useful for frequency sweeping or modulation. An external phase-locked loop (PLL) can make them accurate frequency sources as well.

An early approach to triangle-wave generation coupled a Miller ramp generator with a bistable MV or Schmitt trigger (Fig. 11.67). The bipolar square wave from the switch is integrated by the op-amp, producing a triangle wave. For a symmetric (50% duty-ratio) output, the magnitudes of the square-wave levels must be equal for equal triangle-wave slopes. The frequency can be adjusted, as shown, by varying the amplitude of the square-wave input to the integrator. Like ramp generators using an op-amp integrator, this approach is both speed- and precision-limited. For a VCF range of three decades and a full-scale square-wave amplitude of 10 V, at zero scale (the low-frequency end) the square-wave amplitude is 10 mV. For 1% waveform symmetry, the integrator input offset error must be less than 100 μV.

Speed-wise, for a high-frequency limit of 1 MHz, triangle-wave amplitude of 10 V and triangle nonlinearity of less than 1%, the slope magnitude is

$$\frac{dv_{TW}}{dt}=\frac{V_{SQ}}{RC}=\frac{V_{TW}}{T/4}=\frac{10\text{ V}}{250\text{ ns}} \tag{11.218}$$

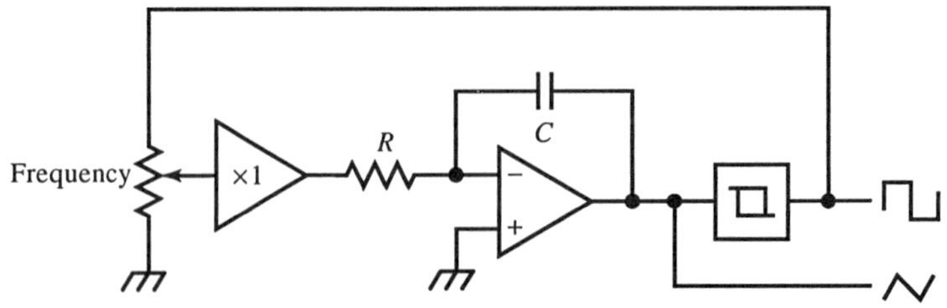

FIG. 11.67 Low-performance triangle-wave generator (TWG).

where V_{SQ} and V_{TW} are square-wave and triangle-wave amplitudes and T their period. Since V_{SQ} is also 10 V, then $RC = 250$ ns. The integrator op-amp gain requirement for the specified nonlinearity is calculated from (11.169):

$$\text{nonlinearity} = \eta = \frac{T/2}{2(1+K)RC} \Rightarrow K = \frac{T}{\eta 4RC} - 1 \qquad (11.219)$$

or K must be at least 999 at 1 MHz, an almost 1 GHz unity-gain bandwidth.

To achieve better performance, current-source ramp generation is used with a bipolar current source (Fig. 11.68a). Two matched current sources are

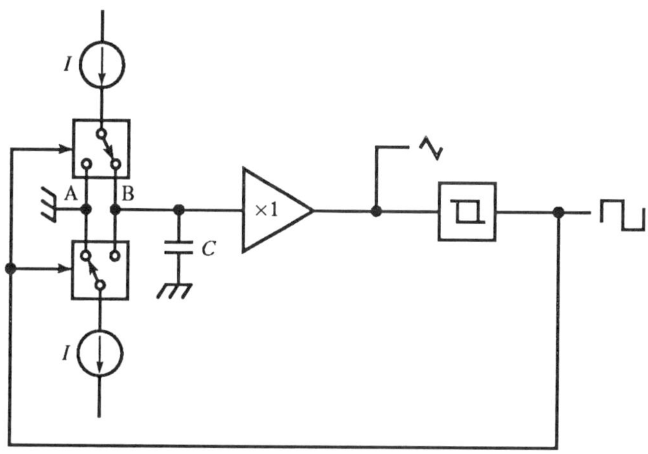

(a)

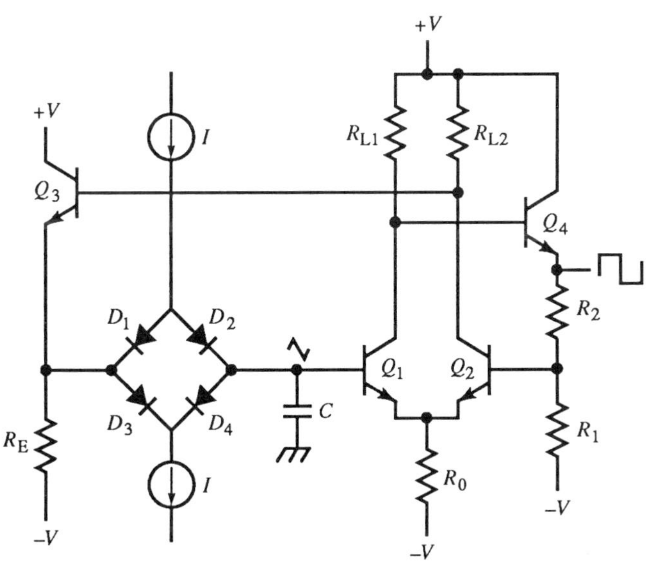

(b)

FIG. 11.68 Balanced current-source TWG block diagram (a) and high-speed realization (b).

used to achieve bipolar charging of C. The positive (source) and negative (sink) currents are alternately switched into C each half-cycle. This scheme is much faster and more precise than the first one but requires matching of current sources for time symmetry and symmetric hysteresis thresholds for voltage symmetry.

A more concrete circuit realization (Fig. 11.68b) is a fast TWG, capable of over 100 MHz oscillation. The hysteresis-switch and current-switch driver Q_1–Q_4 can be implemented with a single ECL NOR gate. The diode bridge is switched by Q_3 and is limited mainly by diode shunt capacitance. Sometimes a Faraday shield is placed between the left and right halves of the diode bridge to decouple the square-wave node v_{E3} from the triangle-wave node at C. Since the input resistance of the Schmitt trigger is not high, appreciable nonlinearity results. If we add a good buffer, its additional delay slows the loop. Speed and precision are, as usual, in conflict.

A variation on the bipolar current supply uses only one current source for charging C. This eliminates the matching problem between sources. The Signetics NE565 FG and NE566 PLL both have TWGs designed this way (Fig. 11.69). Q_1 and Q_2 form a current source or V/I converter with input v_F, the VCF input. It generates I through R, the timing resistor. Switching of I is

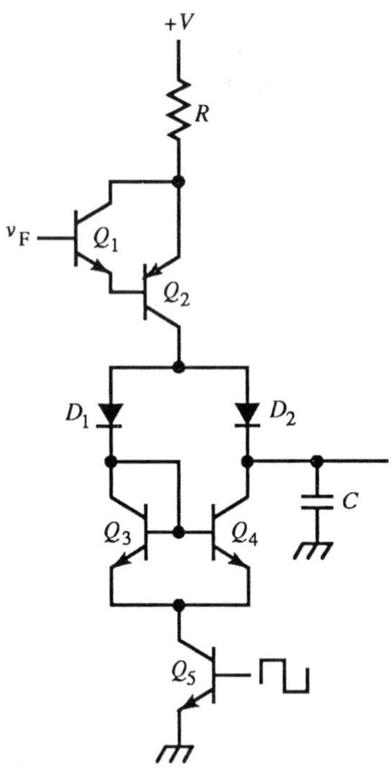

FIG. 11.69 Balanced-current topology in the Signetics NE565 and NE566 uses current mirror to generate negative current.

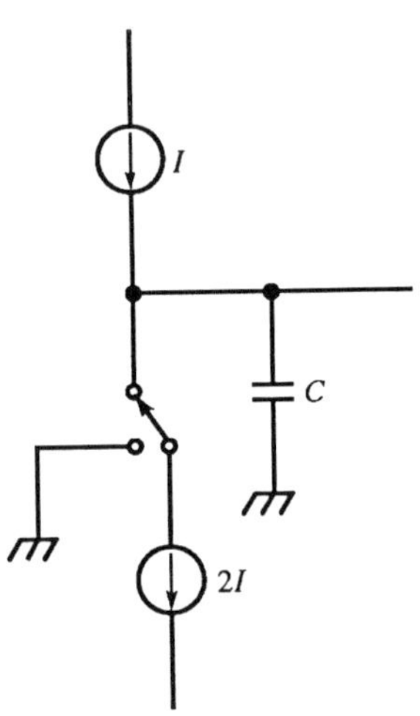

FIG. 11.70 I-$2I$ triangle-generator topology has only one current switch.

controlled by Q_5. When it is off, D_2 conducts I, and C charges (positive). When Q_5 is on, $v_{C3} < v_{C4}$, and D_1 conducts I through Q_3, where it is replicated in Q_4 as $-I$. Thus, C is discharged with the same magnitude of current as it is charged. Symmetry depends on current-mirror matching.

It is not necessary to switch both current sources. To simplify switching, the I-$2I$ scheme is used (Fig. 11.70). Only one current switch is required, but now the I and $2I$ sources must be matched at an accurate ratio of 2.

A more recent approach to better symmetry is to control one current source based on the triangle-wave slope generated by the other. In Fig. 11.68a, instead of connecting node A to ground, it charges another capacitor of value C. Its triangle-wave is inverted relative to the one at node B. The two waveforms are then summed. Ideally, the sum is zero, but any difference is an error voltage that is applied to one of the current sources to correct the slope of the waveform it is generating to match the other. This scheme increases symmetry enough at low currents to extend the VCF range an extra decade.

A conceptually similar approach is taken in the Exar XR2206 FG IC, shown in simplified form in Fig. 11.71. The triangle-wave is developed differentially across C and requires a diff-amp pick-off. Q_1–Q_4 are the timing-current

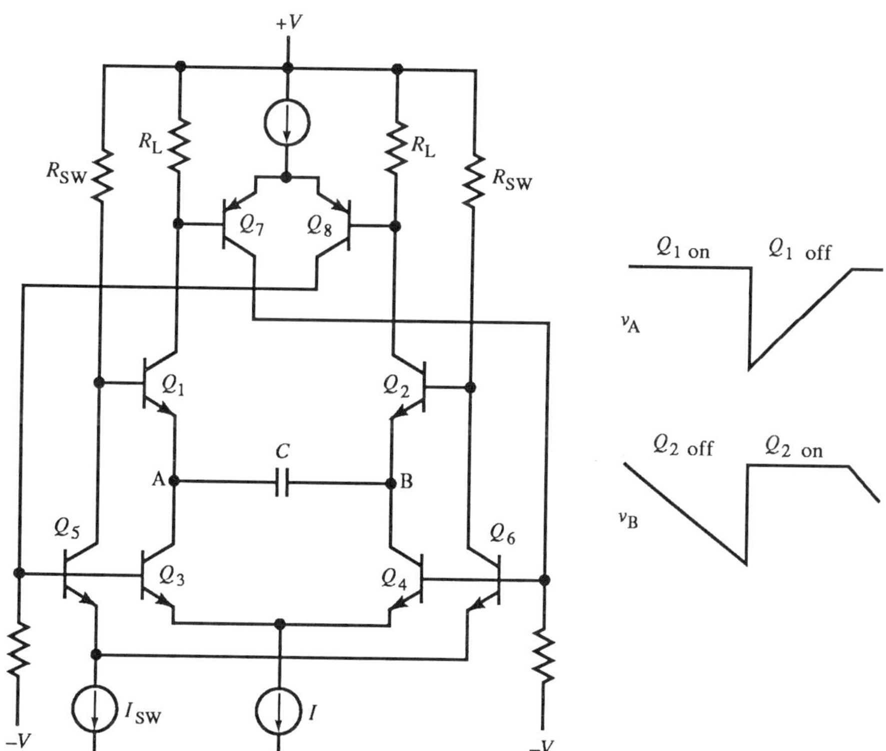

FIG. 11.71 Differential TWG with current-switch BJTs as part of Schmitt trigger. Only one timing current source is required.

switches, but there is only one timing-current source. Q_1, Q_2, Q_7, and Q_8 are the hysteresis switch. The *b–e* junctions of Q_1–Q_2 sense the waveform and switch on when v_E is reduced sufficiently. On one half-cycle, Q_1 and Q_4 conduct; on the other half-cycle, Q_2 and Q_3 conduct. I_{SW} sets the hysteresis thresholds. If the R_{SW} are matched, the levels are symmetric. A disadvantage in this scheme is that the switching voltage is added to the capacitor ramp voltage at nodes A and B, as shown. It is difficult for a diff-amp to common-mode reject these fast switching edges, and some "glitches" appear in the triangle-wave output.

The performance of a TWG loop depends on the subsystems within the loop. The timing capacitor must be of sufficient quality (plastic), and for multiple frequency ranges, a matched set is usually required. The triangle-wave buffer must be fast, have low input-voltage offset and bias current, and high input resistance. The current switches must be fast and have low current leakage when off. Transistors are generally superior to diodes in both leakage and switching characteristics and are used as switches as shown in the TWG loop of Fig. 11.72. A minimum of two switches is required in a balanced two-source scheme, as shown. The switching scheme here consists of complementary diff-amps. Two BJTs switch the current-switch BJTs.

The timing-current generators also must be capable of operating accurately over as many decades of current as the VCF range because output frequency is proportional to timing current. Complementary V/I converters, such as in Fig. 11.16, are commonly used to supply both polarities of current. These

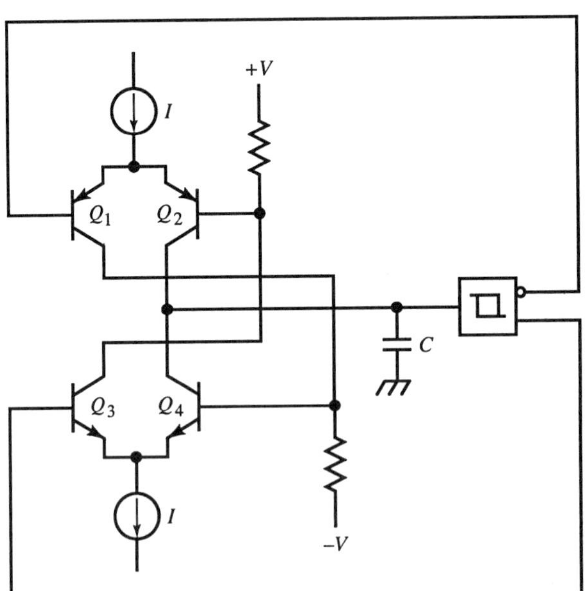

FIG. 11.72 High-performance BJT current switch with diff-amp current switching.

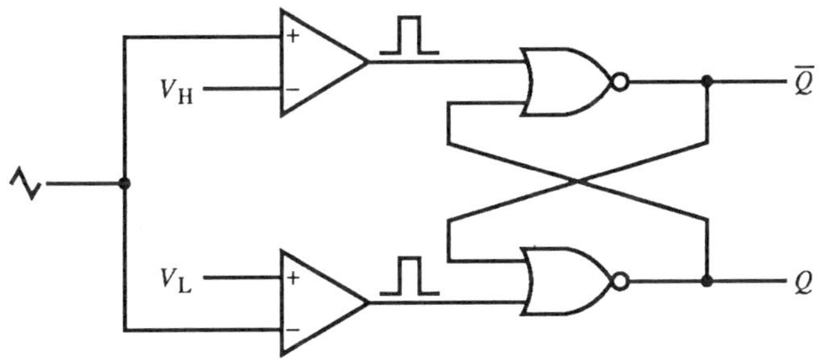

FIG. 11.73 Typical hysteresis switch: two comparators and RS flop.

converters must be driven by precision circuitry that establishes symmetric voltages at their inputs. These circuits need only the bandwidth required of the FM VCF signal. Sometimes BJTs are used as the current-source transistors in V/I converters instead of FETs for their higher output impedance. When a source is supplying the timing current through a diode bridge, the triangle-wave is at its output. Whatever parasitic output capacitance the current source has affects the timing. The additional isolation offered by a BJT switch over a diode minimizes this problem.

The final subsystem is the hysteresis switch. Schmitt triggers of the regenerative MV kind are sometimes used, but their thresholds are often not accurate enough. The dual-level comparator circuit of Fig. 11.73 is more accurate. The comparator outputs are asserted (high) only momentarily as the ramp crosses a threshold at a peak. The RS flop, made of NOR gates, is set to alternating states. This is not the fastest circuit since it consists of several stages of processing and can be used up to about 20 MHz.

Not only must the thresholds be symmetric, a hysteresis switch must have little delay because the total loop delay determines the maximum operating frequency. As loop delay time t_d becomes an appreciable fraction of the period T, the ramp increases in magnitude beyond the threshold before the slope changes. The triangle-wave amplitude thus begins to increase with frequency.

In Fig. 11.74, the effect is graphed. V_M is the low-frequency triangle amplitude and hysteresis threshold. But due to delay, it increases as $v_M(f)$. For a ramp slope of m,

$$v_M = V_M + mt_d \tag{11.220}$$

where the slope is

$$m = \frac{v_M}{T/4} = \frac{4v_M}{T} \tag{11.221}$$

Substituting into (11.220), we get

$$v_M = V_M + \frac{4v_M}{T} \cdot t_d = \frac{V_M}{1-(4t_d/T)} = \frac{V_M}{1-(\omega/\omega_d)}, \qquad \omega_d = \frac{\pi}{2t_d} \tag{11.222}$$

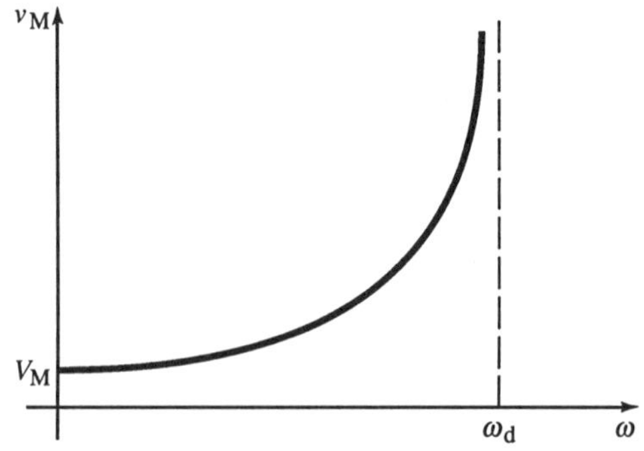

FIG. 11.74 Triangle-wave amplitude versus frequency for TWG loop delay of $\omega_d = \pi/2t_d$. V_M is the comparator threshold and low-frequency amplitude.

With the approximation,

$$\frac{1}{1-x} \cong 1 + x, \qquad x \ll 1 \tag{11.223}$$

then we have

$$v_M \cong V_M\left(1 + \frac{\omega}{\omega_d}\right), \qquad \omega \ll \omega_d \tag{11.224}$$

This equation also results from assuming the low-frequency slope of

$$\text{low-frequency } m = \frac{4V_M}{T} \tag{11.225}$$

in (11.220). The period increases over its low-frequency value by $4t_d$. From (11.224), ω_d is like a zero break frequency for v_M; but from the exact expression (11.222), v_M is vertically asymptotic at ω_d. As ω approaches ω_d, t_d dominates T. A t_d of 10 ns breaks at

$$f_d = \frac{1}{4t_d} \tag{11.226}$$

or 25 MHz.

Because $v_M(\omega)$ is nonlinear, a nonlinear compensator is required to make $v_M(\omega) = V_M$. Since transfer-function compensation is based on linear analysis, another approach is required. This is an adaptive control problem since a parameter V_M must be varied to achieve ideal compensation. The amplitude error is

$$\Delta v_M = v_M - V_M = \frac{V_M(\omega/\omega_d)}{1-(\omega/\omega_d)} = v_M\left(\frac{\omega}{\omega_d}\right) = v_M\left(\frac{f}{f_d}\right) \tag{11.227}$$

If V_M is replaced by $V_M - \Delta v_M$ or if the triangle wave itself, v_{TW}, is modified to $v_{TW} + \Delta v_M$, then v_M is a constant V_M. Since the TWG frequency is determined by the VCF voltage v_F, we can derive frequency information from it to set the comparator thresholds. The slope of v_{TW} is I/C. If I is generated by a V/I converter by VCF voltage v_F across timing resistor R, then,

$$m = \frac{v_F}{RC} \tag{11.228}$$

For $\omega \ll \omega_d$, (11.225) applies. Equating it to (11.228), we get

$$f = \frac{v_F}{4RCV_M} \tag{11.229}$$

and the adaptive V_M is

$$\text{adaptive } V_M \to V_M - \Delta v_M = V_M - \frac{V_M}{(V_M/v_F)(RC/t_d) - 1} \cong V_M - v_F\left(\frac{t_d}{RC}\right) \tag{11.230}$$

This is an instance of *model-reference adaptive control* and is based on *a priori* knowledge of the circuit model, (11.229). When R or C are switched to change frequency ranges, the model changes. Switching of RC in (11.230) is required for the model to represent the circuit.

Another approach to TWG-loop compensation is to place in the loop a time advance to cancel t_d. In the s-domain, the delay is e^{-st_d} and a compensating advance is e^{st_d}. Since time advances are not realizable in physical (causal) systems, this form of compensation can only be approximated. The power-series expansion of the time delay is

$$e^{-st_d} = 1 - st_d + \frac{s^2 t_d^2}{2} - \frac{s^3 t_d^3}{6} + \cdots \cong 1 - st_d \cong \frac{1}{st_d + 1} \tag{11.231}$$

The single-pole approximation to a time delay suggests that a phase-lead compensator with zero at $1/t_d$ and higher-frequency pole provides approximate compensation.

A rational approximation to the time advance is the Padé approximation. A first-order approximation is

$$e^{st_d} \cong \frac{st_d/2 + 1}{-st_d/2 + 1} \tag{11.232}$$

It is a nonminimum-phase transfer function. A second-order Padé approximation is

$$e^{st_d} \cong \frac{s^2 t_d^2 + 6st_d + 12}{s^2 t_d^2 - 6st_d + 12} \tag{11.233}$$

It can be factored, which results in

$$z_{1,2} = \left(\frac{1}{t_d}\right)(3 \pm j\sqrt{3}), \qquad p_{1,2} = \left(\frac{1}{t_d}\right)(-3 \pm j\sqrt{3}) \tag{11.234}$$

The pole-zero placement is symmetric about the origin and represents an all-pass filter. This is consistent with the delay function since it effects only a shift in time with no amplitude change. The pole and zero pairs have an angle of

$$\phi = \cos^{-1}\frac{\sqrt{3}}{3} \cong 55°$$

Besides being rational approximations, these delay compensators are linear whereas the circuit is nonlinear. Frequency-domain analysis assumes sinusoids, not triangle waves. For $\omega \ll \omega_d$, the nonlinearity is not too severe and can be approximated as linear, and the phase error in the rational approximations is minimal.

The versatility of the FG is partly due to its wide VCF range. For frequency-response or Bode magnitude plots, a logarithmically swept VCF directly produces a log-frequency plot on an oscilloscope display. A simple way of producing a logarithmic sweep is shown in Fig. 11.75. On each cycle of the FG output, the MMV is triggered. It gates on a current source for a fixed time, transferring a fixed charge to the VCF capacitor C_F. This increments v_F by a fixed amount, causing the output frequency to increment. As the frequency increases, the rate of increase of v_F increases along with it and is exponential. The resulting frequency sweep, when displayed, is stepwise logarithmic.

Quantitatively, let the MMV time-out be $\tau < T = 1/f$. During a given time interval Δt, n periods of the output occur. Then

$$\frac{\Delta v_F}{\Delta t} = \frac{n(I_F\tau/C_F)}{nT} = f\left(\frac{I_F\tau}{C_F}\right) \tag{11.235}$$

In the limit, for small Δv_F or infinite n

$$\lim_{n\to\infty}\frac{\Delta v_F}{\Delta t} = \frac{dv_F}{dt} = \left(\frac{I_F\tau}{C_F}\right)f \tag{11.236}$$

As a voltage-to-frequency (V/F) converter (VFC),

$$f = k_F v_F \tag{11.237}$$

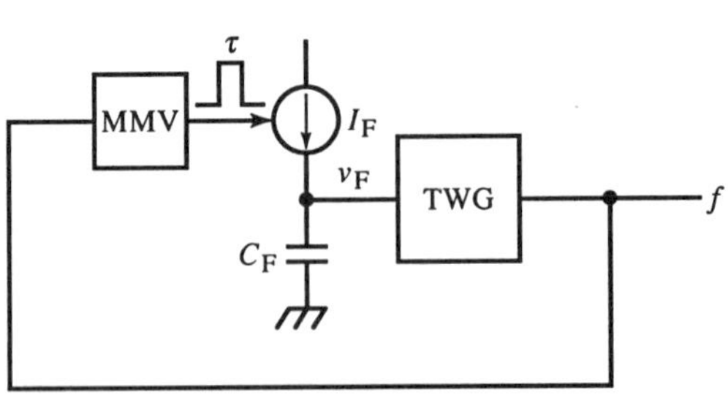

FIG. 11.75 Log-sweep FG uses output frequency of FG to change VCF input voltage proportionally.

Substituting into (11.236) and solving gives

$$v_F = v_F(0)\, e^{(I_F\tau/C_F)k_F t} \tag{11.238}$$

The sweep-rate constant is $(I_F\tau/C_F)k_F$, and the sweep time is

$$t_{SWP} = \frac{\ln(f/f(0))}{(I_F\tau/C_F)k_F} \tag{11.239}$$

where $f(0)$ is the starting frequency and f the ending frequency. The VFC constant k_F can be found from the TWG parameters. For triangle-wave amplitude of V_M, the slope is

$$\frac{V_M}{T/2} = \frac{I}{C} = \frac{v_F/R}{C} \Rightarrow f = \left(\frac{1}{2RC}\cdot\frac{1}{V_M}\right)v_F \Rightarrow k_F = \frac{1}{2RC}\cdot\frac{1}{V_M} \tag{11.240}$$

Besides VCF, the TWG current sources can be individually controlled for variable symmetry or duty-ratio. A *voltage-controlled symmetry* (VCS) input allows pulse-width modulation (PWM) of the output waveforms.

A third waveform available on FGs is the sine wave. This function is not generated by the TWG loop but by a sine converter circuit. Most commonly, this is a multiple diode clamp that performs a piecewise-linear waveform shaping of the triangle wave. The better shapers use diode bridges for symmetry and diode drift cancellation. The initial part of the triangle and sine waves has the same slope (Fig. 11.76). For sine amplitude of V_A, the slopes are equated:

$$\frac{V_M}{T/4} = V_A\omega \Rightarrow \frac{V_M}{V_A} = \frac{\pi}{2} \cong 1.59 \tag{11.241}$$

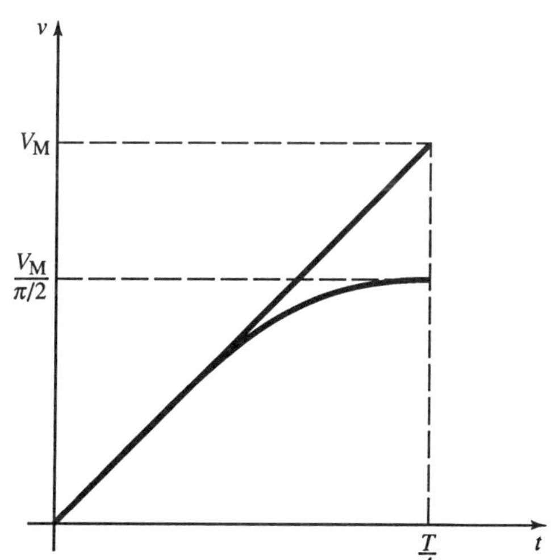

FIG. 11.76 Triangle-to-sine conversion amplitude ratio for equal slopes of triangle and sine wave at $v = 0$.

The most significant anomaly of these sine-shapers is the triangle-wave peaks in the sinusoid. At the peaks, the derivative of the triangle wave is discontinuous and is difficult to remove entirely from the sine wave. Typically, three-break-point sine-shapers produce less than 0.25% THD in the audio range.

Another approach is to use the hyperbolic tangent function of the BJT diff-amp as an approximate sine converter. This approach has somewhat more distortion than the multiple-clamp circuit. MOS diff-amps have a quadratic transfer function and also approximate a sine output with adjusted parameters.

A more elegant approach is to return to the translinear cell concept. By stacking diodes with diff-amp pick-offs, arbitrary power series expansions can be realized, as in Fig. 11.77 for three terms of a sine expansion. The Taylor-series expansion is

$$\sin \frac{i_1}{i_2} = \sin x = x - \frac{x^3}{6} + \frac{x^5}{240} \tag{11.242}$$

The truncation error is less than 0.07%. For two terms, the error is 0.14%, still favorable relative to the other approaches. The sine converter in Fig. 11.77 operates over an input range of x from 0.5 to 2.

The two-term sine shaper of Fig. 11.78, with a topology similar to a Gilbert gain cell, passes the linear term through Q_5–Q_6 while Q_1–Q_2 develop the

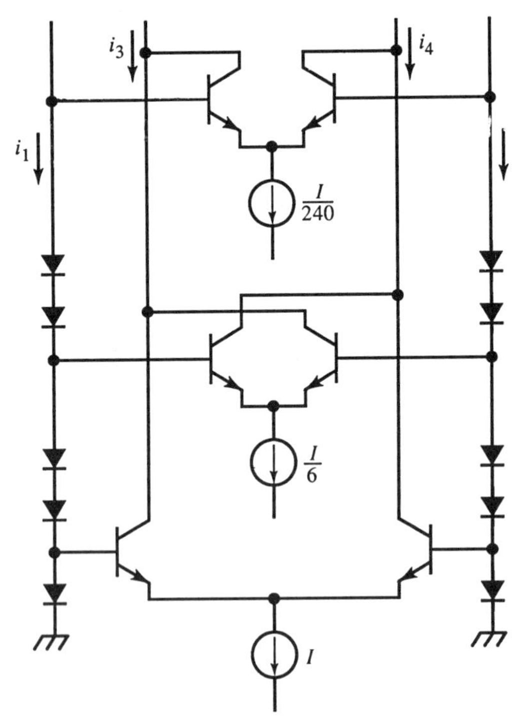

FIG. 11.77 Triangle-to-sine converter based on Gilbert gain cell topology.

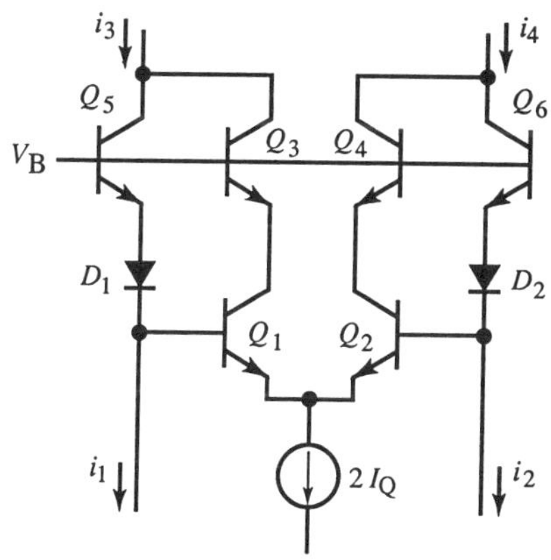

FIG. 11.78 Two-term sine power-series generator based on generalized translinear cell.

second term of the output. Together,

$$i_O = i_I\left[1 - \frac{2I_Q/I_I}{1+(i_I/2I_I)^2}\right] \qquad (11.243)$$

where

$$i_I = i_1 - i_2, \qquad i_O = i_3 - i_4, \qquad 2I_I = i_1 + i_2, \qquad 2I_Q = i_{C1} + i_{C2} \quad (11.244)$$

The ratio $2I_Q/I_I$ is adjusted for minimum THD of less than 0.1% when the ratio is about 1.5.

11.14 Precision Rectifiers or Absolute-Value Circuits

A full-wave rectifier performs the absolute value function:

$$|x| = \begin{cases} x, & x \geq 0 \\ -x, & x < 0 \end{cases} \qquad (11.245)$$

A half-wave rectifier performs the function

$$f(x) = \begin{cases} x, & x \geq 0 \\ 0, & x < 0 \end{cases} \qquad (11.246)$$

Half-wave rectification is mainly applied as an ac-to-dc conversion technique in low-power power supplies or AM demodulators. Rectifiers can be designed to output either polarity; the positive outputs of (11.245) or (11.246) could instead be $-|x|$ or $-f(x)$. For signal processing involving mathematical functions, $|x|$ is more commonly needed.

A single diode can half-wave rectify (Fig. 11.79a). Two diodes are used for full-wave rectification with a ground-based differential input (b) and a

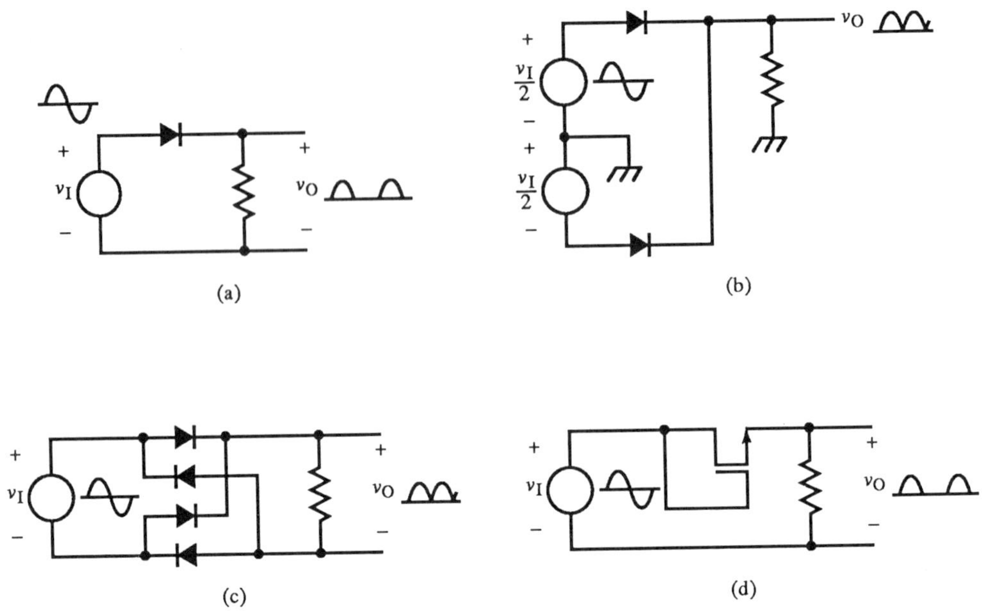

FIG. 11.79 Rectifier circuits: (a) diode half-wave; (b) diode full-wave with ground-referenced differential input; (c) diode bridge full-wave with floating input; (d) MOSFET half-wave synchronous rectifier.

diode bridge for a floating input (c). A more recent addition to rectifier circuits is the *synchronous rectifier*. It uses active devices, usually MOSFETs, that are switched by the input itself. These circuits are associated with power conversion and are commonly found in power supplies, both linear and switched. The precision clamps of Fig. 11.33 are half-wave rectifiers.

Absolute-value circuits can be designed many ways. The common constraint is that the gain magnitude for $v_I < 0$ be the same as for $v_I \geq 0$, so that $-A_{v-} = A_{v+}$. The most common of these circuits is in Fig. 11.80. It has similar frequency responses for positive and for negative inputs. On the positive half-cycle, D_1 conducts, and op-amp A operates in the inverting configuration.

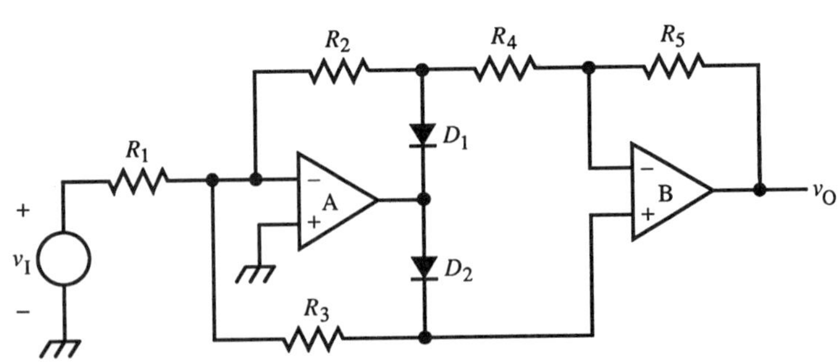

FIG. 11.80 Standard op-amp absolute-value circuit.

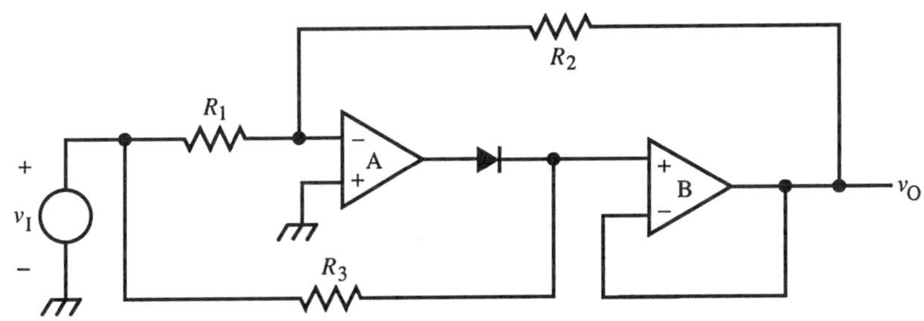

FIG. 11.81 Absolute-value circuit with one diode.

Op-amp B converts this output to a single-ended signal v_O. The gain expressions are

$$A_{v+} = \left(-\frac{R_2}{R_1}\right)\left(-\frac{R_5}{R_4}\right), \qquad A_{v-} = \left(-\frac{R_3 \parallel (R_2 + R_4)}{R_1}\right)\left(\frac{R_5}{R_2 + R_4} + 1\right) \quad (11.247)$$

Equating A_{v+} to $-A_{v-}$, the constraint on resistor values is

$$R_3 = R_5, \qquad R_2 = R_4 \quad (11.248)$$

Since R_1 is unconstrained, it can set the gain. Op-amp B must have a fast large-signal response to follow the discontinuities in the waveform it amplifies.

The circuit in Fig. 11.81 uses only one diode. Op-amp A functions for negative inputs as the first stage of gain in the forward path, cascaded with op-amp B, a ×1 buffer. The gain is

$$A_{v-} = -\frac{R_2}{R_1}$$

For positive inputs, op-amp A is disconnected from B, and v_I is applied directly to the input of B through R_3. The value of R_3 does not affect gain; its purpose is to limit current from op-amp A when the diode conducts. Thus, positive inputs bypass A and are merely buffered by B with a gain of $A_{v+} = 1$. The constraint is simply that $A_{v-} = -1$:

$$R_1 = R_2 \quad (11.249)$$

The positive-gain path involves only one op-amp and has a faster response than the negative-gain path for negative inputs.

A similar kind of circuit with only one op-amp (Fig. 11.82) requires matched diodes D_1 and D_2 and current sources as a trade off for fewer components. In IC form, this is attractive. The positive-gain path is through D_1 and the op-amp, with a gain of unity. D_2 also conducts, causing the inverting op-amp input to also follow the input signal. For negative inputs, D_1 is off, D_3 is on to satisfy the current source, and the op-amp inverts with a gain of $-R_2/R_1$. D_2 conducts all the time to balance D_1. The constraint is (11.249).

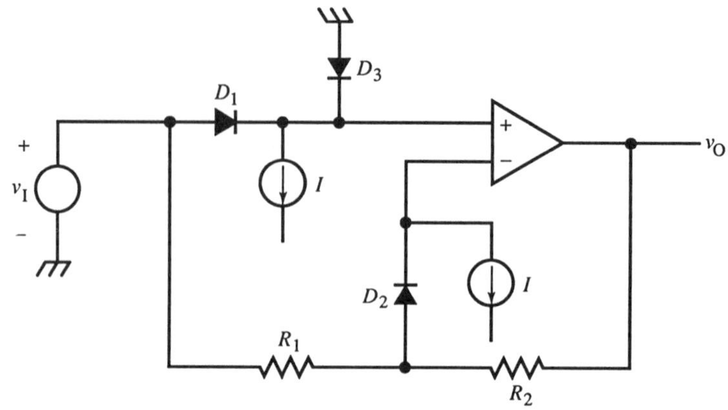

FIG. 11.82 Absolute-value circuit with one op-amp and matched diodes and current sources.

Another variation on this theme, which allows adjustment of gain at the output by means of R_4, is that of Fig. 11.83. The gains are

$$A_{v+} = \frac{R_4}{R_2} + 1, \qquad A_{v-} = -\left(\frac{R_4 + R_2}{R_1}\right) \tag{11.250}$$

and (11.249) again applies. Since R_4 is not constrained, it can be used to set the gain. D_2 clamps op-amp A input to virtual ground for positive signals so that R_1 does not affect A_{v+}.

The absolute-value circuits of Figs. 11.84 and 11.85 have high-impedance inputs. They use the same number of components, but in Fig. 11.84 is constrained to unity gain. For it, $A_{v+} = 1$ and

$$A_{v-} = \left(\frac{R_2}{R_1} + 1\right)\left(-\frac{R_4}{R_3}\right) + \left(\frac{R_4}{R_3} + 1\right) = 1 - \frac{R_2 R_4}{R_1 R_3} \tag{11.251}$$

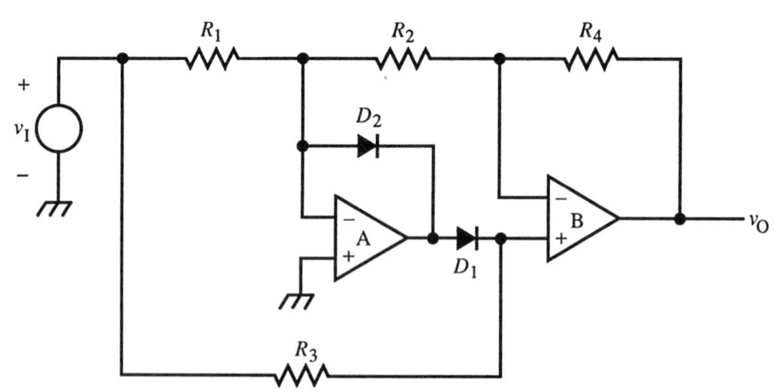

FIG. 11.83 Absolute-value circuit with gain adjustment at output.

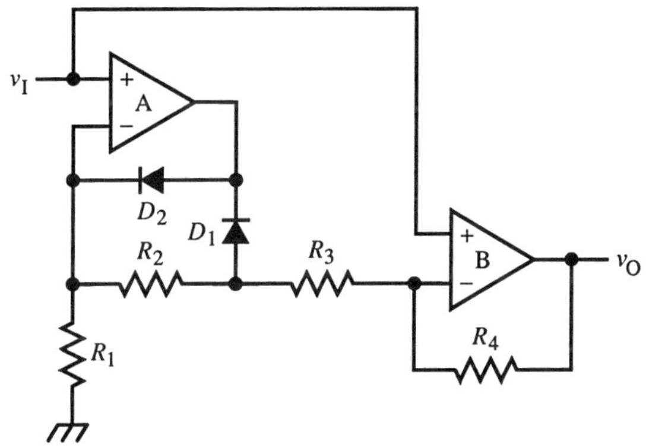

FIG. 11.84 High-impedance input absolute-value circuit limited to unity gain.

The constraint is

$$R_4 = 2\frac{R_1 R_3}{R_2} \tag{11.252}$$

One combination of resistors satisfying (11.252) is

$$R_1 = R_2 = 2R_3 = R_4 \tag{11.253}$$

For Fig. 11.85,

$$A_{v+} = \frac{R_2 + R_3}{R_1} + 1, \qquad A_{v-} = -\frac{R_3}{R_2} \tag{11.254}$$

with constraint

$$\frac{R_1}{R_2} = \frac{(R_3/R_2 + 1)}{(R_3/R_2 - 1)} \tag{11.255}$$

Implicit in (11.255) is the additional constraint that $R_3 > R_2$.

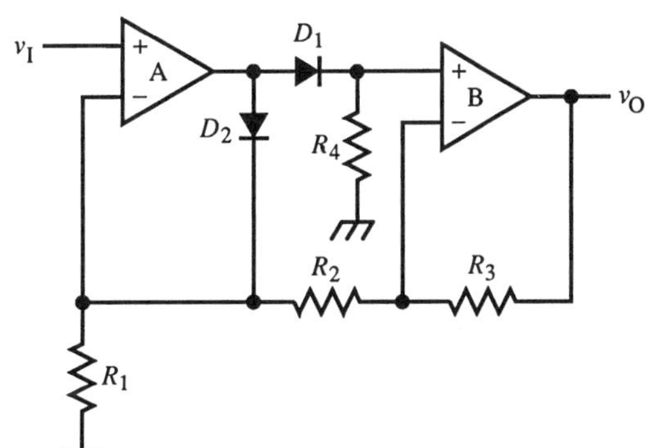

FIG. 11.85 High-impedance input absolute-value circuit with arbitrary gain.

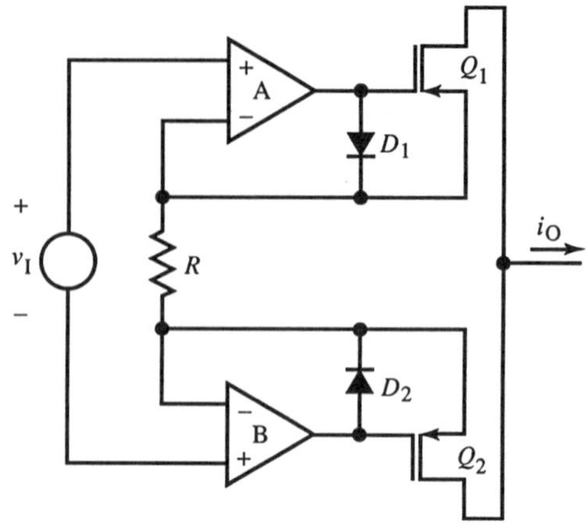

FIG. 11.86 Absolute-value transconductance circuit using FETs.

Not all absolute-value circuits must be designed by matching gains of positive- and negative-gain paths. The circuit of Fig. 11.86 resembles an instrumentation amplifier, but its output current is unipolar. For $v_I > 0$, D_1 and Q_2 conduct. D_1 reverse-biases Q_1. The negative-gain path is through D_2 and Q_1. The circuit is symmetric and has output current

$$i_O = \frac{|v_I|}{R} \tag{11.256}$$

Finally, Fig. 11.87 shows a circuit with a rectifier section similar to Fig. 11.80 but with gain set by R_5. The positive-gain path is through A, then B; the negative-gain path is through R_3 and B. For negative inputs, D_2 conducts through R_6, a current-limiting resistor, forcing the noninverting input to virtual ground. R_4 and R_2 then shunt the input of B, but this does not affect the gain.

Some op-amps have inherently unipolar outputs and can be fashioned into precision rectifiers. When current mirrors, CMOS inverters, and other elemental circuits are also used, the collection of absolute-value circuits becomes extensive.

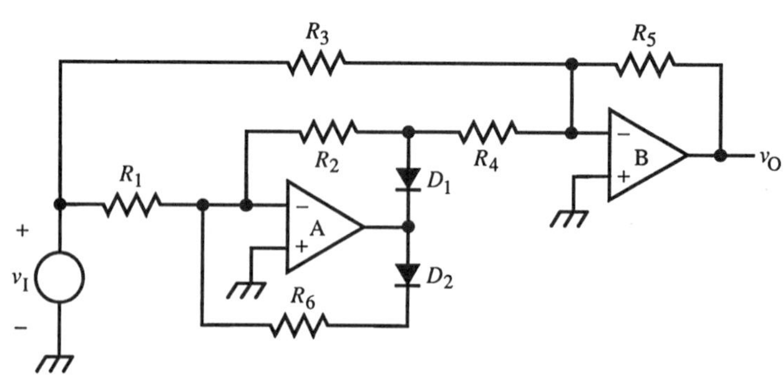

FIG. 11.87 Absolute-value circuit.

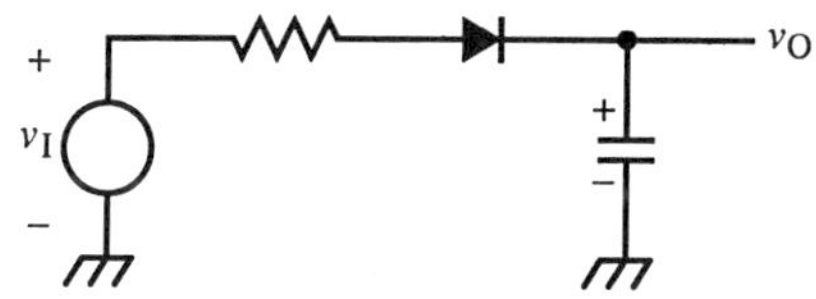

FIG. 11.88 Diode peak detector.

11.15 Peak Detectors

Finally, we consider a class of circuits that extract the extrema of waveforms. These *peak detectors* are essentially rectifier circuits with a memory. A simple example is the rectifier–filter combination of power supplies (Fig. 11.88). The rectifier conducts to charge the filter capacitor whenever its output exceeds the capacitor voltage. The capacitor charges to the peak input voltage.

Fast, simple detectors of maxima (positive peaks) and minima (negative peaks) are shown in Fig. 11.89, using a CC that charges C. Slower, more precise peak detectors (Fig. 11.90) are reset through FETs that discharge the capacitors. Ideally, after C has been charged to the peak voltage, it retains its charge indefinitely. (In this respect, peak detectors are similar to S/H circuits of Chapter 12.) The capacitor must have low leakage (high insulation resistance) to minimize the discharge rate. It must also have low dielectric absorption so that when reset, it retains 0 V until recharged by the input signal. These requirements suggest a plastic capacitor.

This applies also to its load, including the reset FET, and to the op-amp bias current. Load leakage is minimized by using a high-impedance buffer with low input bias current. When it drives the input op-amp, as in Fig. 11.91, its offset-voltage error is compensated by being in the loop. The feedback loop also effectively reduces the time constant $r_{out}C$ by the loop gain, where r_{out} is the resistance in series with the diode. Fast peaks are thus detected more

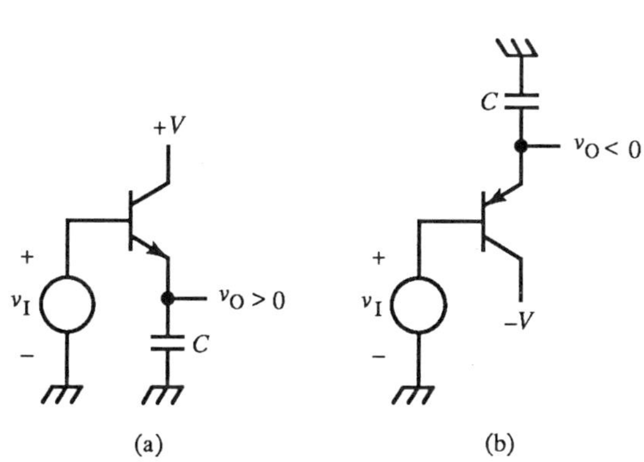

FIG. 11.89 Fast peak detectors using CC: (a) positive and (b) negative.

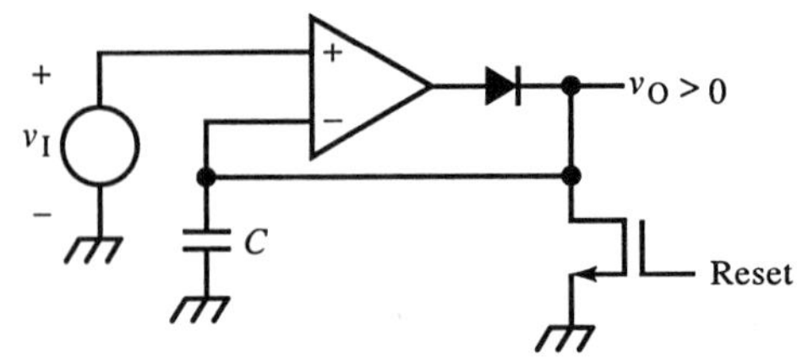

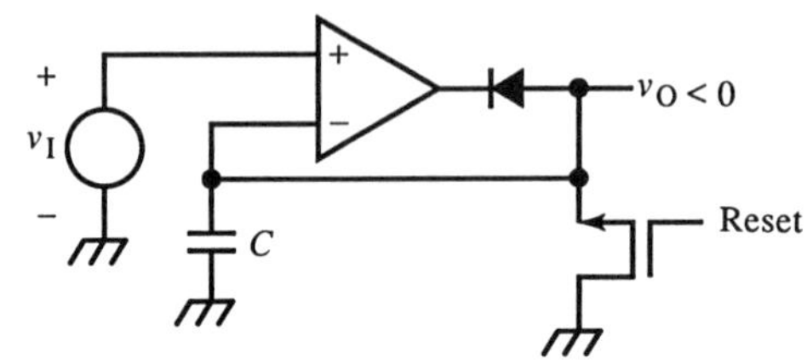

FIG. 11.90 Op-amp peak detectors with reset function: (a) maximum and (b) minimum.

accurately. To avoid overcharging, the feedback response must not be underdamped or overshoot occurs. Op-amp B must be faster than A to minimize loop delay and avoid overshoot. That is, a single dominant pole in the loop due to A yields a damped response. Because of the output loading of C and the additional pole it causes, op-amp A usually must be frequency compensated.

If op-amp A has limited output current, a CC buffer can replace the diode. This increases charging current by the β of the transistor and slew-rate increases by β.

D_2 and R are added to keep op-amp A from being driven into saturation when D_1 is off. R limits D_2 current. This enables the op-amp to respond more quickly since its output now follows the input signal. This also keeps its output

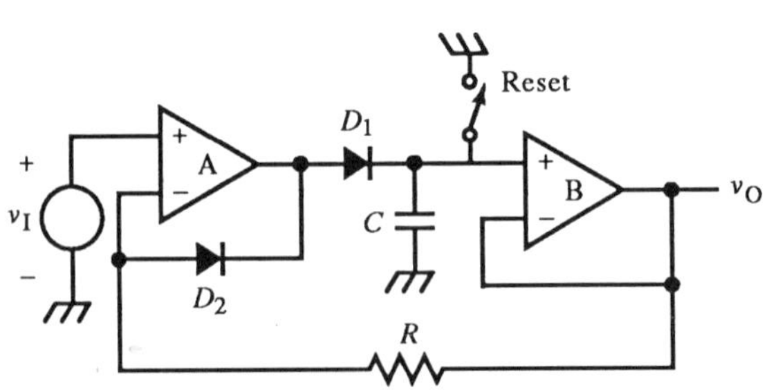

FIG. 11.91 Peak detector with feedback and output buffering.

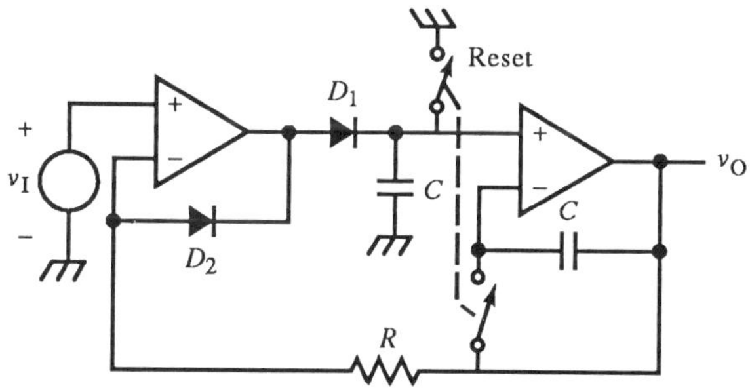

FIG. 11.92 Circuit of FIG. 11.91 with differential capacitor compensation.

from quickly switching to its saturation limits when D_1 cuts off, thus minimizing transient feedthrough to the hold capacitor via D_1 shunt capacitance.

To further minimize bias-current error, a similar capacitor can be placed around op-amp B (Fig. 11.92). Both capacitors are charged by the bias current, resulting in the same Δv across each. The error voltage on the feedback capacitor subtracts from the hold-capacitor voltage, thus compensating its error. The feedback capacitor, however, must also be reset; two reset switches are required.

Finally, the diode must have a low reverse saturation current to minimize leakage when off. Reverse current varies with reverse voltage and is minimized by minimizing the voltage across the nonconducting diode. This can be done by, again, using the versatile technique of bootstrapping (Fig. 11.93). D_3 and R_2 have been added to Fig. 11.91. Since v_O follows the capacitor voltage when holding, R_2 applies this voltage to the anode of D_3, reducing its voltage to zero. D_1 blocks this node from the signal-varying output of op-amp A, but its leakage is not critical for low values of R_2. When C is charging, R_2 isolates the diode node from the output of op-amp B. With this approach, special low-leakage diodes can be avoided in most applications. An alternative is to replace D_1 of Fig. 11.91 with a JFET. Its gate-source junction typically has lower leakage than discrete diodes.

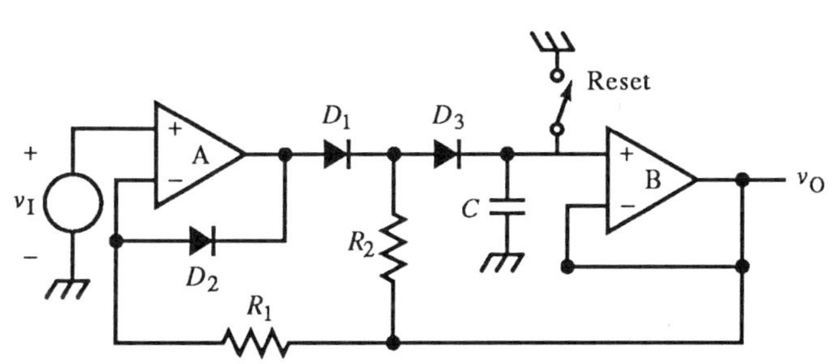

FIG. 11.93 Bootstrapped diode reduces leakage by minimizing its reverse voltage.

This bootstrapping technique of *leakage decoupling* can be applied to the reset switch also. Two switches are placed in series, with a resistor from the output connected to them. Both switches are driven by the Reset signal.

Bootstrapping can also decrease acquisition time if input impedance is not critical. In Fig. 11.91, we add a resistor in series with v_I and a bootstrap diode from output to the op-amp noninverting input so that it conducts from output to input. Then the first fast peak charges C. Through the bootstrap diode, the output of op-amp A is now driven to this voltage, which is near the peak. The next fast peak has less voltage over which to slew the output to further charge C.

Most of the charging time occurs when op-amp A output is near the peak voltage because it is not required to supply the large slew-rate-limited currents that a large voltage difference causes. Feedback then increases response time if $r_{out}C$ corresponds to frequencies at which the loop gain is still high.

The hold capacitor size is chosen as a trade off between hold time and acquisition time. For fast peaks, a small C is preferred for faster charging. But a smaller capacitor develops hold error at a higher rate than a large C. Therefore, C is chosen as a compromise between acquisition and hold-time requirements. A two-stage peak detector mitigates the trade off. The first stage is optimized to be fast whereas the second stage has a long hold time.

References

E. James Angelo, Jr., *Electronics: BJTs, FETs, and Microcircuits*, McGraw-Hill, 1969. Ch. 15.

Alvin B. Phillips, *Transistor Engineering*, McGraw-Hill, 1962. p. 131.

Paul E. Gray and Campbell L. Searle, *Electronic Principles: Physics, Models, and Circuits*, Wiley, 1969. Ch. 4.

"Voltage-reference basics," *EDN*, 1 Apr. 1981. pp. 64, 65.

Ron Knapp, "Selection criteria assist in choice of optimum reference," *EDN*, 18 Feb. 1988. pp. 183–192.

Jerald Graeme, "Manipulate current signals with op amps," *EDN*, 8 Aug. 1985. pp. 147–158.

Robert A. Pease, "Improve circuit performance with a 1-op-amp current pump," *EDN*, 20 Jan. 1983. pp. 85–90.

Jerald G. Graeme, "Low-cost quad op amps boost circuit performance," *EDN*, 3 Sep. 1987. pp. 213–222.

Jim Williams, "Current-source alternatives increase design flexibility," *EDN*, 1 Sep. 1982. pp. 169–174.

Rudy Stefenel, "Positive pulse triggers 555 integrated-circuit timer," *Electronics*, 14 Jul. 1982. p. 169.

Mark Kolber and Frank J. McNerney, "Reduce capacitor leakage resistance for stable, long time delays," *Electronic Design*, 23 Jul. 1981. p. 173.

John J. O'Farrell, "555 timer triggers on millivolt signal," *EDN*, 5 Jan. 1989. pp. 208–209.

J. F. Gibbons and H. S. Horn, "A Circuit with Logarithmic Transfer Response over 9 Decades," *IEEE Trans. Circuit Theory*, Vol. CT-11, No. 3, Sep. 1964. pp. 378–384.

Jim Williams, "Expand linear circuit functions with nonlinear design schemes," *EDN*, 12 May 1982. pp. 153–158.

Charles Kitchin and Lew Counts, "Multifunction IC provides diverse math functions," *EDN*, 22 Aug. 1985. pp. 175–180.

Derek Bray, "The Use of Bipolar Semiconductor Junctions in Linear Circuit Design," Interdesign Monochip Application Note APN-33, 1983.

Art Kapoor and Derek Bowers, "Diverse circuits exploit matching in quad-transistor IC," *EDN*, 6 Mar. 1986. pp. 223–232.

Nonlinear Circuits Handbook, Daniel H. Sheingold, ed. Analog Devices, Inc., Norwood, Massachusetts, 1974.

Jerald Graeme, "Peak detector advances increase measurement accuracy, bandwidth," *EDN*, 5 Sep. 1974. pp. 73–78.

William S. Shaw, "Triangular-wave generator spans eight decades," *Electronics*, 8 May 1972. p. 104.

Dennis Feucht, "Exploring function-generator design problems and solutions," *EDN*, 5 Jun. 1975. pp. 37–43.

Gert Globas *et. al.*, "A 250-MHz Pulse Generator with Transition Times Variable to Less than 1 ns," *Hewlett-Packard Journal*, Vol. 26, No. 1, Sep. 1974. pp. 1–7.

Günter Krauss and Rainer Eggert, "A Moderately Priced 20-MHz Pulse Generator with 16-Volt Output," *Hewlett-Packard Journal*, Vol. 25, No. 7, Mar. 1974. pp. 10–15.

Barrie Gilbert, "Circuits for the precise synthesis of the sine function," *Electronic Letters*, Vol. 13, No. 7, 18 Aug. 1977. pp. 506–508.

John W. Fattaruso and Robert G. Meyer, "Triangle-to-Sine Wave Conversion with MOS Technology," *IEEE JSSC*, Vol. SC-20, No. 2, Apr. 1985. pp. 623–631.

G. Klein and H. Hagenbeuk, "Accurate triangle-sine converter," *Electronic Engineering*, Nov. 1967. pp. 700–704.

G. Klein and H. Hagenbeuk, "An Accurate Triangular-Wave Generator with Large Frequency Sweep," *Electronic Engineering*, Jun. 1967. pp. 388–390.

Vincent G. Bello, "Design of a Diode Function Generator Using the Diode Equation and Iteration," *IEEE Trans. Circuit Theory*, Mar. 1972. pp. 213–214.

Robert G. Meyer *et. al.*, "The Differential Pair as a Triangle-Sine Wave Converter," *IEEE JSSC*, Jun. 1976. pp. 418–420.

Robert Baedtke, "Triangle Waveform Generator Having a Loop Delay Compensation Network," Tektronix patent.

CHAPTER 12

Digitizing and Sampling Circuits

12.1 Electrical Quantities Both Encode and Represent Information

An electrical quantity in time $x(t)$ is a *signal* when it encodes information. The information is interpreted according to a representational theory, such as logic theory for digital signals or transforms based on analogy for analog signals. The theory of representation is independent of the encoding scheme. In communications theory, encoding is called *modulation.* What the modulating signal represents is independent of its encoding. A thermometer output, for example, can be encoded in analog or digital form but represents temperature regardless.

Another way to think about encoding information is that two levels of representation are employed. The encoding scheme is a representation at the electrical level, and the encoded information represents a quantity that is independent of electricity. This "higher" level of representation has to do with the application. Consequently, electronics is useful in domains that have nothing to do with electronics because both signals and their processing operations have meaningful interpretations for the application.

Information can be encoded as discrete or continuous functions of either an electrical quantity (usually voltage or current) or of time. The information to be encoded and the encoding scheme can be either discrete or continuous. The compatibility of an encoding scheme with the encoded information is a design consideration. For example, discrete functions are often best represented by digital encoding. Engineers sometimes differ over the relative merits of discrete versus continuous encoding and processing of information. The difference between discrete, or *digital*, and continuous, or *analog*, encoding is so important that each constitutes a major subdiscipline within electronics.

A digital signal is discrete in both x and t. Binary encoding is by far the dominant digital encoding of x, where $x \in \{X_L, X_H\}$. These two values or "levels" are named *low* (X_L) and *high* (X_H) and represent binary logic states of *true* and *false*; or in Boolean algebra, 0 and 1. Whether the low level represents *true* or *false* depends on the polarity of the logic; a low level is false in positive logic and true in negative logic. Digital encoding can have more than two levels. The number of levels equals the modulus or base of the number representation. For example, decimal numbers can be encoded in a 10-level scheme. As the modulus increases, the representation approaches a continuous form.

An example is the output of DACs. For an 8-bit DAC, 256 discrete levels may adequately approximate a continuous function in some applications. The DAC could, however, be considered an encoder of base-256 numbers. In this sense, continuous signals are of infinite modulus, and analog engineers are actually digital engineers who specialize in infinite-base encoding.

Discrete functions can also be encoded in time. Frequency-shift keying, a kind of binary FM used in modems, is one way. More common to computer electronics is synchronous and asynchronous serial encoding of alphanumeric characters in ASCII. Many other purely digital encoding schemes make digital encoding and communications a specialty in itself.

Continuous functions can also be encoded purely in time as the width of a (binary) pulse (pulse-width modulation) or by its position relative to another event in the signal (pulse-position modulation) or simply by the pulse frequency, as is the output of voltage-to-frequency converters.

Finally, signals that are continuous in x and discrete in t are *sampled signals*. These signals are of great importance in association with analog-to-digital (A/D) and digital-to-analog (D/A) conversion and with sampled-data systems in general, systems that contain discrete-time signals, such as a motor servo controller with a digital position encoder or any system with sample-and-hold circuits.

12.2 Digital-to-Analog Converters

Digital-to-analog (D/A) converters (DACs) convert digital input codes to output voltages or currents. The transfer curve for a unipolar 3-bit DAC (Fig. 12.1) has discrete voltages at the discrete (integer) values of the digital input code *d*. The digital code is an ordered set of bits that represent integers. Various number representations are possible, but the most common are shown for

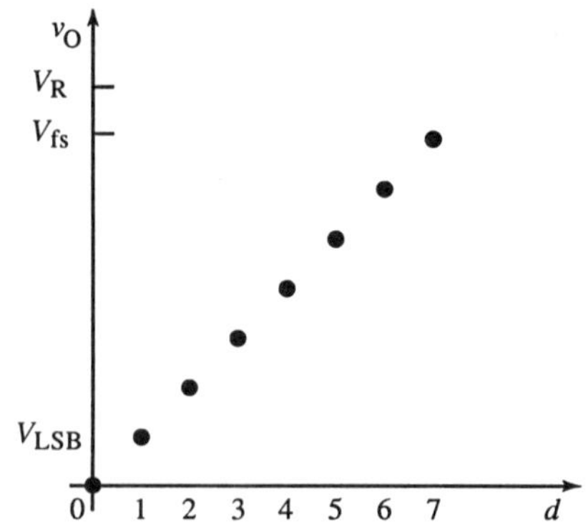

FIG. 12.1 Unipolar DAC transfer characteristics.

three bits in the table:

integer	offset binary	two's complement	sign-magnitude	
3	111	011	011	+fs
2	110	010	010	
1	101	001	001	
0	100	000	000 ↑ 100 ↑	zs
−1	011	111	101	
−2	010	110	110	
−3	001	101	0^+ 111 0^-	
−4	000	100	↑	−fs
	↑ inverted MSB (sign) ↑		1⇒ −	

These are signed (bipolar) representations of integers. The most common number representations are two's complement and offset binary. They differ only in the polarity of their sign bit. Binary-coded decimal (BCD) is also sometimes used, in which the first 10 binary numbers represent a decimal number.

In the table, the positive full-scale (fs) value is 3, and the negative fs value is −4, one greater in magnitude. This asymmetry results from assigning a state to zero. Sign-magnitude coding is symmetric, but it has two zero states. In general, for n bits, there are 2^n states. The transfer characteristic for a unipolar n-bit DAC is

$$v_O = V_R\left(\frac{d}{2^n}\right) \tag{12.1}$$

where V_R is the DAC reference voltage. The fs voltage is less than V_R because the

$$\text{maximum } d = 2^n - 1 \tag{12.2}$$

Accordingly,

$$V_{fs} = V_R\left(\frac{2^n - 1}{2^n}\right) = V_R - \frac{V_R}{2^n} = V_R - V_{LSB} \tag{12.3}$$

That is, the fs output is less than V_R by V_{LSB}, the *quantum* voltage: the voltage difference corresponding to a difference of one input state. Since V_{LSB} is the smallest output voltage difference of the DAC, it is also its resolution, its minimum Δv_O.

The DAC input often represents a continuous function, but because it is discrete (or *quantized*), for values between integers the DAC output remains constant. In Fig. 12.1, v_O is zero over the interval $[0, 1)$. (This is the *least-integer function*). At 1^-, infinitesimally below one, $v_O = 0$ V, though the correct value is infinitesimally less than V_{LSB}. The output is in error by V_{LSB} at 1^- and has no error at zero. The magnitude of the error can be split so that the error range is $\pm\frac{1}{2}V_{LSB}$ by offsetting v_O by $\frac{1}{2}V_{LSB}$ (Fig. 12.2). Then the DAC output fs magnitudes are also equal.

Quantization error causes *quantization noise*, which is a sawtooth function that cycles between $-V_{LSB}/2$ and $+V_{LSB}/2$ between each state ($\Delta d = 1$). The rms value of this noise v_n is

$$\text{rms } v_n = \sqrt{\frac{1}{\Delta d}\int_{-\Delta d/2}^{+\Delta d/2}\left(\frac{V_{LSB}}{\Delta d}\cdot d\right)^2 d(d)} = \frac{V_{LSB}}{\sqrt{12}} \cong 0.3\, V_{LSB} \tag{12.4}$$

A signal-to-noise ratio definition for n bits is

$$\text{SNR} = \frac{V_R}{\text{rms } v_n} = \frac{2^n \cdot V_{LSB}}{V_{LSB}/\sqrt{12}} = \sqrt{12}\cdot 2^n \cong 3.46 \cdot 2^n \tag{12.5}$$

In decibels, this is

$$\text{SNR(dB)} = 20\log(\sqrt{12}\cdot 2^n) = 20\log\sqrt{12} + 20n\log 2 \cong 10.8 + 6.02n \tag{12.6}$$

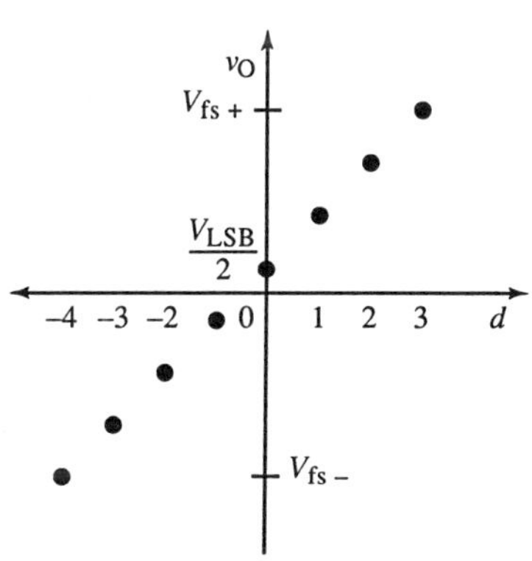

FIG. 12.2 DAC transfer characteristic, offset to reduce quantizing error.

The dynamic range is about $6n$ dB, and quantization noise is about 10.8 dB independent of the number of bits. Since each additional bit increases the range by ×2, an octave, this is 6 dB/octave, the slope of a Bode plot zero.

A different characterization of the SNR is as the ratio of rms signal to rms noise for a sinusoidal signal. The rms value of a sine of amplitude V is $(\sqrt{2}/2)\,V$. Then

$$\mathrm{SNR} = \frac{\text{rms sine}}{\text{rms noise}} = \frac{(2^n/2)\cdot(\sqrt{2}/2)\,V_{\mathrm{LSB}}}{V_{\mathrm{LSB}}/\sqrt{12}} = 2^n\left(\frac{\sqrt{6}}{2}\right) \cong 1.23\cdot 2^n \tag{12.7}$$

In dB scaling, this is

$$\mathrm{SNR(dB)} = 20\log\left(\frac{\sqrt{6}}{2}\right) + 20\log 2^n \cong 1.76 + 6.02n \tag{12.8}$$

The dynamic range remains the same under this definition of SNR, but the signal is less relative to the noise. This explains why the constant term of 1.8 dB is less than in (12.6).

In actual DACs, the step size of quantum V_{LSB} is not constant, which affects the linearity of v_{O}/d. A measure of this nonlinearity is the *differential linearity error* (DLE), or *differential nonlinearity*, the amount a step differs from V_{LSB}:

$$\mathrm{DLE} = \Delta v_{\mathrm{step}} - V_{\mathrm{LSB}} \tag{12.9}$$

If DLE exceeds V_{LSB}, the transfer curve is nonmonotonic, decreasing in output value with increasing d. This behavior can wreak havoc in control system applications. DACs also have offset and scaling (gain) errors, but these are nulled by external adjustment; the DLE cannot be.

DACs often output functions of time, $v_{\mathrm{O}}(t)$; their dynamic response is important. This is characterized by the settling time to within $\frac{1}{2}V_{\mathrm{LSB}}$ of error. A dynamic anomaly of DACs is that when a large number of bits change in d, the effects of individual bits on the output are not exactly synchronized. At the output, momentary pulses or "glitches" appear until all the bits settle. This phenomenon is especially evident at midscale, when d changes from 011 to 100 (for a 3-bit DAC). The change in most-significant bit (MSB) must be cancelled by the combined changes of all the other bits, to within V_{LSB}, the correct Δv_{O}.

When glitches are unacceptable, as in CRT display systems, the DAC is followed by a deglitcher. These are either hf limiters or samplers with a delay. In delayed samplers, the DAC output is allowed to settle. Then it is sampled, and this value is output. The sampling control signal is delayed from the clock that changes DAC input states.

DAC designs are mainly either BJT (bipolar) or CMOS. Both achieve binary weightings of voltage or current for each bit by a resistive network. Unless the number of bits is few (≤ 4), these are R-$2R$ or *resistive ladder networks.* Otherwise, a set of binary-weighted resistors suffice. A 4-bit binary-

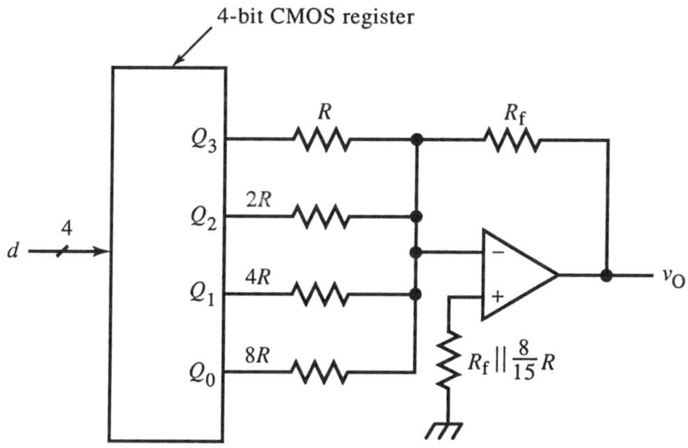

FIG. 12.3 Discrete CMOS 4-bit DAC with binary-weighted resistors.

weighted resistor DAC is shown in Fig. 12.3, voltage-driven by a CMOS register. The output is a function of each of the bits b_i of d:

$$v_O = -R_f\left(\frac{v_R}{R}\cdot b_3 + \frac{V_R}{2R}\cdot b_2 + \frac{V_R}{4R}\cdot b_1 + \frac{V_R}{8R}\cdot b_0\right) = -\left(\frac{R_f}{R}\right) V_R(2^{-i}b_{3-i}),$$

$$i = 0, 1, 2, 3 \qquad (12.10)$$

where b_0 is the LSB and V_R is the CMOS register supply voltage, the DAC voltage reference. (CMOS digital outputs accurately approach the supply rails.) The resistors must have sufficient precision to minimize DLE. The resistor requiring the most precision is at the MSB, R, since it must be within

$$\frac{\Delta R}{R} = \pm\frac{1}{2}\cdot\frac{R}{R_{LSB}} = \pm\frac{1}{2}\cdot 2^{-(n-1)} = \pm 2^{-n} \qquad (12.11)$$

A four-bit DAC must have a tolerance on R of 6.3%. A 5% resistor suffices. For 8 bits, the tolerance is 0.4%. This is difficult to achieve in monolithic form when the resistor values have such a wide range. A standard alternative is the *voltage-switching* R-$2R$ network of Fig. 12.4, another 4-bit DAC.

In general, the binary weighting,

$$W(d) = \sum_{i=0}^{n} 2^{-i} b_{n-i} \qquad (12.12)$$

is the heart of DAC function. This weighting is achieved in the R-$2R$ network. Beginning with b_0, if it is 1, Q_O output is V_R; if 0, then the output is 0 V. For $b_0 = 1$, the Thévenin equivalent circuit is shown in Fig. 12.5. At each stage of an R-$2R$ network, the input resistance is $2R$ and the voltage of the previous stage is halved. From the input end of the network, the b_0 voltage is consequently halved four times. At the output, the op-amp is driven by a source

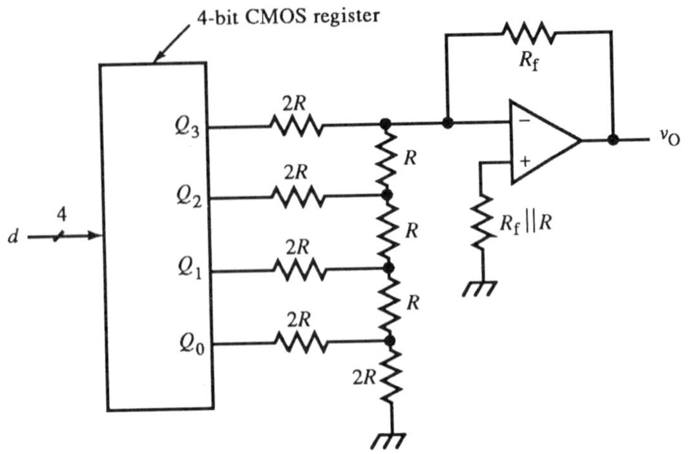

FIG. 12.4 Discrete CMOS 4-bit DAC with R-$2R$ network.

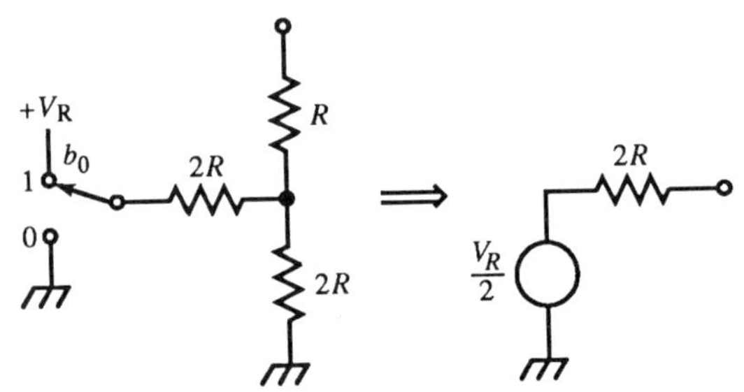

FIG. 12.5 R-$2R$ network equivalent circuit at LSB.

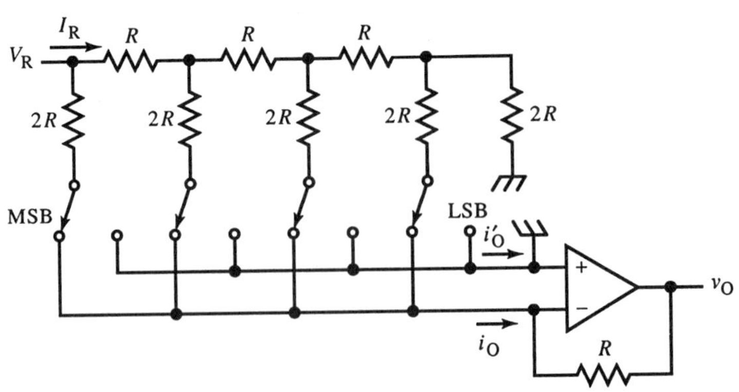

FIG. 12.6 CMOS DAC.

resistance of R (another series R was not added to the network to make it $2R$) and voltage of $W(d)V_R$.

CMOS DACs are typically designed as in Fig. 12.6, with CMOS switches at the output and V_R at the input. This reversal does not change the operation except that the DAC outputs must be kept at ground (or virtual ground, as shown) to avoid errors in output current. This *current-switching* R-$2R$ network is still voltage-driven, and the output is $-WV_R$ with op-amp feedback resistor R. IC DAC's often include this resistor to ensure its match with those in the network. The outputs are complementary currents that sum to a full-scale current,

$$i_O + i'_O = I_{fs} = \frac{V_R - V_{LSB}}{R} \tag{12.13}$$

Switch resistance must be minimum (or binary-ratioed) for minimum network error. MOSFET switch areas are scaled to achieve equal voltage drops across all switches.

CMOS DAC output impedance changes with d. The two extremes are with all zero and all one bits (Fig. 12.7) for an AD7520, a 10-bit DAC with $R = 10\ \text{k}\Omega$ and leakage current I_L of 200 nA. Response compensation for op-amp input capacitance can only be based on an average or worst-case input state. Since output resistance varies extremely, op-amp bias-current compensation is also suboptimal.

Typical BJT DAC design is based on a current-switching R-$2R$ network (Fig. 12.8). An input op-amp establishes a reference current I_R in one of several BJTs with emitters connected to an R-$2R$ network. The emitter areas are ratioed with the current each conducts, to maintain the same V_E for all BJTs.

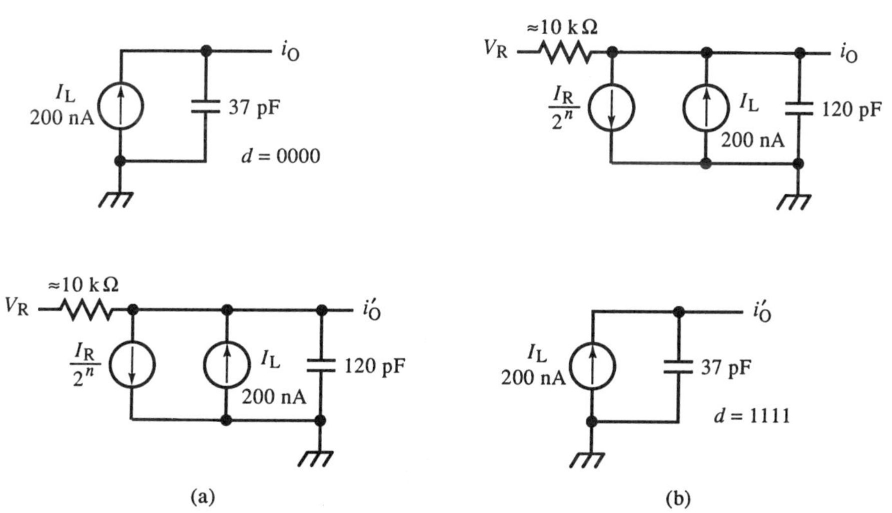

FIG. 12.7 Outputs of CMOS DAC for inputs of all zeros (a) and all ones (b).

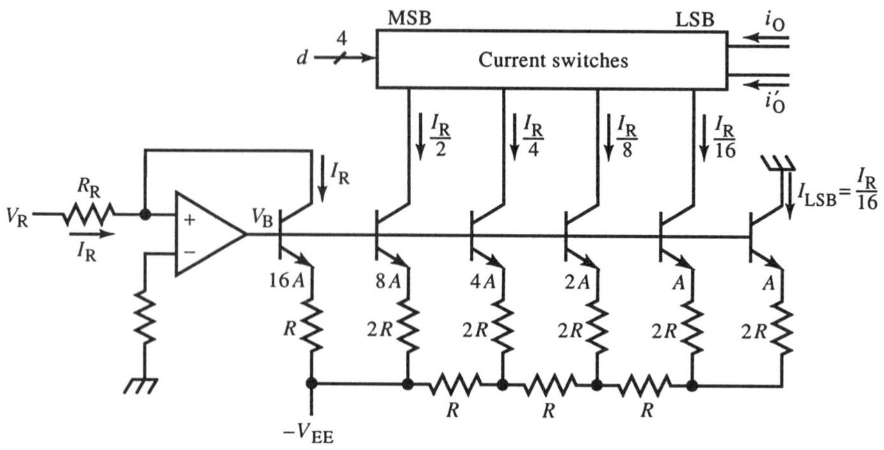

FIG. 12.8 BJT (bipolar) DAC with R-$2R$ binary current generator.

The BJT collectors drive current switches. These can be diff-amps with logic-compatible inputs. Their collectors are connected to either the i_O or i'_O outputs. CMOS DAC outputs have no voltage compliance, but the BJT current outputs from collectors need not be held at a fixed voltage.

The R-$2R$ network is not switched but is a multiple binary current divider. The voltage in the series-R string doubles at each successive stage toward the termination at the LSB. The reference current is established by the op-amp circuit as V_R/R_R. The output is

$$i_O = W(d)I_R \tag{12.14}$$

and complementary output

$$i'_O = I_{fs} - i_O = \left(\frac{2^n - 1}{2^n}\right) I_R - i_O \tag{12.15}$$

Equation (12.13) applies here. The relationship between I_R and I_{fs} is the same as for the CMOS DAC:

$$I_{fs} = \left(\frac{2^n - 1}{2^n}\right) I_R = I_R - I_{LSB} \tag{12.16}$$

The complementary output current is also related to $W(d)$ as

$$i'_O = \overline{W(d)} \cdot I_R \tag{12.17}$$

where

$$\overline{W(d)} = \sum_{i=0}^{n} 2^{-i}\bar{b}_{n-i}, \qquad \bar{b}_{n-i} = \text{logical complement of } b_{n-i} \tag{12.18}$$

That is, $\overline{W}$ is the complementary weighting, the result of the bitwise negation (or one's complement) of d.

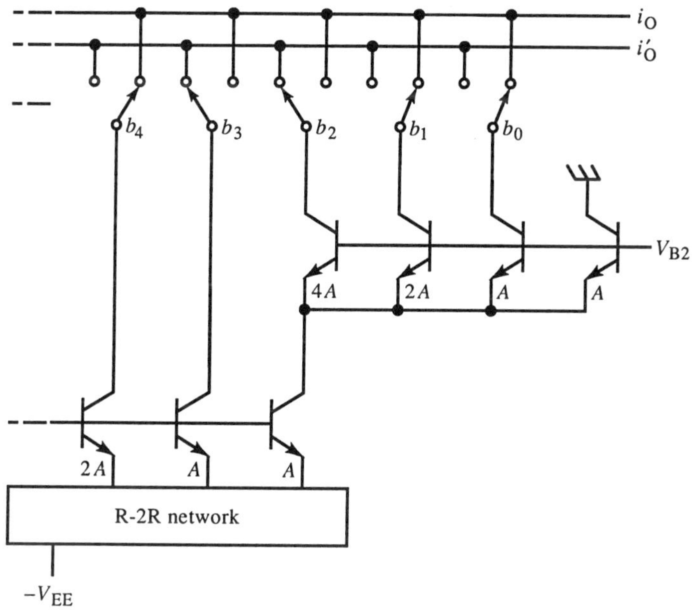

FIG. 12.9 Scaled-emitter technique for DAC LSBs.

Monotonicity among the LSBs, achieved with a *scaled-emitter* technique (Fig. 12.9), is used with the ladder network. The LSB terminating current of the ladder network, instead of being grounded, is fed to a second branch of emitter-scaled BJTs that switch the three LSBs. Monotonicity is ensured by branching.

For DACs with a large number of bits, the branching idea can be realized in a different topology. The *segmented* DAC, shown in Fig. 12.10 for a two-bit segmentation, does current weighting with two networks. The input network is driven by a reference current as before. The two MSBs are decoded by a segment decoder. They switch four equal currents either to output or to the branching input of an R-$2R$ network of the remaining m LSBs. A segment decoder successively switches more segments with larger MSB codes to i_O as the m LSBs divide the current from one segment (s_1 in the figure) between i_O and i'_O. The remaining segment currents go to i'_O. For $m = 2$, when $d \leq 0011$, all segments except s_0 are switched to i'_O. When $d = 0100$, s_0 switches to i_O, and s_1 switches to drive the branch DAC, as shown. The remaining switches stay on i'_O.

Although the branching effect assures monotonicity, the transfer curve linearity can be much worse than the DLE. But in many applications, the DLE is all that matters for linearity. The match of the segment currents does not determine DLE, only the overall linearity. The segmented DAC uses fewer resistors since each segment requires only one resistor, not two as in a ladder network.

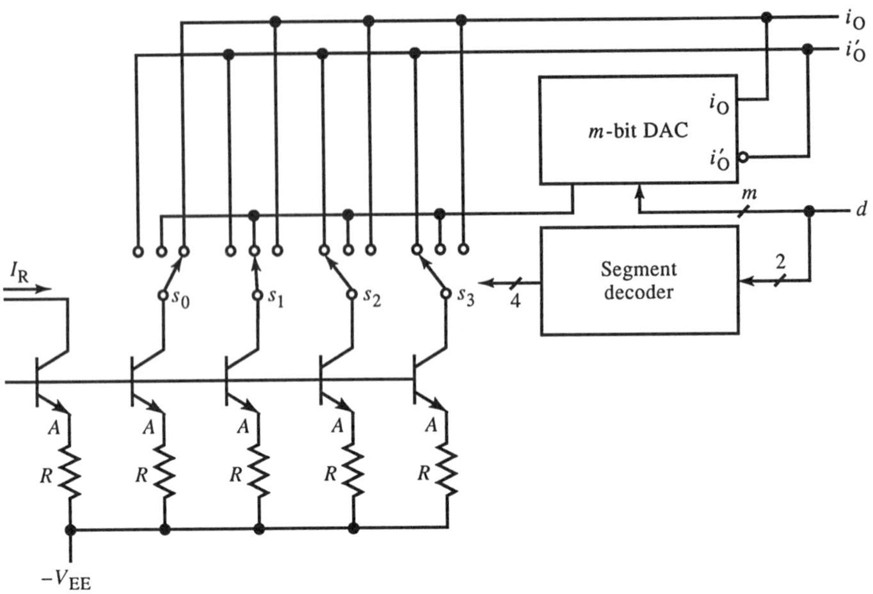

FIG. 12.10 Segmented DAC. One of four segments supplies current to m-bit DAC for m LSBs.

A very simple DAC design is the serial-output DAC, easily realized by one filtered output line of a computer. A PWM generator, either in hardware or software, drives a low-pass filter with break frequency far below the PWM frequency. The average output voltage is proportional to the duty-ratio. The disadvantage of this scheme is that it is slow and inherently noisy due to ripple from the filtered pulse. But if the pulse amplitude is accurate and the transitions are fast, a high-resolution output is achievable in direct trade-off with response time.

The ripple amplitude varies with pulse duration, which depends on the duty-ratio D. For high or low D, ripple is least and is highest at $D = 50\%$. In steady-state, the ripple extends from v_L to v_H around the average, DV. The ripple amplitude,

$$\Delta v = v_H - v_L$$

is derived from the decaying exponential, when the pulse is low:

$$\frac{v_H - v_L}{v_H} = \frac{\Delta v}{v_H} = e^{-(1-D)T/\tau} \cong \frac{\Delta v}{DV} \qquad (12.19)$$

The approximation assumes that ripple is small relative to average output voltage and that $v_H \cong DV$. Also, τ is the filter time constant. Solving (12.19) for the ratio of PWM frequency to filter break frequency for 1 LSB of ripple, we obtain

$$\frac{f_{PWM}}{f_{bw}} \cong -\frac{2\pi(1-D)}{\ln(1-2^{-n}/D)}, \qquad \Delta v = V_{LSB} \qquad (12.20)$$

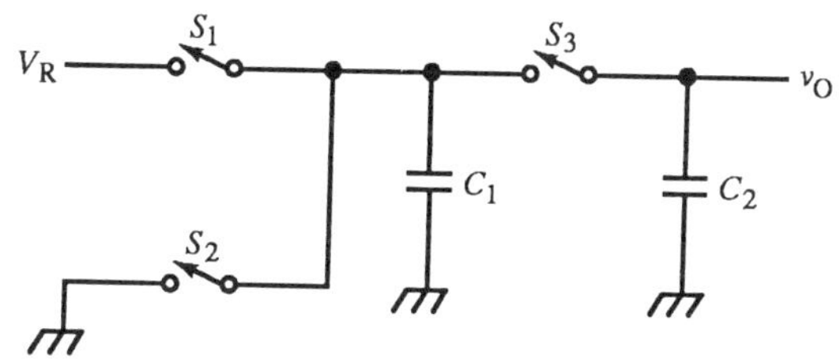

FIG. 12.11 A switched-capacitor serial-output DAC. $C_1 = C_2$.

For four bits, f_{PWM} must be $23.5 f_{bw}$ at $D = 0.5$ and $401 f_{bw}$ for 8 bits. At 10% duty-ratio for 4 bits, the frequency ratio is only 5.8 and is 8.7 for 90%. (At the extremes of D, the approximation fails. At $D = 2^{-n}$ and $D = 1$, the ratio is 0.) Other pulse waveforms from rate-multipliers or statistically biased digital pseudo-random noise have less ripple for the same clock-to-filter frequency ratio but are harder to generate.

A serial-input DAC is shown in Fig. 12.11, with three switches and two equal capacitors $C_1 = C_2$. A serial digital input begins with the LSB. The DAC operates in two phases for each successive bit. On phase 1, switch S_3 is open, and a serial input bit closes either S_1 (for $b = 1$) or S_2 (for $b = 0$), charging C_1 to

$$q_i = Cb_i V_R \tag{12.21}$$

On phase 2, S_1 and S_2 are open, and S_3 is closed. S_2 contains the net charge from previous cycles. On the kth bit on phase 1, this charge is

$$q_{i-1} = Cv_{i-1} \tag{12.22}$$

On phase 2, S_3 closes, and

$$v_i = \frac{q_i + q_{i-1}}{2C} = \left(\frac{V_R}{2}\right) b_i + \frac{v_{i-1}}{2} \tag{12.23}$$

For n bits, the voltage is

$$v_n = \left(\frac{V_R}{2}\right) b_n + \left(\left(\frac{V_R}{4}\right) b_{n-1} + \left(\left(\left(\frac{V_R}{8}\right) b_{n-2} + \cdots\right)\right)\right) \tag{12.24}$$

This iterative equation reduces to the closed form of

$$v_n = \left(\sum_{i=1}^{n} 2^{-i} b_{n-i}\right) v_R = W(d) V_R \tag{12.25}$$

After n bits, v_n is the converted voltage. It must be stored separately for output during the next conversion. The number of bits of monotonic conversion is limited by capacitor matching and switch leakage.

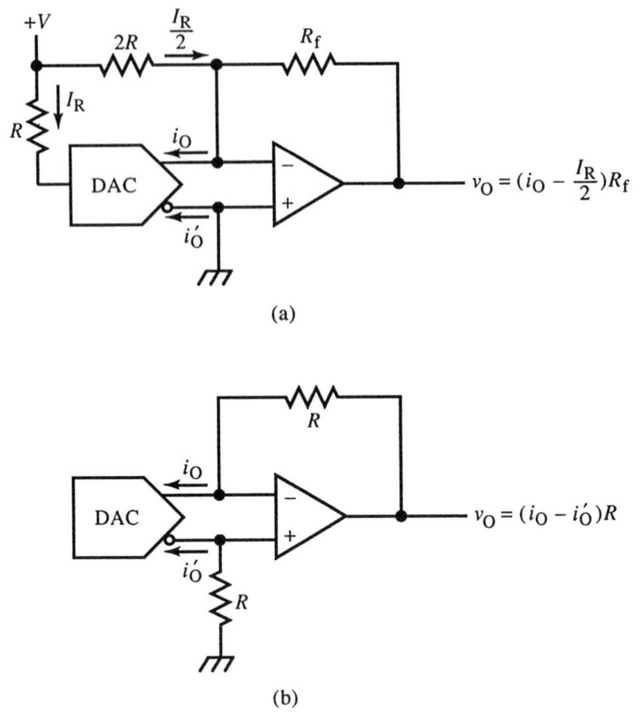

FIG. 12.12 DAC interface circuits: (a) offset binary and (b) symmetric outputs.

12.3 Digital-to-Analog Converter Circuits

In Fig. 12.12, digital-to-analog converters (DACs) are used as components in circuits. The DAC schematic symbol is used, with a small circle at i'_O to indicate the complementary output, after the convention of logic symbols. In Fig. 12.12a, the op-amp output is bipolar and is offset by $I_R/2$ by a resistor of $2R$, where R is the current-reference resistor. Without this offset, v_O is unipolar, ranging from 0 V to $I_{fs}R_f$. If we shift i_O down by $I_R/2$, v_O at negative fs is one V_{LSB} greater in magnitude than positive fs. $I_R/2$ corresponds to the midscale or zero state of d. The output is

$$\text{offset-binary } v_O = \left(i_O - \frac{I_R}{2}\right) R_f = \left(W - \frac{1}{2}\right) I_R R_f \tag{12.26}$$

Some output values for a 4-bit d are tabulated:

d	$W - 1/2$
1111	7/16
1000	0
0111	−1/16
0000	−1/2

A two's complement coding of d produces the same results when the MSB is inverted.

In Fig. 12.12b, the output range is symmetric about zero. The op-amp is driven differentially by the DAC output so that

$$v_O = i_O R - i'_O R = (i_O - i'_O)R = (W - \overline{W})I_R R = (2i_O - I_{fs})R \quad (12.27)$$

The expression for v_O in terms of W follows from (12.14) and (12.17). Compared with the offset-binary ouput, the symmetric-offset output range and step size are twice as large because i'_O is used. The last expression of (12.27), compared with (12.26), has twice the gain ($2i_O$) and a comparable offset difference of

$$\frac{I_{LSB}}{2} = \frac{I_R}{2} - \frac{I_{fs}}{2} \quad (12.28)$$

This leaves the symmetric v_O with a $V_{LSB}/2$ positive offset relative to the offset-binary output. Some 4-bit output values are the following:

d	$W - \overline{W}$
1111	15/16
1000	1/16
0111	−1/16
0000	−15/16

The extreme states have outputs of equal magnitude while zero is offset by $V_{LSB}/2$. An inverted output results from exchanging the DAC outputs.

A DAC bipolar current source with a current mirror is shown in Fig. 11.17. Another circuit that does not require a current mirror, Fig. 12.13, has similar topology to a Howland current source but is simpler in operation. The load current is

$$i_L = \frac{v_O - v_L}{R_S} - i'_O = \frac{i_O R_1 - i'_O R_2}{R_S} - i'_O = \frac{i_O R_1 - i'_O(R_2 + R_S)}{R_S} \quad (12.29)$$

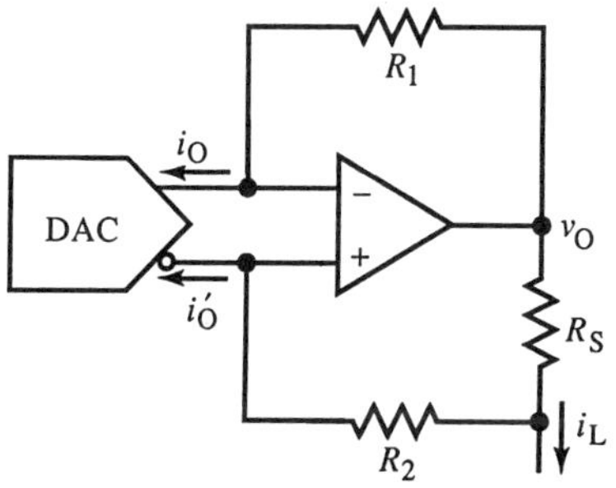

FIG. 12.13 DAC output bipolar current source.

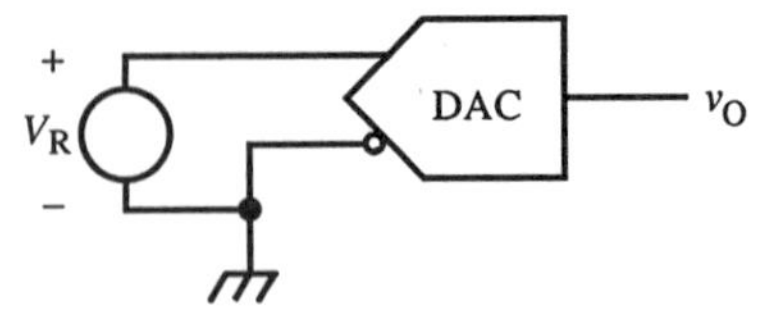

FIG. 12.14 Reverse DAC configuration.

This reduces to

$$i_L = \left(\frac{R_1}{R_S}\right)(i_O - i'_O), \qquad R_1 = R_2 + R_S \tag{12.30}$$

The differential current output from the DAC is converted into a bipolar single-ended current output.

Since CMOS DACs are switched ladder networks, they can be used "backward" as in Fig. 12.14. The reference voltage is applied across the i_O terminals, and the voltage output is taken from where the reference voltage is normally applied. This scheme is similar to that of Fig. 12.4 without the op-amp. Because the CMOS switches are driven from the supply voltage V_{CC}, minimum switch resistance (and linearity error) results by keeping V_R well below V_{CC}.

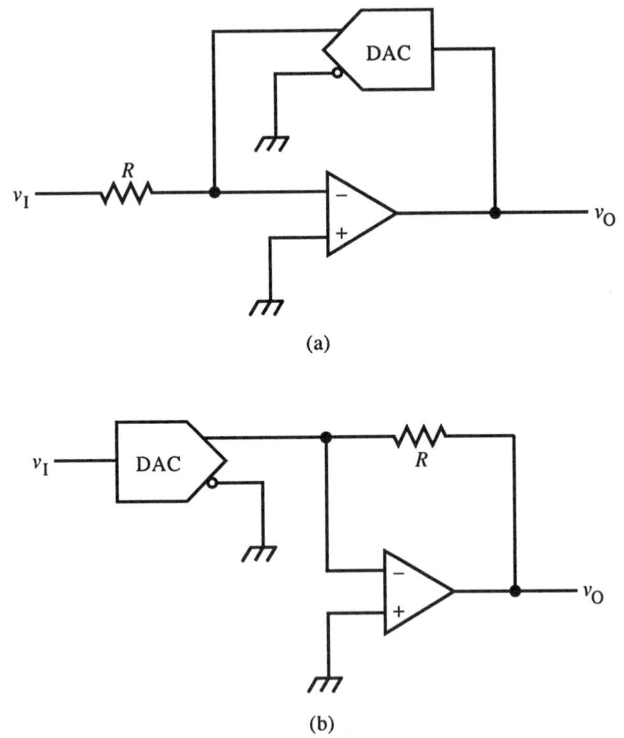

FIG. 12.15 PGA using DAC in feedback path (a) and as input attenuator (b).

In Fig. 12.15, DACs are combined with op-amps to provide programmable gain. The PGA in (a) has a high gain but can have significant voltage offset errors, whereas in (b) the op-amp gain is limited to $\times 2$. In both circuits, R is included in the DAC IC and matches and tracks the ladder resistances. The noninverting configurations are similar in concept. For applications in which the digital input is a dynamic signal and not merely a scale factor, it is multiplied by v_I; the DAC multiplies a digital by an analog quantity.

12.4 Analog-to-Digital Converters: Parallel Feedback

The inverse function of D/A conversion is *analog-to-digital* (A/D) conversion, performed by A/D converters (ADCs). We consider here four categories of ADCs, which include many variations. Approximate ranges for conversion rate and precision are given:

ADC type	conversion rate	precision, bits
integrating	0.1 Hz–10 Hz	14–20
cyclic (serial)	1 kHz–100 kHz	10–16
parallel-feedback	50 kHz–5 MHz	8–10
parallel (flash)	5 MHz–500 MHz	4–8

Parallel-feedback converters are based on a concept similar to that of placing a function block in the feedback loop of an op-amp to achieve the inverse function. Figure 12.16 shows two realizations of the *ramp* converter. In (a), the digital form, a counter driven by a clock generates a digital sawtooth output. It drives a voltage-output DAC that outputs the ramp in analog form. When it crosses v_X, the comparator output clocks the register and holds the digital count. When the counter overflows, the DAC output resets to its minimum value, and the comparator output goes low, completing the cycle. The comparator output is also an end-of-conversion signal indicating valid register data.

In Fig. 12.16b, the same concept is realized with an analog current-source ramp generator. The counter overflow turns on the reset switch, discharging the capacitor at the end of the conversion cycle. The analog circuit is subject to errors in ramp slope relative to the clock frequency. The digital form in (a), although not having these timing errors, must have an accurate DAC reference voltage.

A third realization of the ramp converter (Fig. 12.17) makes use of a microcomputer (μC) (or any computer) and minimal additional hardware: an n-bit DAC and a comparator. The μC must have one digital input bit from the comparator and n output bits to drive the DAC. The software algorithm for ramp conversion uses software variable, VX, to hold the digitized value

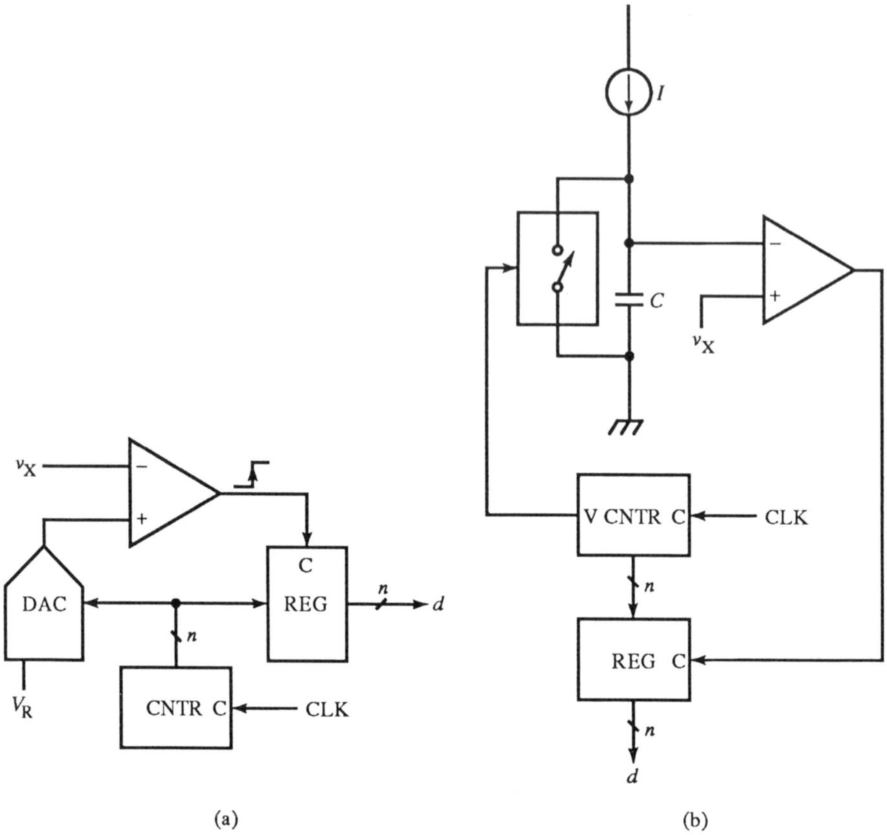

FIG. 12.16 Ramp ADC: (a) digital and (b) analog realizations.

of v_X, and OUT to hold the DAC output value. The procedure is

0. Ramp ADC
1. Set OUT to zero: OUT ← 0.
2. Input the IN bit.
3. If IN = 0, then VX ← OUT; go to 1.
   Else increment OUT: OUT ← OUT + 1.
4. Output OUT; go to 2.

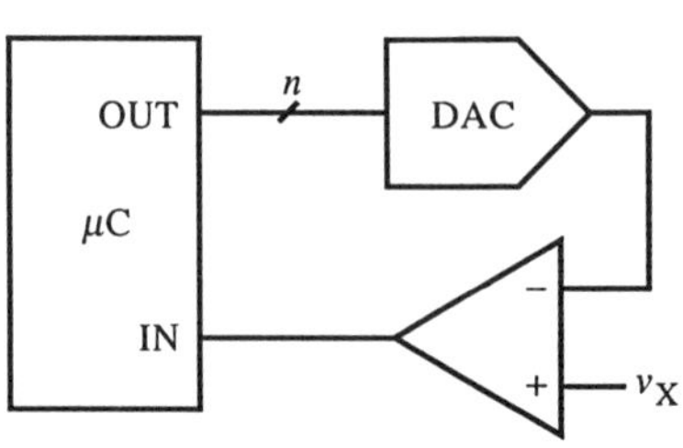

FIG. 12.17 General μC-based parallel-feedback ADC.

Conversion time is usually limited by the μC, but for many applications, it is fast enough; and the few additional components are an advantage. The ramp ADC is a poor technique and is seldom used. The conversion time varies but can take up to 2^n clock periods.

Parallel-feedback converters have a generalized topology (Fig. 12.18). The type of logic block used determines the type of converter. The ramp ADC uses a simple counter. A slightly better ADC is the *tracking* converter. Its logic is a bidirectional (up/down) counter. As v_X changes, the comparator output causes the counter to count up if the DAC input d is low and down if it is high. The counter servos the DAC to minimize input error at the comparator. Since the counter counts either up or down, the error is always ±1 LSB. For a dc input, a converged counter dithers by one state around the correct value; the comparator output alternates logic levels each clock cycle.

The tracking ADC is an improvement over the ramp ADC because it can be used to follow an input signal, digitizing it as it occurs (that is, in *real time*). For small input changes, the counter must change only a few states. This is done in a few clock periods, and conversion is fast. For large input changes, such as a square-wave step, the converter shows slew-rate limitations and a longer conversion time. The DAC output slew-rate, if limited by the counting rate, is

$$\text{tracking ADC slew-rate} = \frac{V_{fs}}{2^n} \cdot f_{CLK} \tag{12.31}$$

The clock frequency is limited by the loop delay time: the DAC settling time, comparator delay time, and counter clock-to-output time. For a sinusoidal input,

$$v_X(t) = \frac{V_{fs}}{2} \sin \omega t \tag{12.32}$$

its maximum slew-rate is $\omega V_{fs}/2 = \pi f V_{fs}$. Equating to (12.31) and solving for the maximum fs sine frequency gives

$$\text{maximum fs sine } f = \left(\frac{1}{\pi 2^n}\right) f_{CLK} \tag{12.33}$$

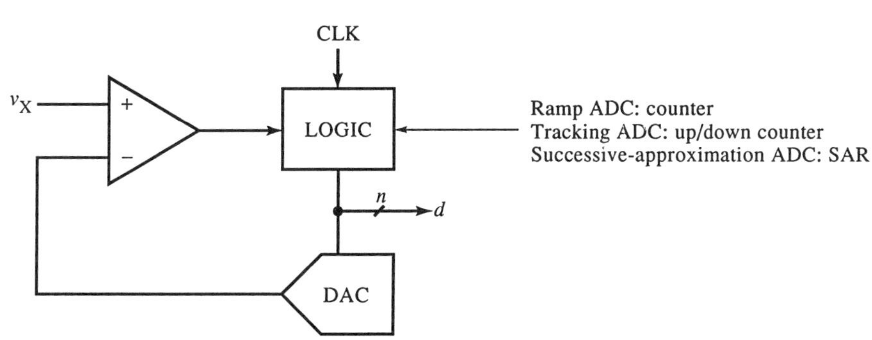

FIG. 12.18 General parallel-feedback ADC. The logic block determines the converter type.

The tracking converter can be implemented with the same hardware as the ramp converter. Figure 12.17 also applies generally to parallel-feedback converters. Instead of hardware logic, the software logic distinguishes among parallel-feedback ADC types. A tracking ADC procedure, based on the same software variables as the ramp ADC, is

0. Tracking ADC.
1. Output OUT.
2. Input IN.
3. If IN = 0, then decrement OUT: OUT ← OUT − 1.
   Else, increment OUT: OUT ← OUT + 1.
4. Set VX to OUT: VX ← OUT.
5. Go to 1.

This procedure is not more complicated than that for the ramp converter but has the performance advantages of the tracking ADC.

The tracking ADC is useful as a *track-and-hold* (T/H) circuit. The digital output follows the input signal until the clock is gated off or the count clocked into another register. Then the input value, in digital form, is held indefinitely; no analog hold circuit can do this. As we shall see about sampling circuits, and as we saw for peak detectors in Section 11.15, a capacitor can accurately maintain its charge for only a limited time.

The circuit of Fig. 12.18 compares DAC voltage to v_X at the comparator input. Current-output DACs require an additional I/V converter stage. In Fig. 12.19, a current-output DAC of either bipolar or CMOS type forms a voltage difference with v_X by dropping $i_O R$ in series with it. The comparator now senses this difference against 0 V. This *current-mode* comparison works with bipolar inputs. The inputs of the comparator must be reversed from voltage-mode comparison, or the complementary current output of the DAC must be used instead.

A third parallel-feedback converter is the *successive-approximation* (SA) converter, a very common conversion technique and the most popular of the parallel-feedback converters. It takes $n+1$ clock cycles to convert n bits using

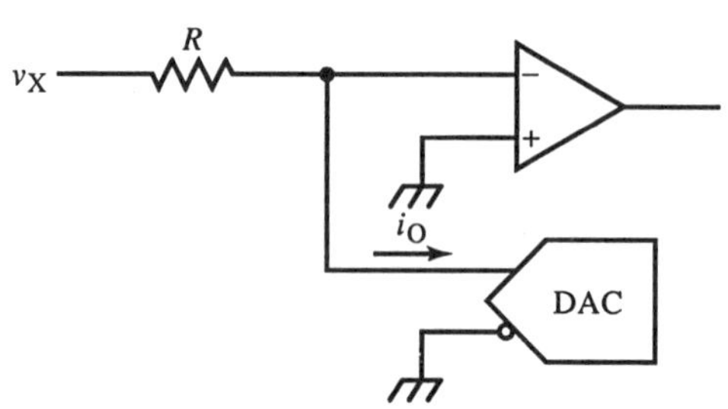

FIG. 12.19 Current-mode comparison at input of comparator.

a bitwise algorithm. It determines one bit per block cycle after an initialization cycle.

In Fig. 12.18 its logic block is a *successive-approximation register* (SAR). This register can be realized by an n-bit shift register (SR) and n latches. [A latch is a kind of flop with a level-sensitive clock input. When the clock is asserted (high), its output follows its input; when the clock is unasserted, the output remains with the value at the falling edge of the clock.] At the start of conversion, the SR bits are cleared and the MSB set. The latch feeds this digital midscale value to the DAC. If v_X is larger than midscale, the comparator output is high. When the clock goes low, the MSB is latched. The next clock edge shifts the 1 bit in the SR to the $n-1$ bit position, and the cycle is repeated. In effect, beginning with the MSB, n decisions are made, each of which narrows the range of possible values for v_X by half. The convergence rate of this procedure is on the order of log n, and the conversion time is independent of v_X.

The generic hardware of Fig. 12.17 is again used to implement a μC-based SA ADC. The procedure is only slightly more complicated than previous ones but is usually well worth the speed increase. Besides the IN and OUT address locations, the software model is shown in Fig. 12.20. SAR is a variable that emulates the SAR latch. Variable SR emulates the shift register, which has an additional "carry-bit" stage that is included in the shift loop, as shown. This formulation suggests the efficiency of assembly-language programming because most μCs have a carry bit and a "rotate right" instruction that includes the carry bit (C). Both software variables can be μC registers. In the following procedure, bitwise logic operations of AND, OR, and NOT (logic negation) are used and are μC instructions. For μCs without a NOT instruction, X is complemented by using the exclusive-OR (EOR or XOR) instruction with binary 1111 . . . (all binary ones, a two's-complement −1, or hexadecimal FFF . . .) and X.

0. Successive-approximation ADC.
1. Clear SR and SAR: SR ← 0; SAR ← 0.
   Set C to one: C ← 1.
2. Rotate SR right.
3. If $C = 1$, then return.
4. Output SR OR SAR to OUT: OUT ← SR OR SAR.

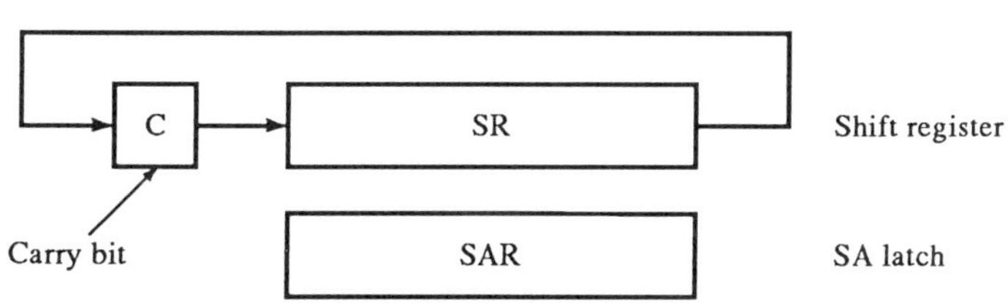

FIG. 12.20 Software registers for successive-approximation ADC algorithm.

5. Input from IN.
6. If IN = 1, then go to 2.
7. Else, set SAR to SAR AND $\overline{\text{SR}}$: SAR ← SAR AND (NOT SR). (Alternative: SAR ← SAR AND (SR EOR 1111 . . .).
8. Go to 2.

The 1 bit, initially in C, is shifted right, into SR, one bit per iteration. When it gets back to C (step 3 checks this), the procedure is done. Step 4 sets the SR 1 bit in the SAR. If the comparator (IN) is high, v_X is still greater than the SAR value, and this test bit is left set. If IN is low, the set bit made the SAR value too large, and it is cleared in step 7. Each bit, beginning at the MSB, is tested and then left set or cleared in SAR.

A speed enhancement for SA converters is to increase the clock rate after the first or second bit is determined. These bits have the most range and require the most slew time of the loop hardware. The less-significant bits cause less comparator voltage change and can be determined more quickly, allowing an increased clock frequency at the expense of more digital hardware.

The ramp and SA converters do not function correctly unless v_X is constant during conversion. For dynamic inputs, a sample-and-hold (S/H) must precede the ADC.

12.5 Integrating Analog-To-Digital Converters

A second category of ADC integrates v_X and outputs its average value over the conversion period. The *dual-slope* ADC of Fig. 12.21a is an instance. The input to an op-amp integrator is switched between input $v_X < 0$ and positive voltage reference V_R. The integrator output zero-crossing is detected by a comparator, and the count of a free-running counter is clocked as the digitized output. The conversion starts when the counter is reset and v_X is switched into the integrator. The ramp output has a slope of $-v_X/RC$ and ramps up until the counter overflows. For an n-bit counter, this phase lasts 2^n clock cycles or T amount of time. In the second phase, the reference is integrated instead. Since its polarity is opposite that of v_X, the slope changes polarity (Fig. 12.21b). When the integrator output crosses zero, the comparator latches the count. The second phase lasts for t_X time. The converter then begins another cycle.

Since the change in integrator output voltage Δv_O is the same for both phases,

$$\Delta v_O = \frac{v_X}{RC} \cdot T = \frac{V_R}{RC} \cdot t_X \Rightarrow t_X = \left(\frac{v_X}{V_R}\right) T \tag{12.34}$$

For a constant-frequency clock, the counts relate to the times by

$$N = f_{CLK} \cdot \Delta t \tag{12.35}$$

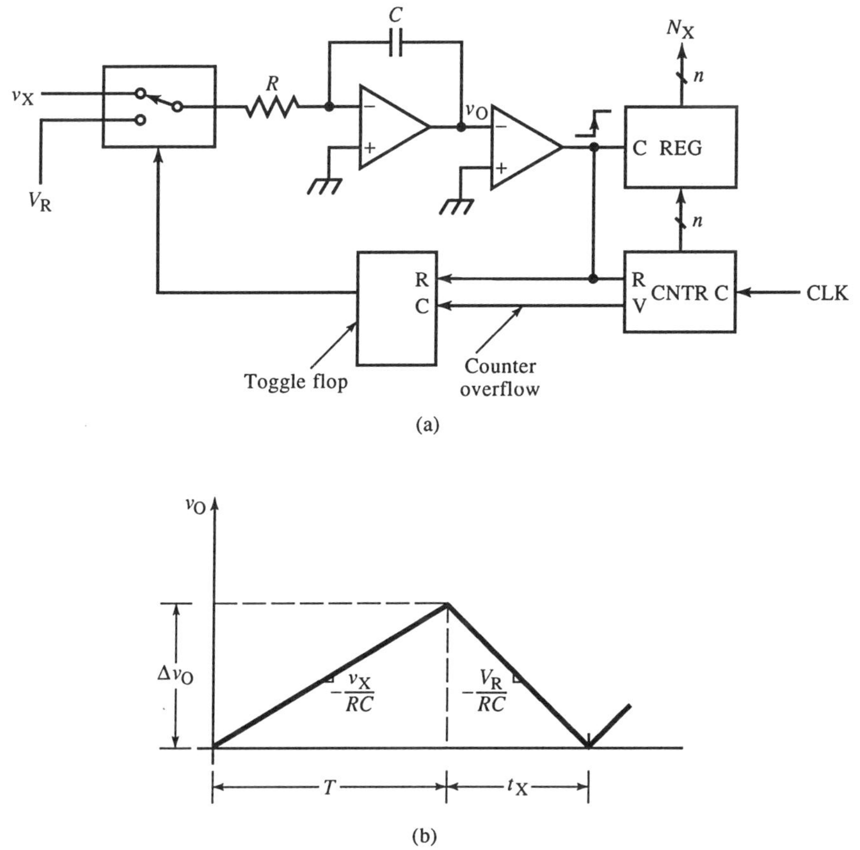

FIG. 12.21 Dual-slope integrating ADC topology (a) and dual-slope waveform (b).

Therefore, the output count is

$$N_X = \left(\frac{v_X}{V_R}\right) 2^n \tag{12.36}$$

Dual-slope converter accuracy is not dependent on long-term drift in R, C, or f_{CLK}, only V_R. What this analysis assumes is a perfect op-amp and comparator. Their input offsets and delay times degrade converter accuracy.

Most actual dual-slope converters correct for offset by introducing a third auto-zero phase before phase 1. In addition, for digital voltmeters (DVMs), a high input impedance is desired, and a buffer amplifier is added before the integrator. An alternative is the noninverting integrator (Fig. 12.22). If the DVM ground is "floating" (not connected) to the measured source, then the ×1 buffer provides high input impedance as it supplies the charging current for C through R.

Bipolar inputs require another reference, $-V_R$. Reference selection is determined by the comparator output at the end of phase 1. Another design

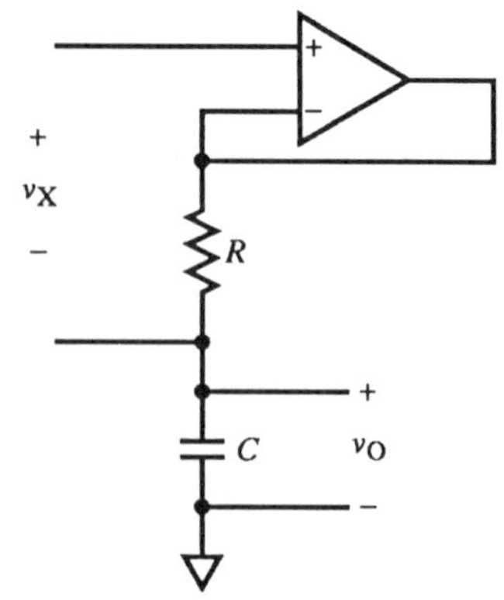

FIG. 12.22 A high input-impedance integrator for a floating circuit.

option for bipolar inputs is to exchange the input terminals by switching. This scheme, however, has difficulty with v_X near zero. Offsets can cause the readings for $+v_X$ and $-v_X$ to have different magnitudes. More significantly, when offsets dominate the input, the converter can integrate with the wrong (shallow) slope. When the reference is integrated, it is of the same polarity, and v_O never crosses zero. To avoid switching in the wrong polarity of reference, hysteresis around zero is sometimes added. But all of this is avoided with two references.

Another input circuit is a V/I converter and a current reference. This eliminates R from the integrator and could also eliminate the op-amp in some designs.

The accuracy of the dual-slope ADC is extended by the *triple-slope* ADC. An additional comparator senses that v_O is approaching 0 V and switches in a smaller reference and another counter. The slope magnitude decreases for this next phase and the time duration is extended. The extra counts contribute additional LSBs.

At somewhat less speed, the simpler *modified dual-slope* converter (Fig. 12.23a) uses only one switch and integrates the input during both phases. In phase 1, the negative reference $-V_R$ is integrated along with the input. If we assume $V_R > |v_X|$, the integrator output has a positive slope. When it reaches comparator threshold voltage V_C, the reference is switched off, and v_X integrates until the counter overflows at T. The integrator voltage, v_O, at this time depends on v_X. The next conversion cycle thus begins at a different initial v_O.

The conditions for convergence of v_O (and a steady digitized value) are found by solving for $v_O(i)$ where i is the cycle index. For the new cycle,

$$v_O(i+1) = v_O(i) + ut_X(i) + d[T - t_X(i)] \tag{12.37}$$

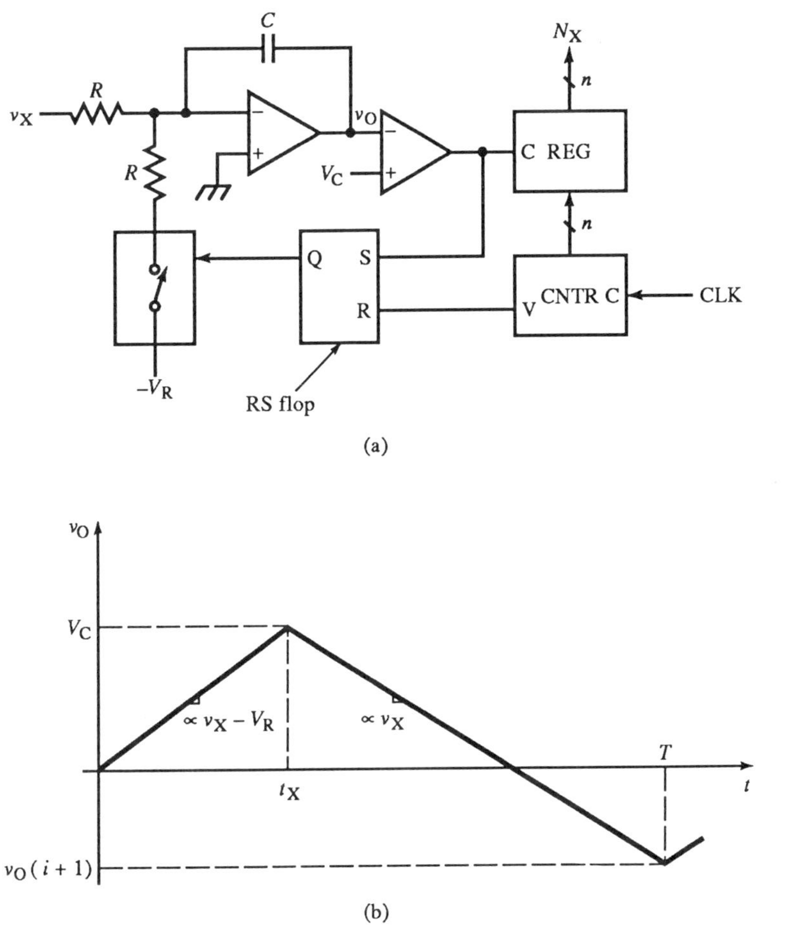

FIG. 12.23 Modified dual-slope ADC topology (a) and dual-slope waveform (b). The input is integrated for both slopes.

where the slopes are

$$u = -\frac{v_X - V_R}{RC} = \frac{V_R - v_X}{RC}, \qquad d = -\frac{v_X}{RC}; \qquad v_X,\ V_R > 0 \tag{12.38}$$

Also, from Fig. 12.23b,

$$t_X(i) = \frac{V_C - v_O(i)}{u} = \frac{V_C - v_O(i)}{V_R - v_X} \cdot RC \tag{12.39}$$

Substituting for t_X in (12.37) gives

$$v_O(i+1) = \left(\frac{d}{u}\right) v_O(i) + \left[V_C\left(1 - \frac{d}{u}\right) + dT \right] = a v_O(i) + b \tag{12.40}$$

This difference equation is solved by expanding several iterations, beginning with $i = 0$. The resulting recursion equation for $i+1 = n$ is

$$v_O(n) = a^n v_O(0) + \left(\frac{1-a^n}{1-a}\right) b, \qquad |a| < 1 \tag{12.41}$$

and is attained by using the geometric-series formula

$$\sum_{k=0}^{N-1} z^k = \frac{1-z^N}{1-z}, \qquad |z| < 1 \tag{12.42}$$

The series converges only when $|z|$ decreases with increasing k. For the converter, the convergence condition is

$$\left|\frac{d}{u}\right| < 1 \Rightarrow -d < u \Rightarrow \frac{v_X}{RC} < \frac{V_R - v_X}{RC} \Rightarrow v_X < \frac{V_R}{2} \tag{12.43}$$

That is, v_X must not exceed half the reference voltage V_R. Or, in time, $t_X < T/2$. The converged (or steady-state) value of v_O can be found by letting n go to infinity in (12.41) or by setting

$$v_O(i+1) = v_O(i)$$

in (12.40) and solving for v_O:

$$\text{steady-state } v_O = V_C + \left(\frac{ud}{u-d}\right) T \tag{12.44}$$

where the second term is always negative, as required for $v_O < V_C$. With v_O, we can now find t_X from (12.39):

$$\text{steady-state } t_X = \left(\frac{-d}{u-d}\right) T = \left(\frac{v_X}{V_R}\right) T \tag{12.45}$$

But this is the same as (12.34), and the digital output is expressed by (12.36). The modified dual-slope converter has the same transfer characteristic as the dual-slope ADC, though its dynamic response is first-order and takes a few cycles to converge.

In this realization of the modified dual-slope ADC, the Rs must match. V_C need not be accurate, only stable during convergence. Both Rs can be eliminated by driving the integrator with a V/I converter for v_X and replacing R and $-V_R$ with I_R. The switch must then be a current switch. This can be accomplished by letting the flop output divert I_R through a diode. For low leakage, a transistor is used instead. The RS flop consists of two cross-coupled NOR gates. The other two gates in a quad NOR-gate IC implement the clock generator.

Since t_X must be kept less than $T/2$, the fs t_X is set at $T/4$ by adding two additional bits to the counter (for $n+2$ bits total). This wastes 50% of the available integration time but is easy to implement (by a dual flop IC) and gives the converter a near-100% overrange capability—an additional half-digit. Besides the register and counter, the total parts count is less than a dozen

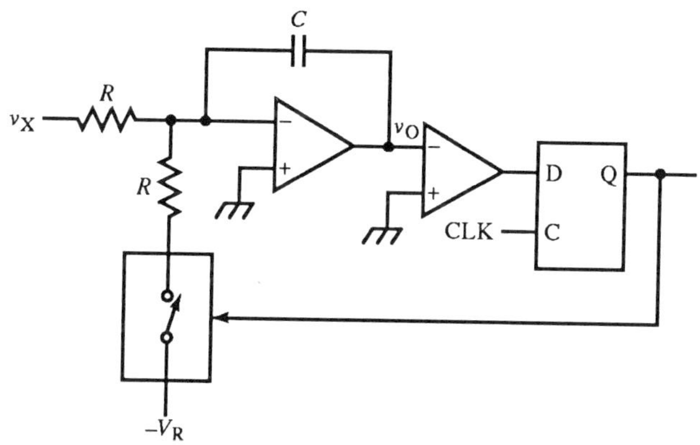

FIG. 12.24 Quantizing ADC. The logic block changes from the RS flop of Fig. 12.23 to a D flop.

to implement a three-digit DVM. (A featureless converter such as this is usually called a digital panel meter (DPM) instead.)

Dual-slope ADCs require a large v_O range to achieve precision. An idea that is the digital analog of the virtual ground is realized in the *charge-balancing* (or quantizing or delta-sigma) ADC of Fig. 12.24. The circuit topology is very similar to the modified dual-slope ADC, but it works differently. The big circuit difference is that the flop driven by the comparator is clocked, a D-type flop instead of an RS flop.

On a given cycle of the clock, the reference is switched in or out of the integrator to keep v_O near ground. The comparator output gives the sign of the error. In other words, v_O is nulled by discrete-time feedback. The number of clock cycles that the flop was high, N_X, over the total number of conversion counts N, indicates v_X.

The transfer characteristic is calculated by constructing the charge-balance equation for the total charge from v_X and $-V_R$ input to the integrator. For $v_O = 0$, they must be equal, or

$$Q_X = Q_R \tag{12.46}$$

These charges are the sums of the per-cycle charges:

$$q_X = \left(\frac{v_X}{R}\right) T_{CLK}, \qquad q_R = \left(\frac{V_R}{R}\right) T_{CLK} \tag{12.47}$$

The total charge of each depends on the number of cycles each is integrated. Then

$$Q_X = q_X N = \left(\frac{v_X}{R}\right) NT_{CLK}, \qquad Q_R = q_R N_X = \left(\frac{V_R}{R}\right) N_X T_{CLK} \tag{12.48}$$

Then substituting into (12.46) and solving for the output, we obtain

$$N_X = \left(\frac{v_X}{V_R}\right) N = \left(\frac{v_X}{V_R}\right) 2^n \tag{12.49}$$

for an n-bit conversion-time counter. This result is, again, the same as for the previous converters.

The charge-balancing circuit we have analyzed is also used as a modulator for serial digital telecommunications (in CODEXs) and speech processing.

An advantage of the integrating ADC is its measurement of the average v_X. By integrating, it has inherent noise rejection and does not need a S/H circuit. The noise rejection capability is quantified by beginning with a dc V_X with sinusoidal noise added:

$$v_X = V_X + V_N \sin \omega_N t \tag{12.50}$$

The integrator averages v_X over the conversion period T, so that

$$\text{avg } v_X = \frac{1}{T}\int_0^T v_X\, dt = V_X + \frac{V_N}{\omega_N T}(1 - \cos \omega_N T) \tag{12.51}$$

The normal-mode rejection (NMR) of the noise is

$$\text{NMR} = \frac{\text{input noise}}{\text{output noise}} = \frac{V_N}{(V_N/\omega_N T)(1 - \cos \omega_N T)} = \frac{\omega_N T}{1 - \cos \omega_N T} \tag{12.52}$$

Since $\omega_N = 2\pi f_N$, and $1 - \cos 2x = 2 \sin^2 x$, then

$$\text{NMR} = \frac{\pi f_N T}{\sin^2(\pi f_N T)} \tag{12.53}$$

In the decibel scale this is

$$\text{NMR(dB)} = 20 \log \text{NMR} = 20\{\log(\pi f_N T) - \log[\sin^2(\pi f_N T)]\} \tag{12.54}$$

At $f_N = n/T$, for whole-number n, NMR is infinite. In practice it is typically about 60 dB of rejection. An exact number of noise cycles fit the integration interval T, and the sum of the areas of their positive and negative half-cycles cancel (Fig. 12.25a). As f_N varies from n/T, the half-cycles of noise at the ends of the interval are truncated and contribute some fraction of a half-cycle. Figure 12.25b shows the worst case, in which an entire extra positive half-cycle is integrated at $f_N = 1.5/T$. For rejection of power-line noise, f_N is often chosen to be a multiple of the power-line frequency.

More significant are the NMR minima of $\pi f_N T$. They occur midway between the maxima, at $f_N = 1.5n/T$. As f_N increases, according to (12.54), the NMR minima increase at 20 dB/dec. At noise frequencies of about n times $1/T$, about n cycles of noise occur during T. The more half-cycles, the less each contributes to the integrated total. Thus, a fraction more of a half-cycle contributes less error the higher f_N is. Note that NMR is the reciprocal of the integrator frequency response, which rolls off at 20 dB/dec with periodic notch filters.

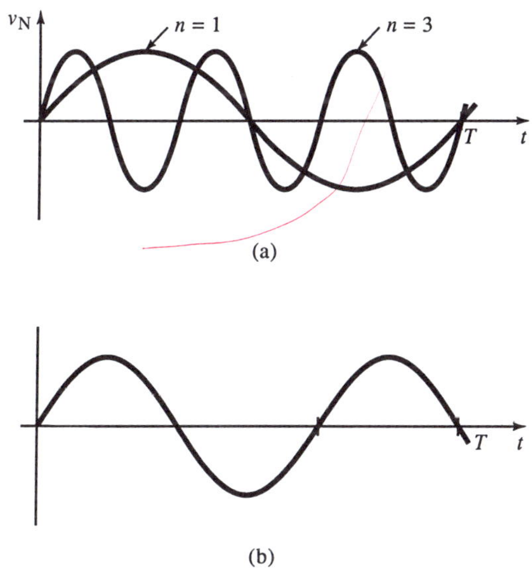

FIG. 12.25 Integration filters out harmonics of integration period (a). The worst case (b) is the net addition of an entire half-cycle of noise.

12.6 Voltage-to-Frequency Converters

A special kind of integrating ADC converts input voltage or current to a pulse frequency. It is a kind of linear VCO or FM modulator with digital output. The topology of the *voltage-to-frequency* (V/F) converter (VFC) is similar to previous integrating ADCs (Fig. 12.26a). As with parallel-feedback converters, the topological variations among integrating ADCs is in the logic block driven by the comparator. For the *asynchronous VFC*, a MMV replaces the flop of the charge-balancing converter.

The operation resembles the modified dual-slope ADC. When the integrator output v_O goes below V_C, the comparator output goes high, triggering the MMV and turning on the reference switch. The MMV time-out is t_h, the time that the output pulse v_f is high. During t_h, v_O ramps up with a slope of u. When the MMV times out, the reference is switched out, and $v_X > 0$ causes v_O to ramp down with slope d. Slopes u and d are the same as those in (12.38). The change in v_O over one cycle is

$$\Delta v_O = ut_h = \frac{V_R - v_X}{RC} \cdot t_h \tag{12.55}$$

From this,

$$t_l = -\frac{\Delta v_O}{d} = \left(\frac{V_R}{v_X} - 1\right) t_h \tag{12.56}$$

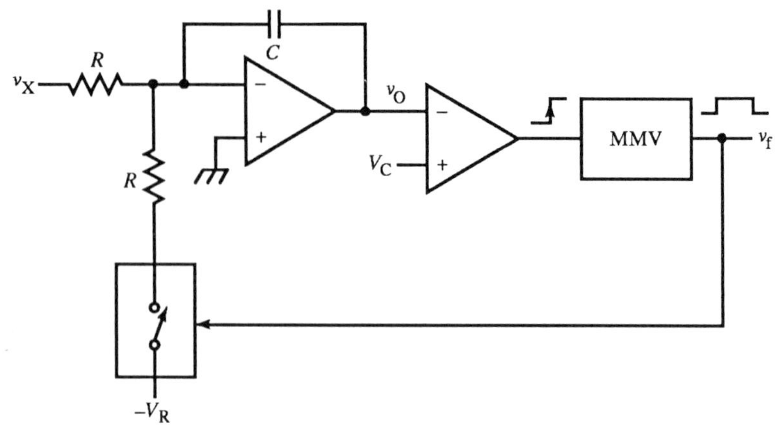

(a)

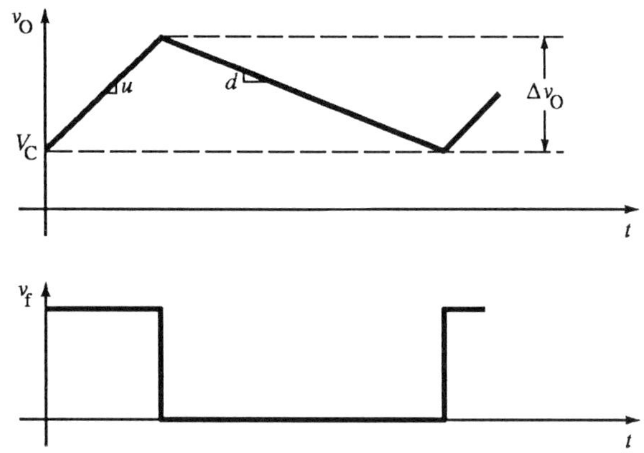

(b)

FIG. 12.26 Asynchronous VFC. The logic block is a MMV (a). Integrated waveform and digital output pulse timing (b).

The output period is the sum of the half-cycles, or

$$T = t_h + t_l = \left(\frac{V_R}{v_X} - 1\right) t_h + t_h = \frac{V_R}{v_x} \cdot t_h \tag{12.57}$$

Finally, the output frequency is

$$f = \frac{1}{T} = \left(\frac{v_X}{V_R}\right) \frac{1}{t_h} \tag{12.58}$$

This formula is similar to that of previous integrating ADCs except that it depends on t_h, the MMV time-out duration, instead of a counter overflow period. Since t_h is typically set by an *RC* circuit, asynchronous VFC accuracy is limited by it. The accuracy also depends on the matching of the R, but the analysis could have been based on an input current i_X and reference current I_R instead. The resistors are implementation dependent and not fundamental to the operating principle.

The LM331 is an 8-pin VFC IC (Fig. 12.27a). Instead of using an op-amp integrator, it avoids op-amp error by integrating with a shunt RC that is maintained at v_X. The shunt RC voltage v_O must be kept small to avoid

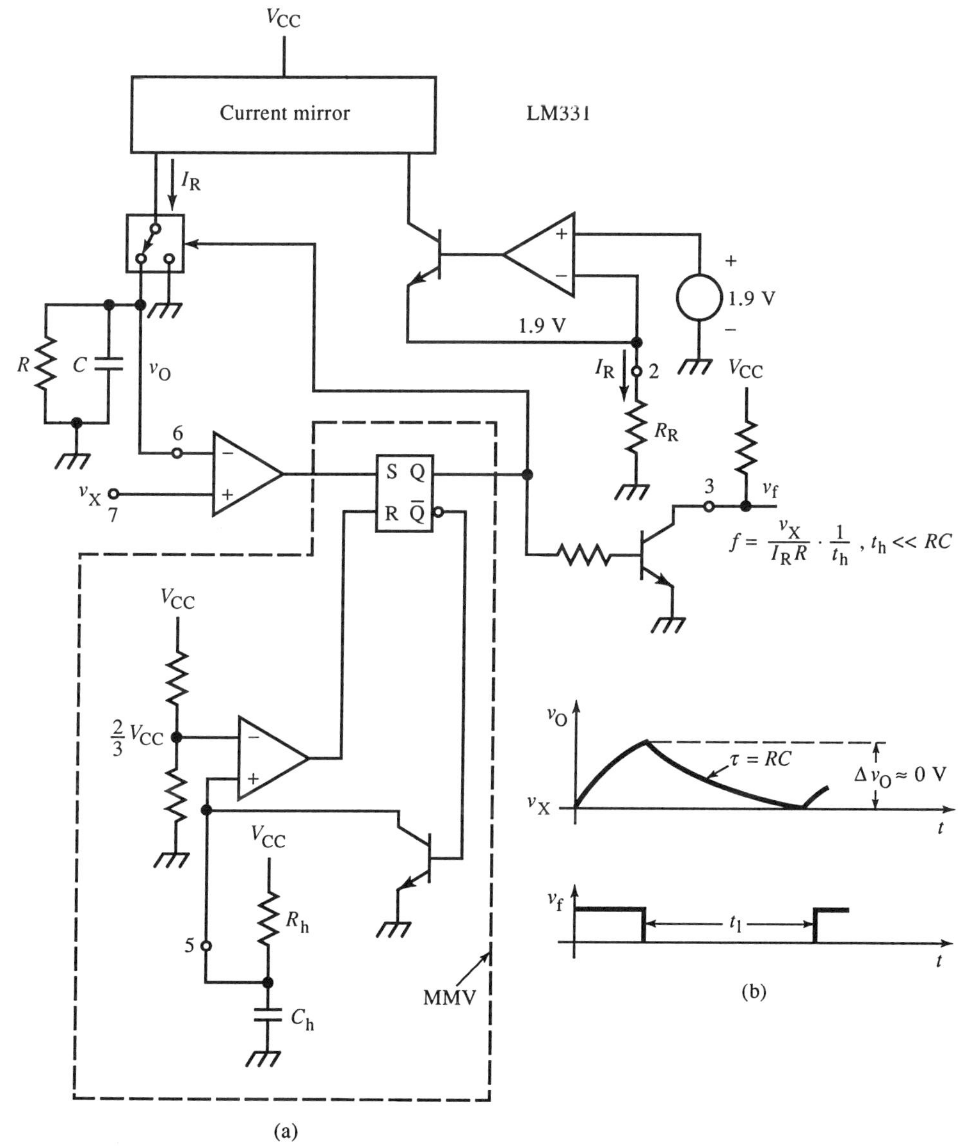

FIG. 12.27 LM331 8-pin VFC IC (a) uses charge-balancing at v_O (b). Δv_O must be kept small to avoid nonlinearity.

nonlinearity. If the exponential waveforms of v_O (Fig. 12.27b) have a time constant RC that is much larger than t_h, they are approximately linear. By keeping $v_O \cong v_X$, the LM331 performs charge balancing at v_O. The charge through R over T must be the reference charge during T, or

$$\frac{v_O}{R} \cdot T \cong \frac{v_X}{R} \cdot T = I_R t_h, \qquad t_h \ll RC \tag{12.59}$$

Solving for $f = 1/T$ gives

$$f = \left(\frac{v_X}{RI_R}\right)\frac{1}{t_h}, \qquad t_h \ll RC \tag{12.60}$$

This result is also valid for a charge-balancing VFC with a linear integrator but without the constraint. The typical fs frequency is 10 kHz at an output duty-ratio of 50%. Unlike the modified dual-slope ADC, no convergence condition exists, but as t_l approaches zero, the fs frequency asymptotically approaches $1/t_h$ of 20 kHz. Since the MMV timing is based on a threshold voltage of $\frac{2}{3}V_{CC}$, then

$$t_h = R_h C_h \ln 3 \simeq 1.1 R_h C_h \tag{12.61}$$

Also, $I_R \cong 1.9\,V/R_R$.

A more precise analysis, calculated from the exponential v_O, yields a period of

$$T = RC \ln\left[\left(\frac{I_R R}{v_X}\right)(e^{-t_h/RC} - 1) + 1\right] \tag{12.62}$$

For $t_h \ll RC$,

$$(e^{-t_h/RC} - 1) \cong 0$$

We now apply the approximations

$$\ln(1+x) \cong x, \qquad x \cong 0; \qquad e^x \cong 1 + x, \qquad x \cong 0 \tag{12.63}$$

to (12.62):

$$T \cong RC\left[\left(\frac{I_R R}{v_X}\right)\left(\frac{t_h}{RC}\right)\right] = \left(\frac{I_R R}{v_X}\right) t_h, \qquad t_h \ll RC \tag{12.64}$$

This period is consistent with (12.58). For applications in which a compressed scale for v_X is desired, the nonlinearity of this converter can be advantageous, thereby invoking the addage "if you can't fix it, feature it."

The VFC is most sensitive to noise at zs, when the downslope d is shallowest, causing comparator output jitter among crossings of its threshold, and thereby jittering f. However, the VFC is an integrating type of ADC because a frequency measurement requires counting v_f over a known period. This counting function is the digital equivalent of integration. The longer the count interval, the more the input is averaged, the greater the precision, and also the slower the conversion rate. For faster conversion at the same precision, the fs frequency must be increased. By changing count intervals, we can make speed–precision trade offs without a converter change.

The drift in MMV t_h can be averted by using a digital timer with an accurate clock. Then t_h would, on the average, be accurate. Since the clock is asynchronous with the comparator output, the timer has phase jitter and the time-out varies up to a complete clock cycle. Elaborate schemes have been devised to synchronize a digital counter with an asynchronous trigger to produce an accurate time-out. One simpler scheme combines an analog ramp generator with a counter. The ramp slope is set to V_C / T_{CLK}, where V_C is a comparator threshold. The trigger starts the ramp. It runs up until the active clock edge occurs. The ramp output is held constant until the counter overflows. (More likely, it is a down counter that underflows.) The ramp is restarted. When it crosses V_C, the comparator signals the time-out. The counter counts one less cycle than is required for the time-out because the ramp generator adds a cycle. Its slope error affects the time-out as an error in only one clock period.

Instead of substituting a clocked timer for the MMV, the *synchronous* (or clocked) VFC operates similarly to the charge-balancing ADC except that the reference is turned on for only one clock cycle at a time. The comparator output switches the reference only at the active clock edge (Fig. 12.28). The D-flop input is gated to enable its output to be high for the cycle, thus generating the output pulse. In commercial synchronous VFCs, the flop output triggers a MMV that sets the output pulse width.

The dual-slope waveform of v_O is synchronous with the clock only at discrete values of v_X. For v_X between these quantum levels, the average level of v_O slowly drifts due to accumulating phase error (Fig. 12.29). The comparator edge drifts relative to the clock, causing the reference on-time to change linearly. This causes the average level of v_O to ramp up or down. When the phase between comparator and clock outputs drifts by a full clock cycle (or 2π radians of phase), the comparator and clock are again in sync; v_O has drifted to a quantum level where the phase error is zero. Since comparator

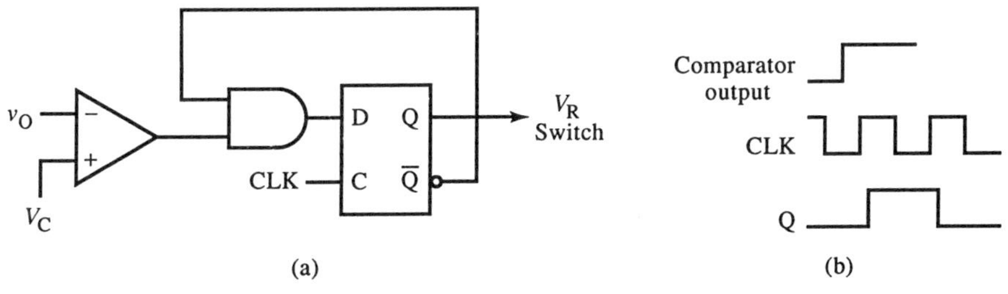

FIG. 12.28 Synchronous VFC logic block. The output also drives the reference switch and is allowed to be high for only one clock period at a time.

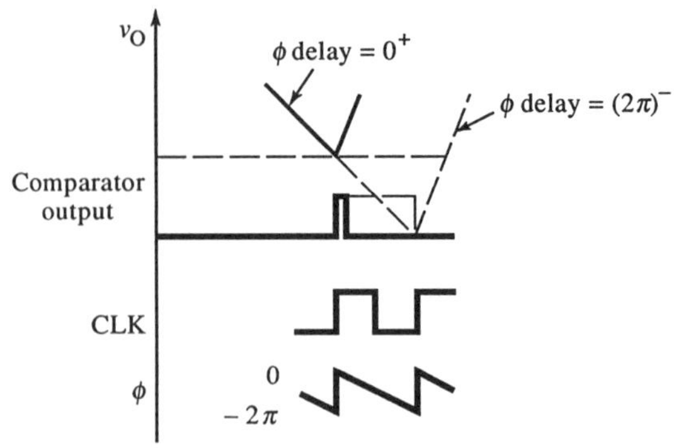

FIG. 12.29 Timing waveforms for synchronous VFC show effect of asynchronous comparator input. Phase ϕ drifts linearly from cycle to cycle, but average error (over many cycles) is zero.

and clock edges can coincide, the output can be indeterminate for some time, causing frequency jitter. (See Section 11.9.) A kind of trigger-generator circuit is required for high performance.

12.7 Parallel and Recursive Conversion Techniques

The fastest ADCs are *parallel* or *flash* converters. They have a resistive-divider string of 2^n resistors for an n-bit converter. Each resistor drops V_{LSB} and sets the reference input on one of 2^n latching comparators that drive an encoder. A clock stores the data as 2^n decisions are made simultaneously: $2^n - 1$ for n bits of conversion plus one for overrange detection. No S/H function is required. Because the circuit complexity grows exponentially with the number of bits, these converters trade off cost, simplicity, and lower power for speed. Also with complexity comes a loss of precision because many parts must meet design tolerances.

Parallel ADC power is reduced by CMOS implementation. Switched-capacitor comparators designed from CMOS logic inverters (see Fig. 9.35) reduce power over BJT comparators and can be easily auto-zeroed. But for many applications, the optimum criteria are less complexity and more precision at somewhat reduced speed. This has led to conversion topologies that use m-bit parallel ADCs to digitize $n > m$ bits by iteration.

The *multistage* or *subranging* flash converter of Fig. 12.30 has two stages of flash ADCs. The first ADC converts m bits. These MSBs drive a DAC. Its output is subtracted from the input. This remainder or residue is a fraction of one V_{LSB} of the first converter. It is the difference between v_X and the m-bit quantized v_X. The second ADC converts this remainder for the remaining $n - m$ LSBs. If its input range is the same as ADC1, then each V_{LSB} (each

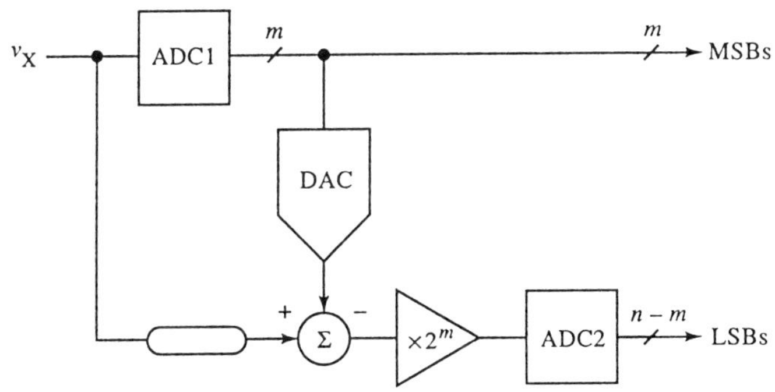

FIG. 12.30 Two-stage parallel ADC feedforward topology.

step) of ADC1 spans the input range of ADC2, and the remainder must be multiplied by 2^m for proper scaling. Consequently, ACD1 must have n-bit accuracy in the placement of its voltage levels or steps. Also, to avoid misalignment in time, or phase error, subtraction from v_X requires that v_X be delayed by the same amount as the path delay of ADC1 and the DAC.

This idea can be taken further. To save on ADCs and DACs, the *recursive subranging* ADC has a feedback topology (Fig. 12.31) instead of the feedforward topology of the multistage flash ADC. In effect, it is a parallel-feedback converter with an m-bit comparator (the ADC) instead of the usual one-bit comparator. It requires n/m iterations or cycles for n-bit conversion. For each iteration, beginning with the m MSBs, the ADC output is stored in m bits of an n-bit output register. The multiplexer (MUX) directs the bits. The PGA gain is increased by 2^m each iteration. This ADC technique requires a hold circuit for v_X.

The multistage idea can be taken to its limits by converting one bit per stage (Fig. 12.32). In this *n-stage* flash ADC, each ADC is a comparator designed to have accurate output levels of 0 V and $V_R/2$ V. Instead of iterating in time, this design iterates hardware stages. It needs no hold on v_X since v_X ripples through the stages, being processed as it goes, much like a distributed amplifier. Since it is a bitwise converter, it implements the SA algorithm in

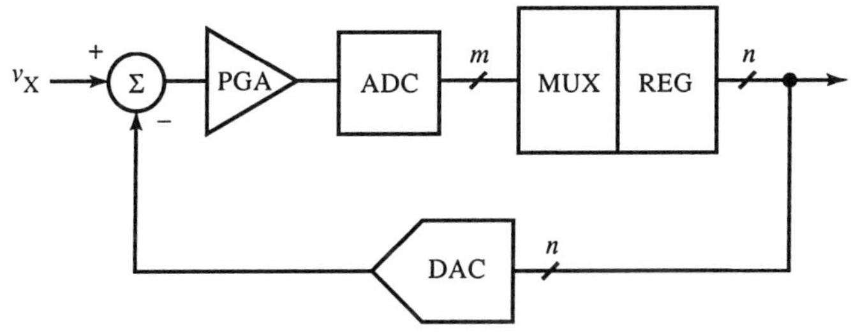

FIG. 12.31 Recursive subranging ADC, a feedback form of Fig. 12.30.

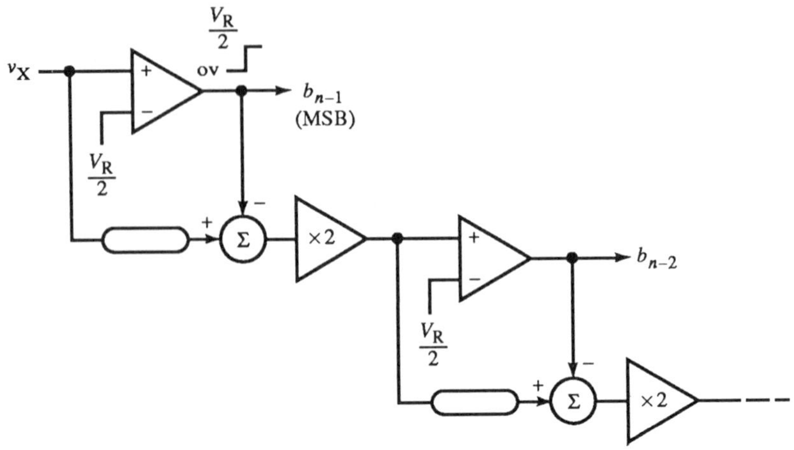

FIG. 12.32 An n-bit n-stage ADC with 1-bit ADCs (comparators) at each stage.

space (hardware), instead of in time as the SA ADC does:

$$v_i = 2\left(v_{i+1} - b_{i+1}\left(\frac{V_R}{2}\right)\right) \tag{12.65}$$

where $v_n = v_X$.

This same idea has been used in John Fluke Co. DVMs, called the *recirculating-remainder* or *cyclic* converter (Fig. 12.33). It follows a similar recursive equation:

$$v_i = 2|v_{i+1}| - V_R \tag{12.66}$$

where $v_n = v_X$. In the n-stage flash ADC, the remainder passed to the next stage is always positive. Here, the error is bipolar; its sign determines the bit. It is made positive by $|v|$, amplified by two and then subtracts V_R. The block diagram in Fig. 12.33 can be repeated, like the n-stage flash, or a S/H can hold the output for recirculation n times.

The serial bit output is ordered MSB first, but the encoding is in *Gray code.* This code is commonly used in mechanical shaft position encoders

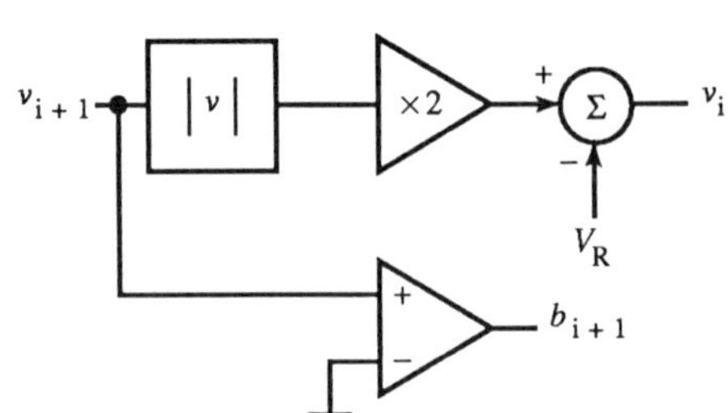

FIG. 12.33 One stage of a recirculating remainder ADC.

because only one bit changes between adjacent states. If the bit outputs are misaligned, an error of only ±1 LSB occurs. Gray-code encoders are used in fast flash converters for the same reason; any time-skew among output bits between two successive outputs results in at most 1 LSB of error. Gray code is converted to offset binary by the formula

$$b_i = b_{i+1} \oplus g_i \tag{12.67}$$

where b_i are output offset-binary bits, g_i are input Gray-code bits, and $\oplus$ is the exclusive-OR logic operation.

12.8 Time-Domain Sampling Theory

The explanation of A/D conversion assumes that a voltage at one point in time is converted. For a dynamic signal, some means of sampling a voltage at an instant and holding this voltage constant is essential to the conversion process. Even flash converters require that all comparators sense v_X at the same instant. Delays in the latching clock and the input among comparators causes this time instant to be instead a time interval t_a, called the *aperture uncertainty* or *aperture jitter.* Besides this, there is delay from the clock edge to when the input is actually sampled, or *aperture delay.*

Aperture jitter limits the maximum sine frequency of v_X that can be digitized. For a frequency f, all comparators must settle within one V_{LSB} or $2^{-n}V_{fs}$ for n bits. The sine slew-rate is $2\pi f V_{fs}$ for worst-case. Then t_a must be less than the time taken to slew 1 LSB, or V_{LSB}; that is,

$$t_a < \frac{V_{LSB}}{\text{maximum slew-rate}} = \frac{2^{-n}V_{fs}}{2\pi f V_{fs}} = \frac{1}{2^{n+1}\pi f} \tag{12.68}$$

The maximum sine frequency for a given aperture jitter is thus

$$\text{maximum sine } f = \frac{1}{2^{n+1}\pi t_a} \tag{12.69}$$

An 8-bit converter with 100 ps aperture jitter has a maximum digitizing bandwidth of about 6 MHz. By its nature, aperture jitter is a statistical quantity, leading to rms values of the quantities calculated with it.

ADCs that require their input to be held constant over their full conversion period must be preceded by a sampling circuit that then holds the sampled value constant. These are *sample-and-hold* (S/H) circuits. DACs are inherently digital hold circuits. They hold the sampled output constant and effect a *zero-order hold* (ZOH). A S/H variation is the track-and-hold (T/H) circuit. Its output follows the input in the tracking mode. S/H theory also applies to T/H circuits.

S/H circuits are based on an underlying theory that has general application to discrete-signal (or sampled-data) systems. We begin its development in the

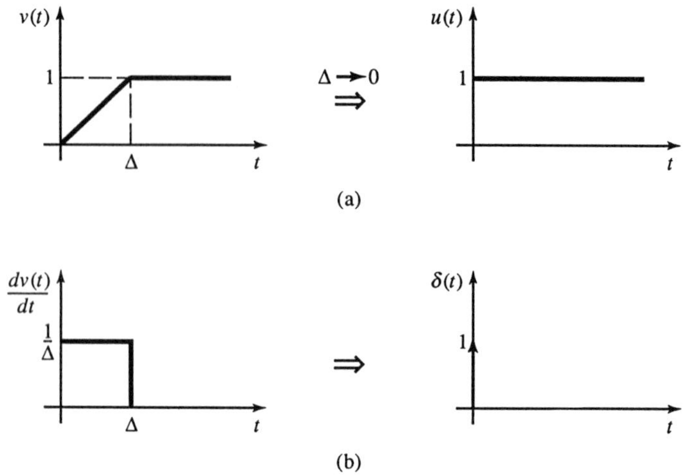

FIG. 12.34 Unit step function (a) and impulse function (b) result from a limiting process.

time domain with the step and impulse functions (Fig. 12.34). The step function is derived by taking the limit of $v(t)$ in (a) as $\Delta \to 0$. Then for $t < 0$, $u(t) = 0$, but at $t = 0^+$, it is 1. Similarly, the impulse function (in a limiting sense) is derived in (b) as the derivative of $v(t)$. As $\Delta \to 0$, the width of the rectangular pulse goes to zero, but the amplitude goes to infinity. The area remains constant in the limiting process and is the "weight" or "strength" of the impulse. In the limit,

$$\lim_{\Delta \to 0} \int_0^\Delta \frac{1}{\Delta}\, dt = 1 \tag{12.70}$$

The unit impulse has unit strength at $t = 0$ and is zero elsewhere so that

$$\int_{-\infty}^{+\infty} \delta(t)\, dt = 1 \tag{12.71}$$

Now multiply $\delta(t)$ by a continuous function $v(t)$ in the integral. Since δ is nonzero only at zero, $v(0)$ effectively weights $\delta(t)$, and

$$\int_{-\infty}^{+\infty} v(t)\, \delta(t)\, dt = v(0) \tag{12.72}$$

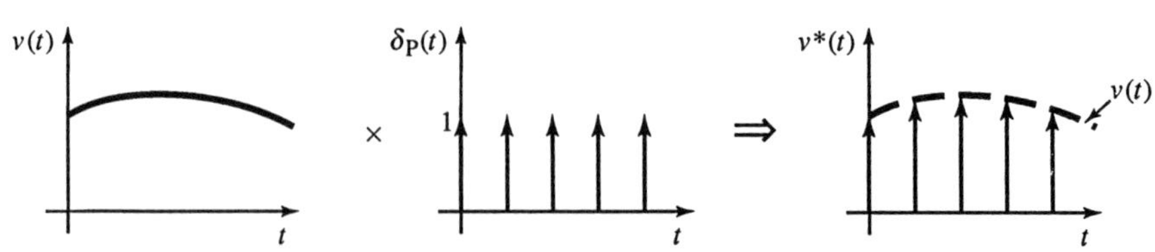

FIG. 12.35 A signal $v(t)$ is impulse modulated or sampled by multiplying it by an impulse train.

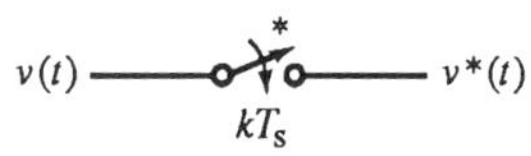

FIG. 12.36 Schematic representation of sampling switch. It closes momentarily every T_s and transforms continuous $v(t)$ into discrete $v^*(t)$.

More generally, if δ is shifted in time by kT, then

$$\int_{-\infty}^{+\infty} v(t)\,\delta(t-kT)\,dt = v(kT) \tag{12.73}$$

This can also be expressed as an integral with t as upper bound:

$$\int_{C}^{t} v(\tau)\,\delta(\tau-kT)\,d\tau = v(kT), \qquad t > kT \tag{12.74}$$

The impulse function is central to sampling theory. A periodic sequence (or "train") of impulses conveniently characterizes the sampling process. A repetitive $\delta(t)$ with period T_s is the sum of an infinite number of time-shifted impulses spaced T_s apart, or

$$\delta_P(t) = \sum_{k=-\infty}^{\infty} \delta(t-kT_s) \tag{12.75}$$

When $v(t)$ is multiplied by $\delta_P(t)$, a sampled form of $v(t)$, or $v^*(t)$, results:

$$v^*(t) = v(t)\delta_P(t) = \sum_{k=0}^{\infty} v(kT_s)\,\delta(t-kT_s) = v(kT_s), \qquad k = 0, 1, \ldots \tag{12.76}$$

The resulting function is nonzero only where the impulses occur, with strengths determined by $v(t)$ (Fig. 12.35). The amplitudes of the impulses, though infinite, graphically represent their strengths, which are determined by $v(t)$. This is the behavior of the *ideal sampler*, a switch that closes only for an instant (Fig. 12.36).

Two equivalent graphic representations of $v^*(t)$ are shown in Fig. 12.37. According to (12.76), the discrete $v(t)$ for $t = kT_s$ in (a) is equivalent to the

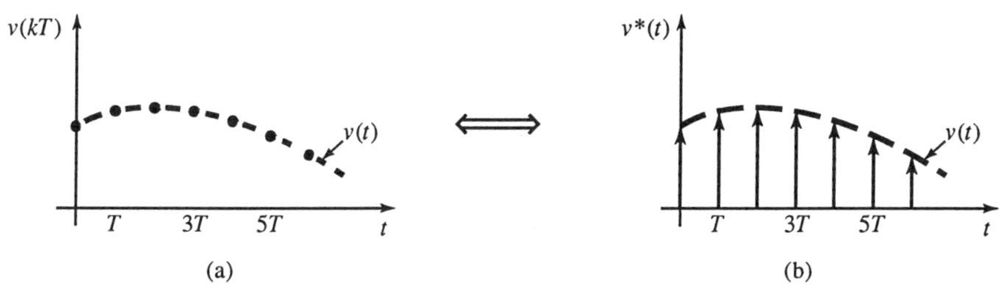

FIG. 12.37 Equivalent representations of a sampled function: (a) $v(kT)$ is discrete $v(t)$, and (b) $v^*(t)$ is weighted impulse train. $v(kT) = v^*(t)$.

weighted-impulse representation of (b):

$$\text{sampled } v(t) = v^*(t) = v(kT_s)$$

Consequently, $\delta_P(t)$ can also be interpreted as a sequence of unit samples.

12.9 Frequency-Domain Sampling Theory

In the frequency domain, sampling is *impulse modulation*; $v(t)$ amplitude-modulates the impulse train. The Laplace transform and Fourier series reveal another perspective on sampling and lead to important design criteria.

To derive the Laplace transform of $v^*(t)$, we begin with $\delta(t)$ and adapt (12.72) to give

$$\mathscr{L}\{\delta(t)\} = \int_0^\infty \delta(t)\, e^{-st}\, dt = e^0 = 1 \tag{12.77}$$

Second, the Laplace transform of δ_P is found from (12.75) using (12.74):

$$\mathscr{L}\{\delta_P(t)\} = \sum_{k=0}^{\infty} e^{-skT_s} = \Delta_P(s) \tag{12.78}$$

where T_s is the sampling period. One period of a function $f_1(t)$, such as a single cycle of a sinusoid, can be made repetitive as the series

$$f(t) = f_1(t) + f_1(t - T_s) + f_1(t - 2T_s) + \cdots = \sum_{k=0}^{\infty} f_1(t - kT_s) \tag{12.79}$$

Given $\mathscr{L}\{f_1(t)\} = F_1(s)$, then

$$\mathscr{L}\{f(t)\} = F_1(s) + F_1(s)\, e^{-sT_s} + F_1(s)\, e^{-s2T_s} + \cdots = \sum_{k=0}^{\infty} F_1(s)\, e^{-skT_s} \tag{12.80}$$

Applying the formula,

$$\sum_{k=0}^{\infty} z^k = \frac{1}{1 - z}, \qquad |z| < 1$$

to (12.79) gives

$$\mathscr{L}\{f(t)\} = \frac{F_1(s)}{1 - e^{-sT_s}}, \qquad |e^{-sT_s}| < 1 \tag{12.81}$$

This is now applied to δ_P:

$$\mathscr{L}\{\delta_P(t)\} = \Delta_P(s) = \frac{1}{1 - e^{-sT_s}}, \qquad |e^{-sT_s}| < 1 \tag{12.82}$$

Finally, the Laplace transform of $v^*(t)$ is

$$\mathscr{L}\{v^*(t)\} = V^*(s) = \sum_{k=0}^{\infty} v(kT_s)\, \Delta_P(s) = \sum_{k=0}^{\infty} v(kT_s)\, e^{-skT_s} \tag{12.83}$$

This infinite series of exponentials in s makes $V^*(s)$ nonalgebraic and is unwieldy for systems analysis. It does, however, resemble the Laplace transform of $v(t)$ for $t = kT_s$. It is simplified by a change of variable,

$$z \equiv e^{sT_s} \tag{12.84}$$

Solving for s, we get

$$s = \frac{1}{T_s} \ln z \tag{12.85}$$

and substituting for s in (12.83) yields

$$V^*(s)\big|_{s=(1/T_s)\ln z} = \mathscr{Z}\{v(t)\} = \sum_{k=0}^{\infty} v(kT_s) z^{-k} \tag{12.86}$$

The operator $\mathscr{Z}$ is the *Z transform.* The Z transform of $v(t)$ is written as $V(z)$, with the understanding that this is not $V(s)$ with z substituted for s. Note that z is a shifting variable; z^{-k} shifts $v(kT)$ by k periods. The Z transform is used in sampled-system analysis the way that the Laplace transform is used with continuous functions. The s-domain offers a continuous view of discrete signals and the z-domain a discrete view of continuous signals.

We now express $v^*(t)$ using the Fourier series. Repetitive $v(t)$ with frequency ω_s can be expressed as the sum of sinusoids at integer multiple frequencies (or *harmonics*) of ω_s:

$$v(t) = \underbrace{\frac{a_0}{2}}_{\text{dc term}} + \underbrace{\sum_{n=1}^{\infty} a_n \cos n\omega_s t}_{\text{even harmonics}} + \underbrace{\sum_{n=1}^{\infty} b_n \sin n\omega_s t}_{\text{odd harmonics}} = \sum_{n=-\infty}^{\infty} c_n e^{jn\omega_s t} \tag{12.87}$$

where

$$a_n = \frac{2}{T_s} \int_{-T_s/2}^{T_s/2} v(t) \cos(n\omega_s t)\, dt, \qquad b_n = \frac{2}{T_s} \int_{-T_s/2}^{T_s/2} v(t) \sin(n\omega_s t)\, dt \tag{12.88}$$

and for the complex Fourier series,

$$c_n = \frac{1}{T_s} \int_{-T_s/2}^{T_s/2} v(t)\, e^{-jn\omega_s t}\, dt \tag{12.89}$$

The two representations are equivalent, and are related by

$$\|c_n\| = \tfrac{1}{2}\sqrt{a_n^2 + b_n^2}, \qquad \vartheta_n = \tan^{-1}\left\{\frac{b_n}{a_n}\right\} \tag{12.90}$$

Even functions of time have no sine terms; odd functions have no cosine terms. Some $v(t)$ can be made odd by subtracting a dc offset. The odd function is then transformed and the offset is added as a dc term.

In actual samplers, the sampling signal is an approximation to an impulse train. It has finite amplitude and time duration. The effect this has on sampling

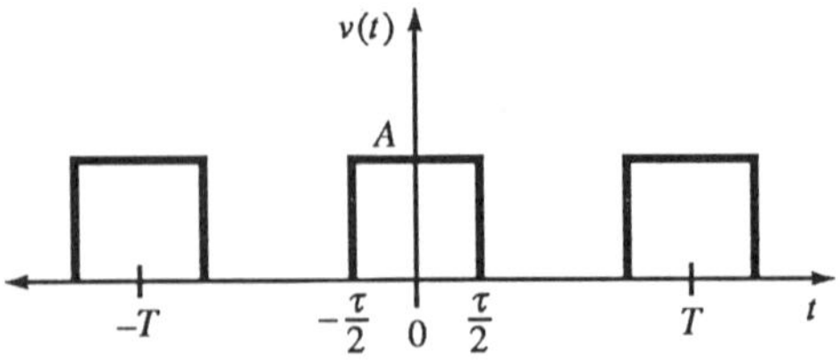

FIG. 12.38 Pulse train. As $\tau \to 0$, it approaches an impulse train.

can be found by assuming the sampling signal to be a pulse train with amplitude A and pulse width τ (Fig. 12.38). Let $\omega_n = n\omega_s$. Since $v(t)$ is centered around $t = 0$, it is an even function, and

$$a_n = \frac{2A}{T_s} \int_{-\tau/2}^{\tau/2} \cos \omega_n t \, dt = A\tau \cdot \frac{\sin(\omega_n \tau/2)}{\omega_n \tau/2} \tag{12.91}$$

where the discrete a_n have the continuous *envelope* of the form

$$\operatorname{sinc} x \equiv \frac{\sin x}{x}$$

shown in Fig. 12.39 for $x = \omega_n \tau/2 = n\pi(\tau/T_s)$. Instead of an impulse, sinc x is the result of finite-width sampling pulses.

As $\tau \to 0$, the pulse train approaches an impulse train. The separation of $a_n(\omega)$ decreases in frequency. If instead we let T_s increase, then the effect is the same; harmonic frequency separation decreases. As $T_s \to \infty$, the a_n merge into a continuous sinc function with a continuous frequency spectrum:

$$\lim_{T_s \to \infty} \Delta\omega_n = \lim_{T_s \to \infty} \left(\frac{2\pi(n+1)}{T_s} - \frac{2\pi n}{T_s} \right) = \lim_{T_s \to \infty} \frac{2\pi}{T_s} = 0 \tag{12.92}$$

As $T_s \to \infty$, the function becomes aperiodic, and the Fourier series becomes the *Fourier transform*:

$$\mathscr{F}\{v(t)\} = V(j\omega) \equiv \int_{-\infty}^{+\infty} v(t)\, e^{-j\omega t} \, dt \tag{12.93}$$

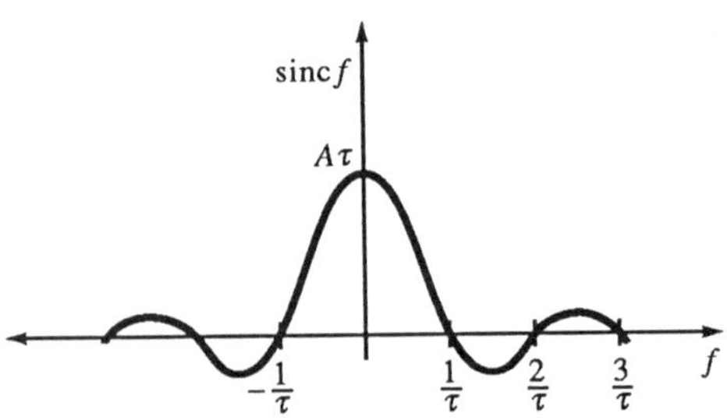

FIG. 12.39 Frequency spectrum of single pulse of Fig. 12.38 centered about origin is a sinc f $[(\sin f)/f]$ function. As τ decreases, sinc broadens to a constant.

In the limit, (12.87) undergoes these changes:

$$\sum_{k=-\infty}^{\infty} \to \int_{-\infty}^{+\infty}, \qquad c_k \to V(j\omega), \qquad k\omega_s \to \omega, \qquad \Delta t \to dt \tag{12.94}$$

Except for the lower limit of integration, the Fourier transform is a special case of the Laplace transform when $s = j\omega$. The unit step and impulse functions have no Fourier series, but they have Fourier transforms.

As T_s increases (or τ decreases), the sinc response broadens until, in the limit, it is constant over all frequencies. Thus, the frequency response of an impulse is independent of frequency, as (12.77) shows.

The frequency spectrum for δ_P is

$$c_n = \frac{1}{T_s}\int_{-T_s/2}^{T_s/2}\left(\sum_{k=-\infty}^{\infty} \delta(t-kT_s)\right) e^{-jn\omega_s t}\, dt = \frac{1}{T_s} \tag{12.95}$$

This spectrum is also flat for all frequencies with a constant amplitude of $1/T_s$. It differs from the spectrum for $\delta(t)$ in that it is discrete. The Fourier series of the impulse train is

$$\delta_P(t) = \sum_{k=-\infty}^{\infty} \delta(t-kT) = \frac{1}{T_s}\cdot \sum_{n=-\infty}^{\infty} e^{jn\omega_s t}, \qquad \omega_s = \frac{2\pi}{T_s} \tag{12.96}$$

The waveforms of δ_P in both the time and frequency domains are shown in Fig. 12.40. This representation of δ_P leads to a different expression for $\mathscr{L}\{\delta_P(t)\}$, from that of (12.83):

$$\mathscr{L}\{\delta_P(t)\} = \int_{-\infty}^{\infty} v(t)\left(\frac{1}{T_s}\sum_{n=-\infty}^{\infty} e^{jn\omega_s t}\right) e^{-st}\, dt \tag{12.97}$$

Since the index n is independent of t, the summation can be removed from the integral:

$$\frac{1}{T_s}\sum_{n=-\infty}^{\infty}\int_{-\infty}^{\infty} v(t)\, e^{jn\omega_s t}\cdot e^{-st}\, dt = \frac{1}{T_s}\sum_{n=-\infty}^{\infty}\int_{-\infty}^{\infty} v(t)\, e^{-(s-jn\omega_s)t}\, dt \tag{12.98}$$

The resulting integral is the Laplace transform, $V(s-jn\omega_s)$. Thus

$$V^*(s) = \frac{1}{T_s}\sum_{n=-\infty}^{\infty} V(s-jn\omega_s) \tag{12.99}$$

This expression for V^* has a geometric interpretation in the s-domain. The

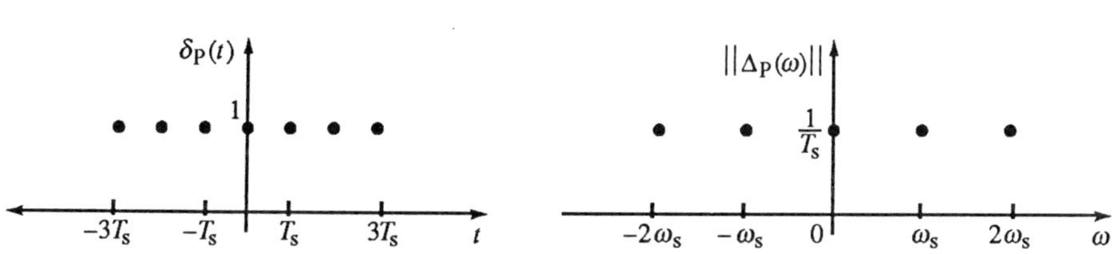

FIG. 12.40 Time- and frequency-domain plots of impulse train.

transform of $V(s)$ is periodic in ω_s so that

$$V^*(s) = V^*(s - j\omega_s)$$

$V(s)$ repeats along the $j\omega$ axis at intervals of $j\omega_s$.

In (12.96), $\delta_P(t)$ is expressed as a series of complex sinusoids with amplitude $1/T_s$ and frequencies of $n\omega_s$. The frequency spectrum of $v(t)$ is convolved (or *heterodyned*) in the frequency domain with the spectrum of $\delta_P(t)$ (Fig. 12.41). (Multiplication in one domain corresponds to convolution in the other.) The sine and cosine terms in $v(t)$ multiply by the terms of δ_P to produce sum and difference frequencies according to the trigonometric formulas:

$$\cos\alpha \cdot \cos\beta = \tfrac{1}{2}\cos(\alpha - \beta) + \tfrac{1}{2}\cos(\alpha + \beta) \qquad (12.100)$$

$$\cos\alpha \cdot \sin\beta = \tfrac{1}{2}\sin(\alpha + \beta) - \tfrac{1}{2}\sin(\alpha - \beta) \qquad (12.101)$$

The frequency-domain plots are the magnitude envelopes of the complex Fourier coefficients, the amplitudes of the harmonics. For $V^*(\omega)$, the spectrum of $v(t)$ is centered around harmonics of ω_s. So the effect of sampling is to generate frequency-shifted copies (or bands) of $V(\omega)$ centered around harmonics of ω_s.

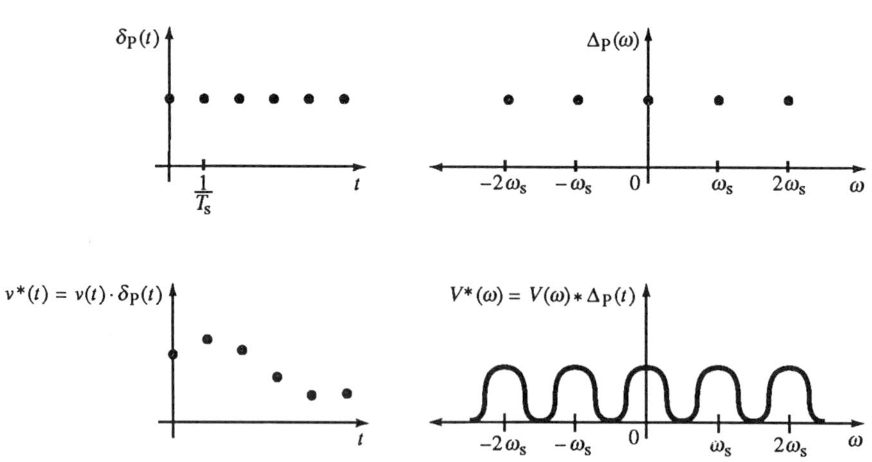

FIG. 12.41 Time- and frequency-domain plots of sampled $v(t)$. Discrete samples in t correspond to periodic spectra in ω, centered about harmonics of the sampling frequency.

12.10 The Sampling Theorem (Nyquist Criterion)

The *sampling theorem* gives a criterion for recovery of $v(t)$ from $v^*(t)$. If ω_s is not larger than twice the highest frequency in $V(\omega)$, then the frequency-shifted bands of $V(\omega)$ overlap (Fig. 12.42) and cannot be separated by filtering. The *Nyquist criterion* for recoverability of the original continuous signal is

$$\omega_s > 2\omega_h \tag{12.102}$$

where ω_h is the highest frequency component of $V(\omega)$. The original signal is recoverable from its sampled form when the highest frequency component is less than the *Nyquist frequency*, $\omega_s/2$. In Fig. 12.42, the band $V_1(\omega)$ is a replica of $V(\omega)$ centered at ω_s. It has frequency components below ω_s that overlap with the positive frequency components of $V(\omega)$. These are negative frequencies in $V(\omega)$ shifted up in frequency by ω_s.

The significance of negative frequency components in $V(\omega)$ is that they are inverted (180° phase-shifted) from their corresponding positive counterparts. Since the frequency spectrum of $V(\omega)$ is symmetric around $\omega = 0$, it is an even function and $V(-\omega) = V(\omega)$. The phase, however, is an odd function and is negative for $\omega < 0$; for negative n, the angle of c_n, from (12.89), is $\vartheta = -n\omega_s t$. Then $\vartheta(-n) = -\vartheta(n)$.

In Fig. 12.42, $V(\omega)$ and $V_1(\omega)$ are symmetrical around the Nyquist frequency. In effect, V has been folded over at $\omega_s/2$. The larger ω_h is, the further back toward lower frequencies the folding extends. These folded frequency components from V_1 are *alias* frequencies in $v^*(t)$ and have a frequency of ω_a relative to ω_s.

The significance of an alias frequency in the time domain is that a sequence of samples has more than one frequency interpretation. In Fig. 12.43, $V(\omega)$ has one frequency component at $\omega = \frac{3}{4}\omega_s$. The samples also fit a sinusoid of $\omega = -\frac{1}{4}\omega_s$, an alias frequency within the band of $V(\omega)$. The alias sinusoid is inverted relative to that of V_1 because its frequency is negative.

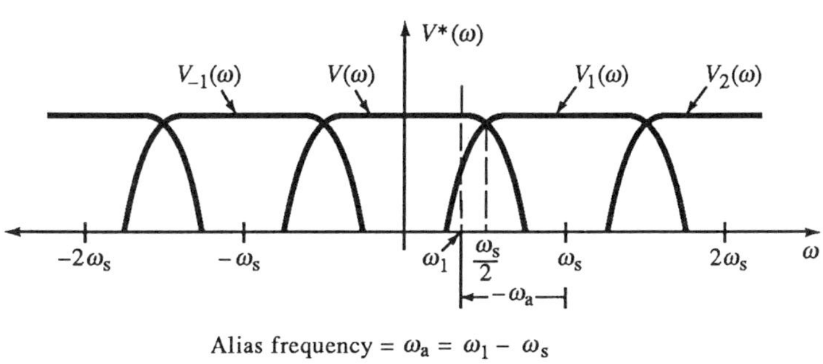

FIG. 12.42 Overlapping spectra of $V(\omega)$ due to undersampling. An alias frequency from $V_1(\omega)$, ω_a, at ω_1 of $V(\omega)$ is indicated.

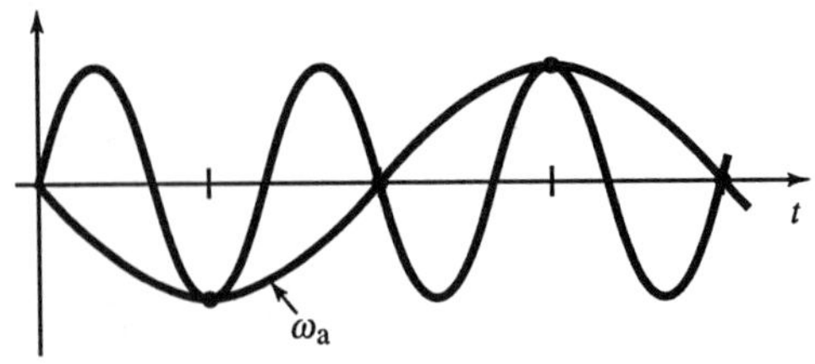

FIG. 12.43 Aliasing in the time domain. The discrete samples fit sinusoids of two frequencies. The alias is inverted, being a negative frequency.

More generally, if ω_1 of $V(\omega)$ is sampled at ω_s, then from Fig. 12.42,

$$\omega_1 = \omega_s - (-\omega_a) = \omega_s + \omega_a \tag{12.103}$$

and

$$\text{alias frequency} = \omega_a = \omega_1 - \omega_s \tag{12.104}$$

In Fig. 12.43, sinusoids of both ω_1 of V and ω_a of V_1 fit the sample points. The discrete samples of $v(t)$ are too few per cycle to eliminate ω_a; $v(t)$ is *undersampled.* The sampling theorem requires more than two samples per cycle for recovery of $v(t)$. Such a $v(t)$ is *oversampled.*

Recovery of $V(\omega)$ from $V^*(\omega)$ for oversampled signals is achieved by a low-pass filter (LPF) that passes only $V(\omega)$. The ideal filter, $H(\omega)$, is shown in Fig. 12.44a. It has an immediate cutoff just above ω_h. The ideal maximum-bandwidth filter has a cutoff at the Nyquist frequency.

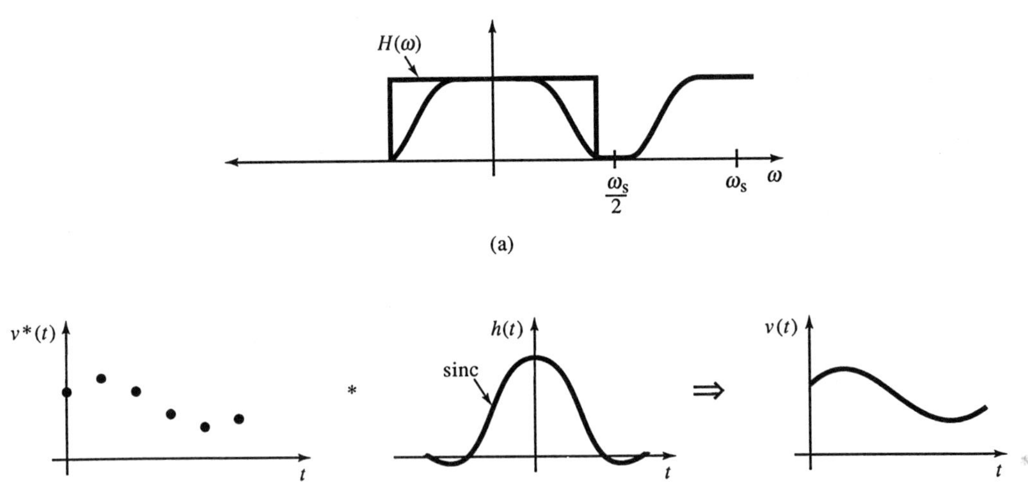

FIG. 12.44 The ideal low-pass antialiasing filter $H(\omega)$ is a spectral "pulse" (a) corresponding to a sinc-function convolver or interpolator in t (b), where $*$ is the convolution operator.

In the time domain, this filter function transforms into a sinc function (Fig. 12.44b). Since nonzero sinc values extend to $t = -\infty$, it is noncausal and can only be approximated by realizable circuits. The pulse shape of the ideal LPF transforms into a sinc function in t just as a pulse in the time domain does in ω. $H(\omega)$ is multiplied by $V^*(\omega)$ in ω to recover $V(\omega)$; in t, $h(t)$ is convolved with $v^*(t)$ to produce $v(t)$. For bandlimited $v(t)$,

$$v(t) = \sum_{k=-\infty}^{\infty} (kT_s)\,\mathrm{sinc}\left(\frac{\omega_s}{2}(t - kT_s)\right), \qquad -\frac{\omega_s}{2} < \omega < \frac{\omega_s}{2} \tag{12.105}$$

The sinc function acts as an interpolator, filling in the missing values of $v(t)$.

Our final derivation is the spectrum of a zero-order hold. This is the frequency response of a S/H. In the s-domain, a ZOH can be regarded as an integrator of weighted impulses, producing $\hat{v}(t)$ in Fig. 12.45a. This is the typical waveform from a S/H or DAC. This integrator signal is periodic at the sampling rate. An integrator in s is $1/s$. A periodic integrator is constructed by integrating for T_s, or

$$\text{ZOH} \qquad H_0(s) = \frac{1}{s} - \frac{1}{s} \cdot e^{-sT_s} = \frac{1 - e^{-sT_s}}{s} \tag{12.106}$$

In the time domain, this is a unit step turned off T_s later, or

$$\text{ZOH} \qquad u(t) - u(t - T_s) \tag{12.107}$$

The Laplace transform of (12.107) is (12.106). The frequency response of H_0 is found by letting $s = j\omega$. Then

$$H_0(j\omega) = \frac{1 - e^{-j\omega T_s}}{j\omega} = T_s\,\mathrm{sinc}\left(\frac{\omega T_s}{2}\right) \cdot e^{-j\omega T_s/2} \tag{12.108}$$

Then

$$\|H_0(j\omega)\| = T_s \left|\mathrm{sinc}\left(\frac{\omega T_s}{2}\right)\right|, \qquad \angle H_0(j\omega) = \frac{-\omega T_s}{2} \tag{12.109}$$

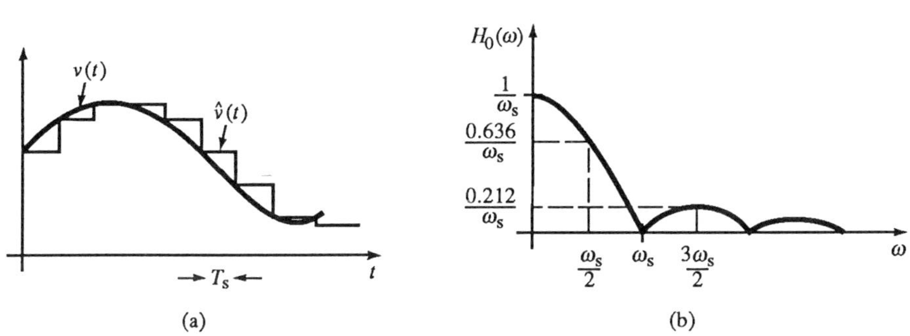

FIG. 12.45 DAC output (or ADC input), $\hat{v}(t)$ in (a) is zero-order hold response to $v(t)$. Zero-order hold frequency-response magnitude $H_0(\omega)$ is |sinc| function (b).

Once again, the sinc function appears. The magnitude plot of the frequency response is shown in Fig. 12.45b. The phase response is linear and only time-shifts the output. The phase delay can be seen in Fig. 12.45a by noting that a best-fit of $v(t)$ to $\hat{v}(t)$ requires $v(t)$ to be shifted to the right (delayed in time) by half a step, or by $-T_s/2$, as (12.109) predicts. Ideal recovery of $v(t)$ from $\hat{v}(t)$ requires an inverse sinc filter, or *sinc compensator.* This compensator can be implemented in either digital or analog form. It is digital if it precedes a DAC or follows an ADC and analog if it follows a DAC or precedes an ADC.

12.11 Sampling Circuits

Sample-and-hold (S/H) or track-and-hold (T/H) circuits are switched between the sample or track state and hold state by a digital control line. Ideally, the input voltage at the instant of switching to HOLD is retained as a constant at the output of the S/H. T/Hs are similar to S/Hs; in the nonheld state, the output follows the input. In a S/H this is not necessarily so, though most S/Hs are actually T/Hs. The sampling impulse of sampling theory corresponds to the active edge of the HOLD signal.

The speed of a S/H is determined by the *acquisition time*, the time from when sampling or tracking of the input begins to when a settled, held output is available. This time has two terms. The first is from the time when tracking begins to the time when the hold capacitor follows the input signal. A large initial difference between v_I and v_C requires slewing time before tracking is accurate. The second term is the setting time at v_C when the hold state begins. In addition to acquisition time, aperture delay and jitter (as discussed in Section 12.8) also apply to S/H circuits.

Several errors are associated with S/H circuits, and their design considerations are closely related to those of peak detectors. Errors occur in the sampling process or in the hold state. The first are dynamic sampling errors. *Digital delay* causes the effective sampling instant to be delayed. For a rising input signal, this translates into a voltage error of

$$v_\varepsilon = \left(\frac{dv_I}{dt}\right) t_d \tag{12.110}$$

where the digital delay time t_d is multiplied by the signal slew-rate. The second cause of error is *analog advance.* If the input signal is delayed instead, an effective negative delay occurs in sampling since the signal lags behind where it should be when sampling occurs. A rising input signal is below where it should be and a negative error occurs. It is equivalent to sampling the signal in advance of the actual sampling instant.

The dominant cause is a voltage lag on the hold capacitor; its charging always lags somewhat behind the source. This is largely due to charging-source

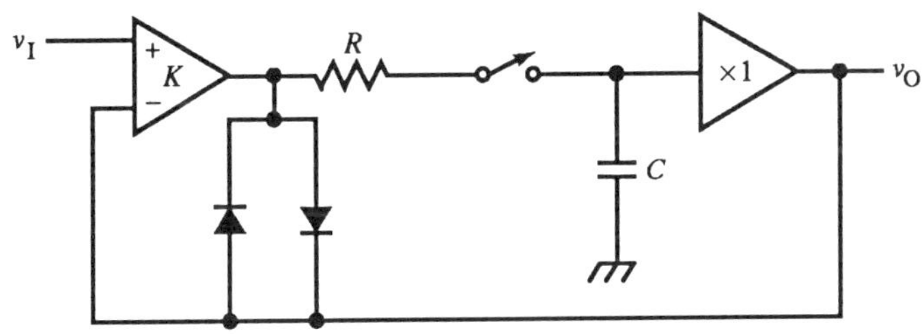

FIG. 12.46 Feedback S/H reduces charging time constant RC by $K+1$.

resistance R (Fig. 12.46). By closing the loop with an op-amp input, we reduce the charging time constant RC by $K+1$. The diodes around the op-amp keep its output from saturating when in the hold state. Signal advance is the major cause of delay error and is compensated by delaying the sampling command.

A third dynamic error is due to stray capacitance C_s between the hold capacitor C and the sampling command line (Fig. 12.47a). When this line switches, it causes charge to flow through C_s into C. If the capacitor voltage v_C is plotted against a range of dc inputs, the plot is linear. Its slope represents a hold gain. As the input voltage v_I increases, the step of extra voltage on C grows in size because the voltage between the hold line v_H and v_C varies

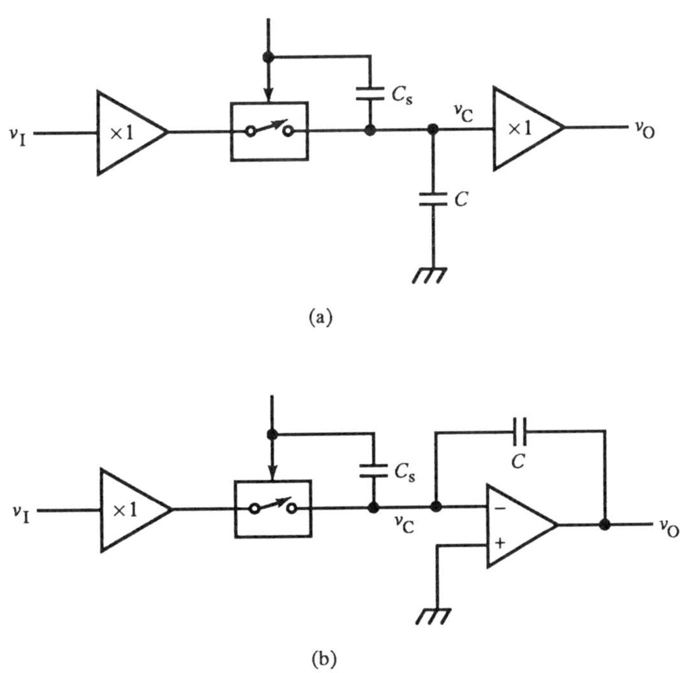

FIG. 12.47 Parasitic capacitance C_s from sampling control line to hold capacitor node causes a hold-gain error (a). In (b) v_C is at virtual ground and voltage variation across C_s due to v_I is eliminated.

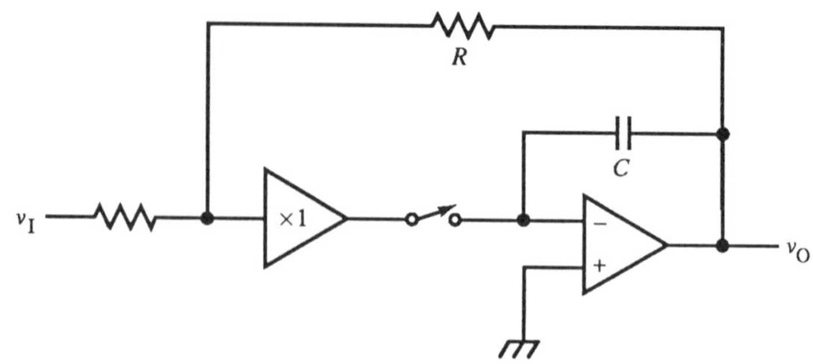

FIG. 12.48 Leakage decoupling technique. R maintains zero voltage across switch in hold mode.

linearly with v_I. As the difference between the sample level, V_S, of v_H, and v_C increases, C_s is charged more, and this charge is transferred to C when v_C changes to the hold state. The hold step, or *pedestal*, thus increases with v_I.

A circuit that avoids this problem is that of Fig. 12.47b. The hold capacitor is the feedback C of the op-amp. The op-amp isolates v_C from the switch node by holding it at virtual ground. Then the voltage across C_s is independent of v_I and the same amount of charge is transferred to C on switching. The charge on C_s is $C_s V_S$. The hold gain varies somewhat as C_s varies with voltage as do semiconductor junctions.

The hold capacitor dielectric absorption must be low to avoid recovery effects during the hold state. Its leakage causes static sampling error during the hold state. Any other leakage paths for capacitor charge contribute to leakage error. The buffer amplifier and sample switch must be low in leakage. A leakage compensator is shown in Fig. 12.48. The leakage decoupler R has the same function as in peak detectors (see Section 11.14). It keeps the voltage across the switch near zero, thus minimizing leakage through it.

Another hold-step compensator (Fig. 12.49) places another switch similar to Q_1 in series with it. This additional switch is shorted, but its C_s (C_{GS} for a MOSFET) connects to the same node. It is driven with an opposite polarity edge so that its stray charge cancels that of Q_1.

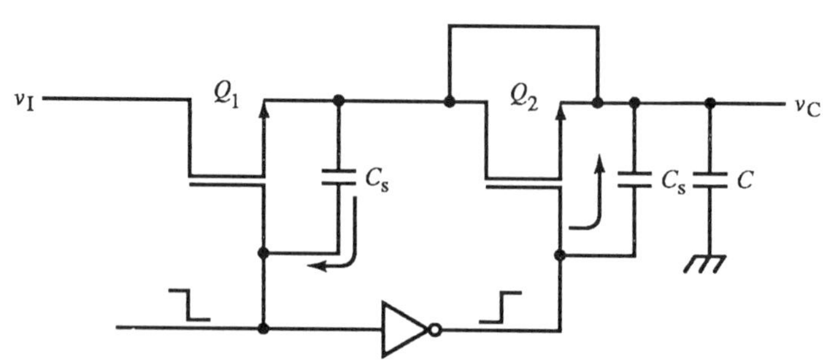

FIG. 12.49 C_s charge compensator uses equal and opposite C_s of Q_2 to cancel charge from Q_1.

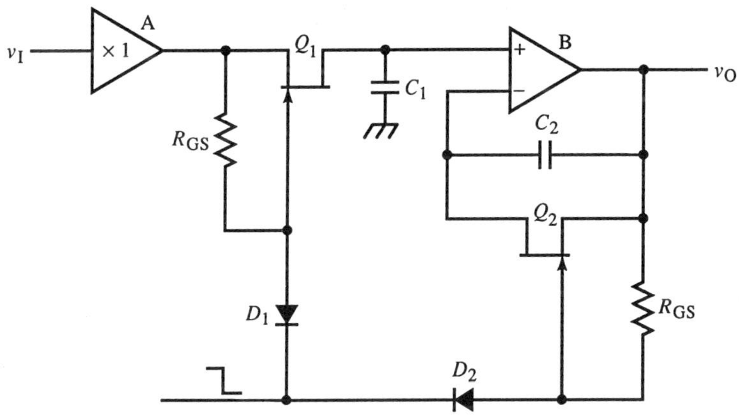

FIG. 12.50 A JFET-switched S/H with op-amp bias-current compensation capacitor $C_2 = C_1$. C_2 must be reset during track mode.

In Fig. 12.50, a JFET switch Q_1 passes the signal through its source, connected to the buffer A. When the control line goes low, Q_1 cuts off. D_1 conducts a small amount of current through R_{GS} to keep the gate reverse-biased. At the same time, Q_2 is also cut off by a similar circuit. The capacitor C_2, equal to C_1, is a bias-current compensator for the op-amp (as with the peak detectors). As the hold capacitor C_1 charges with I_B, so does C_2. The differential voltage is cancelled at the output. Of course, offset current is not compensated.

Finally, very fast S/H have the additional error of signal leakage through shunt switch capacitance C_s during hold. In Fig. 12.51 is the well-used sampling bridge. The diode bridge has two C_s in series for each leg of the bridge, or an equivalent of one C_s from input to output. It is current-switched for speed. Sampling bridges of this kind have commonly been used in sampling oscilloscopes. The practical limitation in their switching time is often the switching speed of their drivers.

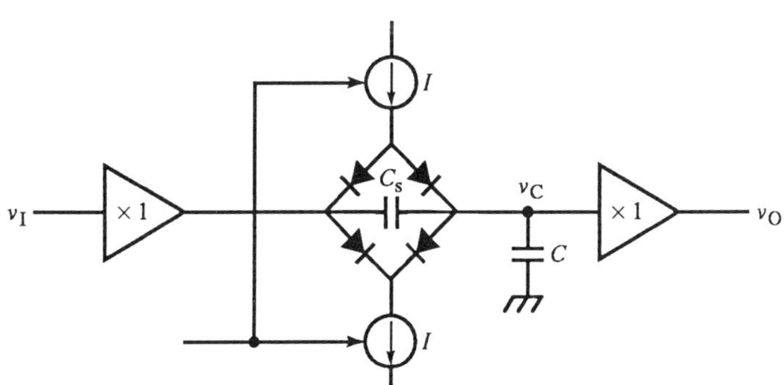

FIG. 12.51 Diode bridge sampler. C_s is minimized to keep signal from affecting v_C during hold mode.

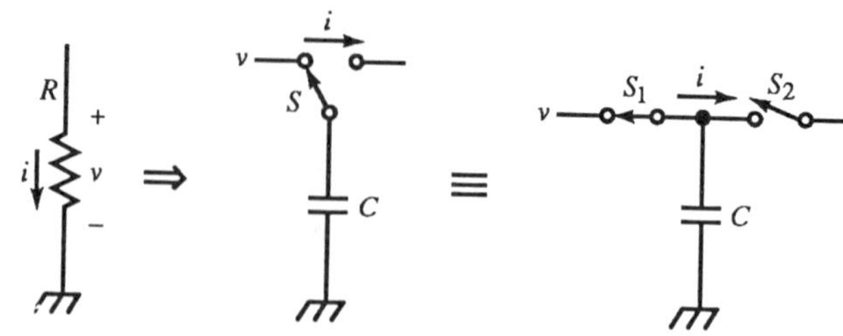

FIG. 12.52 Switched-capacitor equivalent grounded resistor.

12.12 Switched-Capacitor Circuits

Switched-capacitor circuits replace resistors with capacitors and switches. In ICs, *diffusion resistors*, made by connecting to the ends of a diffused area, are not optimal since their values are hard to control, and they have large areas (their relative values are much better; they match well). Large-value resistors take up so much area that they are often impractical. When accuracy is not important, a kind of resistor made of a thin layer of, say, n material between two p layers—a pinch resistor—can be made large but with ±20% accuracy. NiCr (nichrome) resistors are very good but costly to make and trim.

This limitation makes switched-capacitor resistors an attractive alternative. The equivalent resistance is shown in Fig. 12.52. It is a SPDT switch S and a capacitor. The SPDT switch is equivalent to two SPST switches, synchronized as shown. When S is switched to the input, it charges to the input voltage v with a charge of Cv. When it switches to an output held at ground, it delivers this charge. The output is typically the virtual ground of an op-amp (Fig. 12.53). If the switching rate is f_s, then the charge delivered per unit time, or current, is

$$i = Cf_s v \Rightarrow r = \frac{1}{Cf_s} = \frac{T_s}{C} \tag{12.111}$$

The equivalent resistance follows directly and is subject to the Nyquist criterion

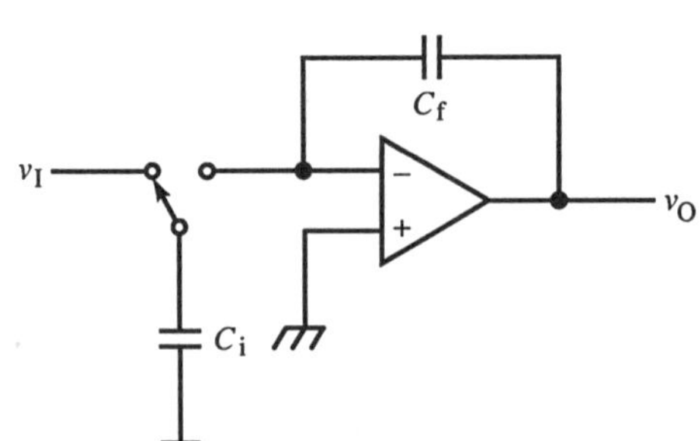

FIG. 12.53 Switched-capacitor replacement of R_i of op-amp circuit.

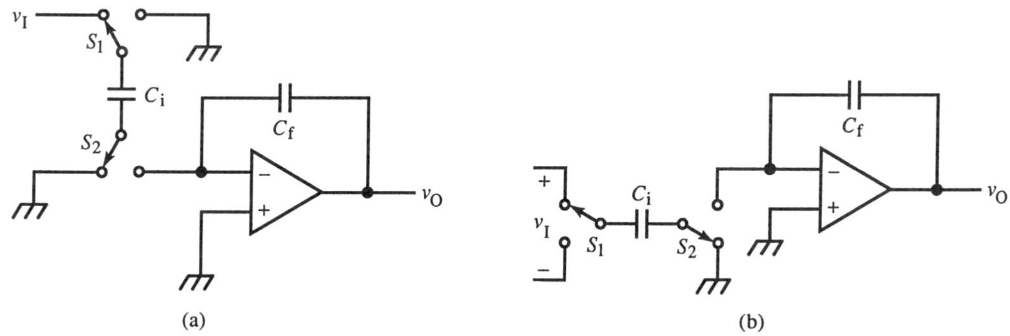

FIG. 12.54 Inverting (a) and differential-input (b) switched-capacitor circuits.

due to switching. The bandwidths of switched-capacitor circuits must be well within the Nyquist frequency for accurate equivalence.

The switching scheme of Fig. 12.54a inverts v_I. C_i charges with switches in the position shown. When switched, charge flows out of the op-amp input to ground. In effect, the two-switch circuit is a negative R. If we drive the grounded side of S_1 with an input instead, as in (b), the two terminal voltages of v_I subtract upon switching. The differential voltage v_I determines the charges.

12.13 Closure

The world of digital electronics merges with analog electronics in digitizing and sampling circuits, but the merged areas—mainly ADCs, DACs, and switched-capacitor and sampling circuits—do not involve logic design. Instead, the underlying theory is an extension of that for continuous functions. The mathematics is similar; difference equations replace differential equations. Since sampled-data circuits also include commutating and switched-capacitor filters and digital signal processing, the full story, including dithering, FFTs, DSP filters, and windowing are subjects for other books.

References

Tim Regan, "CMOS DACs work backwards to provide voltage outputs," *Electronic Design*, 16 Sep. 1982. pp. 117–122.

Roy Sprowl, "PWM DAC simplifies output filtering," *EDN*, 18 Aug. 1988. p. 220.

Bob Peterson, ed., "Inherently monotonic DAC-IC design eliminates costly resistor matching," *EDN*, 5 Mar. 1979. pp. 46, 49.

Steve Kirby, "Deglitcher circuit refines d-a-converter output," *Electronics*, 21 Apr. 1983. p. 151ff.

M. Lobjinski and R. Bermbach, "Switched-capacitor technique improves a-d conversion," *Electronics*, 2 Jun. 1982. pp. 149, 151.

Doug Rife, "High accuracy with standard ICs: an elegant ADC's forte," *EDN*, 28 Apr. 1982. pp. 137–144.

"What's a recursive subranging ADC?," boxed section, *Electronic Design*, 23 Feb. 1989. p. 122.

David Potson and Craig Swing, "Clamping Op Amp Improves ADC," *Electronic Design*, 13 Oct. 1988. pp. 124–125.

"What's a delta-sigma a-d converter?," boxed section, *Electronic Design*, 14 Apr. 1988. p. 50.

Thomas J. Mego, "Resolve 22 bits easily with charge-balance ADCs," *Electronic Design*, 25 Jun. 1987. pp. 109–114.

Albert Gookin, "A Faster Integrating Analog-to-Digital Converter," boxed section, *Hewlett-Packard Journal*, Vol. 28, No. 6, Feb. 1977. p. 9.

Larry DeVito, "Synchronous V/F converter aids linearity in data acquisition," *EDN*, 16 Oct. 1986. pp. 183–190. (See 13 Nov. 1986 *EDN* for Part 2.)

Jim Williams, "Design techniques extend V/F-converter performance," *EDN*, 16 May 1985. pp. 153–164.

T. Anderson and B. Trump, "Clocked v-f converter tightens accuracy and raises stability," *Electronic Design*, 6 Sep. 1984. pp. 235–244.

Steve Connors, "Voltage-to-frequency converters: A/D's with advantages," *EDN*, 5 Jun. 1974. pp. 49–54.

J. W. Fennel, "Feedback Improves Sample-and-Hold Performance," *EDN*, 1 Jul. 1969. pp. 63–64.

Eugene L. Zuch, "Keep track of sample-hold from mode to mode to locate error sources," *Electronic Design*, 6 Dec. 1978. pp. 80–87.

Michael A. Wyatt, "MOSFET switch compensates sampler," *EDN*, 2 Mar. 1989. p. 176.

Peter Henry, "JFET-input amps are unrivaled for speed and accuracy," *EDN*, 14 May 1987. pp. 161–169.

Al Little and Bob Burnett, "S/H amp-ADC matrimony provides accurate sampling," *EDN*, 4 Feb 1988. pp. 153–166.

Jerry Neal and Jim Surber, "Track-and-holds take flash converters to their limits," *Electronic Design*, 3 May 1984. pp. 381–388.

Kerry Lacanette, "Universal switched-capacitor filter lowers part count," *EDN*, 3 Apr. 1986. pp. 139–147.

Index